CHIRONOMIDS

From Genes to Ecosystems

EDITOR
PETER CRANSTON

Best wishes
Pete Cranston

CSIRO
AUSTRALIA

National Library of Australia Cataloguing-in-Publication

Chironomids : From genes to ecosystems / editor, Peter Cranston.

Includes bibliographic references and index.
ISBN 0 643 05656 4

1. Chironomidae. 2. Chironomidae -- Ecology.
I. Cranston, P.S. (Peter S.). II. CSIRO.

595.771

This book is available from:
CSIRO Publications
PO Box 89 (314 Albert Street)
East Melbourne, Victoria 3002
Australia

Tel. (03) 418 7217 Int: +(613) 418 7217
Fax (03) 419 0459 Int: +(613) 419 0459

Managing Editor: Kevin Jeans
Designer: Anita Adams
Production Manager: Jim Quinlan
Layout: Anita Adams
Line Drawing on front cover: Karina McInnes

Contents

• Part 3: Chironomidae in Biological Monitoring

• Part 4: Austral chironomid studies

• PART 5: CHIRONOMIDAE AND HUMANS

• PART 6: ECOLOGY AND BEHAVIOUR OF CHIRONOMIDAE

• Part 7: Chironomid morphology and systematics

INTRODUCTION

The Chironomidae popularly are called "non-biting midges" to distinguish them from their biting relatives. Although the impact of these flies excludes human-feeding and disease transmission, chironomids are still the subject of much study. At the suborganismal level, the presence of giant chromosomes in the larval salivary glands has encouraged much molecular and cytological research. The dominance of the larvae in aquatic ecosystems means that they are unavoidably encountered by every freshwater ecologist. Adult flies can impinge on our activities by their nuisance numbers outdoors or through their ability to thrive in domestic water distribution systems. Furthermore, their taxonomic diversity has encouraged many systematists to study Chironomidae in all life-history stages.

Scientists involved in all aspects of chironomid studies met in Canberra, Australia, in January, 1994, for their 12th International meeting to discuss current research. Financial support was provided by the Commonwealth Environment Protection Agency and CSIRO Division of Entomology. The value of chironomids and other aquatic biota in the monitoring of waterway health was emphasised with reviews from the northern hemisphere and case histories from the southern, including Australia and New Zealand. However, as our title indicates, the contemporary studies reported represent a range from the genetic through the autecological to the ecosystem scale.

This volume is dedicated to the memory of Lars Brundin, whose wide-ranging studies of the Chironomidae emphasised the need for, and illustrated the benefits of, a global perspective. His work in the then poorly-known southern hemisphere fauna not only enlightened his colleagues about the diversity of chironomids in this part of the world, but more importantly allowed Brundin to elaborate a modern biogeography in which distribution patterns were interpretable through rigorous phylogenetics. The austral contributions to this volume owe much to Brundin's pioneering studies in the south. However, Brundin's influence on the topics discussed in Canberra is by no means restricted to the austral - his interests in classification of water bodies and monitoring change and his life-long belief that systematics provides the framework for all other studies are reflected by many of the contributors to this book.

The following new taxa are described in this book:

Tanypodinae:	*Conchapelopia insolens* Murray,
Buchonomyiinae:	*Buchonomyia brundini* Andersen and Sæther,
Diamesinae:	*Mapucheptagyia* Willassen, *Mapucheptagyia brundini* Willassen,
Orthocladiinae:	*Corytibacladius* Oliviera, Messias and Santos, *Corytibacladius gercinoi* Oliviera, Messias and Santos, *Eukiefferiella brundini* Boothroyd and Cranston, *Parakiefferiella minax* Ferrington and Sæther, *Pirara* Boothroyd and Cranston, *Pirara matakiri* Boothroyd and Cranston, *Pirara australiensis* Boothroyd and Cranston
Chironominae:	*Chironomus maddeni* Martin and Cranston in Madden *et al.*, *Sublettea wilesi* Ashe and O'Connor

The following new combination is made:

Parakiefferiella harrisoni (Freeman), from *Pseudosmittia*, by Ferrington and Sæther.

Peter S. Cranston, Canberra

1. Lars Brundin, in England, 1981

2. Attenders at 12th International Symposium on Chironomidae, Canberra, January 1994

In Memory of Lars Brundin

30 May 1907 - 18 November 1993

On the 18th of November 1993, at the age of eighty-six, the rich and full research life of Lars Brundin came to an end when he passed away peacefully in his study in Stockholm.

Lars Brundin established the basis for scientific chironomid research and his lifelong work has made a fundamental impression on modern chironomid taxonomy, systematics and phylogeny. He was among the first to understand Hennig's ingenious principles of phylogenetic systematics and to apply these principles to his research in a convincing and practical way. It was he who initiated a new epoch of causal biogeography.

Born on May 30th 1907 in Gasslanda, Lars Brundin grew up in the Aneboda-Vaxjo region of the wooded lake district in the heart of southern Sweden. As a student at the University of Lund his inclination towards biology / natural history, already strongly defined at the age of nineteen, was apparent by the publication of his first scientific paper on the native beetle fauna. When the biological station was opened in 1927 at Abisko, a region of Lapland which at that time was not known to tourists and largely unknown to biologists, the twenty year old Lars Brundin was its first guest. Even then a trait in Brundin's character, which distinguished him during his scientific career, was already apparent in his practice of purposefully and persistently seeking out new research territory.

In 1934 Brundin completed his university studies with his thesis "Die Coleopteren des Tornetraskgebietes - ein Beitrag zur ökologie und Geschichte der Käferwelt in Schwedisch-Lappland". In 1937, by then a well known and respected coleopteran specialist and an experienced field biologist, Brundin obtained a position at the Institute for Freshwater Research in Drottingholm. He seems to have had some difficulty in adapting to freshwater biological research and five years passed before he published his first specialist paper "Zur Limnologie jämtländischer Seen". In the meantime he also published numerous articles on beetles, particularly on the Staphylinidae, and continued to do so until 1954. It took almost ten years of Brundin's professional working life to find the path he believed would allow him to use his abilities fully to pursue his scientific goals and inclinations. That direction involved the chironomids, and in 1945 he commenced work on the imagines.

As an outsider in dipteran studies, Brundin sought with an open mind new and definite possibilities to cope with the ever-present problems of methodology and taxonomy. He obtained his chironomid material using funnel traps, developed by Russian colleagues, which he improved in design. This procedure enabled him to make the first quantitative and ecological correlations of imago and pupa collected at the same time. Often the larval skin was also found allowing Brundin to become acquainted with the complete metamorphosis of individual species.

His imaginal material was preserved and examined in alcohol while the hypopygium only was cleared in KOH if necessary. Brundin's extraordinary talent for artistry - he was proficient in water-colours - enabled him to draw the morphological features to a very high standard that had not been seen previously in chironomid research. The taxonomic experience stemming from his two decades of research on the difficult staphylinid beetles enabled him to initiate a new epoch of chironomid research.

For several years Brundin continued his intensive studies on the benthic fauna of various lakes in southern Sweden, based on grab samples and submerged funnel traps. Within a short time he became a limnologist who could define the ecological requirements of the larvae of a number of chironomid species. His first taxonomic study "Zur Kenntnis schwedischer Chironomiden" was published in 1947, the second "Über die Metamorphose der sectio Tanytarsariae connectentes" was published in the following year. In the next year, in 1949, Brundin produced the work that was to become an authoritative volume - the bible - particularly for those working on lake benthos. With the publication of this 914 page volume on "Chironomiden und andere Bodentiere der südschwedischen Urgebirgsseen", new standards were set for lake research.

Brundin's success in including chironomids in limnological research to the extent that he did, and with the proven thoroughness and high standard of his work, must be judged from the background of the hopelessly confused taxonomical situation of chironomid science which prevailed at that time. Specific identification of the larval and pupal stages was possible only for a few groups, largely due to Thienemann's work, and identification of the adults remained unsatisfactory using Edwards (1929) key to the British Chironomidae. The taxonomic chaos prevailing at that time is evident in the published volumes of "Tendipedidae" in "Die Fliegen der Paleärktischen Region". After Edwards' unexpected premature death, Goetghebuer, already elderly and in failing health, saw himself as the sole curator of this large and difficult group of Diptera, and also of Kieffer's work, which was none too elaborate.

While Brundin continued to publish papers on the Coleoptera until 1954, his real work concentrated on the Chironomidae. He extended his collecting area into the mountain lakes of northern Scandinavia and the Alps. His detailed study on the genera *Protanypus*, *Prodiamesa* and *Monodiamesa*, published in 1952, is testimony to his deep involvement in problems of chironomid systematics and biogeography. Brundin turned his attention to address the question of the composition of the benthic chironomid fauna of temperate lakes in the southern hemisphere.

Everyone who knew Brundin could easily imagine that Thor Heierdal's adventurous and sensational research exploits may have inspired him to organise the Swedish limnological expedition to South America. He was accompanied on this expedition during 1953-54 by his companions Dr. H. Loffler, Vienna and Dr. K. Thomasson, Uppsala.

Interpretation of the benthic faunal material obtained in the lakes of South America was unsatisfactory for Brundin because he was unable to determine if the chironomids living in the profundal region of Andean lakes were endemic there or related to the Boreal species. However, material collected at the same time in the Andean rivers and streams proved much richer in species and promised to provide more information on phylogenetic and biogeographic aspects than did the lake material. Since the Orthocladiinae were the predominant component of streams, the European representatives of which were at that time difficult to identify and evaluate systematically, Brundin postponed detailed examination of the South American material until he could survey phylogenetic relationships of the European groups. In 1955 and 1956 he collected Orthocladiinae in the running waters of Scandinavia with drift nets, the so-called "Brundin nets". With these nets he was able to collect, in a few excursions, the essential elements of the northern European chironomid fauna as pupae, exuviae and emerging or emerged adults. On the basis of that material and of that already obtained from his studies on lake benthos, he published in 1956 the paper "Zur Systematik der Orthocladiinae". This work had a pivotal role in determining the further development of chironomid research. At the same time his pioneering study "Die bodenfaunistichen Seentypen und ihre Anwendbarkeit auf die Südhalbkugel, zugleich eine Theorie der produktionsbiologischen der glazialen Erosion" was published. This certainly was a cornerstone for the discussion of lake typology based on the profundal chironomid communities.

In 1952 Brundin was awarded a lectureship in entomology at Stockholm University, followed by his appointment in 1957 as Professor and Director of the Entomology Section of the Naturhistoriska Riksmuseet in Stockholm. With this nomination Brundin was given the independence without which his further studies could not have been carried out so successfully. During a five-month expedition in 1957 he made drift net collections in lotic waters of North America and in 1959 he collected in Andean mountain streams from Bolivia to Tierra del Fuego. The results of his second expedition to South America was beyond all his expectations. The exceptional position of the chironomid fauna in the Chile-Patagonia biogeographic region was striking in the strong representation of the subfamily Podonominae, only weakly represented in the northern hemisphere, and also by the various endemic groups of Diamesinae and Orthocladiinae.

Since Brundin was a systematic worker, he felt obliged to investigate relations within the family Chironomidae, based largely on pupal exuvial material already collected, and to correlate them with expected continental distribution patterns. Hennig's (1950) work "Gründzüge einer Theorie der phylogenetischen Systematik" provided the necessary theoretical basis for this undertaking.

Already, as a young scientist, Brundin had admired Wegener's ideas on the formation of the continents and oceans and was acquainted with Jeannel's ideas on the origin of continental faunas, the distribution of beetles and insect systematics. Later he was deeply impressed by Croizat's voluminous and creative "Panbiogeography". Brundin's own ideas on the relationships between the southern hemisphere faunal elements had already matured when, in 1960, Hennig's work "Die Dipterenfauna von Neuseeland als systematisches und tiergeographisches Problem" was published. In that contribution, Hennig tried to examine, with his phylogenetic reasoning, the faunal relationship of the southern continents based on the Brachycera. The result was unconvincing, since this Dipteran order was not really suitable for such studies. At the same time, however, Brundin obtained drift net material from streams in New Zealand in which he found many chironomid types familiar to him only from his earlier studies in the southern Andes in South America. With the background of his substantial and thoroughly assimilated theoretical, systematical and biogeographical knowledge, Brundin saw the great opportunity offered by the Chironomidae - in contrast to the Brachycera - to interpret southern hemisphere distribution patterns of related species of plants and animals. He collected further material himself in autumn 1961 in Australia (including Tasmania), New Zealand, Fiji, Tahiti and the Himalayas which confirmed his expectations.

He had almost finalised his systematic morphological study on the Podonominae in October 1963 when he received from Paul Freeman an unusual plesiomorph podonomine specimen from southern Rhodesia. One month later he was himself searching for additional material in south Africa, and with much success. Among other taxa he found all life-history stages not only of new Podonominae genera from southern Africa but also those of the new subfamily Aphroteniinae which he had already recognized in Chile and Australia.

In a painstaking and meticulous manner he gradually assembled the mosaic of faunal correlations of the southern continents and presented his results to the scientific world for discussion. In 1963 Brundin published his ideas on the "Limnic Diptera in their bearings on the problem of transantarctic faunal connections". In 1965 the forum was enlarged with his publication (in the journal *Evolution*) of the treatise "On the real nature of transantarctic relationships". Both papers presented a strong challenge to prevailing authorities on biogeography and evolutionary biology, such as P.J.Darlington, E. Mayr and G.G. Simpson, who were intolerant of Alfred Wegener's ideas and unwilling to analyse Hennig's theory of phylogenetic systematics. Therefore, Brundin's seminal work on "Transantarctic relationships and their significance as evidenced by chironomid midges" published in 1966 by the Swedish Academy of Science, must have been regarded as provocative.

This monographic work on the subfamilies Podonominae, Aphroteniinae and Heptagyiae from the southern hemisphere, prepared to Brundin's typically high standard, provided the fundamental basis for a new comprehensive evaluation of faunal origins of the southern continents following the split-up of Gondwana during the Mesozoic Era. That publication was important, not only for chironomid science, but also for phylogenetic and biogeographic research. Although during the sixties some geophysicists had already undermined the ideas of Wegener's opponents, Brundin's ideas were still in opposition to the prevailing dogma. It was really unusual that evidence from the study of small aquatic insects should provide more convincing arguments for the interpretation of circum-antarctic distribution patterns than had been proposed prior to that time by botanists dealing with plants and zoologists dealing with vertebrates. Brundin inaugurated a new epoch in causal biogeographic research. In the years following he continued to promote the results of his phylogenetic, biogeographic and systematic researches at congresses and symposia. He became an interpreter and a non-compromising advocate of Hennig's theory of phylogenetic systematics.

Throughout this time Brundin continued to obtain and work on supplementary material from the southern hemisphere to further clarify distribution patterns and phylogeny of the Chironomidae. In 1967 he collected again in Australia and New Zealand, in 1969-70 in Chile and Argentina and again in 1971 in Australia and New Zealand. In the year of his retirement in 1973 he organised the 5th International Chironomid Symposium in Abisko. He witnessed the way in which his ideas were to become incorporated and accepted in the general areas of zoological systematics, phylogeny and biogeography. He was honoured by numerous international scientific institutions.

Brundin rounded off his systematic work on chironomids with description of podonomine species from Lebanese amber (1976), established the subfamilies Buchonomyiinae (together with Ole Sæther, 1978) and Chilenomyiinae (1983) as well as producing keys to the genera of Holarctic Podonominae (1983, 1986, 1989). Even in his advancing years Brundin remained active in research and discussion in the Riksmuseet in Stockholm or at expert meetings. During summer months he took reflective holidays at his parental property at Gasslanda, where he had earlier started his zoological research work with beetles and later continued with chironomids.

His love and enjoyment of nature and sport, his open-minded appreciation of fine arts, geography and ethnology and also history and politics had a significant influence on his work and life and that of his family right up to the end.

We, the members of the chironomid research community, may regard Lars Brundin as an exemplary person and scientist and for those of us who knew and worked with him, he was a critical teacher, an honest and sincere colleague and an outstanding friend.

Professor E.J. Fittkau

THE PUBLICATIONS OF LARS BRUNDIN

1926 Coleoptera från Småland och Blekinge. *Entomologisk Tidskrift*, **47**, 86-91.

1929 Eine neue *Gnypeta* aus dem Torneträskgebiet (Col., Staphylinidae). *Entomologisk Tidskrift*, **50**, 14-15.

1930 Några för landet nya coleoptera från Torneträskområdet. *Entomologisk Tidskrift*, **51**, 48-52.

1931 Insektfaunan inom Abisko Nationalpark. I. Coleoptera. - *Kungliga Svenska Vetenskapsakademiens Skrifter i Naturskyddsärenden*, **16**, 8-36.

1931 Insektfaunan inom Abisko Nationalpark. II. Lepidoptera. - *Kungliga Svenska Vetenskapsakademiens Skrifter i Naturskyddsärenden*, **16**, 37-64.

1934 *Die Coleopteren des Torneträskgebietes. Ein Beitrag zur Ökologie und Geschichte der Käferwelt in Schwedisch-Lappland.* Inaugural Dissertation,. University of Lund, Pp. 1-43b.

1937 Om insamling av akalbaggar i fjällen. *Entomologbladet*, **1**, 44-51.

1939 Resultaten av under perioden 1917-35 gjorda fiskinplanteringar i svenska sjöar. *Meddelanden fran Statens Undersöknings-och. Försökanstalt för Sötvattensfisket*, **16**, 3-41.

1940 Studien über die *Atheta*-Untergattung *Oreostiba* Ganglb. (Col. Staphylinidae). *Entomologisk Tidskrift*, **61**, 56-130, 18 Taf.

1940 Zwei neue nordische *Atheta*-Arten aus der *islandica*-Gruppe (Col. Staphylinidae) - Vorläufige Mitteilungen. *Entomologisk Tidskrift*, **61**, 131-133.

1942 Zur Limnologie jämtländischer Seen. *Meddelanden fran Statens Undersöknings-och. Försökanstalt för Sötvattensfisket*, **20**, 1-104.

1943 Monographie der palaearktischen Arten der *Atheta*-Untergattung *Hygroecia* (Col. Staphylinidae). *Annalen des (K.K.) Naturhistorischen (Hof) Museums. Wien*, **53 (1942)**, 129-300.

1943 Zur Kenntnis einiger in die *Atheta*-Untergattung *Metaxya* M. & R. gestellter Arten (Col. Staphylinidae). *Acta Universitatis Lundensis*, N.F. **(2) 39**, Nr. **4**, 1-38, 7 Taf.

1945 Eine neue Aleocharinen-Gattung aus Nordeuropa. *Opuscula Entomologica*, **1945**, 100-7.

1946 Gäddkläckningsförsök på olika bottensubstrat. *Svensk Fiskeri Tidskrift*, **1**, 1-4.

1947 Zur Kenntnis der schwedischen Chironomiden. *Arkiv för Zoologi*, **39 A**, No. **3**, 1-95, 25 Taf.

1948 Über die Metamorphose der Sectio Tanytarsariae connectentes (Diptera, Chironomidae). *Arkiv för Zoologi*, **41 A**, No. **2**, 1-22, 7 Taf.

1948 *Microdota*-Studien (Col., Staphylinidae). *Entomologisk Tidskrift*, **69**, 8-66, 13 Taf.

1949 Chironomiden und andere Bodentiere der südschwedischen Urgebirgsseen. Ein Beitrag zur Kenntnis der bodenfaunistischen Charakterzüge schwedischer oligotropher Seen. *Report. Institute for Freshwater Research, Drottningholm*, **30**, 1-914.

1950 —. Autorreferat. *Archiv für Hydrobiologie*, **30**, 175-181.

1950 Myggor och Knott (Midges and gnats). *Svenska Djur*, 518-532.

1951 [Review] Thienemann, A., Verbreitungsgeschichte der Süsswassertierwelt Europas. Die Binnengewässer Bd. 18, Stuttgart 1950. *Oikos*, **3(2)**, 268-275.

1951 The relation of O_2-microstratification at the mud surface to the ecology of the profundal bottom fauna. *Report. Institute for Freshwater Research, Drottningholm*, **32**, 32-42.

1952 *Acrotona*-Studien (Gattung *Atheta*, Col. Staphylinidae). *Entomologisk Tidskrift*, **73**, 93-145.

1952 Zur Kenntnis der Taxonomie und Metamorphose der Chironomiden-Gattungen *Protanypus* Kieff., *Prodiamesa* Kieff. und *Monodiamesa* Kieff. *Report. Institute for Freshwater Research, Drottningholm*, **33**, 39-53.

1953 Die palaearktischen Arten der *Atheta*-Untergattung *Dimetrota* Muls. et Rey (Col., Staphylinidae). Eine systematische Studie. *Arkiv för Zoologi*, **(7) 5**, 369-434.

1954 Neue palaearktische Arten der Gattung *Atheta* C. G. Thoms. (Col. Staphylinidae). *Norsk Entomologisk Tidsskrift*, **9**, 1-17.

1955 Limnological study in western South America and Mexico of the hydrography, hydrochemistry, plankton and bottom fauna of the Andine lakes. *Yearbook. American Philosophical Society*, **1955**, 125-8.

1956 Zur Systematik der Orthocladiinae (Dipt. Chironomidae). *Report. Institute for Freshwater Research, Drottningholm*, **37**, 5-185.

1956 Die bodenfaunistischen Seetypen und ihre Anwendbarkeit auf die Südhalbkugel. *Report. Institute for Freshwater Research, Drottningholm*, **37**, 186-235.

1958 The bottom faunistical lake type system and its application to the southern hemisphere. Moreover a theory of glacial erosion as a factor of productivity in lakes and oceans. - *Internationale Vereinigung für Theoretische und Angewandte Limnologie Verhandlungen*, **13**, 288-97.

1960 In memoriam August Thienemann. *Entomologisk Tidskrift*, **81**, 144-6.

1961 August Thienemann in memoriam. Erbe und Vermächtnis August Thienemanns. *Oikos*, **12**, 310-9.

1962 [Review] Fittkau, E. J., Die Tanypodinae (Diptera, Chironomidae). Berlin 1962. *Entomologisk Tidskrift*, **83**, 242-4.

1962 Insects of Macquarie Island. Diptera, Chironomidae. *Pacific Insects*, **4**, 945-54.

1963 Limnic Diptera in their bearings on the problem of transantarctic faunal connections, in *Pacific Basin Biogeography* (ed. J. L. Gressitt), Bishop Museum, Honolulu, pp. 425-34.

1963 In memoriam Einar Wahlgren. *Entomologisk Tidskrift*, **84**, 313-4.

1965 On the real nature of transantarctic relationships. *Evolution*, **19**, 496-505.

1965 [Nekrolog] Richard Frey. *Entomologisk Tidskrift*, **86**, 68.

1965 [Nekrolog] Rolf Krogerus. *Entomologisk Tidskrift*, **86**, 69-70.

1966 Transantarctic relationships and their significance, as evidenced by chironomid midges. With a monograph of the subfamilies Podonominae, Aphroteniinae and the austral Heptagyiae. *Kunglica Svenska Vetenskapsakademiens Handlingar*, **11**, 1-472 + 30 plates.

1967 Håkan Lindberg - In memoriam. *Entomologisk Tidskrift*, **88**, 115.

1967 Insects and the problem of austral disjunctive distribution. *Annual Reviews of Entomology*, **12**, 149-68.

1968 Application of phylogenetic principles in systematics and evolutionary theory, in *Current Problems of Lower Vertebrate Phylogeny, Nobel Symposium 1967* (ed. T. Orvig , **4**, 473-95.

1969 Biogeografisk metodik och livets historia på sydkontinenterna. *Fauna och Flora*, **64**, 240-53.

1970 Diptera: Chironomidae of South Georgia. *Pacific Insects Monograph*, **23**, 276.

1970 Antarctic land faunas and their history, in *Antarctic Ecology*,(ed. M.J. Holdgate), Academic Press, London, pp. 41-53.

1972 Phylogenetics and biogeography. *Systematic Zoology*, **21**, 69-79.

1972 Evolution, causal biology, and classification. *Zoologica Scripta*, **1**, 107-20.

1973 Nutidens fauna speglar kontinenternas delning. *Forskning och Framsteg*, **1973**, 17-9.

1973 [Review of] Erwin, T. L., A Reclassification of Bombardier Beetles (Carabidae, Brachinidae). Quaestiones Entomologicae, 6, 4-215, 1970 and Whitehead, D. R., Classification, Phylogeny, and Zoogeography (Coleoptera: Carbidae: Scaritini). Quaestiones Entomologicae, 8, 131-348, 1972. *Systematic Zoology*, **22 (2)**, 203-6.

1974 Fifty years' limnic zoogeography. *Mitteilungen der Internationalen Vereinigung für Theoretische and Angewandte Limnologie*, **20**, 287-300.

1974 (Editor). *Proceedings of the 5th International Symposium on Chironomidae, Abisko, Swedish Lapland 7. - 9. August 1973. Entomologisk Tidskrift Supplement*, **95**, 1-232.

1975 Phylogenetische Biogeographie als Glied der Evolutionsbiologie. *Verhandlungen der Deutschen Zoologischen Gesellschaft*, **67**, 372-80.

1975 Circum-Antarctic distribution patterns and continental drift. Biogéographie et liaisons intercontinentales au cours du mésozoique. 17th International Congress of Zoology, Monaco 1972. *Mémoirs du Musée National d'Histoire Naturelles, A, Zoologie*, **88**, 19-28.

1976 A Neocomian chironomid and Podonominae-Aphroteniinae (Diptera) in the light of phylogenetics and biogeography. - *Zoologica Scripta*, **5**, 139-60.

1978 & O. A. Sæther. *Buchonomyia burmanica* sp. n. and Buchonomyiinae, a New Subfamily amaong the Chironomidae (Diptera). *Zoologica Scripta*, **7**, 269-75.

1979 Entomologiska Föreningen i Stockholm 1879-1979. *Entomologisk Tidskrift*, **100,** 111-20.

1980 Methods and principles of phylogenetic biogeography (abstract), in *Chironomidae - Ecology, Systematics, Cytology and Physiology* (ed. D. A. Murray). Proc. 7th International Symposium on Chironomidae, Dublin, August 1979, Pergamon Press, Oxford etc. pp. 91-92.

1981 Croizat's Panbiogeography versus Phylogenetic Biogeography, in *Vicariance Biogeography, A Critique* (eds. G. Nelson & D. E. Rosen), Columbia Univ. Press, New York, pp. 94-138 (+ response, 151-8).

1983 *Chilenomyia paradoxa* gen. n., sp. n. and Chilenomyiinae, a new subfamily among the Chironomidae (Diptera). *Entomologica Scandinavica*, **14,** 33-45.

1983 Two new aphrotenian larval types from Chile and Queensland, including *Anaphrotenia lacustris* n. gen., n. sp. (Diptera, Chironomidae). *Entomologica Scandinavica*, **14,** 415-33.

1983 The larvae of Podonominae (Diptera, Chironomidae) of the Holarctic region - Keys and diagnoses, *Entomologica Scandinavica Supplement*, **19,** 23-31.

1986 The pupae of Podonominae (Diptera: Chironomidae) of the Holarctic region - Keys and diagnoses, *Entomologica Scandinavica Supplement*, **28,** 19-30.

1986 Evolution by orderly stepwise subordination and largely nonrandom mutations. *Systematic Zoology*, **35,** 602-7.

1988 Evolutionsprocessen. Försök till en överblick. *Fauna Flora, Stockholm*, **83,** 97-106.

1988 Phylogenetic biogeography, in *Analytical Biogeography* (eds. A. A. Myers and P. S. Giller), Chapman & Hall, London and New York, pp. 343-69.

1989 Varför bli entomolog? *FaZett*, **1 (1988),** 2-6

1989 The adult males of Podonominae (Diptera, Chironomidae) of the Holarctic region - Keys and diagnoses. *Entomologica Scandinavica Supplement*, **34,** 23-36.

The Indicator Concept in Freshwater Biomonitoring

Richard K. Johnson

Department of Environmental Assessment,
Biodiversity Section,
Swedish University of Agricultural Science,
Box 7050, S-750 07 Uppsala, Sweden

Introduction

The indicator concept is an idea which is well founded, but seldom tested in the ecological literature. The idea that organisms reflect their environment is not new, in fact, some of the earliest writings mention observations between organisms and their environment. The Egyptians believed for example that frogs and toads arose from silt deposited by the flooding Nile, while the Greek philosopher Aristotle taught that worms were born of slime (Time Magazine, 11 October, 1993, p. 57). In more recent years, the use of the indicator approach, in particular the use of macroinvertebrates, has received considerable attention in biomonitoring programs (e.g. Rosenberg and Resh, 1993).

Ideally, reliable indicators are those taxa that have narrow and specific tolerances, while conversely, organisms that have wide tolerances are often less informative and poor indicators of environmental quality (Johnson *et al.*, 1993). Indicators may be used to understand the response, adaptation, and recovery of ecosystems and their components to natural and anthropogenic disturbances. In particular, ideal indicators should provide an indication of status and early-warning indications of change, either at the ecosystem, population, or genetic level, as well as provide insight into potential causal mechanisms. Stressed aquatic systems often show: (i) a reduction in species richness, with the disappearance of sensitive species, (ii) a predominance of pollution-tolerant species, and (iii) a change in the number of individuals within a species (i.e. communities may become monospecific) (Rapport, 1991). Schindler (1987) showed, for example, how stress can be inferred from changes in community structure, or more specifically, that changes in community composition to small and fast reproducing species are often the earliest indications of stressed aquatic communities. Biomonitoring of communities, with emphasis on taxonomic richness and composition, is considered one of the most sensitive tools for detecting alterations in aquatic ecosystems (Cairns and Pratt, 1993). Physiological and morphological abnormalities in fish and invertebrates may, however, precede species loss, thereby constituting an even earlier warning of stress (Rosenberg and Resh, 1993).

Although the concept that organisms may be used as indicators of their environment is basically a sound one and has proven fruitful over the years, the indicator concept is often abused. Problems with using the indicator concept have stemmed from the panacea-approach of selecting "universal" indicators, or the use of index/score systems in areas for which they were not designed. Other caveats in using the indicator approach include the lack of formalization of specific questions. Although numerous studies advocate the utility of the indicator approach, the degree of precision and level of error associated with the use of biotic indices in evaluating stress is poorly understood. Power, as discussed below, pertains to the ability to detect change, and is intimately associated with indicator variability and the magnitude of effect to be detected. In this paper I discuss a number of factors which confound and constrain the use of biological indicators in biomonitoring studies, as well as provide a number of suggestions to better formalize the use the indicator approach in aquatic studies.

A short historical review of the indicator concept in Europe

Benthic ecologists were pioneers in understanding the concept of how indicator assemblages relate to ecosystem quality, and during the last century, and in particular in the last two decades, a number of biotic indices have been constructed to evaluate stress. In natural communities the number of individuals belonging to each species often varies widely, usually with a few taxa being very numerous, while the majority are represented by few or even single individuals. This pattern is often expressed mathematically using log-normal curves of community structure, where stress often results in the loss of some species and the predominance of functionally similar taxa (e.g. deposit-feeders).

Building on this premise, the use of biotic index/score systems assumes that polluted sites generally contain fewer species than unimpacted sites, and that along a pollution gradient species will tend to be removed according to their relative susceptibility to the pollutant. Basically, three types of indices are used: (i) indices which incorporate several indicator levels, such as periphyton and invertebrates, into a composite community assessment (e.g. the Saprobien index), (ii) indices which consist of a single indicator group such as macroinvertebrates or fish, but consist of both information on species diversity and ecological tolerance of indicator taxa, and (iii) indices which are solely mathematical expressions of community structure (based on taxa richness, evenness and abundance). Diversity indices have been frequently used in North America, while saprobic and biotic score systems are more commonly used in Europe (Cairns and Pratt, 1993). Reviews of biological indicator systems presently being used for lotic systems in Europe were recently given by Metcalfe (1989) and De Pauw *et al.* (1992). Given below are a selected number of examples to illustrate the evolution of the indicator concept in aquatic sciences.

Lotic systems

The now classic work of Kolkwitz and Marsson (1909) is considered as the first formalized use of biological indicators in evaluating pollution stress (Fig. 1). These authors developed the idea of saprobity (saprobes are organisms of decomposition) in rivers as a measure of the degree of contamination by organic-sewage matter, and the resulting changes in community structure. The Saprobien system approach has undergone several, often simplifying, modifications over the years, and is mainly used in Central Europe. For example, a version of this index was recently accepted as a standardized method for biological assessment of lotic systems in Germany (Friedrich, 1990). Friedrich (1990) modified the Saprobien index by revising the list of indicator taxa to include only relatively common benthic, heterotrophic organisms, some 90 microorganisms and 157 macroinvertebrate taxa for which taxonomic keys are readily available, and including an indicator weight which was determined for each species.

Originating in part from the Saprobien system, the Trent Biotic Index (TBI) (Woodiwiss, 1964) was developed for assessing pollution effects in the Trent River in England in the late 1950's, and appears to the be the origin of most modern biotic indices (Metcalfe, 1989). The index is qualitative, as macroinvertebrates are collected using a handnet from all available habitats and the samples are pooled. TBI is based on the sensitivity of six groups of key organisms, with rankings based on their sensitivity to oxygen concentrations. For example, the presence of Plecoptera and Ephemeroptera taxa, which need higher levels of oxygen, results in higher scores than the presence of *Tubifex* and *Chironomus*. Scores range from 0 for gross

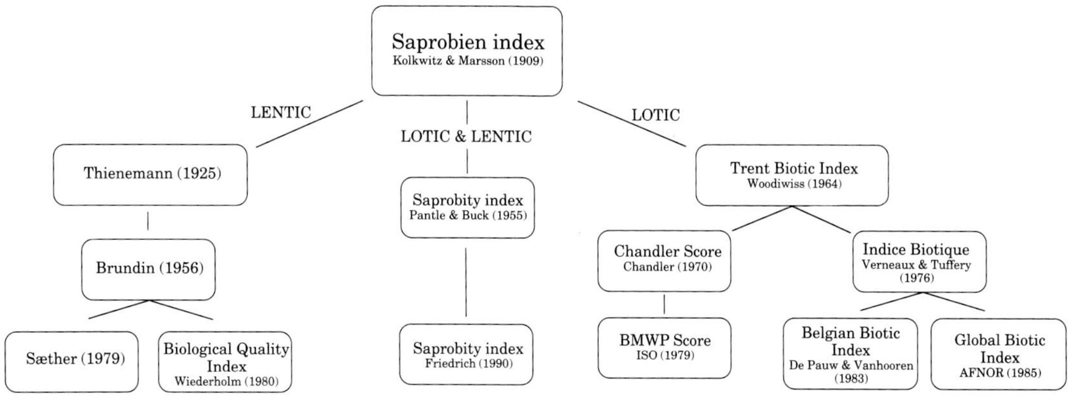

Figure 1. Schematic diagram of a selected number of biotic index/score systems showing the historical development of the indicator concept in Europe.

organic pollution where none of the indicator taxa are present to 10 for unpolluted conditions where macroinvertebrate taxa indicative of clean water, such as Plecoptera, are present.

The Trent Biotic Index has been modified and used extensively in Europe (Metcalfe, 1989). For example, the Biological Monitoring Working Party Score (BMWP), a modification of the TBI, has been developed for use in national river surveys in the UK (e.g. ISO 179). Relying on qualitative handnet sampling, the system consists of three relatively simple steps: (i) all macroinvertebrate families which are collected from a site are listed, (ii) scores are allocated to the families according to their known levels of tolerance, and (iii) the scores for all the families present are summed to give a cumulative site score. Recent work in the UK has dealt with predicting BMWP site scores using predictive modelling of the probability of occurrence of taxa (e.g. Wright *et al.*, 1984 and see below).

One of the major shortcomings of the Trent Biotic Index is that macroinvertebrate abundance is ignored. Not surprisingly, several modifications of TBI include the relative abundance of indicator taxa. For example, the Chandler Biotic Score which was developed for use in Scotland (Chandler, 1970), the Global Biotic Index which was recommended as a standardized method in France (AFNOR, 1985), and the Belgian Biotic Index (De Pauw and Vanhooren, 1983) all use semi-quantitative or quantitative methods of sampling. Other modifications include the number of taxa and level of taxonomic identification used in the assessment, and the tolerance or indicator weights given to the taxa.

An EC working group recently summarized the use of biological index/score methods being used in the assessment of lotic systems in Europe (De Pauw *et al.*, 1992). Eleven countries are included in the review, and nine different indicator methods are presently being used in national and regional monitoring programs (Table 1). Seven countries (Belgium, Denmark, France, Germany, Ireland, Luxemburg, and the UK) have accepted biotic index/score methods as a national standard, and in two additional countries (Italy and Netherlands) assessment methods have been standardized in regional monitoring programs. Sampling is often qualitative or semi-quantitative using a handnet. Although a few assessment systems require species-level identification, there appears to be a tendency for quick assessment techniques using genus- and/or family-level identifications.

Table 1. Biological index/score methods used in the assessment of lotic systems in Europe. Modified from De Pauw et al. (1992).

Country	Method	Sampling	Lowest taxonomic level	Standard
Belgium	Belgian Biotic Index	Qualitative	Genus	National
Denmark	Saprobic Index	Qualitative	Species	National
France	Global Biotic Index	Quantitative	Family	National
Germany	Biol. Effect. Org. Load and Saprobic Index	Qualitative	Species	National
Ireland	Quality Rating System	Qualitative	Species	National
Italy	Extended Biotic Index	Qualitative	Genus	Regional
Luxemburg	Biotic Index	Qualitative	Family	National
Netherlands	Quality Index	Qualitative	Species	Regional
Portugal	Belgian Biotic Index	Qualitative	Genus	-
Spain	Biological Monitoring Working Party (modified)	Qualitative	Family	-
United Kingdom	Biological Monitoring Working Party	Qualitative	Family	National

Lentic systems

Classification methods used in the assessment of lentic systems are also derived from the Saprobien system of Kolkwitz and Marsson (1909). Analogous to the use of the Saprobien system to classify the degree of organic pollution in lotic systems, profundal communities, in particular chironomid species assemblages, have long been used for determining the degree of pollution and trophic classification (eutrophication) of temperate lakes (Johnson *et al.*, 1993). Nutrient enrichment and eutrophication of lakes often results in predictable changes in profundal communities. The development of lake classification schemes using benthic macroinvertebrates is thoroughly summarized in Brinkhurst (1974). In brief, Thienemann (1913) was the first to introduce the concept of using chironomid taxa for classifying European lakes. Originally two lake types were classified; relatively shallow "Baltic" lakes with low hypolimnion oxygen concentrations and dominated by *Chironomus* larvae, and deeper, colder "alpine" lakes with relatively high hypolimnion oxygen concentrations and dominated by *Tanytarsus* larvae. This classification system was later revised to include oligotrophic, eutrophic, and dystrophic lake types (Thienemann, 1925). Brundin (1949) stressed the importance of the geographical distribution of chironomid taxa in lake classification schemes. Four lake types were given in his classification system, and emphasis was placed on indicator assemblages as opposed to single taxa. Lake types ranged from ultraoligotrophic with *Heterotrissocladius subpilosus* (Kieff.) as indicator taxon to strongly eutrophic systems with *Chironomus plumosus* L. and dystrophic lakes having *Chironomus tenuistylus* Brundin (Brundin, 1956). Brundin (1956) also demonstrated the large-scale applicability of the chironomid classification system in areas outside of Europe. A multivariate approach, using classification/ordination of profundal chironomid communities of 45 Swedish lakes, tended to support these more subjective classification of lake types (Johnson, 1989).

More recently, the Thienemann (1925) - Brundin (1956) classification system has been modified by Wiederholm (1980a,b) and Sæther (1979). These authors proposed modifications primarily for developing classification systems suited for use in biomonitoring studies. The Biological Quality Index (BQI) of Wiederholm (1980a) consists of five indicator species, with values ranging from a score of one for eutrophic lakes with exclusively *Ch. plumosus* larvae, to a score of five for oligotrophic lakes with exclusively *H. subpilosus* larvae; a site is scored with a zero if no indicator taxa are present. Sæther (1979) criticized the use of simplified models in describing trophic levels, and proposed a lake classification system for Palaearctic and Nearctic communities with 15 lake types based on the presence of chironomid indicator taxa. However, Aagaard (1986) using a data set of 50 Norwegian lakes found a strong correlation between the two classification systems, and a number of studies have shown the value of these methods for classifying lake type (e.g. Kansanen *et al.*, 1984; Meriläinen, 1987; Gerstmeier, 1989; Meriläinen and Hamina, 1993).

FACTORS THAT CONFOUND THE USE OF BIOTIC INDICES IN BIOMONITORING STUDIES

The ability to detect change using biological indicators is often confounded and constrained by a number of factors, two of the most common are probably the lack of addressing well-formalized questions (i.e. concise statements of what is being tested) and weaknesses in study design. The use of indicators in biomonitoring often entails a comparison of an observed index/score value with an expected (reference) value. Thus, strictly speaking, the use of the indicator concept in biomonitoring studies involves the testing of a null (no change) hypothesis (H_0). Although often some level of significance (α) is reported in which H_0 is rejected, researchers seldom determine the probability of Type II (β) errors and statistical power, even though knowledge of statistical power may be critically important in interpreting the data. For example, assume that a biomonitoring program is aimed at determining the effects of acidification (or liming) on benthic communities. A number of samples (sites and macroinvertebrate indicator metrics) are taken at two or more sites or time intervals and the null hypothesis, that "no differences exist in indicator metrics between sites or times" is tested. A statistical analysis of the data shows no difference between population means, i.e. H_0 is not rejected. The next question to address is whether or not the design was robust or powerful enough to find a difference if a difference indeed exists (e.g. Cohen, 1988; Lipsey, 1990). Here is where the use of power analysis comes into play.

Power is simply the probability that a statistical test will reject the null hypothesis when H_0 is, in fact, false. When a statistical analysis fails to reject the null hypothesis decisions are often based on the assumption that H_0 is true. However, the conclusion that H_0 is true is not justified unless the probability of making a Type II error is sufficiently low. Although for historical reasons ecologists are often more concerned with Type I errors (conventionally set at 0.05), in biomonitoring studies it may be considered more prudent to err on the side of false positives, in other words, to accept a greater probability of Type I errors (i.e. the detection of change when no change has occurred), than to not detect change when change, in fact, occurs (e.g. Underwood, 1993).

A number of factors affect study design power, most notably: (i) the α level selected, (ii) statistical assumptions, (iii) sample size and variability, and (iv) the magnitude of effect to be detected (Lipsey, 1990). The probability of Type I errors (α level selected) and statistical assumptions are two factors which are often relatively easy to manipulate to increase study design power. For example, increasing α from 0.05 to 0.1 greatly increases the power to detect differences between means, and selecting parametric tests using graduated scores, over less powerful nonparametric tests which use categorical data, and one-tailed directional tests as opposed to two-tailed nondirectional ones, increase power.

Factors which are more difficult to manipulate, but which should be carefully considered in the study design phase are effect size and indicator variability. For example, the first question to be addressed is what magnitude of effect will one be able to detect with the selected indicator, or simply, how reliable or valid is the selected indicator to the stressor? An indicator must be a valid measure of the change to be detected, hence taxonomic resolution is a frequently discussed topic. Secondly, how sensitive is the selected indicator to natural variability? Ideally, indicators should have low within-site variability but high among-site variability, i.e. signal to noise ratios should be high.

Taxonomic resolution

One of the major debates concerning the use of benthic macroinvertebrates as indicator organisms for water quality is taxonomic resolution (see discussion in

Resh and McElravy, 1993). Though the appropriate level of identification will depend on the question being asked and the aims and objectives of the study, it is often considered that the *species* is the appropriate level of taxonomic resolution for most biomonitoring and conservation studies.

In situations where a gross pollution gradient exists, generic or higher levels of taxonomic identification may be adequate for identifying pollution-induced changes (Woodiwiss, 1964), while conversely species-level identifications may be necessary at moderate pollution levels. Family-level identification may, for example, be adequate for monitoring the improvement of relatively stressed systems, as the family will be scored as present when one or more taxon of the family is recorded. However, if the goals of the monitoring program include the assessment of relatively pristine sites, much degradation (loss of taxa richness) may occur before a family is scored as absent from the site. Moreover, although gross pollution may affect whole taxonomic groups of macroinvertebrates (Hynes, 1960), intermediate pollutant levels often only affect certain species, and much overlap in tolerance occurs at the generic-level of taxonomic resolution. For example, of the 89 genera of macroinvertebrates with known water quality tolerances to organic pollution that were reviewed by Resh and Unzicker (1975), 61 had species that fit into more than one general tolerance category. Hence, biomonitoring studies may often require species-level or, in some instances, even subspecies-level identifications (Resh, 1979).

Indicator variability

The objectives of many biomonitoring studies are satisfied by the collection of replicate samples which are then analyzed to determine if mean values of a particular metric differ significantly between sites or times. Not surprisingly, one of the most frequently discussed topics in biomonitoring studies pertains to the number of replicate samples that should be collected to reach a desired level of precision. Although estimating the number of replicate samples needed to detect a given difference between mean values is related to precision, this topic has received little attention (cf. Resh and McElravy, 1993). Similarly, the spatial and temporal variability of biotic indices, as well as the degree of precision and level of error associated with the use of biotic indices is poorly understood.

Indicator variability may be broken down into component parts such as: (i) measurement error, e.g. sampling precision, (ii) within-site habitat and temporal (within-a-year and among-year) variability, and (iii) among-site variability (Johnson *et al.*, unpubl.). As mentioned above, an indicator should exhibit low within-site variability but high among-site variability; within-site variability may be considered as noise and should be reduced as much as possible. Habitat and temporal stratification, and the selection of robust indicators are common techniques used to minimize within-site variability while maximizing among-site variability.

Estimates of indicator variability and precision

Six commonly used benthic indices were selected to evaluate indicator variability across various spatial and temporal scales in Swedish reference lakes. Twenty-six, ecoregion-representative reference lakes are situated from the deciduous forests of southern Sweden to the alpine regions of the north (Johnson *et al.*, 1990). For profundal and sublittoral habitats, total density, taxa richness, Shannon diversity (Shannon, 1948), a modified Biological Quality Index (BQI2), and the oligochaete/chironomid index (Wiederholm, 1980a,b) were selected, whereas for littoral habitats the number of Ephemeroptera taxa was used in place of BQI2. Estimates of indicator variability were calculated using macroinvertebrate samples collected from Swedish reference lakes over 5-years, 1988 to 1992. Five replicate samples were randomly taken using an Ekman sampler from quadrats (ca. 200 x 200 m) situated over the profundal and sublittoral, and six samples were taken from the wind-exposed littoral of each lake by standardized kick-net sampling. To minimize the effects of acid-stress on indicator variability, only lakes with a pH > 5.5 were used in the analyses (n = 17). Following the approach of Resh and McElravy (1993), the percent difference which can be detected between two mean values was calculated for these selected benthic indices.

Indicator precision and risk of error were calculated using an algorithm for determining the number of replicate sample units needed to detect a difference between means (Sokal and Rohlf 1981) as:

$$n \geq 2\left(\frac{\sigma}{\delta}\right)^2 \left\{t_{\alpha[v]} + t_{2(1-P)[v]}\right\}^2$$

where n= the number of replicates; σ = the true standard deviation; δ = the smallest true difference that is desired to detect; ν = the degree of freedom of the sample standard deviation; α = significance level; P = the desired probability that a difference will be found to be significant; $t_{\alpha[\nu]}$ and $t_{2(1-P)[\nu]}$ = values from a two tailed t-table with n degrees of freedom and corresponding to probabilities of α and 2(1 - P), respectively. This procedure uses a Students *t*, and requires a priori knowledge of the standard deviation or the coefficient of variation (CV = standard deviation ÷ mean) of the indicator variable. The user selects the degree of difference in the means to be detected (δ) and the significance for Type I (α) and Type II (β or 1 - P) errors.

Number of replicate samples

The effect of sample size on the ability to detect change in mean indicator values was estimated by using median indicator variability (CV) of habitat stratified samples collected from Swedish reference lakes in 1992. In calculating the percent difference which could be detected between mean values, the number of systems being tested was kept constant at two, while the number of replicate sample units varied from 2 to 10, and α and β levels were set at 0.05 and 0.20, respectively.

The percent difference in mean values (PDIM) which could be detected with the selected biotic indices decreased markedly from two to three replicate samples, and above five replicates little reduction in the minimum detectable differences occurred (Fig. 2). Of the six indicators studied, BQI2, Shannon diversity, number of taxa, and number of Ephemeroptera taxa often showed the lowest minimum detectable differences. At five replicate sample units, coefficients of variation of the selected indices ranged from 0 % for BQI2 to 50 % for macroinvertebrate density of profundal habitats, from 1.5 % for BQI2 to 55 % for macroinvertebrate density of sublittoral habitats, and from 12 % for Shannon diversity and to 45 % for O:C index of littoral habitats. These CVs resulted in PDIMs for profundal habitats of < 1 % for BQI2 to 102 % for density, for sublittoral habitats from 3 % for BQI2 to 112 % for density, and for littoral habitats from 23 % for Shannon diversity to 91 % for O:C index.

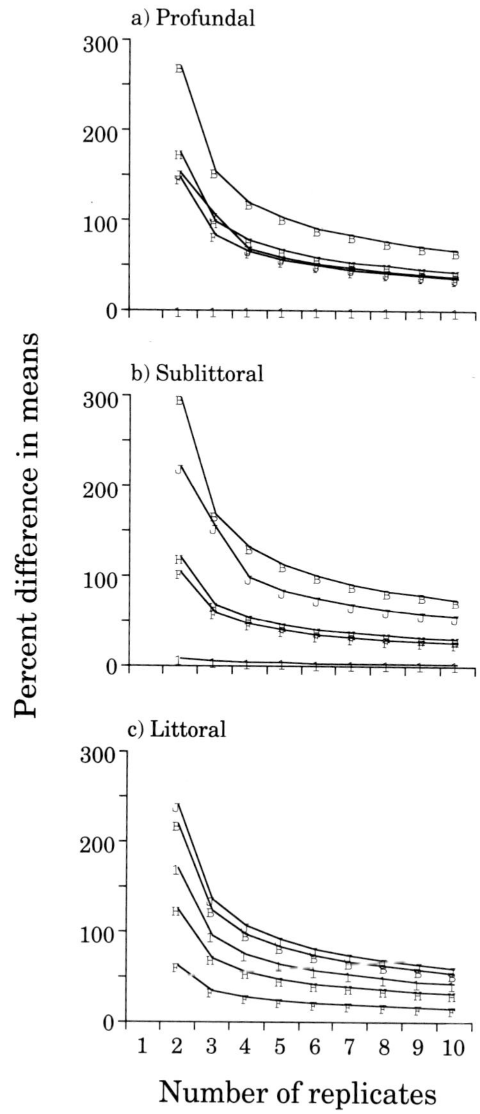

Figure 2. The effect of sample size on the difference in means that can be detected using data from Swedish reference lakes (see text). PDIMs were calculated from the equation in Sokal and Rohlf (1981) for finding the sample size required for a test. Median values of indicator variability (coefficients of variation) were used for three lake habitats, profundal, sublittoral, and littoral. Total macroinvertebrate density (■, taxa richness (▲), Shannon diversity (◆), oligochaete:chironomid ratio (O:C) (●), and BQI2 (profundal and sublittoral habitats) or number of Ephemeroptera taxa (littoral) (×).

Within-a-year variability

Seasonal indicator variability was calculated using median, minimum and maximum indicator CVs of zoobenthos collected 4x annually in 1988; spring (2-weeks following ice-out), early (June) and late (August) summer, and autumn (October - November) sampling (Johnson *et al.*, 1989). Sublittoral habitats showed the lowest seasonal variability, with median indicator variability ranging from 2.4 % for BQI2 to 29 % for O:C (Fig. 3). Profundal and littoral habitats showed higher levels of indicator variability. Median coefficients of variation ranged from 3.8 % (BQI2) to 52 % (O:C) for profundal habitats and from 9.7 % (Shannon diversity) to 44 % (O:C) for littoral habitats. The lower variability found for benthic indices of sublittoral habitats resulted in a greater ability of detecting change between sample means (Table 2). For example, three benthic indices of sublittoral habitats are able to detect differences between mean values of < 50 %, namely, BQI2 (4.9 %), Shannon diversity (12 %) and number of taxa (30 %). Conversely, for profundal habitats, only BQI2 (7.7 %) and Shannon diversity (48 %) were able to detect differences in means values of < 50 %. For littoral habitats, Shannon diversity (20 %) and number of taxa (41 %) were able to detect differences in means values of < 50 %.

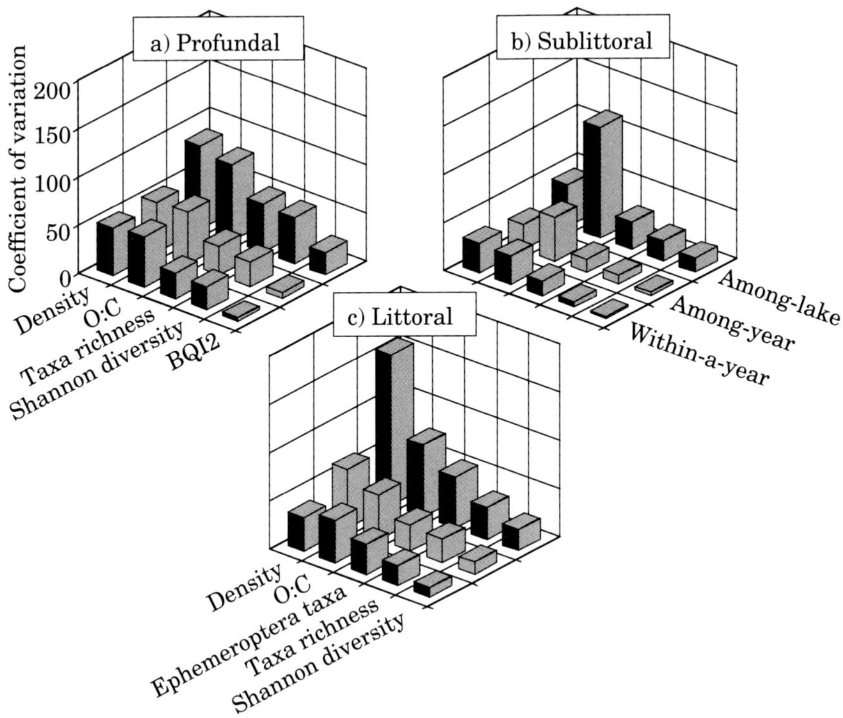

Figure 3. Spatial and temporal variability of selected biotic indices in Swedish reference lakes. Within-a-year variability (coefficients of variation) were calculated for samples collected in 17 lakes during 1988 (see text), among-year CVs were calculated using average lake values for 5-yr (1988 - 1992), and among-lake CVs were calculated using data from 55 (profundal) and 26 (sublittoral and littoral) Swedish lakes.

Table 2. Percent differences between two means (PDIM) detectable with a 95 % chance of being correct at a 5 % level of significance, for six benthic measures and three habitats. PDIMs were calculated using an ANOVA design (Sokal and Rohlf 1981), with five (profundal and sublittoral) or six (littoral) replicate samples, and median, maximum, and minimum coefficients of variability for within-a-lake seasonal and among-year and among-lake indicator variability. Note density values have not been transformed.

	Density	Taxa richness	Shannon diversity	O:C	BQI2	Number of Ephemeroptera taxa
Within-a-year						
Profundal						
Median	99	53	48	105	7.7	
Maximum	31	113	212	238	143	
Minimum	8.7	< 1	15.6	< 1	< 1	
Sublittoral						
Median	63	30	12	59	4.9	
Maximum	236	103	68	249	41	
Minimum	5.7	< 1	< 1	< 1	< 1	
Littoral						
Median	71	41	20	89		63
Maximum	204	92	79	137		164
Minimum	38	20	6.6	< 1		< 1
Among-year						
Profundal						
Median	105	61	52	109	13	
Maximum	208	101	129	254	61	
Minimum	52	34	17	32	< 1	
Sublittoral						
Median	53	28	18	91	6.8	
Maximum	178	108	47	317	20	
Minimum	5.5	6.8	3.4	13	1.7	
Littoral						
Median	123	49	26	95		55
Maximum	281	78	60	162		143
Minimum	36	< 1	8.6	95		55
Among-lake						
Profundal						
Median	175	101	97	159	47	
Sublittoral						
Median	87	57	41	232	30	
Littoral						
Median	315	67	44	150		108

Among-year variability

The effect of increasing the number of years on the ability of the analysis to detect changes in means was examined by using sample sizes of five (profundal and sublittoral habitats) or six (littoral habitats) replicate samples, but using the median indicator variability over a 5 year (1988 to 1992) period. Median values of indicator variability were often only slightly higher among years compared with within-a-year, seasonal values for

1988 alone. Thus increasing the number of study years resulted in little improvement in the ability of detecting changes between mean index values. Similar to within-a-year variability, sublittoral habitats showed the lowest among-year indicator variability, resulting in higher resolution of detecting significant differences between mean indicator values. In sublittoral habitats, BQI2, Shannon diversity, and number of taxa had CVs of 3.4 %, 8.9 %, and 14 %, resulting in PDIMs of 6.8 %, 18 % and 28 %, respectively. Of the indices calculated using profundal and littoral macroinvertebrates, only BQI2 (13 %) for profundal habitats and Shannon diversity (26 %) and number of taxa (49 %) for littoral habitats were able to detect less than 50 % changes in means.

Among-lake variability

Among-lake variability was calculated from a 55 lake data set for profundal habitats, and 26 reference lakes for sublittoral and littoral habitats. In calculating among-lake indicator variability, sample size was five for profundal and sublittoral habitats and six for littoral habitats. Median indicator variability was used to calculate the percent difference which could be detected between mean values of selected indices. Among-lake indicator variability was greater than within-lake seasonal and among-year variability. Coefficients of variation ranged from 23 % (BQI2) to 87 % (density) for profundal habitats, from 15 % (BQI2) to 115 % (O:C) for sublittoral habitats, and from 33 % (Shannon diversity) to 156 % (density) for littoral habitats. The lower indicator variability found for sublittoral habitats resulted in a greater detection of differences between mean values between lakes. PDIMs of less than 50 % were found for BQI2 (30 %) and Shannon diversity (41 %) indices, whereas for profundal and littoral habitats, only BQI2 (47 %) of profundal habitats and Shannon diversity (44 %) of littoral habitats were able to detect differences of less than 50 %.

As shown above, two aspects of study design, namely, the number of replicate samples taken and stratification of sample units by habitat resulted in increased power. Although the number of replicate samples needed to detect significant differences between mean values varied greatly between two and ten sample units, after five replicate samples little reduction occurred. Stratifying sampling within well-defined habitats, on the other hand, greatly reduced within-site indicator variability, resulting in increased power (lower PDIMs). Conversely, increasing the number of sample years from one to five had little influence on indicator variability and study design power. These findings show that careful consideration should be given in the study design phase to methods which increase indicator signal to noise ratios or power to detect change.

Indicator change in stressed versus nonstressed systems

The magnitude of effect which can be detected using benthic indices is clearly dependent on the variability of the indicator used (Table 2). If minimum values of indicator coefficients of variation are substituted for median values, a greater detection in differences between mean values is possible, conversely, if maximum CV values are used the ability to detect a significant change decreases markedly. However, as stressed systems often showed increased variability, benthic indicator CVs will most probably increase with moderate stress, confounding the detection of significant differences between mean values. Here it is important to ask: What is the effect size in mean indicator values that is expected to occur with impact?

For several of the benthic indices reviewed above it was estimated that a change in mean value of 50 % could be detected with five replicate samples. To estimate the degree of change (effect size) of benthic indices in stressed systems, indicator values of nonstressed Swedish reference lakes (lakes with pH > 5.5, n = 17) were compared with acid-stressed (pH < 5.5, n = 9) lakes (Fig. 4). No difference was found for macroinvertebrate densities (ind. m^{-2}) of pH stressed lakes compared to lakes with pH > 5.5 (log density was tested by a two-tailed ANOVA, P > 0.05). Profundal and sublittoral densities of pH stressed lakes averaged (± 1 SD) 3579 ± 4009 and 2358 ± 1695, respectively, compared with 1025 ± 889 and 1902 ± 822 for lakes with pH > 5.5. The somewhat higher, but nonsignificant, densities of pH stressed lakes are due to the high numbers of *Chaoborus flavicans* Meigen and *Chironomus* spp. taxa found in these acid-stressed lakes.

Several benthic indices were, however, significantly lower in acid-stressed lakes compared with nonstressed lakes. For example, taxa richness and Shannon diversity of sublittoral and littoral habitats were significantly lower in pH stressed lakes. For sublittoral

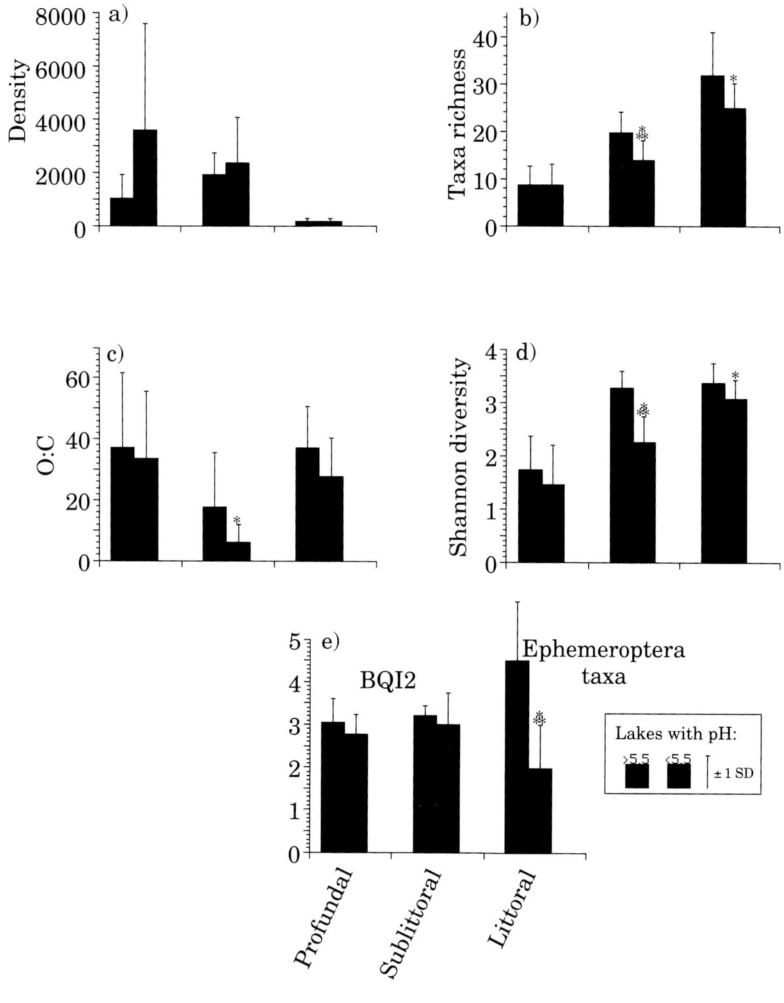

Figure 4. Indicator values for acid-stressed (pH < 5.5, n = 9 lakes, columns to the left) and nonstressed (pH > 5.5, n = 17 lakes, columns to the right) Swedish reference lakes. Mean (± 1 standard deviation) values for the years 1988 to 1992. Group differences were tested by ANOVA. * P < 0.05, ** P < 0.01, *** P < 0.001.

habitats, taxa richness decreased by 30 %; from 20 ± 4.4 taxa in lakes with pH > 5.5 to 14 ± 4.1 taxa in pH stressed lakes (ANOVA, P < 0.001). Similarly, the number of taxa found in littoral habitats of pH stressed lakes (25 ± 5.1) was lower than the number of taxa in lakes with pH > 5.5 (32 ± 9.2) (ANOVA, P < 0.05). Shannon diversity decreased by 30 % in sublittoral (ANOVA, P < 0.0001) and 9 % in littoral (ANOVA, P < 0.05) habitats. For sublittoral habitats diversity ranged from 3.26 ± 0.324 to 2.25 ± 0.492 and for littoral from 3.36 ± 0.37 to 3.06 ± 0.35, for nonstressed and acid-stressed lakes, respectively. For littoral habitats, the number of Ephemeroptera taxa varied greatly between nonstressed and acid-stressed lakes (4.5 ± 1.37 and 2.0 ± 1.01, respectively, ANOVA, P < 0.0001), a decrease of 56 %.

These findings lend support to the power analysis performed above. For example, indicator variability was shown to be lowest for sublittoral habitats, indicating a relatively high power of detection of change. Of the five indices examined for sublittoral habitats, three showed significantly lower values in acid-stressed compared with nonstressed reference lakes. BQI2, which often showed the lowest variability both spatial and temporally, did not decrease significantly in acid-stressed compared with nonstressed lakes. However, this finding is not too surprising, as the index was constructed for assessing the effects of nutrient enrichment on lentic systems, and several of the indicator organisms used in the index are not considered as indicators of acidification stress.

Predictive modelling

Sample size and indicator variability are two factors which have been shown to strongly affect study design power. Power may also be further increased by using a predictive approach, where indicator metrics are individual taxa, as opposed to more simplified and often less informative biotic indices. Biotic index/score systems are usually pollution-specific since they are often based on the presence/absence or abundance of indicator organisms, and as reviewed above, the use of biotic indices to monitor water quality changes when causal mechanisms of the disturbance are known, or fairly well-understood has proved fruitful.

It is obvious, however, that before a biotic index can be developed for a specific type of pollution, considerable knowledge concerning how a pollutant affects the biota is necessary. When species-specific sensitivities or tolerances are not available, or when knowledge of the stressor is poorly understood, the detection of stress may be best approached using a less-focused strategy. A multivariate approach is often selected for understanding complex systems, where confounding factors may be obscuring pollution-induced effects. Examples are given below for the prediction of taxa changes along relatively well-known natural and pollution gradients. The examples given below are by no means a comprehensive review of multivariate statistical procedures, but a few examples to show the utility of using these approaches in biomonitoring studies.

Multivariate models

Although linear or multiple regression models are often successfully used to predict macroinvertebrate densities or biomass, a shortcoming of these approaches is that they can be very time consuming. For example, the prediction of taxon occurrences using these procedures may be very tedious if a large number of taxa are involved. Furthermore, when a number of gradients strongly influence the presence or abundance of indicator organisms, a multivariate approach is often necessary to illucidate anthropogenic-induced effects. In contrast to univariate analyses, multivariate procedures often consider each species to be a variable and the presence/absence or abundance of each species to be a result of spatial or temporal explanatory factors. Hence, subtle changes in species composition or abundance across sites or times can be more readily detected using multivariate methods (Norris and Georges 1993). It is not too surprising that as environmental problems often involve multiple variables or gradients, the use of multivariate techniques in environmental monitoring studies is increasing (Johnson *et al.*, 1993).

Model construction

Wright *et al.* (1988, 1989) recently presented a novel approach using discriminant function analysis (DFA) and predictive models of benthic macroinvertebrates as indicators in biomonitoring. RIVPACS (River InVertebrate Prediction And Classification System) is a program which uses the probability of capture of taxa, predicting the occurrence of macroinvertebrate fauna at a given site from a small number of environmental variables (Moss *et al.*, 1987). Group classification is performed using TWINSPAN (Two-Way INdicator SPecies ANalysis, Hill, 1979) of stream macroinvertebrate communities, and then using environmental data, multiple discriminant analysis is used to develop predictive models which discriminate between groups of sites. The probability that new sites are members of established groups is calculated, and then subsequently the probability of taxa occurrence at a new site is established.

Adopting this approach, classification and ordination techniques and discriminant function analysis were used to develop models to predict taxa occurrence in Swedish reference lakes (Johnson *et al.* unpubl.). Briefly, predictive models were constructed as follows: (i) TWINSPAN classification of Swedish reference lakes was used to group lakes by profundal benthic macroinvertebrate community types, (ii) stepwise partial, constrained CCA (canonical correspondence analysis, ter Braak, 1989) ordination was used to select a number of explanatory variables, and (iii) DFA was used to develop models to predict taxa occurrence in Swedish lakes (see Moss *et al.*, 1987 for explanations and the algorithms used).

Classification

TWINSPAN is a polythetic, divisive, hierarchical classification technique that is frequently used to analyze benthic macroinvertebrate data (Norris and Georges, 1993). The TWINSPAN program classifies both sites (e.g. lakes/streams) and taxa, constructing an ordered two-way table to exhibit the relationship between sites and taxa as clearly as possible. A partitioning of the data is performed by reciprocal averaging, and the taxa quantities are weighted so as to emphasize the influence of "indicator taxa".

TWINSPAN classification of mean autumn densities (October or November) of profundal macroinvertebrate communities of 26 Swedish reference lakes sampled from 1988 to 1992 resulted in four terminal lake groups (Fig. 5) (Johnson *et al.*, unpubl.). Group 1, consisting of four lakes, was formed at the first TWINSPAN division. Indicator taxa for this lake group consisted of taxa indicative of oligotrophic conditions, e.g. *Micropsectra* spp., *H. subpilosus*, *Paracladopelma* spp. and *Spirosperma ferox* (Eisen). This group consists of relatively large (surface area = 2.5 ± 2.17 km^2), deep (depth 29.7 ± 9.2 m), well-buffered (pH = 6.48 ± 0.793) and clear-water (absf = 0.044 ± 0.037) lakes (mean ± 1 SD).

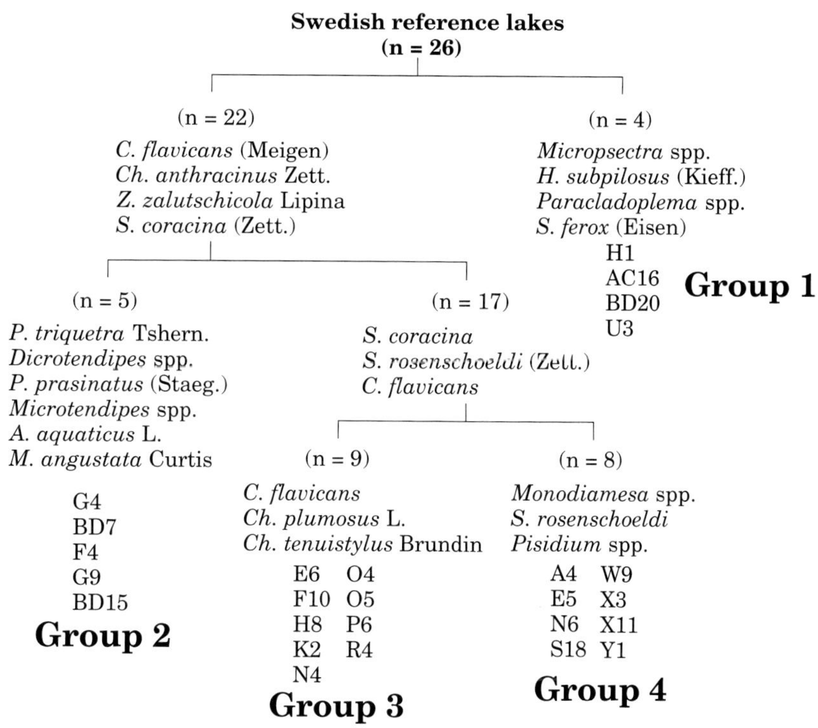

Figure 5. TWINSPAN classification of profundal macroinvertebrates from 26 Swedish reference lakes. Overall mean autumn densities for the years 1988 to 1992 were used in lake classification. Taxa given are "indicator taxa", i.e. strong preferential taxa for the lake groups, and lakes are given as a lake code.

At TWINSPAN division two, the remaining 22 lakes formed lake group 2, and subsequently groups 3 and 4 at division three. Lake group 2 (n = 5) had indicators of relatively shallow lakes such as the chironomids *Paratrichocladius triquetra* Tshern., *Dicrotendipes* spp., *Pseudochironomus prasinatus* (Staeg.) and *Microtendipes* spp., and *Asellus aquaticus* L. and *Molanna angustata* Curtis. Lakes of this group are relatively shallow (depth = 5.4 ± 0.38 m, range 4.9 - 5.74) and dystrophic (absf = 0.114 ± 0.028). Lake group 3 (n = 9) had *C. flavicans*, *Ch. plumosus* and *Ch. tenuistylus* as indicator species, and consisted of relatively small (surface area = 0.32 ± 0.389 km^2), acidic to well-buffered (mean pH = 5.57 ± 0.682, range 4.42 - 6.72) and clear to brown-water (absf = 0.111 ± 0.089, range = 0.008 - 2.52) systems. Similarly, lakes of group 4 (n = 8) are relatively small (surface area = 0.85 ± 0.588km^2), however, lakes of this group are relatively well-buffered (pH = 6.48 ± 0.226, range 6.2 - 6.76). Indicator taxa for this lake group included *Monodiamesa* spp., *Stictochironomus rosenschoeldi* (Zett.) and *Pisidium* spp. Hence, profundal macroinvertebrate communities appear to discriminate adequately among lake types.

Ordination

Similar to classification, ordination procedures are frequently used to reduce the complexity of data sets by representing site and taxa relationships in a low-dimensional space (Gauch, 1982). However, in ordination procedures taxa and sites are often fitted together with explanatory variables to exhibit relationships. Indirect and direct gradient analysis are two forms of ordination; see Norris and Georges (1993) for a recent review of the use of ordination procedures in biomonitoring studies using macroinvertebrates. In the former, axes are constructed from the variation of the total community data, and then the variation is subsequently interpreted in terms of environmental gradients. Conversely, in direct gradient analysis, taxa abundance or the probability of occurrence is described directly as a function of environmental variables.

Canonical correspondence analysis is a form of direct gradient analysis which simultaneously shows the relationship between taxa/sites and environmental gradients. The ordination axes are derived as linear combinations of environmental variables, and individual taxa are related to these axes assuming a unimodal response. Arrows in the ordination depict the direction of change of environmental variables in the ordination, with their length being proportional to their influence.

Profundal macroinvertebrate communities of 26 Swedish reference lakes were analyzed using partial, constrained CCA (Johnson *et al.*, unpubl.). Partial CCA allows for the effects of covariables to be partialled out during the analysis. In this case, time (year) was the covariable, as macroinvertebrate communities over a 5-yr period (1988 - 1992) were analyzed. Significance of each environmental variable was assessed by means of a Monte Carlo permutation test for significance of the first (constrained) ordination axis. Alpha was set at 0.05, and the option of species scores as weighted mean sample scores, with down-weighting of rare species, was selected.

Using forward selection, CCA identified seven environmental variables that explained significant proportions of the variance of the profundal taxa. In the first step, lake surface area was selected and accounted for 19 % of the total variance explained by all environmental variables. Subsequently, depth (15 %), latitude (7 %), water color (6 %), pH (6 %), maximum temperature (5 %), and minimum oxygen concentration (2 %) were selected, and cumulatively accounted for 75 % of the explainable variance of the weighted averages of the taxa. The covariable time (among-year variation) accounted for only 1.1 % of the total variation. A substantial amount of the total variation was not accounted for by the selected environmental variables. Only 27.1 % of the total variation was accounted for using these seven environmental variables. Part of this unexplained variability may be due to attempting to fit unimodal models to linear taxa responses. Thus CANOCO was rerun, using redundancy analysis (RDA), which is a form of multivariate multiple regression. RDA improved the amount of explainable variation by only 10 % (a total of 37.3 % of the variability was accounted for by the seven environmental variables), hence a large amount of profundal macroinvertebrate variability is left unaccounted for in these lakes.

Although CCA and RDA ordinations explained only a low percentage of the total variability, such ordination diagrams may be quite informative (Gauch, 1982). Species - environment correlations were high (axis 1 = 0.903 and axis 2 = 0.907), and the cumulative percentage of variance of the species data was 9.8, 17.7, 21.3, and 24.1 %, respectively, for the first four axes. Lake surface area (r = 0.81) and latitude (r = 0.72) were strongly correlated with taxa axis 1, and maximum temperature (r = -0.78) and depth (r = 0.74) strongly influenced axis 2 (Fig. 6). TWINSPAN indicator taxa such as *Micropsectra* spp., *S. ferox* and *H. subpilosus* which are indicators of oligotrophic conditions were positively aligned with axis 1, while the chironomids *Dicrotendipes* spp., *P. prasinatus* and *Microtendipes* spp., and the trichopteran *M. angustata* were negatively correlated with lake depth and positively correlated with maximum temperature.

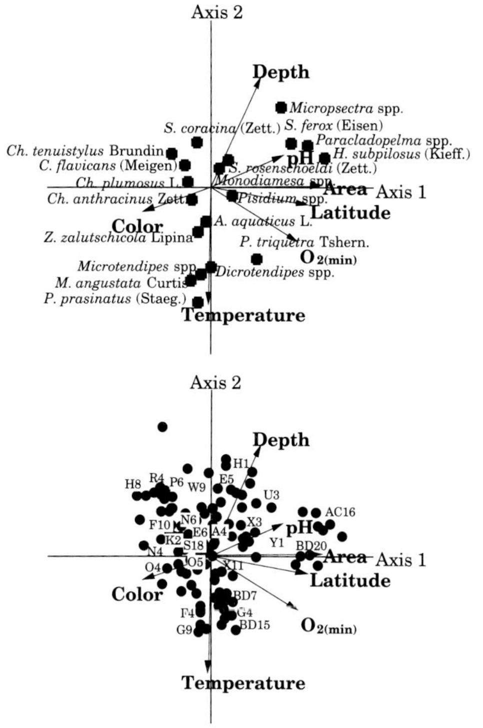

Figure 6. Canonical correspondence analysis (CCA) ordination of TWINSPAN indicator taxa (a) and 26 Swedish reference lakes (codes) (b) with seven environmental variables. Partial, constrained CCA was performed with year as a covariable. Taxa given are TWINSPAN indicator taxa, and lakes are given as a lake code.

DFA

Discriminant function analysis is a multivariate approach to pattern recognition and interpretations that is frequently used to construct predictive models using environmental variables and macroinvertebrate groups. In DFA, discriminant functions are chosen as independent linear combinations of the original environmental variables that best discriminate between the discrete groups. The selection of a parsimonious subset of environmental variables that can be used to adequately predict biological metrics is one goal of a predictive approach. However, often of greater interest is the development of predictive models that can be used to predict group membership of the sites or taxa which were previously established, or the prediction of group membership of new sites or taxa.

Discriminant function analysis of the four groups formed by TWINSPAN analysis of profundal macroinvertebrates of Swedish reference lakes showed distinct differences in water physico-chemical values between lake groups. Ninety-two percent of the lakes were classified "correctly" using the seven environmental variables selected in the CCA ordination. Two of the 26 reference lakes were "misclassified" when profundal communities were compared with environmental variables. Lake Humsjön, was placed in group 3 according to profundal macroinvertebrates, but placed in group 4 according to physico-chemical values. This lake has a mean pH of 6.7, which is more similar to group 4 (mean pH = 6.5) than group 3 (5.6) lakes. Posterior probabilities for group 3 and 4 lakes were 0.101 and 0.890, respectively. Lake Stensjön was also "misclassified", although not as strongly, posterior probabilities for group membership were 0.489 for group 2 and 0.386 for group 3.

As the DFA models constructed using 26 Swedish reference lakes seem to adequately describe lake types, these models may then be used to predict site or taxa occurrence of new sites. Using DFA, the probability that a site would be a member of each group is proportional to the number of members of that group multiplied by the exponential of minus one half the distance from the group mean. The probability of occurrence at the site of each species *k* is calculated as the sum over all groups of the probability that the site belongs to a group *j* multiplied by the proportion of members of group *j* which species *k* occurs (sensu Moss *et al.*, 1987).

Macroinvertebrate taxa occurrence was predicted for limed reference lakes using the predictive models constructed from 26 reference lakes. The biology and water chemistry of fourteen limed reference lakes have been studied since 1989 (Appelberg and Aldén, 1992). Similar to Swedish reference lakes, macroinvertebrates were sampled in autumn, with five replicate Ekman samples taken from the profundal region of each lake. Table 3 shows the predicted probability of occurrence of profundal macroinvertebrate taxa as well as the frequency of observed taxa in 14 limed reference lakes (Johnson *et al.*, unpubl.). The majority of limed reference lakes (n = 11) were placed in TWINSPAN group 1 by DFA using seven environmental variables. As mentioned above, group 1 lakes were characterized as being relatively large, deep, well-buffered, clear-water lakes. Thirteen taxa were predicted to occur with ≥ 75 % probability. Only seven of these taxa were found in > 50 % of the lakes, and of the 21 taxa predicted to occur with a probability ≥ 50 %, only 10 taxa occurred in > 50 % of the lakes (observed to expected ratio = 0.476). The seemingly poor agreement between predicted and observed taxa occurrence may be partly explained by the effects of liming forest lakes in southern Sweden. Several of the taxa which were predicted to occur are indicators of large oligotrophic lakes (e.g. *H. subpilosus*, *Micropsectra* spp., *Paracladopelma* spp. and *Monodiamesa* spp.), and are less common in forest lakes of southern Sweden.

A somewhat better agreement between predicted and observed taxa was found for lakes classified in TWINSPAN group 4. Group 4 lakes are relative small (mean surface area = 0.70 km^2), with a mean pH of 6.51. Of the ten taxa which were predicted to occur with ≥ 50 % probability, eight of these taxa were found

Table 3. The predicted probability of occurrence of selected taxa in limed reference lakes compared with observed values (frequency of lakes where the taxon occurred). Discriminant function analysis was performed with seven environmental variables and TWINSPAN group classification of 26 Swedish reference lakes (see text). MP = median probability of taxon occurrence and % O = percent of lakes where taxon was found occurring.

TWINSPAN group 1	MP	% O	TWINSPAN group 4	MP	% O
Oligochaeta	99	100	*Procladius* spp.	75	100
Procladius spp.	75	100	*S. coracina*	75	100
Tanytarsus spp.	75	75	*Tanytarsus* spp.	75	100
Pisidium spp.	75	67	Oligochaeta	75	100
S. ferox	75	58	*Pisidium* spp.	75	0
H. apicalis	75	58	*C. flavicans*	66	100
Nematoda	75	50	*Monodiamesa* spp.	66	0
Protanypus spp.	75	42	Hydracarina	57	100
Micropsectra spp.	75	33	*Z. zalutschicola*	56	100
Turbellaria	75	25	*S. rosenschoeldi*	56	50
H. subpilosus	75	25	*Ch. anthracinus*	47	50
Monodiamesa spp.	75	17	*H. apicalis*	47	0
Paracladopelma spp.	75	8	*Polypedilum* spp.	38	0
S. coracina	50	92	Nematoda	38	0
S. rosenschoeldi	50	67	*Micropsectra* spp.	28	0
Hydracarina	50	58	Turbellaria	28	0
Psectrocladius spp.	50	25	*T. tubifex*	28	0
H. marcidus	50	25	*S. ferox*	19	0
Polypedilum spp.	50	17	*H. marcidus*	19	0
Gammarus spp.	50	0	*Psectrocladius* spp.	19	0
T. tubifex	50	0	*Protanypus* spp.	9	0
C. flavicans	25	58	*Gammarus* spp.	0	0
Z. zalutschicola	0	33	*H. subpilosus*	0	0
Ch. anthracinus	0	17	*Paracladopelma* spp.	0	0

occurring in the two limed reference lakes (O:E = 0.80). For example, *Sergentia coracina* (Zetterstedt) and *Tanytarsus* spp. were predicted to occur with a probability of 75 %, both of these taxa were found in the limed lakes.

A predictive approach using macroinvertebrates is a relatively new and promising direction in biomonitoring studies. These new techniques often de-emphasize the expertise of the individual investigator, and allow for the calculation of probability estimates of taxa occurrences as well as measures of tolerance of individual taxa along environmental gradients (Johnson *et al.*, 1993). The predicted taxa occurrence may also be used to calculate biotic indices or scores such as BMWP. This is presently being done in the UK, where observed BMWP scores are compared to predicted scores in national river surveys (Wright *et al.*, 1988). Moreover, these predictive techniques may also be used to estimate the power of a study design. Predictive techniques may be used, for example, to simulate environmental impacts, and then predictive models may be used to estimate whether adequate power was obtained to detect the expected magnitudes of effect. Finally, a more widespread use of predictive modelling and power analysis in biomonitoring studies would greatly enhance our understanding and interpretation of the effects of anthropogenic stress on aquatic systems.

Acknowledgments

This paper is dedicated to the memory of Professor Lars Brundin. I would like to thank Dr Peter Cranston for the invitation to attend the 12th International Symposium on Chironomidae, and the Australian Commonwealth Environmental Protection Agency for providing the financial funding. I also thank Dr Torgny Wiederholm and Dr Björn Söderbäck for providing useful suggestions on an earlier draft of the manuscript, and Lars Eriksson for doing the macroinvertebrate identifications and enumerations.

PART 1

GENETICS OF CHIRONOMIDAE

Cytologically identified Chironomus species from lakes in North Dakota and Minnesota, USA

M.G. Butler[1], I.I. Kiknadze[2], J.K. Cooper[1], and M. Siirin[2]

[1]Zoology Department, North Dakota State Univ., Fargo, ND 58105, USA2

[2]Institute of Cytology and Genetics, 630090 Novosibirsk, Russia

Introduction

Chironomus larvae are often the most abundant benthic macroinvertebrates in highly productive waterbodies. The need for species-level identification of *Chironomus* larvae in ecological studies is confounded by the low degree of morphological diversity within this genus (Lindeberg and Wiederholm, 1979). Fortunately, there is a long history of cytotaxonomic work on *Chironomus*, based on the polytene chromosomes of the salivary glands and other tissues. Compared to the Nearctic, the *Chironomus* fauna of Europe and the western Palearctic is relatively well-known, due to extensive work by Keyl, Wülker, Michailova, Kiknadze, and others (references below).

Townes (1945) reported 19 species of *Chironomus* in the Nearctic, based on adult morphology, but some of Townes' taxa are now known to comprise multiple, karyologically-distinct species. Although some earlier studies involved cytogenetic analysis of single species of *Chironomus* (e.g. Rothfels and Fairlie, 1957; Acton, 1959), systematic investigation of the North American *Chironomus* fauna based on larval karyotypes really began with the report at the Second International Symposium on Chironomid Research in Helsinki, by Wülker, Sublette, and Martin (Wülker *et al.*, 1968). This group has published a series of papers revising the taxonomy of Nearctic *Chironomus* species, based on both morphological and cytogenetic characters (Wülker *et al.*, 1971; Sublette and Sublette, 1974a, 1974b; Wülker and Martin, 1974; Martin and Wülker, 1974; Martin *et al.*, 1974; Wülker *et al.*, 1991). Wülker and Butler (1983) compared karyotypes of three European and three North American species with larvae of the *salinarius*-type. Wülker (1991) suggested from karyological evidence that two Fennoscandian *Chironomus* species, *C. tenuistylus* and *C. longistylus*, have holarctic distributions, based on collections from bog lakes in Alaska and Wisconsin. Jon Martin (University of Melbourne) maintains an unpublished list describing over 60 *Chironomus* karyotypes known from North America, although only 25 of these are reported from lakes.The goals of this report are to (1) document the diversity of *Chironomus* species found to date in lakes in North Dakota and west-central Minnesota, and (2) to report preliminary findings from cytogenetic comparison of these Nearctic populations with karyotypes known for Palearctic populations of the same, or closely related, species.

Methods

All specimens were collected during the winter or early spring of 1993 by sampling with a Ponar grab, usually near the deepest point in each lake, and washing the contents through a 0.5 mm mesh. *Chironomus* larvae were sorted live and fixed in a 3:1 mixture of absolute ethanol and glacial acetic acid. Karyotype squashes were made from salivary gland chromosomes stained with orcein, according to the method of Kiknadze *et al.* (1991). Table 1 gives the locations and sampling depths of the study lakes, and numbers of *Chironomus* larvae examined from each.

Results

At least ten karyotypes, distinct at the species level, have been found among the 110 individual *Chironomus* larvae examined cytologically. Five of these karyotypes are similar to karyotypes of species known from the Palearctic, although all show some structural differences relative to their European counterparts. Five other karyotypes appear to represent uniquely North American species, four of which have been reared to the adult male (*C. cucini*, *C. staegeri*, *C. decorus*-gr. sp.1, and *Chironomus* sp. r). With the exception of *C. tentans*, the other karyotypes have not been associated definitely with the adult stage.

C. nr. *plumosus*

Karyotypes from a total of 15 specimens from five lakes were examined (Table 1). Although these karyotypes are similar to that of the palearctic *C. plumosus*, there are differences in the banding sequences in arms A, D, and E. In arm A of these American larvae, two sequences were observed, one identical to the standard sequence in European *C. plumosus*, the other differing by a simple inversion. Similarly, arm D of the American larvae exhibits two sequences, one identical to a sequence known in European populations and another differing by a complex inversion. The former arm D sequence includes a nucleolus in some of the American material, whereas a nucleolus is never found in this arm in European *C. plumosus*.

In the Palearctic, *C. plumosus* is part of a complex of closely related species which differ in subtle cytological, morphological, and ecological details (Devai *et al.*, 1983; Ryser *et al.*, 1983; Michailova and Fischer, 1986b; Kiknadze, 1987; Kiknadze *et al.*, 1991). Michailova and Fischer (1986a) have described a nearctic species, *C. vancouveri*, as a member of this complex that they consider distinct from the European *C. plumosus*. Hybrids produced by artificially crossing European *C. plumosus* with *C. vancouveri* were viable, but the eggs had variable hatching success and the hybrid offspring showed nearly complete asynapsis of homologues. Our material of C. nr. *plumosus* from North Dakota and Minnesota differs from the published karyotype of *C. vancouveri* by inversions in arms A and B, and shows a high level of inversion heterozygosity.Michailova and Fischer (1986a) reported no inversion polymorphism for *C. vancouveri*, but their material appears to have been derived from a single collection from one lake. A more representative picture of karyotypic variability among C. nr. *plumosus* populations across North America is needed to determine whether Nearctic and Palearctic forms warrant species-level distinction.*Chironomus entis* Schobanov is another species in the *C. plumosus*-group that is morphologically very similar to *C. plumosus* (Belyanina *et al.*, 1992), but has a distinctly different karyotype. These species can be distinguished by differences in chromosome arms E and G (Kerkis *et al.*, 1989; Kiknadze *et al.*, 1991). We have found larvae with karyotypes similar to the Palearctic *C. entis* in three lakes, always sympatrically with C. nr. *plumosus* populations. The American C. nr. *entis* differs from the Palearctic karyotype by unique banding sequences in arms A, C, and D. For both *C. plumosus* and *C. entis* , a broader picture of chromosomal variability among Nearctic populations is called for to elucidate relationships between representatives on different continents.

C. nr. *anthracinus*

Chironomus anthracinus Zetterstedt is a common species in mesotrophic to moderately eutrophic Palearctic lakes (Sæther, 1979), and has long been considered to occur in the Nearctic as well (Townes, 1945; Wülker *et al.*, 1968; Sæther, 1979). J. Martin and W. Wülker (pers. comm.) consider the Nearctic karyotype closest to the "true *C. anthracinus*" to be one that Martin lists as "*Chironomus* sp. q", known from lakes in Ontario, New Hampshire, Wisconsin, and California. Martin also lists two other Nearctic karyotypes as being "near *C. anthracinus*". We found larvae in three North Dakota lakes with a karyotype identical to the Palearctic *C. anthracinus* in arms A, E, and F, but with

Table 1. Locations and sampling depths for lakes from which *Chironomus* larvae were examined cytologically. The number of karyotype preparations assigned to each taxon is given. 1=C. nr. *plumosus*, 2=C. nr. *entis*, 3=C. nr. *anthracinus*, 4=C. *annularis*, 5=C. *(Camptochironomus) tentans*, 6=C. *cucini*, 7=C. *staegeri*, 8=C. *decorus* sp.1, 9=C. *decorus*-group, 10=*Chironomus* sp r.

Lake	County	Sampling Depth (m)	Taxon									
			1	2	3	4	5	6	7	8	9	10
North Dakota												
Braddock Dam	Emmons	4										
Dead Colt Creek Dam	Ransom	10	1	2								
Kota Ray Dam	Williams	7									3	
McVille Dam	Nelson	6			1	1			4			
Larimore Dam	Grand Forks	9	3		4	1						
Red Willow Lake	Griggs	6	3									
Silver Lake	Sargent	3	7	3								
Warsing Dam	Eddy	6			2	6	5					
Lake Williams	Kidder	7				3					3	
Minnesota												
Anderson Lake	Clearwater	6							10	11		
Bad Axe Lake	Hubbard	10							5	3		
Lake Christiana	Douglas	1	2	1	7		47					
Long Lake	Clearwater	20						2				
Turtle Lake	Becker	20								6		18

unique banding sequences in arms B, C, and D. Also, arm B of this North Dakota species bears a large nucleolus, which is not present in Palearctic C. *anthracinus*. The North Dakota material is most similar to Jon Martin's unpublished "species i: C. nr *atrella*", collected from lakes in Ontario and California.

C. nr. *annularius*

Although there has been some confusion regarding proper nomenclature (Lindeberg and Wiederholm, 1979), the karyotype of a Palearctic species referred to as *Chironomus annularius* Meigen is illustrated and described by Keyl (1962), Michailova (1989), and Kiknadze *et al.* (1991). We have found a related karyotype in one Minnesota and three North Dakota lakes (Table 1). These larvae show a high level of chromosomal polymorphism, but differ from the Palearctic C. *annularius* by homozygous inversions in arms A, B, C, and D as well as by the presence of a nucleolus in arm A. The karyotype matches Jon Martin's "species 3d: C. nr *annularius*", collected from lakes in Indiana and Wisconsin.

C. *(Camptochironomus) tentans* Fabricius

Chironomus tentans is common in shallow, organically enriched waterbodies across the Holarctic (Acton and Scudder, 1971; Topping, 1971). We found larvae of the Nearctic form of this species in two of the North Dakota lakes sampled in 1993 (Table 1). Only in Warsing Dam were karyotypes examined, but the distinctive dark stripe on the clypeus permitted identification of this species in Braddock Dam. Following chemical eradication of fish from Minnesota's Lake Christina in 1987 (Hanson and Butler, 1993), a very dense population of this species developed for several years. A laboratory culture was established from egg masses collected at Lake Christina in 1992, and karyotypes were examined for 47 larvae from this culture. This species has been well-studied cytologically in Europe (Beerman, 1955; Acton, 1957; Kiknadze *et al.*, 1991). Several cytogenetic analyses of North American populations have focused on genetic control of sex determination (Acton, 1959; Thompson, 1971; Martin and Lee, 1981), while Acton (1959, 1962) and Acton and Scudder (1971) have studied in detail the differences between European and Canadian karyotypes of this species. Based on Acton's work, as well as analysis of C. *tentans* populations in Siberia (Kiknadze *et al.*, 1991; Kiknadze *et al.*, in prep), from Minnesota's Lake Christina (Kiknadze *et al.*, in prep), and from Warsing Dam, North Dakota, it is clear that Nearctic populations differ greatly from those found in the Palearctic. North American karyotypes have unique banding sequences in all chromosome arms, and bear nucleoli in arms B and C (Acton's chromosomes 2 and 1, respectively); Palearctic C. *tentans* have nucleoli in arms B and D (chromosomes 2 and 3).

C. *cucini* Webb

This Nearctic species has a larva of the *salinarius*-type, typically lacking lateral or ventral abdominal tubuli. One pair of very short ventral tubuli may be present on the posterior of abdominal segment 8 (Wülker and Butler, 1983). The species is an indicator of oligotrophic-mesotrophic conditions, as it occurs in the profundal sediments of moderately productive lakes that maintain an oxygen regime capable of supporting coregonid and salmonid fishes (Hare and Carter, 1976; Butler and McMillan, 1990), often occurring sympatrically with *Sergentia coracina* (Zett). We found this species only in the profundal of Minnesota's Long Lake, the least productive lake in this survey, where it was the predominant benthic macroinvertebrate.

The karyotype of C. *cucini* is characterized by a chromocenter of centromeric heterochromatin to which all arms are attached, as illustrated by Martin (1979). Wülker and Butler (1983) describe the entire karyotype and provide a key to distinguish this species from other *salinarius*-type *Chironomus* whose karyotypes are known.

C. *staegeri* Lundbeck

The morphology and karyotype of this species are described by Wülker *et al.* (1971), who placed it in a morphologically and cytologically well-defined group of three species within the *thummi*-complex. These authors characterize C. *staegeri* as a widely distributed species found in lentic habitats ranging from deep lakes to shallow pools. Martin and Wülker (1971) provide evidence for genetic differences between populations in deep and shallow habitats.

The karyotype of C. *staegeri* consists of three chromosomes pairs, owing to fusion of arm G with arm E. We found larvae of this species to be the predominant profundal macroinvertebrates in two of the five Minnesota lakes sampled, Bad Axe Lake and Anderson Lake. It was also identified from a North Dakota impoundment, McVille Dam.

C. *decorus*-group

Townes (1945) recognized C. *decorus* (Johannsen) as a widespread, abundant, and highly variable species in North America. Martin (1979) noted that as many as 15 cytologically distinct species may have adults of the *decorus*-type. Wülker *et al.* (1991) proposed a "karyosystematically defined *decorus*-group", consisting of species in the *thummi*-complex (arm combination AB CD EF G) and characterized by certain key inversions in arms A and F. Two European species in this group (C. *obtusidens* and C. *acutiventris*) and two undescribed species from Lake Michigan have larvae of the *fluviatilis*-type (Lenz, 1954-62), characterized by straight, pointed ventral tubuli on the eighth abdominal segment (Wülker *et al.*, 1983).

In our material we frequently found larvae characterized by relatively straight ventral tubuli that tapered evenly to a rounded tip, much like Lenz's illustration of the *fluviatilis* larval type. At least two karyotypes have been obtained from these larvae, both fitting the "karyosystematic *decorus*-group" of Wülker *et al.* (1991). C. *decorus*-gr sp. 1 (Fig. 1) has been found abundantly in the profundal of three Minnesota lakes (Table 1), where it is co-dominant with C. *staegeri* in two lakes and with *Chironomus* sp. r (see below) in Turtle Lk. Morphologically similar larvae were found in eight North Dakota lakes, but examination of karyotypes from one of these lakes (Lake Williams)(Fig. 2) indicates species-level differences relative to C. *decorus*-gr sp. 1 in the Minnesota lakes. Preliminary analyses of other *decorus*-group karyotypes found in the North Dakota lakes indicate that several sibling species may occur among these *Chironomus* larvae with *fluviatilis*-type ventral tubuli.

Chironomus sp r

As early as 1968, Wülker *et al.* mention the existence of a *Chironomus* species collected from lakes in Ontario and Wisconsin, notable in having only two pairs of chromosomes. Jon Martin has listed this undescribed species as "*Chironomus* sp r" in his unpublished listing of North American karyotypes, based on specimens collected from Clarke Lk in Algonquin Provincial Park in

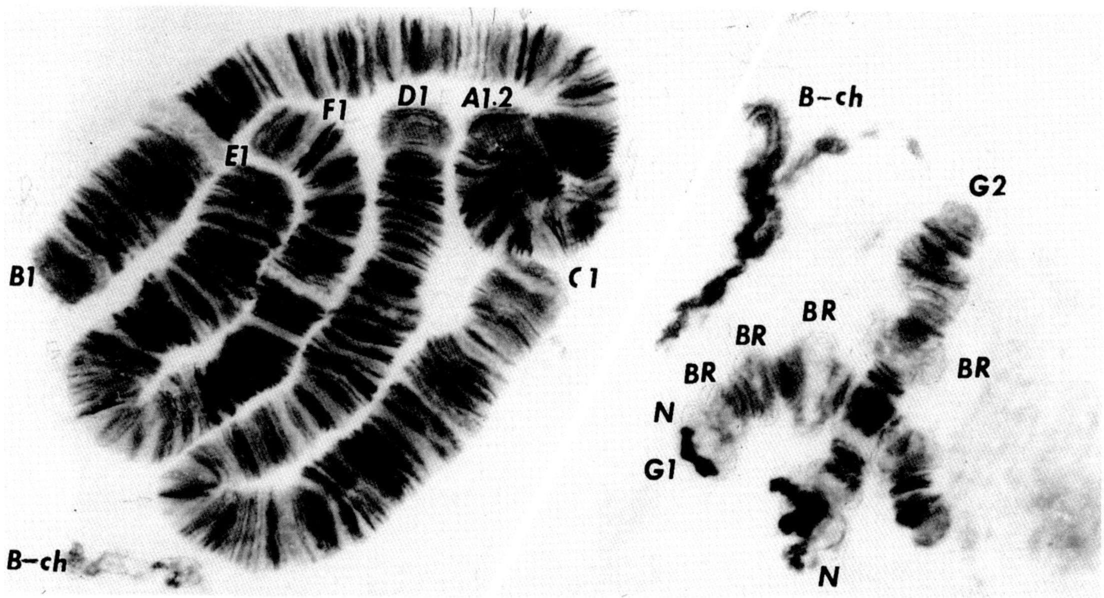

Figure 1. Karyotype of *Chironomus decorus*-gr. sp 1, from Anderson Lake in Minnesota. Normal band sequences in homozygous arms labeled B1, C1, etc. A1.2 indicates a heterozygous inversion in arm A; G2 identifies an alternative sequence in unpaired arm G. BR = Balbiani ring. B-ch = B- chromosome material. N = nucleolus.

Ontario, and Freibauer Lk, Bayfield Co., Wisconsin. We found this species to be the dominant profundal chironomid in Turtle Lk, Becker Co., Minnesota. Other specimens with the same karyotype have since been obtained from Spearhead Lk, Beltrami Co., Minnesota, and from Lk St. Joseph, Quebec (collected by L. Hare).

This karyotype is a unique modification of the *thummi*-complex arm combination in that arm G has fused to arm A, producing a chromosome consisting of GAB, while a second chromosome is formed by the union of CD to EF (Fig. 2). Possible mechanisms and consequences of these chromosomal rearrangements will be discussed in detail in a full description of this species, which we have reared to the adult stage.

DISCUSSION

With the exception of *C. tentans*, all the species found in this study belong to the *thummi*-complex or modifications thereof. This complex is reported to be widespread in North America, (Martin *et al.*, 1974), and seems to be predominant in deeper lentic habitats. The findings reported here are a biased sample of the *Chironomus* fauna that may exist in the Minnesota-North Dakota region, in that we have sampled only in lakes, and at depths of at least 3m. In addition, only a small and non-random sample of *Chironomus* larvae collected from each lake have been studied cytologically. This means that our results present an incomplete view of the *Chironomus* fauna that exists in each lake. This is especially true for the North Dakota lakes. In the deeper, dimictic Minnesota lakes, the profundal fauna appears to be predominantly *C. staegeri* and *C. decorus*-gr. sp. 1 (Anderson and Bad Axe Lakes) or *Chironomus* sp. r and *C. decorus*-gr. sp. 1 (Turtle Lake). More extensive sampling and more uniform cytological analysis is needed to determine if there are typical associations among these species, and what relationships their distribution may have to limnological conditions in different lakes.

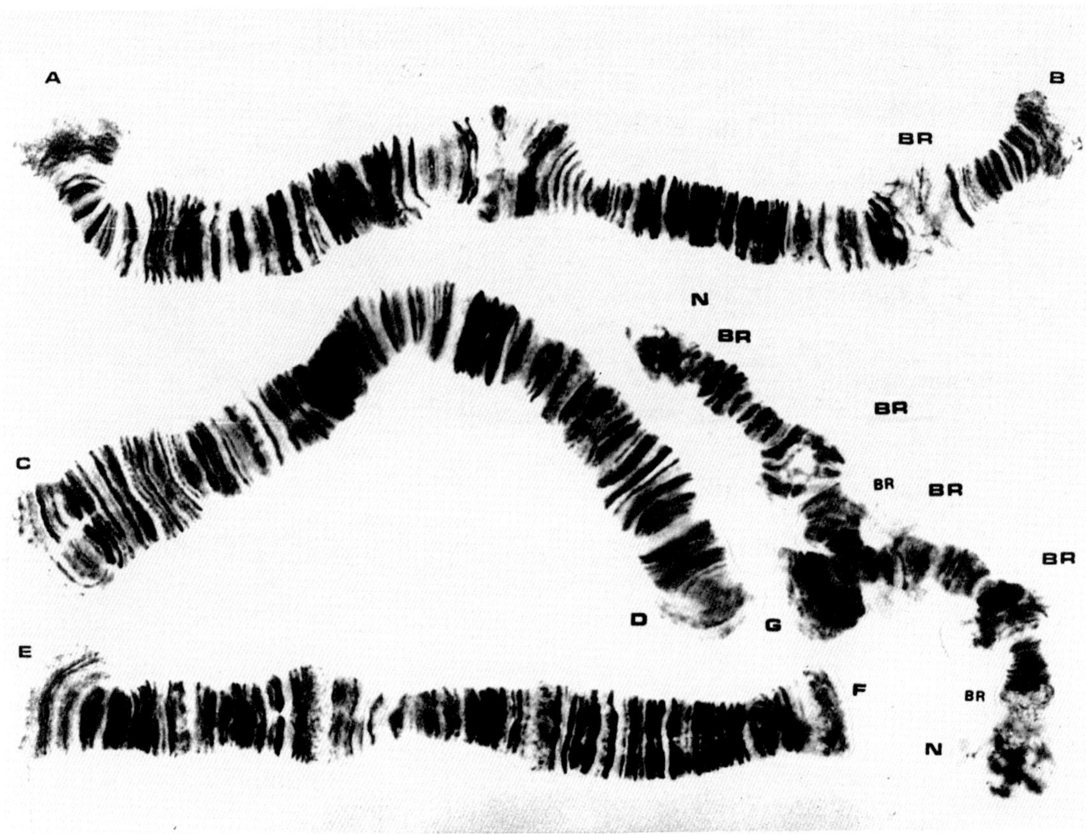

Figure 2. Karyotype of *Chironomus decorus*-group sp 2 from Lake Williams in North Dakota. Arms and other structures labeled as in Fig. 1.

The relatively gross karyological differences among many of the *Chironomus* species encountered in this study indicate that it is feasible to obtain much greater taxonomic resolution within this genus than has been attempted in most limnological studies and water-quality surveys. Existing schemes for evaluating trophic status of Nearctic lakes on the basis of their chironomid fauna typically distinguish only two morphological types of *Chironomus* larvae: *salinarius*-type versus *plumosus*-type (Sæther, 1979; Courtemanch, 1987). This may obscure potential environmental differences among lakes at the eutrophic end of the trophic spectrum, where *Chironomus* is typically the predominant genus.

Another issue raised by this work is appropriate application of the biological species concept to taxa with holarctic distributions. At what point should sibling species pairs found on different continents be recognized as distinct species? Genetic differences between Palearctic and Nearctic populations of any holarctic species can be expected to exceed variation within a continent, and some of these differences may be reflected in chromosome structure. At what degree of cytological difference should different species names be applied? Acton (1959) believed that one of the four chromosomes of C. *tentans* had achieved species-level distinctness between European and Canadian populations, but that overall the chromosomal differences represented only "incipient speciation" (Acton, 1962), in part because European and Canadian populations could be made to hybridize. On the other hand, Michailova and Fischer (1986a,b) have erected a new

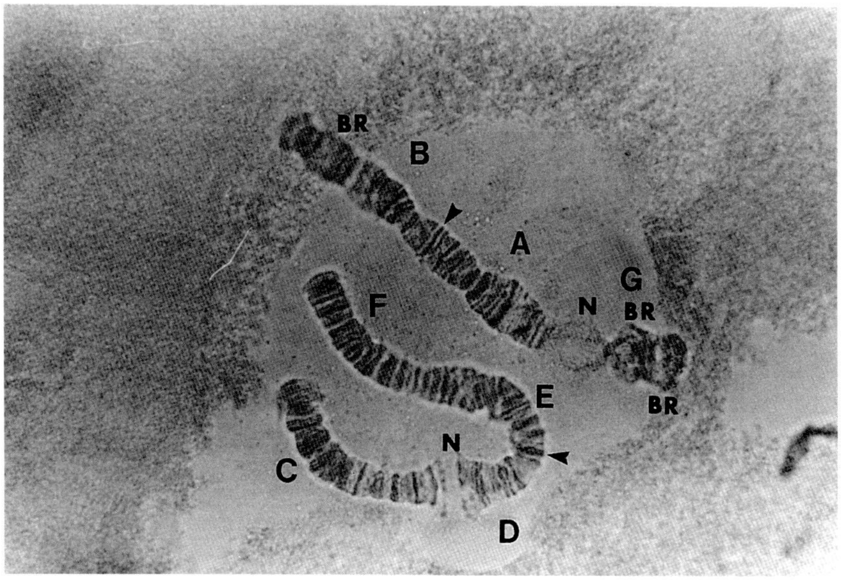

Figure 3. Karyotype of *Chironomus* sp r from Turtle Lake, Minnesota. Arms and other structures labeled as in Fig. 1.

species (*C. vancouveri*) for a Nearctic population closely related to the European *C. plumosus*, despite the fact that hybridization is possible in the laboratory.

Because populations on different continents will not interbreed in nature, the issue of appropriate nomenclature may be less significant than the question of potential ecological differences between geographically distant populations. Use of a specific name implies certain behavioral and ecological characteristics. If sibling species have similar behaviors, habitat requirements and ecological roles on different continents, use of the same name may be beneficial. On the other hand, understanding the degree and direction of genetic change in populations across their range is of value to our understanding of evolutionary processes. Further comparison of closely related chironomid taxa on different continents by morphological, cytological, and molecular techniques can contribute to this end.

Acknowledgements

This work was supported by funding from the North Dakota State Department of Health through the US Environmental Protection Agency's Clean Lakes Program, and a Grant-in-Aid from the North Dakota State University Research and Consulting Committee. We thank Dr. Jon Martin, University of Melbourne, for providing unpublished information on karyotypes of Nearctic *Chironomus*. We also appreciate the critical review of an earlier version of this text by both Dr. Martin and Prof. W. Wülker, and we thank them both for valuable discussion of issues raised during this study.

Basic patterns in the chromosome evolution of the genus *Chironomus* (Diptera): Polytene chromosomes of the three Indian species *C. plumatisetigerus*, *C. calipterus* and *Chironomus* species.

By Sumitra Saxena,

S.G.T.B. Khalsa College, University of Delhi,
Delhi, India

Introduction

The genus *Chironomus* is very useful for investigations into chromosome evolution because it possesses good quality polytene chromosomes. As a result of this, the banding patterns can be compared easily and homology of the banding patterns of the chromosomes can be recognised in most species. In spite of these favourable conditions, papers on this subject have only appeared in the last 35 years. Thus in 1959 Keyl and Keyl published the first results of a detailed investigations on the German *Chironomus* species, followed by Keyl (1962) who carried out a major study of 22 European species of the genus. Later an international team organised by J.E. Sublette extended these investigations to other countries and continents e.g. North America (Wülker 1968, 1980, Wülker and Martin 1974, Wülker, Martin and Sublette 1971), Australia (Martin 1969, 1971, 1974, 1979) Japan (Martin and Sublette 1972), South Africa (Wülker 1980) and now I have carried out similar studies in India.

Species of the genus *Chironomus* in general possess four polytene chromosomes, consisting of three metacentrics and a short acrocentric chromosome, with seven arms (A-G). Some species may have three or two chromosomes. The standard species studied by Keyl (1962), *Chironomus piger*, had the arm combination AB, CD, EF & G. However, not all the species have this combination and differ due to the occurrence of whole arm translocations. As a result, based on different arm combinations caused by whole arm translocations, the species are presently grouped into eight major complexes (Martin *et al.* 1974) e.g. pseudothummi, thummi, maturus, parathummi, lacunarius, Camptochironomus, carus and columbiensis (Wülker *et al.*, 1989). Of these the pseudothummi- and thummi-complexes are most widespread all over the world and most of the species belong to these two complexes.

It is of interest to note that in the system of translocation types, the thummi- and pseudothummi-complexes are connected to each other by a single reciprocal translocation of the arms AB, EF of *thummi* to give AE, BF combination of *pseudothummi* and that the maturus-complex has a unique position in that it can be connected with both the thummi- and pseudothummi-complexes by only one reciprocal translocation.

Inside each complex, species can usually be related and distinguished by simple inversion steps, i.e. paracentric inversions, and their phylogenies can be contructed. Diagrams of the phylogenic relationships have been published by Keyl (1962) for arms A, E & F of the German species; by Wülker and Klotzli (1973) for the same arms in the lacunarius-complex; by Martin (1979) for arms F of all karyotpically known species of the genus *Chironomus* and by Wülker (1980) for arms A, E & F of three *Chironomus* species collected from three different continents. A large number of species in the thummi- and pseudothummi-complexes can be derived from one or the other which suggests that one of these may be the ancestral complex.

Keyl (1962), based on his study on the German *Chironomus* species, identified some banding patterns which were common in species of two or more complexes, such as thummi and pseudothummi, as compared to other patterns which occurred in species of only one complex. Therefore the former banding patterns were considered to have existed in the hypothetical stem species before the separation of the two complexes by a reciprocal translocation and can be regarded as basic or more plesiomorphic patterns. Based upon such findings Keyl tried to reconstruct the banding patterns of the hypothetical or ancestral species which crossed the border between the thummi- and pseudothummi-complex. Thus according to Keyl (1962) this ancestral species should have had in arm F standard pattern of C. *thummi piger*, in arm A structure type 12-4 of C. *pseudothummi* and the subsequent inv 10 - $2^{d,}$ and in arm E the standard form, inv 10^b - 3^f and 5 - 10^b.

Support to the validity of Keyl's speculations was provided by Wülker (1980) who studied three *Chironomus* species living in three different continents of the world (e.g. California, Kenya and Sweden), belonging to thummi- (two species) and pseudothummi- (one species) complexes, and were found to have retained the basic patterns of the ancestral species of the two complexes postulated by Keyl. These results are summarized in Figure 1.

Similar results have also been obtained by me for the three Indian species, and this paper provides therefore additional evidence to the validity of Keyl's speculations of existence of the basic patterns. All the three species belong to the pseudothummi-complex and have retained at least three of six basic patterns postulated by Keyl for the hypothetical ancestor between the thummi- and pseudothummi-complexes.

Materials & Methods

All three species were collected from India. *Chironomus plumatisetigerus* was collected from Yamuna River, Delhi, C. *calipterus* from Lodi Gardens, Delhi and *Chironomus sp.* from Mussouri Hills. Details will be given in separate papers (in preparation). Methods of chromosome preparation are the same as described for *Kiefferulus tainanus* (= *Nilodorum biroi*) (Saxena & Martin, 1985). For identifications of the arms A-G and for chromosome mapping of the arms A, E & F, methods described by Keyl (1962) for the European species have been used.

Results

The salivary gland chromosomes of the three chironomid species studied viz. C. *plumatisetigerus*, C. *calipterus* and *Chironomus* species indicate a sufficient similarity in banding pattern to those of the European species to permit identification of the arms under the scheme proposed by Keyl (1962). Thus all three species have the arm combination BF, DC, EA & G and belong to the pseudothummi-complex. Only the three arms A, E and F, of the three Indian species have been compared with the European species, particularly of the *pseudothummi* and the *thummi* complexes, since the standard maps of the other three arms (Devai *et al.* 1989) only became available subsequent to this work and no basic sequences have yet been designated.

Although all three Indian species belong to the pseudothummi-complex, they have revealed the existence of some banding patterns which are common to both the complexes, i.e. thummi and pseudothummi, thereby providing the second proof to Keyl's speculations, after Wülker (1980), about existence of the basic banding patterns. Figure 1 therefore provides an updated picture of the basic patterns given by Wülker (1980).

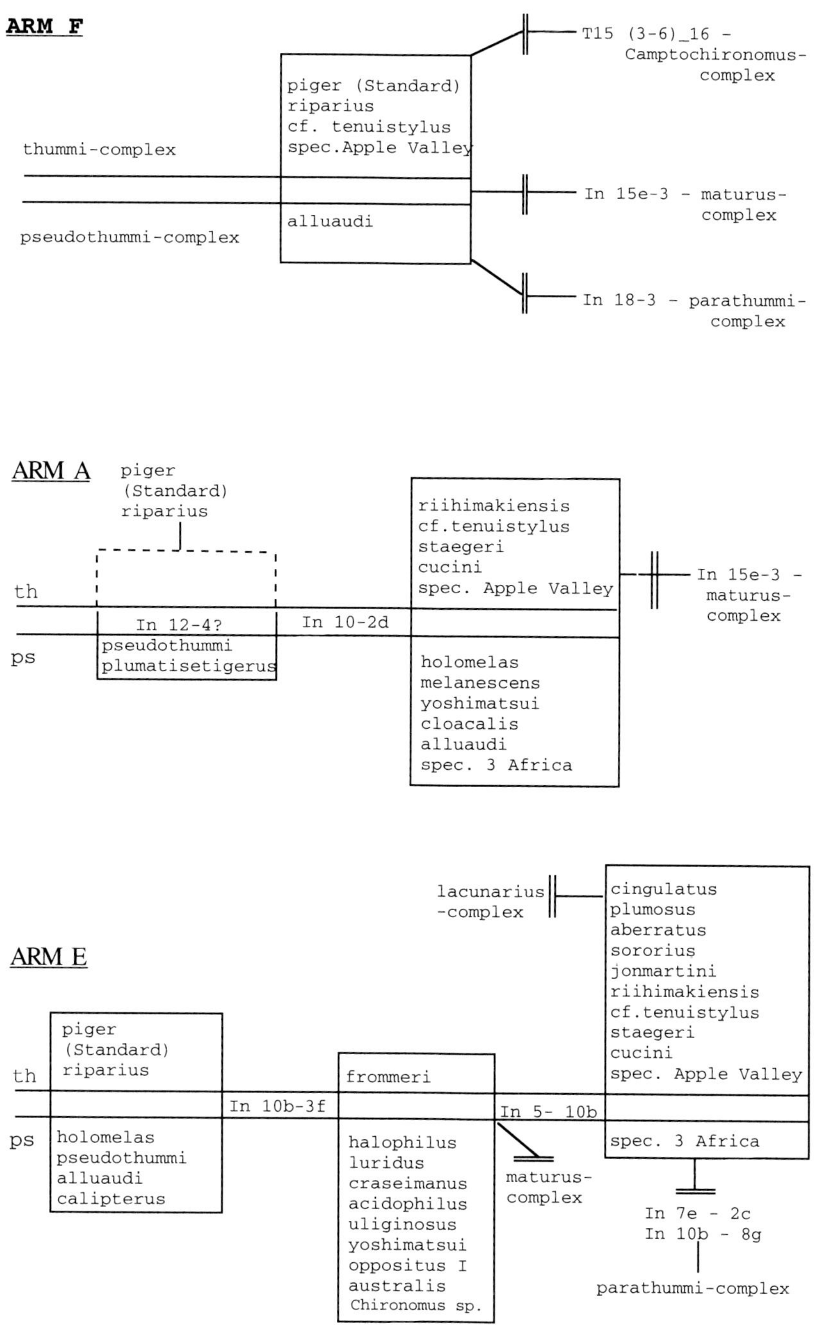

Figure 1. Diagram showing the occurrence of the basic patterns in the thummi- and pseudothummi- complexes in recent *Chironomus* species. Modified from Wülker (1980).

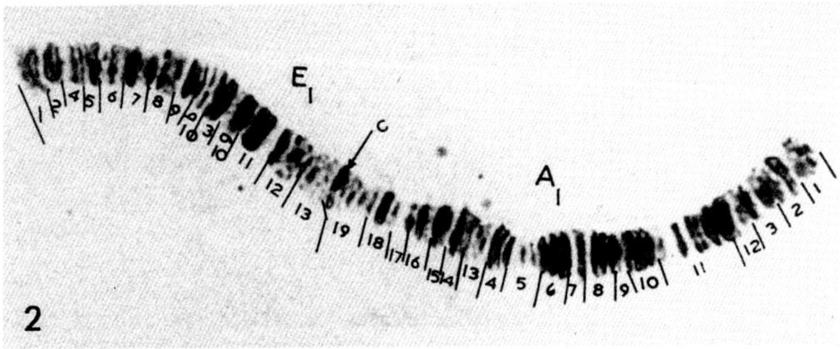

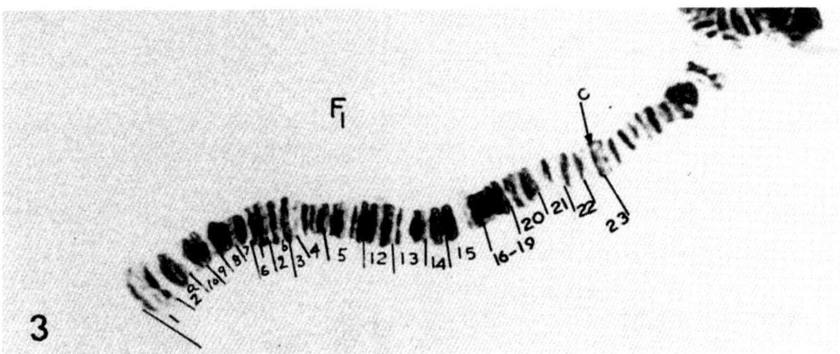

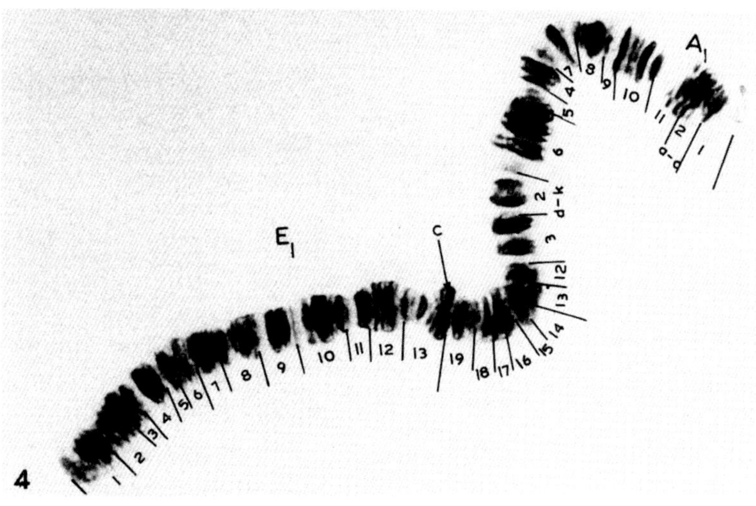

Figure 2. Sequences of arms E and A of *Chironomus plumatisetigerus* under the scheme of Keyl (1962). x950.
Figure 3. Sequence of arm F of *Chironomus plumatisetigerus* under the scheme of Keyl (1962). x950.
Figure 4. Sequences of arms E and A of *Chironomus calipterus* under the scheme of Keyl (1962). x950.
Figure 5. Sequence of arm F of *Chironomus calipterus* under the scheme of Keyl (1962). x950.
Figure 6. Sequence of arm E of *Chironomus* sp. under the scheme of Keyl (1962). x950.
Figure 7. Sequence of arm A of *Chironomus* sp.under the scheme of Keyl (1962). x950.
Figure 8. Sequence of arm F of *Chironomus* sp. under the scheme of Keyl (1962). x950.

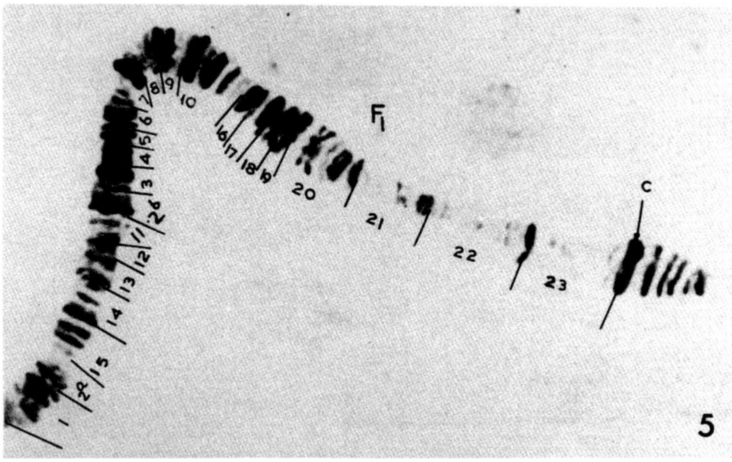

5

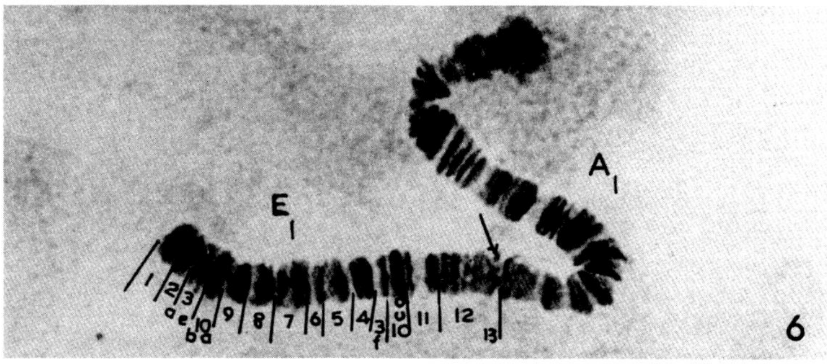

6

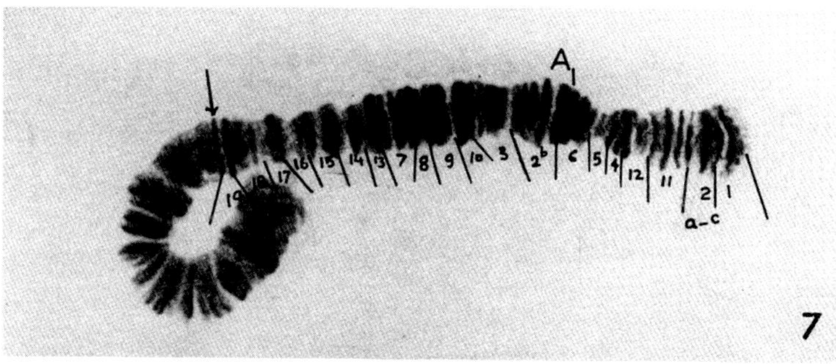

7

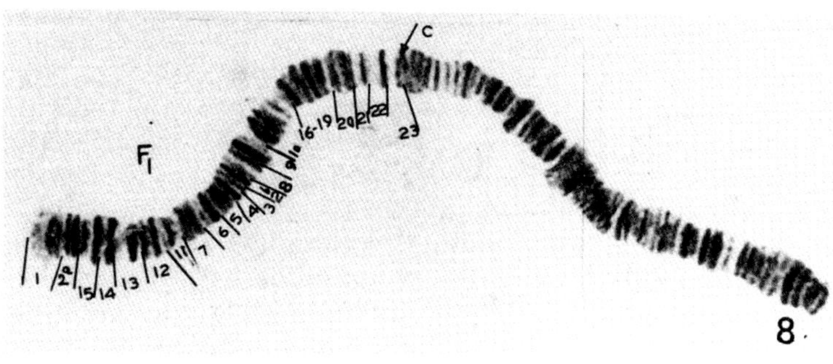

8

Description of the arms A, E and F of C. plumatisetigerus:

Arm A (Fig.2)

The bands 4, 5, 6, 7, 8, 9, 10, 11 and 12 can be made out easily and compared with those in C. *thummi piger*. Also the bands 16, 17, 18 and 19 can be compared easily. The conspicuous stretch of bands 4, 5, 6 and 7 is visible in the middle of the arm. The banding sequence in C. *plumatisetigerus* differs from the standard sequence in *thummi piger* by the inversion 12-4. Thus the sequence corresponds with that in C. *pseudothummi* and in this way C. *plumatisetigerus* is the second species after C. *pseudothummi* itself in the literature to have this banding pattern of the hypothetical ancestor postulated by Keyl.

1 2 3 4 5 6 7 8 9 10 11 12 13 14 15 16 17 18 19 — C. *thummi piger*

1 2 3 12 11 10 9 8 7 6 5 4 13 14 15 16 17 18 19 — C. *plumatisetigerus*, C. *pseudothummi*

Arm E (Fig. 2)

The sequence in the arm E can be readily compared with that in C. *thummi piger*, since the bands 5, 6, 7, 8, 9, 10, 11, 12 and 13 are sufficiently distinct. The arm on the whole has the same banding sequence as that in C. *thummi piger* and also the Indian species C. *calipterus* differing, however, in the shift of the band group 3^{a-f} in between 10^{b} and 10^{c}.

1 2 3^{a-f} 4 5 6 7 8 9 10 11 12 13 — C. *thummi piger*, C. *calipterus*

1 2 4 5 6 7 8 9 10^{a-b} 3^{a-f} 10^{c-d} 11 12 13 — C. *plumatisetigerus*

Arm F (Fig 3)

The band groups 1, 2, 3, 4, 5, 6, 7, 8 and 9, 10 and then 11, 12, 13, 14, 15, 16, 17, 18, 19 and 20 and 23 can be identified easily. The banding sequence can be derived from that in C. *thummi piger* by two inversions. Thus the first inversion extends from 2^{b}-10 giving an intermediate sequence. The second inversion is a small inversion having the break points at 11 and 5. Thus this inversion extends from 5-11.

1 2^{a-b} 3 4 5 6 7 8 9 10 11 12 13 14 15 16 17 18 19 - 23 — C. *thummi piger*

1 2^{a} 10 9 8 7 6 5 4 3 2^{b} 11 12 13 14 15 16 17 18 19 - 23 — C. *yoshimatsui*, C. *februarius*, C. *oppositus* I, C. *tepperi*, C. *duplex*

1 2^{a} 10 9 8 7 6 11 2^{b} 3 4 5 12 13 14 15 16 17 18 19 - 23 — C. *plumatisetigerus*

Description of arms A, E *and* F *of* C. calipterus :

Centromeres very heterchromatic.

Arm A (Fig 4)

The arm, on the whole, is not easy to compare. The only bands which can be made out with certainty are the bands 4, 5, 6, 7, 8, 9, 10, 11, 18 and 19. The arm seems to differ from the standard sequence by inversion of the segment 7-11 and its displacements towards the proximal end. As a result the bands 4-6 have been shifted toward the middle of the arm. Also the bands 2^{d}-3 are present behind the band 6.

1 2^{a-k} 3 4 5 6 7 8 9 10 11 12 13 14 15 16 17 18 19 C. *thummi piger*

1 2^{a-c} 11 10 9 8 7 4 5 6 2^{d} 3 12 13 14 15 16 17 18 19 C. *calipterus*

The sequence can also be derived from that in C. *holomelas* I and C. *melanescens* through the inversions of the segments 10-11, 6-4 and of the segment 12-2^{d} which has been inserted in between the bands 6 and 13.

1 2^{a-c} 10 11 12 3 2^{k-d} 9 8 7 6 5 4 13 14 15 16 17 18 19 C. *holomelas* I
C. *melanescens*

1 2^{a-c} 11 10 9 8 7 4 5 6 2^{d} 3 12 13 14 15 16 17 18 19 C. *calipterus*

Arm E (Fig 4)

The bands 1, 2, 3, 4, 5, 6, 7, 8, 12 and 13 can be compared with certainty. The species has the same sequence as the standard form of C. *thummi piger*. Thus C. *calipterus* is the fourth species after C. *pseudothummi*, C. *holomelas* and C. *alluaudi* from the pseudothummi-complex to retain this basic pattern of the stem species for Arm E:

1 2 3 4 5 6 7 8 9 10 11 12 13 14 C. *thummi piger*
C. *pseudothummi*
C. *holomelas*
C. *alluaudi*
C. *calipterus*

Arm F (Fig 5)

This arm is fairly easy to compare. The bands 2-10 and 16-23 are very distinct. Arm F differs from the standard species C. *thummi piger* and also C. *alluaudi*, C. *tenuistylus* and *Chironomus* species Apple Valley, by the inversion and shift of the segment 11-15.

1 2 3 4 5 6 7 8 9 10 11 12 13 14 15 16 17 18 19 - 23 C. *thummi piger*
C. *alluaudi*
C. *tenuistylus*
Chironomus sp. Apple Valley

1 2^{a} 15 14 13 12 11 2^{b} 3 4 5 6 7 8 9 10 16 - 23 C. *calipterus*

The sequence thus is also very similar to that in C.*holomelas* I, C. *melanescens*, C. *dorsalis* and C. *halophilus* ,and differs by the shift of the segment 15-11 only.

1 2 3 4 5 6 7 8 9 10 15 14 13 12 11 16 17 18 19 - 23 — C. *holomelas* I, C. *dorsalis*, C. *halophilus*, C. *melanescens*

1 2^{a} 15 14 13 12 11 2^{b} 3 4 5 6 7 8 9 10 16 17 18 19 - 23 — C. *calipterus*

As compared to the sequence in the Indian species, C. *plumatisetigerus*, the sequence in C. *calipterus* differs by the shift and inversion 15-12 and 6-10.

1 2^{a} 10 9 8 7 6 11 2^{b} 3 4 5 12 13 14 15 16 - 23 — C. *plumatisetigerus*

1 2^{a} 15 14 13 12 11 2^{b} 3 4 5 6 7 8 9 10 16 - 23 — C. *calipterus*

Description of the arms A E and F of Chironomus species:

Centromeres not heterochromatinised.

Arm A (Fig 7)

On the whole, this arm is fairly easy to compare. The band group 4 5 6 7 8 9 10 11 17 18 and 19 can be identified with absolute certainty. The banding pattern differs from the standard sequence in C. *thummi piger* by the shift and inversion 10-7, shift 2^{d} - 3 and 11-12 and can be derived in two steps.

1 2^{a-k} 3 4 5 6 7 8 9 10 11 12 13 14 15 16 17 18 19 — C. *thummi piger*

1 2^{a-k} 3 4 5 6 11 12 .10 9 8 7 13 14 15 16 17 18 19 — hypothetical sequence

1 2^{a} 11 12 4 5 6 2^{d-k} 3 10 9 8 7 13 14 15 16 17 18 19 — *Chironomus* species

The species also shows relations with C. *holomelas* I, C. *melanescens*, C. *yoshimatsui* (Japanese species) and the Indian sequence can be derived from the sequence of these species by an inversion and shift of the segment 6-4, inversion 3-2^{d} and shift of 10.

1 2^{a-c} 10 11 12 3 2^{k-d} 9 8 7 6 5 4 13 14 15 16 17 18 19 — C. *holomelas*, C. *melanescens*, C. *yoshimatsui*

1 2^{a} 11 12 4 5 6 2^{d-k} 3 10 9 8 7 13 14 15 16 17 18 19 — *Chironomus* species

Arm E (Fig 6)

This arm is easy to compare and has retained the second banding pattern 10^b - 3^f of the hypothetical stem species for arm E, thereby differing from the standard sequence in C. *thummi piger*, C.*pseudothummi*, C. *holomelas*, C. *alluaudi* and C. *calipterus* by the inversion 10^b - 3^f. In this way the sequence corresponds with that in C. *acidophilus*, C. *luridus*, C. *crassimanus*, C. *uliginosus* and C. *halophilus*, and also with C. *yoshimatsui* (Japanese species), C. *oppositus* and C. *australis* (Australian species).

1 2 3^{a-f} 4 5 6 7 8 9 10^{a-g} 11 12 13 — C. *thummi piger*, C. *pseudothummi*, C. *holomelas*, C. *allauaudi*, C. *calipterus*

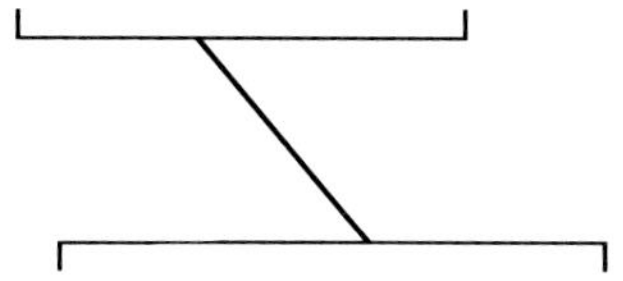

1 2 3^{a-e} 10^{b-a} 9 8 7 6 5 4 3^f 10^{c-g} 11 12 13 — *Chironomus* species, C. *luridus*, C. *acidophilus*, C. *oppositus*, C. *australis*, C. *crassimanus*, C. *uliginosus*

Arm F (Fig 8)

The arm is not very easy to compare. Only the bands 2-7, 16-23 and 12-15 can be made out easily. The sequence appears to differ from the standard sequence by the shift and inversions of the segments 11-15 and 2^b-7.

1 2 3 4 5 6 7 8 9 10 11 12 13 14 15 16 17 18 19 - 23 — C. *thummi pigert*

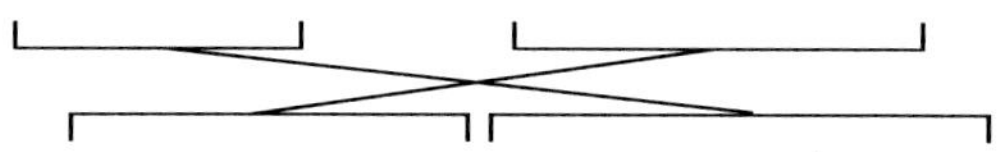

1 2^a 15 14 13 12 11 7 6 5 4 3 2^b 8 9 10 16 17 18 19 - 23 — *Chironomus* species

The sequence can also be derived from that in C. *holomelas* I, C. *melanescens*, C. *dorsalis* and C. *halophilus* through the inversion of the segment 2^b -7 and shift of the segment 15-11 in between 2^a and 7.

1 2 3 4 5 6 7 8 9 10 15 14 13 12 11 16 17 18 19 - 23 — C. *holomelas* I, C. *dorsalis*, C. *halophilus*, C. *melanescens*

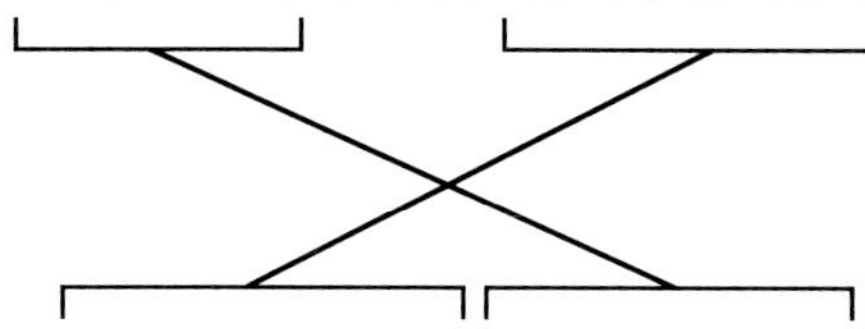

1 2^a 15 14 13 12 11 7 6 5 4 3 2^b 8 9 10 16 17 18 19 - 23 — *Chironomus* species

Discussion

All the three Indian species studied, viz. *C. plumatisetigerus*, *C calipterus*, and *Chironomus* species, have the arm combination BF, DC, EA & G and thus they belong to the pseudothummi-complex. While details of the arms A-G for the three species are being described elsewhere under Martin's (1969) system of mapping, in this paper comparisons have been made of the arms A, E & F with the European species under the scheme proposed by Keyl (1962). Comparisons of these three arms have revealed the existence of some common banding patterns of the species of thummi- and pseudothummi-complexes in contrast to other banding patterns which only occur in species of one of these complexes. The occurrrence of these basic banding patterns in the three Indian species provides additional support to the hypothesis of Keyl (1962) regarding the existence of basic patterns in the hypothetical stem species before the separation of the two complexes i.e. thummi and pseudothummi from each other by a reciprocal translocation. So far only one direct proof has been provided to this hypothesis, and that is by Wülker (1980). Figure 1 therefore provides an updated picture of the basic patterns given by Wülker (1980), with the inclusion of the three Indian species.

Based upon his studies, Keyl (1962) tried to formulate the sequences which a common ancestor must have carried in arm A, E & F. Thus this ancestor would have arm F of *C. piger* i.e. the standard form, arm A of *C. pseudothummi* i.e. the structure type 12-4, and arm E as in *C. piger* i.e. the standard form or inv10^b - 3^f of *C. crassimanus* or inv5-10^b of *C. plumosus* (i.e. three sequences for arm E). Of the six sequences suggested by Keyl for the ancestral species, at least three sequences have been found to exist in one or more arms in the living form in the Indian species studied. Thus *C. plumatisetigerus* has been found to have the structure type 12-4 of arm A of the ancestral species and is thus only the second species after *C. pseudothummi* itself known to possess this banding pattern (Fig. 1). Discovery of this pattern from India not only removes the doubts of Wülker (1980) about the disappearance of the pattern 12-4 for arm A, but also proves that our faunistic data are incomplete and that *C.pseudothummi* does not belong to any separate branch of evolution from the other pseudothummi-group of species but has evolved from the same stem species.

The Indian species, *C. calipterus*, has retained the second basic pattern for arm E i.e. the standard form of *C. piger* and *C. riparius* of the thummi-complex and in this way it is the fourth species from the pseudothummi-group of species after *C. holomelas*, *C. pseudothummi* and *C.alluaudi* to possess this banding pattern of the ancestral species (Fig 1). Similarly in arm E, *Chironomus* species has retained the third basic pattern i.e. inv10^b - 3^f of *C. crassimanus*, which is also possessed by *C. frommeri* from the thummi-complex and *C. halophilus*, *C. luridus*, *C. acidophilus*, *C. uliginosus*, *C. yoshimatsui*, *C. oppositus* and *C. australis* of the pseudothummi-complex (Fig 1).

For arm F, *C. calipterus* differs from the hypothetical ancestor by a small shift of the segment 11-15 and for arm E *C. plumatisetigerus* differs from the standard sequence of *C. piger* by the shift of the band group 3^{a-f} only.

Since three of the six basic patterns postulated by Keyl for the hypothetical ancestor of thummi- and pseudothummi-complex have been reported for the three Indian species, all of which belong to the pseudothummi-complex, it is possible that the pseudo-thummi-complex is older than the thummi-complex. This further supports the opinion of Keyl (1962) that the thummi-complex is more recent in origin than the pseudothummi-complex because interspecific chromosomal differences exist as intraspecific inversion polymorphisms only in the thummi-complex species. Also, since all the three studied species belong to the same geographical region, i.e. India or oriental region, it might have been the centre of evolution of the genus *Chironomus*. As not only thummi- and pseudothummi-complexes, but even maturus- (Wülker and Martin, 1974) and parathummi-complexes are also connected, and even the comparison to the Camptochironomus-complex contains identical or closely related patterns (Martin, 1979), this has led Martin *et al.* (1974) to the conclusion that all these complexes may be derived from a common ancestor at the same evolutionary period.

Acknowledgments

I wish to express my gratitude to Dr. Peter Cranston, Entomology Division, CSIRO, Canberra, Australia, for inviting me to the XIIth International Conference on Chironomidae; to Dr. Jon Martin, University of Melbourne, Victoria, for his expert guidance; and to Prof. C.M.S.Dass, University of Delhi, India, for providing laboratory facilities and supervision. I am also grateful to DST, CSIR, and UGC, Delhi, India, for providing financial assistance towards my travel and maintenance in Canberra.

Comparison of a Gene for a Possible Sex-influenced Protein in *Chironomus* and *Kiefferulus*

Z.Z.Chen, S.A.Fabb[1], *Jon Martin and B.T.O.Lee*

Genetics Department, University of Melbourne, Parkville, 3052, Australia

[1]Present Address: Division of Medical & Molecular Genetics, Guy's Hospital, London SE1 9RT UK.

Introduction

In a review of insect storage proteins, Thomson (1975) noted that in *Chironomus* the haemoglobins were the major storage proteins, in contrast to other insects which lacked haemoglobins but possessed a number of storage proteins carrying out different functions. He therefore suggested that the haemoglobins might be a polyphyletic grouping of proteins which had independently acquired the ability to transport oxygen. Subsequent studies of the major haemoglobins have indicated that, while they can be divided into the monomeric and dimeric forms which have been separated for about 300 myr, they have a common origin (Goodman *et al.* 1988).

However, it is still possible that Thomson is correct and some unrelated proteins also serve the double function of storage protein and oxygen transporter. This study provides some data that may support Thomson's hypothesis.

A sex-influenced protein (S.I.P.) is present in *Chironomus tentans* and a number of other *Chironomus* species (Fabb *et al.* 1990). This protein is essentially confined to females and produced in the latter half of the fourth larval instar. It appears to have some of the properties of a haemoglobin although ultimately taken up into the developing eggs. Analysis using native and denaturing PAGE indicated that it was a dimeric protein of M_r about 36 kd, probably homodimeric with the subunits of M_r =17.8 kd (Fabb *et al.* 1990). This size is roughly in accord with those of haemoglobins. Although displaying some properties similar to the described haemoglobins, there are sufficient differences to suggest that it might be an unrelated protein in the category envisaged by Thomson (1975). There is also a report in the literature of a haemoglobin being taken up into eggs of *C. thummi* (= *C. riparius*) (Travis & Schin 1976, Trewitt *et al.* 1986), but since the larvae were not sexed it cannot be determined if it was sex-influenced or corresponds to the protein we are studying.

The present study aimed to clarify the relationship of the S.I.P. of *C. tentans* to previously characterized haemoglobins. We also wished to determine whether an homologous gene occurred in the sister genus *Kiefferulus*, since immunological tests had not demonstrated the protein outside the genus *Chironomus* (Cheong 1984). The comparison of such homologous genes would be a start in the evaluation of the molecular relationships between the two genera.

MATERIALS

Two species were used in this study:

Chironomus tentans belonging to the East Canadian race, from a laboratory stock derived from Winnipeg, Manitoba, Canada, provided by Dr. B. Townsend.

An undescribed *Kiefferulus* species from a laboratory stock derived from a population at the M.M.B.W. Sewerage Farm, Werribee, Victoria. This species has been referred to elsewhere as *K.* 'Werribee species' but will be called here *K.* sp.

RESULTS

An antibody was produced against the *C. tentans* S.I.P. and this was used to probe an expression library of *Eco* RI-cut genomic DNA in λgt11, since no cDNA library was available and it seemed that if it was a haemoglobin it would lack introns (Vinogradov *et al.* 1992). Although this may not have been a valid assumption (see below), a positive clone (SF#13) with an insert of about 1.9 kb was obtained. Subsequent attempts to confirm that this clone contained the gene for the S.I.P., by demonstrating reaction of the antibody with the expressed b-galactosidase fusion-protein from the lgt11, were unsuccessful and it was concluded that the protein was being rapidly degraded.

SF#13 was sequenced from both strands of the cloned insert. Since it was subcloned from a vector in which it had been expressed, the sequence had to be in-frame with the **lac**Zgene of the λgt11. A *Pvu* I digest of the insert permitted the orientation of the gene to be established (Fig.1). Translation of the sequence indicated an ORF of 86 amino acids in-frame with the **lac**Z. This is shorter than expected from the size of the protein subunit and, since the 3' end appears to be complete including a polyadenylation site, it was concluded that the 5' end of the gene must be missing.

SF#13 was used as a probe on a Southern transfer of genomic DNA from *Kiefferulus* sp. The results (not shown) indicated that a DNA with considerable homology existed in single copy in the genome of *K.* sp. SF#13 was therefore used to probe a genomic library of *K.* sp. DNA in λGEM-11. Four positives were obtained, all from the same region of the genome

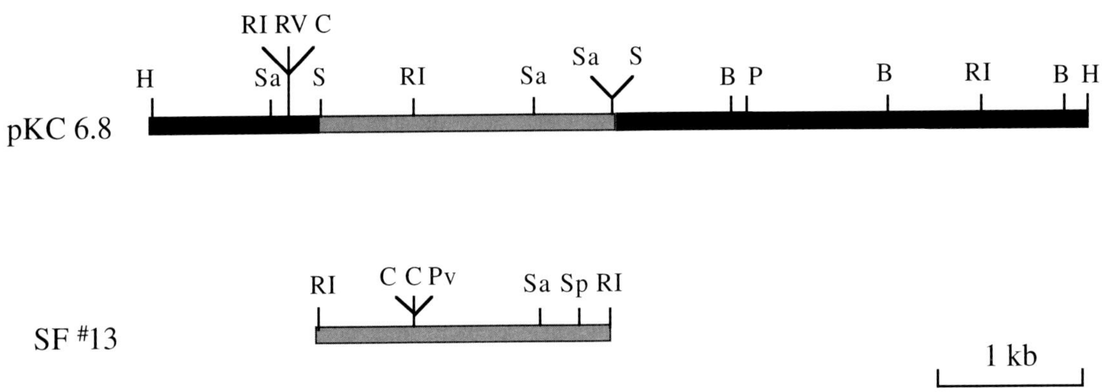

Figure 1. Restriction map of plasmid clones pKC6.8 from *K.* sp., and SF#13 from *C. tentans*. The stippled regions are homologous fragments.
Restriction enzyme sites B, *Bgl* II; C, *Cla* I; H, *Hind* III; P, *Pst* I; Pv, *Pvu* I; RI, *Eco* RI; RV, *Eco* RV; S, *Sal* I; Sp, *Sph* I.

since they could be aligned into a single restriction map. One of these clones with a 6.8 kb insert, with homology to SF#13 near the middle of the clone (Fig. 1), was chosen since this offered the possibility of obtaining a more 5' of the gene. The region was subcloned into pUC 19 for further analysis and called pKC6.8. A length of 3386 bp, straddling the region of homology to SF#13, was sequenced from both strands.

The ORF of the *Kiefferulus* gene was longer but mainly at the 3' end where it extended a further 30 codons. At the 5' end there was a stop codon only 6 codons further upstream. This is less than expected on the basis of the subunit size of the protein in *Chironomus*. Since a Northern blot suggested that the *Kiefferulus* gene was only expressed in the adult, it was possible that the 5' end differed from that of *C. tentans*. The 5' end of the *C. tentans* gene is assumed to possess a haem-pocket, to give the oxygen binding properties reported in earlier studies (Cheong 1984, Fabb *et al.* 1990). An oxygen binding capacity would only be required in the larva, and would not be necessary in a gene expressed only in the adult, as appears to be the case in *K.* sp. To check this possibility, a pair of 20 base oligonucleotides which spanned the end of the homology to SF#13 were constructed and used as primers for PCR amplification of *C. tentans* DNA. A 346 bp PCR fragment was obtained and sequenced, providing a 2131 bp sequence of the *C. tentans* gene. A dot plot comparison (Dotty Plot, version 1.0c by Don Gilbert, Indiana University) of the sequences from the two species showed that there was significant homology (Fig. 2). The homology was not confined to just the ORFs but was along the whole length of the overlap between the two clones. This homology was largely retained when the stringency was increased to 23 bases, with 572 points of homology along the whole length of the overlap (data not shown).

When the ORF of the extended sequence from *C. tentans* was checked, it proved to be very similar to that of *K.* sp., terminating at the same 5' position due to a stop codon (Fig. 3). Upstream of this stop codon there is no sequence that corresponds to the 3' consensus sequence of an intron, while there are sequences with homology to appropriate regulatory sequences in the 5' upstream region of the *Kiefferulus* clone (Fig. 3), although there is variation from the consensus sequences. The TATA box is not exact and may be the sequence TATGAT at –65 to –60. Alternatively there could be a modified sequence such as that reported for the *Drosophila E74*

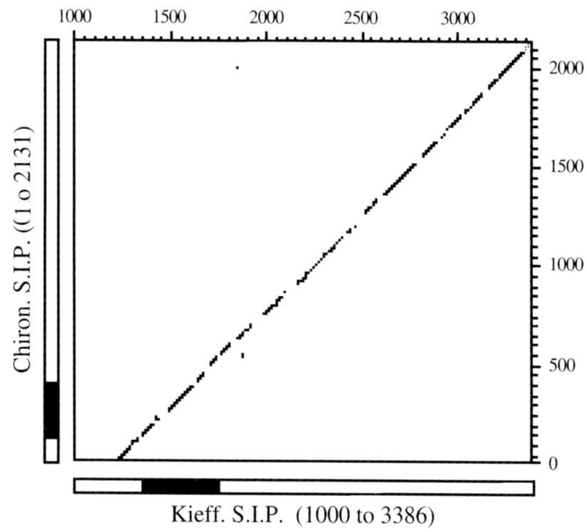

Figure 2. Dot plot comparison (see text) of the DNA sequences of the clones from *Chironomus tentans* and *Kiefferulus* sp. (excluding the first 1000 bp of *K.* sp. which has no homology to the *C. tentans* clone). The axes are the number of bp from the 5' end of the clone, based on the orientation of the ORF. The bar diagrams indicate the location of the ORF (solid bar) in the clones. A dot is entered if there is agreement of 20 bases in a frame of 25 bases. Note that significant homology exists along the whole of the clones, not just in the ORF. The total number of dots is 1404.

gene (Jones *et al.* 1991), in which case it might be included with a CAAT box-like sequence at about –108 to -98 (Fig. 3). About 450 bases upstream from the putative start codon is a sequence with homology to the consensus sequence for ecdysone receptor binding (Segraves & Richards 1990). These sequences, except for the ecdysone receptor binder, which is not within the sequence available, are also present in *C. tentans*. Using the ATG codon indicated in Fig. 3 as the start of the ORF in each species, the putative protein product is very similar. There is about 78% AA identity overall but with over 85% in some regions (Fig. 4). In one region of 44 AA there are only two AA changes, each due to a single base substitution. There are nine other base changes in the region but they are silent substitutions. Both species share the leucine rich region near the N-terminal end, with 14 leucine residues in *Kiefferulus* and 10 plus an isoleucine in *Chironomus*.

AATCTTCAAT TATTTTACCT GGATGAATTT GAACATTTTG TTGCTGCTGC
TGCTGCTGCT CAGCATGACC AAAGATCATG GAAAGCTCCT CACTTGAGTT
GTACAATGAT TCGGTATCGG ATGGACCGAC CTTTGGTAAT ACAATTGGTT
TGATTTGAAC TGGCGTTGAT GGCTCCTTTG ATGGTGTATT GTATTGCGAT
GCTTCAGATG CTGTCTCACT CACGGCTTCT GAATTCTGTC GACTCAACGA
TATCTTTTGC TCTAATTTTG AGATGCTCTC TTTGACACTC ATCTTTTGCG
GTTGTTGAGC CGCGTCAGCT TCACTCTTGA CAGCTTCCAA ATTTTCATTA
ATTTTATCGA TGTAATTCAA GACAACTTCT TTGTTTTGAC GCATTGATTT
AATGGTGATT TATGATCATT CTCATTTGTA GCGGCTGAAA TGGCTTTCTG.....
ATCGACAGTT TCTTTCAGAA ATTCTATGGC ATCTTTGTAT ATG
met

Figure 3. The first 490 bp of the 5' upstream region of the gene from *K.* sp., on the assumption that translation starts at the ATG codon at the end of the sequence shown. Possible TATA/CAAT boxes are underlined (see text) and a putative ecdysone receptor binder is double underlined. The stop codon in-frame with the ORF is indicated by dotted underlining.

Chironomus				Met	Ala	Ser	Leu	Tyr	Met	Ala	Thr	Glu	Ser
Kiefferulus				Met	Ala	Ser	Leu	Tyr	Met	Ala	Ala	Glu	Ser
Ala	Gln	Ser	–	Leu	Phe	Thr	Leu	Leu	Leu	Leu	Leu	Ile	Leu
Val	Gln	Ser	Leu	Leu	Leu	Leu	Leu	Leu	Leu	Leu	Leu	Leu	Leu
Leu	Leu	Leu	–	–	–	–	Arg	–	Phe	Ser	Ile	Asp	Glu
Leu	Leu	Leu	His	Cys	Cys	Ser	Arg	Cys	Ser	Ser	Phe	Asp	Asp
Leu	Ser	Phe	Asp	Glu	Leu	Ser	Thr	Asp	Val	Ser	Arg	Asn	Asp
Val	Ser	Phe	Asp	Glu	Leu	Ser	Thr	Asp	Val	Ser	Arg	Asn	Asp
Glu	Leu	Val	Val	Val	Ile	Asp	Leu	Ser	Ser	Gly	Glu	Phe	Ser
Glu	Leu	Val	Val	Val	Ile	Asp	Leu	Ile	Ser	Gly	Glu	Phe	Ser
Ser	Tyr	Ser	Ile	Ser	Asp	Asp	Ser	Phe	Asp	Met	Val	Thr	Ile
Ser	Tyr	Ser	Ile	Ser	Asp	Asp	Ser	Phe	Asp	Met	Ala	Thr	Ile
Asn	Gly	Ser	Ser	Phe	Gly	Arg	–	–	–	–	–	–	Asn
Asn	Gly	Ser	Cys	–	Gly	Arg	Lys	Ser	Gly	Leu	Met	Thr	Asn
Gln	Ala												
Val	Ala	Val	Ile	Ile	Val	Gly	Leu	Ile	Ser	Asp	Asp	Asp	Ala
Ser	Thr	Leu	Val	Leu	Leu	Phe	Asp	Ile	Leu	Ser	Asn	Phe	Pro
Ser	Val	Ile	Ser										

Figure 4. Comparison of the deduced amino acid sequences of the sex-influenced proteins in *Chironomus* and *Kiefferulus*, aligned to give the greatest homology. Identical amino acids are underlined.

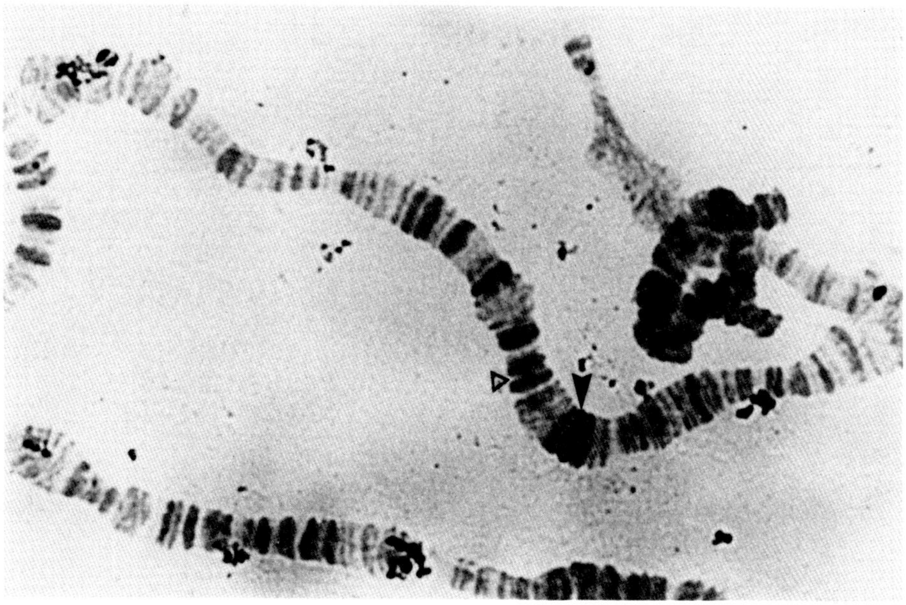

Figure 5. *In situ* localization of SF#13 to the left arm (=arm D) of the third chromosome of the East Canadian race of *C. tentans* (arrowhead). Open triangle indicates the location of the haemoglobin HbVIIB gene cluster (see Hankeln *et al.* 1993, Fig. 3).

This leads to a very hydrophobic region of the type that may be associated with lipid-binding, so that the lipid present in the native protein (Cheong 1984) may bind at this site. Alternatively, such a hydrophobic region could be associated with transport of the protein out of the cell.

These sequences show no significant homology to any in the protein sequence databases searched (Brookhaven linear (NRL), Doolittle, Genbank, Melbourne, PIR annotated (NBRF), PRF of Japan, PSD/Kyoto, Swissprot) and, in particular, they show none to the known *Chironomus* haemoglobins. Indeed, there is no clear evidence of a haem-binding site in the putative protein.

This suggests that either we do not have the complete gene or that the gene we have, although sex-influenced in its expression, is not the gene for the protein identified by the antibody. One way to answer this problem was to sequence the N-terminal amino acids from the *C. tentans* protein, which could be purified in relatively large quantities. Dr. Tim Tetaz, Baker Medical Research Institute, sequenced 10 of the first 11 amino acids: — Pro Ala Tyr Val Glu Leu Asp Gln Thr Glu. These do not correspond to any part of the putative protein sequence of either species and so do not clarify the situation. Such clarification must await further studies, such as the screening of a cDNA library which is now available. It is worth noting, however, that the AA sequence also does not correspond to the sequence of any known chironomid haemoglobin. Therefore it seems likely that Thomson (1975) was correct in his hypothesis that the oxygen transporting proteins in *Chironomus* may include some of polyphyletic origin.

Whether or not this is the gene we set out to clone, it must be a gene with a conserved function in view of the high degree of homology between the two genera. Also it would represent another chironomid gene, in addition to the haemoglobins (Vinogradov *et al.* 1992), with no introns. In order to determine the cytologenetic location of the gene, SF#13 and pKC6.8 were used as probes for *in situ* hybridization to the polytene chromosomes.

Both clones gave identical results. The gene is located at about 8B on arm D of *C. tentans* (Fig. 5) (It is difficult to be certain of the exact band identification since no fully labelled photographic maps exist for *C. tentans*). This is close to the location of the dimeric haemoglobin VIIB gene cluster (Fig. 5), which here differs from the relative position on the arm shown by Hankeln *et al.* (1993, Fig. 4) because the sequences in the North American populations differ from those in Europe. In *Kiefferulus* sp. it is located at 7D9 at the distal end of chromosome IR, as it is in *K. intertinctus* (Phillips *et al.* 1994, Fig. 2b). The dimeric haemoglobin and histone clones of Hankeln *et al.* (1993) also bind to this arm indicating that IR is the homologue of arm D.

The changes within the ORF of this gene in the two species allows us to make a first estimate of the time since the *Camptochironomus* group of *Chironomus*, and *Kiefferulus* separated from a common ancestor. Two estimates can be made, using the synonymous substitution per site (Sharp & Li 1989) and the nucleotide divergence of silent positions of fourfold degenerate codons assuming a nucleotide substitution rate of 1.1%/myr as in *Drosophila* (Bodmer & Ashburner 1984). Synonymous substitution gives an estimate of 24-48 myr, which is probably a minimal estimate because of the gaps in the optimized alignment. The silent nucleotide substitution rate gives an estimate of 50 myr. Since the two estimates are relatively similar it can be concluded that *Kiefferulus* and the *Camptochironomus* group last shared a common ancestor about 50 mya.

Acknowledgements

We are grateful to Dr. T. Tetaz, Baker Medical Research Institute for the N-terminal sequencing of the purified sex influenced protein.

A Mutation in *Chironomus tentans* Fab. that Leads to Gynandromorph and Mosaic Formation

Jon Martin

Genetics Department, University of Melbourne, Parkville, 3052, Aust.

Introduction

In the early part of 1991 some unusual adults were noted in a stock of *Chironomus tentans* . The initial observation was of a female with male antennae (Fig. 1). This may not have been the first mutant in the stock but this phenotype is the most easily recognized with the naked eye. Subsequently more abnormal individuals were identified and have continued to arise, but always in this one stock, suggesting that a mutation of some kind was involved. However the phenotype was quite variable. At one extreme were the gynandromorphs - anteroposterior of both forms, i.e as well as the female with male antennae, there was a male with female antennae (Fig. 2) and occasionally bilateral gynanders (Fig. 3). At the other extreme were females with just an anal point , or perhaps with the antennae slightly masculinized; or in the case of males, with one side of the terminalia feminized (see Figs. 24, 16 & 30 in Sublette & Martin, this volume). The details of the morphological features of the antennae and hypopygia of these abnormal individuals have been studied by Sublette and Martin (this volume), while this paper investigates other aspects of the sexual phenotype and attempts to gain some insight into the nature of the genetic change involved.

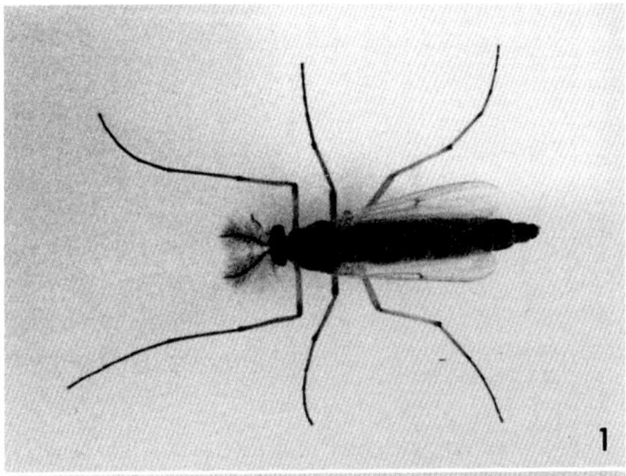

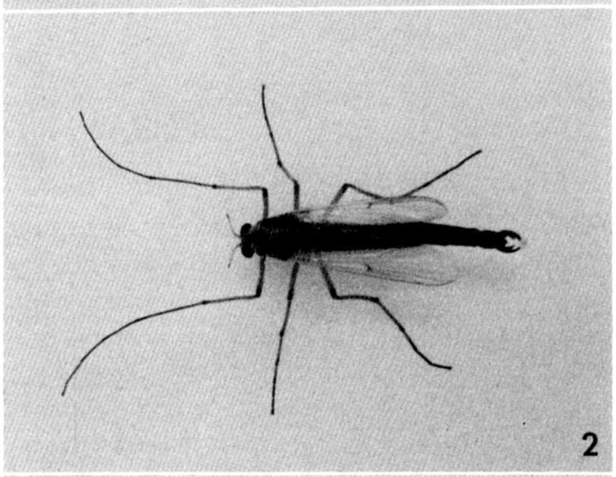

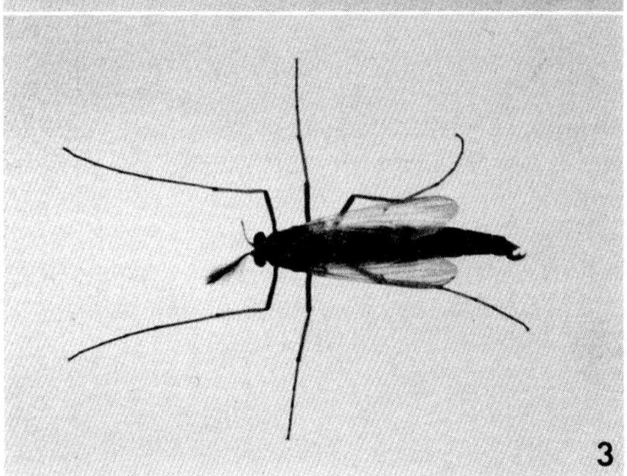

Figure 1. Female anteroposterior gynandromorph with female genitalia but male antennae.

Figure 2. Male anteroposterior gynandromorph with male genitalia but female antennae.

Figure 3. Bilateral gynandromorph with female structures on the left and male structures on the right of the individual.

Material and Methods

The stock of Eastern race of *Chironomus tentans* Fab. from the vicinity of Winnipeg, Manitoba, Canada was obtained from Dr. Bruce Townsend in 1982. It has been maintained in the laboratory in Walter's medium (Walter 1973) with alcohol-washed Kleenwipe tissues and, in the last two years with a thin layer of stone dust, as substrate. They were fed weekly with 0.4g trout starter , supplemented with TetraMin flakes.

Mating of individual pairs for genetic crosses was carried out in 100x25mm glass vials. The females were left in the same vial with a small amount of water added in which to lay eggs. The egg masses were examined for fertility and normality of embryonic development before being set up in individual cages.

Adults for study were killed by chilling in a deep freeze, then the body parts were dissected and the thorax put into a microtiter plate containing buffer for subsequent electrophoresis for malate dehydrogenase (Martin & Lee 1988). In some cases the abdomen was dissected fresh in saline, in others it was put into 85% ethanol with the other dissected body parts. These were subsequently cleared in 10%KOH and mounted in euparal. Measurements were made using an eyepiece micrometer in a dissecting microscope. The measurement of wing length was modified to allow for the fact that the stem vein was often torn off during dissection. Wing length was therefore measured from the humeral crossvein to the tip of the wing. Wing width was measured across the position of the RM crossvein. Lengths of tibial, metatarsal and second tarsal leg segments were measured from the distal end of the segment to the point of insertion with the previous segment. AR for both sexes was measured as the length of the distal segment relative to the combined length of the other segments.

Results and Discussion

One of the first problems that arises in respect of this work is how to refer to the basic types for the purposes of discussion. Previously recorded gynandromorphs in *C. tentans* have mostly been the result of mermithid parasitism (Sæther & Galloway 1980). Here the presence of the male hypopygium represents an individual which is genetically male. Therefore for this work, antero-posterior gynandromorphs or mosaics with a predominantly male hypopygium were called 'abnormal males', while those with a predominantly female hypopygium were called 'abnormal females'. As will become obvious below, this is not correct but it does provide a convenient means of categorizing the specimens.

Dissection of the abdomen of normal and abnormal adults indicated that there was considerable variability in the degree of development of the reproductive organs in the abnormal individuals. Abnormal individuals varied from an apparently normal set of reproductive organs consistent with the hypopygium, through a mixture of male and female tissues with some organs reduced or undeveloped, to virtually lacking all internal reproductive organs. Some normal females also lacked an ovary on one side. No more detailed description of this variation will be given here.

There is a further way that genetic sex of these specimens may be determined:In this stock the dominant male determining (MD) locus is very tightly linked to an electrophoretically detectable enzyme locus - malate dehydrogenase 1 (*Mdh-1*) This is the fastest migrating of the two electrophoretically detectable *Mdh* variants in *Chironomus* (Martin & Lee 1988). Thus in this stock, all males are heterozygous for *Mdh-1* and all females are homozygous at this locus. There are problems however as the enzyme begins to break down in adults, so that the protein may migrate further or give rise to additional, breakdown bands. These additional bands can cause a problem in some specimens in determining whether the *Mdh-1* genotype is homozygous with breakdown products, or a mosaic of the homozygous and heterozygous genotypes in the same individuals. However the *Mdh-1* genotypes do shed some light on the situation. Not all individuals with the same major phenotype were identical, in MDH phenotype e.g. some of the female anteroposterior gynandromorphs (the ones with male antennae and female terminalia) were homozygous and some were heterozgous. This is significant as it suggests that individuals with the same apparent phenotype have a different MD genotype. Thus these phenotypes do not constitute a uniform grouping.

The question therefore arises as to whether the morphological phenotypes within each group were as uniform as first appeared. The sex of the head can be categorized by the antennae, and of the abdomen by the hypopygium, but this tells nothing about the parts in between.

Therefore a more detailed study of the wings and legs was undertaken to see if those of normal males and females were sufficiently dimorphic to be used as a means of classifying sexual phenotype. Since ratios can vary with environment, normal individuals should preferably come only from the same stock. Measurement of a small number of specimens from other localities confirmed that this was the case. Therefore the only specimens that could be used as controls were the normal sibs of the abnormal specimens, and other specimens from the same stock. If the hypothesis that other parts of the body could be differentially affected was correct, these sibs could include abnormal individuals in which the antennae and genitalia were not affected. The results suggest that some were indeed abnormal, but these could be recognized by the large difference in value between left and right side of the body.

A number of characters have previously been recognized as differing between males and females. In some *Chironomus* species the male has a beard on the foretarsus, which is absent in the female. No beard is present in *C. tentans* (Townes 1945). The number of sensilla chaetica on the mid and hind metatarsi differ quite markedly between the sexes, but these are hard to count, depending upon how the leg segment is lying, and therefore could be misleading in an intersexual phenotype. The measurements chosen, after some preliminary investigations, were the wing ratio (i.e. ratio of length to width); and the ratio of metatarsal length to length of tarsus 2 (met/tar 2) of each leg. The AR was also measured. This is over 3 in males, and about 0.3 in females. However only a couple of individuals with intersexual antennae were present in the sample so it was not used for practical purposes.

Wing ratio: Although it is apparent that males have a shorter narrower wing compared to females (ratio of 3.41 in ♂, cf. 2.94 in ♀), this ratio is not clearly dimorphic in the sexes since the values overlap. The overlap shown in Figure 4 is actually due to one wing only in each of two male specimens. In both of these the ratio of the other wing was much higher (3.4 and 3.6 respectively), suggesting that these specimens may have been mosaics with one female wing. The ratio permitted most wings to be clearly designated female or male, but those with values close to the overlap values were classified as uncertain.

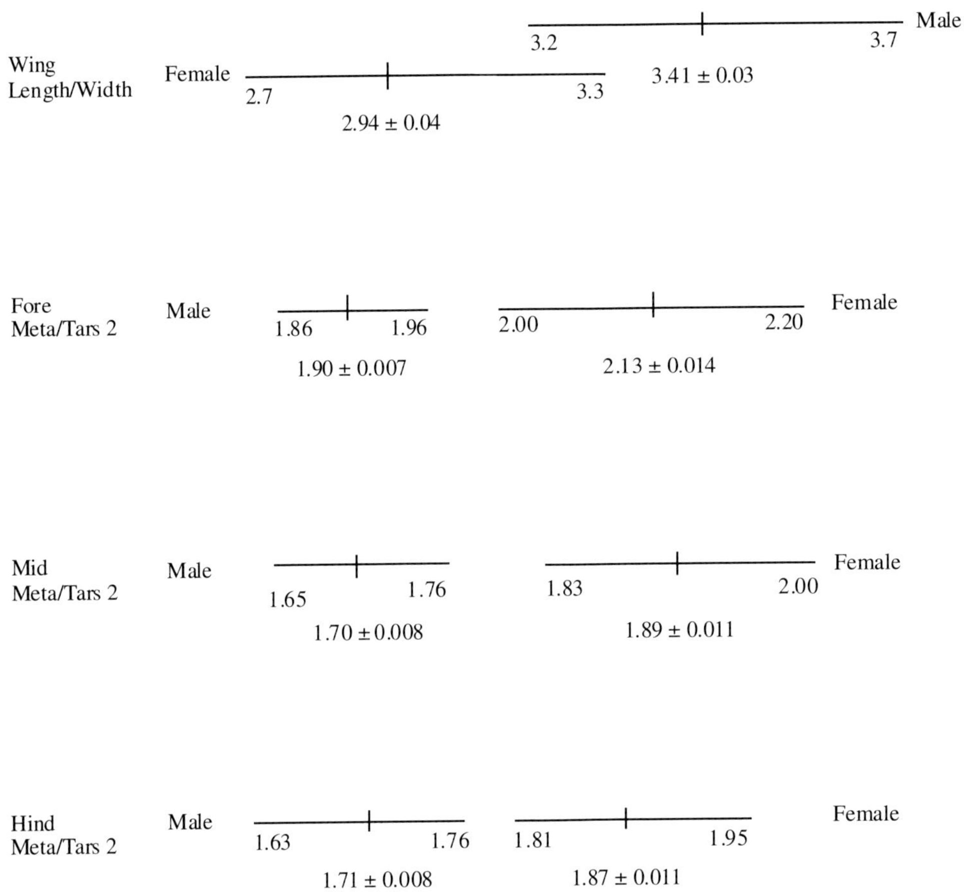

Figure 4. Distribution of values for the wing and leg ratios of normal males and females of *C. tentans*. Vertical lines represent the means, which are given underneath along with the standard error of the mean. The manner in which the ratios are obtained is detailed in the text.

Meta/Tars 2: This ratio clearly separates females and males on each pair of legs (Fig. 4) with no overlap.

The first thing that became obvious from this analysis was the large difference that can occur between the right and left hand measurements of a specimen, even where the obvious gynandromorphism is anteroposterior - e.g. a female gynandromorph had a wing ratio of 2.97 on one side and 3.62 on the other side. This asymmetry is interpreted as indicating that the tissue of one wing was female, while that of the other was male.

The analysis further emphasized the difference between specimens in the major gynandromorph classes (Fig. 5). Specimens which were heterozygous for *Mdh-1* ('genetic males') tended to have male leg and wing ratios, while those homozygous for *Mdh-1* ('genetic females') tended to have female ratios. This of course is a generalization but the main point to be noted is that many of these specimens are not really gynandromorphs, but mosaics with quite different amounts of male and female tissue.

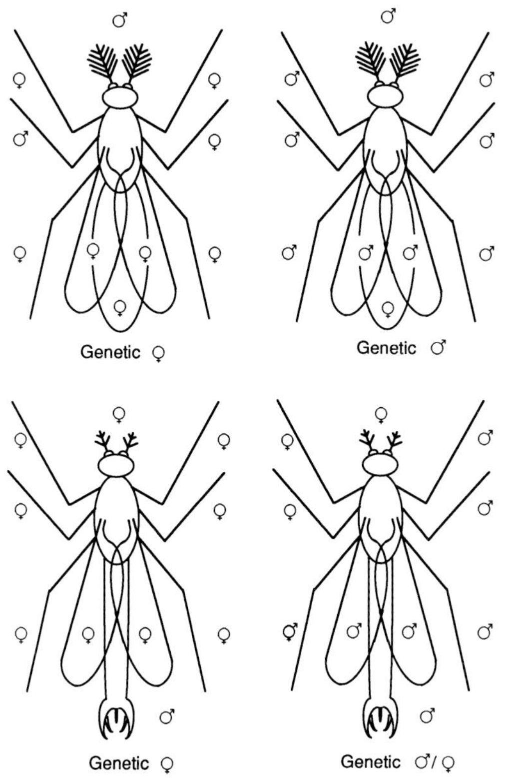

Figure 5. Cartoons of extreme phenotypes of anteroposterior gynandromorph showing the different extent of male and female tissue. Genetic ♀ = homozygous for *Mdh-1*, Genetic ♂ = heterozygous for *Mdh-1*, and Genetic ♂/♀ = *Mdh-1* phenotype with ♀ band the strongest, but also showing two faster bands as if some ♂ tissue also present.

In order to determine the nature of the mutation, genetic crosses are necessary. Hence it was necessary to determine whether the abnormal individuals would mate and, if so, whether they were fertile. The mating behaviour of the abnormal individuals is similar to that reported for gynandromorphs in *Drosophila*, in that it is determined by the sex of the head (Hall 1978). Specimens with male antennae behaved as males and would attempt to copulate with other specimens despite the fact they had a female terminalia. This in fact made them easy to mate, for their attempts to copulate with males made it relatively easy for the males to copulate with them. None of these individuals displayed the avoidance behaviour of tucking the abdomen under themselves, shown by normal females. However ease of mating was not reflected in ease of obtaining eggs. Only one of these females has laid eggs - and then only about three. Dissection has shown that some of them at least have well developed ovaries filled with apparently normal eggs. Therefore it is not certain whether they are infertile or simply lack the necessary egg laying behaviour.

The gynandromorphs with female antennae act as females and therefore it is very difficult to get them to attempt to copulate with a female. However when it can be achieved, by continually knocking them into contact with a virgin female, they are often fertile. Therefore data on the genetic nature of the mutation can only be obtained from crosses of male gynander x normal female or from normal x normal crosses which produce abnormal offspring.

None of these crosses fit any normal genetic ratio - abnormal offspring are generally less than 10% of offspring, although there is also a high degree of infertility (up to 40%) with an additional 5% embryonic lethality. Infertility level can be misleading because this stock now has variable rates of infertility, probably due to inbreeding effects from the 12 years in culture. However crosses which did not produce abnormal offspring did not show embryonic lethality, so that characteristic is taken to be a feature associated with the production of abnormal offspring.

The basic conclusion from this data is that the primary genetic mutation does not directly produce the gynandromorphs or mosaics. The detailed genetic hypothesis will be explained in detail in a later paper (Martin & Lee, in preparation), but it seems overall that the results can be explained on the basis of a mutation that increases the rate of mitotic recombination. This would require that all the gynandromorphs or mosaics had in fact started life as genetic males, but one or more recombination events lead to some cells, and the resulting tissues, becoming female, as has been reported in houseflies (Hilfiker-Kleiner *et al.* 1993). The *MDH-1* locus, because of its close linkage to the male determining locus, also gets recombined to give the homozygous female genotype, which can be detected when the thoracic tissue has been converted to female. This hypothesis seems capable of explaining all abnormal individuals obtained so far, even those that are about 90% female with just abnormal antennae or perhaps an anal point.

Acknowledgements:

I am grateful to Dr. B.T.O.Lee and Prof. R.Nöthiger for useful discussions of this problem. I am also grateful to Dr. Lee for critical comments on the manuscript; to Sarah and Stuart Pike for technical assistance in maintaining the stock; to Quentin Lang for preparing the photographs and figure 5; and to Cheryl Grant for preparing figure 4.

PART 2

ENVIRONMENTALLY-INDUCED DEVELOPMENTAL ABNORMALITIES

THE *IN SITU* RELATIONSHIPS BETWEEN SEDIMENT CONCENTRATIONS OF MICROPOLLUTANTS AND MORPHOLOGICAL DEFORMITIES IN *CHIRONOMUS* GR. *THUMMI* LARVAE (DIPTERA, CHIRONOMIDAE) FROM LOWLAND RIVERS (BELGIUM): A SPATIAL COMPARISON

L. Janssens de Bisthoven, C. Huysmans and F. Ollevier

Laboratory for Ecology and Aquaculture,
Catholic University of Leuven, Naamsestraat 59,
3000 Leuven, Belgium.

INTRODUCTION

A direct relationship between morphological deformities of the head capsules of benthic larval Chironomidae and heavy metal and pesticide pollution of aquatic sediments was hypothesised by Hamilton and Sæther (1970). Such a relationship was confirmed later in field studies, with deformed larvae occurring in sites polluted with heavy metals or pesticides, other organic xenobiotics, radionuclides or a mixture of these (e.g. Hare and Carter, 1976; Köhn and Frank, 1980; Warwick, 1980; Cushman, 1984; Wiederholm, 1984; van Urk *et al.*, 1985; Warwick *et al.*, 1987; Pettigrove, 1989; Warwick, 1990; Dermott, 1991, van Urk *et al.*, 1992). In reviewing deformities in chironomid larvae, Warwick (1988) stressed the many advantages of their use as biological indicators for heavy metal and pesticide pollution in aquatic systems. Although there are many circumstantial semi-quantitative data indicating a relationship between sediment pollution and chironomid deformities, no attempt has been made so far to quantify this field-derived relationship. The purpose of this study was to quantify *in situ* relationships between morphological deformities in chironomid larvae and a wide range of micropollutants in fluvial sediments from different lowland rivers and watersheds in Belgium (Flanders). We emphasise a global approach of inter-site comparisons of raw total and transformed total sediment concentrations and deformity data. These are summarized as mean deformity strength, variance of deformity strength and percentage deformed larvae for five head capsule structures of *Chironomus* gr. *thummi* larvae. The approach in this spatial comparison of deformity variables and pollutants is essentially governed by principal factor analysis, correlation and linear regression analysis. Some drawbacks resulting from the linearity assumption are discussed.

Materials and methods

Fourth instar *Chironomus* gr. *thummi* Kieffer larvae were collected in 1987 and 1992 with a 250µm mesh size handnet during at least 20 minutes along a 20 m river embankment from a total of 36 sites in Flemish lowland rivers (Belgium) (Fig. 1, Table 1). At the same sites and the same periods, the upper 5 cm sediment layer had been sampled with a PVC Eckman grab or a plexi-glass hand-corer by various government agencies for chemical analysis of heavy metals, organic compounds and total organic carbon (TOC) (Table 2). These lowland rivers flow through heavily populated areas with a high degree of urbanisation, industrialisation and intensive agriculture and are meso- to hypereutrophic. Larvae of C. gr. *thummi* collected from the Dijle and the Dommel basins were reared in the laboratory to imagines, which were identified as *Chironomus riparius* Meigen, according to Pinder (1978). However, as (1) rearing of imagines could not have been systematically done and (2) some reared imagines from other river basins could not be identified to the species level, the species is referred to in this study as belonging to the *thummi* species-group. The head capsules of about 30 larvae per site (see Table 1) were mounted in Euparal balsam after digestion in hot 10% KOH, cleansing in distillated water and dehydration in an ethanol series. A micro-dissection of the head capsules was performed on slides in order to exhibit optimally all buccal structures and antennae for latter light micocopical observation. For the designation of the morphological structures we refer to the nomenclature of Sæther (1980). Each departure from the normal pattern in the head capsule structures was considered as a deformity. However, concerning the toothed buccal structures, abnormalities resulting from mechanical wearing or mounting artefacts were not included. These could readily be distinguished from real deformities by their structural characteristics: i.e. the edge of abnormalities resulting from mounting damage or mechanical wearing is irregular, so that the double chitinous wall of such structures (unequal wearing of both walls) becomes apparent. Moreover, mechanical wearing is not localised, but symmetrically spread over the teeth of a buccal structure. Real deformities, however, are often localised and asymmetrical, they exhibit a smooth edge, the double wall is invisible and often show a lack of pigmentation at the base of a gap. If there remained doubt concerning the origin of the abnormality, a conservative approach was taken and the abnormality was excluded from the analysis. For the antennae, lack of 2nd, 3rd, 4th and 5th segments was taken into account, although a proportion of these anomalies probably resulted from mounting damage.

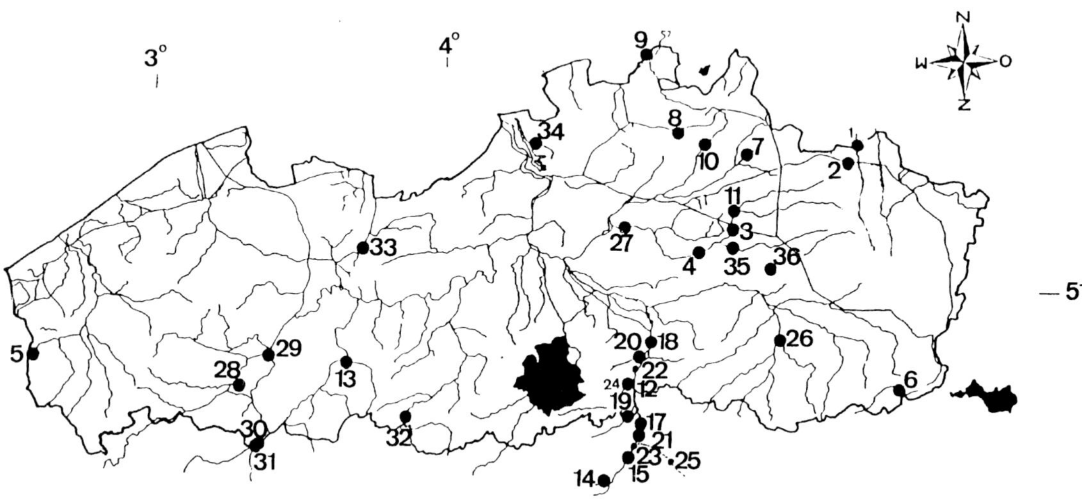

Figure 1. Hydrological map of Flanders with the sampling sites (numbers). For the river and locality names, see Table 1.

Table 1: Sampling sites: numbering, river, locality, Lambert-coördinates (Belgian km grid system), sampling dates, number of mounted larvae and site codes (only for Dijle watershed).

NR .RIVER	LOCALITY	Lambert coördinates	SAMPLING DATE	number of mounted larvae	CODE
1. Dommel	Hageven (Neerpelt)	224.05/218.10	03 September 1992	30	
2. Eindergatloop	Neerpelt	223.50/214.30	06 August 1992	30	
3. Grote Nete	Geel	193.18/200.90	06 August 1992	20	
4. Grote Nete	Westerlo	185.66/196.10	06 August 1992	30	
5. Ijzer	Roesbrugge	027.25/180.32	11 August 1992	30	
6. Jeker	Tongeren	231.19/162.63	10 August 1992	30	
7. Kleine Nete	Retie	197.86/216.73	06 August 1992	15	
8. Laakbeek	Beerse	181.65/222.31	26 February 1992	30	
9. Leiloop	Hoogstraten	175.20/241.73	10 September 1992	30	
10. Aa	Turnhout	187.82/220.28	26 February 1992	30	
11. Molse Nete	Geel	194.42/204.38	06 August 1992	19	
12. Ijse	Neerijse	167.85/166.98	02 March 1992	30	
13. Zwalm	Nederzwalm	101.31/175.59	11 August 1992	20	
14. Dijle	Loupoigne	155.21/142.80	01 October 1987	52	LOD
15. Dijle	Basse Wavre	169.20/157.80	01 October 1987	21	BWD
16. Dijle	Grands-Prés	169.20/162.80	01 October 1987	19	GPD
17. Dijle	St. -Agatha-Rode	168.80/163.70	01 October 1987	22	SAD
18. Dijle	Leuven	174.00/175.80	01 October 1987	31	LED
19. Laan	St.-Agatha-Rode	168.40/164.20	01 October 1987	18	SAL
20. Voer	Leuven	172.90/173.25	01 October 1987	28	LEV
21. Dijle	Florival	169.22/163.80	01 October 1987	7	FLD
22. Dijle	Korbeek-Dijle	169.25/169.60	01 October 1987	7	KOD
23. Dijle	Limal	169.20/160.10	01 October 1987	0	LID
24. Ijse	Neerijse	167.85/166.98	01 October 1987	0	NEI
25. Train	Archennes	171.45/160.01	01 October 1987	0	ART
26. Gete	Zoutleeuw	202.79/175.48	11 August 1992	0	
27. Grote Nete	Lier	165.80/202.80	11 August 1992	0	
28. Gaverbeek	Harelbeke	075.50/170.62	11 August 1992	0	
29. Gaverbeek	Waregem	082.80/177.55	11 August 1992	0	
30. Zwarte Spierebeek	Spiere-Helkijn	076.47/156.70	11 August 1992	0	
31. Grote Spierebeek	Spiere-Helkijn	079.09/157.17	11 August 1992	0	
32. Dender	Geraardsbergen	114.82/161.36	11 August 1992	0	
33. Kan. Gent-Terneuzen	Gent	106.35/200.24	11 August 1992	0	
34. Groot Schijn	Antwerpen	148.77/221.72	11 August 1992	0	
35. Grote Laak	Laakdal	193.30/196.60	06 August 1992	0	
36. Winterbeek	Tessenderlo	201.70/191.11	06 August 1992	0	

In that case however, there was no way of discriminating between real deformities and artefacts. The choice of the structures analyzed, the zonations quantified and the number of points allocated was based on previous quantification attempts and stepwise optimization of the quantification system, as well as on the quantification framework proposed by Warwick (*pers. comm.*, 1985, 1991): Warwick's quantification system gives a certain 'deformity weight' to a particular midge population, based on a summing up of point values from different kinds of deformities in different structures, analysed in a detailed way. The system allows inter-site (i.e. inter-population) comparisons and is open for addition of other structures, other deformities, and even other midge species. The method used in our study is purely a working instrument and does not provide at this stage a tool for routine biomonitoring. It has the essential aim of demonstrating simultaneous increases of deformities with increases of pollutant concentrations: morphological deformities of mentum, mandibles, pecten epipharyngis, premandibles and antennae were quantified according to a points allocating system (Table 3).

Table 2: The screened pollutants, their abreviations and their affiliation to particular datasets and transformations. Only datasets containing deformity data are given. All transformations were performed on data from 9 sites of the Dijle basin (sites 14 to 22).

HEAVY METALS	abreviation	dataset	transformation
cadmium	Cd	22 sites (nr. 1 to 22)	yes
lead	Pb	"	"
chromium	Cr	"	"
copper	Cu	"	"
zinc	Zn	"	"
iron	Fe	9 sites (nr. 14 to 22)	"
manganese	Mn	"	"
arsenic	As	13 sites (nr. 1 to 13)	no
nickel	Ni	"	"
mercury	Hg	"	"
POLYCHLORIDE BIPHENYLS			
2,4,4' trichloridebiphenyl	PCB28	22 sites (nr. 1 to 22)	yes
2,2',5,5' tetrachloridebiphenyl	PCB52	"	"
2,2',4,5,5' pentachloridebiphenyl	PCB101	"	"
2,3',4,4',5 pentachloridebiphenyl	PCB118	"	"
2,2',3,4,4',5' hexachloridebiphenyl	PCB138	"	"
2,2,'4,4',5,5' hexachloridebiphenyl	PCB153	"	"
2,2',3,4,4',5,5' heptachloridebiphenyl	PCB180	"	"
2,2',4,5' tetrachloride biphenyl	PCB49	13 sites (nr.1 to 13)	no
ORGANOCHLORINE PESTICIDES			
aldrin	aldrin	22 sites (nr. 1 to 22)	no
dieldrin	dieldrin	"	"
endrin	endrin	"	"
isodrin	isodrin	9 sites (nr. 14 to 22)	"
hexachloridecyclohexane alfa	aHCH	22 sites (nr. 1 to 22)	yes
hexachloridecyclohexane epsilon	eHCH	9 sites (nr. 14 to 22)	"
hexachloridecyclohexane gamma	gHCH	22 sites (nr. 1 to 22)	"
heptachloride	hepta	"	"
heptachloride epoxide	epox	"	no
hexachloridebenzene	HCB	9 sites (nr. 14 to 22)	yes
pp'-dichloridediphenyltrichlorideethane	pp'DDT	22 sites (nr. 1 to 22)	"
op' - dichloridediphenyldichlorideethylene	opDDE	9 sites (nr. 14 to 22)	no
pp' - dichloridediphenyldichlorideethylene	pp'DDE	22 sites (nr. 1 to 22)	yes, on 9 sites
pp' -dichloridediphenyldichlorideethane	pp'DDD	"	no
metoxychloride	metox	9 sites (nr. 14 to 22)	"
APOLAR HYDROCARBURES	akws	9 sites (nr. 14 to 22)	"
POLYAROMATIC HYDROCARBURES	PAH	9 sites (nr. 14 to 22)	yes
fluoranthene (b)	fluorant	22 sites (nr. 1 to 22)	"
benzo(b)fluoranthene (b)	benzob	"	"
benzo(k)fluoranthene (b)	benzok	"	"
benzo(a)pyrene	benzoa	"	"
benzo(g,h,i)perylene	benzog	"	"
indeno(1,2,3,c,d)pyrene	indeno	"	"
PHTALATES			
di-ethylphtalate	ef2	13 sites	no
di-isopropylphtalate	ipf2	"	"
di-propylphtalate	prf2	"	"
di-isobutylphtalate	ibf2	"	"
di-butylphtalate	bf2	"	"
di-hexylphtalate	hexf2	"	"
di-iso-octylphtalate	iof2	"	"
di-nonylphtalate	nf2	"	"

Table 3: Quantification framework for allocating points to morphological deformities in *Chironomus* gr. *thummi* larvae. The points per larva are summed up for each structure (paired structures are pooled) and a mean and variance are calculated per midge population.

structure	deformity type	point allocation
mentum	- 1 tooth	3
	- 1/2 tooth	2
	minus >1 tooth in 4 external lateral teeth	10
	minus >1 tooth in 2 inner lateral teeth	20
	minus >1 tooth in trifid medial tooth	30
	Köhngap of size equal to 1 tooth of corresponding zonation	10
	1 1/2	15
	2	20
	3	40
	4	80
	>5	100
	+ 1 tooth	30
	+ 2 additional structures	20
	+ 3 additional structures	40
	twisted or splitted tooth	10
mandibles	- 1 tooth	3 p
	- 1/2 tooth	2 p
	minus >1 tooth in inner teeth	10
	minus >1 tooth in inner teeth and dorsal tooth	20
	+ 1 tooth	30
	blunted to unrecognizable	50 to 100
Premandibles	1 shortened or twisted tooth	10
	2 shortened or twisted teeth	30
	- 1 tooth	40
	- 2 teeth	90
	+ 1 tooth	40
	shaft more than 1/2 shortened	30
Pecten epipharyngis		
	overlapping teeth (1 pair)	10
	overlapping teeth (2 pairs)	20
	overlapping teeth (3 or more pairs)	40
	asymmetry	20
	widening (at least 2x normal width) of 1 tooth	10
	widening of 2 teeth	20
	widening of 3 teeth	40
	widening of 4 teeth or more	80
	less than 10 teeth	10
Antennae	- 1 upper segment	5
	- 2 upper segments	10
	- 3 upper segments	20
	- 4 upper segments	40
	- 5 segments	considered as artefact
	filamentous segment	10
	Ring organ moved to upper half of 1st segment	10
	Blade reduced or lacking	10
	+ 1 additional structure	10

The deformed structures of each larva of a site were allocated a number of points, which increased in a geometric series (factor 2 or 3), according to the degree of deformity. For each structure from each larva, the allocated points were summed up; points of paired structures were pooled. For each population (per sampling site), a mean and variance to this mean was calculated for the deformity points of each structure. Moreover, the percentage deformed larvae for each structure was noted. Thus, the three parameters used in this analysis (mean, variance and percentage) give complementary information about the distribution of deformities in a midge population. Although the raw deformity data of individual larvae are most often not normally distributed, the percentages, means and variances per population tend to approach normality and stability of variance, according to the central limit theorem (Snedecor and Cochran, 1980). In order to preserve the geometric point allocating scale, and since we were interested also in the few extremely deformed larvae of a population (outlying values), data were not transformed prior to statistical analysis. More detailed analysis on the frequency distribution of the deformities, the different types of deformities, some morphometric information pertaining to sublethal growth impairment and temporal patterns are not considered at this stage. For a limited number of sites, the frequency of granulometric size fractions was assessed by sieving and sorting and with a sedimentation method based on Stokes' law in Gullentops cylinders.

The relationships between the 15 deformity variables and a wide array of micropollutant concentrations were assessed by means of correlation analysis. When interactions between pollutants were to be taken into account, multiple linear regression analysis was used. The variability within the larvae data-set and within the micropollutant data-sets was illustrated with Principal Factor Analysis (centroid method with equimax normality rotation, referred to as 'factoranalysis') in biplots with factor loadings (arrows) representing the variables (either deformity or pollution variables) and factor scores (named or numbered dots) representing the sites (midge populations). In a second approach, a number of heavy metals from smaller data-subsets were weighed to different components of the sediments as follows: metal/total iron, metal/total manganese, metal/clay, metal*total iron/total manganese*TOC . The choice of these formulae are based on studies by Luoma and Bryan (1978) and Hall and Bindra (1979). However we could use only total iron and manganese concentrations instead of more loosely bound concentrations as reported in these studies. The organic xenobiotics were weighed on total organic carbon content of the sediment as org. xenob./TOC, based on the strong affinity of such components for organic matter (van de Guchte and Van Leeuwen, 1988). All computations were performed with the STATISTICA-CSS package.

Results

Deformities

In 22 from the 36 sampled sites, samples contained individuals of *Chironomus* gr. *thummi* instar 4 larvae (Table 1). Percentages of larvae that were deformed ranged from 0% to more than 60% (mentum, mandibles, antennae). Mean number of points ranged from 0 to 24 (mentum) and 45 (total) point units. Variance to mean deformity strength ranged from 0 to 1280 (mentum) and 3850 (total) point units. Within each structure, percentages, means and variances were positively correlated with each other ($n=22$, $r>0.4$, $P<0.05$). However, the 5 structures did generally not correlate amongst each other for their percentages, means and variances (i.e. at population level). The raw point values (i.e. at individual level) show a weak correlation between mentum and mandible deformities ($n=267$, $r=0.26$, $P<0.05$).

A Factoranalysis performed on the means, variances and percentages of the deformity data (Fig. 2) shows that antennae have more weight on the global analysis, when looking at percentages, while mandibles have more weight when looking at the means and variances. Sites tend to cluster around particular structures in a different way when looking at either means, percentages or variances.

Granulometry and pollutants in the Dijle basin sites (12 sites) and the occurrence of *Chironomus* gr. *thummi* larvae

From a sub-dataset containing 12 sites from the Dyle watershed, the absence of *Chironomus* larvae can be predicted by the granulometric composition of the sediment, as is demonstrated in Fig. 3: sites with few (sites 21 and 22) or no larvae (sites 23, 24 and 25) cluster around the 125µm fraction, while sites with more

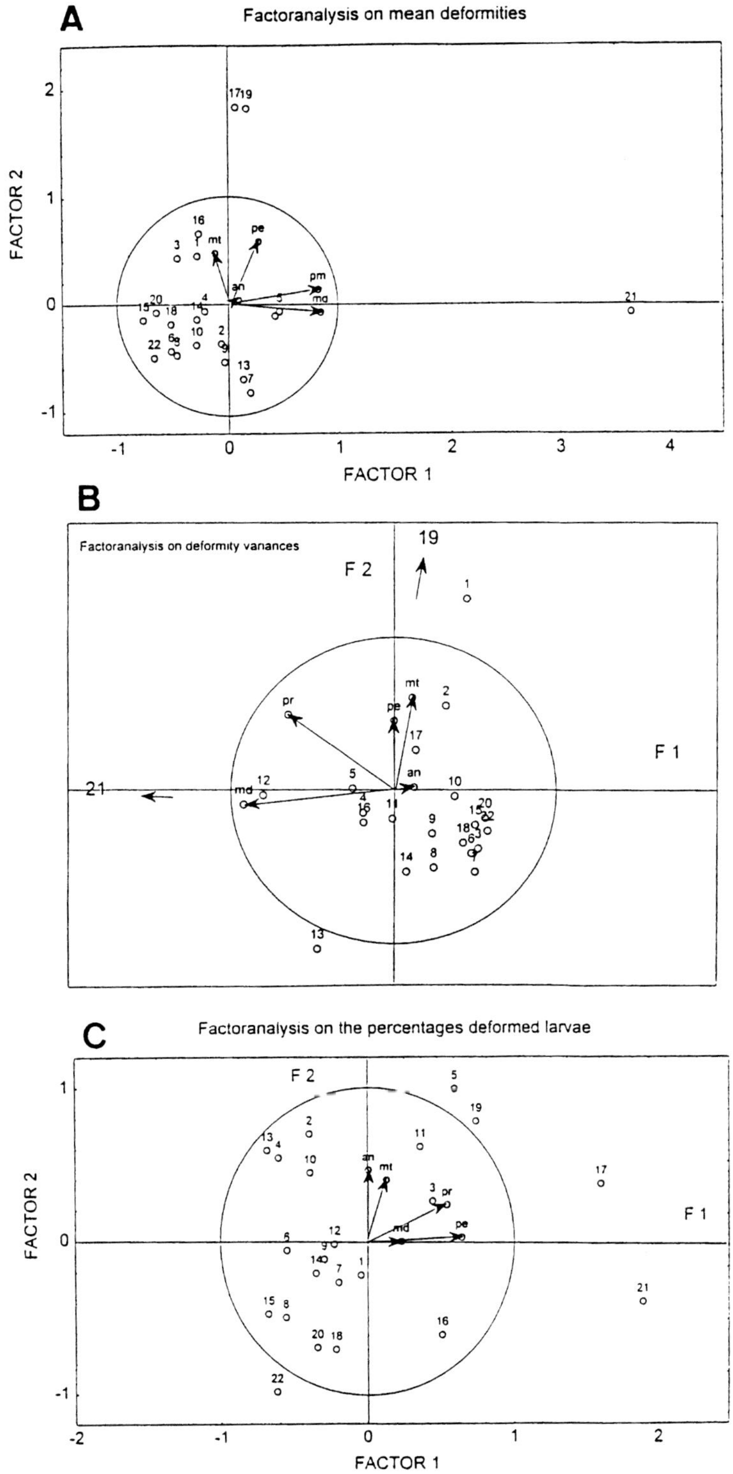

Figure 2. Factor analysis on the means (A), variances (B) and percentages (C) of the deformity data from *Chironomus* gr. *thummi* populations from 22 sites in Flanders. Dots represent sites. For site names see Table 1, arrows represent deformity variables. Sites marked with an arrow are located outside the graph range. F 1 represents 20 to 29 % of the total variability; F 2, 10 to 16 % of the remaining variability.

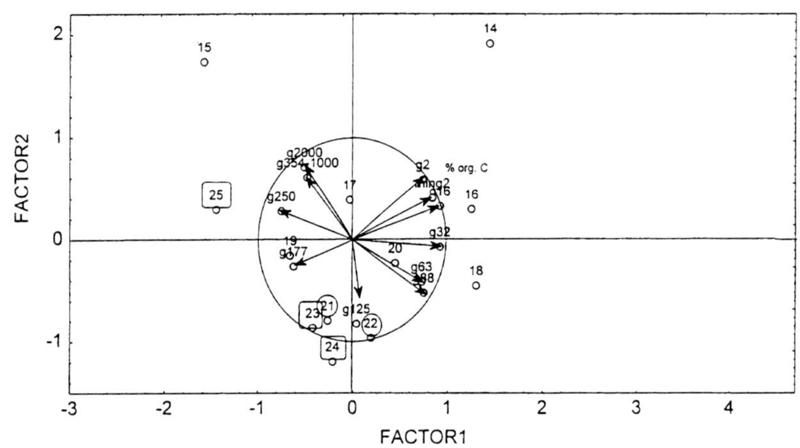

Figure 3. Factor analysis on the granulometric size fractions (arrows) of the sediments from 12 sites (numbered dots) of the Dijle watershed. For site names, see Table 1. The squared sites are sites where no larvae were found. The circled sites are sites where only a few larvae were found. F 1 represents 50% of the total variability; F 2, 26% of the remaining variability.

larvae cluster around finer sediment fractions and organic carbon content. When looking at the pollutants in the same sites (Fig. 4) it is remarkable to note that the sites without larvae cluster around the clay and chromium content of the sediment on the one hand and around the PCB 52 and PAH content on the other hand.

Pollutants in the global analysis (36 sites) and the occurrence of Chironomus gr. thummi larvae

Almost all sites without larvae show outlying values of micropollutant concentrations: sites 28, 30, 34, 35 and 36 are well discriminated from the other larvae containing sites by their heavy metal contents (Fig. 5); site 34 for its pp'-DDE content; sites 28, 30, 31, 32, 34 and 36 for their organochlorine pesticide contents or their phtalate contents (30 and 31). In general, pollutants from the same 'families' are strongly positively intercorrelated ($n=22$, $r>0.45$, $P<0.05$). However, chromium and aldrin show no correlations at all. There are a number of positive correlations between pollutants of different 'families' as well: e.g.: heavy metals with PCBs and phtalates with nickel and op-DDE.

Pollutants and deformities

Iron, arsenic, nickel, manganese, lead and zinc all show a number of positive correlations with either means, variances or percentages of deformity data for the different structures of the head capsules (Table 4). Cadmium and copper are negatively correlated to deformities of either mentum, mandibles or antennae. Iron and zinc correlate negatively with the variance of premandible and antennal deformities. The organochlorine pesticides alfa-HCH, pp'-DDT, op-DDE and aldrin correlate positively with antennae and pecten deformities and with the variance of mentum deformities. Five phtalates correlate positively with deformities of all 5 structures. One PAH, benzo (a) pyrene, correlates positively with the variance of antennal deformities. Endrin, pp'-DDD, pp'-DDT, indenopyrene and hexachlorobenzene correlate negatively with percentage larvae with deformed antennae.

Weighing of the lead values to either iron, manganese, clay or a combination of these generated much more positive correlations with deformities than lead taken as a raw value. This was also true for zinc/clay as compared to zinc alone. However, some transformations generated negative correlations with deformities as well, as is shown in Table 4. Transformation of the organic xenobiotics to total organic carbon content of the sediments prooved beneficial for obtaining positive correlations for alfa-HCH, pp'-DDT, benzo (a) pyrene and PCB 52.

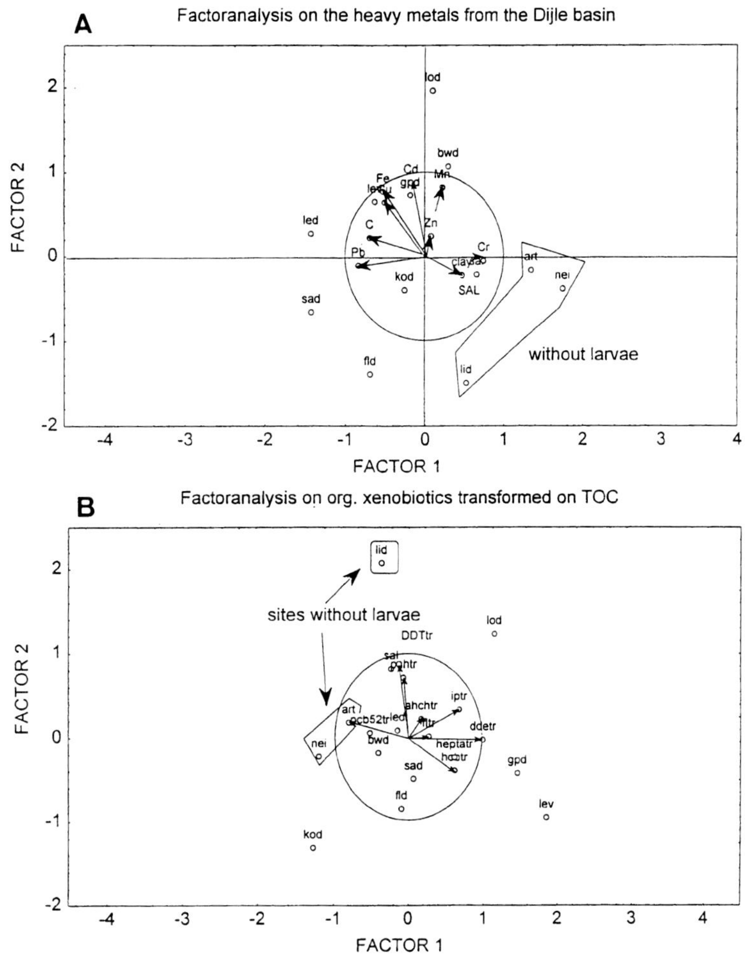

Figure 4. Factor analysis on the heavy metal (arrows) concentrations in the sediments of the 12 Dijle watershed sites (dots) (A) and the concentrations of organic xenobiotics transformed to TOC (B). The sites without larvae are indicated in the graphs. F 1 represents 28 % of the total variability; F 2, 18 % of the remaining variability.

Most positive correlations between heavy metals and deformities concern mandibles, premandibles and pecten epipharyngis (Table 5). Negative correlations are often related to deformity variances. For the organic xenobiotics, the situation looks somewhat different: most positive correlations are to be found for mentum and antennae and to a lesser extent, premandibles, while negative correlations are almost entirely due to percentage larvae with deformed antennae.

A number of significant deformity-pollutant relationships were displayed in linear regressions, with the deformity variable as dependent and the pollutant as independent variable. This, in order to illustrate why a certain relationship appears to be either positive or negative in the global analysis of 22 sites and in a more limited analysis of 9 Dijle sites: when looking at the 22 sites as a whole, the overall relationship between pp'-DDT and the percentage larvae with deformed antennae was

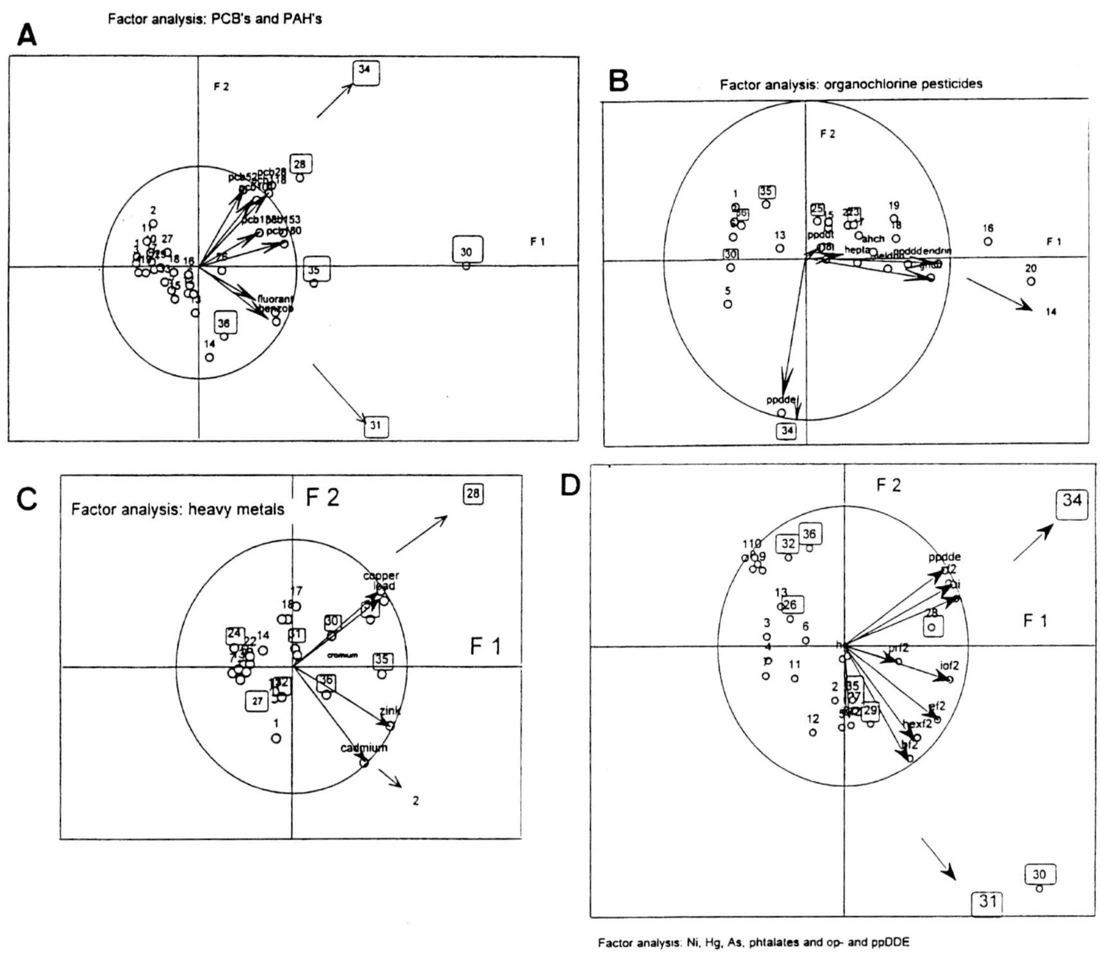

Figure 5. Factor analysis on the PCBs and PAHs (A), organochlorine pesticides (B) and heavy metals (C) of 36 sites from lowland rivers in Flanders. The squared sites are sites where no larvae were found. D represents a factor analysis on the pollutants which were measured in 24 sites. Ni, Hg, As, phtalates, op-DDE and pp'-DDE.

negative (Fig. 6). However, this relationship is shaped by one cluster of sites having a range of 20 to 80% of larvae with deformities and with no detectable concentration of pp'-DDT. It is likely that the deformities encountered in these sites were induced by other stressors. Moreover, a local increase in antennal deformities with increasing pp'-DDT concentration is noted over a range of 0.1 to 0.4 µg/kg DW sediment and the site which contained a maximum of pp'-DDT had also a relatively high TOC content. This last could offset the effect of pp'-DDT on the larval morphology by binding the pesticide and thus lowering the bioavailability.

Some relationships between deformities and pollutants are characterized by an evenly ditribution of the sites over the deformity and pollutant ranges as it is the case for nickel and hexf2 with percentage deformed menta (Fig. 7). However, in some cases the regression is strongly determined by one or a few outlying sites, as is the case for alfa-HCH and pp'-DDT with mentum mean deformities or lead and zinc with mandible mean deformities (Fig. 8).

Table 4: Significant Pearson correlation coefficients (P<0.05) between pollutants and deformity variables . Raw values: n=22, unless * n=9, ** n= 13 ; transformed values: n=9.

structure		mentum			mandibles			premandibles			pecten			antennae			total		
+-	poll.	x	va	%	x	va	%	x	va	%	x	va	%	x	va	%	x	va	%
+	Fe																		
+	As		0.68 **																
+	Ni			0.68 **															
+	Mn														0.83 *				
+	Pb						0.74 **				0.53	0.43	0.57						
	Pb/Fe				0.85	0.78	0.89	0.67											
	Pb/Mn				0.87	0.81	0.93	0.85	0.84	0.88							0.74		
	Pb*Fe/Mn*C				0.81	0.72	0.94	0.75	0.74	0.80			0.74				0.76		0.76
	Pb/clay				0.92	0.95	0.71	0.87	0.71	0.81									
+	Zn																		
	Zn/clay				0.72	0.79		0.71											
-	Cd						0.75 *												
	Cd/Fe																		
	Cd/Mn													0.74	0.72				
-	Cu																	0.42	
	Cu/Mn	0.72													0.83				
	Cu/clay			0.71															
-	Zn/Mn														0.73				
-	Fe								0.69*										
	Fe/Mn														0.76				
+	aHCH		0.68 0.47*								0.43	0.43	0.53						
	aHCH/C		0.93															0.82	
+	ppDDT/C		0.69											0.81	0.72	0.72			
+	opDDE							0.62**		0.61**							0.56**		
+	aldrin									0.51									
+	ibf2																		0.76**
+	bf2																		0.73**
+	prf2													0.62**			0.57**	0.67**	
+	ef2									0.66**									
+	hexf2			0.55**		0.61**													
+	benzoa/C											0.68							
+	PCB52																		
	PCB52/C															0.45* *			
			0.74															0.67	
-	endrin															0.49			0.44
-	pp'DDD															0.46			
-	pp'DDT															0.43			
-	indeno															0.44			0.43
-	HCB														0.67*				

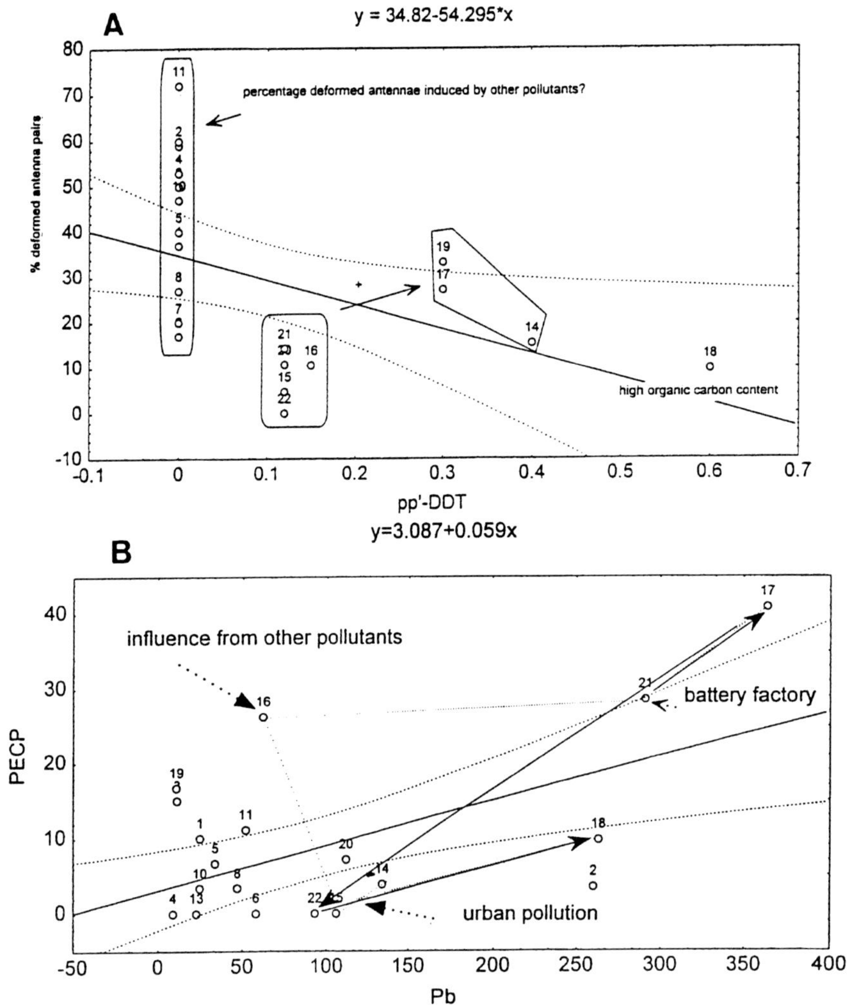

Figure 6. A, significant negative relationship between a pollutant (pp'-DDT) and a deformity variable (% larvae with deformed antennae). B, significant positive relationship between lead and % pecten deformities (PECP) for 22 sites, which is strongly influenced by a lead gradient in the Dijle river in only 7 sites.

In these cases, this relationship is characterized by extremely low and extremely high values of deformities and pollutant concentrations with no values in between: The pollutant concentration-dependency of the morphological response is only apparent over a wide pollutant range. Cushman (1984) demonstrated an increase in mentum deformities of *Chironomus tentans* larvae in earthen ponds, which were highly contaminated with coal tar (containing PAHs and other undefined toxic components), while he noted no differences between ponds which were only slightly different in their coal tar contamination.

From Fig. 6 it is clear that the overall positive correlation between lead and the percentage pecten deformities is due to a clear lead gradient in the Dijle river, provocated by a battery factory. Site 16 however has a peculiar position on this gradient: the pecten deformi-

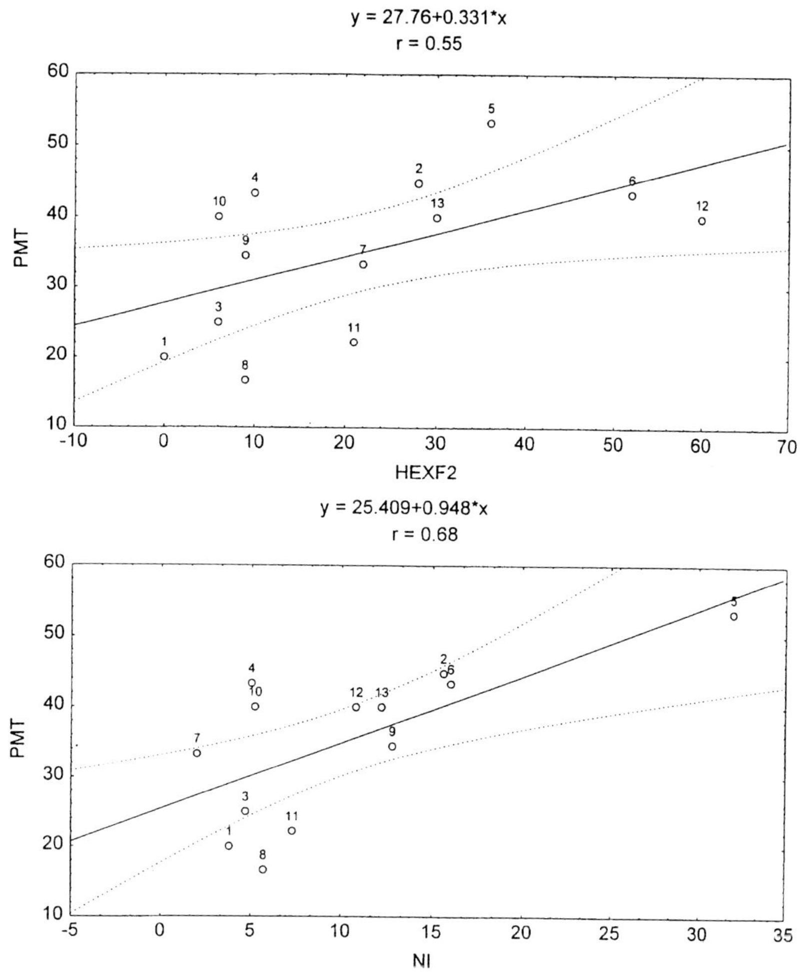

Figure 7. Positive relationship between a pollutant and a deformity variable, where the sites are evenly distributed over the deformity and the pollutant range (PMT = % mentum deformities).

ties observed in larvae from this site are probably induced by other stressors than lead alone, such as organochlorine pesticides which are present at this site (see Fig. 5). This lead gradient in the Dijle is more clearly represented when taking the Dijle sites alone for transformed values of lead and percentage mandible deformities (Fig. 9A). When the sites are ranked according to their lead concentrations, we find a different response for the different structures to lead stress (Fig. 9B): mean pecten deformities gradually increase with increasing lead concentrations and therefore generate a significant correlation with lead (see Table 4). On the other hand, antennal and mentum deformities have a maximum well before the lead peak, and drop suddenly when the lead contamination in the aquatic sediments takes more serious proportions at the battery factory. At the following site, where lead contamination is even more severe, antennal and mentum deformities become again stronger, while premandible and mandible deformities decrease in strength. At the site

Table 5: Number of significant Pearson correlations ($P<0.05$) between deformities in mentum (MT), mandibles (MD), premandibles (PR), pecten epipharyngis (PE), antennae (AN) and the sum of deformities from all structures (TOT) for *Chironomus* gr. *thummi* larvae and concentrations of heavy metals and organic xenobiotics in the aquatic sediments. GTOT is the sum of the number of correlations for a particular row, excluding TOT.

pollutant	sign	parameter	MT	MD	PR	PE	AN	TOT	GTOT
heavy metals	+	mean	0	5	5	1	0	2	11
"	+	variance	1	5	3	2	1	0	12
"	+	%	1	5	3	2	0	1	11
"	-	mean	1	0	0	0	1	0	2
"	-	variance	0	0	1	0	4	1	5
"	-	%	1	1	0	0	0	0	2
org. xenob.	+	mean	3	0	1	1	2	2	7
"	+	variance	5	1	0	1	2	3	9
"	+	%	1	0	3	1	2	2	7
"	-	mean	0	0	0	0	0	0	0
"	-	variance	0	0	0	0	1	0	1
"	-	%	0	0	0	0	4	2	4

where lead suddenly increases, mandibles and premandibles exhibit an optimal mean deformity response, however before the actual lead maximum, followed by a drop at the lead maximum. It is clear that such bell-shaped morphological response will not be detected by linear regression analysis.

Discussion

The incidence of deformed larvae and the intensity of deformities in Belgian lowland rivers are high and the deformities are encountered in all 5 assessed structures (mentum, mandibles, premandibles, pecten epipharyngis and antennae) and this, in all populations studied. Both (high) incidence of deformities and kind of deformed structures (köhngaps, additional or fewer teeth in mentum and mandibles, splittting of medial mentum tooth, lacking of antennal segments, twisting or shortening of premandibles, broadening or splitting of pecten teeth) are comparable with results from Warwick and Tisdale (1988) for *Chironomus plumosus* larvae from stressed sites in Tobin lake (Canada), Warwick *et al.* (1987) for *Chironomus* spp. from Port Hope Harbour, Lake Ontario and Köhn and Frank (1980) for *Chironomus* sp. in Teltowkanal, Germany. We however never encountered any deterioration (thickening or incomplete moult) of the larval skin, as reported by Hamilton and Sæther (1971) and antennae with non-homologous structures or very deformed segments, as reported in Warwick (1985) were extremely scarce.

Within all 5 structures, percentage deformed larvae, mean deformity strength and variance to this mean are positively intercorrelated; as a general rule, the more larvae deformed for a particular structure, the more likely that a larva will be found with a strong deformity, with subsequent increase in mean and variance. At population level, no correlations exist between the deformity means, variances and percentages between the different structures, except between pecten, premandibles and antennae. This means that sites with e.g. a high mean deformity in the menta will not automatically yield high means for antennae or mandibles etc. The individual level (raw point values) gives some positive correlation between the three strongest buccal structures: mentum, mandibles and premandibles; larvae which are deformed in their mentum are likely to be deformed for at least one mandible and/or premandible. Pecten and antennal deformities seem to occur independently from the other structures.

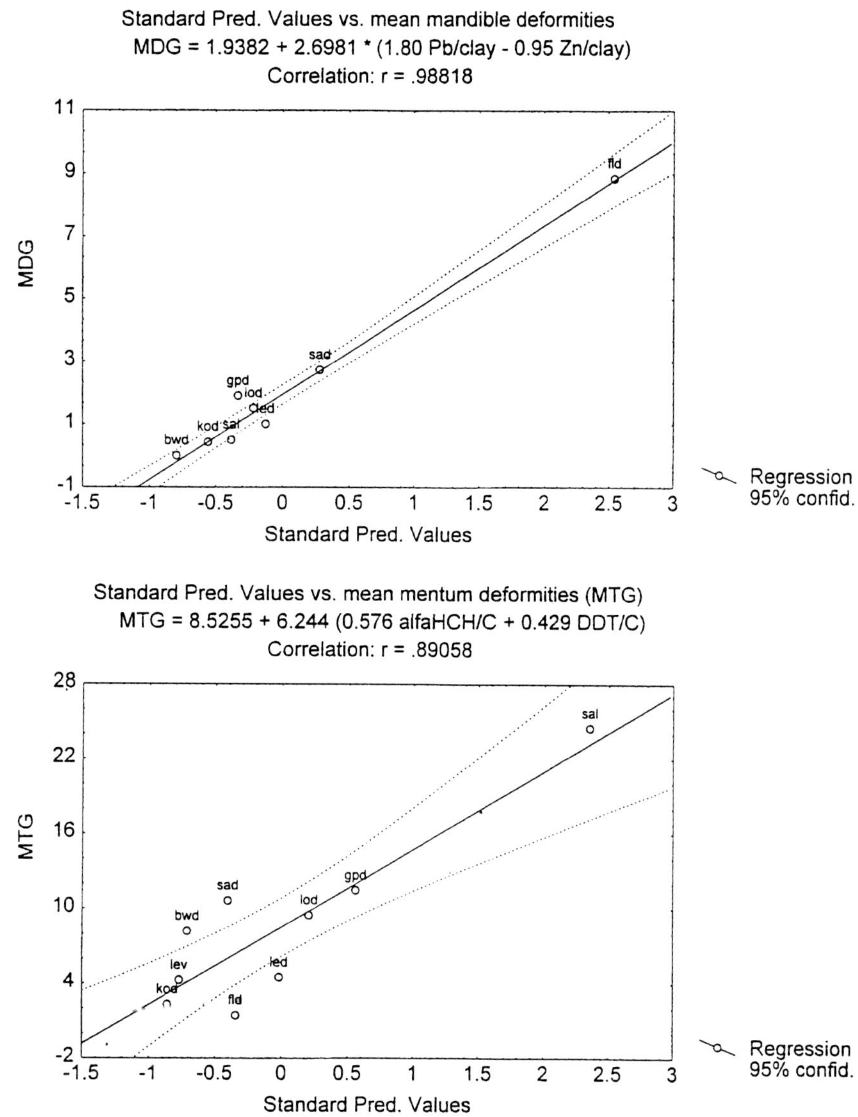

Figure 8. Positive relationship between some pollutants and a deformity variable, where the sites are clustered at the left side of the regression and where only one site with extreme values of deformity and pollutants is strongly influencing the overall regression.

The absence of *Chironomus* larvae in certain sampling sites is likely to be due to the granulometry of the sediments and the lack of enough organic carbon, or (for 7 other sites from another dataset), to extremely high values of pollutant concentrations in the sediments. Only 5 sites (26, 27, 29 and 33, see Table 1 for site-names) were found to harbour apparently viable pollutant concentrations, though it might be possible that their substrate composition was not adequate for *Chironomus* (as this could not be assessed for these sites).

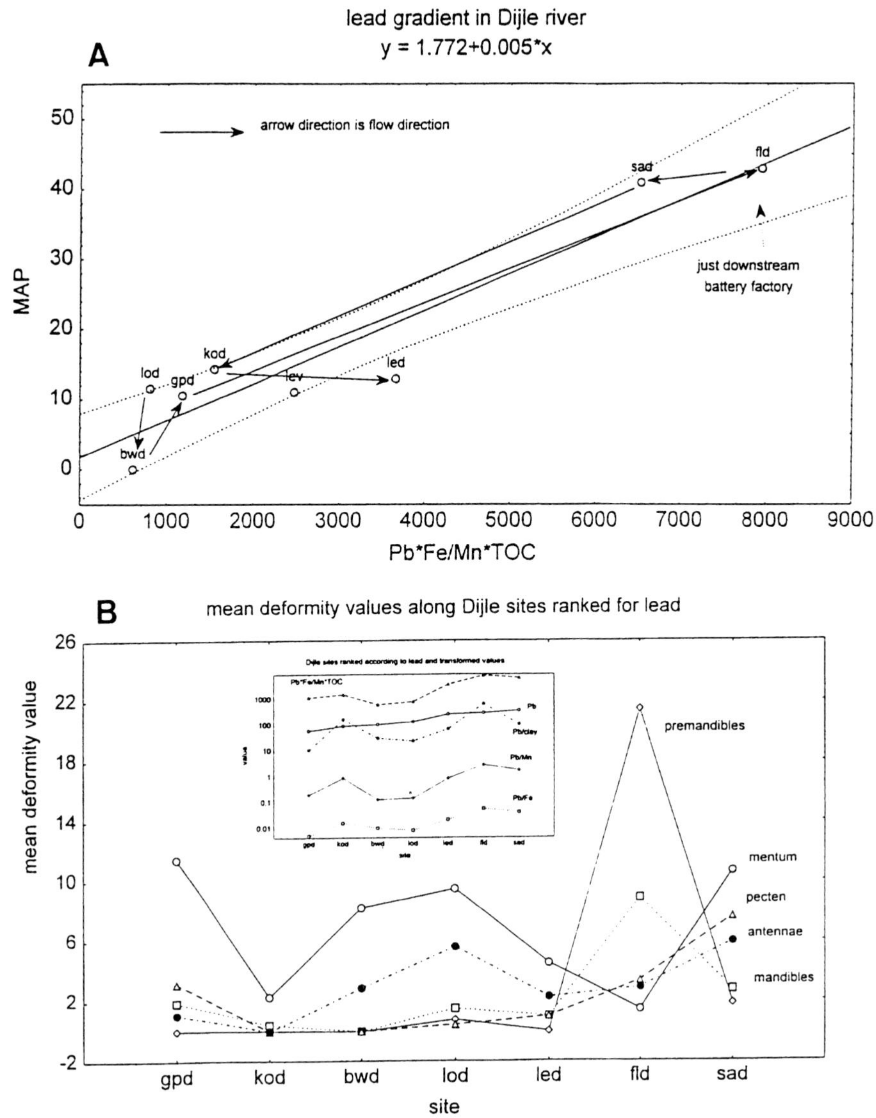

Figure 9. A, Significant linear regression between transformed values of lead concentrations and % deformed mandible pairs. B, deformity response (means) in function of increasing lead concentrations in the Dijle river. Internal graph gives the lead gradient and the corresponding transformed values on a log scale.

Some aquatic sediments are apparently too toxic even for the extremely tolerant *Chironomus* larvae. Amongst the sites where no larvae were found, at least one and often more than one site had pollutant concentrations higher than the recommended 'limit concentration' above the accepted quality criteria concentrations for The Netherlands (van de Guchte and Van Leeuwen, 1988).

Although concentrations of 10 heavy metals and 40 organic xenobiotics were screened in river sediments and matched against 15 deformity variables for 9 to 22 sites (depending on the kind of pollutants involved in the analysis), the lotic ecosystems under scrutinity remain largely 'black boxes', where other stressors might play an important role (e.g. temperature and radionuclides). A combination of (i) many correlations within the pollutant dataset and within the deformity variable dataset, (ii) the occurrence of opposite spatial gradients of different pollutants (a heavy metal from a point pollution source could steadily decrease in downstream direction, while pesticides from diffuse pollution sources could steadily increase along the same stretch of river) and (iii) the accidental incorporation of artefacts (some antennal deformities) created a high degree of redundancy in the overall analysis. Moreover, though pollutants locked in a sediment sink are well-known to represent a time integration of pollutional events (e.g. Reynoldson, 1987), it has been frequently demonstrated that only a fraction of the measured total concentration is available for accumulation in benthic midge larvae (e.g. Luoma, 1989). Despite this background noise, which probably accounts for the low number of significant correlations between deformities and pollutants, still some relevant correlations could be detected. Consequently, adequate transformations of pollutant concentrations according to established formulae in order to approach a virtually more bioavailable fraction, generated even more positive correlations. Most negative correlations were found between pollutants and deformity variances, which can be interpreted as: the higher pollutant concentrations, the more larvae will be found with similar strength of deformities. However, in some cases, the opposite prevailed.

For the untransformed heavy metals, iron, lead, zinc, arsenic and nickel exhibited positive relationships with deformities.

For the transformed heavy metals (note that this could not be performed with As, Ni and Hg) lead and zinc showed a remarkable relationship with mandible and premandible deformities. The existence of a clear lead gradient in the Dijle river contributed strongly to the number of significant correlations between lead and deformities found in the global analysis, involving all 22 sites.

The fact that we did not discriminate between different deformity types of the mentum, such as Köhngaps, split medial tooth and addition of lateral teeth, could have depleted the number of positive correlations between mentum deformities and pollutants. Janssens de Bisthoven *et al.* (1992) demonstrated significantly higher concentrations of lead and copper in larvae with a clear Köhngap, compared to larvae with a normal mentum from a Dijle site. This finding has subsequently been confirmed at other sites and with other heavy metals (*unpublished*) and the same *modus operandi* has been used by Dickman *et al.* (1992) to demonstrate a higher bioaccumulation of 15 PAHs from a coal tar contaminated site in *Chironomus anthracinus* larvae whith deformed menta as compared to normal larvae. Both these literature data and the results in present study demonstrate the relatively low pollutant-specificity of mentum deformities in benthic midge larvae.

Members of the phtalate group are positively correlated with mandible, mentum and antennal deformities. op-DDE and aldrin correlate with premandible deformities. pp'-DDT transformed to total organic carbon is positively correlated with antennal deformities, but when used as raw values, negatively correlated, together with pp'-DDD, to the percentage larvae bearing antennal deformities. Madden *et al.* (1992) showed experimentally a strong concentration-response relationship between DDT and mentum deformities and a much weaker one for antennal deformities. However, with the herbicide Dactal@, the relationship with mentum and antennal deformities over a certain concentration range showed a bell-shaped curve. After reexamination of the material resulting from experiments by Hamilton and Sæther (1971), Warwick (1988) could demonstrate a negative relationship between DDE and antennal deformities. However, this reexamination also showed that Aldrin, dieldrin, DDD and DDT all eperimentally could induce deformities that were similar to deformities found in contaminated Northern Canadian lakes. In the present field study, positive correlations between aldrin, DDE, DDT and deformities could also be shown. It is clear from these literature data and from our results that organochlorine pesticides such as DDT and its derivates have some teratogenic effects on buccal structures and antennae of *Chironomus* larvae. However, this effect on the larval morphology seems to be confined to a very narrow pollutant concentration range and probably only occurs under narrow ecological conditions. Any departure from these apparently result in a decrease of particular deformities, often with an increase of other deformities.

The whole process is not well understood and resulted into the formulation of a 'quantal dose-response hypothesis' by Warwick (1988): with increasing pollutant environmental contamination, it is suggested that midge populations might exhibit a shift in deformity response from e.g. antennae to pecten to mentum etc. A negative correlation between a pollutant and a deformity could then be explained as an event occurring at the right side of a bell-shaped response curve; i.e. when the deformity response of a particular structure is decreasing over a certain range of increasing pollutant concentration, while another structure appears to be more and more deformed (at population level, e.g. in terms of percentage deformed larvae). In this study, the observation of a sequential shift of deformity response to different structures along a lead gradient, and the presence of some negative relationships between pollutants and deformities in the overall analysis give arguments to support this hypothesis.

The presence of (i) a background morphological anomaly rate wich is not related to pollution stress, as has been shown in experimental blanco conditions by Madden *et al.* (1992) and in current experiments (*unpubl.*), (ii) of natural variation in morphology (especially for pecten epipharyngis, unpublished data) and (iii) the absence in this study of an uncontaminated site where *Chironomus* gr. *thummi* could be found in sufficient numbers, remain important obstacles in the quantification process of the pollutant-deformity relationship.

In conclusion, this study supports the Hamilton-Sæther hypothesis: morphological deformities in field populations of *Chironomus* larvae are related to environmental stress, provoked by both heavy metals and organic xenobiotics such as PCBs, phtalates and organochlorine pesticides. The relationship between pollutants and deformities is either weakly to strongly linear, 'graded' (i.e. a strong deformity response only for extreme pollutant values) or bell-shaped, depending on the kind of pollutant and deformity involved. The deformities are rather pollutant-aspecific, though mandible and premandible deformities seem to be more related to heavy metals, while mentum and antennal deformities, are rather related to organic xenobiotics. For the further development of a useful biomonitoring instrument, we recommend to direct future research efforts on the bioaccumulation-deformity relationship in field studies as well as in experimental chronic exposure tests.

Acknowledgements

We are indebted to many persons from various national and federal agencies for providing us with sediment analysis data. We particularly would like to thank Dr. G. Van Hooren, Dr. D. Ringelé ('Dienst Hydrobiologie' of the 'Instituut voor Hygiene en Epidemiologie' (I.H.E.) in Brussels (Belgium)), Lic. R. Vannevel (Vlaamse Milieu Maatschappij, V.M.M.), and Prof. Em. Dr. F. Gullentops (Lab. for Geomorphology and Recent Sediments, K.U.Leuven). We also would like to acknowledge Dr. B. Van Hattum (Institute for Environmental Studies, Free University of Amsterdam) for useful advice on statistical problems and Dr. B. Goddeeris (K.B.I.N., the Royal Belgian Institute for Natural Sciences, Brussels, Section Freshwater Biology) for sharing his expertise on Chironomidae.

THE IMPACT OF PESTICIDES USED IN RICE AGRICULTURE ON LARVAL CHIRONOMID MORPHOLOGY

V. Pettigrove[1], *W. Korth*[2], *M. Thomas*[2] *and K.H. Bowmer*[2]

[1] Catchment and Drainage, Melbourne Water, Yarra Region Private bag 1, Mitcham Vic. 3132, Australia.

[2] CSIRO Division of Water Resources, Griffith Laboratory, Private Bag No. 3, Griffith N.S.W. 2650, Australia.

INTRODUCTION

Few water quality monitoring programs have monitored pesticide contamination of Australian inland waters. Analyses are expensive, and many pesticides are difficult to detect as they degrade rapidly. The morphological condition of indicator species is a potentially cheap and effective means of detecting and assessing the impact of toxicants on aquatic systems. The occurrence of structural abnormalities in chironomid larvae have been used frequently as an indicator of environmental stress, as chironomids are sensitive to environmental changes, have a relatively short life cycle, and inhabit niches that makes them likely to contact toxicants (Warwick, 1988).

The link between the occurrence of structural abnormalities in chironomid larvae and exposure to toxic stress has been substantiated in relation to copper (Kolsawat and Knight, 1987; Van De Gutche and Van Urk, 1989; Janssens de Bisthoven *et al.*, 1992), lead (Michailova and Belcheva, 1990; Janssens de Bisthoven *et al.*, 1992), cadmium and zinc (Van De Gutche and Van Urk, 1989), ultraviolet irradiation of eggs (Percy *et al.*, 1986), the organochlorines DDE (Warwick, 1985), DDT and the herbicide Dacthal (Madden *et al*, 1993), and have been suggested in relation to a pesticide mixture (Hamilton and Saether, 1971), industrial pollution (Dermott, 1991) and 'Roundup' and atrazine (Dickman *et al.*, 1989). This technique could become a cheap and effective screening tool for identifying stress in aquatic systems (Nowak *et al.*, 1993).

This study was initiated following observations of a high frequency of abnormal ligulae in *P. paludicola* Skuse larvae collected from the Murray River, near Euston and from the Darling River, near Burtundy, New South Wales (Pettigrove, 1989). As these sites received waters from agricultural areas, it was possible that the causative agents were pesticides used in irrigation agriculture. This present work investigates the morphological response of several chironomid taxa that inhabited two rice bays which were treated with a cocktail of four pesticides that are widely used in rice agriculture (malathion, chlorpyrifos, molinate and bensulfuron). This study aims to establish whether the frequency of structural abnormalities is higher in chironomid populations in the rice bays than compared to reference sites. If so, to describe the delay between impact and morphological response, to determine how the frequency of abnormalities change over time, and to better understand of the processes involved in the development of an abnormality. The implications of these findings are discussed in relation to the potential of using this technique as an indicator of pesticide contamination.

Materials and Methods

The two rice bays (Bays 1 and 2) are located approximately 10 km south of Griffith, New South Wales. Taxa that were abundant in these rice bays were also collected from three local reference sites. These reference sites were a seepage area adjacent to Bay 1, a supply canal at Brays Dam and a drainage channel downstream of the Griffith Sewage Treatment Plant.

Rice Bay Treatment

The rice bays were treated with the insecticides malathion (Maldison) and chlorpyrifos (Dursban) and the herbicides molinate (Ordram) and bensulfuron (Londax). The concentrations of chlorpyrifos-ethyl, malathion and molinate in the floodwater of the two rice fields were monitored at daily intervals for up to 24 days after application on the rice bays. Bensulfuron was not measured as it is known to be relatively non-toxic to invertebrates such as bees (LC_{50} for bees > 12.5 g/bee) and vertebrates (Royal Society for Chemistry, 1992). The concentrations of the pesticides present in the sediment of the bays were not measured.

Chemical Analysis

Water samples were collected in amber glass bottles, buffered (pH 6.9, 2.5 mM Na_2HPO_4 / KH_2PO_4 and cooled to -4C for transport and storage. Subsamples (500 mL) were spiked with Fenchlorphos (added as an internal standard) and extracted using methylene chloride (3 x 50 mL). The three methylene chloride extracts were pooled, dried (Na_2SO_4) and concentrated by evaporation (nitrogen) to a final volume of 1.0 mL. Extracts were analysed by gas chromatography - mass spectrophotometry (GC - MS) using a Varian 3300 gas chromatograph coupled to a Hewlett Packard 5970 Mass Selective Detector (operated in a multiple ion monitoring mode) and a 5% phenylmethyl silicone column (25 m x 0.2 mm, film 0.25 m, Hewlett Parkard). One L was directly injected using a spitless injection (splitter off 1 min.) and an injector temperature of 200°C. The column temperature was ramped from 50°C (isocratic 1 min.) to 160°C at 20°C/ min., the temperature was held at 160°C for 4 min. and then ramped at 5°C/ min. to a final temperature of 250°C. The pesticides of interest were quantified using Fenchlorphos as an internal standard. The minimum detectable concentration was 0.05 µg/L for molinate, chlorpyrifos and malathion.

Biological Field Sampling

The chironomid larvae were collected with a 300 µm mesh handnet. Samples were collected at around weekly intervals for 14 weeks after the pesticide applications (1 Nov. 1991 to Mar. 1992). Additional samples were also collected on 19 Feb. and 12 Mar. 1992 from Bays 1 and 2 respectively, before the bays were drained and the rice harvested. We were unable to collect samples from these bays prior to the pesticide applications. However, the only chironomid species present in other local bays prior to the pesticide applications was *Chironomus tepperi* (Skuse), which did not recolonise any of the bays that were examined after the pesticide applications. All samples were preserved in 70 to 80% denatured ethanol.

Laboratory Processing

Many chironomid taxa were collected from the rice bays, however *Polypedilum nubifer* Skuse, *Procladius paludicola*, *Chironomus "februarius"* Martin (a nomen nudum) and *Dicrotendipes* SWL sp.1 were selected for investigation as they were reasonably abundant at one or both rice bays and at least one of the reference sites. Chironomid larvae were sorted from the samples until 30 *P. nubifer* or *Dicrotendipes* SWL sp.1 larvae were collected from the samples or until the sample had been processed for an hour. The remainder of the sample was stored in 80% ethanol. Occasionally more of these samples were processed if it was necessary to help substantiate any trends. The larvae were mounted on microscope slides using the method of Warwick (1985), except that the larvae were mounted in Hoyers medium. Neither the larvae or pupae of *Dicrotendipes* SWL sp.1 could be keyed to any of the Australian species described by either Epler (1988) or Cranston (1990).

Some variation in interpretation exists as to what constitutes an abnormal structure (Warwick, 1988; Janssens de Bisthoven and Ollevier, 1989; Madden *et al.*, in press). A structure was assessed as being abnormal if it varied from its normal configuration and its condition was not a result of damage during slide preparation (*cf* Warwick, 1985; Warwick and Tisdale, 1988). Those structures that have been mechanically damaged either before they were collected or during the collection and processing of the samples were not considered to be abnormal. A mentum or ligula was assessed as being abnormal if it was asymmetrical and/or the number of teeth differed to the normal configuration. The pecten epipharyngis was assessed as being abnormal if it was bilaterally asymmetrical. The antennae were considered to be abnormal if the number of antennal segments differed from the normal configuration. Antennal organelles were not examined. A Tanypodinae larvae was considered to have abnormal paraligulae if bilateral asymmetry occurred in the number of spines between its paraligulae. The instar of each larva was determined by measuring the width of the mentum (Chironomidae) or ligula (Tanypodinae).

The frequency of structural abnormalities was only examined in fourth instars. The severity of abnormalities will be discussed elsewhere.

Results

Physico-chemistry

Data on pesticide applications for the experimental rice Bays 1 and 2 are presented in Table 1. This information was based on records from the aerial operators and the landholders.

The insecticides malathion and chlorpyrifos were applied for the control of *Chironomus tepperi*. The herbicides molinate and bensulfuron were applied for the control of barnyard grass and broadleaf weeds respectively.

Chlorpyrifos-ethyl concentrations were found to reach around 5 µg/l within a few hours after spraying and declined below detection limits (0.05 µg/L) within 14 and 11 days after application at Bays 1 and 2 respectively. The 5 µg/L concentration of chlorpyrifos measured after spraying is much lower than the theoretical value of 40 µg/L, assuming no loss from the water and that a 10 cm water depth occurs in the bays. The clayey substratum of the rice bays would have a good adsorptive capacity to remove chlorpyrifos from the water (Rawn *et al.*, 1978) and it is likely that the chlorpyrifos would have persisted much longer in the sediment than in the open water (Brock *et al.*, 1992).

Table 1: A list of pesticides applied to the rice bays in the 1991/92 season. All chemicals were aerially applied unless otherwise stated.

Date	Bay 1	Bay 2
Crop grown in 19/92 season	Rice	Fallow
Oct. 1-18		Molinate dripped into bay with supply water
Oct. 11		Rice aerially sown with malathion (300 g a.i./ha)
Oct. 14	Rice aerially sown with malathion (300 g a.i./ha)	
Oct. 16		Chlorpyrifos (40 g a.i/ha)
Oct. 17	Malathion (40 g a.i./ha)	
Oct. 19	Molinate (3.6 kg a.i./ha) Chlorpyrifos (40 g a.i./ha)	
Oct. 28		Bensulfuron (40 g a.i./ha) Chlorpyrifos (40 g a.i./ha)

The molinate concentrations in Bay 1 (where it was aerially applied) exceeded 3000 µg/L, three days after application and was still around 200 µg/L, 22 days after treatment. The molinate concentration in Bay 2 (where it was dripped into the water supply) was still around 350 µg/L two days after application ceased and was still around 200 µg/L, 22 days after treatment. The malathion concentrations were 36 and 38 µg/L in Bays 1 and 2 respectively, one day after application and fell below detection limits (0.05 µg/L) by 14 and 10 days after treatment in Bays 1 and 2 respectively.

Colonisation of the rice fields by chironomids

No chironomids were present in either bay for at a least 2 weeks after the pesticide applications. The cumulative number of taxa that colonised Bays 1 and 2 after the pesticide applications is presented in Figures 1 and 2 respectively. A total of 17 chironomid taxa were collected from the rice bays during the following months after the pesticides ceased to be applied. The first larvae found in the rice bays after the pesticide applications (using a 360 µm mesh handnet) were second and third instars.

The composition of the chironomid fauna was similar in both bays, as the fauna was dominated by *Polypedilum nubifer*, *Chironomus februarius*, *Ablabesmyia notabilis* Skuse, *Procladius paludicola*, *Dicrotendipes* SWL sp.1, *Cricotopus* SRV sp. 8 and *Larsia albiceps* (Johannsen). Some noticeable differences did occur in the relative abundance of some taxa between these two bays: *P. nubifer* was abundant only in Bay 1 and *Dicrotendipes* SWL sp.1 was only abundant in Bay 2. Few pupae were found in the bays over the four months that the chironomid fauna was monitored. In particular, no pupae and few mature (*ie* those with enlarged thoracic segments) fourth instars of *P. nubifer*, *C. februarius* or *P. paludicola* were found despite the large numbers of larvae examined.

The frequency of abnormalities in the fourth instars of *P. nubifer* and *Dicrotendipes* SWL sp.1 collected from Bays 1 and 2 are presented in figures 3 and 4 respectively. The frequency of abnormalities in the fourth instars of *P. nubifer* did not significantly vary over the 97 days they were present in Bay 1 (χ^2test, P > 0.1). Furthermore, the residuals from this chi-square test did not indicate any underlying trend (Mosteller and

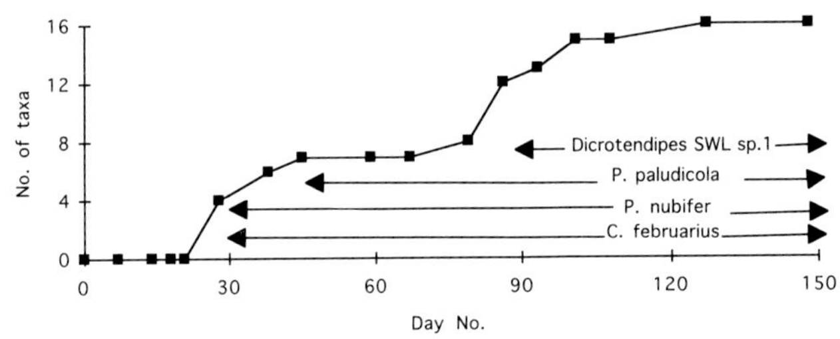

Figure 1. The cumulative number of taxa present in the Rice Bay 1 following the application of pesticides.

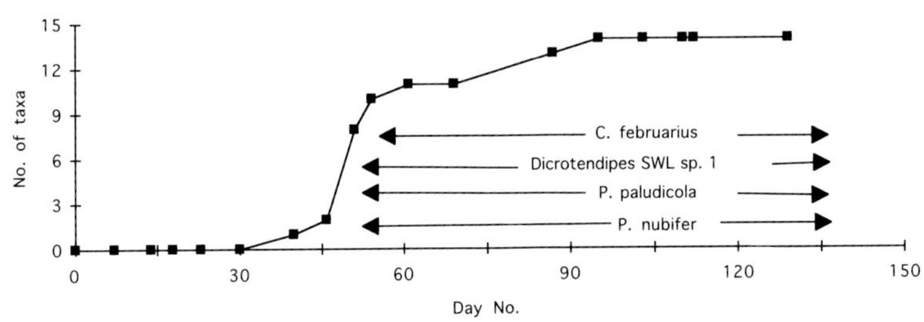

Figure 2. The cumulative number of taxa present in Rice Bay 2 following the application of pesticides.

Table 2: The number and type of abnormalities in fourth instars of *P. nubifer* collected from the rice bays and three reference sites.

Site	No. Larvae	Total no. abnormal	No. abnormal pecten epipharngyis	No. abnormal menta	No. other abnormal
Bay 1	150	63	61	1	1
Bay 2	11	4	4	-	-
Brays Dam	0	0	-	-	-
Seepage Area	32	3	3		
Sewerage Outfall	44	12	10	2	-
Total	237	81	78	3	1

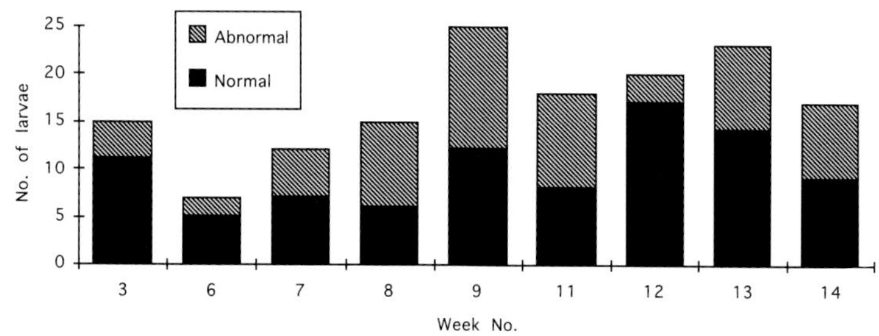

Figure 3. The frequency of abnormalities in *P. nubifer* fourth instars collected from bay 1.

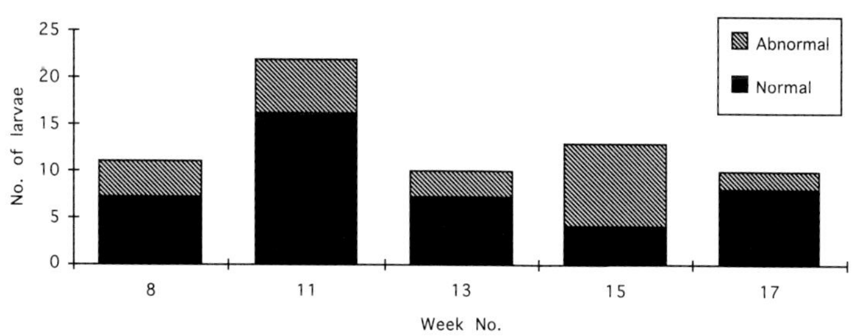

Figure 4. The frequency of abnormalities in *Dicrotendipes* SWL sp. 1 fourth instars collected from Bay 2.

Tukey, 1977). Similarly, the frequency of abnormalities in *Dicrotendipes* SWL sp.1 did not vary over the 97 days they were present in Bay 2, nor were any underlying trends evident in the residuals (χ^2 test, $P > 0.1$).

Spatial comparison of the frequency of abnormalities

Eighty-one of the 237 (34.2%) fourth instars of *P. nubifer* collected from the rice bays and the reference sites had a structural abnormality (Table 2). 78 larvae (32.9%) had an asymmetrical pecten epipharyngis, 2 larvae (0.8%) had an abnormal mentum and two larvae (0.8%) had an abnormal antenna. The frequency of abnormalities in the fourth instars of the *P. nubifer* populations from the rice bays (41.6%) was significantly higher than that detected in the seepage area and the Griffith Sewerage Outfall (19.6%) (χ^2 test, P < 0.01).

Table 3: The number and types of abnormalities in fourth instars of *Dicrotendipes* SWL sp.1 larvae collected from the two rice bays and three reference sites.

Site	No. of larvae	No. abnormal
Bay 1	7	1
Bays 2	66	24
Brays Dam	48	3
Seepage area	0	0
Sewerage Outfall	2	0
Total	123	28

Twenty-eight of the 123 (22.7%) fourth instars of *Dicrotendipes* SWL sp.1 examined had a morphological abnormality (Table 3). All of these abnormalities were bilateral asymmetries in the number of spines on the pecten epipharyngis. The frequency of abnormalities in the Bay 2 populations (36.4%) was significantly higher than that present in the Brays Dam population (6.3%) (χ^2 test, P < 0.005).

Twenty-five of the 149 fourth instars (16.8%) of *P. paludicola* collected from the rice bays and the reference sites had a structural abnormality (Table 4). Nineteen larvae (12.8%) had bilaterally asymmetrical paraligulae and six larvae (4.0%) had an abnormal ligula. The frequency of abnormalities in *P. paludicola* was significantly higher in the rice bays (23.0%) than that present in the reference site populations (8.1%) (χ^2 test, P < 0.05).

Nineteen of the 121 (15.7%) of fourth instars of *C. februarius* collected from the rice bays and reference sites had a morphological abnormality (Table 5). Sixteen larvae (13.2%) had an asymmetrical pecten epipharyngis and three larvae (2.5%) had an abnormal mentum. The frequency of abnormalities was similar in Bay 1 (25%), Bay 2 (16%) and in the seepage area (15%), whereas it was only 4% in the Griffith Sewerage Outfall. The frequency of abnormalities in *C. februarius* did not significantly vary between the rice bays and reference sites' populations (χ^2 test, P > 0.1)

Other Taxa

No abnormalities were detected in the ligula, paraligulae, antennae, mandibles or hypopharngeal pecten of 51 *Larsia albiceps* larvae collected from the rice bays. 33% of *Ablabesmyia notabilis* larvae collected from the rice bays had an asymmetrical hypopharyngeal pecten. Unfortunately, insufficient numbers of *A. notabilis* larvae were collected from any of the reference sites to enable a statistical comparison of the frequency of abnormalities between the various populations.

Table 4: The number and type of abnormalities identified in fourth instars of *P. paludicola* collected from the two rice bays and three reference sites.

Site	No. of larvae	No. Abnormal	No. abnormal paraligulae	No. abnormal ligulae
Bay 1	61	13	8	6
Bay 2	26	7	7	-
Brays Dam	28	1	1	-
Seepage Area	10	2	2	-
Sewerage Outfall	24	2	1	1
Total	149	25	19	7

Table 5: The number and type of abnormalities identified in fourth instars of *C. februarius* collected from the two rice bays and three reference sites.

Site	No. of larvae	No. abnormal	No. abnormal pecten epipharngis	No abnormal antennae
Bay 1	24	6	6	-
Bay 2	31	5	4	1
Brays Dam	1	1	-	1
Seepage area	41	6	5	1
Sewerage Outfall	24	1	1	-
total	121	19	16	3

None of the remaining taxa collected from the rice bays were collected in either sufficient numbers from the rice bays or from the reference sites to facilitate any statistical comparisons of the frequency of abnormalities in relation to their exposure to toxicants.

DISCUSSION

The pesticide analyses were restricted to the rice bays where four pesticides were known to be applied. The selection of the reference sites was not ideal but they were the only sites sampled, apart from the rice bays, where one or more of the four indicator taxa were abundant. Pesticide analyses were not taken from the reference sites. It was likely that the larvae collected from the reference sites were exposed to some toxic pollution. It is unlikely that the Griffith Sewerage Outfall contained any significant amounts of pesticides; however elevated levels of metals may have been present. The irrigation supply canal at Brays Dam may have received some spray drift from pesticides aerially applied to the rice fields that are located within the vicinity of the canal; it could have also received small amounts of pesticide contaminated drainage water from the rice fields. The seepage area (adjacent area to Bay 1) may have been contaminated by spray drift from the aerial application of pesticides onto Bay 1. However, the chironomid assemblages were present at all reference sites before, during and after the pesticides were applied to the rice bays whereas it took about 2 weeks before the rice bays were recolonised after the pesticide treatments. As the pesticides degraded, the chironomid fauna would have been exposed to a sub-lethal dose of pesticides once they began to successfully recolonise the bays. The technique of investigating the frequency of abnormalities in chironomid larvae must meet several criteria if it is to be a good indicator of pesticide contamination when used in routine biomonitoring programs. Morphological abnormalities should quickly appear in chironomid larvae following exposure to toxic stress, so that any stress experienced by a population from toxic contamination may be detected soon after an impact. Furthermore, the frequency of abnormalities in a population exposed to toxic contamination should remain above background levels for many months after the impact, so that the occurrence of these impacts can be detected by routine biomonitoring programs which may examine the fauna from a site only three to four times per year. The results of this study are encouraging for the use of this technique in routine biomonitoring as it has shown that a high frequency of abnormalities was present in the epipharyngeal pecten of *P. nubifer* and *Dicrotendipes* SWL sp.1 and in the paraligulae of *P. paludicola* larvae as soon as fourth instars were detected in the rice bays, and that a high frequency of abnormalities were expressed in these populations for at least three months after the impact.

The consistently high frequency of abnormalities that were detected in the epipharyngeal pecten of *P. nubifer* and *Dicrotendipes* SWL sp.1 rice bay populations may be due to retarded larval development, which may also be a response to the sub-lethal stress caused by exposure to the pesticides. It was likely that pupation and emergence was suppressed in the rice bay populations as no pupae or even fourth instar larvae with enlarged thoracic segments were detected.

The available information on the life histories of these species confirms that they should have developed into pupae within the four months they were present in the rice bays. The growth rate of chironomid larvae is dependent on water temperatures (Pinder, 1986). Edward (1986) reported that *P. nubifer* adults emerged from 65 to 80 days after hatching when water temperatures ranged from 8 to 17°C but took only 20 to 30 days when water temperatures ranged from 14 to 30°C. Field observations from the Murrumbidgil Swamp and Lake Merrimajeel, which are distributaries of the Lachlan River, New South Wales, indicate that the developmental time for *P. nubifer* (from egg to adult) ranged from 45 to 145 days (Maher and Carpenter, 1984). *In situ* measurements of the water temperatures at Bays 1 and 2 ranged from 12 to 22°C in the morning to 17 to 30°C in the afternoon. These temperatures should have been sufficient to enable the *P. nubifer* larvae in the rice bays to develop into pupae well within the three months they were present in the bays, if they were not stressed by pesticide contamination.

The reproduction and behaviour of *P. nubifer* is known to be affected by chronic levels of stress. Emergence success decreased in larvae exposed to high concentrations of cadmium (Hatakeyema, 1987) or copper (Hatakeyama, 1988) in water. The herbicide chlornitrofen has been found to delay emergence and emergence success and also affect oviposition success and hatchability of the eggs (Hatakeyema and Shiraishi, 1991). *P. nubifer* larvae exposed to chronic levels of a variety of insecticides could not make a proper tube (Sato and Yasuno, 1979). Other taxa may also have a retarded development in response to low levels of toxicants. Even concentrations of cadmium approximately 5000 greater times than the 48h LC_{50} for fourth instars of *Chironomus riparius* produced significant reductions in larval development, survival and production (Pascoe *et al.*, 1989).

It is important to know the most sensitive developmental stage to toxic contamination when determining the time it takes for a larva to express a morphological abnormality after exposure to toxic contamination. For example, first instars of *P. nubifer* were 11 times more sensitive in 24 hour LC_{50} tests to the herbicide chlornitrofen than fourth instars and were five times more sensitive than second instars (Hatakeyama and Shiraishi, 1991). Similarly, first instars of *Chironomus tentans* are 12 to 27 times more sensitive than fourth instars in 96 hour toxicity tests (Gauss *et al.*, 1985). An extreme example of how larval sensitivity towards toxicants varies between instars is reported by Williams *et al.* (1986) who reported that first instars of *Chironomus riparius* are about 1000 times more sensitive to cadmium than fourth instars (24 hr LC_{50}). Given that the concentration (and toxicity) of the pesticides declined rapidly over time, it is likely that the more sensitive early instars were exposed to the greatest stress from pesticide contamination. The sensitivity of chironomid eggs to toxic contamination is unknown.

The relative toxicity of the water and sediments may have determined which instar was most stressed by pesticides. First instars are usually planktonic and are therefore exposed to material in the water column. Following instars are usually benthic (Pinder, 1986) and are probably exposed to similar levels of hydrophilic materials as the first instars (Muir *et al.*, 1982). They would also be exposed to hydrophobic toxicants associated with the sediment. Therefore, the second instar, which appears to be the second most sensitive instar, may be the most sensitive instar to hydrophobic toxicants. Although the early instars appear to be the most sensitive to toxicants, it is generally more convenient to examine morphological abnormalities in later instars as their larger structures are easier to examine and exist for longer than the early instars.

There is obviously a need to understand the process by which structural abnormalities are produced in the fourth instars, and it is recommended that controlled experiments are conducted. In particular, there is a need to investigate the significance of instar when considering the propensity of a larva to develop a morphological abnormality in response to stress from pesticide contamination. However the current study indicates that the frequency of abnormalities in chironomid population appears to be a useful indicator of stress from pesticides used in rice agriculture and that this technique could be a very useful screening tool for biomonitoring programs.

Acknowledgments

We would like to thank Dr Jon Martin, Ms Linda Anderson Carnahan (USEPA) and Gary Bennison for their comments on the manuscript and Geoff McCorkelle for his assistance in the field.

Pollution Monitoring Using Chironomid Larvae: What is a Deformity?

C.P. Madden[1], A.D. Austin[1] and P.J. Suter[2]

[1]Department of Crop Protection, Waite Campus, University of Adelaide, Glen Osmond, S.A. 5064

[2]Office of the Environment Protection Authority, GPO Box 2607, Adelaide, S.A. 5001.

Introduction

Warwick (1988) defined a deformity in chironomid larvae as any morphological feature that departs from the normal configuration. Examples of many types of deformities have been recorded and illustrated for several chironomid genera (e.g. Warwick, 1985, 1989, 1991; Warwick and Tisdale, 1988). Numerous pollutants have been implicated as possible causal agents but many field studies of deformities in chironomids have been poorly controlled so the attribution of deformities to contaminants remains uncertain (Rosenberg, 1992). However, the recording of high frequencies of deformity in locations known to be seriously polluted are not likely to be random occurrences. Further, very little experimental evidence has been published to show that contaminants do cause deformities (Madden *et al.*, 1992).

The above definition of a deformity necessarily includes variations that occur as spontaneous mutations in individuals as well as deformities that are caused by pollutants, whether they can be attributed to individual chemicals or not. Thus, for chironomid larval deformities to be useful as a monitoring technique, information is required on the extent and type of deformity that occurs naturally at unpolluted sites, as well as the types of deformity that are more likely to occur due to stress from environmental pollution.

Deformities recorded in chironomid larval mouthparts and antennae can be assigned to one of two general categories; absolute deformities (e.g. missing or extra teeth on the mentum or mandibles, or missing antennal segments (Figs 1a, c) and relative deformities (e.g. change in shape or size; Fig. 1b). Breaks that occur in the mentum or mandibles (e.g. Figs 1c,d,and e) are easily distinguished because of the uneven way in which the two fused plates of the mentum are usually broken (described as "jagged edges" by Dickman *et al.*, 1992). As departures from the normal configuration, breaks should be counted as deformities if one narrowly applies Warwick's definition. Breaks are mentioned in very few papers dealing with deformities (Warwick and Tisdale, 1988; Dickman *et al.*, 1992; Janssens de Bisthoven and Ollevier, 1989; Janssens de Bisthoven *et al.*, 1992b). Some authors apparently include breaks as deformities, however this may mask the effect of pollutants on other deformities. As yet, no published comparison has been made between the number of deformities and the number of breaks in a sample of larvae.

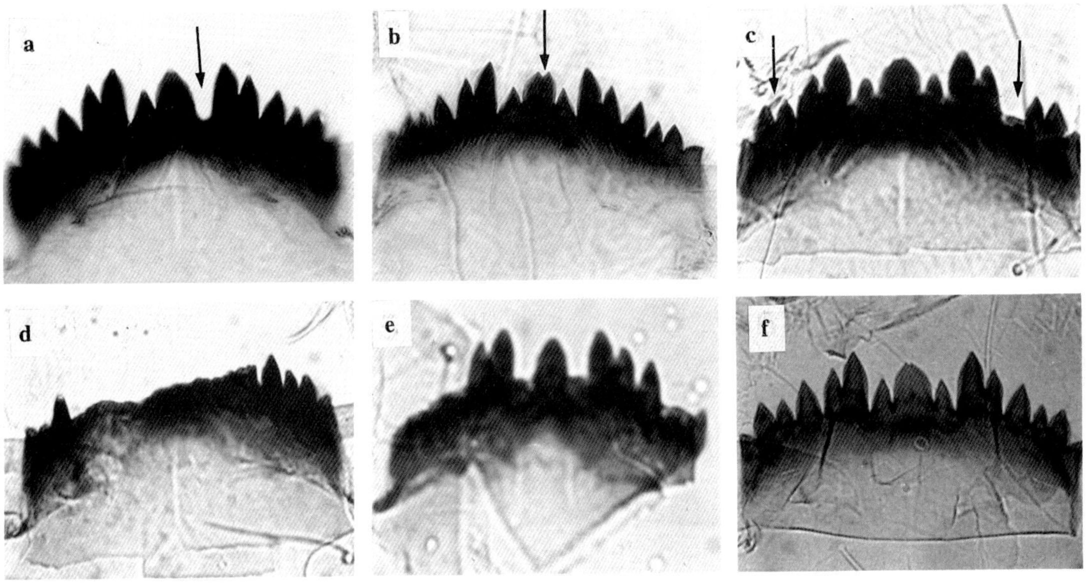

Figure 1. Deformities and breaks in the mentum of *Chironomus maddeni*, (a) gap (arrowed), (b), split median tooth (arrowed), (c), missing lateral tooth on left, broken lateral tooth on right (arrowed); (d), broken teeth; (e), broken teeth on mentum of 3rd instar; (f), normal mentum of 4th instar of the same individual as in Figure 1e.

In this study we set out to investigate the relative merits of scoring deformities and breaks separately when investigating experimentally the effect of pollutants on chironomid larvae in the laboratory, and assessing the frequency of deformities in field populations. The effect of rearing chironomids in artificial solutions as against natural stream water was also examined by comparing the frequency of deformities and breaks in each system.

Materials and Methods

Laboratory cultures were maintained in 2X Martin's solution using the method of Martin *et al.* (1980), and experimental tanks were set up as described in Madden *et al.* (1992). The species maintained in culture and used in laboratory experiments was *Chironomus maddeni* Martin and Cranston (a new species described here; see appendix). Larvae were fed with crushed trout pellets and unbleached tissue paper was used as a medium for shelter building. The laboratory cultures were sampled regularly to record the frequency of deformities as an estimate of the health of the culture and also to provide a measure of the frequency of deformities for comparison with experiments.

All laboratory experiments were conducted in 2X Martin's solution with 48 larvae being sampled randomly from each tank and examined for deformities and breaks. All larval head capsules were mounted in Hoyer's medium. Total deformities as shown in Tables 1, 2 and 3 and Figures 2 and 3 consist of deformities found in all head capsule structures (mentum, mandibles, antennae and epipharyngeal pecten). Breaks represent the proportion of larvae with at least one broken tooth on the mentum. Regression lines and t-tests were calculated using the Macintosh application Statview+Graphics except for the comparison of regression slopes (Zar, 1984) which were calculated manually.

To reduce the effect of variation between egg masses, for one experimental replicate six egg masses were each divided into six approximately equal parts. Each experimental tank then received one part of every egg mass (i.e. the total is equivalent to one whole egg mass per tank).

Table 1. Comparison of frequencies of total deformities and breaks in *Chironomus maddeni* for natural water (from Deep Creek) and artificial water (2X Martin's solution). Data represent means and ranges for nine tanks (see text for further explanation).

	Deep Creek water	2X Martin's
Deformities	$0.19^{a}\pm0.01$ (range 0.13-0.27)	$0.32^{c}\pm0.02$ (range 0.23-0.48)
Breaks	$0.65^{b}\pm0.04$ (range 0.50-0.90)	$0.25^{c}\pm0.05$ (range 0.06-0.50)

[a]Dissimilar superscripts indicate that the means are significantly different using the Student t-test ($p < 0.05$).

Table 2. Comparison of frequencies of total deformities and breaks in *Dicrotendipes conjunctus* larvae collected from the unpolluted Aaron Creek.

Date	Deformities	Breaks
9.xii.90	0.07	0.19
10.iii.91	0.04	0.15
10.vi.91	0.06	0.18
3.xii.91	0.13	0.19
14.iii.92	0.17	0.10
15.vi.92	0	0.42
16.xii.92	0.10	0.16
11.iii.93	0.10	0.17
12.vi.93	0.07	0.20
Mean ± S.E.	0.08±0.02	0.19±0.03

Table 3. Comparison of frequencies of total deformities and breaks in *Chironomus* sp. and *Dicrotendipes conjunctus* larvae from egg masses collected in Cox Creek on 10.ii.89 and raised in the laboratory (data from Madden 1990).

	Deformities	Breaks
***Chironomus* sp.**		
Field	0.14	0.16
Laboratory	0.16	0.52
Dicrotendipes conjunctus		
Field	0	0.05
Laboratory	0.02	0.22

The experiment comparing Deep Creek water with 2X Martin's solution used three tanks each of Deep Creek water and 2X Martin's for each replicate. The water from Deep Creek was tested by the Engineering and Water Supply Department of South Australia using automated gas chromatography and flame photometer for the presence of various pollutants. It was found to be free of pesticides; it had only background concentrations of heavy metals (< 5 µg/L), but had a much higher TDS (total dissolved solids: 564 ppm versus 150 ppm for 2X Martin's solution).

For experiments examining the effects of pollutants, the appropriate concentration of test chemical was added to tanks. Solutions of lead were made up as the lead acetate salt in water. DDT solutions were in acetone (Madden *et al.*, 1992). The experiments using lead and DDT consisted of one control and five test concentrations in each replicate series. Three replicates were conducted for DDT and five for lead.

Field collection methods were as described in Madden *et al.* (1992). The field sites used were Deep Creek and Aaron Creek (two unpolluted sites within the Deep Creek Conservation Park 100 km south of Adelaide) and Cox Creek, a pesticide impacted stream in the Piccadilly Valley vegetable growing area, 15 km east of Adelaide (see Madden *et al.* 1992 for details). The *Chironomus* sampled in Cox Creek belonged to one species but it was not identified.

RESULTS

Laboratory Experiments

Over a period of two years (1990-91) the frequency of deformities in the laboratory cultures ranged from 20% to 46% and the frequency of breaks from 19% to 46%. As these frequencies were considered to be unacceptably high, it was decided to modify the culturing technique. It was thought that the high levels of deformities may have been due to the lack of some micronutrient in the culture medium. Hence, water was collected from Deep Creek Conservation Park, an unpolluted area used as a control for field samples (Madden *et al.*, 1992). To test the hypothesis that the relatively high frequency of deformities in the laboratory culture was due to the medium used for culturing, three replicates of six tanks were set up as described above. The results for this experiment are given in Table 1; the frequencies for which represent the means of the nine individual tanks for each treatment. There was a significant reduction in the frequency of deformities between Deep Creek water and 2X Martin's (d.f. = 8, t = 4.36, P < 0.005), thus confirming the hypothesis. Deformities and breaks in 2X Martin's were not significantly different (d.f. = 8, t = 1.23, P = 0.25). However, an unexpected result in Deep Creek water was that there was a highly significant difference between deformities and breaks (d.f. = 8, t = 9.83, P < 0.0005). Breaks were also significantly increased in Deep Creek water compared with 2X Martin's (d.f. = 8, t = 5.99, P < 0.0005). The results indicate that the frequency of deformities was reduced in the Deep Creek water but surprisingly the frequency of breaks was increased.

To test the hypothesis that deformities and breaks respond differently to pollutants, larvae were exposed to a series of concentrations of DDT and lead. The results for DDT are given in Figure 2. The data indicate a similar trend for deformities and breaks, with the slopes of the two regression lines not differing from each other (d.f. = 8, t = 0.44, P > 0.5) (Zar, 1984). Figure 3 shows the results for lead and, in contrast to the DDT experiment, deformities and breaks indicate opposite trends for which the slopes of the regression lines are significantly different (d.f. = 8, t = 7.59, P < 0.001). If the data for deformities and breaks are combined their regression line is different to that for deformities (d.f. = 8, t = 7.09, P < 0.001), but not the regression line for breaks only (d.f. = 8, t = 1.38, P > 0.20). These results confirm the hypothesis that deformities and breaks respond differently to at least some pollutants.

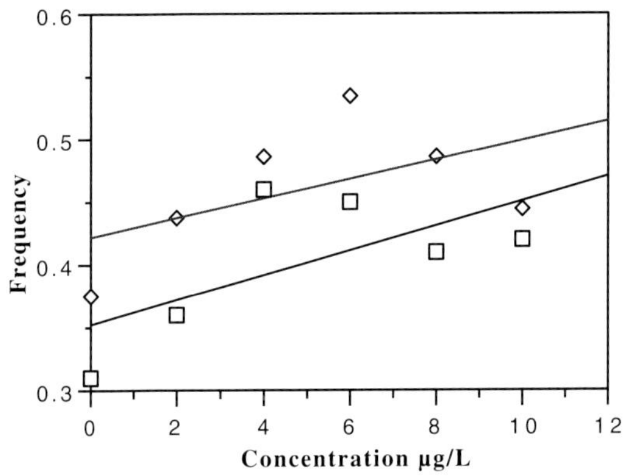

Figure 2. Frequencies of mean total deformities (□) and mean breaks (◇) for *Chironomus maddeni* larvae exposed to various concentrations of DDT.

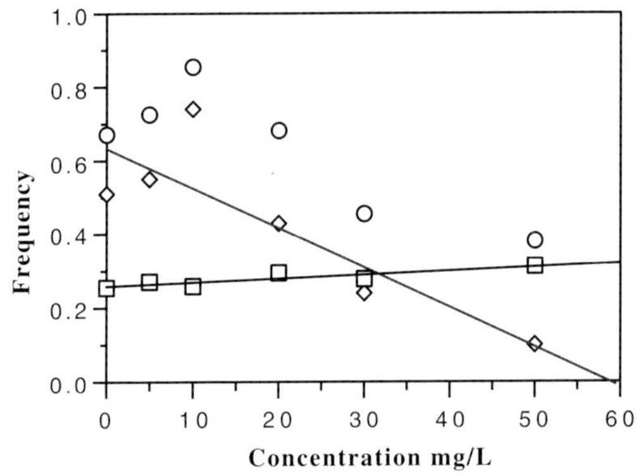

Figure 3. Frequencies of mean total deformities (□), mean breaks (◇), and these data combined (○) for *Chironomus maddeni* larvae exposed to various concentrations of lead. N.B. The dotted line represents the regression for the combined data.

Field Sampling

To test the hypothesis that the frequency of deformities and breaks may differ in the field, *Dicrotendipes conjunctus* larvae were collected from an unpolluted area on Aaron Creek (Madden *et al.*, 1992). *Chironomus* larvae could not be used as they were very rare at this site. The results are shown in Table 2, with a t-test revealing that there was a significant difference between the means for deformities and breaks, pooled across all dates (d.f. = 8, t = -2.61, P < 0.05). The frequency of both breaks and deformities were relatively constant, the only substantial variation being on 15.vi.92 when there were no deformed larvae, whereas breaks increased to 42%. Thus, it can be concluded that, for *D. conjunctus* at this site, the frequency of breaks is greater than that of deformities

A further indication of the variability of the frequency of breaks is shown in Table 3. Larvae of *Chironomus* sp. and *D. conjunctus* were collected from different sites on Cox Creek in the Adelaide region and an egg mass of each species was also collected from the same site and raised in the laboratory in 2X Martin's. *Chironomus* sp. larvae were collected from a more polluted site within the vegetable growing area (Madden *et al.*, 1992), and this is correlated with the higher frequency of deformities for this species. The frequency of deformities for each species was only slightly increased in the egg masses raised in the laboratory. However, the frequency of breaks was greatly increased in the laboratory-reared larvae. Clearly, some stress is placed on the laboratory-reared larvae that result in breaks occurring much more frequently, whereas deformities virtually remain at the same level.

Discussion

The data from both laboratory cultures and experiments and the field show that deformities and breaks often display different trends and give contrasting indications of the effect of environmental conditions on larvae. For example, the results of the experiment shown in Table 1 suggest that the use of Deep Creek water reduces the frequency of deformities in larvae compared with 2X Martin's. However, the frequency of breaks, which was generally low in the laboratory culture, were much higher when Deep Creek water was used as a medium. It should be noted that after two generations in the Deep Creek water, the laboratory culture showed a frequency of breaks that was similar to that observed in 2X Martin's. This indicates (as do the data presented in Table 3) that breaks are possibly a response to stress caused by the sudden change in environmental conditions (i.e. from 2X Martin's to Deep Creek water in Table 1 and from field to laboratory conditions in Table 3).

The results from the laboratory experiments to test the effect of pollutants on the frequency of deformities and breaks were also contrasting. For DDT both sets of data (deformities and breaks) show a slight increase in frequency with concentration and a slight drop at the highest concentration, which may be explained by the increased mortality at this dose. In the case of the lead experiment, however, the trend for breaks is opposite to that of deformities. Deformities increased slightly with concentration (a similar result to that for DDT) but breaks decreased sharply with concentration. When breaks were included as deformities this response was completely masked by the greater, but opposite, trend for the frequency of breaks. Janssens de Bisthoven *et al.* (1992a) found that deformed larvae contained higher metal concentrations than normal larvae. A measurement of the metal concentration in larvae with breaks may also show a difference between normal and deformed larvae, though such a comparison would not be easy as some larvae exhibit both breaks and deformities. The important result from the above experiments is the difference in response found between deformities and breaks to the increasing concentration of chemical. Deformities in each case increase with concentration as one would expect, intuitively. However, breaks increased with the concentration of DDT but decreased with the concentration of lead.

The field data from Aaron Creek (Table 2) shows that the frequency of deformities is, as expected at an unpolluted site, very low. The frequency of breaks is higher and is similar to that recorded in the laboratory culture using 2X Martin's solution (see Table 1) and hence, must be considered a less useful indicator of the severity of pollution.

Previously few authors have referred to breaks in mouthparts when recording deformities in chironomid larvae and opinions differ on how they should be treated. Warwick (*pers. comm.*) does not regard breaks as deformities; Dickman *et al.* (1992) described a break as a "slight deformity" and did not include them as deformities; while Janssens de Bisthoven and Ollevier (1989) and Janssens de Bisthoven *et al.* (1992b) recorded "mechanical wear", a result of abrasion due to the hard substrate in which larvae were raised that includes, but is not identical to, breaks. Janssens de Bisthoven and Ollevier (1989) found that mechanical wear was reduced in the laboratory culture when the Kiezelgür substrate was organically enriched. This would seem to contrast with the result obtained here where the use of Deep Creek water (which undoubtedly is organically richer than 2X Martin's) initially increased the frequency of breaks. It can be argued that breaks indicate a weakening of the mentum but the available data do not allow for the patterns of contaminant response to be explained. In an experiment using cadmium, Janssens de Bisthoven *et al.* (1992b) recorded heavy mechanical wear, and a trend for maximum percentage wear to occur at higher concentrations (9 and 27 ppm). An experiment conducted by us using cadmium at concentrations of up to 100 µg/L (data not shown) produced very similar results to those in Figure 3 for lead. The frequency of breaks decreased as the concentration of cadmium increased.

The presence of breaks may interfere with the scoring of deformities by reducing the chance of identifying, for example, deformities in median teeth. If the median tooth is broken, as in Figure 1d, it is impossible to determine if there was a deformity of the type shown in Figure 1b. Thus, a high frequency of broken teeth may cause a reduction in the number of observed mentum deformities in a sample of larvae. However, Figure 1c shows that it is still possible to see mentum deformities if only one or a few teeth are broken.

Whether or not the mentum can function properly is also relevant when discussing breaks. It would be expected that the mentum in Figure 1b would be affected only very slightly when compared to a normal mentum. However, the menta in Figures 1d and e are missing many teeth and their performance should be substantially impaired. However, larvae of the type illustrated in Figure 1d were found alive in cultures (and in many cases they successfully pupated) and so, presumably, drastically altered menta still allow adequate feeding to ensure survival. Figure 1e illustrates a broken mentum of a third instar larva which survived despite many broken teeth, with Figure 1f showing the mentum (with no broken teeth) of the fourth instar of the same individual. This not only indicates that larvae can survive despite the deleterious effect of broken teeth but also that breaks are purely a physical effect given that the fourth instar developed a normal mentum. Interestingly, in our cultures, breaks were very rare in newly moulted larvae but deformities occurred with the same frequency as in other larvae. Similarly, Janssens de Bisthoven and Ollevier (1989) observed that the frequency of mechanical wear increased with the age of an individual in a specific instar.

The conclusion to be drawn from this work is that breaks are a common occurrence in the menta of chironomid larvae, both in laboratory cultures and the field, and that they should be recorded separately when examining larvae for deformities. However, the patterns of response to environmental variation are often quite different between these two types of abnormality. Deformities show greater response to contaminants than breaks and, hence, should be more useful as an indicator of pollution. We, therefore, advocate that breaks should not be included as deformities but recorded separately, because there is a chance of erroneously concluding that there is no response to a pollutant if breaks are regarded as deformities. Breaks may prove to be most useful in determining whether the conditions in which laboratory experiments are being undertaken are comparable to those in the field, and that new stresses are not being introduced into experimental systems.

Acknowledgments

We would like to thank the Land and Water Resources Research and Development Corporation and the Engineering and Water Supply Department of South Australia for funding the Postgraduate Research Grant to one of us (CPM) that allowed this work to be undertaken. Eveline Bartowsky kindly read a draft of the manuscript.

APPENDIX

Description of *Chironomus maddeni* sp. nov

Jon Martin[1] *and Peter S. Cranston*[2]

[1]Genetics Department, University of Melbourne, Parkville, 3052, Australia

[2] CSIRO Division of Entomology, GPO Box 1700, ACT 2601, Australia

Introduction

In the preceding study (Madden *et al.*, 1995), the investigated species of *Chironomus* evidently is undescribed. This judgement is based upon comparison with the cytology of all known species of Australian *Chironomus* by the senior author of this appendix contribution. Here we provide details of the cytology and morphology of all stages and provide a formal description of *Chironomus maddeni* as a species new to science.

Chironomus maddeni n.sp.

Type-locality: **South Australia,** Glen Osmond, pool in grounds of Waite Institute, University of Adelaide.

Description. Conforms in all stages with the generic diagnoses of Cranston *et al.* (1989; adult male), Pinder and Reiss (1986; pupa) and Pinder and Reiss, (1983; larva). All measurements in µm unless stated (as mm).

Adult Male (n=2) (Fig. 4a,b)

Body length 5.3-5.6 mm, wing length 2.4-2.5 mm, colour yellow to pale brown with mid-brown scutal vittae, scutellum, and pre-episternum. Abdomen pale, with brown, broadly diamond-shaped transverse area in basal half of tergites I-IV. Legs pale, unbanded.

Head. Without frontal tubercles, with 25-27 temporal setae in a continuous multiserial row from frontals to postoculars, 19-21 clypeals. Antenna with apical flagellomere 850-920µm long, basal 10 flagellomeres 320-355 long, Antennal Ratio 2.55-2.60. Palp segment 2-5 lengths: 50-65, 140-150, 180-205, 240-315.

Thorax with quite prominent scutal tubercle. Thoracic setation: acrostichals about 8, biserial, dorsocentrals 10-12 anteriorly uniserial, posteriorly biserial, 5 prealars, 12-14 scutellars. Wing with Venarum Ratio 1.03. Vein setation: R with 31, R_{2+3} 27-29, R_{4+5} 31-34; squama with 14-16.

Leg lengths and proportions as follows:

	Fe	Ti	Ta1	Ta2	Ta3	Ta4
PI	1045-1080	865-900	1510-1620	765-790	570-580	500-505
PII	1115-1225	1010-1045	610-615	325-360	250-320	175-180
PIII	1260-1295	1225-1295	900-935	505-540	360-395	215-220

	Ta5	LR	BV	SV	BR
PI	215-250	1.75-1.80	1.69-1.70	1.22-1.26	2.0-2.5
PII	105-110	0.58-0.60	2.92-3.20	3.47-3.70	3.0-4.0
PIII	140-145	0.72-0.73	2.72-2.76	2.75-2.76	3.0-4.0

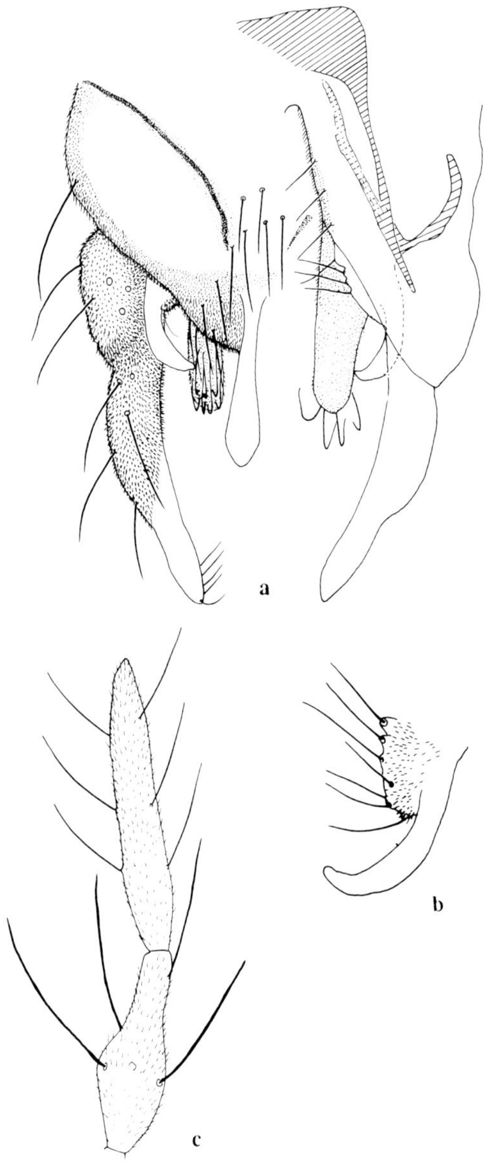

Figure 4. *Chironomus maddeni* sp. nov., adult: (a) male hypopygium; (b) superior volsella, (c) female antenna apex.

Sensilla chaetica: PII with 5-6 on subapex of Ti_1, PIII with 7-8 on subapex of Ti_1.

Hypopygium: see Figure 4a, b.

Adult female (n=2)

Body length 5.3-5.4mm, wing length 2.8-2.9mm, colour as male; abdomen less well marked than in male.Head. Without frontal tubercles, with 22+ temporal setae in a continuous multiserial row from inner verticals, 26 clypeals. Antenna (Fig. 4c) with apical flagellomere 400-420 long, basal 5 flagellomeres 156-158 long, Antennal Ratio 0.35-0.37. Palp segment 2-5 lengths: 52-60, 140-155, 190-210, 315-335.

Thoracic setation: antepronotals absent, acrostichals c. 17, biserial, dorsocentrals 27-28 biserial, 6 prealars, 14-21 scutellars. Wing with Venarum Ratio 1.02-1.09; setation: R with 35-40, R_1 32-35, R_{4+5} 78-80, squama with 13-16.

Leg lengths and proportions as follows:

	Fe	Ti	Ta1	Ta2	Ta3	Ta4
PI	1190-1260	972-1010	-	-	-	-
PII	1290-1295	1190-1260	650-685	325-360	250-255	180
PIII	1405-1440	1440-1475	1010-1045	540-576	430-435	250

	Ta5	LR	BV	SV	BR
PI	-	-	-	-	-
PII	145	1.74-1.83	3.38-3.48	3,63-3.94	4.0-5.5
PIII	145	1.40-1.46	2.79-2.84	2.75-2.85	3.5-4.0

Sensilla chaetica: PII with 50-60 uniserial, absent only on basal 10% of Ti_1, PIII with 60-70 uniserial, absent only on basal 10% of Ti_1.

Genitalia: typical for *Chironomus* s.str.

Pupa (n=10) (Fig. 5d-g).

Pale brown, with basal ring and tracheal bases, abdominal apophyses, posterolateral spur and bases of anal lobe setae all darker brown. Cephalic tubercles (Fig. 5d) mid-brown, 122-157 long subapically bearing 28-25 long frontal seta. Thorax with dense brown rugulosity to midpoint (Fig. 5e). Basal ring oval, 120-132 long, 58-98 wide, with elliptical fused double tracheal bases.

Abdomen (Fig. 5f) with very restricted shagreen, present only along apophyses on TV and TVI. Pedes spurii B quite strong on II. Pedes spurii A on IV, much weaker on V, absent on VI. Tergal spinule pattern as in Figure 5f. Hook row with 60-75 hooks, extending about 55-60% of the width of tergite II. Conjunctives all bare. Paratergite IV with narrow spines, remainder bare. Setation ('O', D, V and L) typical for genus. Posterolateral spur (Fig. 5g) simple, narrow, tapering to fine point, sometimes with one or two small fine spines at base. Anal lobe with about 150 multiple ranked setae in fringe (not drawn).

Fourth-instar larva (n=10) (Fig. 5a-c).

Medium sized, length about 15-18mm (female), 13-14mm (male). Head capsule with mentum mid-brown, mandibular teeth brown, remainder of mandible yellow, gular region slightly darkened towards posterior margin, very slight darkening of frontoclypeus and dark brown to black occipital margin. Ventral tubules relatively short (c. 0.65-1.3mm), not coiled, anterior pair usually slightly shorter than the posterior pair; no lateral projections on eighth abdominal segment. Similar to larvae of C. *oppositus*-group.

Measurements. Head capsule length 580-660. Antenna (Fig. 5c) segments 100-112, 26-31, 8-9, 9-11, 6-8; Antennal Ratio 1.9-2.2; Ring organ 33-42 from base of 1st segment; style 39-44 bifid; Lauterborn organs and peg 5-6, about half length of third antennal segment. Mandible (Fig. 5b) 202-225, with yellow-brown outer tooth, dark apical and two inner teeth, mola yellow. Mentum (Fig. 5a) 160-170 wide, with typical triple-toothed median part and six pairs of lateral teeth of which the innermost pair are rather more closely appressed than are the remainder; ventromental plate 108-117 with 29-35 striae across median part.

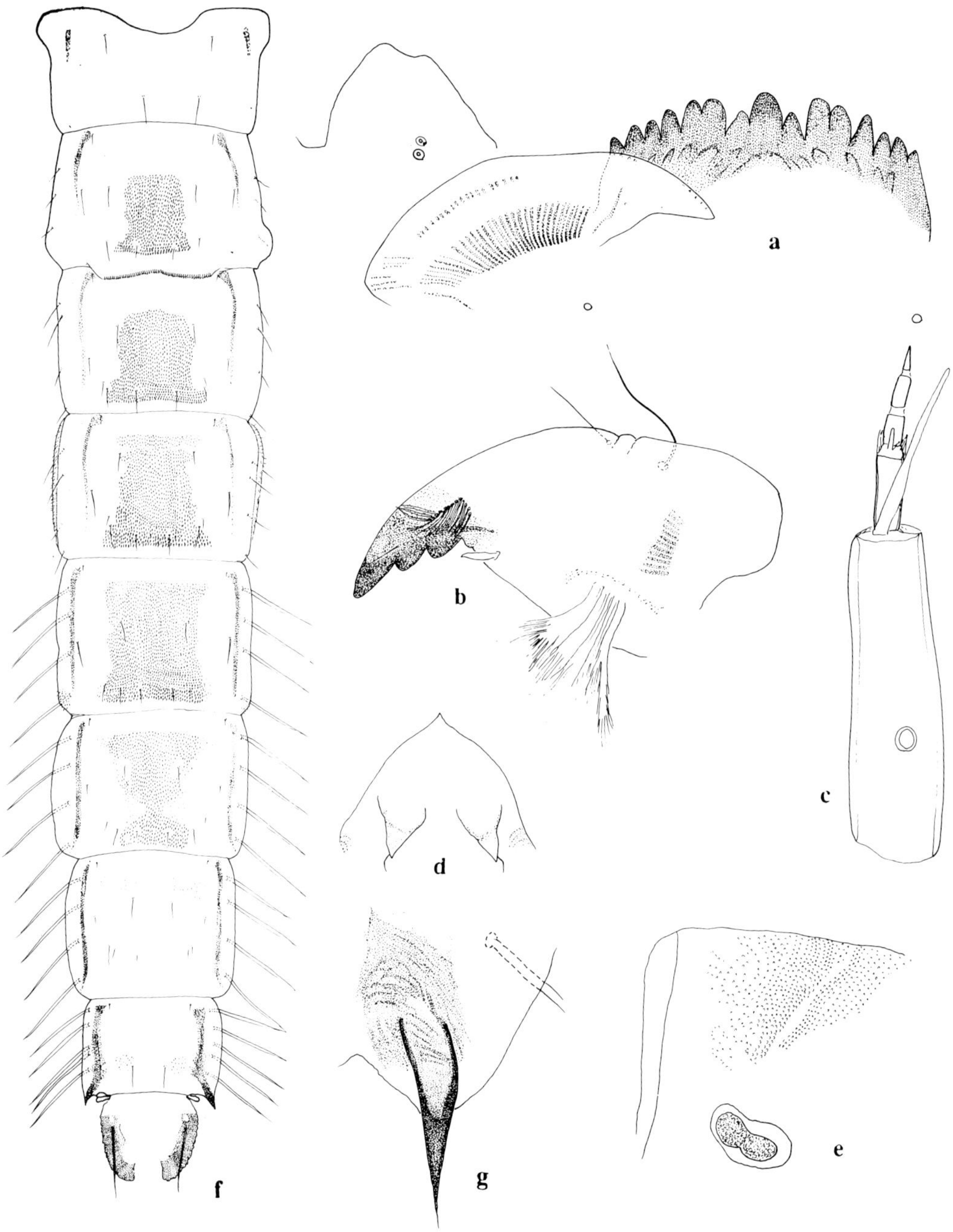

Figure 5. *Chironomus maddeni* sp. nov. larva and pupa. (a)-(c) Larva: (a) mentum, (b) mandible, (c) antenna; (d)-(g) Pupa: (d) cephalic tubercles and frontal setae, (e) anterior thorax, (f) tergites (anal lobe fringe not figured), (g) posterolateral comb.

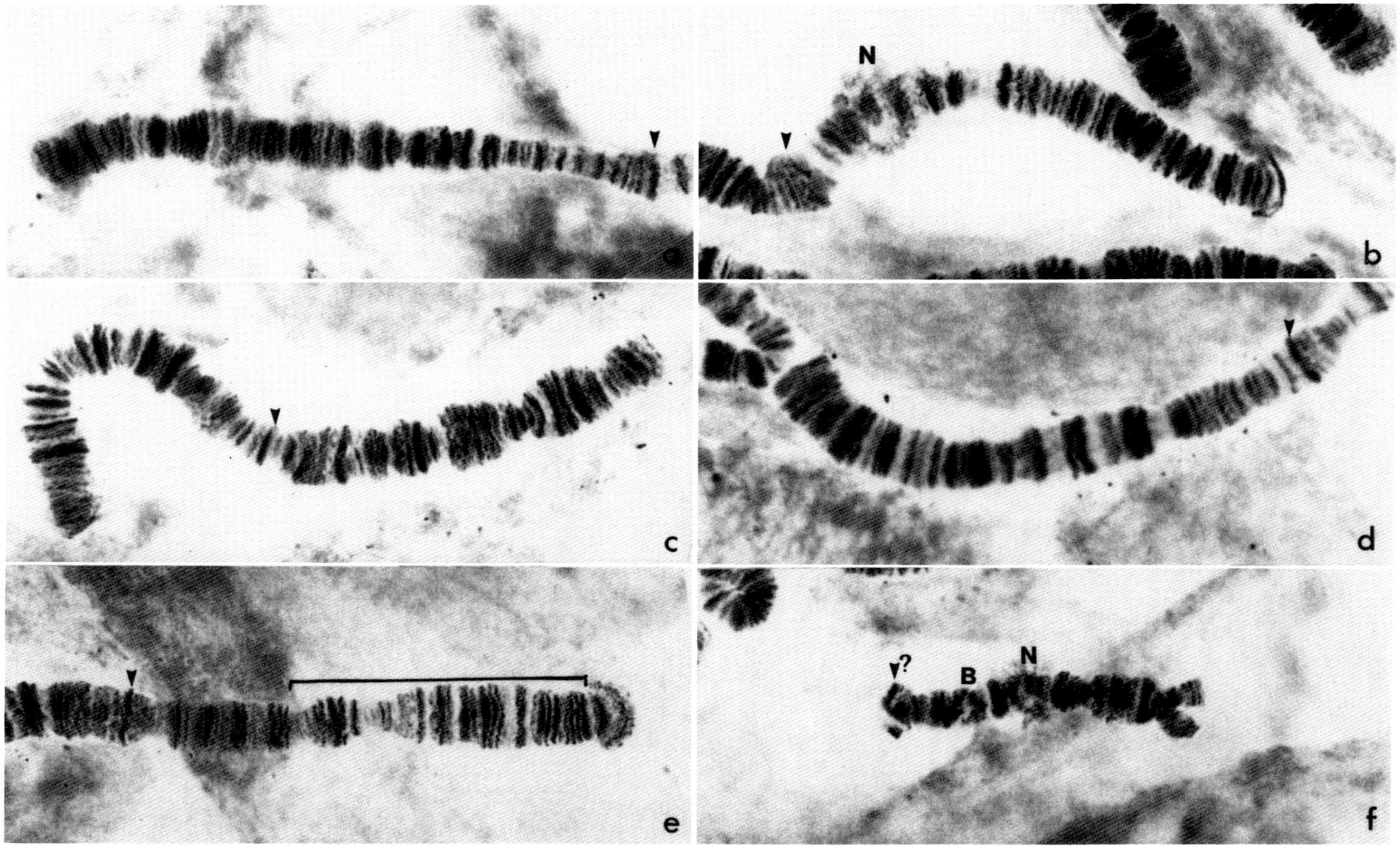

Figure 6. *Chironomus maddeni* sp. nov., polytene chromosomes: (a) Arm B, (b) Arm F, (c) Arms D (left) and C (right), (d) Arm E, (e) Arm A, (the bracket above the arm indicates the approximate extent of the heterozygous inversion observed in one individual), (f) Arm G. The postulated centromere is at the left end. Centromeres are indicated by arrow heads; N - nucleolus, B - Balbiani ring.

Larval cytology (Fig. 6a-f)

Four polytene chromosomes (Fig. 6) with the arm combination of the pseudothummi-group, i.e. BF, CD, AE, G. Large nucleolus developed near the centromere of arm F, with a smaller one developed medially in arm G. Only one inversion polymorphism, in arm A, noted in the material available for study.

Arm B (Fig. 6a): Sequence as B2 of C. *oppositus* (Martin 1969; Martin *et al.*, 1978), i.e. large puff of 2D is near the distal end with the dark bands of 2E towards the telomere.

Arm F (Fig. 6b): Sequence as F1 of C. *oppositus* (Martin 1969; Martin *et al.*, 1978), including the large nucleolus at 4C.

Arms D and C (Fig. 6c): Banding pattern does not appear to correspond to any previously described sequence.

Arm E (Fig. 6d): Sequence as E1 of C. *oppositus* (Martin 1969; Martin *et al.*, 1978).

Arm A (Fig. 6e): Differs from the previously described sequences by an inversion of approximately the region 14D1-15C8 of sequence A4. This results in the ball of 15E having a similar position and orientation to that of A5. However it can be distinguished from that sequence because the distal region remains similar to A4. One specimen was heterozygous for an inversion of the region from about 15D4-15E.

Arm G (Fig. 6f): Differs from arm G of C. *oppositus* by the nucleolus being near the middle of the arm, with an obvious Balbiani ring between the nucleolus and the postulated centromere.

Material examined

Holotype male slide-mounted in Euparal, ex-**South Australia,** Glen Osmond, pool in grounds of Waite Institute, University of Adelaide (Madden), reared in culture in Melbourne University Genetic Department by Jon Martin, Australian National Insect Collection (ANIC).

Paratypes, 1 male, 2 females, 10 pupal exuviae, 10 larvae, all mounted in Euparal, same data as holotype.

Comments

Cytologically this species is similar to those of the broader C. *oppositus*-group (i.e. including the undescribed species C. *alternans* b and C. *alternans* c (Martin and Lee, 1984)). It can be distinguished from C. *oppositus* by the lack of a subterminal nucleolus in arm G, and from C. *alternans* b and C. *alternans* c by the orientation of group 15E of arm A, since both the latter species have sequence A4.

Although this species is distinguishable on the basis of cytology, it has not been possible to find unambiguous character states to differentiate from related species in the C. *oppositus*-group. However, characters that may prove of value in future differentiation include the following:

Adult: "normal" male genitalia, unpatterned wing and legs, short bristle ratio (BR) on the fore leg, longer on mid and hind legs; presence of acrostichal setae and absence of frontal tubercles.

Pupa: this stages provides less diagnostic information than in many other genera. The tergal spinule pattern, presence of spines on paratergite IV, thoracic rugosity and pigment distribution may prove significant in species differentiation.Larva: the short, uncoiled ventral tubules with the anterior pair usually slightly shorter than the posterior pairand lack of lateral projections on eighth abdominal segment typify the *oppositus*-group. Within this group the number of striae in the basal part of the ventromentum (29-35), and the distribution of pigment on the head capsule may allow separation from other members of the group.

Fluctuating Asymmetry of Chironomid Larvae as an Indicator of Pesticide Contamination in Freshwater Environments

Geoffrey M. Clarke[1], *Angela H. Arthington*[2] *and Bradley J. Pusey*[2]

[1] CSIRO Division of Entomology, GPO Box 1700, Canberra, ACT 2601, Australia.

[2] Centre for Catchment and In-Channel Research, Griffith University, Nathan, QLD 4111, Australia

Introduction

One of the most widespread organisms used in the analysis of water quality have been genera of the family Chironomidae (Diptera) which display a wide range of observable sub-lethal responses to environmental stresses. These include reduction in larval growth (Kosalwat and Knight, 1987), changes in patterns of adult emergence (McCahon and Pascoe, 1991) and increased frequency of structural deformities in the antennae and head capsule of larvae (e.g., Hamilton and Sæther, 1971; Hare and Carter, 1976; Wiederholm, 1984; Warwick, 1985; 1989). The incidence of larval deformities appears to be a widespread phenomenon occurring in a wide variety of chironomid genera, both free-living and tubiferous in the benthos, in response to a variety of stresses (Wiederholm, 1984; Warwick, 1989). A considerable database on types of deformities correlated with different types of stress has been assembled and a number of index systems have been developed for use in biomonitoring programs (Warwick, 1985). This technique has gained wide acceptance among environmental agencies throughout the world, primarily because it is sensitive, relatively simple and inexpensive, and has been used often in environmental assessment programs.

Despite the popularity of this approach there are a number of inherent problems: identification and classification of deformities is very subjective, difficulty in distinguishing between developmental-derived deformities and physical changes resulting from wear and tear, and the need to know the underlying normal variation in character expression in unstressed environments. Although a number of attempts have been made to develop standardised methods of characterising deformities and quantifying data using elaborate index system, deformity analysis still remains essentially a qualitative approach.

An alternative approach is that of developmental stability analysis. Essentially developmental stability analysis and deformity analysis reflect the same underlying developmental changes, i.e. departures from the normal genetically determined development pathway which results in aberrant phenotypes. However, developmental stability analysis does not suffer from the same problems as deformity analysis. The primary technique of developmental stability analysis is measurement of departures from symmetry in normally bilaterally symmetric characters (either meristic or morphometric). Increases in asymmetry relative to background levels (determined by control populations) is an indication of developmental stress. Measurements are totally objective and statistical methodology for handling data sets are well established.

Fluctuating asymmetry is defined as those random differences which occur between left and right sides of a normally bilaterally symmetrical organism (Van Valen, 1962). It has been used as an indicator of the level of developmental stability and in defining the influence of both environmental and genetic stress on development (reviewed by Palmer and Strobeck (1986) and Zakharov (1989)). The underlying theory of fluctuating asymmetry analysis proposes that the presence of environmental or genetic stress during ontogeny may reduce the efficiency of normal developmental processes, this being reflected by an increase in the level of fluctuating asymmetry (Adams and Niswander, 1967; Clarke *et al.*, 1986).

Relationships between fluctuating asymmetry and environmental stressors have been shown in a number of organisms and assessment of asymmetry has been shown to be applicable across a wide variety of animals, for the determination of the effects of different types of environmental stresses (see Clarke, 1994 for review). Much of the research to date has concentrated on target species, specifically trying to address the impact of a particular stressing agent on a species of interest, e.g. mercury levels on a target fish species, for which the fluctuating asymmetry approach has proved extremely sensitive and useful (see Palmer and Strobeck (1986); Zakharov (1989) and Parsons (1990) for reviews). While such information is obviously valuable, there is also a need for a more generalised system to be able to assess biologically the impact of a potential stressor on a particular environment, e.g. the biological impact of agricultural chemicals on the quality of freshwater and saline aquatic systems. The fluctuating asymmetry approach has the potential to be used as such a bioindicator of environmental quality, particularly when applied to invertebrate populations (Clarke, 1993*a*, 1994 5*a*,*b*).

Fluctuating asymmetry has also been shown to provide an early-warning system for potential detrimental impacts on conventional life-history parameters (fitness), e.g. survival, fecundity, and fertility, in a number of species (Clarke, 1995a). Thus fluctuating asymmetry may have potential for determining impacts at sufficiently low stress intensity to allow remedial action to be taken before local populations or species are endangered or lost.

This project aimed to investigate the potential of using fluctuating asymmetry of larval mouthparts of *Chironomus* (Diptera: Chironomidae) as such a bioindicator system for identification of the biological impact of the organophosphate pesticide chlorpyrifos on this species. Chlorpyrifos is widely used to control a variety of insect pests of agricultural systems as well as flies and other household pests. It is one of a suite of pesticides detected in water samples from freshwater bodies in the vicinity of cropping lands in several areas of Australia. Chironomid midges were chosen as the test organism as they are widespread and abundant within freshwater systems and would be one of the organisms of choice in any freshwater biomonitoring program. Fluctuating asymmetry analysis has previously been successfully applied to *Chironomus* spp. in the assessment of organic pollution of freshwater systems (Clarke, 1993*a*).

This project took advantage of an established experimental artificial channel system (mesocosm) which allows manipulation of pesticide concentration and delivery mechanisms under simulated field conditions (Pusey *et al.*, 1994; Ward *et al.*, 1994). It must be emphasised that the experimental design utilised for this study was not specifically designed to investigate potential impact of chlorpyrifos on asymmetry in *Chironomus*. Rather it was established to assess the effects of chlorpyrifos on invertebrate species composition, abundances and community structure (Ward *et al.* 1994). The current study took advantage of collections made for the previous community experiment and must be considered a pilot study investigating the potential use of asymmetry as an indicator of pesticide stress.

Both acute and chronic exposure of chlorpyrifos, utlilising this mesocosm system, have been shown to impact freshwater macroinvertebrate communities, particularly larval Chironomidae (Pusey *et al.* 1994; Ward *et al.* 1994).

Materials and Methods

The Artificial Channels

The original chronic dose experiment was conducted in an artificial channel system, details of which can be found elsewhere (Pusey *et al.*, 1994; Ward *et al.*, 1994).

The channels were initially established for use in an acute dose experiment where treatments were randomly allocated to six channels. The same channel allocation were used in the chronic dose experiment.

Experimental Design

The experiment involved a chronic application of chlorpyrifos which was delivered constantly for 21 days. Treatments were randomly allocated to six artificial channels (see above). Channels 1 & 5 were control channels dosed with the carrier solution (1,1,1-trichloroethane and xylene) at the concentration of that in the high dose channels. Low dose channels were channels 2 & 4 and they received $0.1\mu g.l^{-1}$ chlorpyrifos plus the carrier solution; high dose channels were 3 & 6 and they received $5\mu g.l^{-1}$ of chlorpyrifos plus the carrier solution. The availability of only 6 channels precluded the use of true controls in the experiment (i.e. channels with no carrier solution).

Sampling

A Surber sampler (20 x 20cm) was used to sample each channel with four samples taken from each channel on each sampling date. Thirty-two different positions were used and these were ordered randomly over the eight sampling times:

July 2 - 9 days before treatment
July 11 - day of dosing of channels
July 15 - after 3 days of dosing
July 22 - after 10 days of dosing
Aug. 1 - after 21 days of dosing
Aug. 8 - 7 days after dosing ended
Aug 22. - 21 days after dosing ended
Sep. 19 - 49 days after dosing ended

Chironomid larvae were made available for asymmetry analysis after the event. *Chironomus* spp. were selected because the genus showed a strong numerical response to the pesticide. *Chironomus* spp. were not present in all channels from all collection dates (see results). For samples in which *Chironomus* were present, *Chironomus cloacalis* was the dominant species in terms of abundance with relatively few *Chironomus februarius*. These two species are almost impossible to distinguish based on larval morphology.

Asymmetry Analysis

Chironomid larval heads were removed and placed in a 10% potassium hydroxide solution overnight to clear the muscle. Cleared heads were then neutralised in glacial acetic acid and placed in 80% alcohol prior to mounting. Heads were mounted from isopropanol, dorsal side up in Euparal on glass microscope slides. A glass coverslip was placed on the preparation and gentle pressure applied to spread the mandibles enabling a clear view of the pecten epipharyngis, the character chosen for analysis. The number of teeth on either side of the midline on the pecten was counted at x400 magnification using phase contrast microscopy.

Individual sample sizes were small from each surber sample so samples from different locations within the same channel within a sampling date were pooled for subsequent analyses. Fluctuating asymmetry is statistically represented by a normal distribution of signed L–R differences around a mean of zero. Statistical procedures used for asymmetry analysis are well defined (Palmer and Strobeck, 1986, 1992; Palmer *et al.* 1993). For each sample, asymmetry values were calculated as the sum of the absolute differences between left and right sides divided by the sample size, i.e.

$$\Sigma | L_i - R^i | / N$$

Differences in asymmetry values between and among channels and sampling periods were tested using factorial analyses of covariance and single-classification analyses of variance. All analyses were performed using the statistical software package SYSTAT for Windows version 5.0 (Systat Inc., 1992). All scoring was done blind without knowledge of corresponding doses for each channel.

Results

Chironomus larvae were not present in samples taken on July 11, July 15, July 22 (i.e. during the dosing period) for all channels, including controls. All channels had *Chironomus* present 49 days after dosing ended (19/09/91). High dose channel 3 did not contain *Chironomus* between the initial pre-dose sample (02/07/91) and the final sample (19/09/91) and the other high dose channel 6 had no *Chironomus* at the completion of dosing (01/08/91), few individuals 7 days following the completion of dosing (08/08/91) and then no individuals until the final collection (19/09/91) (see Table 1).

Table 1. Mean asymmetry values of Chironomus collected from artificial channels between July and September 1991.

Channel	02/07/91	01/08/91	08/08/91	22/08/91	19/09/91
1	0.52 ± 0.08 (48)	0.39 ± 0.09 (28)	0.53 ± 0.06 (101)	0.58 ± 0.10 (38)	0.53 ± 0.09 (36)
2	0.39 ± 0.07 (59)	—	0.46 ± 0.14 (13)	0.56 ± 0.11 (41)	0.73 ± 0.09 (63)
3	0.39 ± 0.06 (59)	—	—	—	0.65 ± 0.15 (23)
4	0.36 ± 0.07 (69)	0.50 ± 0.13 (26)	0.56 ± 0.09 (54)	0.48 ± 0.09 (46)	0.60 ± 0.10 (40)
5	0.30 ± 0.10 (23)	0.39 ± 0.10 (23)	0.46 ± 0.07 (86)	0.36 ± 0.09 (45)	0.51 ± 0.09 (35)
6	0.46 ± 0.07 (61)	—	0.57 ± 0.30 (7)	—	0.59 ± 0.10 (34)

Sample sizes are shown in parentheses

Table 2. Mean asymmetry values of Chironomus collected from artificial channels (channels pooled within dosage) between July and September 1991.

Dose	02/07/91	01/08/91	08/08/91	22/08/91	19/09/91
Control	0.45 ± 0.06 (71)	0.39 ± 0.07 (51)	0.50 ± 0.05 (187)	0.46 ± 0.06 (83)	0.52 ± 0.06 (71)
Low	0.38 ± 0.05 (128)	0.50 ± 0.13 (26)	0.53 ± 0.08 (67)	0.52 ± 0.07 (87)	0.68 ± 0.07 (103)
High	0.43 ± 0.05 (120)	—	0.57 ± 0.30 (7)	—	0.61 ± 0.09 (57)

Sample sizes are shown in parentheses

Table 3. Two way analysis of covariance with position along channel as covariate and channel-within-dosage for mean asymmetry value of Chironomus collected from artificial channels between July and September 1991. DF = 1057

Source of Variation	DF	Mean Square	F	p
Position along channel	1	0.2400	0.6612	0.4163
Sampling Date	4	1.5186	4.1833	0.0023
Dosage	2	0.2360	0.6502	0.5222
Channel-within-dosage	3	0.5466	1.5030	0.2122

Table 4. Results of single-classification analyses of variance testing for differences in mean asymmetry value among channels and among doses for each collection date for Chironomus collected from artificial channels between July and September 1991.

	Among Channels			Among Doses		
Date	DF	F	p	DF	F	p
02/07/91	5	0.8149	0.5397	2	0.5124	0.5995
01/08/91	2	0.3271	0.7320	1	0.6630	0.4181
08/08/91	4	0.2357	0.9181	2	0.0986	0.9061
22/08/91	3	1.1622	0.3259	1	0.3916	0.5323
19/09/91	5	0.7548	0.5833	2	1.3131	0.2701

Table 5. Results of single-classification analysis of variance testing for differences in mean asymmetry value among collection dates for each channel and dosage for Chironomus collected from artificial channels between July and September 1991.

Channel	DF	Mean Square	F	p
1	4	0.1519	0.4246	0.7909
2	3	1.2190	2.9211	0.0356
3	1	1.1390	3.6084	0.0611
5	4	0.4611	1.1935	0.3144
5	4	0.2488	0.7487	0.5599
6	2	0.1965	0.5544	0.5762
Dose	DF	Mean Square	F	p
Control	4	0.1770	0.5106	0.7280
Low	4	1.3348	3.3596	0.0101
High	2	0.7151	2.1377	0.1209

Table 6. Results of regression analyses testing for relationship between mean asymmetry value and collection date for each channel and dosage for Chironomus collected from artificial channels between July and September 1991.

Channel	R^2	Mean Square	F	p
1	0.001	0.0991	0.2788	0.5979
2	0.045	3.3767	8.1542	0.0048
3	0.043	1.1390	3.6084	0.0611
5	0.014	1.2890	3.3591	0.0681
5	0.004	0.3093	0.9349	0.3347
6	0.011	0.3783	1.0773	0.3018
Dose	R^2	Mean Square	F	p
Control	0.001	0.2314	0.6698	0.4135
Low	0.029	4.7706	12.0536	0.0006
High	0.023	1.4122	4.2435	0.0408

Mean asymmetry values for each channel and dose are given in Tables 1 and 2 respectively. Distributions of signed left-minus-right differences are given (Fig. 1) for channels 5 and 2 for two collection dates to illustrate variance patterns typical of asymmetry analyses. The distributions for control channel 5 for the two dates are similar revealing low asymmetry value and variance. The channel 2 distribution for the 22/08/91 sample differs from the pre-dose sample (02/07/91) showing a greater variance and greater asymmetry value. These distributions illustrate that increased asymmetry value for a sample is not the result of the presence of a few highly asymmetric individuals within the sample.

Two-way analysis of covariance, with position along channel as the covariate, revealed differences in mean asymmetry value among collection dates (Table 3). There was no significant effect of dose on asymmetry value. Position within each channel did not contribute significantly to variation among asymmetry values and the insignificant channel-within-dosage term indicates no significant differences between replicate channels.

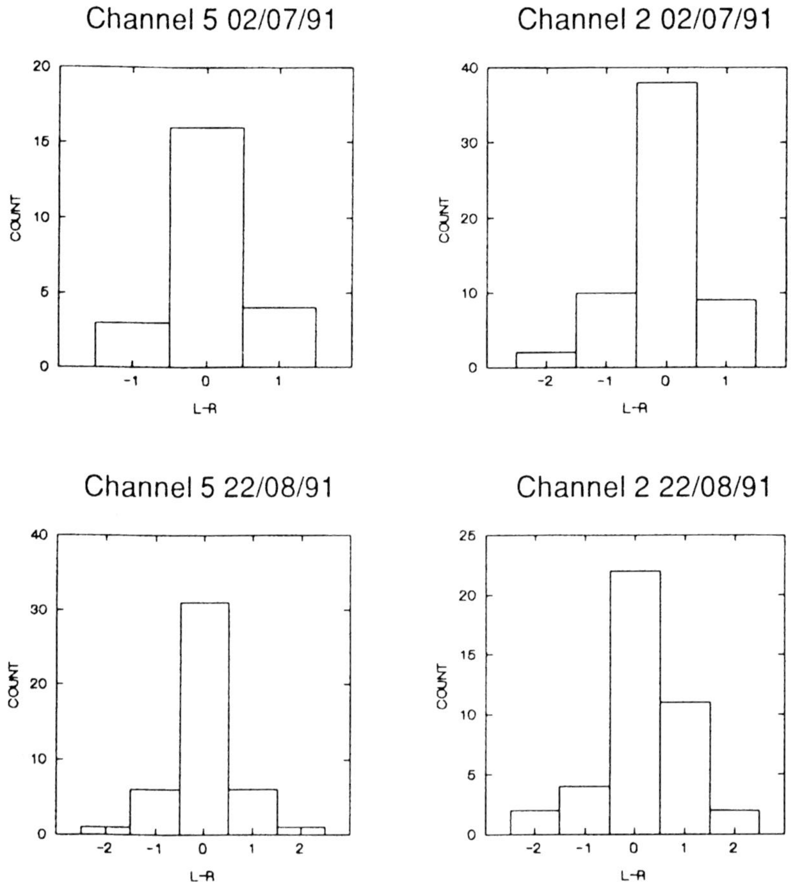

Figure 1. Distribution of signed left-minus-right differences for Channels 2 and 5 for two collection dates (02/07/91 and 22/08/91).

There were no significant differences in mean asymmetry values among channels within each of the collection dates (Table 4) and with a single exception (Channel 2) no significant differences among collection dates for each channel treated independently (Table 5). Single-classification ANOVA revealed significant differences between collection dates for Channel 2. Pairwise Tukey highest-significant-difference (HSD) tests revealed the only significant comparison for this channel to be between the initial and final samples (i.e. 02/07/91 vs 19/09/91). Pooling replicate channels into control, low and high dose categories also revealed no significant differences between doses within a given date (Table 4). There were significant differences between collection dates for the pooled low dose samples (Table 5); again Tukey HSD tests revealed the differences to be due to differences between initial and final samples. All other comparisons between dates for either control or high dose were not significant. There was a significant positive relationship between asymmetry values and collection date for channel 2 (Table 6, Fig 2a) and between asymmetry values and collection date for the pooled low dose (Table 6, Fig 2b) and high dose groups (Table 2) as revealed by regression analyses.

a

b

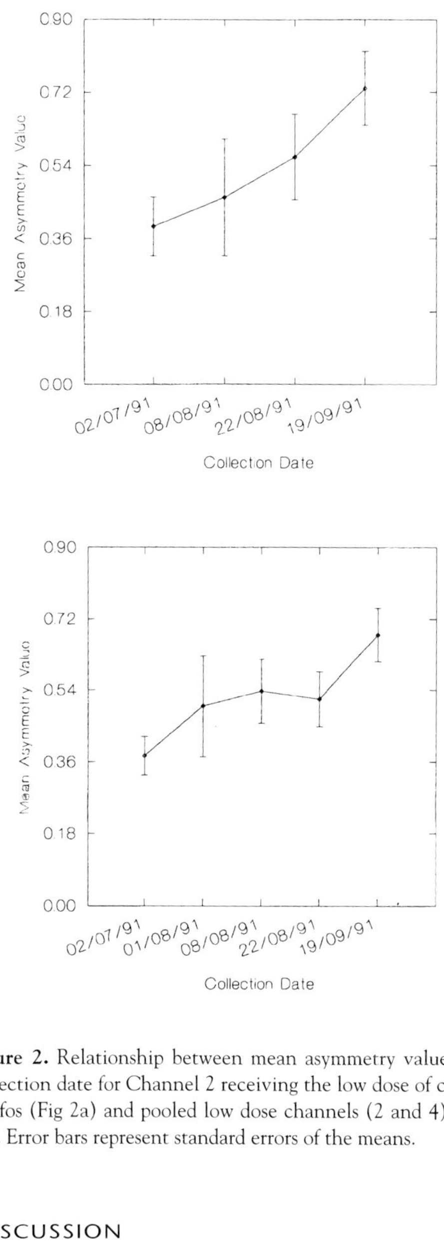

Figure 2. Relationship between mean asymmetry value and collection date for Channel 2 receiving the low dose of chlorpyrifos (Fig 2a) and pooled low dose channels (2 and 4) (Fig 2b). Error bars represent standard errors of the means.

DISCUSSION

The high dose of chlorpyrifos (5µg.l^{-1}) was obviously lethal to the two *Chironomus* spp. utilised in this study as evidenced by their general absence from high dose channels until 7 weeks after dosing ceased. As such, asymmetry analysis is irrelevant for this dosage as more obvious parameters of impact are available, i.e. mortality (see Ward *et al.* 1994).

The apparent lack of response in asymmetry measures for the low dose channels compared with the control channels may have a number of explanations. The low doses of chlorpyrifos used in this study may not have an effect on larval development in *Chironomus cloacalis* (*februarius*). It is likely that *Chironomus* larvae were exposed to pesticide via residues present in water, sediments and sedimentary organic matter. Chlorpyrifos has a tendency to partition from the aqueous to organic phase in the environment. It is strongly lipophilic (n-octanol/water partition coefficient of 66,000 at 23°C; Racke 1993). It also has a high soil and sediment sorption coefficient (K_{oc} = 13,600; Racke 1993). Furthermore the half-life of chlorpyrifos in sediments is comparatively long, especially where organic matter is present. Chlorpyrifos has been detected in sediments up to 76 days after application to water by aerial spraying (Marshall and Roberts 1978). Thus *Chironomus* larvae may be expected to have experienced pesticide exposure via the sediments directly for some time after dosing ceased. Chlorpyrifos levels in water and sediments were measured in each channel (top and bottom) at each sampling occasion. Unfortunately, all laboratory analyses using gas chromatography were unusable due to contamination problems. In the previous acute dose experiment, chlorpyrifos was applied at the same concentrations as those used in the current study for only 6 hours. At the end of the dosing period, chlorpyrifos concentrations were slightly higher at the top of the channels than at the bottom. Traces of chlorpyrifos were detected in water 10 days after cessation of the dosing period. Therefore in the current experiment, one would expect that sedimentary residues would be present 21 days after dosing, and they could be at quite high concentrations, possibly even higher than the actual doses delivered, since the chemical tends to absorb to sediment particles and organic matter containing body fats.

The egg to 4th instar larval development time in this species is approximately 12 days at 25°C (P.S. Cranston, *pers. comm.*). Thus, individuals present in the 22/08/91 sample would have colonised the channel at least a week after dosing had ended, and thus may be expected to have received reduced exposure to the pesticide. Similarly individuals from the 08/08/91 sample would have been exposed for 4-5 days of the dosing period. Channel 4 individuals from the 01/08/91 sample (completion of dosing) will have been exposed to the pesticide in water during their entire development.

This sample does show elevated asymmetry compared with both control channels although the differences are not significant. It is likely that there is only a small window of opportunity during development when the stressing agent may have an impact on the character under investigation, as such exposure time may be a critical determining factor in this experiment (Clarke 1994c). To investigate the possibility of such a critical window of opportunity would require more frequent sampling and shorter time intervals than that conducted in the current experiment. Sampling interval was selected in this study to detect population changes and effects on total invertebrate structure. There was no opportunity to check on factors such as sampling interval, sample size, number of samples per channel, as the experiment progressed. Again it must be stressed that this can only be regarded as a pilot study with regard to asymmetry analysis.

Samples taken from different locations within each channel were scored individually for asymmetry but in order to increase sample sizes were pooled prior to analysis. Such pooled samples may contain a mixture of individuals exposed to different levels of chlorpyrifos. If the concentration of chlorpyrifos varies along the length of the channel then it might be expected that individuals taken from the lower end of the channel were exposed to lower concentrations of chlorpyrifos than those taken from the top of the channel. Such heterogeneous samples would tend to even out the magnitude of asymmetry within and among samples thus reducing the ability to detect significant differences among samples. The fact that samples were not taken from the same distance within each channel for each collection date, the small number of individuals collected at each distance and the fact that chlorpyrifos concentration data are not available, precludes further elaboration of this explanation.

Control channel 1 displayed consistently high levels of asymmetry throughout the course of the experiment compared with control channel 5 suggesting that water quality in this channel may be detrimental to *Chironomus* larval development. Again, Ward *et al.* (1994) report significant differences between channel within dosage for a number of water quality parameters. However, excluding Channel 1 from the analyses does not change the overall result although comparisons between low dose channels and control channel 5 for the 22/08/91 sample (21 days after dosing was completed) approach significance (p = 0.0897). There is a possibility that the carrier solution, present in both control channels, may have a detrimental impact of development of chironomid larvae. The lack of a true (carrier free) control precludes any investigation of this potential.

The significant positive regressions of asymmetry against time for channel 2 suggest that water quality has deteriorated over time in this channel compared with the other channels. Examination of the water quality parameters measured by Ward *et al.* (1994) shows no differential changes in these parameters between Channel 2 and other channels. Without concentration measures of chlorpyrifos within each channel it is not possible to determine if this relationship is associated with chlorpyrifos. It appears as though the differences between the initial and final samples are responsible for the significant positive relationships (Fig. 2). Chlorpyrifos may be expected to still be present in sediments at this sampling date. Also Ward *et. al.* (1994) noted higher temperature, conductivity and turbidity in all channels at this final collection date suggesting the possibility of additional factors which may contribute to the observed increased asymmetry levels.

Despite the lack of conclusive results in the current study the fluctuating asymmetry approach still has considerable potential as a biomonitoring tool for assessment of water quality and impact of anthropogenic substances and habitat change (see Clarke, 1994 for review). The major failing of the current experiment we believe was not due to a lack of sensitivity or applicability of the technique but rather inherent problems with the ability of the experimental design to address the issues under investigation. As this was a pilot study it was not possible to modify the design to suit asymmetry analysis. With adjustment to the experimental design the artificial streams could be used to follow up several issues raised by this study. For example, one channel could be as a true control (no carrier solution); taking more samples more frequently; and monitoring pupal or adult emergence throughout the experiment to determine the developmental time and cycle taking place. The results of this pilot study indicate that the two doses used may represent two extremes of impact on chironomid development. The high dose ($5 \mu g.l^{-1}$) is obviously lethal and the low dose ($0.1 \mu g.l^{-1}$) may be having a slight impact. Future studies should include additional low doses. Controlled field (artificial stream) and laboratory-based experiments could also be undertaken to establish relationships between asymmetry and fitness parameters such as fecundity and fertility under control and pesticide contaminated conditions.

Studies using the same character as that used in this study in both *Chironomus salinarius* and *C. plumosus* in Russia have shown the technique to be sensitive and useful in defining changes in water quality as a result of chemical pollution (primarily methanol) (Clarke, 1993*a* and unpublished data). Preliminary results from controlled laboratory aquaria experiments have revealed significant increases in asymmetry after exposure to chlorpyrifos using larval and adult characters in *Chironomus tepperi* (Clarke and Hillman, unpublished data).

As with all biomonitoring approaches the availability of suitable relevant reference sites is critical for meaningful analysis. Under certain circumstances it may be possible to measure asymmetry of museum specimens collected from the area under investigation prior to any detrimental impact, thus providing a temporal control. The feasibility of this approach has been shown in a study of Baltic seals (Zakharov and Yablokov, 1990). The major value of the fluctuating asymmetry approach lies with its increased sensitivity over more conventionally used parameters thus providing an early-warning system for environmental deterioration. The relationship between fluctuating asymmetry and fitness parameters shown in a number of species provides a means of assessing potentially negative changes in fitness before they occur thus allowing for remedial action. Additionally the technique is not limited to use in a single indicator species but is applicable for use in a variety of species thus allowing more widespread application than species-specific parameters. Perhaps most importantly (together with other biologically-based approaches) it provides biological information on the potential impact of a variety of stressors and thus has more relevance then physico-chemical analysis of either water or organisms.

The technique, like any other, is not without its limitations. There is a need for relatively large sample sizes (>30) for valid statistical analyses. Also the technique is non-specific, i.e. asymmetry levels may be affected by a variety of factors within a single population. Thus in many cases it will not be possible to identify the causative agent from what may sometimes be a complex of stressing agents (which is true of nearly all biomonitoring techniques). However, fluctuating asymmetry is seen primarily as a cheap, simple and sensitive early-warning system useful in alerting environmental managers etc. of potential environmental problems, i.e. a first line of attack. If the fluctuating asymmetry approach reveals some environmental deterioration then alternative techniques can be used to attempt to identify the causative agent and assess the severity of the impact. The important message in all of this is that no technique should be taken in isolation and that meaningful environmental assessment is only possible via a coordinated multidisciplinary approach.

Of particular interest is the relationship between fluctuating asymmetry and the more widely used deformity analysis in chironomids. Deformities of the type analysed in chironomids fall into a more general class of abnormalities commonly referred to in the developmental stability literature as phenodeviants (Lerner 1954, Rasmuson, 1960). In general terms there is a relationship between fluctuating asymmetry and frequency of phenodeviants. The relative sensitivities of the two types of measurement appears to be dependent on the type of characters under investigation (Clarke, 1993*b*). It would thus appear likely that a relationship between asymmetry and deformities of mouthpart characters would exist. An independent re-examination of the material used in this study, scoring deformities, has the potential to reveal any such relationship and address the relative sensitivity of the two approaches.

Acknowledgments

Thanks to Drs. Peter Cranston (CSIRO Entomology) and Jon Martin (Genetics Dept., University of Melbourne) for identifying the larvae, and to Mick Smith, Renate Sadler and Karina Hansen-McInness for slide mounting the material and to Selina Ward for field sampling and laboratory sorting of surber samples. The chronic dose experiment and the study of asymmetry were funded by the Land and Water Resources Research & Developmental Corporation (LWRRDC).

Part 3

Chironomidae in Biological Monitoring

LONG-TERM ASSESSMENT OF WATER AND SEDIMENT QUALITY OF THE RIVER THAMES USING CHIRONOMID PUPAL SKINS

Leslie P. Ruse[1] *and Ronald S. Wilson*[2]

[1]National Rivers Authority, Fobney Mead, Rose Kiln Lane, Reading, Berks, RG2 0SF, UK.

[2]University of Bristol, Department of Zoology, Woodland Road, Bristol, BS8 1UG, UK.

INTRODUCTION

Long-term biological assessment of a river's water and sediment quality is confounded by similar problems to chemical assessment. Sampling and analysis will typically have been performed by different operators while methods of assessment will have varied over the period of interest. Sampling and analysis are complicated further for large, heterogeneous rivers which are predominantly too deep to wade. Sampling equipment and methods are usually restricted to particular types of habitat such as shallow riffles, plant stands or deep water. In biological assessment of water quality the effect of the physical environment also needs to be separated from the effect of the chemical environment.

The chironomid pupal exuvial technique proposed by Wilson and McGill (1979) provides a method of long-term river quality assessment that largely overcomes the problems referred to above. Pupal exuviae (skins) discarded by emerging adults will float for about two days before sinking (Coffman, 1973; McGill 1980). Studies by Wilson and Bright (1973) and McGill (1980) indicate that the majority of skins do not travel far before becoming trapped behind obstacles in the river, most drift less than 100m in a 10m wide river. Samples of floating debris from the margins of a river are assumed to contain an integrated collection of pupal skins accumulated over the previous two days approximately. These skins belong to chironomids that could have emerged from any of the available habitats a short distance upstream of the sampling point. Owing to the passive collection of pupal skins there appears to be little opportunity for the operator to affect the sample contents. In contrast to many biological sampling techniques, it is claimed that skimmed collections of pupal skins remain representative of the local population, whether derived from the shallow headwaters or the profundal lowland reaches.

Many chironomid species will be available to colonise existing habitats along the whole course of a river. The species that succeed in maintaining a population at a river site, and the proportion of their numbers relative to other species, will reflect the physical and chemical quality of the sediments and the overlying water.

The results of four surveys between 1977 and 1992 are presented here. The first two surveys formed part of a nationwide study to characterise rivers using the pupal exuvial technique (RSW). The subsequent two surveys were collected by a different operator (LPR) for the purpose of providing baseline data prior to proposed flood relief schemes and the construction of a large reservoir to regulate river flows.

METHODS

Thirteen sites along the River Thames in southern England were sampled for chironomid pupal skins in each of four surveys (Fig. 1). Monthly samples were taken from April to September 1977 and repeated during the same months in 1978. The same sites were sampled subsequently on four occasions between September 1986 and July 1987 and on three occasions between May and August 1992. Any combination of monthly samples between May and September were expected to yield at least 80% of the available genera (Ruse and Wilson, 1984). All samples were collected with a 250µm mesh net. Pupal skins were subsampled randomly and identified at least to genus. A minimum of 600 pupal skins were collected from each site during a survey.

Chemical data were retrieved from the National Rivers Authority (NRA) archive. Annual mean and potentially harmful extreme values for BOD_{ATU}, DO, unionised ammonia, TON and orthophosphate were selected because of the relative completeness of their long-term data. Maximum values were taken as extreme for all determinands except DO, for which minimum values were used. Chemical data for each survey covered the twelve months up to and including the last month of pupal skin collection. Where chironomid sampling sites did not coincide with NRA routine chemical sites then chemical data from the nearest appropriate site were used instead. Annual mean flow, from October to September, for chironomid sampling sites during each survey were provided by NRA Hydrological Services. In addition to annually variable measurements, altitude, slope, distance downstream of river source, channel mean depth and width (based on NRA navigation charts) were also recorded for each site.

Data analysis proceeded in two ways. Firstly, physical and chemical quality of each site was assessed from an *a priori* assumption of chironomid substratum preferences and pollution sensitivity. Each taxon was classified as preferring fine (silt, fine sand) or coarse (gravel, rocks, plants) substrata and being tolerant or intolerant of organic pollution (Table 1). Secondly, *a posteriori* data analysis was achieved by calculating Shannon-Weaver diversity indices and by a constrained ordination, canonical correspondence analysis (CCA), using the program package CANOCO version 3.1 (Ter Braak, 1988, 1990a).

Explanation and evaluation of CCA is provided by Jongman, Ter Braak and van Tongeren (1987) and Okland (1990). In constrained ordination, changes in species abundance are directly described as a function of measured environmental gradients by integrating ordination with regression and calibration. In CCA, axes explicitly optimise the fit of species data with the supplied environmental variables. Chironomid data were $\log_{10}(x+1)$ transformed to be normally distributed. Transformed chironomid data were also analysed by unconstrained ordination, detrended correspondence analysis (DCA), to check that unimodal techniques were suitable and to assess the importance of supplied environmental variables. Environmental data were normalised if necessary; mean and maximum concentrations of unionised ammonia, maximum orthophosphate and discharge were all square-root $(x+0.5)$ transformed while maximum TON was $\log_{10}$ transformed. Environmental data were standardised to have zero mean and unit variance. A forward selection procedure (Ter Braak, 1990a) was used to reduce the number of environmental variables supplied to CCA. This removed co-linear variables and avoided the requirement for detrending. Within CCA, chironomid taxa scores were scaled to be weighted averages of the samples scores. This scaling provided an approximate biplot of the correlation between environmental variables, as interpreted for linear ordination methods. With this scaling it was also possible to produce a biplot of approximate *t*-values of the regression of taxa on the environmental variables (Ter Braak, 1990b) so that the significance of the relationships could be assessed. The significance of the observed environment-species relationship for each of the first four ordination axes was tested against ninety-nine random permutations of the original data (Ter Braak, 1992). As in conventional regression analysis, covariables can be used in CCA to account for obvious or unwanted variation. The effects of the residual variation may then be investigated by partial canonical correspondence analysis (PCCA).

Table 1 Total abundance of chironomid taxa collected between 1977 and 1992

Taxon				Taxon			
Apsectrotanypus trifascipennis (Zett)	X	T	13	*Corynoneura* spp.	Y	T	524
Macropelopia spp.	X	T	1	*Epoicocladius flavens* (Mall)	Y	I	78
Macropelopia nebulosa(Mg)	X	T	40	*Limnophyes* spp.	Y	T	368
Procladius spp.	X	T	1231	*Metriocnemus* spp.	Y	T	4
Psectrotanypus varius (Fab)	X	T	3	*Parakiefferiella bathophila* (K)	Y	I	34
Ablabesmyia spp.	X	T	272	*Parametriocnemus spp.*	Y	I	11
Ablabesmyia monilis (L)	X	T	45	*Paraphaenocladius impensus* (Walker)	X	T	3
Arctopelopia spp.	X	I	13	*Paratrissocladius excerptus* (Walker)	X	I	3
Conchapelopia spp.	Y	T	2	*Pseudorthocladius* spp.	Y	T	1
Conchapelopia melanops (Mg)	Y	T	87	*Pseudosmittia* spp.	X	T	39
Larsia spp.	X	I	4	*Thalassosmittia thalassophila* (Goet)	X	T	26
Nilotanypus dubius (Mg)	Y	T	2	*Thienemanniella* spp.	Y	I	407
Paramerina spp.	X	I	4	*Chironomus* spp.	X	T	761
Rheopelopia spp.	Y	I	5	*Cladopelma* spp.	X	T	35
Thienemannimyia spp.	X	I	463	*Cryptochironomus* spp.	X	T	433
Tanypus punctipennis Mg	X	T	14	*Cryptotendipes* spp.	X	T	89
Diamesa spp.	Y	I	2	*Demeijerea rufipes* (L)	Y	I	8
Potthastia spp.	X	I	93	*Demicryptochironomus vulneratus* (Zett)	X	T	1
Sympotthastia zavreli Pag	X	T	5	*Dicrotendipes nervosus* (Staeg)	Y	T	1437
Odontomesa fulva (K)	X	T	2	*Dicrotendipes notatus* (Mg)	X	T	75
Prodiamesa spp.	X	T	6	*Endochironomus* spp.	Y	T	68
Prodiamesa olivacea (Mg)	X	T	448	*Glyptotendipes* spp.	Y	T	800
Brillia flavifrons Johannsen	Y	T	67	*Harnischia* spp.	X	T	592
Cardiocladius fuscus	Y	I	99	*Kiefferulus tendipediformis* (Goet)	X	T	92
Cricotopus (*C.*) spp.	Y	T	7232	*Lauterborniella agrayloides* (K)	Y	I	2
Cricotopus (*C.*) *bicinctus* (Mg)	Y	T	101	*Microchironomus* spp.	X	T	28
Cricotopus (*I.*) spp.	Y	T	923	*Microtendipes* spp.	X	I	1270
Cricotopus (*I.*) *sylvestris* Fab	Y	T	148	*Parachironomus* spp.	Y	T	1251
Cricotopus (*Nost.*) *lygropis* Edw	Y	I	5	*Paracladopelma* spp.	X	I	6
Diplocladius cultriger K	X	I	1	*Paracladopelma camptolabis* gp	X	I	2
Eukiefferiella spp.	Y	I	173	*Paratendipes* spp.	X	I	138
Eukiefferiella claripennis (Lund)	Y	T	286	*Phaenopsectra* spp.	X	I	199
Heterotrissocladius spp.	X	T	1	*Polypedilum* (ss)	X	I	751
Nanocladius spp.	Y	T	2513	*Stenochironomus* spp.	Y	I	19
Nanocladius balticus Palmen	Y	I	153	*Stictochironomus* spp.	X	I	30
Orthocladius (*Eudact*) spp.	Y	I	268	*Xenochironomus xenolabis* (K)	Y	I	384
Orthocladius (*Eudact*) *olivaceus* K	Y	I	3	*Cladotanytarus* spp.	X	I	901
Orthocladius (*Pogo*) *consobrinus* Holm	Y	I	2	*Micropsectra* spp.	X	T	240
Orthocladius (*Euorth*) spp.	Y	I	198	*Micropsectra atrofasciata* K	X	T	693
Orthocladius (ss)	Y	I	2316	*Neozavrelia* spp.	Y	I	140
Paracladius conversus (Walk)	X	I	191	*Paratanytarsus* spp.	Y	I	1226
Paracricotopus niger (K)	Y	T	2	*Rheotanytarsus* spp.	Y	I	890
Paratrichocladius rufiventris (Mg)	Y	T	1328	*Stempellina almi* Brundin	X	I	36
Psectrocladius (*ss*) spp.	X	I	172	*Stempellina bausei* (K)	X	I	20
Psectrocladius (*ss*) *barbimanus* (Edw)	X	I	2	*Stempellinella* spp.	X	I	63
Rheocricotopus spp.	Y	I	565	*Tanytarsus* spp.	X	I	4107
Synorthocladius semivirens (K)	Y	I	2402	*Tanytarsus brundini* Lind	X	I	18
Tvetenia spp.	Y	I	917	*Virgatanytarsus* spp.	X	I	5
Bryophaenocladius subvernalis (Edw)	Y	T	39	*Zavrelia pentatoma* K	Y	I	1

substratum preference
X = fine
Y = coarse

pollution tolerance
I = intolerant
T = tolerant

Significant differences in univariate data were assessed by one-way analysis of variance, percentage data were arcsine transformed.

RESULTS

A total of 41 171 chironomid pupal skins were collected and identified to 98 taxa, mostly genera, over the four surveys (Table 1). Almost 46% of the individuals collected were considered to belong to taxa that were pollution-sensitive. A Shannon-Weaver diversity of 3.34 was obtained for the amalgamated data.

The quality of River Thames sites for each survey was subjectively assessed from a plot of chironomid diversity against the percentage of pollution-intolerant chironomid individuals (Fig. 2). No sites were considered of excellent quality in any survey. Upstream sites at Ashton Keynes, Castle Eaton, Buscot and Newbridge, as well as mid-reach sites at Hurley and Maidenhead, have been consistently good or very good during the last three surveys. Most of the remaining sites were considered of doubtful quality. All four surveys have implicated the river just upstream of Teddington weir to be very enriched, as well as Molesey in 1977 and Whitchurch in 1978. River quality for each site was poorest in 1977, particularly at Castle Eaton.

In Fig. 3 the percentage of fine-sediment dwelling chironomids at each site is presented for each of the four surveys. Assessed by chironomid inhabitants, there was a significant difference in sediment levels between sites ($P < 0.001$). This difference was due to low levels of fine sediment at Castle Eaton, remaining sites showed no significant difference in sediment ($P=0.15$). There was also no significant difference in sediment levels between the four surveys ($P=0.25$).

The gradient length of the first DCA axis was 2.1 standard deviation units, confirming the suitability of the data for unimodal analysis (Ter Braak and Prentice, 1988). Eigenvalues (importance measure) for the first two axes of DCA and CCA (using the refined set of environmental variables including distance) are provided in Table 2. Forward selection revealed that variables remaining constant between surveys were highly correlated. Distance of site from the source explained the greatest amount of chironomid variation while all other immutable variables were superfluous because of their high correlation with distance. Distance was also highly correlated with discharge, which varied annually. Immutable variables were excluded from the analysis while distance from source was used as a covariable. With this strategy the variation in data between surveys could be assessed independently from the influence of longitudinal variation in physical, chemical and biological data. Forward selection was also used to chose between mean or extreme values of each chemical determinand according to which explained more variation in the chironomid data. The refined set of environmental variables supplied in the PCCA comprised annual mean discharge, maximum BOD, minimum DO, mean unionised ammonia, mean TON and mean orthophosphate.

After the effect of longitudinal transition in data was removed, the variation in chironomid data remained highly significantly related to the supplied environmental variables along the first three axes of the PCCA ordination ($P<0.01$ for each axis). A summary of eigenvalues of axes for PCCA is provided in Table 2.

Table 2 Eigenvalues for ordination and constrained analyses

	DCA		CCA		PCCA			
	1	2	1	2	1	2	3	4
eigenvalue	0.18	0.13	0.15	0.11	0.11	0.07	0.05	0.04
cumulative % variance of species data explained	12.5	21.8	10.2	17.8	8.4	14.1	18.1	20.8

Total unconstrained eigenvalues	1.435
Total unconstrained eigenvalues (after fitting covariable in PCCA)	1.302
Total constrained eigenvalues	0.318

The variation in chironomid data explained by the covariable, distance from source, was 9.3%. A further 22.2% of the variation was explained by the supplied environmental variables and the remaining 68.5% was unexplained. In Fig. 4 sites are ordered along the first two PCCA axes according to the relative variation in their biological and environmental data. The overlying arrows represent the relative importance and direction of change of environmental vectors across the ordination plot. The principal axis was best correlated with increasing concentrations of mean unionised ammonia from left to right. The second axis was best explained by variation in annual mean discharge between surveys. The first two axes accounted for 57.6% of explained variation.

In the PCCA biplot (Fig. 4) samples (collated data for one site in one survey) had a strong tendency to associate with other samples from the same survey. Samples for 1977 and 1978 were positively associated with high levels of unionised ammonia (Fig. 5) and discharge (Fig. 6) reflecting their position towards the right of Fig. 4. Samples from 1986/87 were characterised by low levels of ammonia. Samples in 1992 were associated with low ammonia, except at Teddington, and with low flow. In Fig. 4, the outlying samples of Molesey and Teddington from the 1992 survey reflect extremely low flows in comparison to those occurring at these sites in previous surveys. Mean concentrations of TON best explained variation along the third PCCA axis. The effect of high mean values of TON in 1977 (Fig. 7) can also be assessed from Fig. 4. TON was significantly higher in 1977 than subsequent years ($P < 0.001$). Although mean TON was always lowest at Ashton Keynes, it was never significantly so ($P=0.97$).

The significance of the regression coefficients between each chironomid taxon and the variables best explaining the first two PCCA axes are presented in Fig. 8. Remote (non-significant) species have been excluded to improve the resolution of the figure. A chironomid taxon may be significantly related to both variables, for example, *Synorthocladius semivirens* showed a significant inverse relationship with ammonia and a direct relationship with high flows. Two chironomid taxa were indicated as tolerant of un-ionised ammonia, six were intolerant. Seventeen taxa were associated with low flow while nine taxa had a positive relationship with discharge.

Discussion

The four surveys of chironomid pupal skins provide a baseline data set which can be used to assess the effects of future projects, such as flood relief schemes and river regulation on the Thames.

Assessment of river quality from a prior knowledge of chironomid species preferences has shown a consistent pattern of downstream change through the last three surveys. Physical, chemical and biological data indicated that October 1976 to September 1977 was quite distinct from subsequent survey periods. The highest flows occurred over this period following a severe drought which ended in September 1976. Abrupt hydrological changes may have increased the concentrations of toxic ammonia and nitrates entering the river. The analysis of chironomid data, using *a priori* classification of preferences, was sensitive to these changes and can be expected to respond to future fluctuations in discharge and river quality.

The distribution of fine sediment in the River Thames has not varied significantly over the four surveys, as judged from the preferences of chironomid taxa. The methods described in this report are expected to detect significant changes in sediment levels following large scale manipulations of the river's flow. These methods have indicated a progressive rise in soft sediments above Teddington since 1977. An accumulation of poor quality fine sediments will act as a source of oxygen depletion and toxic substances. These sediments may be resuspended by seasonal increases in flow or boat traffic. Present river regulation practices may be storing up water quality problems for the future.

Over 68% of the variation in chironomid data was not explained by the six most influential environmental variables supplied in a PCCA analysis. Despite this, a comparison of eigenvalues between DCA and CCA suggests that the supplied variables were sufficient to explain major non-random variation of chironomid data. Unexplained variation could result from changes in the chemical monitoring sites chosen for each survey and in the number of chemical samples and chironomid skins collected. The analysis would be limited by the variables that were supplied and there will always be a high level of 'noise' from random fluctuations around data measurements. If all sixteen variables were retained in the analysis unexplained variation would have been reduced below 50%. This would have led to a high degree of co-linearity which distorts the ordination.

Among the variables supplied, the most important predictor of the chironomid assemblage of the River Thames was their distance downstream of the source. The PCCA has shown that variation within the chironomid assemblage at particular stations along the river was best explained by mean concentrations of ammonia and by annual variations in flow. It is proposed that the subsets of chironomid taxa presented in Fig. 8 could be used as indicators that changes in ammonia concentrations or flow were having a significant effect.

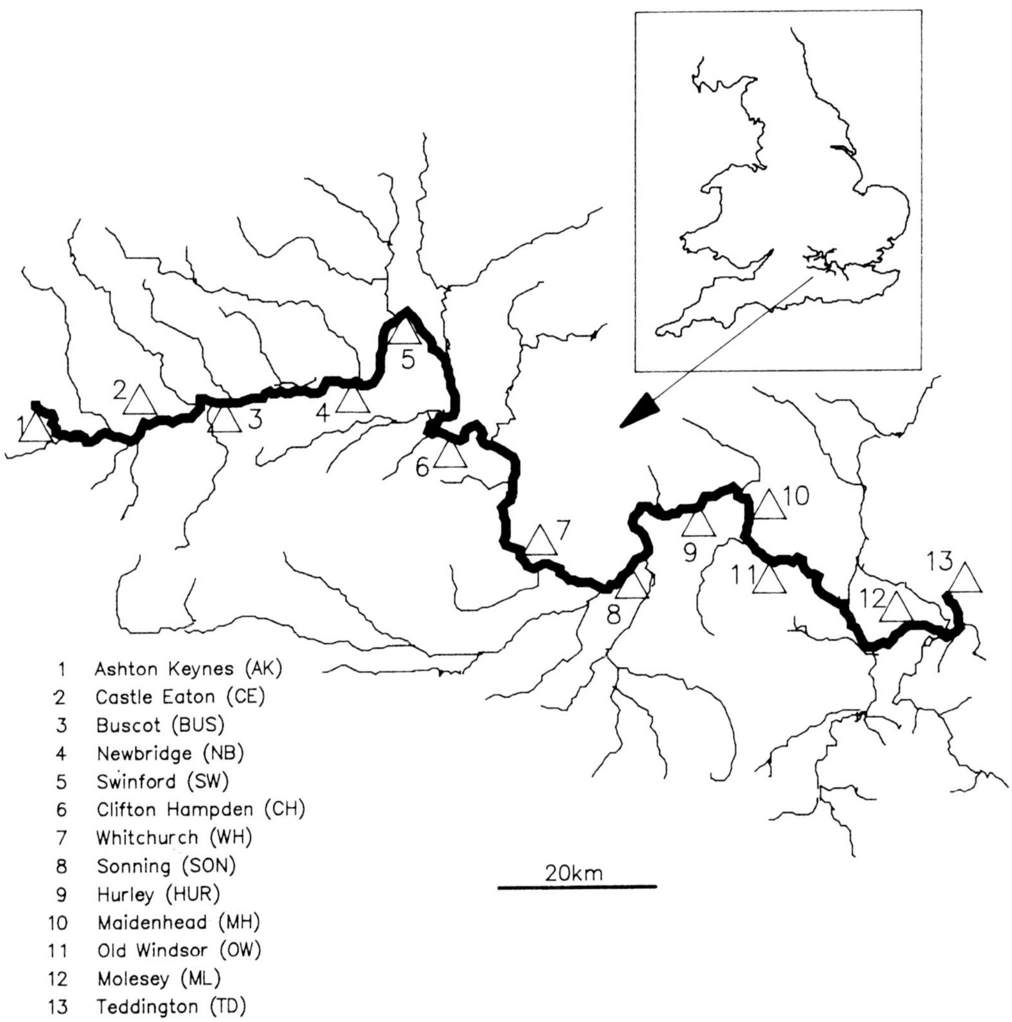

Figure 1. Pupal skin sampling sites on the River Thames

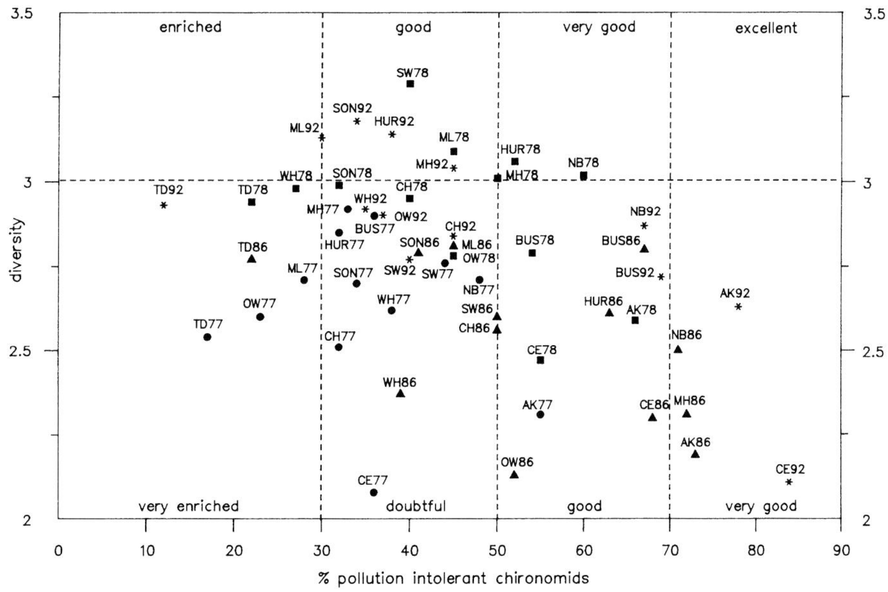

Figure 2. Quality assessment based on diversity and pollution intolerance

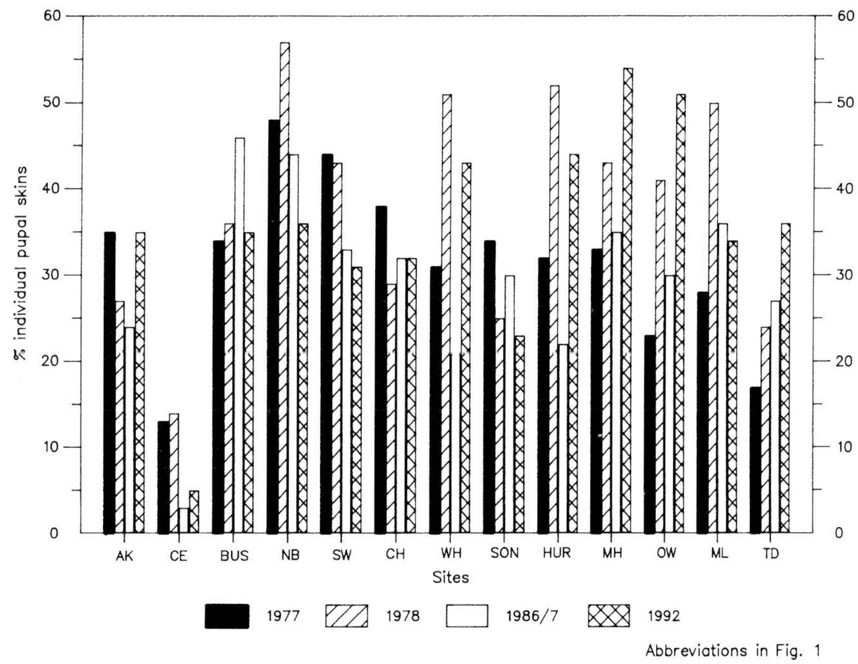

Figure 3. Occurrence of chironomids preferring fine sediments

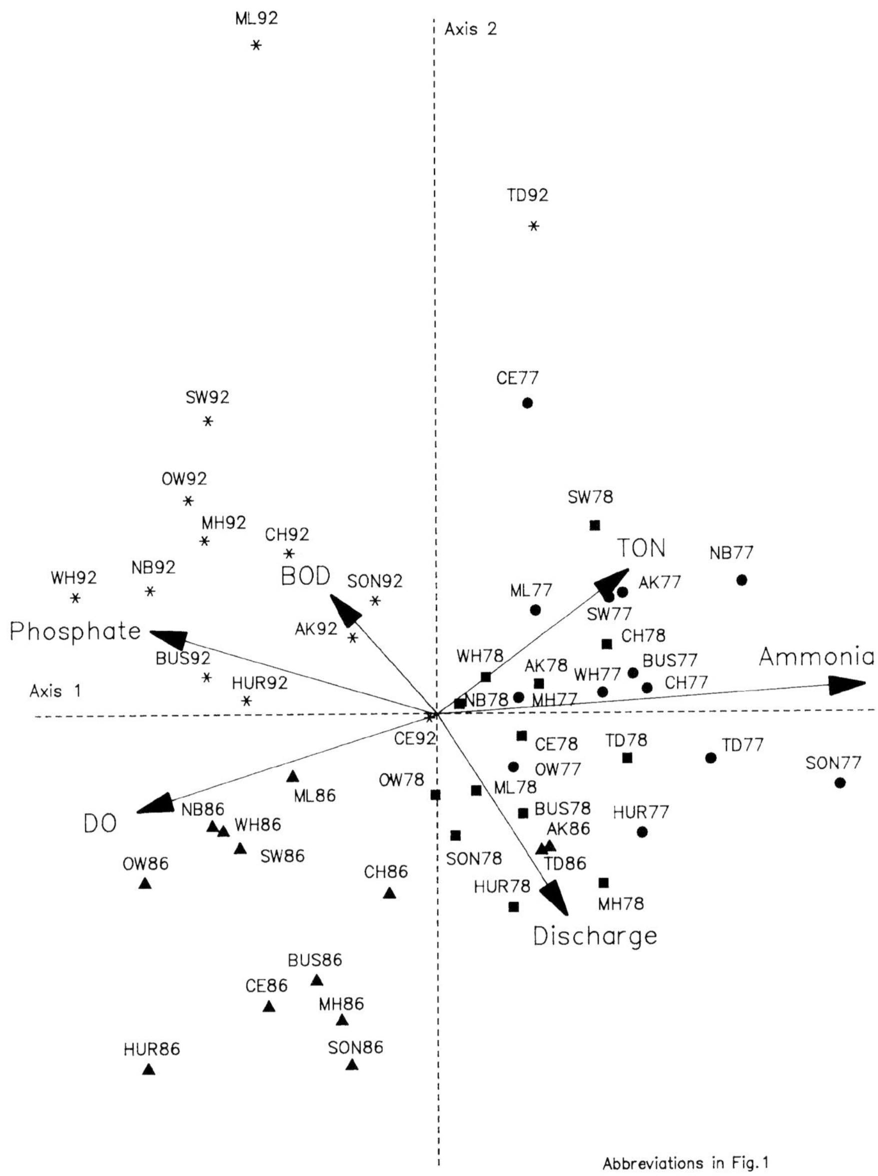

Figure 4. PCCA biplot of samples (distance from source as a covariable)

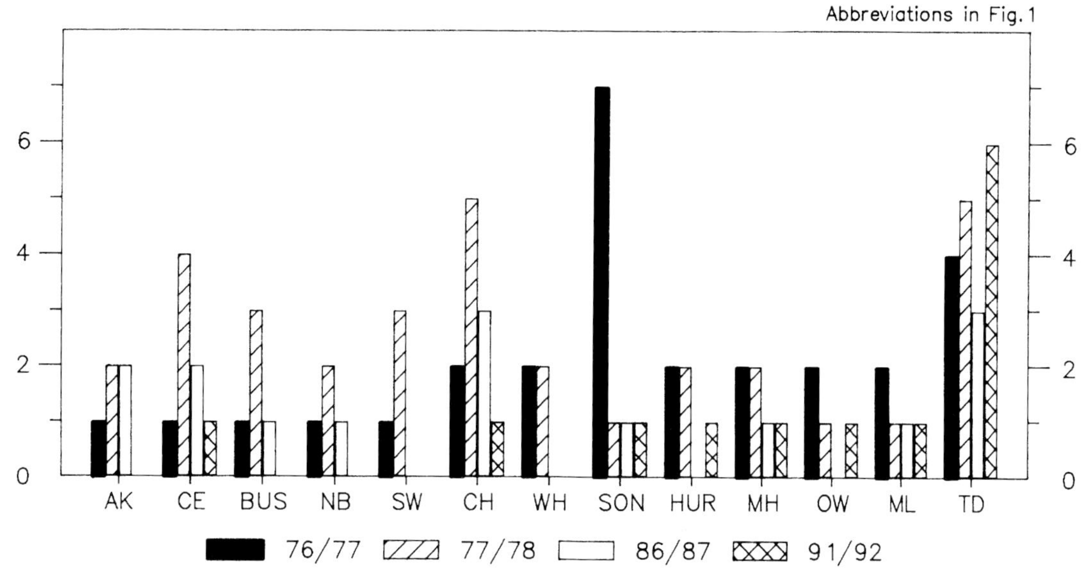

Figure 5. Mean unionised ammonia as N (mgl^{-1})

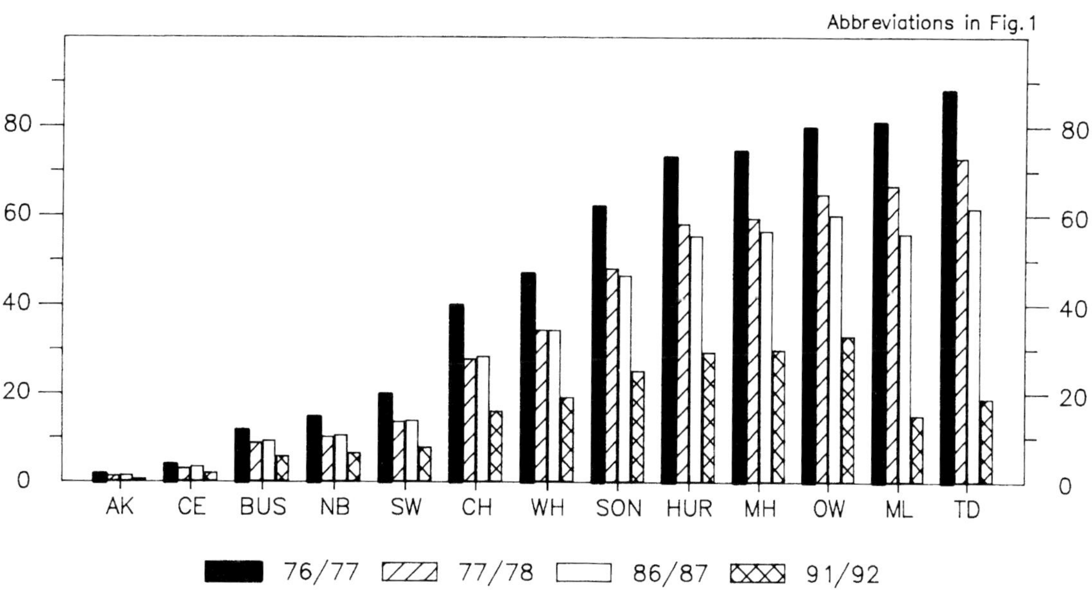

Figure 6. Mean discharge ($m^3 s^{-1}$)

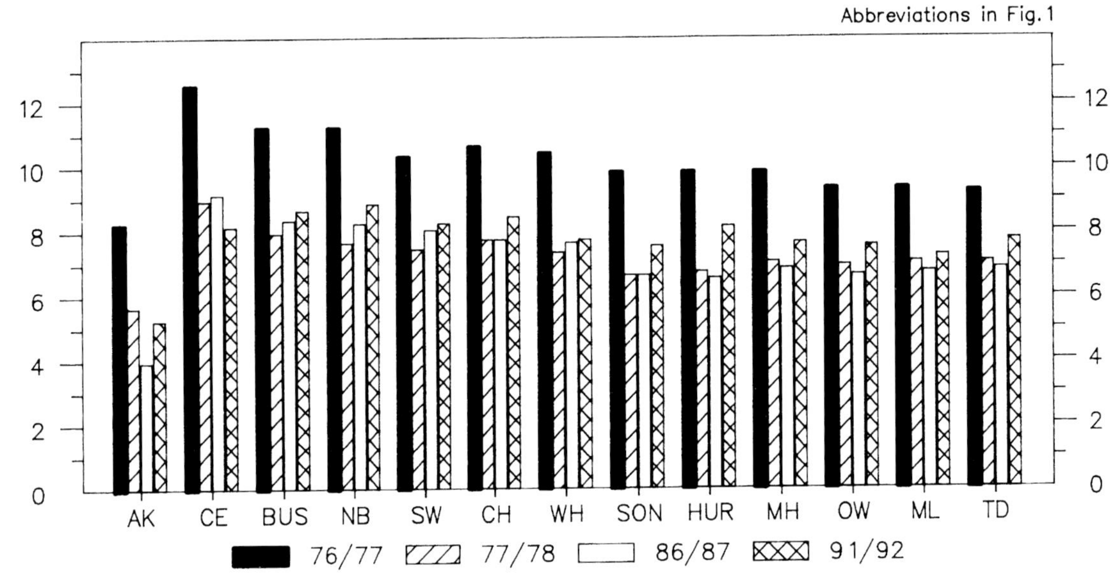

Figure 7. Mean total oxidised nitrogen (mgl^{-1})

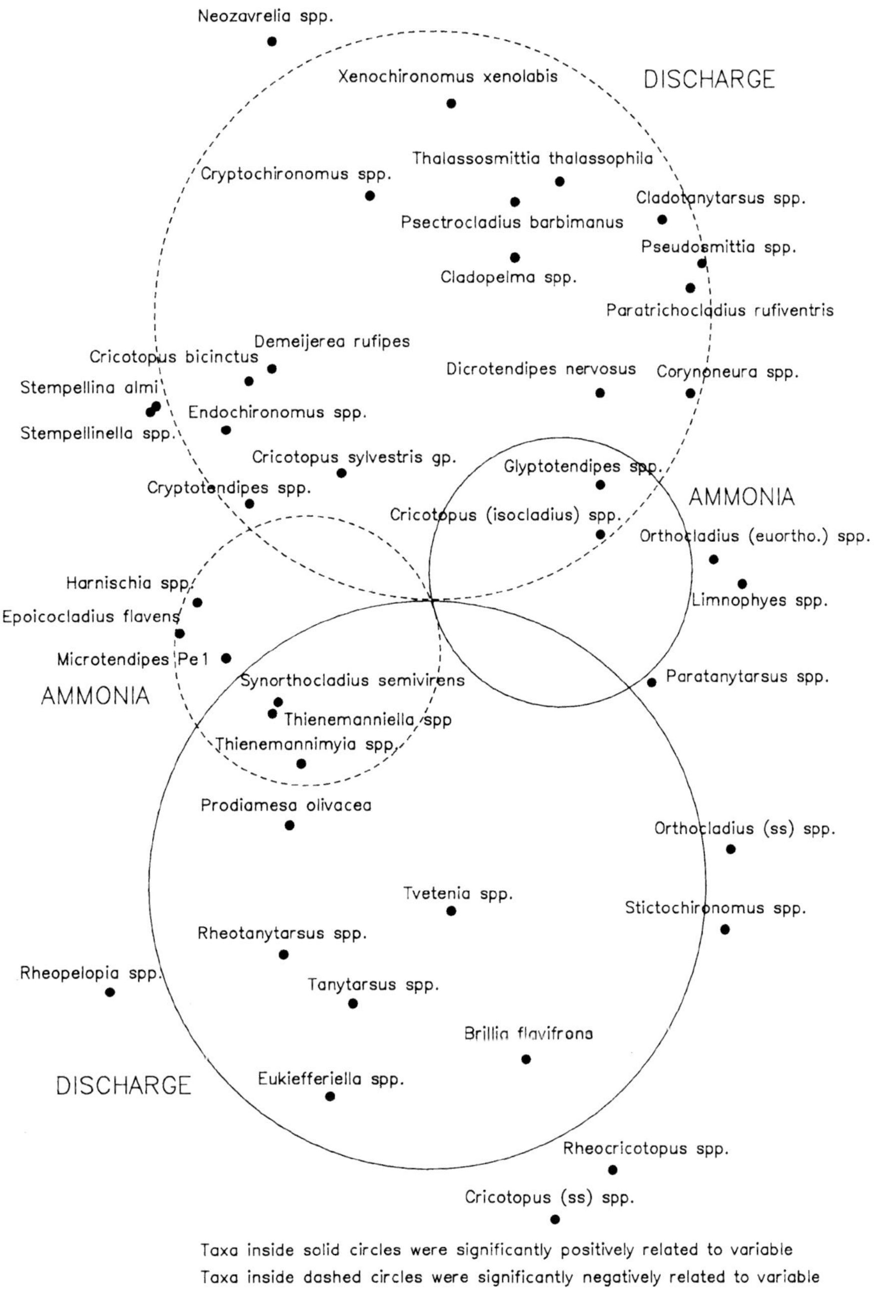

Figure 8. PCCA *t* –value biplot

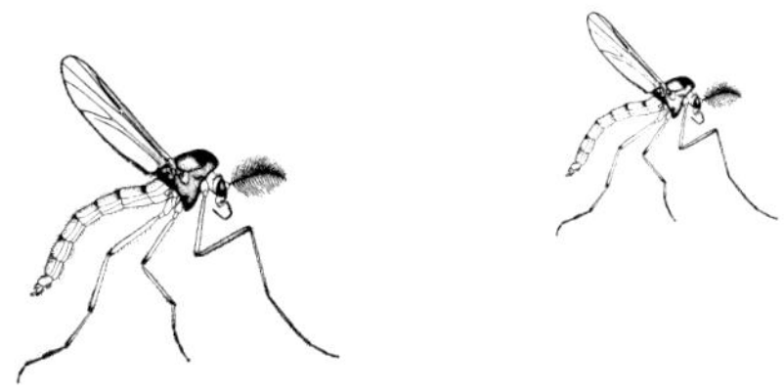

A COMPARISON OF PUPAL EXUVIAE AND LARVAL CHIRONOMIDAE FOR BIOMONITORING OF THE IMPACTS OF AGRICULTURAL PRACTICES ON SURFACE WATER QUALITY

D.R. Barton[1], *D. R. Oliver*[2] *and M.E. Dillon*[2]

[1] Department of Biology, University of Waterloo, Waterloo, Ontario, Canada, N2L 3G1;

[2] Centre for Land and Biological Research Resources, Agriculture Canada, Ottawa, Ontario, Canada K1A 0C6

INTRODUCTION

One of the primary drawbacks to the use of benthic macroinvertebrates for biomonitoring and assessment of water quality in streams is the amount of effort required to process the samples. Sorting of preserved animals from the associated debris which remains even after washing through sieves is extremely laborious. In some situations this problem can be reduced by hand picking in the field (Lenat 1988), but this approach is impractical in more degraded systems where small sized individuals make up the bulk of the fauna. Most of these smaller species are Chironomidae, the pupal exuviae (Coffman, 1973; Wilson and McGill, 1977) of which are relatively easy to collect without much extraneous debris. Such material can be processed much more quickly in the laboratory and offers the further advantage of allowing (in most cases) identification to the species level.

Since 1990, we have been examining the possibility of developing a protocol using benthic invertebrates for monitoring and assessment of the effects of agricultural practices on the quality of surface waters in southern Ontario. As part of that study, we have evaluated several qualitative and quantitative collecting techniques. The best technique should provide maximum information for minimum effort; the initial criterion in this evaluation was the number of recognizable taxa collected with respect to the time needed to process the samples.

This report compares the catches of chironomid taxa in preserved kicknet samples (mainly larvae) and collections of pupal exuviae. The two techniques were used on each of 140 sampling visits during the openwater season of 1991.

Methods

Sampling sites were chosen to include streams draining each of the primary crops and forested basins in the major physiographic regions of southern Ontario. Only first- and second-order (Strahler) streams were selected in order to isolate, as far as possible, the influences of individual agricultural practices associated with pasture (mainly cattle), hay, grains and oilseeds (maize, wheat, oats, soybeans, white beans), tobacco, market vegetables (mainly tomatoes) and fruits (mainly apples, peaches and grapes). About 250sites were sampled during 1990, 1991 and 1993; most were visited two or more times during different parts of the openwater season.

Macroinvertebrates were collected at each site using a "kick" technique; the operator disturbed the stream bottom with a booted foot and netted the dislodged material with a longhandled dip net (200µm aperture mesh). All parts of the stream were sampled - midchannel, margins, aquatic vegetation, woody debris, etc. - with the effort expended in each microhabitat proportional to the relative area which it covered. All material in the net was pooled and preserved in 10% formalin. In the lab, the samples were examined under a dissecting microscope and the first 200 animals seen were removed to 70% ethanol for identification. The remaining material was scanned and, if it was apparent that taxa had been missed, up to 200 more animals were picked (nonselectively).

Pupal exuviae of Chironomidae were collected from backeddies created by woody debris, emergent vegetation or irregularities in the streambanks by skimming the surface film with a handstrainer fitted with 100um netting. Collecting continued until no obviously different exuviae were being found, or for at least 10 minutes. Netted material was preserved in 70% ethanol for later sorting under a dissecting microscope.

After the initial sorting of the samples, chironomid larvae were identified by DRB. Identities of representative pupal exuviae from each sample were confirmed by DRO and MED.

Results and Discussion

Pupal exuviae were found in only 49 of 140 (35%) samples collected from May through midAugust 1991; each of the 140 kicksamples contained chironomid larvae or pupae. Failure to collect exuviae could reflect the effects of weather, the physical structure of the stream channel, seasonality, or operator error (4 individuals were involved). Because we were unable to assess the relative importance of these factors, we will deal here only with those site visits which yielded both larvae and pupal exuviae.Collecting in the field took one person 5–10 minutes for kick samples, and 5–30 minutes for exuvial samples. Processing in the laboratory requires 2.5–5.0 hours for kick samples, and up to 1.5 hours for exuviae. This includes sorting, slidemaking and identification.

The 141 taxa identified from the 49 collections (a complete list is available on request) included 68 genera, of which 22 were represented by two to nine species. The most ubiquitous and abundant genera

Table 1. Mean numbers of specimens and taxa collected from streams draining areas under different crops.

	Forest	Tobacco	Hay	Pasture	Grains
Number of sites	12	4	9	8	16
Exuviae					
Individuals	22.7	42.2	19.6	21.6	17.5
Taxa	7.5	6.8	5.6	7.5	3.4
Larvae					
Individuals	133	134	100	142	154
Taxa	15.9	16.8	11.8	15.3	13.5
Combined Taxa	19.7	19.8	15.0	19.6	14.9

included *Conchapelopia* (3 spp.), *Corynoneura* (2 spp.), *Cricotopus* (9 spp.), *Orthocladius* (6 spp.), *Parametriocnemus* (3 spp.), *Tvetenia* (2 spp.), *Chironomus*, *Cryptochironomus*, *Microtendipes*, *Polypedilum* (6 spp.), *Micropsectra* (6 spp.) and *Tanytarsus* (7 spp.). Twenty-eight taxa (mostly genera) were found only in kick samples, and 43 (mostly species) only as pupal exuviae.

Both the number of specimens and the number of taxa varied widely from sample to sample but, on average, individual kick samples yielded about 5 times as many individuals and twice as many taxa (Table 1, Fig. 1) as did exuvial samples. When sites were grouped according to the landuse immediately upstream, streams draining hay and grain fields appeared to support fewer taxa than did those draining forest, tobacco or pasture. Combining kick and exuvial samples increased the number of taxa by 24 to 28% for forest, hay and pasture streams, but only 10% for grain streams.

While catches in both kick and exuvial samples were relatively similar when grouped by landuse, there was no overall relationship between the numbers of larval and pupal taxa (Fig. 2). There was no evident seasonal pattern in the number of larval taxa collected (Fig. 3). Samples containing the largest numbers of pupal exuvial species were collected during the first half of the study period, but samples with few species were collected throughout the summer (Fig. 4).

About half of the streams sampled have been channelized. These straightened, uniform channels should have fewer eddies where pupal exuviae would accumulate, so exuvial catches would be expected to be lower. This appeared to be the case only in pasture and tobacco streams (Fig. 5), probably because of the growth of emergent macrophytes in grain and hay streams. Grazing cattle tend to eliminate macrophytic growth in ditches. Only one tobacco site was ditched, so it is impossible to assess the validity of this apparent effect of channelization. The apparently greater diversity of chironomids in forested ditches is also based on only one sample.

Figures 6 and 7 summarize the results of kick and exuvial sampling in streams draining different land uses on the basis of the proportion of the catches in various subfamilies and tribes. These results suggest that exuvial collections based on skimming the water surface in marginal eddies is selective for certain groups. Tanypodinae were significantly under-represented in exuvial collections (X^2 = 21.079, df=4, p< 0.01). Exuviae of Diamesinae were very rarely collected, but their small contribution to the larval collections made the difference in catches statistically insignificant. Tanytarsini and Orthocladiinae (X^2 = 11.895, df=4,p<0.02) tended to be over-represented as exuviae. The Percent Similarity of the catches by the two methods was lowest for hay and tobacco streams (69%), but very high (91%) in pasture streams due to large numbers of *Conchapelopia rurika* (Roback) and *Nilotanypus fimbriatus* (Walker) found in two samples.

CONCLUSIONS

Collections of pupal exuviae made by skimming the surface film of backeddies are not appropriate by themselves for purposes of water quality monitoring. Exuviae were found at only 35% of the sites sampled, and even where found averaged only half as many taxa as were found in kick samples. We have some evidence that the nature of the stream channel influences the catch, and suspect that the technique is highly susceptible to variation due to differences among collecting personnel.

For those sites where exuviae were collected, pooling of the kick and exuvial samples increased the estimated species richness by 10 to 28%. These combined estimates represent about a 40% increase in collecting and processing time, so might not seem to be cost effective. We should emphasize, however, that several numerically abundant genera, such as *Conchapelopia*, *Orthocladius*, *Micropsectra* and *Tanytarsus*, appear to include species with very different habitat preferences. The information gained from specieslevel identification suggests that exuvial collections are a worthwhile supplement to kick sampling for routine monitoring purposes.

Acknowledgements

This study was partially supported by the Great Lakes Water Quality Program. We thank M. Farmer, I. Martin and M. Pope for assistance in collecting and processing the samples.

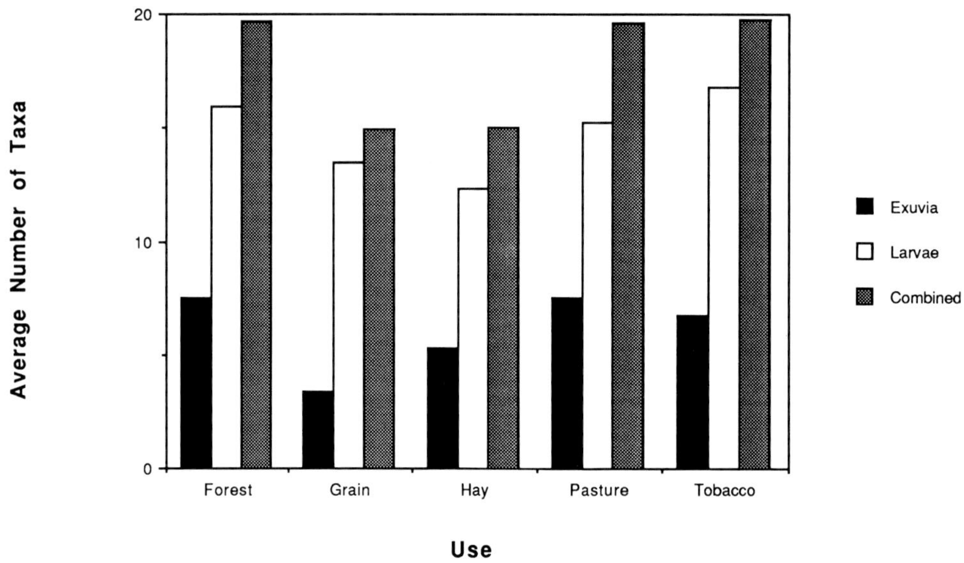

Figure 1. Average numbers of taxa of exuviae, kick (larvae) and combined (exuviae + larvae) samples from streams draining forest and various crops. Number of sites for each land-use are given in Table 1.

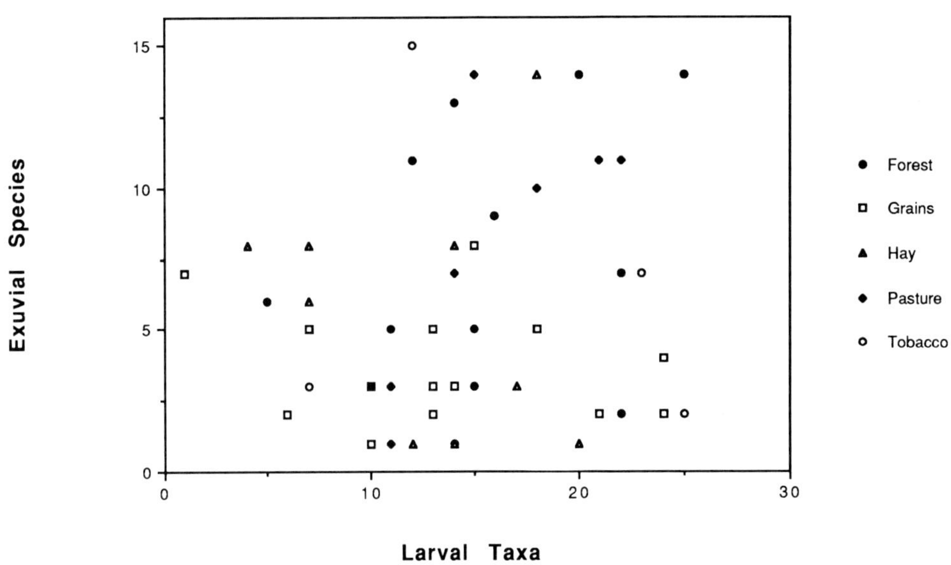

Figure 2. The relationship between the numbers of taxa found in kick samples (mainly larvae) and pupal exuviae collections.

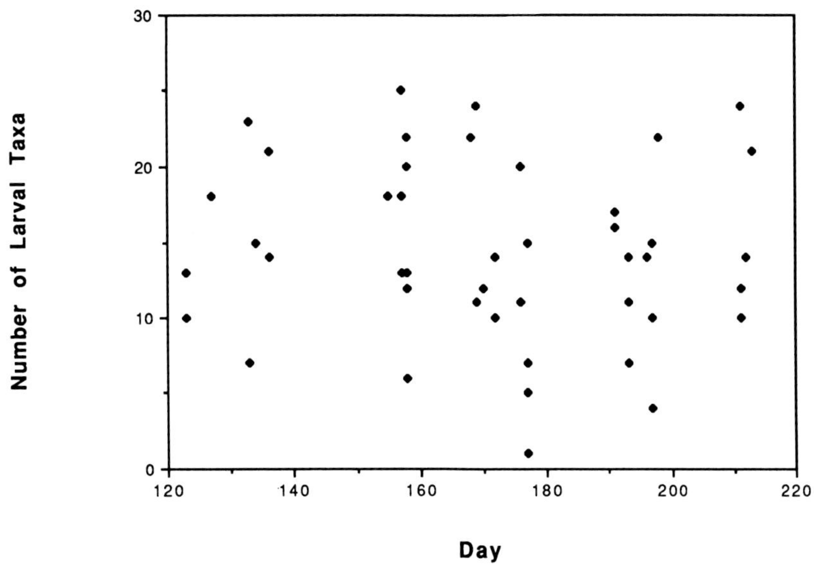

Figure 3. Numbers of larval taxa found in kick samples made on each collecting day in 1991.

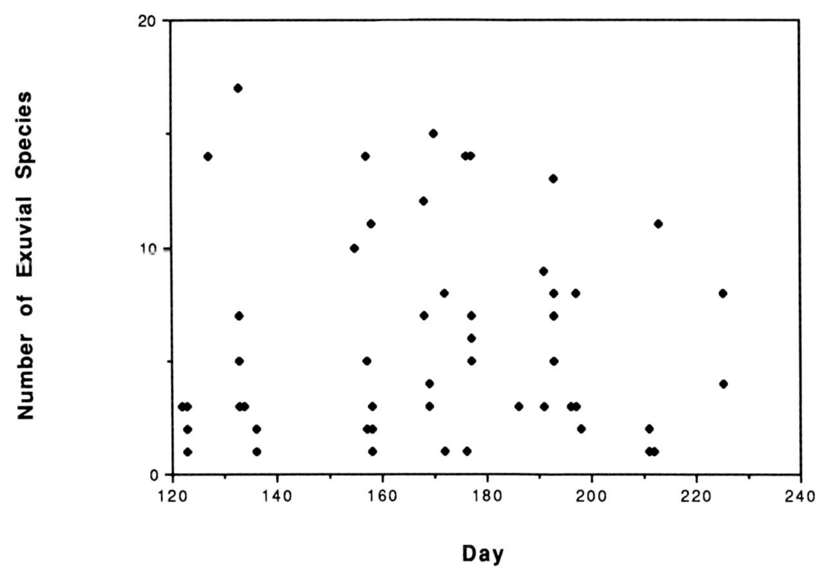

Figure 4. Numbers of exuviae species found in samples made on each collecting day in 1991.

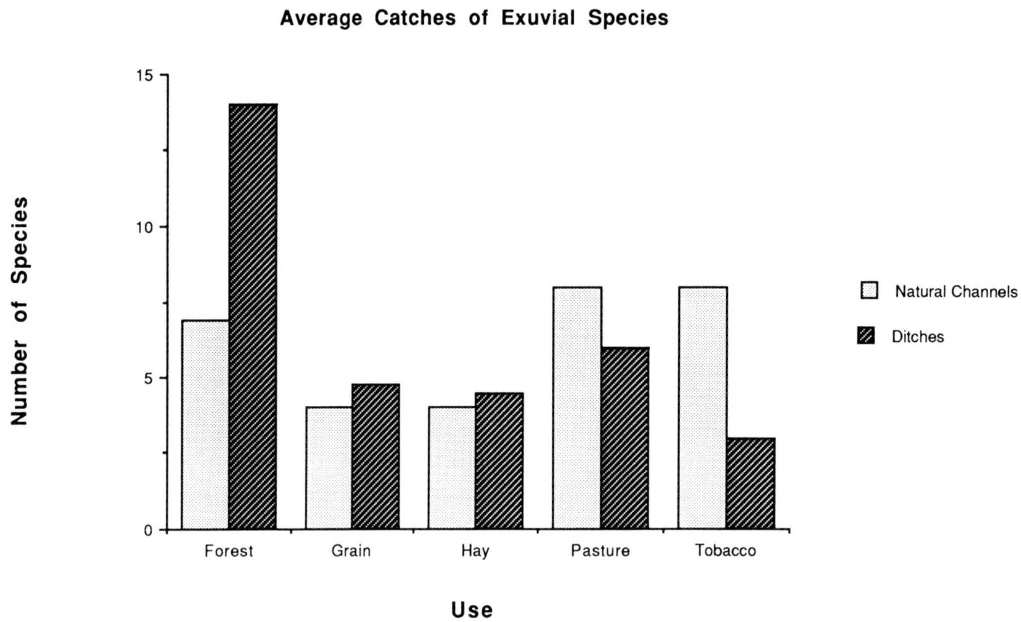

Figure 5. Average catches of pupal exuviae in natural and modified stream channels draining areas under various land-uses. Numbers of sites for each land-use are given in Table 1.

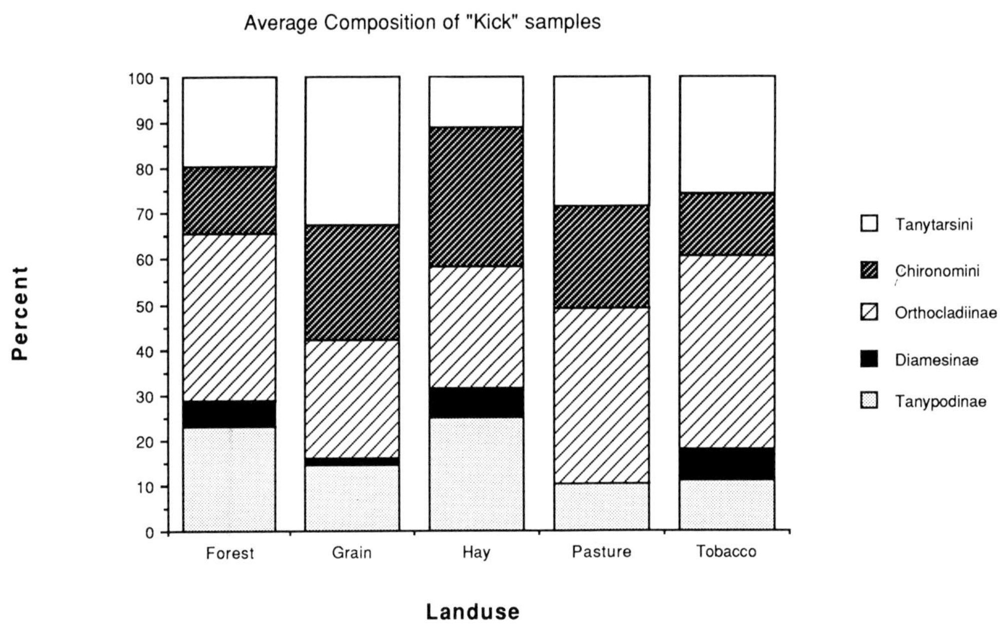

Figure 6. Average percentage composition of the Chironomidae in kick samples from streams draining forest and various crops. Numbers of sites for each land-use are given in Table 1.

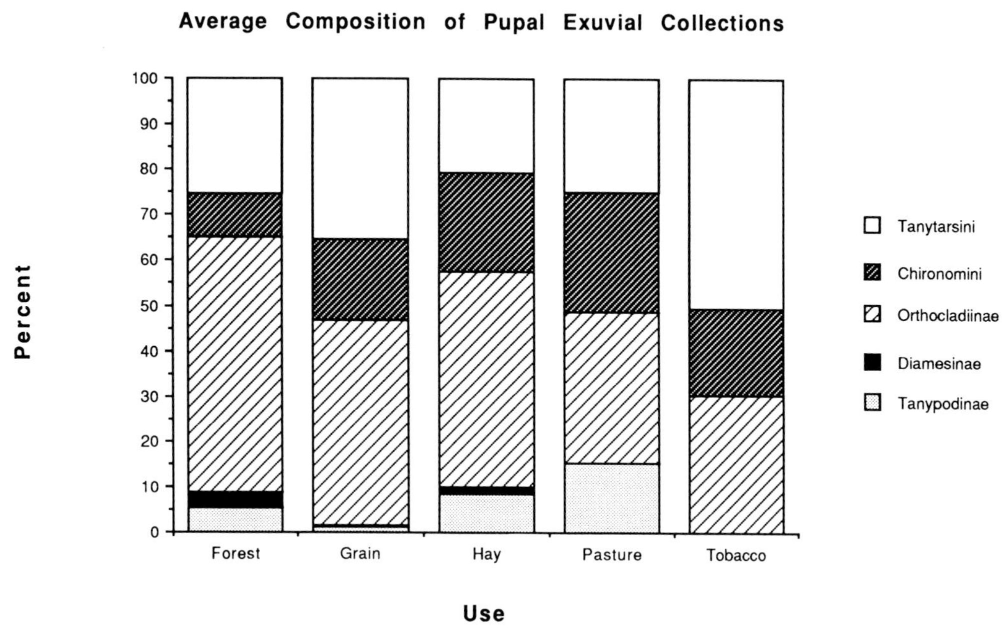

Figure 7. Average percentage composition of the Chironomidae in pupal exuviae samples from streams draining forest and various crops. Numbers of sites for each land-use are given in Table 1.

Flight periods of midges (Chironomidae and Simuliidae) in the River Laxa, N-Iceland

Gisli Mar Gislason, Thora Hrafnsdottir and Arnthor Gardarsson

Institute of Biology, University of Iceland, Grensasvegur 12, IS-108 Reykjavik, Iceland

Introduction

Window traps were designed to monitor population changes in Chironomidae and *Simulium vittatum* Zett. in Lake Myvatn and River Laxa. The larvae, pupae and adults of these insects are the staple food of many water birds and fish in these ecosystems. After a few years operation, the usefulness of window traps was shown, both in revealing the life cycles of these insects (Jonsson *et al.*, 1986) and as indicators of benthic production of *S. vittatum* in the river (Gislason and Gardarsson, 1988).

In this paper, we examine the pattern of flight activity of Chironomidae and *S. vittatum* along the course of R. Laxa. Also we document considerable variation among years in number of flight periods per year over a period of 16 years. We believe that the results may throw some light on the parameters that determine the number of life cycles in the principal species of midges.

Study area

The River Laxa (Fig 1) is the outlet of Lake Myvatn (65°35'N, 17°00'W, 277 m a.s.l.) in the volcanic area of northeast Iceland. The lake is eutrophic, known for its wealth of waterfowl and fish (Jonasson, 1979a). The lake receives water from cold and warm springs by its east shore and is discharged from the western part in the Laxa at the rate of 33 m^3s^{-1} (Rist, 1979). Additional water from tributaries increases the discharge to about 52 m^3s^{-1} on the route to the sea.

The lake is ice covered for 189 days on the average, with break up of ice from early May to 10 June. In 1978 the ice break was on May 15th.

The river is 59 km long, from L. Myvatn to the sea. It is divided into two parts, an upper (33 km) and a lower (26 km). Between the two parts are waterfalls, impassable to fish migrating up the river.

The mean current velocity in the upper part is 1.3 ms^{-1} and 0.7 m^{s-1} in the lower part. Close to the river mouth the river widens, and the average water velocity becomes 0.4 m^{s-1} (Rist 1979). Average monthly water temperatures are around 10-14°C (June - August) in summer and 0.5°C in winter (October - March) (Gislason and Johannsson, 1991). Discharge and concentrations of major ions change little throughout the year (Olafsson, 1979a,b), though year to year fluctuation in major nutrients (phosphorus and nitrogen) and drifting fine particulate organic matter (fPOM) have been recorded (Olafsson, 1979b; Gislason, 1994).

Material and methods

Flying insects were caught in window traps operated from May 4th to October 7th 1978. The trap construction and methods are described by Jonsson *et al.* (1986). Six traps were placed along the river, 0.5, 4, 23, 41, 50.5 and 56 km from the outlet at Myvatn (Fig. 1). The traps were situated in places inaccessible to livestock. The traps were all in exposed positions. At Dragsey (D) the trap was on a bridge crossing the outlet, 2.8 m above the water surface, at Helluvad (He) it was on a mound for a water gauge, 3.2 m above the water surface. Other traps were between 40 - 120 cm above the water surface. At Thvera (T), Holmavad (Ho) and Laxamyri (L) they were on islands in the river but at Nupafoss (N) it was positioned underneath a bridge. All traps were 0.5 to 1.5 m from the water's edge.

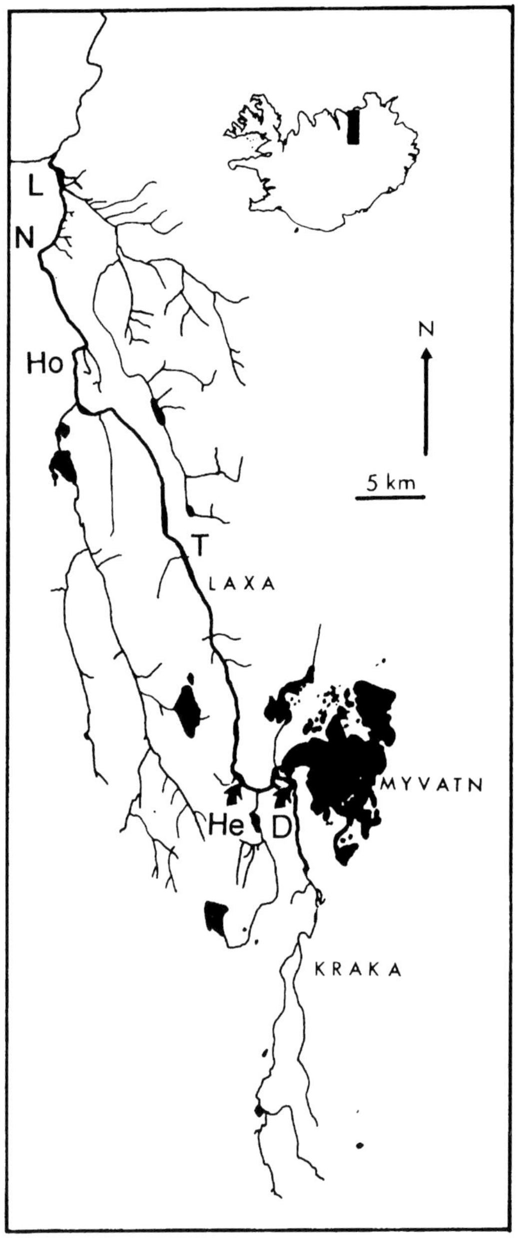

Figure 1. General location of Lake Myvatn and the River Laxa. Location of window traps:
D = Dragsey (outlet),
He = Helluvad,
T = Thvera,
Ho = Holmavad,
N = Nupafoss,
L = Laxamyri.

Samples were collected at 14 days interval on the average in 1978, stored in 70% alcohol and sorted and counted in the laboratory. Most collections were subsampled before further processing. Samples were divided into: miscellaneous insects, *Simulium vittatum* Zett., Chironomidae females, and Chironomidae males which were identified to species.

In correlation analyses on flight activities, log transformed numbers were used. Statistical analyses were not performed on species with a smaller annual catch than 30.

Results

Twenty nine species of Chironomidae were identified from the traps (15 - 25 species from each trap). *Eukiefferiella minor* (Edwards) was the most common Chironomidae in all traps except trap L, where *Tanytarsus gracilentus* (Holmgren) dominated the catch (see Appendix 1). *Simulium vittatum* dominated the catch in trap D.

Emergence of all species began in mid-May to early June (Figs 2-7), except *E. minor*, which had a peak in its flight period in early to mid-May (Fig. 2). *E. minor* and *Micropsectra atrofasciata* Kieffer had two peaks in flight activity at all trap sites (Fig. 2 and 3), *Orthocladius oblidens* (Walker) and *S. vittatum* had two peaks in flight activity at the two uppermost trap sites (D and He) but only a single peak further downstream (Figs 4 and 5). *T. gracilentus* had two peaks at the trap site at the outlet (D) and close to the river mouth (L), but a single peak in between (Fig. 6). *E. claripennis* (Lundbeck) had a single peak in flight period at all trap sites (Fig. 7). The second peak in flight activity of species that had two peaks occurred in late July or August.

Over the period 1977 to 1992, *E. minor* had two flight periods per year (Table 1). Other species had most often two flight periods at site D and either one or two at site He. In 1979 to 1982 and in 1991 *S. vittatum* had only one annual flight period at site He, whereas in other years it had two.

The number of flight periods per year also changed among Orthocladiinae species and *T. gracilentus* (Table 1).

The flight activity of all species, except *O. consobrinus* (Holmgren) and *T. gracilentus* at trap sites downstream of the outlet was independent from flight activities at site D (Table 2). Most other species had a positive association in flight activities from trap site He downstream to site N. *M. atrofasciata* and *T. gracilentus* had correlated activities down to the river mouth (Table 2).

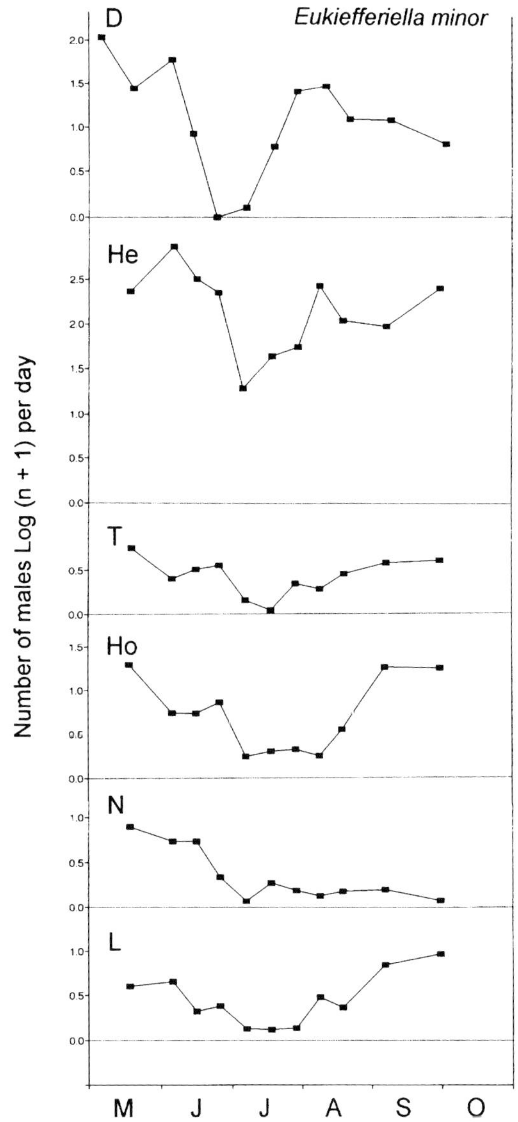

Figure 2. Flight periods of *Eukiefferiella minor* at trap sites along the Laxa.

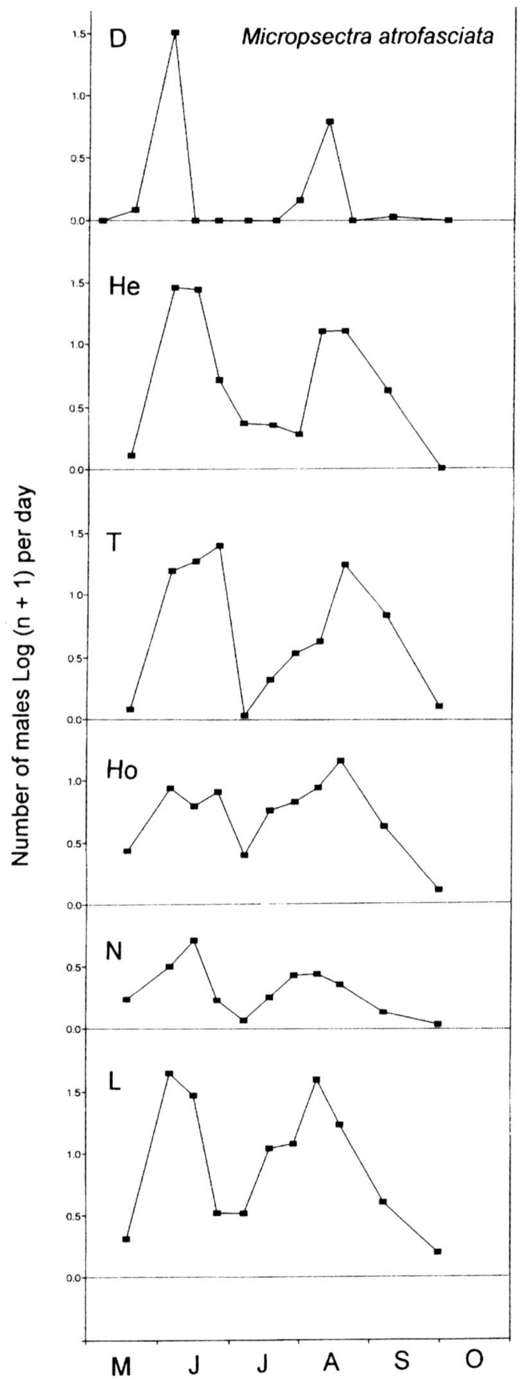

Figure 3. Flight periods of *Micropsectra atrofasciata* at trap sites along the Laxa.

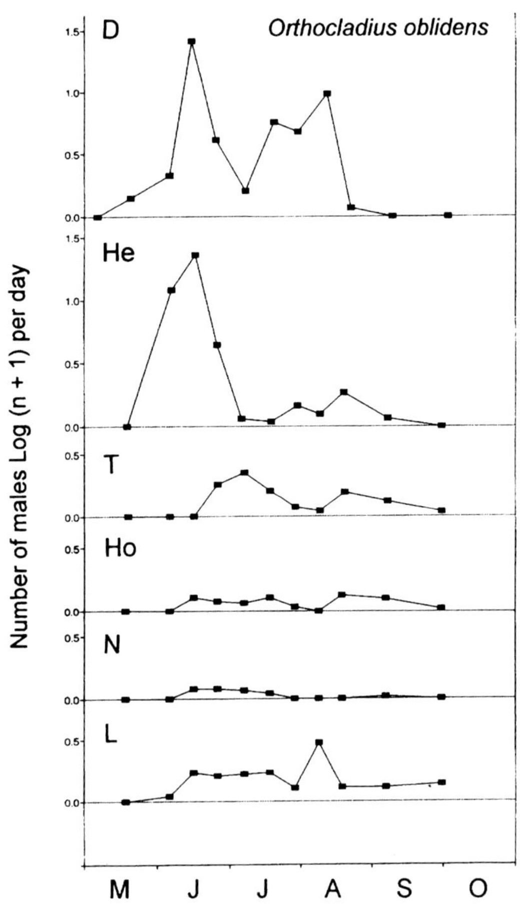

Figure 4. Flight periods of *Orthocladius oblidens* at trap sites along the Laxa.

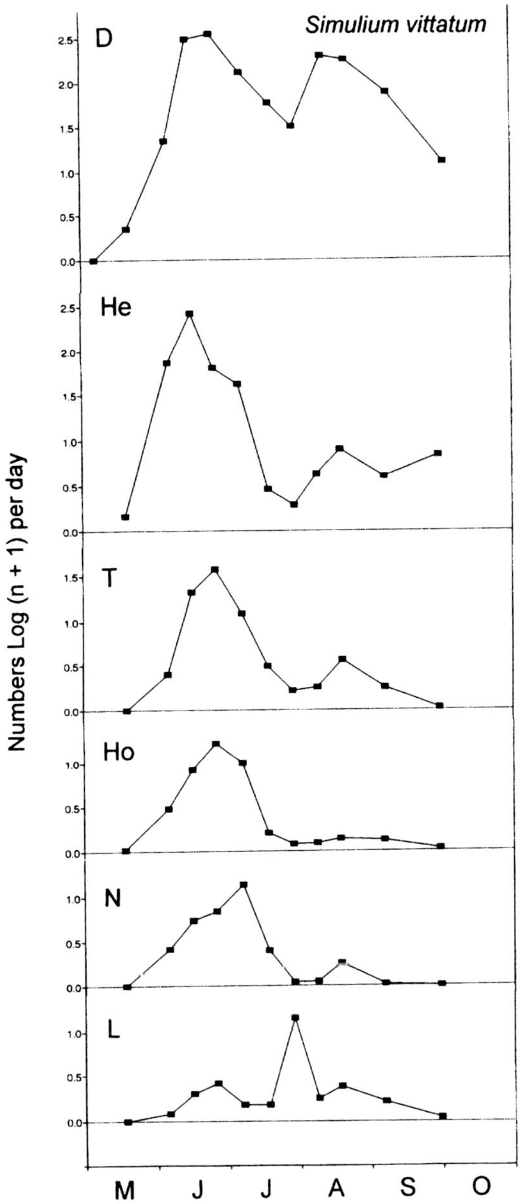

Figure 5. Flight periods of *Simulium vittatum* at trap sites along the Laxa.

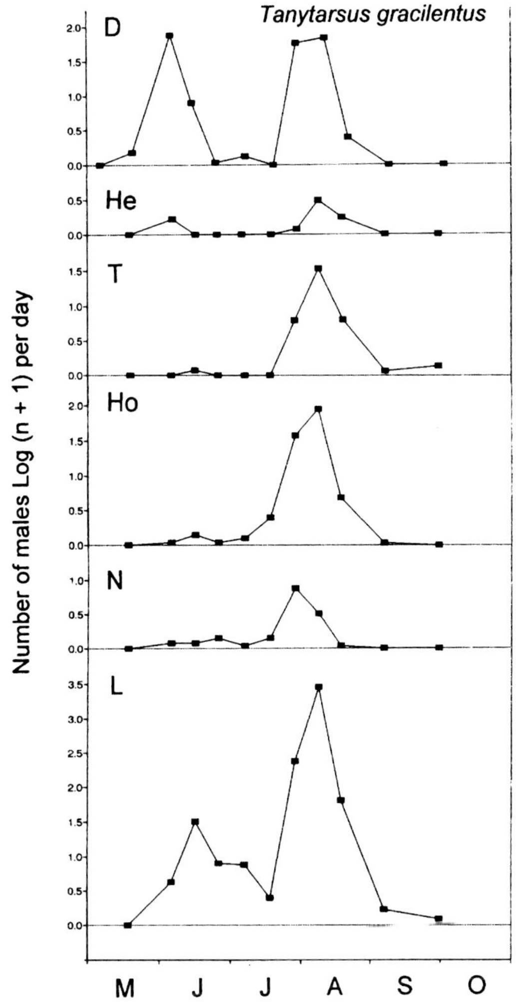

Figure 6. Flight periods of *Tanytarsus gracilentus* at trap sites along the Laxa.

Table 1. Estimated number of flight periods of selected species at Dragsey (D) and Helluvad each year from 1977 to 1992

Year	*Eukiefferiella minor*		*E. claripennis*		*Orthocladius consobrinus*		*O. oblidens*		*Micropsectra atrofasciata*		*Tanytarsus gracilentus*		*Simulium vittatum*	
	D	He	D	He	D	He	D	He	D	He	D	He	D	He
1977	2	2	1	2	2	2	2	2	2	2	2	1	2	2
1978	2	2	1	2	1	2	2	2	2	2	2	2	2	2
1979	2	2	0	1	1	0	2	1	2	1	2	0	2	1
1980	2	2	0	1	2	1	1	1	1	2	2	2	2	1
1981	2	2	0	1	2	1	2	1	2	2	2	1	2	1
1982	2	2	0	1	1	1	1	1	2	2	2	1	2	2
1983	2	2	0	1	2	1	2	1	2	2	1	1	2	2
1984	2	2	0	2	2	1	2	2	1	2	1	0	2	2
1985	2	2	1	1	2	1	2	1	2	2	2	0	2	2
1986	2	2	1	1	2	1	2	1	2	2	2	0	2	2
1987	2	2	1	2	1	1	2	2	0	2	2	1	2	2
1988	2	2	1	2	2	1	2	2	2	2	2	2	2	2
1989	2	2	1	2	2	1	2	2	2	2	1	1	2	2
1990	2	2	1	2	2	2	2	2	2	2	2	2	2	2
1991	2	2	1	1	1	1	2	1	2	2	2	1	2	1
1992	2	2	0	1	2	1	2	1	2	2	2	1	2	2

Table 2. Correlation (Pearson's r) between trap catches, as log(n+1)/day, of common chironomids and *Simulium vittatum* along the Laxa in 1978. Number of observations = 11

Species	D-He	He-T	T-Ho	Ho-N	N-L
Eukiefferiella minor	0.47 NS	0.57 NS	0.89 ***	0.38 NS	0.10 NS
E. claripennis	0.53 NS	0.91 ***	0.94 ***	0.71 *	0.44 NS
Orthocladius consobrinus	0.77 **	0.77 **	—	—	—
O. oblidens	0.53 NS	−0.30 NS	0.50 NS	0.51 NS	0.28 NS
Metriocnemus obscuripes	—	0.43 NS	0.78 **	—	—
Micropsectra atrofasciata	0.55 NS	0.81 **	0.77 **	0.66 *	0.85 ***
Tanytarsus gracilentus	0.68 *	0.84 ***	0.93 ***	0.86 ***	0.73 *
Simulium vittatum	0.54 NS	0.79 **	0.95 ***	0.93 ***	−0.08 NS

NS Not significant * $P < 0.05$ ** $P < 0.01$ *** $P < 0.001$

Discussion

Simulium vittatum has by far the highest production of invertebrates in River Laxa. The growth rate and production of the larvae depend crucially on the amount of drifting organic matter coming from Lake Myvatn (Gislason, 1985; Gislason and Gardarsson, 1988; Gislason and Johannsson, 1991). In most years *S. vittatum* populations near the outlet have two generations per year. The dense populations at the outlet are very effective filter feeders. A decline occurs in the density of drifting organic matter as the water flows further from the lake (Gislason 1994, Gislason et al. 1994). The downstream food limitation brought about in this way explains why S. *vittatum* is limited to one generation in the lower reaches of the river. The present long term study shows that a similar effect, i.e. the decrease to one generation per year, may occur even near the outlet (Table 1). This happens especially in years when *Anabaena*-blooms fail in Myvatn, since the total yearly amount of organic matter drifting out of the lake closely follows the production of Anabaena.

Gislason *et al.* (1994) showed that just as for S. *vittatum*, trap catches of *E. claripennis*, *E. minor*, *O. consobrinus* and *O. oblidens* showed a decreasing downstream trend in R. Laxa, whereas catches of *Metriocnemus obscuripes* (Holmgren), *Micropsectra atrofasciata* Kieffer and *T. gracilentus* did not show any such trends. Compared to S. *vittatum* the chironomid catches were also relatively stable between years at the two trap sites D and He (Gislason *et al.*, 1994).

We suggest that as in S. *vittatum*, annual variation in flight pattern among the Chironomidae in R. Laxa reflects variation in food conditions in the river rather than variation in physical parameters. Analyses of stomach contents have shown that larvae of *Eukiefferiella* spp. and *O. consobrinus* fed more on epilithic algae (*Synedra* sp. and *Epithemia* sp.) than on drifting seston (Gislason *et al.*, 1994). Hence, they should not be affected by high density of S. *vittatum* through competition. Therefore it is not surprising to find that the number of flight periods of S. *vittatum* and the Orthocladiinae species do not show the same pattern of variation among years. The similarity of flight patterns of *T. gracilentus* at the outlet of Lake Myvatn (site D) and at site L, close to the river mouth (Fig. 6) probably reflects habitat selection of the larvae. *Tanytarsus gracilentus* is the most numerous chironomid at Myvatn and the spring generation usually seen in trap catches at the outlet, most likely stems from the dense populations of the lake proper. Larvae of *T. gracilentus* in Lake Myvatn feed on detritus deposited at the lake bottom (Olafsson, 1987). In the river comparable habitats are mainly found in slow flowing parts, especially near the outlet.

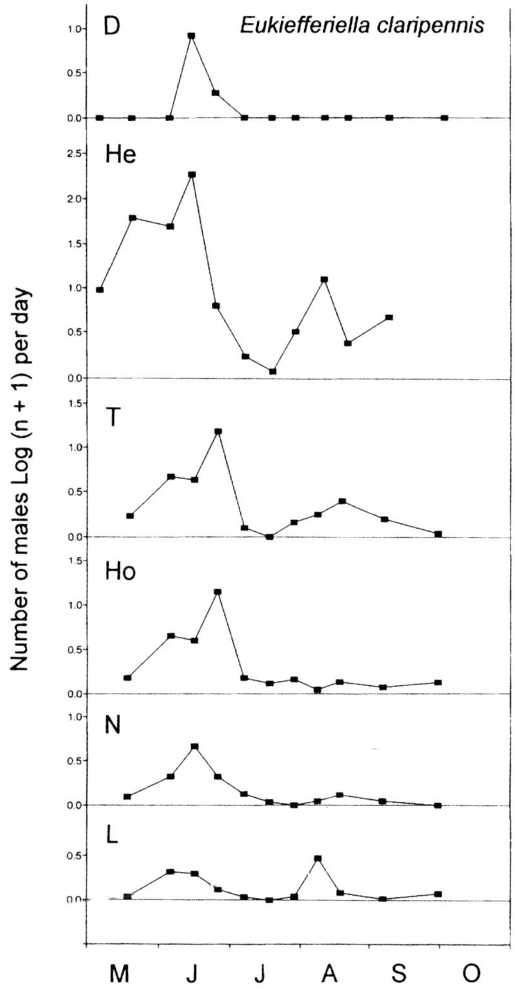

Figure 7. Flight periods of *Eukiefferiella claripennis* at trap sites along the Laxa.

Acknowledgements

We thank the farmers at Haganes, Helluvad, Thvera, Holmavad, Nupar and Laxamyri for allowing the traps on their property. We also thank Erlendur Jonsson, Jon S. Olafsson and Arni Einarsson for their contribution to this work. We are also indebted to several people of the Institute of Biology, University of Iceland, too numerous to mention here, who have contributed to this work. Financial support for parts of this study from the University of Iceland Research Fund and the Icelandic Science Foundation is gratefully acknowledged. Dr. Sigurdur S. Snorrason kindly read and made several helpful comments on the manuscript.

Appendix 1

Numbers of male Chironomidae and total catch of Simuliidae in window traps along the River Laxa 1978

Species Distance from Myvatn (km)	Dragsey 0.5	Helluvad 4	Thvera 23	Holmavad 41	Nupafoss 50.5	Laxamyri 56
Chironomidae						
Tanypodinae						
Macropelopia nebulosa (Meigen)	29	11	1	4	-	6
Procladius islandicus (Goetghebuer)	107	-	1	12	-	39
Ablabesmyia monilis (L.)	5	3	1	1	-	2
Arctopelopia griseipennis (v.d. Wulp)	22	2	-	1	3	2
Podonominae						
Parochlus kiefferi (Garret)	-	5	-	-	-	-
Diamesinae						
Diamesa bertrami Edwards	-	-	1	-	-	3
D. bohemani Goetghebuer	56	482	19	4	5	16
D. latitarsis (Goetghebuer)	22	1686	17	1	-	14
D. zernyi Edwards	9	450	-	-	-	-
Orthocladiinae						
Cricotopus sylvestris (Fabricius)	82	16	12	2	-	167
C. tibialis (Meigen)	1	6	125	40		78
Eukiefferiella minor (Edwards)	2694	30595	356	1497	322	511
E. claripennis (Lundbeck)	83	2975	278	241	77	53
Heterotrissocladius grimshawi Edwards	-	1	-	-	-	-
Orthocladius consobrinus (Holmgren)	1471	851	10	3	1	100
O. frigidus (Zetterstedt)	-	44	7	2	8	1
O. oblidens (Walker)	513	388	48	23	8	69
Psectrocladius barbimanus (Edwards)	10	-	21	10	2	20
P. limbatellus (Holmgren)	-	8	55	1	1	16
Limnophyes difficilis Brundin	-	1	-	-	-	-
Metriocnemus obscuripes (Holmgren)	-	111	89	269	3	1543
Paraphaenocladius impensus (Walker)	-	3	-	-	-	-
Thienemanniella sp.	6	44	11	1	3	156
Chironominae						
Chironomus islandicus Kieffer	19	-	-	-	1	5
Paracladopelma laminata Kieffer	13	28	24	7	8	6
Micropsectra atrofasciata Kieffer	332	934	983	671	160	1604
M. lindrothi Goetghebuer	31	154	1	4	-	29
M. recurvata (Goetghebuer)	-	1	30	10	-	49
Tanytarsus gracilentus (Holmgren)	2249	34	466	1256	106	30500
Simuliidae						
Simulium vittatum Zetterstedt	15774	4935	803	381	298	227

MONITORING CHIRONOMID NUMBERS AT MYVATN, ICELAND: THE FIRST SIXTEEN YEARS

Arnthor Gardarsson, Jon S. Olafsson, Thora Hrafnsdottir, Gisli Mar Gislason and Arni Einarsson[1]

Institute of Biology, University of Iceland, Grensasvegur 12, IS-108 Reykjavik, Iceland; and [1]Myvatn Research Station.

INTRODUCTION

Window traps are currently used at Lake Myvatn, Iceland, to monitor seasonal and yearly variation in numbers of flying insects, primarily Chironomidae and Simuliidae. In this paper we report the main results from the period 1977-92.

This study began as an attempt to monitor year to year variation in food of aquatic vertebrates at Myvatn. The first analyses suggested that window traps were a useful tool in estimating flight activity of insects (Jonsson *et al.*, 1986). Also, trap catches of the blackfly *Simulium vittatum* Zetterstedt were associated with local production of larvae and pupae (Gislason and Gardarsson, 1988). Hence window traps appear to be a simple and inexpensive means of obtaining population indices of flying insects.

By simultaneously recording several chironomid populations and a limited number of environmental variables, we attempt a comparative analysis aimed at explaining variation in numbers. While it is clear that variation thus explained is only a first approximation, we feel that this approach is worth pursuing for several reasons: Long-term monitoring of populations is a necessary background for an array of other studies and for management of the lake ecosystem. In the long run, benthic sampling of larvae may be less cost-efficient than window traps, especially in littoral habitats. Variation in population indices can be used to sort species into ecological groups of species that have similar responses to the environment. Such variation may yield insights into the importance of potential limiting factors and suggest the direction of detailed observation and experiment.

Study area

Myvatn (65°40'N, 17°00'W, altitude 278 m a. s. l.), in the volcanic zone of northern Iceland, is a shallow, eutrophic lake well known for its abundance of midges, fish and waterbirds (Jonasson, 1979b; Gardarsson and Einarsson, 1991). The average air temperature varies between about 10°C in July and –4°C in January-February; the annual precipitation is about 400mm. The lake is ice-covered for an average of 189 days a year, though spring areas and the lake outlet, the River Laxa, remain ice-free in winter.

Apart from Lake Myvatn (37 km^2), the Myvatn basin (Fig. 1) contains numerous permanent small lakes and ponds (11 km^2), about 5 km^2 of temporary sedge marshes and about 1.5 km^2 of streams. Most permanent lakes and ponds have mud bottoms though littoral habitats are often rock or gravel. The bottom mud in the deepest (about 2.5-4.0 m) parts of Lake Myvatn, is covered with the chlorophyte *Cladophora aegagropila* L. (Rabenh.). Stands of rooted macrophytes (mainly *Myriophyllum spicatum* L., *Potamogeton filiformis* Pers., *Ranunculus trichophyllus* Chaix) occur at depths of 0.8-2.5 m in Myvatn. Other permanent lakes and ponds in the basin are characterized by *Myriophyllum alterniflorum* D. C. and emergent *Carex rostrata* Stokes.

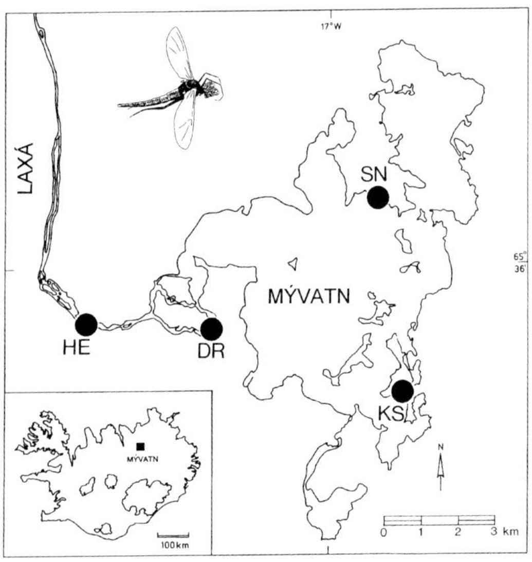

Figure 1. Map of the study area showing locations of window traps.

The benthic distribution of the larvae of the more common chironomid species at Myvatn is described by Lindegaard and Jonasson (1979) and the distribution of some scarcer species by Lindegaard (1979), supplemented by observations of imagines (Jonsson *et al.*, 1986).

Long-term changes at Myvatn are reviewed by Gardarsson *et al.* (1989). Studies of sediment cores cover the history of algae, crustaceans and chironomids through 2,300 years (Einarsson, 1982; Einarsson and Haflidason, 1988). Monitoring studies of aquatic birds began in 1975 and insects in 1977. The production and density of several bird populations at Myvatn is associated with chironomid numbers (Gardarsson and Einarsson, 1994).

Methods

We used a window trap described by Jonsson *et al.* (1986). The trap was made of transparent acrylate and was basically a vertical catching window (20 x 20 cm) mounted over an open box (16 cm deep, 16 x 20 cm wide) containing formaldehyde solution with a few drops of detergent and some ethylene glycol. Frequent refilling was necessary during windy and warm (>20°C) periods which were rather rare. Traps were mounted on aluminium poles in permanent positions on the shores of the lake.

Traps were located at Sydri Neslönd (SN), on the north shore of Myvatn's south basin, at Kalfaströnd (KS) near cold springs south-east of the lake, and at Dragsey (DR) near the headwaters of the River Laxa and 1 km from the lake (Fig. 1). The three traps were separated by 5-6 km. Trap SN was in an exposed position on the shore at 2.8 m above the lake surface. Trap KS was sheltered by low trees; it was 1.4 m above the lake surface and 12 m from the shoreline. Trap DR was in an exposed situation on a bridge, 2.8 m above the river surface. The traps were operated from early May to late September each year.

Samples were collected from the traps at variable intervals, stored in 70% alcohol and sorted and counted in the laboratory. Most collections were subsampled before further processing. Samples were divided into: miscellaneous insects, *Simulium vittatum*, Chironomidae females, and Chironomidae males which were identified to species.

Estimates of particulate organic matter (mainly *Anabaena flos-aquae* (Lyngb.) Bréb.) were made from water samples collected at intervals during summer at the lake outlet in the River Laxa. The method of processing the water samples is described by Gislason (1985).

Standard weather records from Reykjahlid' 4 km northeast of trap SN, were provided by the Icelandic Meteorological Office. The records included mean, maximum and minimum daily air temperatures, precipitation and estimated mean and maximum wind speed. From this we extracted for the relevant periods mean temperatures, total precipitation, and number of windy days with several threshold values.

Weather data were analysed on the basis of assumed periods of growth of larvae of chironomid species and the periods in which the imagines were caught. Only abundant bivoltine species were compared with weather factors. For this purpose the growth period of the spring emergence was set at 21 August - 20 October plus 15 May - 20 June, and of the summer emergence 21 June - 20 August; high winds toward the end of these two periods were excluded since they would have affected the aerial and not the aquatic stage of the generation being caught. The trap periods were set at 21 May to 15 July for the spring emergence and 16 July to 20 August for the summer emergence, eliminating the last month, 21 August through September, when most flight activity has ceased and winds are often comparatively high.

In correlation analyses and multiple regressions we mainly employed the first-order differences of the log-transformed numbers of each generation of bivoltine populations, i.e. $\log(N_t) - \log(N_{t-1})$, or each year when comparing other populations. This excludes some of the autocorrelation inevitably present in long term population data and is more conservative than using the raw transformed figures. Detailed comparison of environmental variables with population change was only attempted for bivoltine populations which have more clearly delimited growth and flight periods than most univoltine populations or those that could not be separated into two flight periods per year.

Results

Over 40 species of chironomids were identified from the traps, but only the most abundant 17 yielded useful quantitative information and are considered here (Table 1). The mean annual catch of chironomids in the three traps was about 290,000 of which 215,000 were taken in trap KS, 69,000 at SN and 8,000 at DR. *Tanytarsus gracilentus* (Holmgren) was the most abundant species, accounting for 84% of the chironomids at KS, 73% at SN and 32% at DR. At both SN and KS *Cricotopus sylvestris* (Fabricius), *Cricotopus tibialis* (Meigen), *Chironomus islandicus* Kieffer, *Orthocladius (Pogonocladius) consobrinus* (Holmgren) and *Orthocladius oblidens* (Walker) were amongst the most abundant species. At SN *Procladius islandicus* (Goetghebuer) and *Metriocnemus obscuripes* (Holmgren) were relatively abundant. At KS the second most abundant species was *Micropsectra lindrothi* Goetghebuer. At DR the second and third most abundant species were *Eukiefferiella minor* (Edwards) and *Orthocladius oblidens*.

The long term stability in numbers varied on a spectrum from *Tanytarsus gracilentus* and *Micropsectra lindrothi* (highest variance, Table 1) to *Orthocladius oblidens* (least variable). A significant long term (i.e. 16 years) trend in numbers was found only in *Ablabesmyia monilis* (L.) which declined at all sites ($P<0.05$). *Simulium vittatum* showed an increasing trend ($P<0.01$).

The numbers of most common species caught in traps SN and KS were significantly correlated each year, with the exception of *Micropsectra atrofasciata* Kieffer and *Ablabesmyia monilis* (Table 2). The correlation was weaker when generations were used instead of years. Correlations between catches in SN and KS, the two traps at the lake, with trap DR were seldom significant, the exception being *Orthocladius (Pogonocladius) consobrinus*.

Tanytarsus gracilentus fluctuated by over 4 orders of magnitude at both SN and KS and was distinctly cyclic (Fig. 2). At SN numbers generally increased through 9 generations from 1977 to spring 1981, then declined to a low 6 generations later, in spring 1984; an increase culminated in spring 1987, after 6 generations; this was followed by a rapid decline (3 generations) to summer 1988 and a continuing increase at least to spring 1992. At KS numbers were already high in 1977 and an irregular high, trending down, continued until 1983. After that numbers closely parallelled those at SN, with the exception of spring 1988 when numbers were still high at KS, one generation after they had crashed at SN.

"Table 1. Mean numbers of Chironomidae (males) and *Simulium vittatum* (total) in three window traps at Myvatn, Iceland, in 1977-92. A: Arithmetic means, and and log-transformed means and variances per year (n=16). B: Log-transformed means and variances per generation of bivoltine species (n = 32). See also Appendix 1.

	All traps	Sydri Neslönd				Kalfaströnd				Dragsey			
	mean	mean	log-transformed		%	mean	log-transformed		%	mean	log-transformed		%
			mean	var.			mean	var.			mean	var.	
A.													
Tanytarsus gracilentus	244413	50525	3.91	1.48	73.02	191366	4.71	0.78	84.40	2522	2.99	0.73	31.62
Micropsectra lindrothi	13669	184	1.71	0.78	0.27	13478	3.69	0.69	5.94	7	0.43	–	0.09
Cricotopus sylvestris	8648	5591	3.44	0.38	8.08	2496	2.97	0.60	1.10	561	2.62	0.14	7.03
Cricotopus tibialis	5384	2854	3.15	0.43	4.13	2518	3.16	0.27	1.11	12	0.78	0.36	0.15
Chironomus islandicus	4546	3837	2.96	0.66	5.55	570	2.16	0.95	0.25	139	1.44	0.69	1.75
Pogonocladius consobrinus	4141	1743	2.96	0.35	2.52	1673	2.96	0.43	0.74	725	2.67	0.20	9.09
Orthocladius oblidens	3810	1386	3.01	0.16	2.00	1189	2.91	0.19	0.52	1235	3.02	0.06	15.48
Eukiefferiella minor	2156	113	1.97	0.08	0.16	219	2.00	0.32	0.10	1824	3.20	0.07	22.87
Procladius islandicus	1493	1219	2.82	0.37	1.76	223	1.94	0.72	0.10	51	1.55	0.19	0.64
Micropsectra atrofasciata	976	419	1.69	1.36	0.61	110	1.60	0.55	0.05	446	2.21	0.72	5.59
Psectrocladius barbimanus	626	190	1.80	0.60	0.28	332	2.10	0.59	0.15	103	1.70	0.38	1.30
Metriocnemus obscuripes	600	593	2.57	0.22	0.86	7	0.47	–	0.00	0	–	–	0.00
Ablabesmyia monilis	410	338	2.15	0.34	0.49	56	1.31	0.61	0.02	16	0.81	0.48	0.20
Macropelopia nebulosa	354	7	0.66	–	0.01	309	1.87	0.89	0.14	38	1.18	0.52	0.47
Psectrocladius limbatellus	221	17	0.65	0.58	0.02	171	1.52	0.98	0.08	32	0.54	0.75	0.41
Heterotrissocladius grimshawi	163	2	0.18	–	0.00	46	0.75	0.93	0.02	114	0.85	1.24	1.43
Smittia aterrima	159	0	0.04	–	0.00	157	1.81	0.48	0.07	1	0.09	–	0.01
other chironomids	470	172	–	–	0.00	148	-	-	0.00	150	–	–	1.88
Total chironomids	292237	69191				215069				7977			
Simulium vittatum	86156	155	2.07	0.13		379	2.25	0.50		85622	4.77	0.19	
B.													
Tanytarsus gracilentus			3.49	1.30			4.14	1.15			2.45	1.30	
Micropsectra lindrothi			1.31	0.81			3.14	1.08			–	–	
Cricotopus tibialis			2.79	0.46			2.79	0.38			0.49	0.30	
Pogonocladius consobrinus			2.58	0.41			2.54	0.61			2.22	0.35	
Eukiefferiella minor			1.65	0.11			1.54	0.53			2.86	0.11	
Micropsectra atrofasciata			1.35	1.20			–	–			1.88	0.67	
Ablabesmyia monilis			1.54	0.73			–	–			–	–	
Smittia aterrima			–	–			1.28	0.75			–	–	
Simulium vittatum			1.50	0.54			1.78	0.63			4.42	0.27	

Table 2. Correlations (Pearson's r) between traps of change in numbers, as $\log(N_t/N_{t-1})$, of common chironomids and *Simulium* at Myvatn in 1977-92.

	SN-KS		SN-DR		KS-DR	
A. Annual totals (n=15)						
Tanytarsus gracilentus	0.56	*	0.37		0.36	
Micropsectra lindrothi	0.82	***	–		–	
Cricotopus sylvestris	0.78	***	0.56	*	0.37	
Cricotopus tibialis	0.54	*	–		–	
Chironomus islandicus	0.63	*	-0.16		-0.02	
Pogonocladius consobrinus	0.84	***	0.87	***	0.68	**
Orthocladius oblidens	0.85	***	0.32		0.47	
Eukiefferiella minor	0.59	*	0.40		0.38	
Procladius islandicus	0.88	***	0.37		0.36	
Micropsectra atrofasciata	0.14		0.21		0.35	
Psectrocladius barbimanus	0.73	**	-0.17		0.24	
Ablabesmyia monilis	0.34		0.34		0.29	
Macropelopia nebulosa	–		–		0.45	
Heterotrissocladius grimshawi	–		–		0.77	***
Simulium vittatum	0.03		0.28		0.45	
B. Generations (n=31)						
Tanytarsus gracilentus	0.45	*	0.22		0.24	
Micropsectra lindrothi	0.34		–		–	
Cricotopus tibialis	0.21		-0.33		-0.06	
Pogonocladius consobrinus	0.36		0.32		0.23	
Eukiefferiella minor	0.38	*	0.34		0.09	
Ablabesmyia monilis	–		0.43	*	–	
Macropelopia nebulosa	–		–		0.38	*
Micropsectra atrofasciata	–		0.06		–	
Simulium vittatum	0.33		0.42	*	0.68	***

* $P < 0.05$, ** $PP < 0.01$, *** $P < 0.001$

Several species fluctuated in phase with *Tanytarsus*, though the magnitude of their fluctuations was generally no more than about 2 orders (Fig. 2). All chironomid species at the lake underwent a simultaneous low in 1988 and most also about 1983. Correlation analyses showed significant association of several species, *Micropsectra lindrothi*, *Orthocladius (Pogonocladius) consobrinus*, *Psectrocladius barbimanus* (Edwards), *Eukiefferiella minor*, *Procladius islandicus* (Goetghebuer), *Micropsectra atrofasciata* and *Ablabesmyia monilis*, with *Tanytarsus gracilentus* at SN (Table 3). At KS only *Micropsectra lindrothi* fluctuated in phase with *Tanytarsus* and at DR only *Ablabesmyia monilis*. At DR *Orthocladius oblidens* and *Eukiefferiella minor* showed significant negative correlations with *Tanytarsus*.

A few species appeared to fluctuate independently of *Tanytarsus*. These included *Cricotopus sylvestris*, *Orthocladius oblidens* and *Metriocnemus obscuripes*, as well as *Simulium vittatum* (Fig. 2). Species in the latter group were relatively stable in numbers and were not significantly correlated with any other chironomids or *Simulium* (analyses not shown).

All species showed significant negative correlations of change in population (N_t/N_{t-1}) with numbers (N_{t-1}), i.e. they were density dependent. The more abundant bivoltine species were examined for delayed density dependence (Table 4) which was very marked in *Tanytarsus*, where the correlation peaked at a lag of three generations. *Micropsectra lindrothi* also showed delayed density dependence with a three year lag period, as did *Orthocladius (Pogonocladius) consobrinus* at SN and *Micropsectra atrofasciata* at DR.

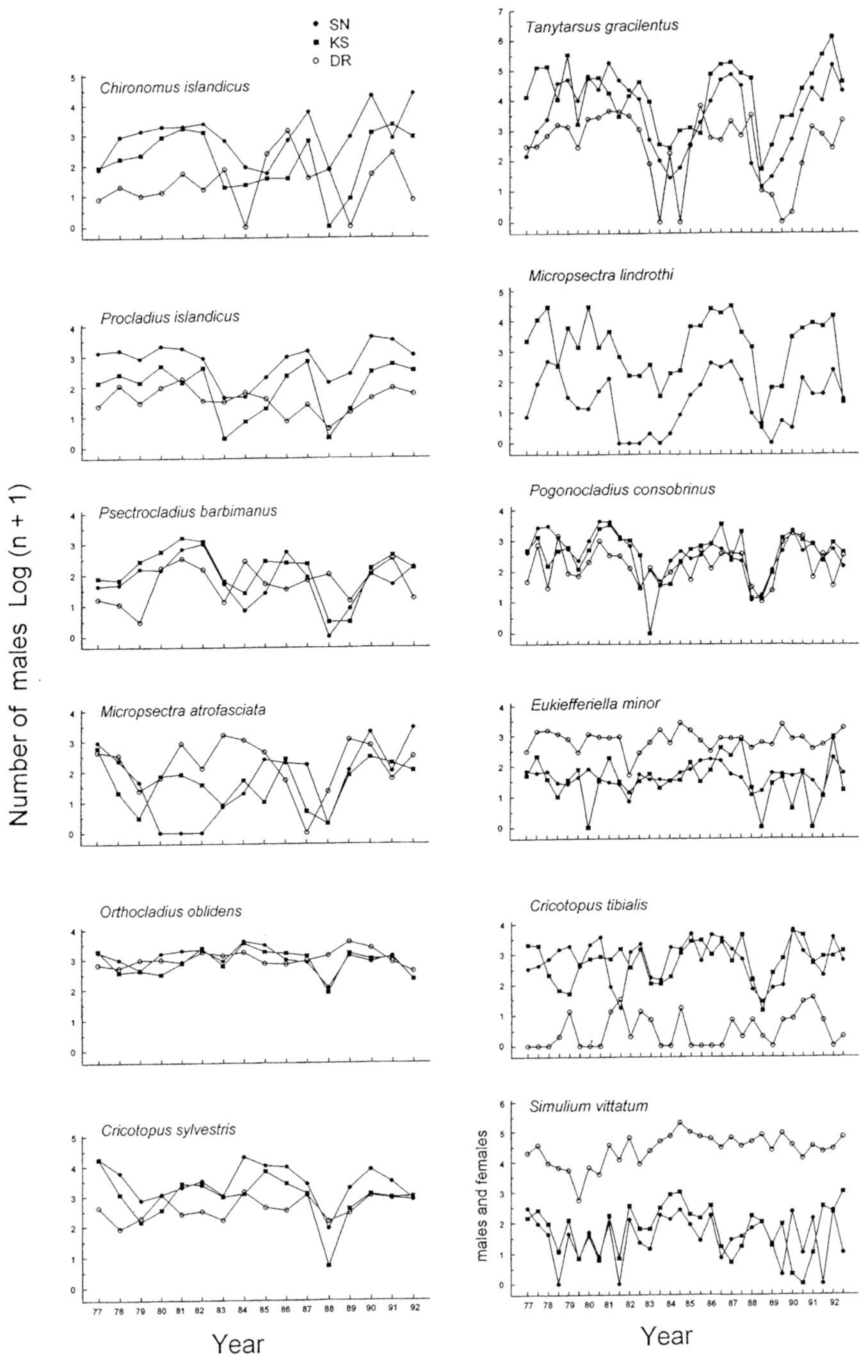

Figure 2. Numbers (log) of the 11 most abundant chironomid species (males) and of *Simulium vittatum* (total males and females) in three window traps at Myvatn in 1977-92. Yearly totals shown in graphs on the left for species that have one flight period or an undetermined flight period. Generation totals on the right for distinctly bivoltine species.

Table 3. Correlations (Pearson's r) between change in numbers, as $\log(N_t/N_{t-1})$, of selected species with *Tanytarsus gracilentus* in the same trap at Myvatn in 1977-92.

	SN		KS		DR	
A. Annual totals (n=15)						
Micropsectra lindrothi	0.65	**	0.44		–	
Cricotopus sylvestris	0.24		–0.16		–0.04	
Cricotopus tibialis	0.35		–0.05		0.00	
Chironomus islandicus	0.66	**	0.26		0.45	
Pogonocladius consobrinus	0.50		0.01		–0.27	
Orthocladius oblidens	0.11		–0.20		–0.63	*
Eukiefferiella minor	0.27		0.08		–0.42	
Procladius islandicus	0.78	***	0.41		0.13	
Micropsectra atrofasciata	0.30		–0.01		–0.48	
Psectrocladius barbimanus	0.80	***	0.39		0.46	
Metriocnemus obscuripes	–0.50		–		–	
Ablabesmyia monilis	0.75	**	0.05		0.31	
Macropelopia nebulosa	–		0.43		0.40	
Heterotrissocladius grimshawi	–		0.26		0.31	
Smittia aterrima	–		–0.12		–	
Simulium vittatum	-0.25		–0.01		–0.03	
B. Generations (n=31)						
Micropsectra lindrothi	0.51	**	0.69	***	–	
Cricotopus tibialis	0.32		0.10		0.03	
Pogonocladius consobrinus	0.38	*	0.17		0.00	
Eukiefferiella minor	0.37	*	0.31		–0.36	*
Micropsectra atrofasciata	0.40	*	–		–0.13	
Ablabesmyia monilis	0.25		–		0.44	***
Macropelopia nebulosa	–		–		0.02	
Smittia aterrima	–		–0.09		–	
Simulium vittatum	0.13		0.26		–0.07	

* $P < 0.05$, ** $P < 0.01$, *** $P < 0.001$

We found little evidence that temperature or precipitation (Fig. 3), as deviations from the mean during the period of assumed larval growth, affected the subsequently observed number of flies. Wind velocity (Fig. 3) fluctuated with about the same periodicity as *Tanytarsus*. Extensive examination of the data by correlation analyses indicated that wind was the only weather factor of any importance to the trap catches. Further analysis by multiple regression, taking the strong density dependent effect into consideration, shows that windy weather in the trap period significantly depressed trap catches of several species in the two exposed traps, SN and DR (Table 5). High winds during the growth period were less clearly associated with decreases, in *Ablabesmyia monilis* and *Macropelopia nebulosa* the effect of wind was positive, and there was no evidence that *Tanytarsus* was directly influenced by wind in the larval period (Table 5). Models using only one of the two wind factors did not markedly change this result.

Population change of *Tanytarsus* (one generation lag) was negatively associated with fine particulate organic matter at all trap stations (Table 6). Negative effects were also found in *Micropsectra lindrothi* at KS and *Ablabesmyia monilis* at SN and DR. Particulate organic matter had a significant positive association with *Eukiefferiella minor* at DR and *Micropsectra atrofasciata* at SN (Table 6).

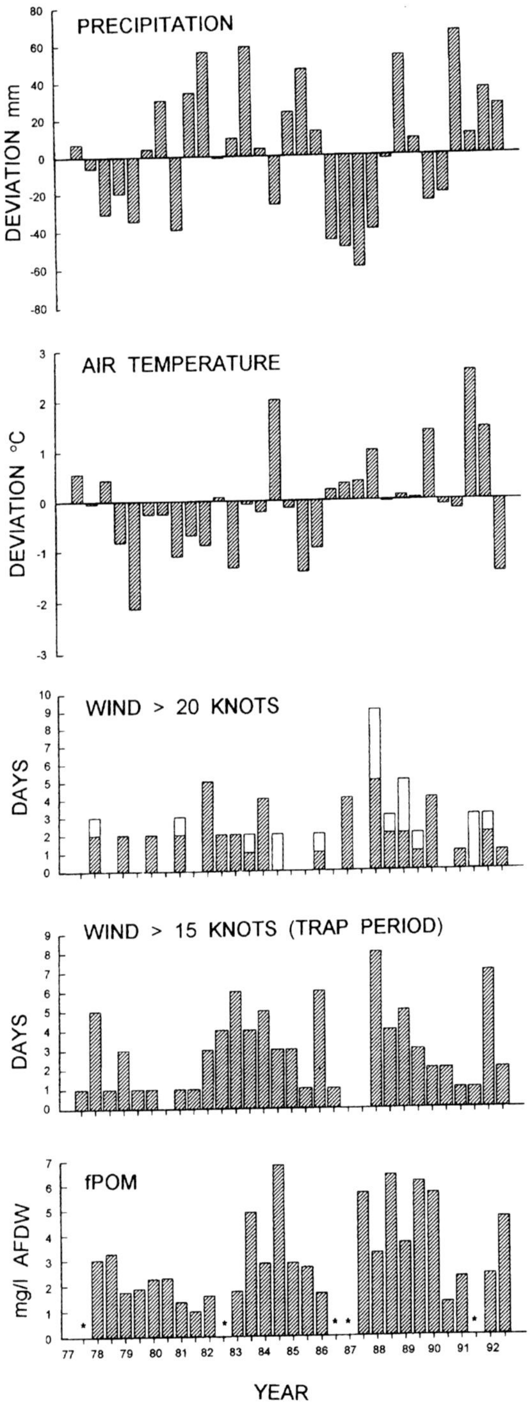

Figure 3. Selected environmental factors at Myvatn in 1977-92. Air temperature and precipitation shown as deviation from the mean of the relevant period of larval growth. Days with wind speed over 20 knots during the growth periods of larvae, white bars indicate days that were excluded from analyses as they affected primarily the aerial stage. Days with winds over 15 knots in the trap periods of bivoltine species. Fine particulate organic matter at the outlet (Laxa) in the larval growth period, missing values are indicated with asterisks.

Table 4. Serial correlations in populations of bivoltine chironomids and *Simulium* at Myvatn in 1977-92. Comparisons are shown as correlation coefficients (Pearson's) of change in numbers at time 1, $\log(N_t/N_{t-1})$ with previous numbers at time t–x, $\log(N_{t-x})$.

	Trap	Lag, generations									
	x:	1		2		3		4		5	
	n	31		30		29		28		27	
Tanytarsus gracilentus	SN	-0.36	*	-0.51	**	-0.63	***	-0.51	**	-0.23	
——	KS	-0.50	**	-0.29		-0.28		-0.25		-0.11	
——	DR	-0.49	**	-0.21		-0.29		-0.29		-0.07	
Micropsectra lindrothi	SN	-0.41	*	-0.42	*	-0.41	*	-0.30		-0.19	
——	KS	-0.49	**	-0.19		-0.54	**	-0.15		-0.30	
Cricotopus tibialis	SN	-0.59	***	-0.39	*	0.02		0.14		-0.24	
——	KS	-0.54	***	-0.43	*	-0.32		0.03		0.25	
——	DR	-0.61	***	-0.26		0.00		-0.07		0.11	
Pogonocladius consobrinus	SN	-0.05		-0.43	*	-0.48	**	-0.13		-0.15	
——	KS	-0.55	**	-0.26		-0.31		-0.18		0.01	
——	DR	-0.69	***	0.01		0.17		0.19		0.25	
Eukiefferiella minor	SN	-0.59	***	-0.21		-0.16		-0.14		0.06	
——	KS	-0.69	***	-0.22		0.31		-0.22		0.08	
——	DR	-0.65	***	-0.19		-0.23		-0.21		0.05	
Micropsectra atrofasciata	SN	-0.35		-0.38	*	-0.20		-0.18		-0.09	
——	DR	-0.43	*	-0.45	*	-0.40	*	-0.30		-0.04	
Ablabesmyia monilis	SN	-0.67	***	0.10		-0.14		0.05		-0.15	
——	DR	-0.59	***	-0.03		-0.09		0.05		-0.17	
Macropelopia nebulosa	KS	-0.75	***	0.21		-0.30		0.37		-0.31	
——	DR	-0.81	***	0.26		-0.19		0.19		-0.05	
Simulium vittatum	DR	-0.47	**	0.06		-0.32		-0.14		-0.02	

* $P < 0.05$, ** $P < 0.01$, *** $P < 0.001$

DISCUSSION

The Lake Myvatn ecosystem is characterized by large scale fluctuations in the biota (Gardarsson *et al.*, 1988). Periods of high numbers of *Tanytarsus* and several other chironomids coincide with clear water. Periods of low numbers of *Tanytarsus* coincide with turbid water, high densities of *Anabaena* and large populations of *Gasterosteus aculeatus* (L.), the three-spined stickleback. The turbid periods are also reflected in high production of *Simulium* in the river Laxa draining Myvatn (Gislason and Gardarsson, 1988). Waterfowl numbers show a lag response to the fluctuation in chironomids (Gardarsson and Einarsson, 1994).

Wind speeds of about 20 knots or more are of particular interest to this study because wind of this velocity (9-10 m/s) has been shown to be sufficent to resuspend the bottom sediment of Myvatn (Johannesson and Birkisson, 1991). The sediment surface is affected in an

Table 5. Summary of the results of multiple regression analyses of change in numbers of bivoltine chironomids as $\log(N_t/N_{t-1})$ versus previous numbers at lag 1, $\log(N_{t-1})$, or lag 3, $\log(N_{t-3})$; number of windy days (force > 15 knots) in trap period; number of days with winds over 20 knots in assumed larval growth period. For lag 1 n = 31, for lag 3 n = 29.

	Trap	R-square (dd)	R-square (final)	Lag	Density dependence (dd) Number at time t-lag		Trap period Wind > 15 knots		Larval period Wind > 20 knots	
					Beta	*P*	Beta	*P*	Beta	*P*
Tanytarsus gracilentus	SN	0.13	0.32	1	-0.47	0.01	-0.45	0.01	-	-
——	SN	0.40	0.44	3	-0.59	0.00	-0.20	0.20	-	-
——	KS	0.25	0.25	1	-0.50	0.00	-		-	-
——	DR	0.24	0.33	1	-0.49	0.00	-0.34	0.07	0.23	0.21
Micropsectra lindrothi	SN	0.17	0.22	1	-0.46	0.01	-	-	-0.24	0.17
——	SN	0.17	0.17	3	-0.41	0.03	-	-	-	-
——	KS	0.24	0.24	1	-0.49	0.01	-	-	-	-
——	KS	0.29	0.33	3	-0.53	0.00	-	-	0.20	0.22
Cricotopus tibialis	SN	0.35	0.35	1	-0.59	0.00	-	-	-	-
——	KS	0.29	0.46	1	-0.54	0.00	-0.27	0.11	-0.21	0.21
——	DR	0.38	0.38	1	-0.61	0.00	-	-	-	-
Pogonocladius consobrinus	SN	0.19	0.25	1	-0.57	0.00	-0.28	0.14	-	-
——	KS	0.30	0.51	1	-0.67	0.00	-0.47	0.00	-	-
——	DR	0.48	0.69	1	-0.72	0.00	-0.47	0.00		- -
Eukiefferiella minor	SN	0.35	0.49	1	-0.56	0.00	0.31	0.07	-0.41	0.01
——	KS	0.47	0.50	1	-0.66	0.00	-	-	-0.16	0.26
——	DR	0.43	0.56	1	-0.63	0.00	-	-	-0.36	0.01
Micropsectra atrofasciata	SN	0.12	0.12	1	-0.35	0.06	-	-		- -
——	DR	0.18	0.32	1	-0.40	0.02	0.37	0.03	-	-
Ablabesmyia monilis	SN	0.45	0.56	1	-0.87	0.00	-0.43	0.02	0.20	0.23
——	DR	0.34	0.49	1	-0.63	0.00	-0.32	0.06	0.42	0.01
Macropelopia nebulosa	KS	0.56	0.63	1	-0.67	0.00	-0.15	0.28	0.31	0.03
——	DR	0.65	0.72	1	-0.80	0.00	-0.26	0.04	0.22	0.08
Simulium vittatum	DR	0.22	0.30	1	-0.48	0.01	-	-	0.28	0.09

Table 6. Summary of the results of multiple regression analyses of change in numbers of bivoltine chironomids and Simulium, as log (N_t/N_{t-1}) versus previous numbers at lag 1, log (N_{t-1}), or lag 3, log (N_{t-3}); versus fPOM (fine particulate organic matter) in assumed larval growth period. For lag 1 n = 27, for lag 3 n = 25.

	Trap	R-square (dd)	R-square (final)	Lag	Density dependence (dd) Number at time t-lag Beta	*P*	Larval period fPOM (AFDW mg/l) Beta	*P*
Tanytarsus gracilentus	SN	0.12	0.22	1	–0.54	0.02	–0.37	0.10
——	SN	0.43	0.43	3	–0.65	0.00	–	–
——	KS	0.31	0.43	1	–0.60	0.00	–0.34	0.04
——	DR	0.24	0.58	1	–0.64	0.00	–0.61	0.00
Micropsectra lindrothi	SN	0.20	0.20	1	–0.45	0.02	–	–
——	SN	0.19	0.19	3	–0.43	0.03	–	–
——	KS	0.30	0.47	1	–0.64	0.00	–0.42	0.01
——	KS	0.32	0.39	3	–0.56	0.00	–0.28	0.11
Cricotopus tibialis	SN	0.37	0.37	1	–0.61	0.00	–	–
——	KS	0.30	0.30	1	–0.55	0.00	–	–
——	DR	0.36	0.36	1	–0.60	0.00	–	–
Pogonocladius consobrinus	SN	0.21	0.21	1	0.21	0.02	–	–
——	KS	0.29	0.29	1	–0.54	0.00	–	–
——	DR	0.48	0.48	1	–0.69	0.00	–	–
Eukiefferiella minor	SN	0.28	0.28	1	–0.53	0.01	–	–
——	KS	0.53	0.53	1	–0.73	0.00	–	–
——	DR	0.33	0.47	1	-0.54	0.00	0.38	0.02
Micropsectra atrofasciata	SN	0.12	0.29	1	–0.35	0.06	0.39	0.04
——	DR	0.31	0.31	1	–0.56	0.00	–	–
Ablabesmyia monilis	SN	0.44	0.57	1	–0.76	0.00	–0.39	0.00
——	DR	0.37	0.50	1	–0.62	0.00	–0.36	0.02
Macropelopia nebulosa	KS	0.39	0.39	1	–0.63	0.00	–	–
——	DR	0.56	0.56	1	–0.75	0.00	–	–
Simulium vittatum	DR	0.20	0.29	1	–0.57	0.02	0.33	0.10

opposite fashion by the tube building activity of *Tanytarsus* at high densities (cf. Gardarsson and Snorrason, 1993). Clearly *Tanytarsus gracilentus* is a key organism in Lake Myvatn and its role must be understood before one can hope to understand the ecosystem.

Paleolimnological studies of Myvatn show that *Tanytarsus* has become less abundant with time, as the lake has become shallower. Two effects of water depth on resources of the larvae were postulated (Einarsson and Haflidason, 1988): (i) decreasing depth increases resuspension of the bottom sediments and reduces their energy and nutrient content; (ii) in a shallow lake less focussing of the sediments (food) takes place toward the central floor of the basin.

Since we failed to find evidence that *Tanytarsus* was influenced by wind in the larval period, it would appear that wind may not have a direct adverse effect on the sediment modifying activities of *Tanytarsus* and associated species. However, the wind was only estimated and more detailed direct measurements should be undertaken before ruling out wave-induced disturbance of the sediment as a cause of population declines. Wind disturbance also causes increased turbidity and may therefore affect the larval populations indirectly through the resources.

Although a detailed, formal analysis has not yet been undertaken, it seems unlikely that vertebrate predators (fish or birds) are able to bring down numbers of *Tanytarsus* from high densities. *Tanytarsus* can increase by an order of magnitude in one generation, and by two orders in a year, an increase which seems to be excessive for any vertebrate predator, or such invertebrate predators as the Tanypodinae. Predation on the other hand may be important in reducing *Tanytarsus* and other chironomids to very low levels and hence may synchronize the low of many species including those that are not otherwise influenced by *Tanytarsus* numbers.

The assumption that predation (or perhaps pathogens) synchronizes the populations of different species, by bringing numbers to very low levels, implies that the subsequent increase of each is an increase into an unrestricted environment to which each species responds at its maximum intrinsic rate of increase.

Most of the species show density dependence with a one-generation lag only and are more stable than *Tanytarsus* (Fig. 4). The latter is far more abundant than any other benthic macrofauna species in Lake Myvatn and seems to respond in a deterministic fashion to its previous population level, suggesting that its interaction with the environment may be central to its own cyclic fluctuation. Such interaction is to be sought in the benthic environment of the larvae which at high densities greatly modify the sediment surface (Gardarsson and Snorrason, 1993). Alterations in the turbidity phase of the lake may also be closely related to the state of the *Tanytarsus* population.

Acknowledgements

We are indebted to the farmers of Sydri Neslönd, Kalfaströnd and Haganes and their families for their willing acceptance of permanent traps on their property, and for their generous hospitality over many years. We also wish to thank Erlendur Jonsson for his contribution to the initial success of this venture and numerous field workers and assistants for their help in collecting the data. Sigurdur S. Snorrason critically read the manuscript. Financial support for parts of this study from the Icelandic Science Fund and the University of Iceland Research Fund is gratefully acknowledged.

Appendix 1. Numbers of Chironomidae (males) and *Simulium vittatum* (total) in three window traps at Myvatn, Iceland, in each year 1977-92. Numbers of each generation of distinctly bivoltine species are recorded separately, as g1 and g2. Rare species are omitted.

		1977	1978	1979	1980	1981	1982	1983	1984	1985	1986	1987	1988	1989	1990	1991	1992
Trap SN (Sydr Neslönd)																	
Tanytarsus gracilentus	g1	143	2346	49311	65355	181304	21941	490	26	340	9199	69795	76	28	482	23856	141818
— —	g2	964	37452	9861	23288	47124	11105	98	57	1818	46630	30210	13	99	4275	9328	19565
Micropsectra lindrothi	g1	6	468	30	12	124	0	1	1	34	399	434	8	0	2	36	228
— —	g2	82	352	13	48	0	0	0	7	74	289	109	2	4	123	37	25
Cricotopus sylvestris		17950	5970	750	1150	1964	3244	1054	20778	10715	9665	2575	86	1838	7926	2995	796
Cricotopus tibialis	g1	335	690	1824	2064	87	1174	170	1665	4748	4383	1398	68	78	6377	571	3516
— —	g2	420	1441	404	3682	16	2203	144	1463	608	3310	667	26	94	1237	199	604
Chironomus islandicus		73	872	1344	1878	1905	2402	645	86	54	659	5930	73	871	19958	780	23860
Pogonocladius consobrinus	g1	400	3008	522	934	4166	682	110	219	256	766	266	11	115	2200	831	533
— —	g2	2698	1274	223	4334	1338	31	36	449	334	544	217	17	464	467	242	148
Orthocladius oblidens		1715	948	439	1532	1962	2104	868	4088	3024	965	798	112	1295	837	1277	216
Eukiefferiella minor	g1	68	66	25	78	28	6	35	31	75	165	52	10	57	46	31	180
— —	g2	59	27	41	37	24	53	35	60	146	146	40	14	54	57	10	56
Procladius islandicus		1307	1541	828	2099	1794	850	44	46	198	915	1410	129	247	3991	3102	998
Micropsectra atrofasciata	g1	238	106	42	0	0	0	0	0	118	171	125	0	9	1694	38	2237
— —	g2	662	114	1	0	0	0	6	18	127	18	41	1	99	317	60	464
Psectrocladius barbimanus		42	46	150	139	658	983	49	6	25	542	82	0	7	99	46	171
Metriocnemus obscuripes		2038	1328	48	554	74	343	149	1392	591	613	554	141	801	559	173	134
Ablabesmyia monilis	g1	70	20	6	18	128	618	0	55	93	54	316	2	13	19	225	45
— —	g2	93	416	162	324	2454	21	59	0	18	90	12	5	24	45	0	0
Macropelopia nebulosa		12	10	12	6	18	37	7	7	0	0	0	2	0	7	2	0
Psectrocladius limbatellus		6	48	0	0	48	120	15	0	30	6	0	0	0	0	3	0
Heterotrissocladius grimshawi		0	8	0	0	0	0	0	0	0	0	0	0	2	23	0	0
Simulium vittatum	g1	295	41	42	48	97	124	13	126	83	167	26	62	19	221	131	241
— —	g2	91	0	6	7	0	21	174	260	25	6	33	98	1	9	0	9
Trap KS (Kalfaströnd)																	
Tanytarsus gracilentus	g1	13202	134876	335469	54273	17527	13915	8889	257	1208	72648	178478	52735	310	2868	70029	1248120
— —	g2	127190	10581	1613	57319	2844	40321	337	984	790	155217	77398	50	2595	23343	317665	38792
Micropsectra lindrothi	g1	2159	27913	5849	29397	4289	154	351	181	6253	25380	31201	1351	60	2806	8424	14137
— —	g2	10921	320	1339	1322	644	151	32	224	6479	18195	3960	3	63	5514	6553	19
Cricotopus sylvestris		16944	1161	144	358	2624	2382	957	1197	6822	2610	1230	4	486	1162	909	945
Cricotopus tibialis	g1	2042	204	49	691	696	364	108	183	2658	972	579	145	240	5752	495	852
— —	g2	1897	64	460	841	1508	1418	107	1082	2886	2490	4348	13	719	4305	838	1262
Chironomus islandicus		83	162	216	851	1672	1239	18	22	36	36	621	0	7	1162	2154	838
Pogonocladius consobrinus	g1	487	150	602	493	3296	1035	0	39	522	810	229	13	93	2028	783	870
— —	g2	1301	460	119	2494	1142	337	36	210	675	3552	2008	12	1256	1056	228	431

Appendix 1. continued.

		1977	1978	1979	1980	1981	1982	1983	1984	1985	1986	1987	1988	1989	1990	1991	1992
Orthocladius oblidens		1833	361	414	315	732	2425	598	3537	1770	1596	1296	76	1762	982	1116	212
Eukiefferiella minor	g1	48	36	34	0	184	13	54	32	129	72	216	18	25	3	0	936
— —	g2	205	9	75	31	30	31	18	33	30	390	724	0	42	63	9	15
Procladius islandicus		133	248	135	479	134	409	1	6	18	216	646	1	23	294	513	317
Micropsectra atrofasciata		579	19	2	69	80	36	7	50	9	258	4	1	74	287	184	102
Psectrocladius barbimanus		75	66	264	541	1600	1231	61	25	273	234	221	2	3	145	414	155
Metriocnemus obscuripes		2	9	0	0	32	0	6	42	0	0	6	0	0	0	18	0
Ablabesmyia monilis		62	162	66	107	224	97	6	6	24	36	9	0	6	1	0	96
Macropelopia nebulosa		150	63	130	376	0	611	1	34	9	162	723	7	27	470	2043	136
Psectrocladius limbatellus		296	60	6	151	512	1052	43	24	48	0	452	0	49	3	0	41
Heterotrissocladius grimshawi		236	12	6	15	0	0	0	0	0	0	0	0	0	191	225	57
Smittia aterrima	g1	58	0	1	15	344	355	0	91	141	126	0	87	59	29	315	0
— —	g2	9	10	1	31	78	0	9	432	12	240	12	16	18	6	0	20
Simulium vittatum	g1	126	91	115	36	165	349	60	759	123	360	4	147	13	0	9	189
— —	g2	16	0	6	0	4	0	0	51	60	0	0	9	3	1	0	27
Trap DR (Dragsey)																	
Tanytarsus gracilentus	g1	292	683	1320	2464	4470	3019	78	169	306	558	1939	3069	6	1	1254	246
— —	g2	301	1566	275	2736	4128	1067	0	0	6471	469	660	9	0	75	708	2018
Micropsectra lindrothi		0	31	0	0	42	0	0	0	0	1	0	0	0	7	24	12
Cricotopus sylvestris		425	82	193	1145	259	324	163	1414	426	339	1116	147	279	1058	846	756
Cricotopus tibialis	g1	0	0	12	0	12	1	6	0	0	0	6	6	0	7	36	0
— —	g2	0	1	0	0	32	12	0	16	0	0	1	1	6	27	6	1
Chironomus islandicus		7	19	9	12	54	15	73	0	234	1404	37	70	0	48	240	6
Pogonocladius consobrinus	g1	46	28	89	201	329	127	132	96	54	127	384	31	24	1749	66	35
— —	g2	719	1443	72	978	318	28	46	200	477	370	373	10	894	1451	378	329
Orthocladius oblidens		653	513	927	946	786	1755	1344	1738	741	692	918	1349	3802	2396	795	400
Eukiefferiella minor	g1	304	1543	780	1119	856	50	600	572	1560	304	803	379	495	790	366	827
— —	g2	1434	1151	282	894	916	269	1575	2738	681	832	833	592	2430	881	508	1823
Procladius islandicus		22	107	28	94	175	33	31	64	39	6	24	3	13	40	84	54
Micropsectra atrofasciata	g1	251	253	16	18	382	116	1098	393	228	45	0	12	216	384	36	87
— —	g2	177	79	6	42	442	9	496	682	201	6	0	9	894	326	19	212
Psectrocladius barbimanus		15	10	2	164	329	146	12	269	51	33	66	99	13	107	324	16
Ablabesmyia monilis		4	5	20	54	36	49	0	32	6	0	42	1	0	0	0	6
Macropelopia nebulosa		11	29	9	14	8	153	31	16	72	0	72	1	0	57	126	1
Psectrocladius limbatellus		4	0	2	0	0	0	7	68	0	246	0	0	192	0	0	0
Heterotrissocladius grimshawi		125	0	0	12	0	107	0	0	6	0	0	0	0	993	516	66
Simulium vittatum	g1	19976	9232	5398	6655	36768	65311	23550	74084	100770	60699	64611	46962	24515	37223	33576	27216
— —	g2	36025	6542	553	3937	12030	8768	50132	207002	72069	30479	33822	78424	90306	13386	22269	67667

Rapid Biological Assessment of Streams in the Blue Mountains, Australia: Characteristics of the Chironomidae (Diptera) Fauna

Rod A. Hardwick, Ian A. Wright, Hugh A. Jones, Bruce C. Chessman and David E. Holleley

Australian Water Technologies,
Science and Environment, P.O. Box 73,
West Ryde, N.S.W., 2114, Australia

Introduction

The Blue Mountains west of Sydney, NSW, are popular for recreation and of high conservation value, the region being proposed in 1989 for World Heritage listing. Development of plateau and ridge-top townships and discharges from a number of sewage treatment plants (STPs) have led to a deterioration in water quality of headwater streams in the region (Currey 1993). As a consequence, the Sydney Water Board developed an effluent management strategy involving the systematic decommissioning of small, local sewage treatment plants and the diversion of effluent to a single, more efficient plant in the lower mountains (Anon., 1990).

Following STP closures, streams currently receiving effluent are expected to show a marked improvement in water quality and ecological health. The extent of this recovery will be a key measure of the effectiveness of the management strategy. As part of a recovery monitoring program, baseline samples of benthic macroinvertebrates were collected in spring 1992, using a rapid bioassessment technique (Chessman, 1994; Growns *et al.*, 1994).

The presence of large numbers of larval chironomids in benthic samples can constitute a major problem in biological monitoring, primarily due to difficulties of species identification. It is common for aquatic biologists to refer to chironomids at the family or subfamily level. In Australia, a further complication is that a substantial proportion of the chironomid fauna (numbering perhaps several hundred species) have not been formally described at any life history stage, and their ecological requirements are unknown.

In the Blue Mountains biological program, chironomid larvae were collected in a non-quantitative fashion along with other components of the macroinvertebrate fauna, and were identified to species level. This paper considers the characteristics of the chironomid fauna in relation to water quality, stream altitude and habitat.

Study area

The Blue Mountains are the portion of the Great Dividing Range to the west of Sydney extending between the Colo River in the north and the Wollondilly River in the south. Consisting of a rugged sandstone plateaux up to 1200 m in elevation, the mountains are drained by two main river systems, the Grose River in the north and Coxs River to the west and south.

To assess the response of Blue Mountains streams to the progressive removal of STP discharges, forty-four sampling sites were chosen, covering all STPs in the upper and lower mountains. Sites selected included sections of streams above and below STP discharge points, and in catchments in undisturbed bushland or affected by urban, industrial or agricultural development. Four categories of sites were distinguished according to influences on water quality: (1) immediately downstream of effluent discharges (n=12); (2) affected by disturbances other than STP discharges (n=9); (3) far downstream of effluent discharges (n=9) and (4) in pristine or near-pristine catchments (n=14).

Materials and methods

Site altitudes (Australian Height Datum) were read from 1: 25000 topographic maps. Macroinvertebrates were sampled by the rapid bioassessment technique described by Chessman (1994). Standardised samples were obtained at the forty-four sites from up to four habitat types depending on the number present at each site. Habitats were defined as riffles, pool edges/backwaters, rocks in pools and submerged wood.

Riffles and edges were sampled with a hand held dip net with 320 by 250mm opening and 250μm mesh. Riffles were sampled by disturbing substrata such as stones and sediments upstream of the net; stones were scrubbed by hands or turned by feet to dislodge attached benthic organisms, for a distance of approximately 10m including areas of maximum and moderate current. Stream edges and backwaters were sampled in areas of little or no current by sweeping the net from open water to the shore over a bank length of about 10m. Pool rocks and submerged wood were sampled by picking off macroinvertebrates with forceps from a number of the respective substrata. Submerged wood was broken open to look for burrowing animals when possible. For each habitat, either about 100 macroinvertebrates were collected, or sampling was terminated after 1 person hour if 100 animals had not been obtained by then. The intention was to collect as wide a range of taxa as possible within these constraints, and rare taxa were deliberately sought in preference to picking large numbers of specimens of common taxa. Generally, if a common species could be distinguished by eye, no more then ten specimens were collected.

All specimens were preserved in 70% ethanol and returned to the laboratory. Chironomid larvae were extracted from the samples, prepared and identified to the lowest taxonomic level possible according to Cranston (1991), or allocated to recognisable taxonomic units (voucher species).

The relationships between the abundances of the 20 most widely distributed chironomid species, altitude and the ordinal classifications of water quality (site category) were assessed using Spearman's partial rank-order correlation, after adjusting for the effect of habitat. In order to examine patterns of species richness and total abundance, a three-way factorial analysis of variance was performed on square-root transformed data using the factors water quality, altitude (divided into three zones of < 330m, 330-660m and > 660m) and habitat. Differences between the levels of significant factors were assessed using Tukey's method of multiple comparisons (Neter *et al.* 1985).

Results

From the 44 localities sampled, 60 species or putative species of Chironomidae were identified from 3929 larvae (Table 1). The majority belonged to the subfamilies Tanypodinae, Orthocladiinae and Chironominae with species of Diamesinae and Podonominae restricted in both distribution and numbers, constituting less then 2% of larvae collected. Aphroteniinae were not collected although they are known to occur locally (Wright, 1992).

Undisturbed sites contained a suite of species that were often absent or reduced in abundance at impacted sites. The latter were numerically dominated by a few species that, although not restricted to such sites, were usually less abundant elsewhere (Table 2). The taxa most common at the impacted sites varied with altitude. For

Table 1. Chironomidae (Diptera) voucher species collected from 44 localities in the Blue Mountains, spring 1992. Codes include Sydney Water Board (SWB), reference numbers and those used by Dr P. Cranston, CSIRO Division of Entomology, Canberra.

Subfamily Tanypodinae

Ablabesmyia notabilis Skuse
Ablabesmyia hilli Freeman
Apsectrotanypus ?maculosus ("SWB 1")
Fittkauimyia disparipes Karunakaran
Larsia ?albiceps
Paramerina ?levidensis ("SWB 1")
Pentaneura "SWB 1"
Procladius "SWB 1"
Pentaneurini genus A ("SWB 1")
Pentaneurini genus D ("SWB 1")
Pentaneurini genus ST1 ("SWB 1")

Subfamily Orthocladiinae

Cardiocladius australiensis Freeman
Corynoneura "SWB 1"
Cricotopus "hillmani"
Cricotopus spp. ("SWB 1")
Eukiefferiella insolida Skuse
Limnophyes "SWB 1"
Psectrocladius "SWB 1"
Parakiefferiella "SWB 1"
Parametriocnemus "SWB 1"
Paratrichocladius "SWB 1"
Rheocricotopus "SWB 1"
Stictocladius "SWB 1"
Thienemanniella "SWB 1"
Orthocladiinae *"Grape th"* ("SWB 5")
Orthocladiinae "*MO5*" ("SWB 6")
Orthocladiinae "SWB 9"
Orthocladiinae "*SO1*" ("SWB 10")
Orthocladiinae "*Wood miner*" ("SWB 13")

Subfamily Diamesinae

Paraheptagyia tonnoiri Freeman

Subfamily Chironominae

Chironomus "SWB 1"
Cladotanytarsus "SWB 1"
Cryptochironomus "SWB 1"
Dicrotendipes cumberlandensis Epler
Dicrotendipes "SWB 1"
Dicrotendipes "SWB 2"
Harnischia "SWB 1"
Imparipecten pictipes Freeman
Parachironomus "SWB 1"
Paracladopelma "SWB 1"
Paratanytarsus "SWB 1"
Polypedilum "*M2*" ("SWB 4")
Polypedilum "*B1*" ("SWB 5")
Polypedilum oresitrophus Skuse
Polypedilum prasiogaster Freeman
Polypedilum watsoni Freeman
Polypedilum (Pentapedilum) "*K1*" ("SWB 6")
Polypedilum "SWB 7"
Rheotanytarsus "SWB 1"
Riethia stictoptera Kieffer
Riethia zeylandica Freeman
Skusella "SWB 1"
Stenochironomus ?watsoni ("SWB 1")
Tanytarsus "SWB 1"
Tanytarsus "SWB 2"
Tanytarsus "SWB 3"
Chironominae "SWB 6"

Subfamily Podonominae

Podonominae "SWB 1"
Parochlus ?bassianus ("SWB 1")
Podonomopsis ?evansi ("SWB 1")

example, such sites located in the upper mountains were predominantly populated by larvae of *Cricotopus* "SWB 1", *Eukiefferiella insolita* (Skuse), *Paratrichocladius* "SWB 1" and in one instance, *Limnophyes* "SWB 1" and Orthocladiinae SWB 6. In the lower mountains, Chironominae species such as *Chironomus* "SWB 1", *Dicrotendipes* "SWB 1" and *Polypedilum (Pentapedilum)* "SWB 1" were more common (Table 2)

Excluding uncommon taxa (fewer than 20 individuals), no larvae were restricted to a single habitat. However, some species were never recorded from a particular habitat category (Table 2).

The number of chironomid species per sample ranged from 0 to 11. Significantly fewer species per sample were

Table 2. Numbers of the 20 most widely distributed chironomid species (found at a minimum of 9 localities) collected in different water quality and altitudinal zones and habitats. Water quality categories are coded as 1: immediately downstream of Sewage Treatment Plant (STP) effluent discharges; 2: affected by disturbances other than STP discharges; 3: far downstream of STP discharges and 4: in pristine or near-pristine catchments. a: significant altitudinal trend; w: significant water quality trend (P<0.05).

Species	Water quality				Altitude			Habitat			
	1	2	3	4	<330m	330m to 660m	>660m	Riffle	Edge	Rock	Wood
Subfamily Tanypodinae											
Ablabesmyia notabilis	0	11	28	5	15	19	10	2	31	7	4
Apsectrotanypus "SWB 1"[a]	10	35	14	42	6	10	85	2	92	7	0
Paramerina "SWB 1"	6	40	30	53	40	69	20	22	80	19	8
Pentaneura "SWB 1"	1	1	4	11	1	13	3	5	9	2	1
Procladius "SWB 1"[a]	15	23	27	4	64	3	2	4	62	0	3
Pentaneurini genus X "SWB 1"[a]	48	3	2	14	2	12	53	20	32	12	3
Subfamily Orthocladiinae											
Cardiocladius australiensis	20	47	70	0	24	47	66	89	3	26	19
Cricotopus "hillmani"	2	1	9	8	1	16	3	15	0	0	5
Cricotopus "SWB 1"[aw]	349	67	1	3	30	110	280	125	51	115	129
Eukiefferiella insolida[aw]	449	20	27	13	20	112	377	195	16	145	153
Paratrichocladius "SWB 1"[w]	97	245	272	38	276	170	206	139	135	217	161
Psectrocladius "SWB 1"[aw]	0	1	1	67	0	63	6	1	60	8	0
Thienemanniella "SWB 1"	5	3	38	3	26	18	5	1	16	11	21
Orthocladiinae "SWB 6"[a]	11	0	4	16	2	13	16	10	11	8	2
Subfamily Chironominae											
Chironomus "SWB 1"[aw]	68	94	35	7	96	29	79	19	147	29	9
Dicrotendipes "SWB 1"[aw]	277	79	11	3	364	3	3	47	59	206	58
Polypedilum "SWB 4"	4	31	43	11	35	30	24	4	21	48	16
P. (Pentapedilum) "K1" ("SWB 6")[w]	134	56	13	4	103	17	77	47	49	61	40
Rheotanytarsus "SWB 1"	37	6	30	48	41	16	64	64	21	22	14
Tanytarsus "SWB 3"	3	65	11	24	38	20	45	11	58	10	24

Table 3. Summary of ANOVA results for number of chironomid species and individuals per sample.

Source of variation	d.f.	Species		Individuals	
		F-value	Prob. (>F)	F-value	Prob. (>F)
Main effects					
Water quality	3	4.7	0.004	19	0.0001
Altitude	2	4.7	0.012	5.6	0.005
Habitat	3	3.7	0.015	0.59	0.622
Interactions					
Water quality x altitude	6	3.8	0.002	1.3	0.278
Water quality x habitat	9	0.78	0.633	2.3	0.024
Altitude x habitat	6	0.38	0.891	0.86	0.529
Water quality x altitude x habitat	15	0.56	0.900	0.85	0.626

collected from pristine and near-pristine sites than from sites immediately downstream of other disturbances than sewage effluent discharges (Tables 3 and 4). The species richness of sites downstream from points of effluent discharge did not differ significantly from that of undisturbed sites. Species richness also varied significantly between habitats (Table 3), with more species collected from pool edges than from pool rocks or wood (Table 4).

Table 4. Mean number of species and 95% confidence intervals for each water quality zone and habitat (numbers back transformed from square root transformed data). Levels which were not significantly different (Tukey's Least Significant Difference) are indicated by vertical bars.

Category	Mean No. of species	95% Confidence Limits	
		Lower	Upper
Water quality			
Near-pristine	3.9	3.3	4.7
Far below STP	4.6	3.9	5.5
Immediately below STP	5.3	4.4	6.2
Other disturbance	6.4	5.3	7.5
Habitat			
Wood	3.9	3.1	4.8
Pool rocks	4.5	3.7	5.4
Riffles	4.7	4.0	5.5
Pool edges	6.1	5.3	7.1

Table 5. Mean number of individuals and 95% confidence intervals for each water quality and altitudinal zone (numbers back-transformed from squareroot transformed data). Levels which were not significantly different (Tukey's Least Significant Difference) are indicated by vertical bars.

Category	Mean No. of individuals	95% Confidence Limits	
		Lower	Upper
Water quality			
Immediately below STP	49	39	59
Other disturbance	32	24	41
Far below STP	19	13	25
Near-pristine	9	5	13
Altitude			
<330m	24	19	29
330-660m	13	9	17
>660m	38	31	46

Chironomid abundance (between 0 and 104 specimens per sample) varied significantly according to water quality category (Table 3). Sites affected by urban development and immediately downstream of STPs yielded similar numbers of individuals (Table 5). The numbers of specimens collected were much lower far downstream from STP effluent discharges, and at undisturbed sites. Although altitude was significantly associated with total abundance (Table 3), there was no clear altitudinal trend in numbers (Table 5). This was probably due to the unbalanced distribution of sites among water quality and altitudinal categories, as the middle altitudes had very few sites immediately downstream of STPs or affected by other disturbances.

DISCUSSION

The patterns in the distribution and abundance of the chironomid larval fauna apparent in relation to water quality, altitude and habitat would have been influenced by the rapid assessment technique that was used. At sites containing a diverse macroinvertebrate fauna, midge larvae would have constituted a small proportion of the sample even if they were abundant, because of the imposed limit of about 100 macroinvertebrates collected in total. Sites with an impoverished macroinvertebrate fauna would have yielded samples with higher proportions of chironomids because certain chironomid species are among the most pollution-tolerant of all stream macroinvertebrates (Hynes, 1960). The proportion of chironomids in such rapid bioassessement samples may therefore indicate ecosystem degradation in a similar manner to the index proposed by Winner *et al.* (1980) for metal pollution.

Similar considerations may have affected estimates of species richness; the apparent greater richness at disturbed sites (Table 4) may have been at least partly a result of more specimens being collected at those sites (Table 5). However, some other studies have found either richer chironomid assemblages at impacted localities (Lenat, 1983; Hardwick, 1992) or at least no obvious reduction in richness (Smith and Cranston, 1995). This cautions against the undiscriminating use of chironomid species richness as an indicator in pollution studies. Lenat (1983) noted that mild pollution sources, stream size and season may cause discrepancies. In the Blue Mountains, factors such as diffuse sources of contamination in the studied streams and variable dilution of effluent by natural streamflow may also be linked to the higher species richness found below discharge points.

The prevalence of pollution-tolerant species at the most disturbed sites was obvious. Orthocladiinae species were often numerically dominant at such sites, particularly in the upper mountains where *Cricotopus* "SWB 1" and *Eukiefferiella insolida* were prevalent. Only one common species, identified as "*Psectrocladius*", indicated sensitivity to water quality impacts (Table 2). In the lower mountains, although Orthocladiinae species were present, Chironominae species such as *Chironomus* "SWB 1", *Dicrotendipes* "SWB 1" and *Polypedilum (Pentapedilum)* "SWB 6" were found to be more prevalent at many of the disturbed sites. However Wright (1992), in a detailed quantitative study of Wentworth Falls STP, in the upper mountains, found the chironomid fauna to be dominated by the Chironominae. His sampling was quarterly, in contrast to the spring sampling for the present paper. It is possible that the dominant chironomid subfamily at organically polluted sites in the Blue Mountains flunctuates seasonally, with the largely cold adapted Orthocladiinae being replaced by Chironominae species in the warmer months. In discussing lentic waterways, Edward (1986) noted that marked seasonal changes in the composition of chironomid faunas in eastern Australia have not yet been clearly demonstrated. Examination of autumn macroinvertebrate samples which were collected from the Blue Mountains in 1993 will allow further investigation of this question.

Species richness varied significantly between habitats, with pool edges having the greatest diversity (Table 3), and several chironomid species were more frequently encountered in a particular habitat (Table 2). The variability between substrata supports the need to sample as many habitats as possible at a site in order to maximise the probability of collecting habitat-specific taxa. This study shows that for bioassessment programs, family-level identification for Chironomidae masks much useful information (cf. Armitage and Blackburn, 1985). Even within genera the constituent species often vary in their responses to different levels and types of pollution (Resh and Unzicker, 1975). The use of recognisable taxonomic units to distinguish difficult taxa in conjunction with a voucher system as suggested by Cranston (1990) has reduced taxonomic difficulties, allowing the inclusion of Chironomidae in species-level studies.

Acknowledgements

We are indebted to Dr. P.S. Cranston for assistance in confirming chironomid identifications, Dr. J. Growns for advice on analysis and the manuscript, and members of the AWT Stream Studies Section for their field work.

"Recovery" of an Acid Mine-Waste Impacted Tropical Stream - The Chironomid Story

Mick Smith and Peter S. Cranston[1]

Division of Botany and Zoology, Australian National University,

and [1]CSIRO Division of Entomology, GPO Box 1700, Canberra, ACT 2601, Australia

Introduction

The introduction of acidic and heavy metal-rich waters from mine workings is a widespread source of aquatic pollution (Glover, 1973; Ramsey and Brannon, 1988). The main detrimental effects of such pollutants on aquatic invertebrates are adverse changes in ionic balance, depressed respiratory efficiency (Winterbourne and Collier, 1987; Hellawell, 1989) and the toxic effects of high concentrations of heavy metals (Wiederholm, 1984).

In Australia, although current legislation limits such discharges, a legacy of past mining with minimal environmental controls has left many polluting adits.

This paper reports changes in the benthic chironomid larval assemblage (as part of a more extensive macroinvertebrate study) of a monsoonal tropical stream following removal (by downstream diversion) of acidic mine effluent. The receiving stream flows seasonally, from the onset of the Wet season (usually December) until about two months into the Dry (usually late May/June) and for some 3 decades it has received continuous drainage from an abandoned uranium mine. Prior to the 1992/3 Wet season, a "recovery" section was created by diversion of the mine effluent downstream.

During a six-week period during the Wet-Dry transition of 1993, 109 species of macroinvertebrates were sampled, of which 36 species belonged to the Chironomidae. We examined the assemblage of Chironomidae at species level, and in relation to the total community, to investigate the reported high species diversity but lack of any species dominance in acid-impacted sections of this tropical stream (Hardwick, 1993). The findings conflict with those reported from several northern hemisphere studies on acidified streams (*e.g.* Simpson, 1983; Zischke *et al.*, 1983). We assessed the utility of the Chironomidae as bioindicators by comparing the results of analyses of chironomid data alone with those derived from the whole macroinvertebrate community of the stream (Smith, 1994). These latter results will be presented elsewhere.

Methods and Materials

Field work was undertaken in the Northern Territory in Rockhole Mine Creek (RMC) (13°30'S 132°30'E) (Fig. 1). This softwater stream arises on an outlier of the sandstone plateau of western Arnhem Land. Detailed descriptions of the geography, climate and major characteristics of the regional streams may be found in Humphrey *et al.* (1990).

RMC is a small, narrow, second-order tributary of the upper South Alligator River, and has a substrate of mainly cobbles and boulders. Monsoon rainforest lines the banks and shades the stream for much of its length. The lower sections of RMC receive an inflow of acidic, heavy-metal rich water, draining from an adjacent abandoned uranium mine adit. Associated with this drainage is an orange-brown ferruginous "floc" which blankets the stream bed (ARRRI, 1989).

The benthic macroinvertebrate community of a reach of RMC has been sampled during the Wet-Dry transition period for three years (1991 to 1993), in 1991 and 1992 by staff of the Alligator Rivers Region Research Institute (Dostine *et al.*, 1992, 1993). Prior to the first sampling event in 1991, the creek was labelled as three sections, A, B and C, with sections A and B located upstream from the usual entry of mine effluent, and section C downstream of this point (Fig. 1).

During 1991, the creek was sampled with the mine effluent entering at its usual point above section C. After two weeks of sampling in 1992, the mine effluent was diverted to enter the stream immediately above section B, and hence exposing section B to effluent for the remainder of that season (Faith *et al.*, in press). In 1993, a downstream diversion was performed so that section C (abbreviated to C93), the section that had been press-polluted for some thirty years, would no longer receive mine effluent and thus would have the opportunity to recover. Two additional sections (D and E) were located downstream of the diversion in 1993 to monitor the polluted sections of the creek. In 1993, creek sections A, B and C were studied in detail and D and E less closely.

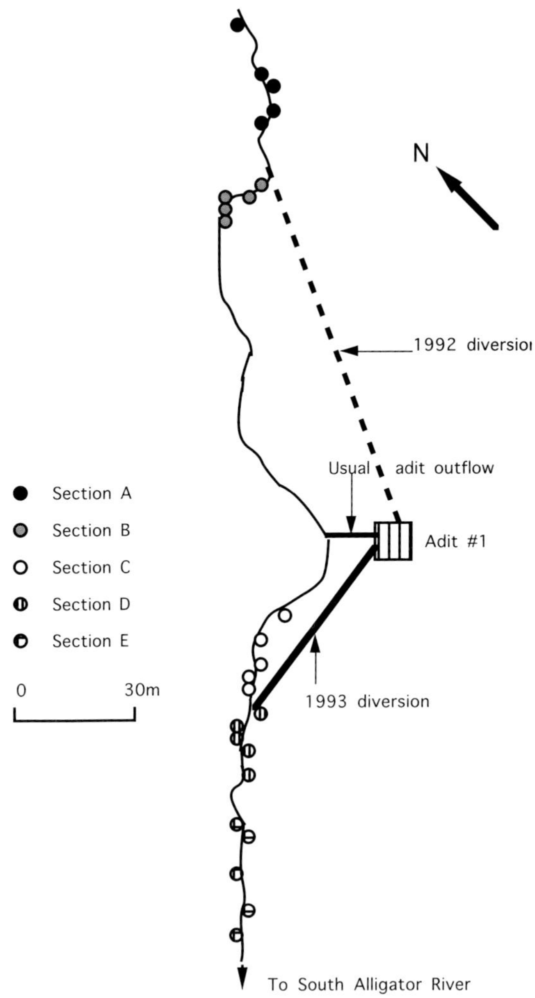

Figure 1. Schematic diagram of Rockhole Mine Creek (RMC) showing location of sections and sites. Creek flows from top to bottom.

During 1993, the sampling period, as in previous years, commenced when the site became accessible and continued until the cessation of creek flow. The duration of this period varied between years: 6 weeks in 1991, 10 weeks in 1992 and 9 weeks in 1993. Comparable sampling weeks (as in Figs. 6-9) have been calculated retrospectively, counting backwards from cessation of flow. For consistency, the five sites within each section (Fig. 1) were the same as those designated by previous workers (Dostine *et al.*, 1992). As in previous years, macroinvertebrates were collected using 16cm long x 9.5cm diameter artificial substrate cylinders, made from plastic mesh filled with gravel pieces of 2-3cm diameter, and placed with the maximum surface area perpendicular to stream flow. Paired replicates of these substrates were placed at three of the five sites in each section. Exposure was for one week, the period calculated by Faith (1990) to be optimal for colonization of substrates (*i.e.* fully colonised, where taxonomic diversity and abundance peak prior to subsequent slow decline). Artificial substrates were recovered, commencing at the most downstream site, by advancing a 250μm net from the downstream direction until the artificial substrate was surrounded, then the sample and net were withdrawn into a 20 L bucket. Elutriation with 5 washes into a 500μm sieve removed macroinvertebrates which were preserved in 70% ethanol for subsequent laboratory identification.

Chironomidae were slide mounted in Hoyer's medium (for speed) or permanently prepared (vouchers, pharate larvae, pupae etc) in Euparal. Identifications using an Olympus CH2 phase contrast microscope followed the key of Cranston (1991). Chironomidae data from 1991 and 1992 were made available by the Office of the Supervising Scientist (identified by P. Dostine and P.S. Cranston). Identifications of other macroinvertebrates followed Wells (1991; Trichoptera), Suter (1992; Ephemeroptera), Hawking (1993, Odonata) and Williams (1980; all other taxa except Chironomidae).

Data were analysed using PATN software (Belbin, 1991). This package incorporates a form of hybrid Multi Dimensional Scaling to derive matrices of dissimilarity, where two identical data points receive a value of zero, and two totally dissimilar points receive a score of one. The Bray-Curtis measure of dissimilarity was used because it is capable of depicting associations without being greatly influenced by noise or biases in the data (Faith *et al.*, 1991). Community structure data (*i.e.* abundance and species richness) were analysed following the procedure of Belbin (1991). Dendrograms and ordinations were used to illustrate the results. The stress value associated with each ordination indicates how well distances displayed in the ordination match their corresponding dissimilarity data, with a stress value below 0.2 usually regarded as being low (Faith *pers. comm.*).

Surface water samples from each section were taken weekly to assess pH, conductivity and alkalinity and each second week to measure heavy metals. All samples were taken between 1300h and 1500h and were transported on ice, then stored overnight at 4°C before analysis next day at 25°C (room temperature). Conductivity was measured with a Metrohm, Model E518 Conductivity Meter and pH with an Orion, Model 720A pH meter. Concentrations of heavy metals were determined using an Iron Coupled Plasma - Mass Spectrophotometer.

RESULTS

Abundance and species richness

During the 1993 sampling period more than 20,000 macroinvertebrates, belonging to 109 different taxonomic categories, were collected, sorted and identified. The number of benthic macroinvertebrates in section C was markedly higher in 1993 (4277) than in 1992 (2610) or 1991 (1785). There were 658, 1486 and 746 chironomids collected from section C for the same three years respectively, and hence, the chironomids represented 15, 57 and 42% of the entire fauna.

Whilst chironomid abundance in section C varied somewhat, the species richness varied little, with 24, 28 and 26 species being recorded from C91, C92 and C93 respectively. In 1993 there were 12 species of chironomid unique to C93 and although 11 of these were rare taxa (< 5 animals), the other, *Fittkauimyia disparipes* (Karunakaran), was reasonably abundant (22 animals).

Dissimilarity dendrograms

The relationships between three different sections (A, B and C) for the three years from 1991 to 1993 are displayed in dendrograms (Fig. 2). The dissimilarity dendrogram derived using all macroinvertebrate taxa showed two main groups comprising uncontaminated (top five) and contaminated (bottom four) sections (Fig. 2a). Within the group of uncontaminated sections there is greater similarity between different sections in any given year, than the same section in different years; *i.e.* the annual effect is more powerful in determining community structure than the seasonal effect. This is not the case with the group comprised of contaminated or previously contaminated (C93) sections. The lower part of Figure 2a shows that the section under scrutiny for evidence of recovery (C93) was well distanced from C91 and C92, and from other sections sampled in 1993; it resembled the manipulated section B92 more than any uncontaminated section. The position of C93 cannot be clearly attributed either to sectional or annual influence, and therefore may indicate recovery. Also of interest (Fig. 2a) is the confirmation of recovery of section B. During 1992 when section B was exposed to the pollutant, its macroinvertebrate community was more typical of a contaminated section than an uncontaminated one. However, in 1993 the macroinvertebrate community of section B has reverted to being more like that of other uncontaminated sections.

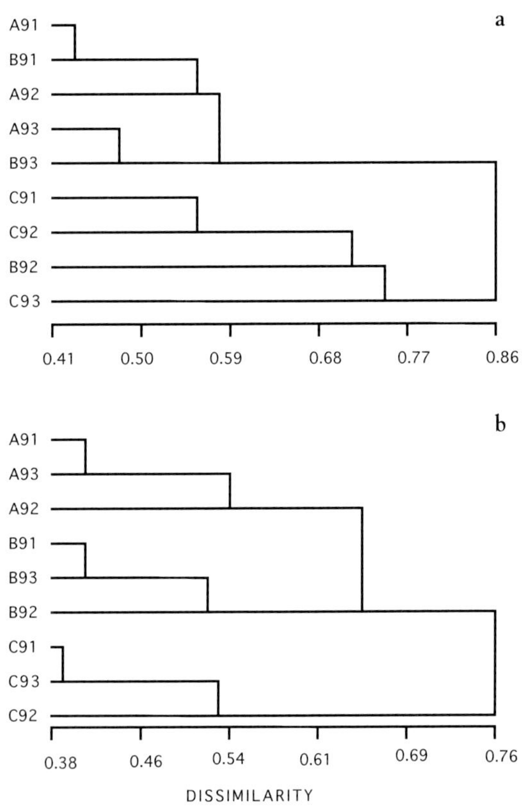

Figure 2. Dissimilarity dendrogram showing the associations between sections A, B, and C of RMC based on comparable sampling over the three years 1991 to 1993 for: (a) the entire macroinvertebrate fauna, and (b) the chironomid assemblages.

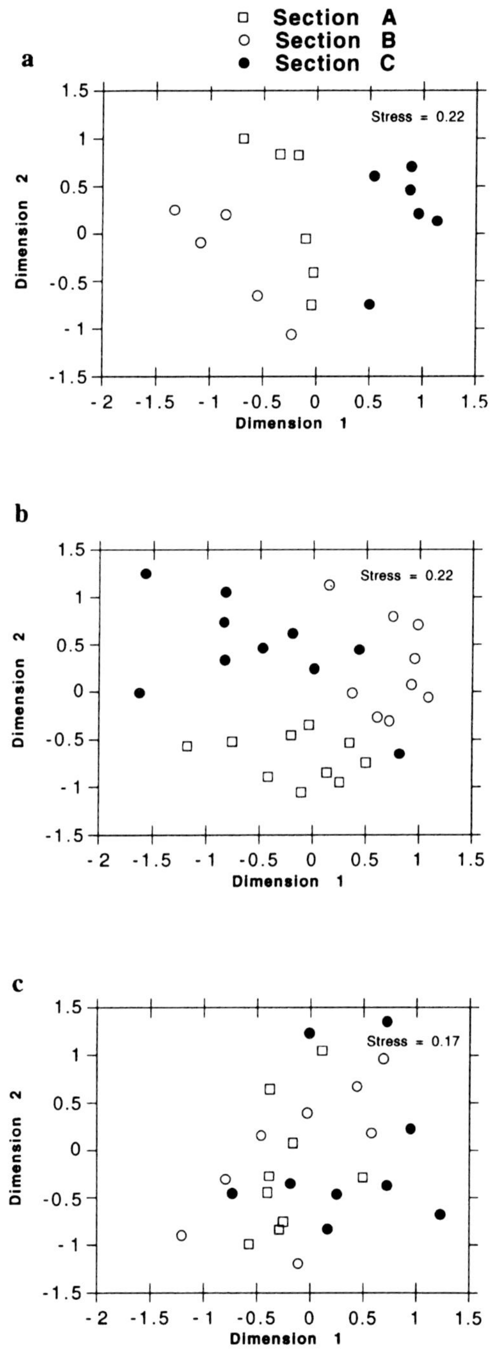

Figure 3. Ordination of Bray-Curtis dissimilarities of all chironomid data for (a) 1991, (b) 1992, and (c) 1993.

The dendrogram based on the chironomid assemblages for the same three sections for the same three years (Fig. 2b) also shows two major groups, but the two groups differ in their components from those based on the entire benthic macroinvertebrate fauna. Chironomid assemblages from any given section of the creek are more similar to one another from year to year than they are to an assemblage from different sections within the same year: the sectional (spatial) effect is more influential than the annual (temporal) effect. The chironomid assemblage of C93, which closely resembles those of long-term (press) polluted sections of the creek (C91 and C92), cannot be taken as indicative of recovery.

Ordinations of dissimilarity

Ordinations of the chironomid data for each year from 1991 to 1993 (Fig. 3) show two notable trends. Firstly, it is evident that the symbols representing chironomid assemblages from any one section cluster more closely with one another than they do with symbols representing either of the other sections. Secondly, in 1991 and 1992, the clusters of points representing the chironomid assemblages of section C cluster together some distance from the symbols depicting the other sections. This is most evident in the 1991 ordination (Fig. 3a) where all the polluted section C data points fall in the positive region of the X axis, whereas all symbols depicting assemblages from pristine sections A and B fall in the negative region of the same axis. In 1992 (Fig. 3b) points representing the chironomid assemblage of the newly-impacted section B now are more proximate to those of polluted section C than of section A. This confirms the trends shown in Faith *et al.* (in press) concerning the total community response to the manipulation. In 1993 (Fig. 3c) no separate clusters are evident, with an intermingling of data points in the ordination space. Such intermingling might be taken as evidence of recovery, but data points from polluted sections D or E would be required to allow any such interpretation.

Because of the experimental manipulation of section B in 1992, and the unusually low rainfall, this year's data have been excluded from the data combined in the same ordination space (Fig. 4) to look for signs of recovery and to highlight the different trends seen when examining the entire macroinvertebrate fauna as opposed to examining the benthic chironomid assemblage alone. These comparisons show that the entire benthic macroinvertebrate community of C93 was more similar to its annual counterparts (A93 and B93) than C91 was to A91 or B91. The benthic macroinvertebrate communities of C93 are again seen to be intermediate in dissimilarity as the symbols depicting their position lie between the uncontaminated and contaminated sections. This, however, is not the case in the ordination of the chironomid assemblages with the chironomids again showing a confusing pattern differing from that of the entire benthic invertebrate fauna.

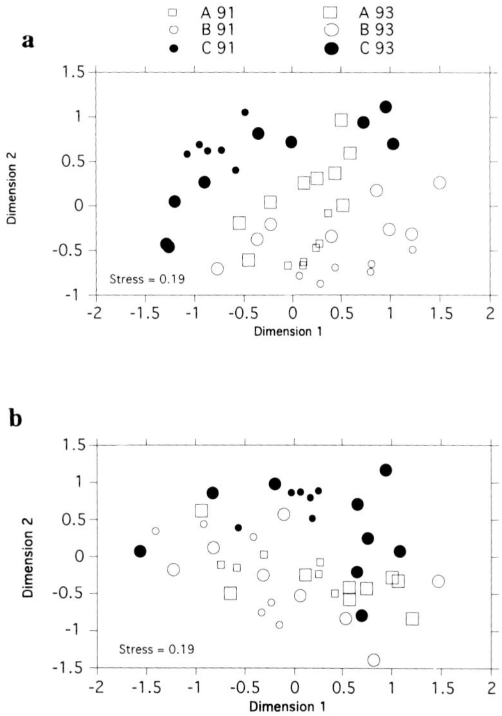

Figure 4. Ordination depicting weekly data points for sections A and C, during 1991 and 1993, based on the community structure of (a) the entire macroinvertebrate fauna, and (b) the chironomid assemblages.

Change in the mean dissimilarities

The pattern of dissimilarity between two different sections also may be compared by plotting their mean dissimilarity values over time. Section A is thus compared with section C, with the entire macroinvertebrate fauna (Fig. 5a) and the chironomid assemblages (Fig. 5b) illustrated [note that the regression lines in Fig. 5a, b represent the mean dissimilarities between sections].

In 1991 and 1992, when section A was uncontaminated and section C was contaminated, the mean dissimilarity between the entire benthic macroinvertebrate communities of these sections (Fig. 5a) increased as the season progressed; in both years the dissimilarity increased at approximately the same gradient. In 1993, however, the mean dissimilarity between sections A and C remains constant for the entire season, indicating that the dissimilarity between the macroinvertebrate communities of these two sections remained constant.

Regression lines of the mean dissimilarities between chironomid assemblages of sections A and C (Fig. 5b) show no consistent pattern prior to the diversion of the mine drainage, nor a constant level of dissimilarity between the two sections. Indeed, the mean dissimilarities for the chironomids vary considerably relative to those derived from the entire benthic fauna, and hence, interpretation is difficult. The step-like change

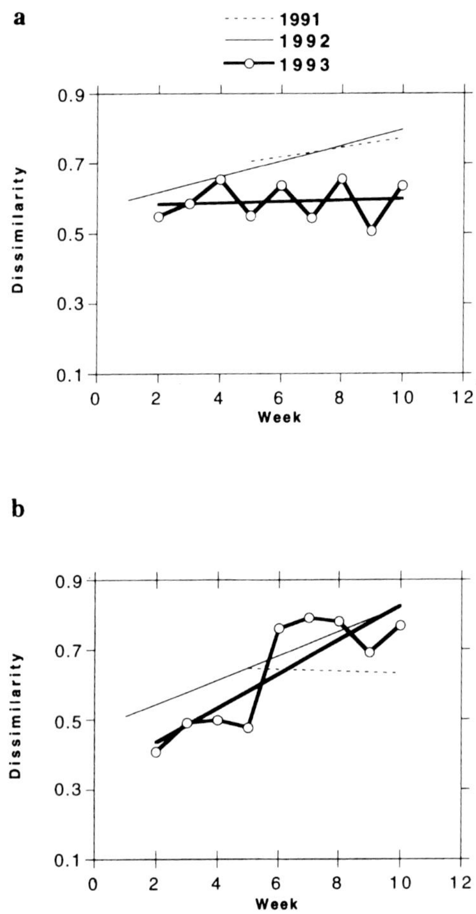

Figure 5. Regression lines of the mean dissimilarities between sections A and C from 1991 to 1993 for (a) the entire macroinvertebrate fauna, and (b) the chironomid assemblages.

between weeks 5 and 6 may indicate that the linear reduction in flow and pH has passed a threshold of physiological significance to some of the community, but we have no way of verifying this.

Changes in abundances of individual species

Abundances of several species of chironomid highlight the different responses to the mine drainage. The abundance of the *Nilotanypus* species (Fig. 6) is far greater in section C during 1993 than in previous years.

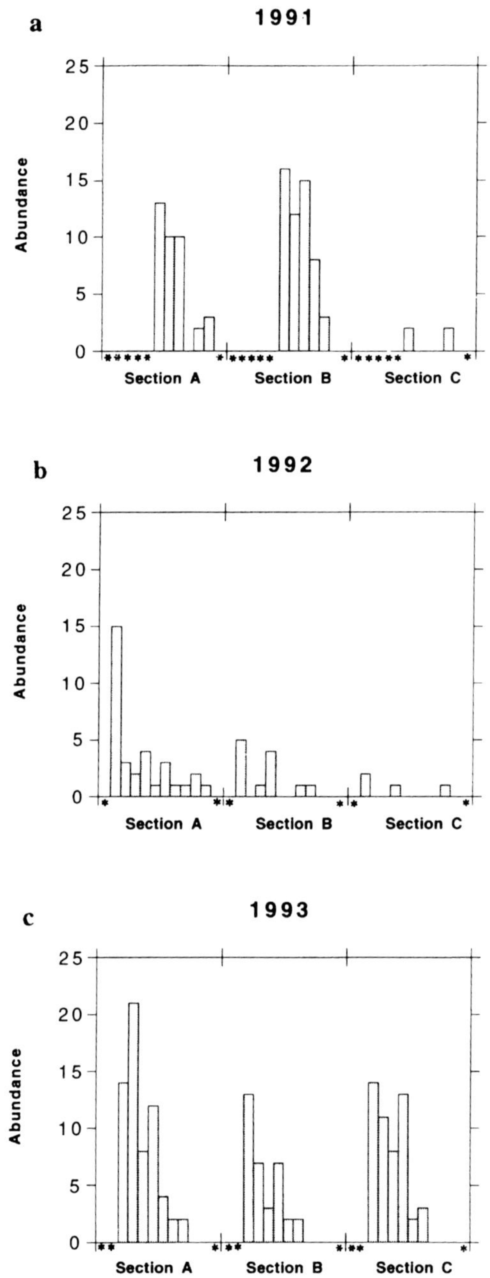

Figure 6. Abundances of *Nilotanypus* sp. in sections A to C of RMC (for the last weeks of flow) for (a) 1991, (b) 1992, and (c) 1993. Each column within each section represents the 3 samples summed per week; * represents unsampled weeks.

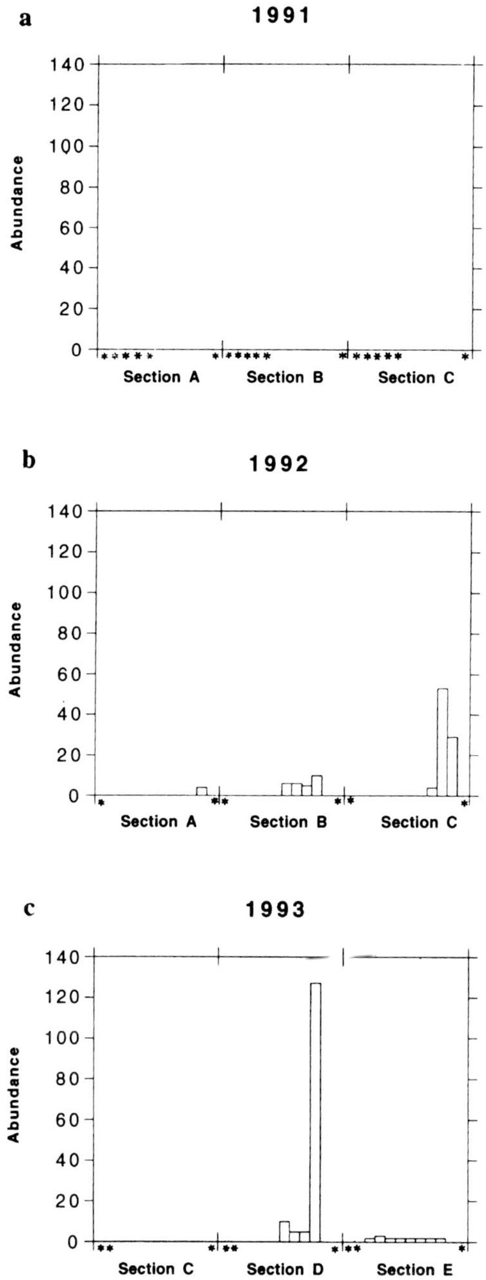

Figure 7. Abundances of *Polypedilum convexum* (for the last weeks of flow) for (a) sections A to C of RMC in 1991, (b) sections A to C of RMC in 1992, and (c) sections C to E of RMC in 1993. Each column within each section represents the 3 samples summed per week; * represents unsampled weeks.

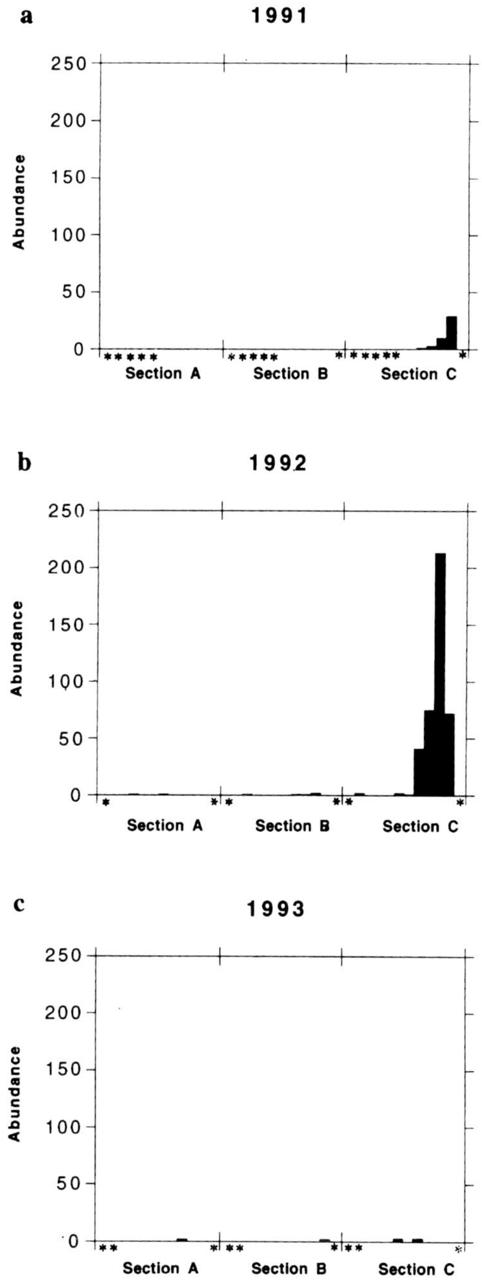

Figure 8. Abundances of *Tanytarsus* "K2" in sections A to C of RMC (for the last weeks of flow) for (a) 1991, (b) 1992, and (c) 1993. Each column within each section represents the 3 samples summed per week; * represents unsampled weeks.

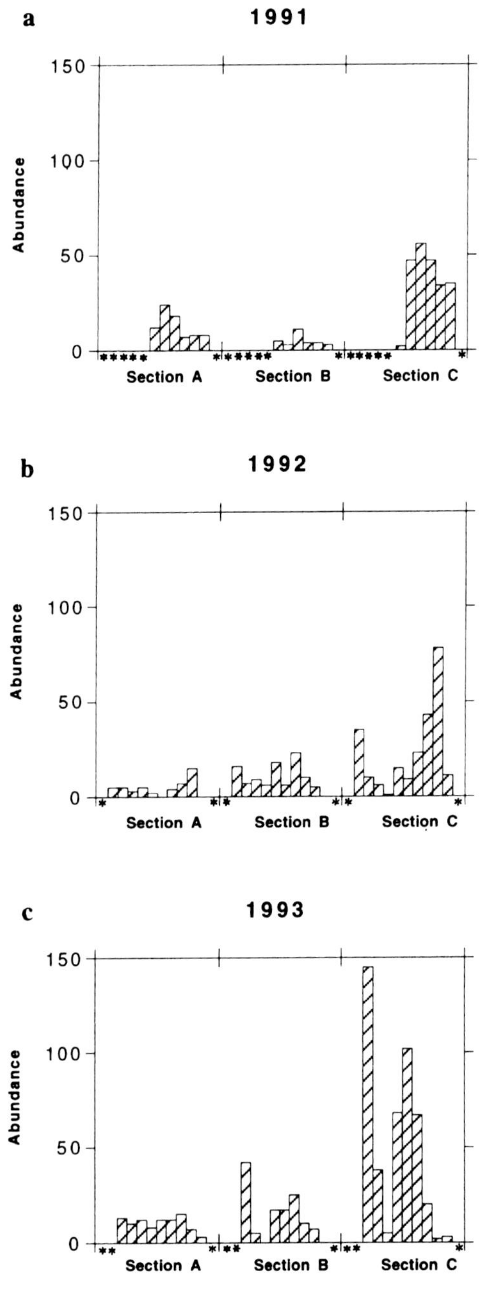

Figure 9. Abundances of *Tanytarsus* "K3" in sections A to C of RMC (for the last weeks of flow) for (a) 1991, (b) 1992, and (c) 1993. Each column within each section represents the 3 samples summed per week; * represents unsampled weeks.

In 1991 and 1992 *Nilotanypus sp.* was rare in section C yet in 1993 it is as common in this section as in sections A and B.

A very different pattern is shown by *Polypedilum (Pentapedilum) convexum* (Johannsen) (Fig. 7). This species was not present during 1991, and in 1992 it occurred only late in the season, predominantly in effluent-exposed sections B and C. In 1993 *Polypedilum convexum* was absent from sections A, B and C and was found only in downstream sections into which the mine effluent was diverted.

Histograms of the abundances of two other species, *Tanytarsus* "K2" (Fig. 8) and *Tanytarsus* "K3" (Fig. 9) illustrate the different responses of two species which are very similar morphologically. *Tanytarsus* "K2" shows a pattern of abundance similar to that of *Polypedilum convexum*; it was present in section C when this section was receiving mine drainage but in 1993, following the diversion of the adit water, numbers were reduced in section C. In contrast, *Tanytarsus* "K3" was present in all sections during all three years: it was more abundant in section C than in other sections during 1991 and 1992, and it occurs in section C in greater numbers in 1993 following the downstream diversion of mine effluent.

Water chemistry

A summary of general water chemistry results is given (Table 1). Available data for C91 have been included in this table to allow comparisons, and hence, comment on recovery. Results in Table 1 reveal the following trends:

- the pH of C93 (measured at a single point) is slightly acid but consistent throughout the season and is much higher than it was in previous years (C91) when the water pH approached the value found in E93;
- the conductivity of C93 (again measured at a single point) increased five-fold as the season progressed, whereas in 1991 the conductivity increased more than fifteen-fold;
- dissolved oxygen concentrations decreased somewhat, in both pristine and polluted areas of RMC; and
- the volume of water being discharged by RMC declined dramatically as the season progressed, whereas the volume of water draining from the mine adit remained constant.

Table 1. Summary of RMC general water chemistry results for 1993 and some data for section C from the final sampling occasion in 1991.

Variable	Location	Wk -3	Wk0	Wk 3	Wk 6	Wk 6 C91
pH	Section A	5.7	5.7	6.0	6.0	6.0
	Section C	5.6	5.5	5.7	5.7	3.9
	Section E	5.3	4.5	4.0	3.7	NA
	Adit	2.9	2.9	2.9	2.9	NA
Conductivity (µS/cm)	Section A	10.1	13.5	13.6	15.7	16.1
	Section C	10.4	16.6	28.2	51.6	185.0
	Section E	14.0	38.5	116.0	180.0	NA
	Adit	920.0	845.0	780.0	760.0	NA
Dissolved Oxygen (mg/L)	Pristine	9.4	8.0	4.8	NA	6.7
	Polluted	8.6	7.8	6.2	NA	6.4
Discharge (L/sec)	RMC	133.0	99.0	8.0	*	*
	Adit	NA	0.3	0.3	0.3	NA

NA - result not available, * - no flow detected.

The concentrations of a series of heavy metals were determined also, and a summary of these results is given in Table 2. From this table it can seen that the concentrations of Fe, Mn and U in E93 are all in excess of the Australian and New Zealand Environment and Conservation Council (ANZECC, 1992) guidelines for the "safe" concentrations of these metals. The concentrations of all three elements are much lower in C93 than in E93. There are no data available for the concentrations of metals in section C during 1991 and 1992, however, it is believed that the concentrations in these sections were very similar to those of E93.

Table 2. Maximum concentrations of some heavy metals in two section of RMC during 1993. The (ANZECC, 1992) guidelines for the "safe" concentrations of these elements is also shown.

Element	ANZECC "safe" concentration	Concentration in C93 (µg/L)	Concentration in E93 (µg/L)
Fe	300	1700	3400
Mn	100	240	500
U	0	120	450

DISCUSSION

The chironomid communities of Rockhole Mine Creek

In contrast to studies on acid lakes, reports on acid and heavy metal-impacted streams rarely have identified the chironomids beyond family level (Lindegaard, 1994). Chironomids are often found to be tolerant to these pollutants (e.g. McKnight and Feder, 1984) but often it is unclear whether the chironomids belong to mixed communities or one or few tolerant species (Yasuno *et al.*, 1985). From family-level abundance data, an increase in the proportion of Chironomidae to other aquatic taxa is evident and ratios derived from such observations have been suggested as indicative of acid and heavy-metal stress (Winner *et al.*, 1980). Certain studies of acidified streams in the temperate northern hemisphere have identified chironomids to an approximate generic level (e.g. Winner *et al.* 1980; Sheehan and Knight, 1985) and others to species level (e.g. Simpson, 1983; Zischke *et al.*, 1983; Yasuno *et al.*, 1985). From these more detailed studies, further generalisations are made concerning the effects of increasing stream acidity on Chironomidae: (i) species richness decreases; (ii) at moderate acidity, a few tolerant species become abundant; (iii) deposition of hydrous metal oxides ("floc") smothers periphyton, increases oligotrophication and reduces grazing Orthocladiinae and Diamesinae (Lindegaard, 1994).

Our studies on RMC confirm the supposed relationship between acidity and total chironomid abundance. Thus the percentage contribution to the abundance of the entire benthic macroinvertebrate fauna made by chironomids is greater in the acid-polluted sections of RMC than in the reference sections. In C93 the reduction in the proportional abundance of chironomids is due to a dramatic increase in Platyhelminths in the recovery section.

Turning to the other generalisations that have been made concerning chironomids and acidity, namely species richness reduction, increased abundance of few tolerant species and reduction in grazers, we can find no substantiation from this study. In three consecutive years, during comparable six-week periods prior to the annual cessation of flow, there is no reduction evident in chironomid larval species richness at sites below the entry of acid mine waste. There appears to be a turnover of species with those intolerant of the contaminated conditions "replaced" by others that can colonise and survive these conditions. Even diversion of the mine effluent into section B in 1992 is not associated with any dramatic reduction in species numbers in this previously unexposed section. This result is unlikely to derive from any sampling limitations since floating pupal exuviae as an estimator of the chironomid assemblage in RMC in 1992 gave a similar pattern (Hardwick, 1993). Although the Chironomidae is renowned for its presence in polluted waters, this is not because the family *per se* has unusually high tolerance, but derives from the speciose "pool" of taxa, within which there are many that tolerate various pollutants. This explains the observation that "chironomid species richness in rivers is not necessarily well correlated with water quality" made by Pinder (1986).

Amongst the taxa that tolerate the pollutant entry, there is no indication of numerical dominance by any one or few species (see Appendix). There are, however, changes in species presence or abundance above and below entry of the discharge: these are discussed below.

Regarding the suggested reduction in grazers, Rasmussen and Lindegaard (1988) in Denmark found a negative correlation between numbers of chironomid taxa and dissolved iron, even in circum-neutral waters, with the primary loss being of grazers. Assessing reduction in periphyton grazing under the acid conditions of RMC is not easily done: this is typical of northern Australian streams, dominated by Chironomini and Tanypodinae, with grazers such as Diamesinae never present and Orthocladiinae rare (Cranston, 1992). Although little is known of the feeding roles of the non-tanypod taxa, our observations (*unpublished*) showed that the larval gut is packed with iron hydroxide "floc" in larvae from the acidified sections. In the oligotrophic RMC, it is possible that bacteria associated with the depositional "floc" in the acid impacted section may provide a larval food source.

Chironomid species responses in Rockhole Mine Creek

As alluded to above, the chironomid community, though comparably rich above and below discharge, varies in composition in response to acid input. For instance, C93 supports 12 species that were found in neither A93 or B93. Although 11 of these were rare taxa (<5 animals), *Fittkauimyia disparipes* Karunakaran was reasonably abundant with 22 individuals. This species, described from Singapore, where the type-locality is a polluted creek (P. Murphy, *pers. comm.*), occurs regularly in polluted sections of RMC.

Amongst the taxa that are present both above and below the discharge, *Nilotanypus* (probably *albiceps* (Johannsen)) typifies the response of relatively intolerant species. In 1991 and 1992, it was rare in the polluted section C, yet with alleviation of the pollutant in C93 it attains comparable densities to those seen in unimpacted sections A93 and B93. The presence of small numbers of this and other apparently intolerant taxa in polluted sections may be due to drift: we lack evidence for completion of the life-cycle in these sections. Whatever "ecological" interactions or changes in water chemistry have allowed *Nilotanypus* sp. to inhabit section C93, their increase in this section indicates recovery towards the similar abundances seen in uncontaminated sections.

Perhaps an even more acidophilic indicator of pollution than *F. disparipes* is *Polypedilum (Pentapedilum) convexum* Johannsen (Fig. 7). This species was described from ditches and a sulphur spring in Sumatra, Indonesia, and, under the name *Polypedilum anale* (Freeman) (a previously unrecognised synonymy), from an acid-impacted stream in South Africa. It frequently inhabits the naturally acidic contents of *Nepenthes* pitcher plants (Cranston *pers. obs.*). In RMC, *Polypedilum convexum* was absent (possibly unrecognised) in 1991, but occurred late in the 1992 sampling period as the acidic input contributed increasingly to the diminishing flow of RMC in pulse-polluted section B92 and long-term polluted C92. In 1993, *Polypedilum convexum* was abundant in the polluted sections D and E below the diverted adit entry but was absent from sections A, B and C.

The decline in the abundance of an contaminant-associated organism can be interpreted as recovery. Indeed, recovery may be better indicated by the absence of such a species, rather than the appearance of one that was previously absent, particularly (as here) where pre-impact data is lacking, because tolerance of pollution explains presence, whereas absence may derive from one of many factors apart from intolerance. As Cairns (1974) stated, "absence of a species is less useful as an indicator of environmental conditions than presence" because absence does not imply causality. Although true for detection studies, the situation is reversed in studies of recovery. The absence of *Polypedilum convexum* in uncontaminated sections and the subsequent appearance when these sections are contaminated suggests that this organism has an affinity with contaminated conditions. The removal of the contaminant and subsequent disappearance of *Polypedilum convexum* in our view indicates that this animal is acidophilic and provides strong evidence for recovery directly attributable to the removal of the adit drainage. Furthermore, this finding highlights the advantages of performing experimental manipulations prior to any recovery study, because they may reveal crucial information to implicate a disturbance and allow interpretation of change as recovery.

The importance of species discrimination in pollution studies (*e.g.* Resh and Unzinger, 1975; Cranston, 1990) is exemplified by two morphologically very similar species of *Tanytarsus*, referred to as "K2" and "K3" (Cranston, 1992) pending formal description. *Tanytarsus* "K2" (Fig. 8) is pollution indicative, with a pattern of distribution resembling that of *Polypedilum convexum* (Fig. 7). In contrast, *Tanytarsus* "K3" (Fig. 9) has been present in all sections in all three years, although it is more abundant in section C93 even after downstream diversion of mine effluent. The fact that *Tanytarsus* "K2" and *Tanytarsus* "K3" show drastically different responses to the contaminant, yet can only be distinguished in the larval stage by a slight change in sclerotisation of the second antennal segment, reinforces the need for species level identification in thorough biomonitoring studies. Such intra-generic variability in responses to acidification as seen here occur in speciose genera in RMC such as *Polypedilum*, *Dicrotendipes* and *Rheotanytarsus* as well as between other *Tanytarsus* species. These different responses by congeners in multi-species taxa, alluded to also by Simpson (1983), is the likely explanation of incongruent opinions on the acid tolerances of chironomid genera (e.g. Scullion and Edwards, 1980; Haines, 1981).

Recovery: macroinvertebrates versus chironomids

In terms of recovery, the entire fauna and chironomid assemblages "tell different stories". The grouping of sections seen in Figure 2a implies that the mine effluent strongly influences the benthic macroinvertebrate fauna. The change in the pattern of grouping between uncontaminated and contaminated sections can only be explained by inferring change caused by the mine effluent. In contrast, the dendrogram of chironomid assemblages (Fig. 2b) implies that these are less affected by the mine effluent. The grouping of the entire fauna is driven by an inter-annual effect, but the grouping of the chironomid assemblages is not. Rather, the chironomids show the pattern expected if spatial variation (zonation of the creek) was the dominant cause behind the relationships between sections.

That chironomid assemblages in each section resemble one another more than those of another section, is to be expected since subtle differences in biotic and abiotic factors of the creek lead to spatial heterogeneity (Marchant *et al.*, 1985). Although communities of a particular lotic section will be characteristic (Vannote *et al.*, 1980; Lindegaard, 1994; Lindegaard and Brodersen, 1995), minor differences in habitat preference will exist. The three sections A, B and C thus are expected to differ in community structure because assemblages of running waters are a dynamic continuum. Furthermore, stochasticity in the annual recolonisation process in RMC may play some role in annual variation in the chironomid assemblage, but further (more than three years') data may be required to test this suggestion.

Analysis of the entire macroinvertebrate community (Fig. 4a) showed the intermediate nature of C93, although equivalent analysis of the chironomid assemblage (Fig. 4b) did not do so. In Figures 2a and 4a, section C93 was seen to be neither typically uncontaminated, nor typically contaminated, but rather, somewhere between these two extremes. The intermediacy of C93 is explained partially by some continued leaching of mine effluent, indicated by the iron-hydroxide "floc" and confirmed by water chemistry data. Secondly, the intermediacy can be ascribed to recovery of the benthic macroinvertebrate community following the downstream effluent diversion.

Identification of recovery requires knowledge of the direction of change: in the absence of pre-impact data we cannot estimate recovery by attainment of a pre-existing condition. Using the physically similar section A as a reference (justified also by ordinations shown in Figures 3 and 4), the direction of change expected in the recovery of section C can be identified. Our observations do not permit us to say whether the changes seen in 1993 represent an end-point in a recovery process: the time period is too short to determine if the dramatic Wet season scouring of RMC is enough to reset the system. Furthermore, the buffering of any system against change (Cairns *et al.*, 1971) is an additional variable. However, although the differences observed in 1993 cannot be attributed directly to the diversion (from this analysis), other results strongly imply that this interpretation is likely.

The plots of mean dissimilarity (Fig. 5) provide strong evidence of recovery in C93, accounting for the sectional, intra-seasonal and inter-annual variation, and for the Dry season concentration of contaminants. Until now, all these factors have confounded comparisons of data collected in different years; however, by analysing the mean dissimilarities between two sections, all the biotic and abiotic (intrinsic) influences can be accounted for. Extrinsic factors can be attributed as causing major change in dissimilarity of community structure. Our interpretation of figure 5 is that the change in the mean dissimilarity between Sections A and C reflects a substantial change in the community structure of C93. Since all intrinsic change has been accounted for, the altered community structure can be ascribed to some extrinsic force. Given that the diversion of adit water is the only known extrinsic change that the fauna in C93 is exposed to, the diversion is implicated as causative. Since the change is in the direction of section C to remaining more like section A for the duration of the season, this indicates recovery in C.

Some chironomids in C93 indicate recovery but the total chironomid assemblage failed to reveal the pattern exemplified by the entire community. In particular two trends were undetected in the chironomid data. Firstly, the entire RMC macroinvertebrate community included two superabundant taxa, a finding typical of intermediate/recovering communities (Rosenberg, 1972). This causes the dramatic decline in the proportion of chironomids, which, although dominant in species richness, were never highly abundant. Secondly, no species of chironomid occurred upstream in sufficient numbers for a reduction in numbers in the polluted downstream section(s) to demonstrate high sensitivity to the mine effluent. In contrast, the remaining macroinvertebrate fauna included an abundant and very sensitive species of leptophlebiid mayfly, which will be reported elsewhere. Such sensitive indicator organisms are useful in detection of pollution but very high sensitivity may render them less useful in studies of recovery. Effective monitoring of recovery would utilise the wide range of sensitivities inherent in a community of organisms. Although the chironomid assemblage seems to lack some informativeness present in the entire macroinvertebrate community, it does provide an array of sensitivities amongst which indicator taxa can be recognised.

Acknowledgements

We are grateful to the Australian National Parks and Wildlife Service (now A.N.C.A.) and the Office of the Supervising Scientist (O.S.S. now E.R.I.S.S.) for the Alligator Rivers Region for financial and logistical support for this research. This study formed part of the requirements of the senior author's Bachelor of Science Honours degree at the Division of Botany and Zoology, the Australian National University. Particular thanks are due to Peter Dostine, Chris Humphrey and Abie Spiers from the Biological Monitoring Section, O.S.S. for the provision of previous years' data and field and laboratory assistance, to Dan Faith (CSIRO Wildlife and Ecology) for help with PATN, and to Penny Gullan and Paul Cooper for their assistance throughout. Paul Cooper, Dan Faith, Penny Gullan, Chris Humphrey, Stuart Halse and John Trueman provided perceptive advice on the manuscript, for which, many thanks.

APPENDIX.

List and abundances of chironomid taxa found during April - June 1993 in Rockhole Mine Creek, Northern Territory. Sections of creek (A-E) follow Figure 1. Taxon names and codes predominantly follow Cranston (1993).

Species	Sect. A	Sect. B	Sect. C	Sect. D	Sect. E	TOTAL
Ablabesmyia sp. 1	0	0	2	2	1	5
Djalmabatista sp.	0	1	0	0	0	1
Fittkauimyia disparipes Karunakaran	0	0	22	13	8	43
Larsia sp.	2	0	13	0	0	15
Telmatopelopia sp.	1	3	22	15	16	57
Nilotanypus sp.	20	11	18	6	1	56
Paramerina parva Freeman	37	39	160	89	43	368
Thienemannimyia sp.	0	0	2	5	0	7
Pentaneurini genus B sp nov.	0	0	2	0	1	3
Procladius paludicola Skuse	0	0	0	2	0	2
Cricotopus "K2"	0	0	1	0	0	1
Corynoneura sp.	2	1	1	0	1	5
Parametriocnemus ornaticornis Kieffer	0	0	1	0	4	5
Rheocricotopus sp.	12	9	5	3	0	29
Thienemanniella (pale long 3rd)	21	7	37	3	3	71
Thienemanniella (dark short 3rd)	0	2	0	0	0	2
Cladotanytarsus "K2 "	0	0	1	0	0	1
Cladotanytarsus "K4 "	0	0	2	0	0	2
Paratanytarsus "K2"	0	1	1	0	0	2
Rheotanytarsus johnstoni Glover	397	30	67	19	13	526
Rheotanytarsus christinae Glover	10	16	3	5	2	36
Tanytarsus "K2"	1	1	4	17	1	24
Tanytarsus "K3"	34	45	157	29	33	298
Tanytarsus "K5"	4	0	26	10	1	41
Tanytarsus "K10"	0	2	3	0	0	5
Tanytarsus "K12"	4	1	6	1	1	13
Tanytarsus manleyensis Glover	0	0	0	0	1	1
Chironomus sp.	2	0	0	0	9	11
Conochironomus sp.	0	1	0	0	0	1
Dicrotendipes lindae Epler	15	5	51	3	1	75
Dicrotendipes sp. 2 (unknown sp.)	4	0	6	0	0	10
Dicrotendipes sp. 3 (5 lateral teeth)	0	1	0	0	0	1
Parachironomus sp.	2	1	0	0	0	3
Polypedilum convexum Johannsen	0	0	0	86	7	93
Polypedilum oresitrophus Skuse	0	2	0	0	1	3
Polypedilum "K2"(nr *seorsus* Skuse)	0	0	1	0	1	2
Polypedilum "K3"	13	17	15	28	14	87
Polypedilum nr "B1"	0	4	0	0	0	4
Riethia cinctipes Freeman	0	0	0	1	0	1
Stenochironomus sp.	0	10	6	14	3	33
INDETERMINATE	4	4	23	1	10	42
	585	214	658	352	176	1985

Part 4

Austral Chironomid Studies

From egg to imago in less than seven days: *Apedilum elachistus* (Chironomidae)

Ulrike Nolte

Universidade Federal de Mato Grosso (UFMT)
(Correspondence address: Box c.p. 3091,
78060-200 Cuiaba - MT, Brazil)

Introduction

While studying the headwaters of a tropical lowland river in central Brazil, densely colonized rock pools adjacent to a spring were encountered. Among the colonizing insects, *Apedilum elachistus* Townes was one of the most abundant species, tending to be more dominant the more ephemeral and simpler in spatial structure such rock pools were. Particularly towards the end of the rainy season, the smallest of the pools - with a maximum volume of 2.8 litres only - was teeming with larvae of *A. elachistus*, and thus the water showed a tinge of red.

Apedilum elachistus has been previously recorded only from North America (south of 42°N) where it seems to be widely distributed (Oliver *et al.*, 1990) but its biology was practically unknown. Epler (1988) gives detailed descriptions of larva, pupa, female and male imago and, in doing so, resurrects the generic status of *Apedilum* in distinguishing it from *Paralauterborniella*. Larvae of *A. elachistus* were collected among submerged vegetation from freshwater and brackish ponds in Florida, the temperature of which ranged from 25 to 39°C, and where imagos emerge all year round (Epler, 1988)

The present paper provides information on abundances of *A. elachistus* in tropical rock pools during one year, and on its generation time and biology.

STUDY AREA AND SAMPLING SITES

The rock pools are located adjacent to one of the main springs of the Rio Bento Gomes, a lowland river which drains into the northern Pantanal (Mato Grosso, Brazil). (Pantanal is the Portuguese word for everglades.)

Most of the more than one hundred springs of this intermittent tropical river arise in savannah woodlands, the so-called cerrado. The degree of the river's cut-off or desiccation strongly depends on local precipitation, which was 1600 mm in 1992/93 (20.vii.92-20.vii.93) near the spring concerned. This rheocrene arises from a drainage area large enough to keep it flowing throughout the year. However, it may dry out in an extremely dry years. The present studies are the first ever undertaken in the headwaters of the Rio Bento Gomes, thus long-term records of environment conditions are not available.

The spring is located 15°45′S, 56°35′W, at approximately 330 m a.s.l. The area with some 20 rock pools, which are either temporarily linked with the spring or completely isolated and just fed by rain and dew, covers about 70 m^2. The bedrock consists of quartzites, therefore conductivity in pool water was low (20-100 µS/cm), and pH ranged from 5.5 to 6.8. The water of the unshaded, shallow pools heated up to 42°C (March), but during winter, temperature dropped to 8°C (July). However, during most of the year it ranged from 24 to 29°C. The pools generally held some silt, sand and detritus, except when torrential downpours washed this material away. Subsequent lighter rainfalls always brought new sediment into the pools, and in the dry season wind-borne material was a constant input. Occasional defecation of small rodents was a further source of organic matter. Although exposed to direct sunlight, visible algal growth (*Spirogyra* sp.) ocurred only towards the end of the rainy season from January to March. Diatoms were present all year round, but always remained scarce. Disturbance frequency (drought, flash flooding) and spatial heterogeneity of the rock pools were found to be the most important factors for the compositon of the colonizing invertebrate communities.

In this paper, densities of *A. elachistus* recorded from the smallest and largest of the pools studied are presented (while results from further pools concerning the entire communities will be published elsewhere). The smallest pool was a simple, unstructured hollow with a maximum surface area of 980 cm^2, then holding 2.8 l of water subject to a wide temperature range. It was a rather short-lived pool, slightly higher located than the spring and the other rock pools thus only fed by rain and dew, and it stayed completely dry from June to September 1992. In the following, this pool is called pool "s" (for simple). The largest pool was a fairly complex system of three excavations which were connected most of the time and always contained some sand and stones. This pool held a maximum of 14.7 l of water, showing a surface area of 3350 cm^2. It was of longer duration since temporarily linked to the spring. In the following, this pool is called pool "c" (for complex).

METHODS

Field data during one year

From April 1992 to May 1993, samples were taken from the rock pools at intervals of 5 to 21 days which resulted in 35 sampling occasions. From all pools and at every sampling occasion samples were taken in the same fashion to keep the series of samples comparable. A small hand net (9 x 14 cm) of 125µm mesh-size was used. To minimize the sampling impact on community succession, samples were taken as gently as possible. Large, mobile species for instance, which were generally less abundant (mostly predators) were observed and counted in the field rather than caught (except for identification). One routine sample was preserved immediately with alcohol, and additional material was taken alive to the laboratory for further observations.

When the dominance of *Apedilum* sp. became apparent, I reared the species and sent larvae, pupae and imagos to Dr Epler (Florida, USA) who identified it as *A. elachistus*.

The data obtained in the field led to the following questions: What is the generation time of this successful colonizer and how is *A. elachistus* adapted to such a highly dynamic, physically controlled habitat?

Life-history studies

The life history of *A. elachistus* was studied by three different methods.

(i) Egg masses were collected from the rock pools and raised up to imagos in the laboratory. For this, individual egg masses were placed in glass dishes (7 cm^2) containing pool water and controlled every hour to record the hatching-point of the larvulae. Subsequently, some detritus was added and the rearings were inspected in two to four hour intervals until pupation. Pupae, again, were checked more frequently, i.e. two times per hour. These rearing series were run during the dry season (August and September 92, July 93) and the rainy season (January, February 1993), and a minimum of sediment and detritus in the petri dishes was used to permit detection of the instars present. Unfortunately, no incubators were available, so rearings had to be run at room temperature which showed a broad variation (20-31°C), in particular during the dry season when nightly temperatures dropped.

(ii) To avoid stress for *A. elachistus* during development, rearings of ten further individual egg masses were run in the laboratory, but without measurements in between and with plenty of sediment and detritus. For these "summary rearings", vials were provided with detritus previously boiled and then allowed to become recolonized by microbes and algae for 14 to 20 days. The summary rearings were also started with newly hatched larvulae and were concluded when the first imago emerged.

(iii) To verify the life history data of *A. elachistus* obtained from the laboratory, a field experiment was conducted from 12-19 February 1993. Rock pool "s" was completely emptied and cleaned and then refilled with filtered (125 µm) spring water. One week later, pool "s" was checked for the oldest instars of *A. elachistus* present. This experiment was conducted during the rainy season to ensure the continuous existence of pool "s" at this time.

Laboratory observations

The biology of *A. elachistus* was observed in the laboratory. Samples kept in petri dishes for several days, were inspected with 6 to 40-fold amplification (dissecting microscope) to study periods of activity and feeding behaviour. The potential drought-resistance of *A. elachistus* was tested by three experiments. (i) Some dry sediment from pool "s" was taken to the laboratory and covered with water to see whether larvae would recuperate via rehydration. (ii) Healthy larvae of all instars were placed in petri dishes filled with sediment and detritus, which were permitted to dry up, so as to refill the dishes with water after 180, 60, 45, 30 and 15 minutes and inspect the larvae's condition. (iii) Freshly laid egg masses, in which embryogenesis was not yet visible, were exposed to drought and flood to observe their viability.

RESULTS

(1) Egg masses, larvae and pupae of *A. elachistus* were present throughout the year. Most striking was the fact that *A. elachistus* was the only chironomid that successfully colonized the simply structured, ephemeral pool "s", where it attained densities of up to 1860 individuals per sample (Feb. 93; Fig.1). Only on five out of 35 sampling occasions, single larvae of *Thienemanniella* sp. and *Polypedilum* sp. were recorded in addition. In the larger and complex structured pool "c", *A. elachistus* was also eudominant, but here the chironomid community was composed of species from at least 14 genera (5 Tanypodinae, 2 Orthocladiinae, 7 Chironominae). In this pool, a maximum of 272 chironomids per sample was found, which included 59 individuals of *A. elachistus* (Aug. 92).

Pool was colonized only by primary consumers, mainly by larvae of *A. elachistus*, *Cloeodes* n. sp. (Baetidae), *Anopheles* sp. (Culicidae) and by adults of a small hydrophilid beetle. Their densities were determined by the frequent droughts in the dry season and flash floodings in the wet season. In May 1992, for instance, pool "s" dried up and refilled in intervals shorter than five days and only 12 to 52 chironomids per sample were recorded (Fig. 1). Much higher densities were attained when the disturbance frequency of the habitat was low. So, 485 (Sep. 92), 1350 (Dec. 92), and 1860 (Feb. 93) chironomids were found per sample when pool "s" remained undisturbed for 38, 28 and 25 days, respectively (Fig. 1). From September to February, the abundance of *A. elachistus* increased evidently faster after pool disturbance which was certainly due to the higher number of potential colonizers ermerging from the numerous water bodies existent in the rainy season. During this time, *A. elachistus* attained highest densities, and Figure 1 illustrates the seasonal existence of the habitats, while chironomids (with the eudominant *A. elachistus*) were present all year. The much lower chironomid densities recorded in pool "c", which

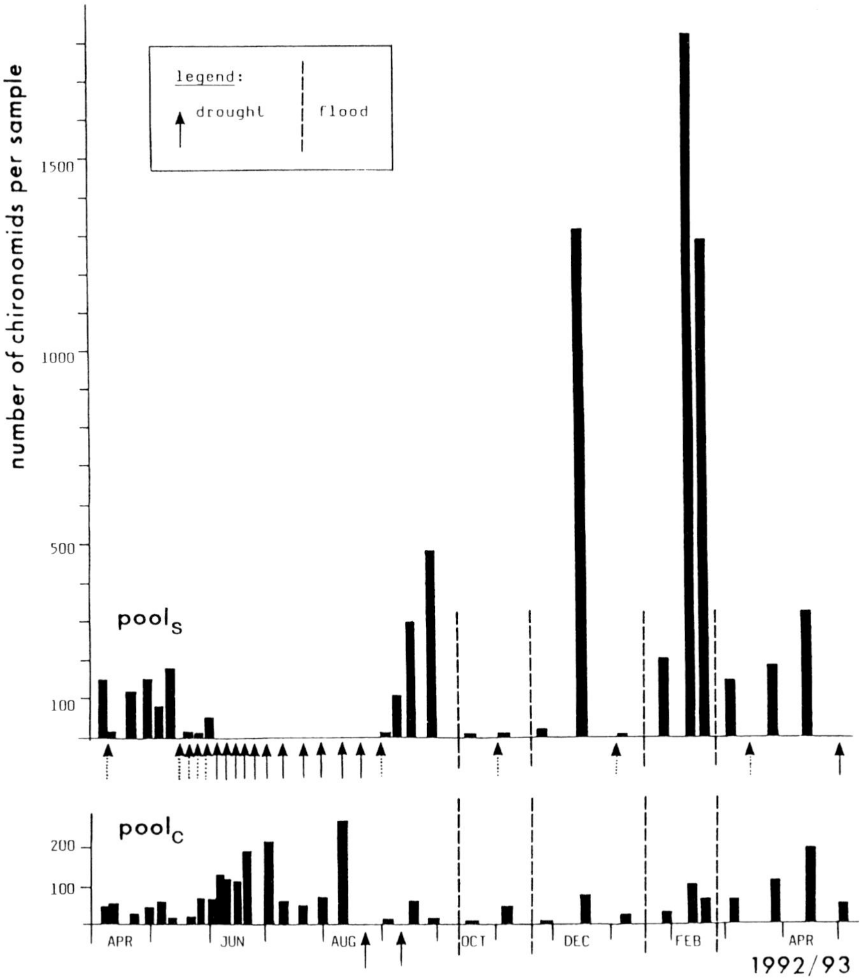

Figure 1. Chironomid densities in two neotropical rock pools during one year. Pool "s": a small and simple hollow colonized by primary consumers only, with *Apedilum elachistus* being the only chironomid. Pool "c": a larger and spatially diversified pool, colonized by many taxa, including predators and at least 14 chironomid genera.

The frequent drought events (indicated by arrows), in particular of pool"s", and flash flooding during the rainy season (dotted lines) characterize the unstable ambient conditions to which the colonizers of the pools were subjected. (The solid arrows show droughts observed directly, while the dotted arrows indicate drought events which were concluded from sediment appearance (compact, cracked layers scaling off the rock).)

housed a complex insect community, reflect the presence of predators such as Coenagrionidae, Belastomatidae and larvae of Hydrophilidae.

(2) To study the life cycle of *A. elachistus*, egg masses were collected from the rock pools during the dry and rainy season and raised up to imagos. In the wet season,

eggs were quite easy to obtain because of their abundance. On 28 March 93, for instance, as many as 17 egg masses were found in the routine sample taken from pool "s". Some of them were close to hatching, others already empty, and others again freshly laid. Out of 26 rearing series which were started with individual egg masses, 14 series ran well and imagos were obtained (Table 1). In the remaining series, larvae died in the first or second instar. In Table 1, development times recorded from the successful series are given for each instar along with the total time span from larvula to imago. On account of the fast development, time spans are given in hours. Temperature ranges are also indicated, and the date when a series was started. Since no incubators were available to keep temperatures constant, and room temperatures were different in the dry and wet season, rearing series are listed according to seasons in Table 1.

A. *elachistus* grew from larvula to imago in 309 ± 17 hours (n=6 egg masses) during the dry season, and in 259 ±24 hours (n=8 egg masses) during the wet season. In the dry season, temperatures were lower (20-26°C), and the total development time was about two days longer than in the rainy season (26-31°C). The data given in Table 1 show that particularly the first larval instar was prolonged, which took almost one third of the total development time in the dry season. It was interesting to observe that the fourth instar took less time than all other larval instars and that its duration was about the same in both seasons. A constant development time of 4 to 4.5 hours was recorded for the pupal instar (Table 1).

Total time spans given in Table 1 do not include egg development because oviposition times were unknown. Only the 51 hours duration, indicated with an asterisk in Table 1, was ensured as this egg mass was laid in the laboratory by a female obtained from the field.

In addition to these frequently controlled rearing series, ten further individual egg masses were raised in the laboratory, but without any measurements until imagos had emerged. These summary rearings were carried out in February 1993, when temperatures ranged from 28 to 31°C. The mean development time from larvula to imago took 187 ±20 hours (n=10), with a minimum time span of 168 and a maximum of 273 hours.

Table 1: Development times [in hours] of *Apedilum elachistus* reared in the laboratory from individual egg masses at different seasons and temperature ranges. The duration of single instars and the total time span from larvula to imago are given.

	egg	LI	LII	LIII	LIV	P	LI - P	T[°C]	date
dry season:	32	96	62	62	58	4	282	20-26	Jul-15
	48	84	67	74	60	4	289	20-26	Jul-15
	51*	102	—141—		68	4	315	20-26	Aug-13
	42	102	72	78	62	4	318	22-26	Sep-04
	36	98	174	82	62	4	320	22-26	Sep-04
	42	100	78	78	66	4.5	327	22-26	Sep-04
mean		97.0	70.6	74.8	62.7	4.1	309	20-26	
±SD:		±6.2	±5.6	±6.9	±3.4	±0.2	±16.8		
wet season:	28	45	59	60	58	4	226	28-31	Feb-12
	48	40	61	67	57	4	229	28-31	Feb-12
	28	42	67	74	58	4	245	26-29	Jan-29
	28	72	59	62	58	4	255	28-31	Feb-19
	31	72	69	58	58	4.5	262	28-31	Feb-12
	36	67	66	72	60	4	269	28-31	Feb-19
	22	96	69	58	58	4.5	286	26-29	Jan-29
	20	53	——238——			4.5	296	26-29	Jan-29
mean		60.9	64.3	64.4	58.1	4.2	259	26-31	
±SD:		±18.1	±4.2	±6.1	±0.8	±0.2	±23.5		

* oviposited in the laboratory by a female collected from the field

Table 2: Development times of *Apedilum elachistus* obtained by different methods.

method used	LI - P [hours ±SD]	LI - P [days ±SD]	n egg mass	T [°C] range
lab rearings with frequent controls				
during the dry season:	309 ±16.8	12.9 ±0.7	6	20-26
during the wet season:	259 ±23.5	10.8 ±1.0	8	26-31
summary rearings, lab, wet season	187 ±20.0	7.8 ±0.8	10	28-31
field experiment, wet season	<166	<7.0	?	25-35

The field experiment revealed that A. *elachistus* took less than 166 hours from oviposition to emergence when temperature ranged from 25 to 35°C. Seven days after rock pool "s" had been emptied, cleaned and refilled, 1302 individuals of A. *elachistus* were recorded in the routine sample (19. Feb., Fig.1), and this already included three pupae and one pupal exuviae.

In Table 2, all data obtained on the development time of A. *elachistus* are listed along with the method used. The data show that generation time is shorter at higher temperatures, and that the frequent manipulations during rearings retarded development time.

(3) *Apedilum elachistus* is a small Chironomini, with fourth instar larvae of 2.39 (±0.16) mm (n=12) and pupae of 2.10 (±0.02) mm (n=8) body lengths. Wing lengths of male and female imagos was 1.05 (±0.07) mm (n=6) and 1.12 (±0.02) mm (n=6) respectively.

In the laboratory, A. *elachistus* was observed to oviposit on the water's surface. The egg masses sank down immediately. Correspondingly, eggs were always obtained from the bottom of the rock pools and never found attached to a substratum such as the rock at the water's edge. The bale-shaped masses of A. *elachistus* were slightly flattened and appeared to consist of two "limbs" between which a second type of gelatin was visible. However, a gelatinous stalk was never present (terminology follows Nolte, 1993). The masses ranged in size from 1.7 x 1.2 mm to 3.3 x 2.5 mm (n=26). The eggs were irregularly arranged within the main gelatin, and 23 to 61 eggs per mass were found, with a mean of 35.6 ±10.1 (n=26). Eggs were elliptical in shape, 309 x 118µm in size (n=30), and retained their white and transparent appearence until larvulae hatched. The egg stage took up to 51 hours (Tab.1), but visible embryogenesis (50 x magnification, 29°C) started only six hours before hatching. Three hours later, first larvulae left the gelatinous mass, and started immediately with tube building. Hatching success per egg mass was very high (close to 100%) but was not measured precisely. Only in one of 41 egg masses examined, were water mites observed to feed on the eggs.

Larvulae were photo-negative, and all larval instars showed highest activities at night. The loose, somewhat unorganized cases of A. *elachistus* were never found to be fixed to the bottom (of a rock pool or petri dish), thus were easily shifted by slight water movements. In particular at night, larvae were observed to leave their tubes spontaneously, either crawling or swimming around, and then looking for an unoccupied case to enter for the subsequent resting period. Pupae which are also good swimmers left their tubes approximately 20 to 15 minutes before emergence.

Gut content analyses and observations of feeding larvae showed that A. *elachistus* is a non-selective primary consumer, ingesting detritus, mineral material and diatoms according to food item availability. Fourth instar larvae were also observed to bite off portions of *Spirogyra* filaments. However, such fragments were relatively seldom found in the guts examined, probably due to the difficulty of handling this bulky alga (assessed by handling time).

The experiments to test potential drought-resistance, revealed that no such adaptation to ephemeral habitats is evolved in A. *elachistus*. Eggs already lost viability when the gelatinous mass became slightly dry, and larvae did not show any rehydration capacity either.

DISCUSSION

With less than seven days from egg to emergence the generation time of A. *elachistus* is the shortest presently known for Chironomidae. Life times reported for African rock pool dwellers range from 12 to 32 days in *Chironomus imicola*, and from 35 to 43 days in *Polypedilum vanderplanki*, data which are "based on chance observations" (McLachlan and Cantrell, 1980). No further development times are known from small water chironomids, and the few studies concerned with biological features of rain pool dipterans are focussed on the spectacular phenomenon of cryptobiosis detected by Hinton (1951) in *P. vanderplanki* (Bishop, 1974; Dodson, 1987; Jones, 1975; MacLachlan and Cantrell, 1980). For the neotropical species *Chironomus sancticaroli* and C. *strenzkei*, generation times of 15 days (19-26°C, Strixino and Strixino, 1982) and 10 days (28°C, Syrjämäki, 1965) repectively, have been reported, and I assess short life cycles such as found for A. *elachistus* to be quite common, in particular among tropical chironomids.

Apedilum elachistus reproduced the year round (Fig. 1) and showed no seasonality. The high abundances recorded from December to March (Fig. 1) rather reflect the seasonality of the rain-dependent habitats than that of A. *elachistus*. Rearings were run sporadically and not throughout the year, thus the annual number of generations A. *elachistus* is likely to produce in the field must be deduced from the data given in Tables 1 and 2. Development times of chironomids are temperature dependent (cf. Oliver, 1971) thus temperatures which prevailed during larval growth have to be taken into account. Because temperature could not be kept constant during the rearings, ranges of temperatures have to be related to the time spans obtained, and this precludes a calculation via a regression coefficient. So, in the following, the number of generations per year are approximated. I am emphasizing that life cycles of chironomids often have proved to be retarded under artificial conditions (as indicated by the present field experiment; discussions with many colleagues, no published data available), i.e. species reared in the laboratory generally take longer to develop than in their natural habitat.

During the wet season, rearings were performed in two fashions, and the mean development time obtained from the frequently handled rearing series (259 hours) was retarded by 39% in comparison to the summary rearings (187 hours, Table 2). Correspondingly, the mean development time observed during the dry season is corrected from 309 to 222 hours. To both, 187 and 222 hours, 50 hours have to be added for the egg stage. These data can then be related to the temperatures in the field. The nearest meteorological station to the rock pools was located at a distance of 70 km, and continuous temperature recordings in 1992 gave the following conditions: On 216 days the daily mean temperature ranged from 26 to 30°C, on 112 days from 20 to 25°C, and on 26 days its was less than 20°C. The combination of these temperature data with the mean time spans obtained from the rearings, result in 32 generations per year. Employing the minimum time spans observed (Tables 1 and 2), leads to 48 generations per year.

Apedilum elachistus shows no evidence of specialization, such as drought-resistance, to frequent habitat perturbation. The small size of 2.4 mm in mature larvae meets the common size range reported from tropical chironomids (Nolte, 1989, Roback, 1966) and thus, is not understood to be an "adaptive response to increases in competition for food and space" which particularly occurs in crowded temporary habitats (McLachlan, 1989). The principle mechanism permitting A. *elachistus* to colonize ephemeral habitats, apparently is the short generation time. But this is thought to be a very common feature among tropical Diptera rather than a specific adaptation. However, as A. *elachistus* was the only successful colonizer among at least 14 chironomid taxa in the nearest neighbourhood (Fig.1), there must be something special in the biology of this species. It tolerated a broad temperature range from at least 8 to 42°C, did not feed selectively, and had little requirements

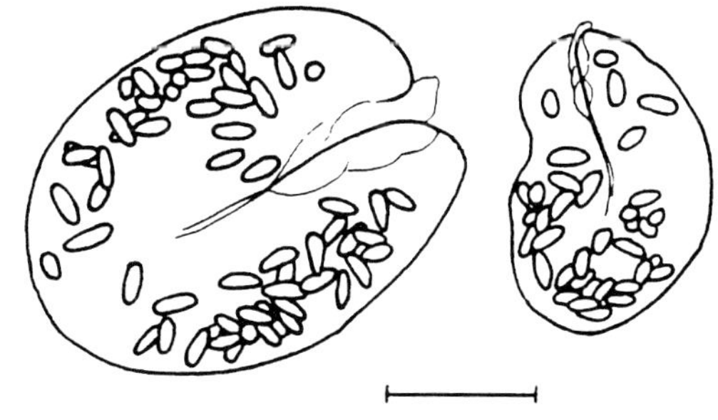

Figure 2. Egg masses of *Apedilum elachistus*; scale indicates 1mm

for spatial habitat structures. This, again, characterizes A. *elachistus* to be a generalist, all the more since it also occurred in adjacent rock pools together with many other species, and is known from submerged vegetation and brackish water (Epler, 1988). Hence it follows that exactly because A. *elachistus* is a generalist, it is capable of living in both harsh and benign habitats, and it nicely illustrates that "unstable and unpredictable ... environments appear to have favoured selection of opportunism" (Winterbourn *et al.*, 1981).

Acknowledgements

I thank J.H. Epler (Crawfordville, Florida) for the identification of A. *elachistus*. W.P. McCafferty (West Lafayette, Indiana) identified the baetid mayfly *Cloedes* n.sp., and the staff of the Pantanal Project (MPI/UFMT) provided some meteorological data. Thanks is due to R. da Silveira Tietböhl for her cautious assistance in the field and in the lab (supported by the Brazilian Council for Sciences and Technology CNPq), and to P. Haase (Leeds, England) who kindly corrected the English manuscript.

Travel funds for attending the conference were provided by the CNPq (453332/93-5 NV).

The response of Chironomid Populations to Flooding and Drying in Flood Plain Wetlands of the Lower River Murray in South Australia

P.J.Suter, P.M.Goonan, J.A.Beer and T.B.Thompson

Australian Centre for Water Quality Research, Private Mail Bag, Salisbury, South Australia 5108.

Introduction

A major influence on the ecology of river flood plain wetlands is the variablity of the hydrology, and in particular, the periods of drying and flooding. A flood pulse is vital for the existence, and interactions of the major biota in a river/flood plain system (Junk *et al.*, 1989). Aquatic fauna respond in a variety of ways to the flooding and drying regimes with some species showing an increased production during the rising hydrograph as the littoral zone traverses the aquatic/terrestrial transition zone (Bayley, 1991) while others species show a negative response to flooding and increase production during the drying phase (Maher and Carpenter, 1984).

In a study of the chironomid populations of two wetlands on the flood plain of the Lachlan River in New South Wales, Maher and Carpenter (1984) showed that *Chironomus tepperi* Skuse and C.*"alternans a"* responded positively to flooding while *Kiefferulus intertinctus* (Skuse), *Polypedilum nubifer* (Skuse) and *Procladius villosimanus* Kieffer had highest densities as water levels fell.

During a two year monitoring study of eight South Australian wetlands on the River Murray flood plain (Goonan *et al.*, 1992) the opportunity to study the responses of the chironomid fauna to flooding was provided by two flood periods. The 1990 flood inundated a considerable area of the flood plain in South Australia, including the eight wetlands being investigated while the 1991 event affected a smaller area of flood plain.

Study Sites

Eight wetlands on the River Murray flood plain from the Calperum region 16-20 Km north of Renmark to 6 km north of Mannum (Fig. 1) were selected as part of a detailed monitoring study of wetlands responses to various hydrological manipulations (Suter *et al.*, 1993). The location and physical characteristics of the eight wetlands are given in Table 1. Full descriptions of the wetlands and the chemical characteristics are given in Suter *et al.* (1993).

Fig. 2 shows the River Murray hydrograph at the S.A. border during the study period from May 1990 to May/June 1992 (M.D.B.C., *unpubl. records*). A major flood with a maximum flow of 113 000 ML/day was recorded at Overland Corner on 26 October 1990, and a smaller flood of 50 700 ML/day was recorded downstream of Rufus River on 16 October 1991. These peak flows correspond to a return period of 1 in 20 and 1 in 3 years, respectively (Ohlmeyer, 1991). All eight wetlands were flooded for at least 2-3 months by the 1990 event, but Clover Lake received no water in 1991 and Berri Evaporation Basin (EB), Disher Creek EB and Ramco Lagoon received only limited inflow of freshwater for less than 2 months.

Figure 1. Map of River Murray floodplain in South Australia with location of wetlands included in Wetlands Management Program.

Table 1. Location and physical characteristics of the eight wetlands included in the Wetlands Monitoring Program

Wetland	Acronym	Site nos.[1]	Location		Area[2] (ha)	Classification[3]	Dominant Vegetation
Clover Lake	CLOL	1-3, 32	34°00'S	140°46'E	140	Intermittent flood plain swamp	*Eucalyptus largiflorens*, *Acacia stenophylla*, *Cyperus gymnocaulos*, *Morgania floribunda*
Lake Merreti	LMER	4-6, 33	34°01'S	140°45'E	390	Permanent regulated flood plain lake	*Eucalyptus camaldulensis*, *E.largiflorens*, *A.stenophylla*, *Muehlenbeckia florulenta*, *C.gymnocaulos*
Lake Woolpolool	LWOO	7-9, 34	34°02'S	140°43'E	330	Seasonal regulated flood plain swamp	*Halosarcia* spp., *Typha* spp.
Disher Creek Disposal Basin	DISC	10-12	34°14'S	140°42'E	260	Permanent regulated flood plain anabranch	*E.largiflorens*, dead *E.camaldulensis*, *M.florulenta*, *Atriplex* spp.
Berri Disposal Basin	BERB	13-15	34°18'S	140°34'E	325	Permanent regulated flood plain swamp	*Suaeda australis*, *Chenopodium* spp., *Halosarcia* spp. *Phragmites australis*, *Bolboschoenus* spp.
Ramco Lagoon	RAML	22-24	34°10'S	139°55'E	91	Permanent regulated flood plain lake	*E.largiflorens*, *M.florulenta*, *C.gymnocaulos*, *P.australis*, *Bolboschoenus* spp., *Paspalum* spp.
Wongulla Lagoon	WONL	25-28	34°43'S	139°33'E	120	Permanent flood plain swamp	*E.camaldulensis*, *Myoporum* sp., *M.florulenta*, *C.gymnocaulos*, *Schoenoplectus validus*, *Vallisneria spiralis*, *Azolla* sp., *Salix babylonica*
Lake Carlet	LCAR	29-31, 35	34°52'S	139°31'E	330	Permanent flood plain swamp	*E.camaldulensis*, *M.florulenta*, *C.gymnocaulos*, *Typha* spp., *P.australis*, *S.validus*, *Ceratophyllum demersum*, *V.spiralis*, *Azolla* sp., *S.babylonica*

1 Site Nos. not in sequential order due to modifications in the number of wetlands included in the final project and the inclusion of additional sites for some wetlands.
2 Based on data in Pressey (1986).
3 Classification after Paijmans *et al.* (1985).

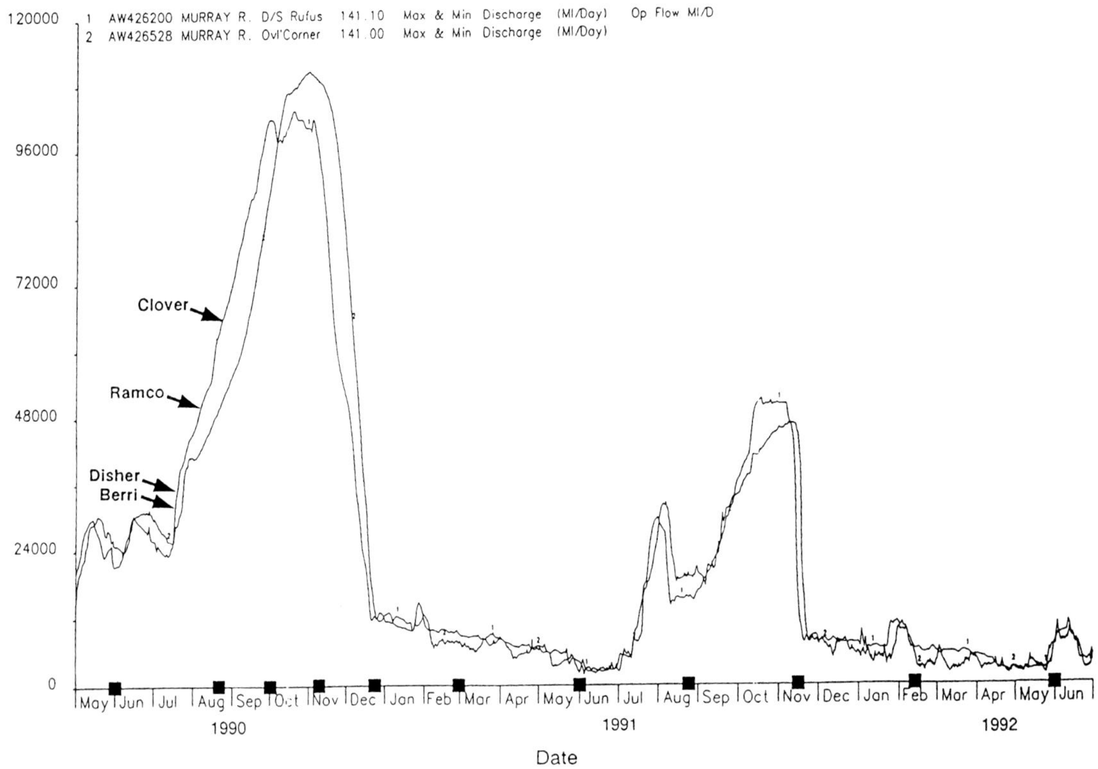

Figure 2. River Murray hydrography in South Australia during the study period from May 1990 to May-June 1992. Sampling periods are indicated on the x-axis (_). Arrows indicate the river flow at which water enters various regulated wetlands and Clover Lake.

Methods

The biota and water chemistry were sampled at each wetland on at least a seasonal basis (every three months), commencing in May-June 1990 and ending in February 1992. During the 1990 flood, samples were collected more frequently (every six weeks) to extend the databases and provide a more detailed understanding of the environmental responses to such a large flood event (Fig. 2).

A sediment sample was collected from the littoral zone at most sites in August 1990 and February-March 1991, and organic matter content, as a percentage of dry weight, was determined from each sample using the method described in Anon (1985).

Aquatic macroinvertebrates were sampled from the littoral zone at each site using 30 second sweep samples with a 200 μm mesh dip net. The number of microhabitats sampled at each site varied from 2-5 during the quarterly surveys, and 1-2 for additional surveys conducted during the 1990 flood. Samples were preserved in 5% formalin and returned to the laboratory for sorting and identification. A minimum of 10% of the sample was sorted using a subsampling technique described by Marchant (1988). Whenever the total number of macroinvertebrates was less than 200, the whole sample was sorted. The remaining sample was always scanned for the presence of rare taxa. Samples from each site were combined to provide a wetland mean.

Chironomids were identified to species where possible, otherwise identifications were made to genus.

RESULTS

A total of 203 taxa of macroinvertebrates were collected from the 29 sites. Of these the Chironomidae were represented by 41 taxa, with 6 Tanypodinae, 9 Orthocladiinae and 26 Chironominae which included 8 Tanytarsini. Fifteen taxa were widespread and commonly occuring in the flood plain wetlands. These included *Procladius paludicola*, *Cricotopus* spp., *Limnophyes*, *Chironomus tepperi*, *Chironomus* EWS sp1, *Cladopelma*, *Cryptochironomus*, *Kiefferulus intertinctus*, *Parachironomus*, *Paratanytarsus*, *Polypedilum nubifer*, *Polypedilum oresitrophus* Skuse, *Tanytarsus semibarbitarsus* Glover and *Tanytarsus* EWS sp4.

In terms of abundance the Chironomidae represented 41.6% of the total fauna collected over the two years. This is an elevated percentage due to the dominance of chironomids at Berri EB where they contributed 91% of the fauna. In the other wetlands the contribution ranged from 16.5% at Lake Woolpolool and Wongulla Lagoon to 37.8% at Lake Merreti.

The chironomid community responded in one of three ways; increased numbers with flooding and a concomitant decline in salinity, increased through the drying phase when salinities increased, or a seasonal response irrespective of the hydrology. The species which had greater than 100 animals per sample over the study period at each wetland are given in Table 2.

Taxa which responded to flooding included *Chironomus tepperi*, *C. cloacalis* Atchley and Martin, *Kiefferulus intertinctus*, *Cricotopus "albitibia"* Walker and *Chironomus* EWS sp1. Of these the first three species showed this response consistently at all wetlands. However, during the first flood event *C. tepperi* did not exceed 100 animals per sample at any wetland except Berri EB where numbers exceeded 1000 animals per sample. *C. tepperi* was not recorded in Ramco Lagoon during the 1991 event. During the second flood event numbers were higher in Lake Merreti (approx. 200 animals per sample), Berri EB (>2000 animals per sample) and Ramco Lagoon (approx. 400 animals per sample). At all other wetlands numbers were similar to the previous year or lower (Fig. 3). *C. cloacalis* was recorded in significant numbers at Berri EB in November 1991 (>2000 animals per sample), Ramco Lagoon from May 1991 to February 1992 with a peak in November (>400 animals per sample) and in Wongulla Lagoon where numbers did not exceed 20 per sample on any occasion.

Kiefferulus intertinctus responded variably at a number of wetlands with <250 per sample recorded at any time. In Lake Woolpolool *K. intertinctus* showed a positive flood response in 1991, but not in 1990. In contrast numbers were high during both flood events in Berri EB, but were only high during the 1990 event at Disher Creek EB and Lake Carlet (Table 2).

The other species only had sufficient numbers in one or a few wetlands to enable any observation of a temporal response. *Chironomus* EWS sp1 showed a peak in numbers at Berri EB during both floods, but never reached sufficient numbers anywhere else to detect a trend. Similarly, *Cricotopus "albitibia"* showed a positive response only in 1991 in Disher Creek EB, Wongulla Lagoon and Lake Carlet but did not show any seasonal or hydrological response in the other permanent, freshwater wetland of Lake Merreti.

Four species (*Polypedilum nubifer*, *Parachironomus* sp., *Tanytarsus semibarbitarsus* and *Procladius oresitrophus*) showed a positive flooding response at Berri EB but only during the second flood event.

Positive responses to the drying phase, and a consequent negative response to flooding were observed in *Procladius paludicola* at all wetlands (Fig 4). Other species which showed this type of response were from temporary or saline wetlands and included *Cladotanytarsus* at Clover Lake; *Tanytarsus semibarbitarsus* at Clover Lake and Disher Creek EB; *Cricotopus "albitibia"*, *Cladopelma*, *Microchironomus* and *Tanytarsus* EWS sp4 in Clover Lake and *Polypedilum nubifer* at Ramco Lagoon. In the permanent, isolated, freshwater wetland of Lake Merreti *Parachironomus* and *Cladotanytarsus* were the only midges showing a positive response to drying. In other permanent, freshwater wetlands, such as Wongulla Lagoon and Lake Carlet, *Cladotanytarsus* was present throughout the year but showed no consistent seasonal or hydrological response.

Seasonal responses as shown by an increase in abundance in spring were mainly observed for *Limnophyes* in the permanent wetlands of Ramco Lagoon, Wongulla Lagoon and Lake Carlet. Although initial examination of the data suggests that the response is related to flooding, it is the Ramco Lagoon result which demonstrates the seasonality. The second flood did not have a major influence on the communities in Ramco Lagoon but there was a similar response to the previous year.

Table 2. The mean number of common Chironomidae sampled from the eight River Murray flood plain wetlands in South Australia from August 1990 to February 1992.

Clover Lake

Species			1990						MONTHS		1991							1992	
	A	S	O	N	D	J	F	M	A	M	J	J	A	S	O	N	D	J	F
Procladius paludicola.	125	0		0	1		0			145			135						
Chironomus tepperi	0	47		0	0		0			3			0						
Cladopelma	40	0		0	0		0			141			147						
Tanytarsus sp4	0	0		0	7		8			90			8			Dry			Dry
Cricotopus albitibia	0	0		1	4		41			40			88						
Tanytarsus semibarbitarsus	10	0		0	0		0			115			27						
Cladotanytarsus	0	0		15	97		199			487			870						
Microchironomus	0	0		0	0		0			137			57						

Lake Merreti

Species			1990						MONTHS		1991							1992	
	A	S	O	N	D	J	F	M	A	M	J	J	A	S	O	N	D	J	F
Procladius paludicola.	0	0		7	0		3			21			53			25			38
Chironomus tepperi	0	0		18	0		0			4			13			220			0
Parachironomus	0	0		3	0		0			56			104			18			13
Cryptochironomus	0	0		3	1		0			34			1			1			58
Cricotopus albitibia	47	0		66	7		4			292			89			58			113
Cricotopus sp.	3	0		542	22		3			0			0			1			5
Cladotanytarsus	40	0		29	33		264			771			392			46			1228

Lake Woolpolool

Species			1990						MONTHS		1991							1992	
	A	S	O	N	D	J	F	M	A	M	J	J	A	S	O	N	D	J	F
Procladius paludicola.	0	3		5	19		0						13			51			
Chironomus tepperi	64	0		0	1		0						55			1			
Kiefferulus intertinctus	0	0		3	2		0			Dry			75			1			Dry
Polypedilum nubifer	0	30		4	21		0						0			65			

Disher Creek Evaporation Basin

Species			1990						MONTHS		1991							1992	
	A	S	O	N	D	J	F	M	A	M	J	J	A	S	O	N	D	J	F
Procladius paludicola.	4	0		0	1		0			97			326			51			113
Chironomus tepperi	7	45		6	1		0			0			0			5			0
Chironomus EWS sp1	336	10		0	0		0			0			1			20			0
Kiefferulus intertinctus	123	3		6	0		1			0			0			7			13
Cricotopus albitibia	3	2		0	0		0			0			0			342			0
Cricotopus sp.	2	27		4	1		0			0			0			33			0
Tanytarsus semibarbitarsus	14	0		0	1		0			56			244			14			53
Cladotanytarsus	2	3		16	27		12			0			0			240			0

Table 2. Cont.

Berri Evaporation Basin

Species	1990							MONTHS			1991						1992	
	A	S	O	N	D	J	F	M	A	M	J	J	A	S	O	N	D	J	F
Procladius paludicola.	138	0		0	20		1			257			272			11723			6555
Chironomus cloacalis	0	0		0	0		0			0			0			2160			0
Chironomus duplex	13	10		0	0		0			0			0			0			0
Chironomus tepperi	0	987		6	3		0			0			0			2493			0
Chironomus EWS sp1	0	147		29	0		0			0			0			143			0
Kiefferulus intertinctus	3	213		21	3		0			0			0			0			0
Parachironomus	0	3		0	0		0			0			3			180			0
Polypedilum nubifer	0	7		0	0		0			0			0			90			0
Tanytarsus semibarbitarsus	20	0		0	20		0			0			0			820			130

Ramco Lagoon

Species	1990							MONTHS			1991						1992	
	A	S	O	N	D	J	F	M	A	M	J	J	A	S	O	N	D	J	F
Procladius paludicola.	0	0		0	13		2			294			190			48			53
Chironomus cloacalis	0	0		0	0		0			28			28			437			13
Chironomus tepperi	0	0		35	0		26			52			99			725			10
Polypedilum nubifer	0	0		0	0		2			62			87			180			77
Limnophyes EWS sp1	120	203		7	0		0			0			229			65			0

Wongulla Lagoon

Species	1990							MONTHS			1991						1992	
	A	S	O	N	D	J	F	M	A	M	J	J	A	S	O	N	D	J	F
Procladius paludicola.	3	0		0	4		13			104			76			28			18
Cricotopus albitibia	7	10		1	1		21			35			39			168			200
Cricotopus sp.	6	96		9	2		1			13			17			11			0
Limnophyes EWS sp1	4	18		5	0		0			6			32			58			0
Nanocladius	10	0		11	44		0			5			13			155			0
Cladotanytarsus	111	5		40	89		71			52			238			207			48
Paratanytarsus	0	0		0	4		0			328			38			54			0

Lake Carlet

Species	1990							MONTHS			1991						1992	
	A	S	O	N	D	J	F	M	A	M	J	J	A	S	O	N	D	J	F
Procladius paludicola.	3	0		0	1		0			27			1			15			5
Chironomus tepperi	0	70		2	1		8			37			5			22			8
Kiefferulus intertinctus	1	10		161	10		0			6			1			0			0
Polypedilum sp.	3	0		131	508		151			484			38			13			30
Cricotopus albitibia	1	0		1	1		12			21			16			166			93
Cricotopus sp.	0	3		35	15		2			12			4			83			25
Limnophyes EWS sp1	12	117		40	1		0			0			18			14			5
Cladotanytarsus	11	13		13	103		34			172			162			225			113
Paratanytarsus	0	0		9	7		0			47			58			40			13

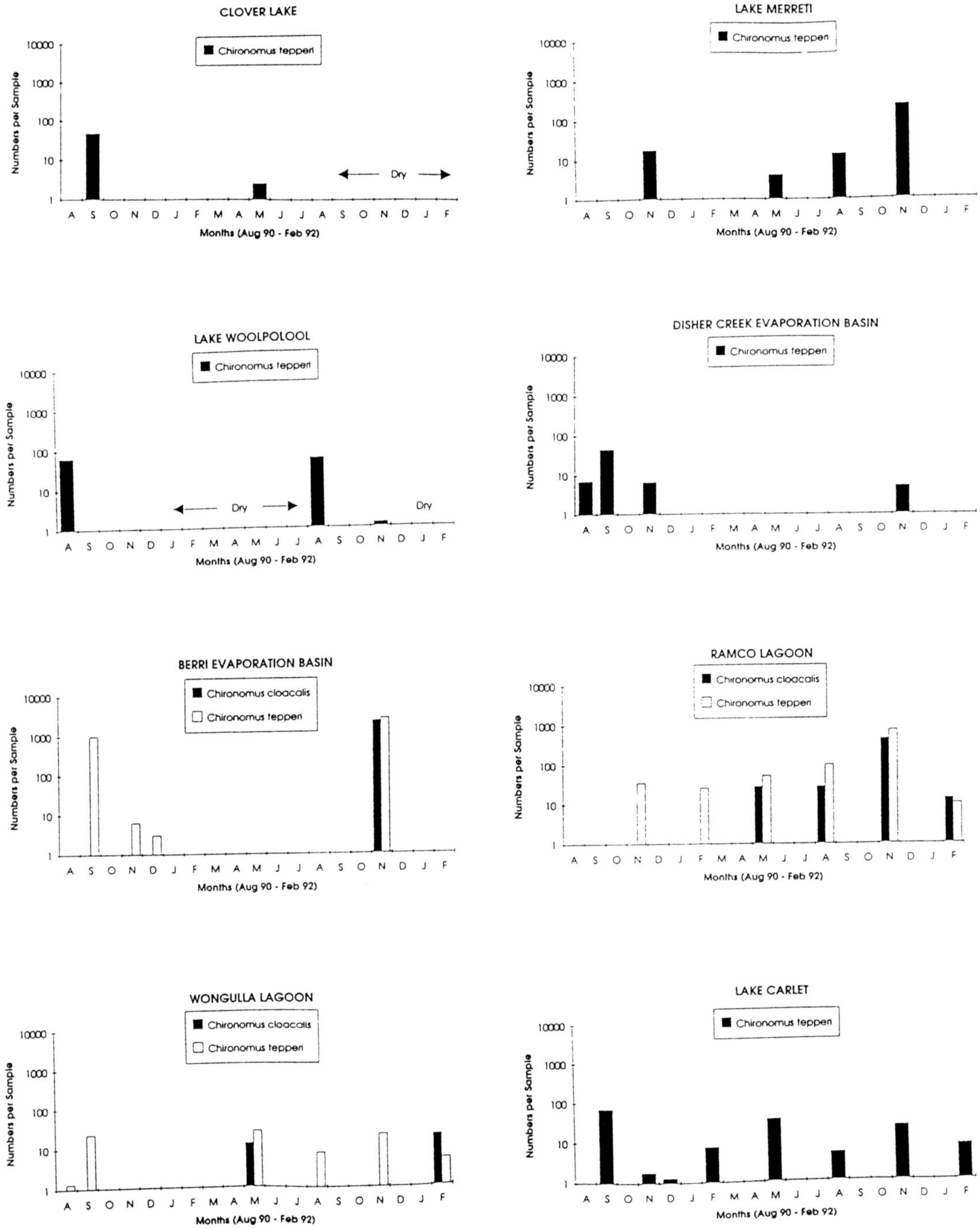

Figure 3. The abundance of *Chironomus tepperi* and *C. cloacalis* in the River Murray floodplain wetlands in South Australia during 1990-1992.

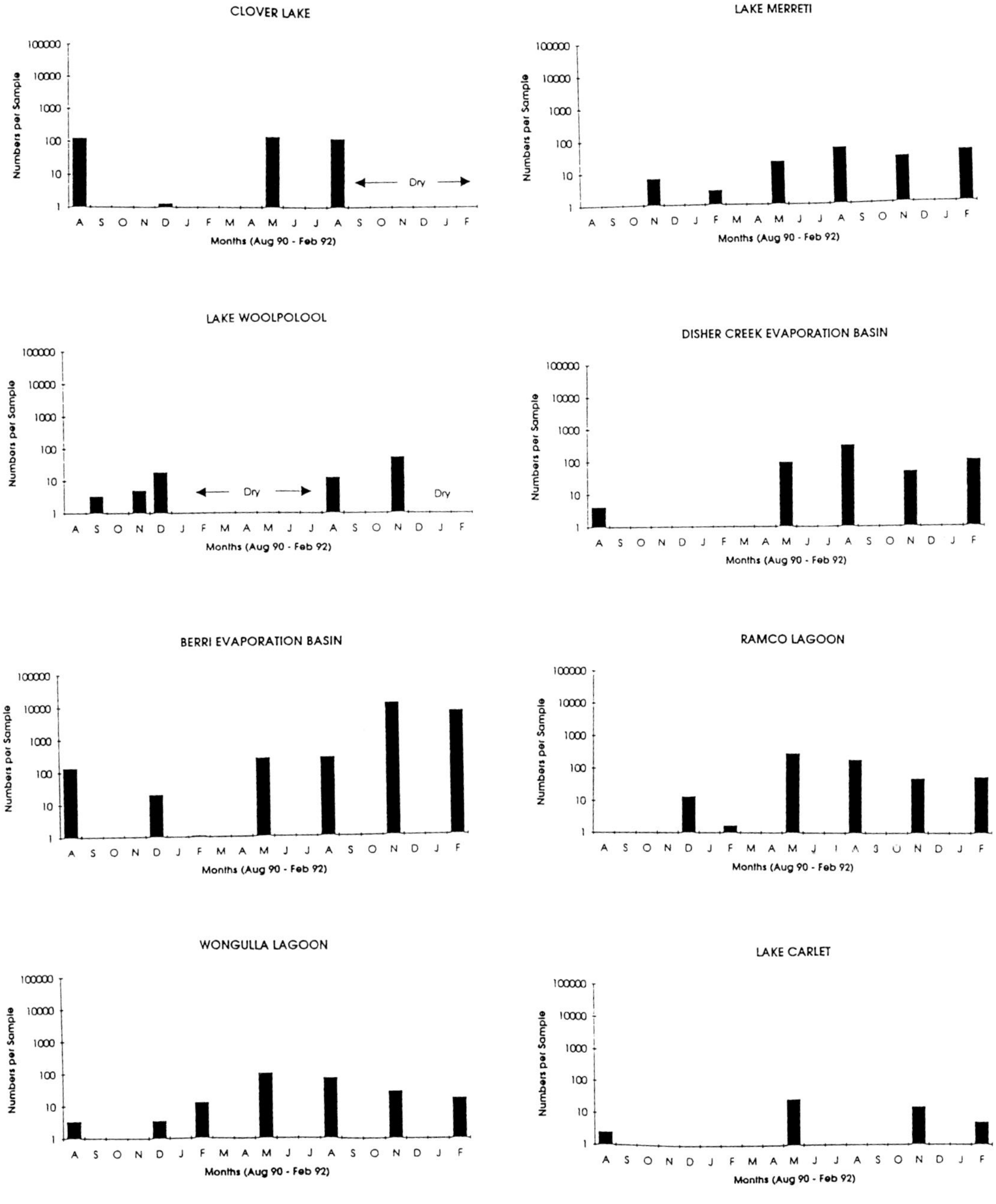

Figure 4. The abundance of *Procladius paludicola* in the River Murray floodplain wetlands in South Australia during 1990-1992.

Nanocladius and *Cladotanytarsus* in Wongulla Lagoon also showed a spring seasonal response. With the exception of the Lake Carlet result, there was an increase in the populations of these species in 1991 compared with 1990.

The organic content of the littoral sediments were generally low (<3%) at most sites but with exceptions at Berri EB and Lake Carlet where the organic content exceeded 5%. In August 1990 Berri EB sites ranged from 1.7 to 9.4% and in February 1991 from 5.7 to 12.7%. At Lake Carlet only one site under willow trees had a high organic content with values of 20.8% and 9.3% recorded in August and February, respectively. When these sites only were considered in terms of *C.tepperi* numbers it was apparent that those with the highest organic content had highest numbers. Site 31 in Lake Carlet contributed almost 90% of *C. tepperi* in 1990 but even though numbers were low in 1991 site 31 contributed 50% of the *C.tepperii* numbers.

Discussion

Three types of responses were observed in the aquatic fauna. These included an increase in numbers in response to flooding, a decline in numbers in response to flooding followed by a positive response to drying and a seasonal response which was unrelated to flooding. During the present study the chironomid fauna had representatives in each of these observed patterns, but the responses were not consistent throughout all wetlands suggesting that there was more to the species response than the hydrology alone.

Positive flooding responses have been recorded for a number of chironomid species, particularly those with a r-selected traits (Bayley, 1991). Murkin and Kadlec (1986) observed non predaceous chironomids colonised rapidly after flooding particularly those species possessing haemoglobin such as *Chironomus* spp. and *Glyptotendipes* spp. Similar responses occur in newly inundated sediments in reservoirs (Paterson and Fernando, 1970; Armitage, 1977; Cantrell and McLachlan, 1977; Sephton *et al.*, 1983). In Australian wetlands, similar observations have been made by Maher and Carpenter (1984) and this study, with positive response to flooding by *Chironomus tepperi*. This species also has rapid growth in rice fields in New South Wales where it causes damage to the crops (Jones, 1968). In addition, this study has shown that other *Chironomus* spp and *K.intertinctus* also respond rapidly to flooding, especially in wetlands which are shallow and which have a high organic content in the sediment. The results also suggest that *C. tepperi* responds with highest population numbers in shallow organic rich wetlands such as Berri EB and the NSW wetlands of Murrumbidgil Swamp and Lake Merrimajeel where organic content was 3.4 to 17.6% and 1.4 and 7.3%, respectively (Briggs and Maher, 1985). Jones (1968) made the observation that "the development of *C. tepperi* as a pest of rice after 30 years of rice growing in southern N.S.W. has been associated with major changes in rice culture. Soils with high organic matter content following pasture rotation...now provide favourable environment....". The relationship between flooding and organic content is consistent with observations made for *Chironomus* spp by Davies and Hawkes (1981), Culp and Davies (1985) and Murkin and Kadlec (1986).

Kiefferulus intertinctus showed no relationship with flooding, drying or season in the NSW wetlands (Maher and Carpenter, 1984), but was related to flooding in the lowland SA wetlands. The difference between the two studies may be the timing of the flood (in Autumn in NSW compared with Spring in SA) and sources of colonisation.

Cricotopus "albitibia" showed a population increase with the floods in the fresh, permanent, and relatively deep wetlands of Wongulla Lagoon and Lake Carlet, both of which are connected to the River Murray. In 1991 this species exceeded 300 animals per sample in Disher Creek EB following the onset of the flood in the River Murray, and may indicate colonisation from the River itself.

Negative response to flooding was noted by Maher and Carpenter (1984) in *Polypedilum nubifer* and *Procladius villosimanus* Kieffer. *Procladius paludicola* showed a similar response in all SA wetlands with greatest abundances in the temporary, saline wetlands, but *Polypedilum nubifer* only showed an increase in abundance as Ramco Lagoon dried and became saline. It is likely that the increase in salinity of the wetlands as they dry provides the appropriate conditions for these species to dominate. Clover Lake, the only temporary, freshwater wetland in the study had a number of species which increased as the wetland dried out.

These included *Cladotanytarsus*, *Cladopelma*, *Microchironomus*, *Tanytarsus* EWS sp4 and *Tanytarsus semibarbitarsus*. The high number of such species in this lake may be related to the temporary nature of the wetland and the absence of a colonising source after the flows in the River Murray receded below 70,000ML/day. In the permanently connected wetlands where some of these species were found there was no discernable responses. This may be due to the differences in hydrology, as well as an influence by predatory fish species from the main channel of the River.

Cryptochironomus griseidorsum and two species of *Dicrotendipes* showed highest numbers in spring/summer in Murrumbidgil Swamp and Lake Merrimajeel (Maher and Carpenter, 1984), but none of these species occured in significant numbers in the SA wetlands to enable comparison. *Limnophyes* showed a seasonal growth pattern in those wetlands closely associated with the River Murray, in particular those connected throughout the year (Wongulla Lagoon and Lake Carlet). During the present study the flood events were in Spring and may have masked some of the Spring/Summer responses noted in the Maher and Carpenter study which examined the responses following an Autumn flood. However, in both studies *C. tepperi* and *Procladius* spp responded similarly irresepective of the timing of the flood event.

The increases in population numbers in most wetlands following the second flood event may be as a result of the ecological changes which occured on the flood plain during the study period. The flood plain had been significantly inundated in 1989/90, some 7 years since the last significant flooding. As a consequence of the long period without inundation the development of riparian and aquatic vegetation was poor. Following the three years of flooding from 1989 - 1992 there was an increase in the growth of aquatic plants and habitat improvement was evident. The chironomid populations clearly showed the increased habitat improvement in each wetland with highest abundances during the second flood event even though the extent of flooding was considerably less than the two previous years.

In conclusion, differences in the chironomid responses between the NSW wetlands and those in SA were probably related to the timing of flooding, salinity variations and the source of colonisation (ie River Murray, a lowland reach versus the Lachlan River, a foothill reach). However, in both the Lachlan River wetlands and those of the lower River Murray the chironomid populations responded to the flood pulses and drawdowns but the physical and chemical conditions also had considerable influence on the species which responded. In shallow, organically enriched wetlands *Chironomus* spp respond rapidly to flooding, and *Procladius* spp increase as wetlands dried out. For some species the responses are consistent irrespective of location whereas others behave differently at different locations. Therefore generalisations on the responses by chironomids to hydrological changes may be inappropriate considering the site specific responses observed in the SA River Murray flood plain wetlands.

Acknowledgements

These results are part of a major monitoring study of the South Australian River Murray flood plain wetlands which was funded by the Murray-Darling Basin Commission through the Natural Resources Management Strategy. We are grateful for the assistance provided by personnel from the Engineering and Water Supply Department, the South Australian Department of Environment and Natural Resources and members of the SA wetlands management committee for their support and constructive input to the project.

Chironomid assemblages of temporal and permanent water bodies in an Australian floodplain forest

T. J. Hillman and D. L. Nielsen

Murray-Darling Freshwater Research Centre, Albury, NSW.

Introduction

Changes in the flood pattern of the River Murray resulting from regulation have influenced the inundation regime in floodplain forests. A long term program has been established to investigate the response of forest trees to a variety of experimental inundation patterns. The work reported here is part of an ongoing study examining the role of inundation patterns in the ecology of floodplain aquatic systems. The aim of the work is to contribute to the design of ecologically sound flow management procedures in a regulated floodplain river.

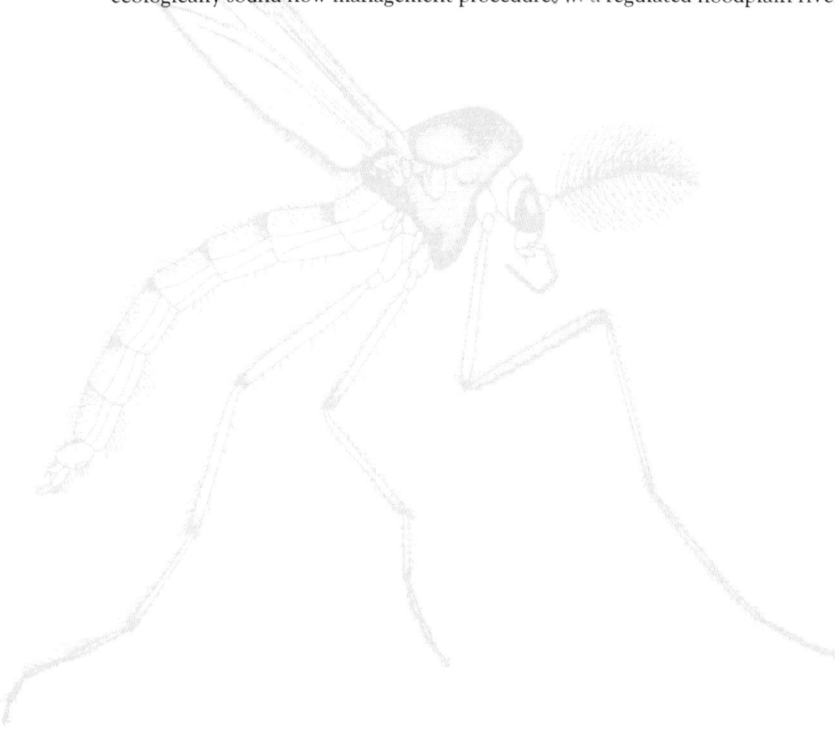

Methods

Chironomid larvae were collected with hand nets from six floodplain water bodies in the Gulpa Forest, New South Wales, S.E. Australia, during the period August 1991 - March 1992. These water bodies were subject to a range of hydrological conditions. Two were permanently inundated; three were filled in late winter for varying periods of time; and one remained dry until summer. Chironomid larvae (and other aquatic invertebrates) were preserved, identified (usually to genus level), and counted.

Similarity analyses of the site-samples and of species-samples were carried out using Jaccard's dichotomy coefficient which calculates the proportion of co-occurrence against total occurrence. The matrices of these coefficients between samples and between taxa were then used in a cluster analysis to rank the sites and taxa according to their similarity. The data were then sorted according to these ranks.

Results

The total number of chironomid larvae found at each site is presented in Table 1. Although care was taken to ensure that sampling effort was comparable, the data must be treated as qualitative with numbers indicating relative abundance only. In further analysis the data are treated in terms of "presence/absence". The number of taxa seen per sampling occasion are presented for each site in Figure 1. The results of sorting according to the ranks of sites and taxa by similarity are shown in Table 2.

Discussion

Because of the limited nature of this preliminary study - particularly in terms of replication - it is not possible to draw strong conclusions regarding the seasonal effects of hydrological pattern.

The data indicate no clearly identifiable "pioneering" taxa although some taxa clearly appear to favour ephemeral systems. Other studies of ephemeral water-bodies (Barclay, 1966; Lake *et al.*, 1989) also indicate that chironomids are not prominent amongst early colonising species although a number of species are adapted to surviving local dessication (Wiggins *et al.*, 1980). In terms of the taxa present, the samples clearly divide into temporary and permanent water bodies; the former further divide into seasonal groups. Thus Group F (Table 2) contains most of the samples from permanent inundated sites. Groups A, B and C are constituted by most of the samples taken during winter-spring from ephemeral sites; Group G represents the same sites in late spring; and Group H those sites in summer. Throughout most of these groupings there is a significant number of common taxa.

With regard to the spatial and temporal distribution of taxa, Group 1 contains rare taxa (outliers in terms of the analysis). Taxa in Groups 2 and 3 were also quite rare being found mainly in ephemeral systems in late spring. Group 5 contained a number of ubiquitous taxa. Several taxa ("MO10", *Paraborniella* sp. and *Corynoneura* sp.) favour ephemeral systems, and most being particularly widespread during the late spring. The aforementioned taxa were absent from summer samples which may have implications for wetlands freshly inundated at that time. Group 6 occurred mainly in permanent water systems, perhaps because they were poor colonisers or (more likely) because of the presence of appropriate habitats at those sites - the diverse macrophyte beds present in the permanently flooded sites did not develop significantly in the ephemeral waters during the study.

The consistent difference between the ephemeral water bodies and the permanent ones from which they were filled, particularly early in the period of inundation, would indicate that transport with the inflowing water was not a common means of colonisation, at least for the majority of taxa.

Conclusions

1) Although some taxa appeared to be more successful in ephemeral waters, none appear especially to favour the early stages of inundation (regardless of season).

2) Seasonal patterns have some influence on the taxonomic composition of chironomid assemblages in temporary water bodies although there was a significant number of taxa present throughout the year.

3) There were assemblages of chironomids which typify permanently inundated systems. These probably reflect differences in habitat diversity between the permanent and ephemeral water bodies in this study.

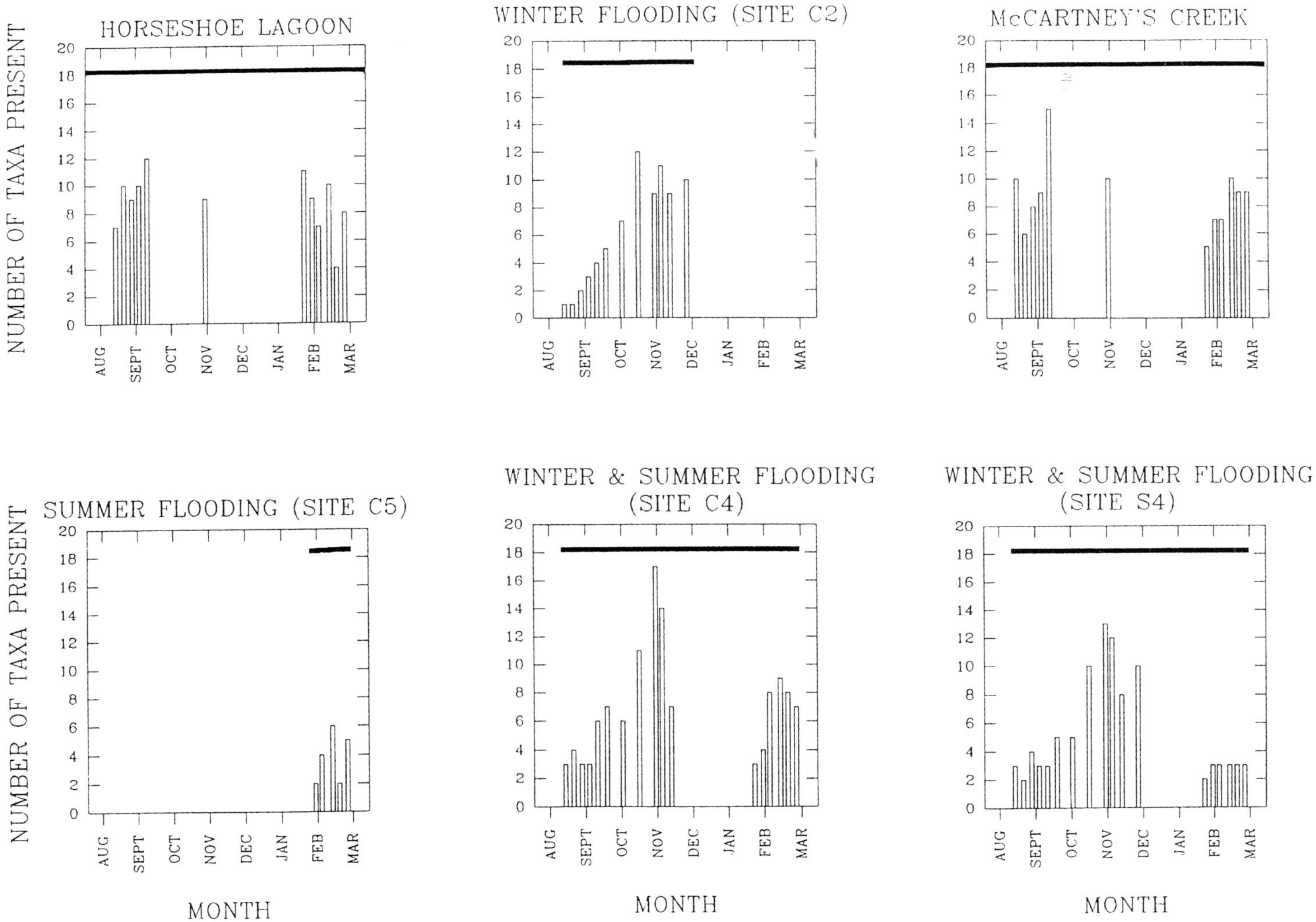

Figure 1. Number of Chironomid taxa collected in composite hand-net samples. Black bar indicates period of inundation

Table 1: Total number of chironomids collected in composite handnet samples

Site	HS	MC	C2	C4	C5	S4
Number of samples	12	12	12	18	6	18
Taxa						
Cladopelma sp	0	0	40	46	0	10
Chironomus sp	10	5	780	315	2100	1199
Cryptochironomus griseidorsum	0	0	0	21	0	0
Dicrotendipes sp	60	30	60	11	68	20
Kiefferulus martini	192	515	480	171	10	810
Kiefferulus sp	37	85	20	96	148	428
Microchironomus sp	0	0	0	63	0	0
Microtendipes sp	0	0	10	0	0	0
Paraborniella sp	10	0	1278	422	0	191
Parachironomus sp	123	126	880	119	0	460
Paracladopelma sp	10	0	0	0	0	0
Polypedilum leei	603	264	190	31	0	0
Polypedilum nubifer	0	0	0	10	0	0
Polypedilum oresitrophus	105	85	0	20	0	0
Riethia sp	1	0	0	0	0	0
Stenochironomus sp	0	0	10	0	0	0
Cladotanytarsus sp	151	359	50	11	0	10
Tanytarsus sp	61	139	690	327	91	290
Paratanytarsus sp	933	582	1020	682	0	180
Ablabesmyia sp	52	11	50	20	0	0
Coelopynia pruinosa	0	0	0	10	0	0
Larsia albiceps	1155	391	900	76	0	321
Paramerina sp	10	30	190	40	0	0
Unknown tanypod sp	10	10	10	0	0	0
Procladius sp	140	190	160	395	22	50
Bryophaenocladius sp	10	0	10	2	0	8
Corynoneura sp	130	136	1360	790	0	1085
Cricotopus sp	131	238	200	642	0	162
Nanocladius sp	111	22	30	10	0	0
Parakiefferiella sp	0	10	0	40	0	10
Paralimnophyes albibasis	130	30	200	41	0	3
"MO10"	10	132	30	843	0	206
Thienemanniella sp	0	0	0	10	0	0

Acknowledgements

Thanks are due to Dr Peter Cranston (CSIRO Division of Entomology) for taxonomic guidance and active participation. Chris Knight, Steve Wilson and Robert Cook are gratefully acknowledged for their help with sampling, sorting, and identification. Ms Marijke Korting prepared the tables and manuscript.

The work was part of a research project supported by the Murray-Darling Basin Commission through a Natural Resources Management Strategy Grant and was coordinated within a larger study of floodplain forest management by State Forests of NSW.

Table 2. Presence (Δ) - Absence (.) data for net samples of chironomid larvae grouped according to similarity (Jaccard coefficient) between samples and taxa.

		A									B							C					D	E
	Site:	S4	C4	S4	MC	C4	C4	C4	C4	HS	C2	C2	C2	S4	C2	C2	S4	S4	C4	S4	C2	C2	C4	S4
	Month:	9	9	9	9	9	9	8	8	9	10	9	9	1	8	9	8	8	8	1	8	8	1	8
	Flooding Duration:	5	4	4	99	6	5	3	2	99	7	6	5	21	3	4	3	2	1	20	1	2	20	1
SPECIES	Sp No.																							
Microtendipes sp	8	.	.	.	.	.	.	.	.	.	.	.	Δ	.	.	.	.	.	.	.	.	.	.	.
Riethia sp	15	.	.	.	.	.	.	.	.	.	.	.	.	.	.	.	.	.	.	.	.	.	.	.
Stenochironomus sp	16	.	.	.	.	.	.	.	.	.	Δ	.	.	.	.	.	.	.	.	.	.	.	.	.
Paracladopelma sp	11	.	.	.	.	.	.	.	.	.	.	.	.	.	.	.	.	.	.	.	.	.	.	.
unknown Tanypod sp	24	.	.	.	.	.	.	.	.	.	.	.	.	.	.	.	.	.	.	.	.	.	.	.
Polypedilum nubifer	13	.	.	.	.	.	.	.	.	.	.	.	.	.	.	.	.	.	.	.	.	.	.	.
Cryptochironomus griseidorsum	3	.	.	.	.	.	.	.	.	.	.	.	.	.	.	.	.	.	.	.	.	.	.	.
Microchironomus sp	7	.	.	.	.	.	.	.	.	.	.	.	.	.	.	.	.	.	.	.	.	.	.	.
Coelopynia pruinosa	21	.	.	.	.	.	.	.	.	.	.	.	.	.	.	.	.	.	.	.	.	.	.	.
Parakiefferiella sp	30	.	.	.	Δ	Δ	.	.	.	.	.	.	.	.	.	.	.	.	.	.	.	.	.	.
Paramerina sp	23	.	.	.	Δ	.	.	.	.	Δ	.	.	.	.	.	.	.	.	.	.	.	.	.	.
Cladopelma sp	1	.	.	.	.	.	.	.	.	.	.	.	.	.	.	.	.	.	.	.	.	.	.	.
Paralimnophyes albibasis	31	.	Δ	.	.	.	Δ	.	.	Δ	Δ	Δ	.	.	.	.	.	.	Δ	Δ	.	.	.	.
MO10 sp	32	Δ	Δ	Δ	Δ	Δ	Δ	Δ	Δ	Δ	.	.	.	.	.	Δ	Δ	.	.	.	.	.	.	.
Paraborniella sp	9	.	.	.	.	Δ	Δ	Δ	Δ	.	Δ	Δ	Δ	Δ	Δ	Δ	Δ	Δ	Δ	Δ	Δ	Δ	.	.
Chironomus sp	2	.	.	.	.	Δ	.	.	.	.	Δ	Δ	Δ	Δ	Δ	Δ	Δ	.	.	.	.	.	.	.
Procladius sp	25	.	.	.	Δ	Δ	Δ	Δ	.	.	.	.	.	.	.	.	.	.	.	.	.	.	.	.
Tanytarsus sp	18	.	.	.	.	.	.	.	.	.	Δ	Δ	.	.	.	.	.	.	.	.	.	.	.	.
Kiefferulus martini	5	.	.	.	.	.	.	.	.	.	Δ	.	.	.	.	.	.	.	.	.	.	.	.	.
Parachironomus sp	10	.	.	.	.	.	.	.	.	.	Δ	Δ	Δ	.	.	.	.	.	.	.	.	.	.	.
Paratanytarsus sp	19	.	.	.	.	.	.	.	Δ	Δ	.	.	.	.	.	.	.	.	.	.	.	.	.	.
Cricotopus sp	28	.	.	Δ	Δ	Δ	Δ	.	.	Δ	.	.	.	.	.	.	.	.	.	.	.	.	Δ	Δ
Corynoneura sp	27	Δ	Δ	Δ	Δ	Δ	Δ	.	.	Δ	.	.	.	.	.	.	.	.	.	.	.	.	.	.
Kiefferulus sp	6	.	.	.	.	.	.	.	.	.	.	.	.	Δ	.	.	.	.	.	.	.	.	.	.
Nanocladius sp	29	.	.	.	Δ	.	.	.	.	.	.	.	.	.	.	.	.		.	.	.	.	.	.
Cladotanytarsus sp	17	.	.	.	Δ	.	.	.	.	.	.	.	.	.	.	.	.	.	.	.	.	.	Δ	.
Polypedilum oresitrophus	14	.	.	.	Δ	.	.	.	.	.	.	.	.	.	.	.	.	.	.	.	.	.	.	.
Polypedilum leei	12	.	.	.	.	.	.	.	.	Δ	.	.	.	.	.	.	.	.	.	.	.	.	Δ	.
Larsia sp	22	.	.	.	.	.	.	.	.	Δ	.	.	.	.	.	.	.	.	.	.	.	.	.	Δ
Ablabesmyia sp	20	.	.	.	.	.	.	.	.	Δ	.	.	.	.	.	.	.	.	.	.	.	.	.	.
Dicrotendipes sp	4	.	.	.	.	.	.	.	Δ	Δ	.	.	.	.	.	.	.	.	.	.	.	.	.	.
Bryophaenocladius sp	26	Δ	.	.	.	.	.	.	.	.	.	.	.	.	.	.	Δ	Δ	Δ	.	.	.	.	Δ
Thienemanniella sp	33	.	.	.	.	.	.	.	.	.	.	.	.	.	.	.	.	.	.	.	.	.	.	.

Table 2. Continued
Presence (Δ) - Absence (.) data for net samples of chironomid larvae grouped according to similarity (Jaccard coefficient) between samples and taxa.

															F											
	Site:	HS	MC	MC	HS	MC	HS	MC	HS	HS	MC	MC	HS	HS	HS	MC	MC	MC	HS	C4	C4	C4	HS	HS	MC	MC
	Month:	2	1	1	2	2	10	9	1	2	2	2	1	8	8	10	8	2	2	2	2	2	8	9	8	8
	Flooding Duration:	99	99	99	99	99	99	99	99	99	99	99	99	99	99	99	99	99	99	22	23	24	99	99	99	99
SPECIES	Sp No.																									
Microtendipes sp	8	.	.	.	.	.	.	.	.	.	.	.	.	.	.	.	.	.	.	.	.	.	.	.	.	.
Riethia sp	15	.	.	.	.	.	.	.	.	.	.	.	.	.	.	.	.	.	Δ	.	.	.	.	.	.	.
Stenochironomus sp	16	.	.	.	.	.	.	.	.	.	.	.	.	.	.	.	.	.	.	.	.	.	.	.	.	.
Paracladopelma sp	11	.	.	.	.	.	.	.	.	.	.	.	Δ	.	.	.	.	.	.	.	.	.	.	.	.	.
unknown Tanypod sp	24	.	.	Δ	.	.	.	.	.	.	.	.	.	.	Δ	.	.	.	.	.	.	.	.	.	.	.
Polypedilum nubifer	13	.	.	.	.	.	.	.	.	.	.	.	.	.	.	.	.	.	.	.	.	.	.	.	.	.
Cryptochironomus griseidorsum	3	.	.	.	.	.	.	.	.	.	.	.	.	.	.	.	.	.	.	.	.	.	.	.	.	.
Microchironomus sp	7	.	.	.	.	.	.	.	.	.	.	.	.	.	.	.	.	.	.	.	.	Δ	.	.	.	.
Coelopynia pruinosa	21	.	.	.	.	.	.	.	.	.	.	.	.	.	.	.	.	.	.	.	.	.	.	.	.	.
Parakiefferiella sp	30	.	.	.	.	.	.	.	.	.	.	.	.	.	.	.	.	.	.	.	.	.	.	.	.	.
Paramerina sp	23	.	.	.	.	.	.	.	.	.	.	.	.	.	.	.	.	.	.	.	.	.	.	.	.	.
Cladopelma sp	1	.	.	.	.	.	.	.	.	.	.	.	.	.	.	.	.	.	.	.	.	.	.	.	.	.
Paralimnophyes albibasis	31	.	.	.	.	.	.	.	.	.	.	.	.	.	.	.	Δ	.	.	.	.	.	Δ	Δ	.	.
MO10 sp	32	.	.	.	.	.	.	Δ	.	.	.	.	.	.	.	.	.	.	.	.	.	.	.	.	Δ	.
Paraborniella sp	9	.	.	.	.	.	.	.	.	.	.	.	.	.	.	.	.	.	.	.	.	.	.	Δ	.	.
Chironomus sp	2	.	.	.	.	Δ	.	.	.	Δ	.	.	.	.	.	.	.	.	.	Δ	Δ	Δ	.	.	.	.
Procladius sp	25	.	.	.	.	.	.	Δ	.	.	Δ	Δ	Δ	Δ	Δ	Δ	.	.	.	.	Δ	Δ	Δ	Δ	Δ	Δ
Tanytarsus sp	18	.	.	.	.	.	.	Δ	.	.	Δ	Δ	Δ	Δ	.	.	Δ	.	Δ	Δ	Δ	Δ	Δ	Δ	Δ	Δ
Kiefferulus martini	5	.	.	.	Δ	Δ	.	Δ	Δ	.	Δ	Δ	.	Δ	Δ	Δ	Δ	Δ	Δ	Δ	Δ	Δ	Δ	Δ	Δ	Δ
Parachironomus sp	10	.	.	.	.	Δ	Δ	Δ	Δ	Δ	Δ	Δ	Δ	Δ	Δ	Δ	Δ	Δ	Δ	.	Δ	Δ	.	.	.	.
Paratanytarsus sp	19	Δ	.	Δ	Δ	Δ	Δ	Δ	Δ	Δ	Δ	Δ	Δ	Δ	Δ	Δ	Δ	Δ	Δ	.	.	.	.	Δ	Δ	Δ
Cricotopus sp	28	.	Δ	Δ	Δ	Δ	Δ	Δ	Δ	Δ	Δ	.	.	.	.	Δ	.	.	.	.	.	.	.	.	.	.
Corynoneura sp	27	.	.	.	.	.	Δ	Δ	.	.	.	.	.	.	.	Δ	Δ	.	.	.	Δ	.	.	Δ	Δ	Δ
Kiefferulus sp	6	.	.	.	.	.	.	Δ	Δ	Δ	.	.	.	.	.	Δ	Δ	Δ	Δ	Δ	Δ	Δ	Δ	Δ	Δ	Δ
Nanocladius sp	29	.	.	.	.	.	Δ	Δ	Δ	Δ	.	Δ	Δ	.	.	.	.	.	.	.	.	.	.	Δ	.	.
Cladotanytarsus sp	17	.	Δ	Δ	Δ	Δ	Δ	Δ	Δ	Δ	Δ	.	Δ	.	.	.	Δ	Δ	.	.	.	.	.	Δ	.	.
Polypedilum oresitrophus	14	Δ	Δ	Δ	Δ	Δ	.	Δ	Δ	Δ	Δ	Δ	.	.	.	.	.	.	.	Δ	.	.	.	.	.	.
Polypedilum leei	12	Δ	Δ	Δ	Δ	Δ	Δ	Δ	Δ	Δ	Δ	Δ	Δ	Δ	Δ	Δ	Δ	Δ	Δ	Δ	Δ	.	Δ	Δ	.	.
Larsia sp	22	Δ	Δ	Δ	Δ	Δ	Δ	Δ	Δ	Δ	Δ	Δ	Δ	Δ	Δ	Δ	.	Δ	Δ	Δ	Δ	Δ	Δ	Δ	Δ	.
Ablabesmyia sp	20	.	.	.	.	.	Δ	Δ	Δ	.	.	.	.	Δ	Δ	.	.	.	.	Δ	.	.	.	.	.	.
Dicrotendipes sp	4	.	.	.	.	.	.	.	.	.	.	.	.	Δ	Δ	Δ	Δ	.	.	.	.	.	.	.	.	.
Bryophaenocladius sp	26	.	.	.	.	.	.	.	.	.	.	.	.	Δ	.	.	.	.	.	.	.	.	.	.	.	.
Thienemanniella sp	33	.	.	.	.	.	.	.	.	.	.	.	.	.	.	.	.	.	.	.	.	.	.	.	.	.

Table 2. Continued
Presence (Δ) - Absence (.) data for net samples of chironomid larvae grouped according to similarity (Jaccard coefficient) between samples and taxa.

		G																	H										
	Site:	C4	C4	C2	C2	C2	C2	C4	S4	S4	S4	C2	S4	S4	C4	C4	S4	S4	C5	C4	C4	C5	C5	C5	C5	S4	S4	S4	S4
	Month:	10	11	11	10	11	11	11	10	11	11	10	11	10	10	10	10	9	1	1	2	2	2	2	2	2	2	2	2
	Flooding Duration:	7	11	12	8	10	11	10	9	10	12	9	11	8	8	9	7	6	2	21	25	4	6	3	5	24	22	23	25
SPECIES	Sp No.																												
Microtendipes sp	8	.	.	.	.	.	.	.	.	.	.	.	.	.	.	.	.	.	.	.	.	.	.	.	.	.	.	.	.
Riethia sp	15	.	.	.	.	.	.	.	.	.	.	.	.	.	.	.	.	.	.	.	.	.	.	.	.	.	.	.	.
Stenochironomus sp	16	.	.	.	.	.	.	.	.	.	.	.	.	.	.	.	.	.	.	.	.	.	.	.	.	.	.	.	.
Paracladopelma sp	11	.	.	.	.	.	.	.	.	.	.	.	.	.	.	.	.	.	.	.	.	.	.	.	.	.	.	.	.
unknown Tanypod sp	24	.	.	.	.	.	.	.	.	.	.	.	.	.	.	.	.	.	.	.	.	.	.	.	.	.	.	.	.
Polypedilum nubifer	13	.	.	.	.	.	.	Δ	.	.	.	.	.	.	.	.	.	.	.	.	.	.	.	.	.	.	.	.	.
Cryptochironomus griseidorsum	3	Δ	Δ	.	.	.	.	Δ	.	.	.	.	.	.	.	.	.	.	.	.	.	.	.	.	.	.	.	.	.
Microchironomus sp	7	.	.	.	.	.	.	.	.	.	.	.	.	.	Δ	Δ	.	.	.	.	.	.	.	.	.	.	.	.	.
Coelopynia pruinosa	21	.	.	.	.	.	.	.	.	.	.	.	.	.	.	Δ	.	.	.	.	.	.	.	.	.	.	.	.	.
Parakiefferiella sp	30	.	.	.	.	.	.	.	.	.	.	.	.	.	.	Δ	.	Δ	.	.	.	.	.	.	.	.	.	.	.
Paramerina sp	23	.	.	Δ	Δ	Δ	Δ	.	.	.	.	.	.	.	.	Δ	.	.	.	.	.	.	.	.	.	.	.	.	.
Cladopelma sp	1	.	Δ	Δ	Δ	Δ	.	Δ	Δ	.	.	.	.	.	.	Δ	.	.	.	.	Δ	.	.	.	.	.	.	.	.
Paralimnophyes albibasis	31	.	.	.	Δ	.	.	Δ	.	.	.	.	.	.	.	.	.	.	.	.	.	.	.	.	.	.	.	.	.
MO10 sp	32	.	.	.	.	.	.	.	Δ	Δ	.	.	.	Δ	Δ	Δ	Δ	.	.	.	.	.	.	.	.	.	.	.	.
Paraborniella sp	9	.	.	Δ	Δ	Δ	.	Δ	Δ	Δ	Δ	Δ	.	Δ	Δ	Δ	.	.	.	.	.	.	.	.	.	.	.	.	.
Chironomus sp	2	Δ	Δ	Δ	Δ	Δ	Δ	Δ	Δ	Δ	Δ	Δ	Δ	Δ	.	Δ	.	.	Δ	Δ	Δ	Δ	Δ	Δ	Δ	Δ	Δ	Δ	Δ
Procladius sp	25	Δ	Δ	.	Δ	Δ	Δ	Δ	Δ	Δ	Δ	.	.	Δ	Δ	Δ	.	.	.	Δ	Δ	Δ	Δ	.	.	.	.	.	.
Tanytarsus sp	18	Δ	Δ	Δ	Δ	Δ	Δ	Δ	Δ	Δ	Δ	Δ	Δ	Δ	Δ	Δ	Δ	.	Δ	Δ	Δ	Δ	Δ	Δ	.	.	.	.	.
Kiefferulus martini	5	.	.	.	Δ	Δ	Δ	Δ	Δ	Δ	.	Δ	Δ	Δ	Δ	Δ	Δ	.	.	.	Δ	Δ	.	.	.	Δ	Δ	Δ	Δ
Parachironomus sp	10	.	Δ	Δ	Δ	Δ	Δ	Δ	Δ	Δ	Δ	Δ	Δ	Δ	Δ	Δ	Δ	Δ	.	.	.	.	.	.	.	.	.	.	.
Paratanytarsus sp	19	Δ	.	Δ	Δ	Δ	Δ	Δ	Δ	Δ	Δ	Δ	Δ	Δ	Δ	Δ	.	Δ	.	.	.	.	.	.	.	.	.	.	.
Cricotopus sp	28	Δ	Δ	Δ	Δ	Δ	Δ	Δ	Δ	Δ	Δ	Δ	Δ	.	Δ	Δ	Δ	Δ	.	.	.	.	.	.	.	.	.	.	.
Corynoneura sp	27	.	.	Δ	Δ	Δ	Δ	Δ	Δ	Δ	Δ	Δ	Δ	Δ	Δ	Δ	.	Δ	.	.	.	.	.	.	.	.	.	.	.
Kiefferulus sp	6	.	.	.	.	.	.	.	.	Δ	Δ	.	.	.	.	.	.	.	.	.	Δ	Δ	Δ	Δ	Δ	Δ	Δ	Δ	Δ
Nanocladius sp	29	.	.	.	.	.	.	.	.	.	.	.	.	.	.	Δ	.	.	.	.	.	.	.	.	.	.	.	.	.
Cladotanytarsus sp	17	.	.	.	.	.	.	.	Δ	.	.	.	.	.	.	Δ	.	.	.	.	.	.	.	.	.	.	.	.	.
Polypedilum oresitrophus	14	.	.	.	.	.	.	Δ	.	.	.	.	.	.	.	.	.	.	.	.	.	.	.	.	.	.	.	.	.
Polypedilum leei	12	.	.	Δ	.	.	.	.	.	.	.	.	.	.	.	.	.	.	.	Δ	.	.	.	.	.	.	.	.	.
Larsia sp	22	.	.	.	.	.	.	.	Δ	Δ	Δ	Δ	Δ	.	.	.	.	.	.	.	.	.	.	.	.	.	.	.	.
Ablabesmyia sp	20	.	.	.	.	.	.	.	.	.	.	.	.	.	.	.	.	.	.	.	.	.	.	.	.	.	.	.	.
Dicrotendipes sp	4	.	.	.	.	.	.	.	.	.	.	.	.	Δ	.	.	.	.	.	.	Δ	Δ	Δ	Δ	.	.	.	.	.
Bryophaenocladius sp	26	.	.	.	.	.	.	.	.	.	.	.	.	.	.	.	.	.	.	.	.	.	.	.	.	.	.	.	.
Thienemanniella sp	33	.	.	.	.	.	.	.	.	.	.	.	.	.	Δ	.	.	.	.	.	.	.	.	.	.	.	.	.	.

Stream drift of Chironomidae: Diel and seasonal patterns in the Acheron River, Victoria, Australia

E.S.G. Schreiber

Cooperative Research Centre for Freshwater Ecology, Department of Ecology and Evolutionary Biology, Monash University, Clayton, 3168, Australia.

Introduction

Chironomids can contribute a relatively large proportion to invertebrate stream drift (Williams, 1989; Sagar and Glova, 1992). Contrary to other commonly drifting invertebrate groups, such as some ephemeropteran taxa, they are often reported as not showing behavioural drift (= the drift that occurs with a diel pattern in densities; refer also to Waters, 1972), or as being the exception to the norm by showing diurnal rather than nocturnal behavioural drift (peaks in drift occuring during the day rather than at night)(Brittain and Eikeland, 1988). However, in many drift studies chironomids are not identified below the family level, and thus we have less information about the drift of chironomids than that of other benthic invertebrates (Ferrington, 1984; Tilley, 1989). This is particularly the case in Australia, where chironomid taxonomy is still undergoing major revision.

As part of a general study on invertebrate stream drift (Schreiber, 1994), I investigated which chironomid taxa drifted and whether diel and/or seasonal drift patterns existed in the Acheron River, a temperate Australian stream.

Materials and Methods

The study site was located in a 4th order section of the Acheron River, Victoria, Australia (145°43'E, 37°38.5'N). Drift samples were collected hourly, for 25 consecutive hours, around New Moon, once every 2 months, from December 1981 to April 1983. At least 2 replicate samples were collected per hour. The drift samplers had a net attached to them with a mesh size of 0.3mm, which has been suggested to be the upper limit for quantitatively collecting chironomid larvae (Ferrington, 1984). Other studies have recommended nets with smaller mesh sizes (Williams, 1985), however, I carried out preliminary tests of the efficiency of nets of different mesh sizes and found that nets of a smaller mesh size became clogged with organic matter, and thus would not have sampled drift with consistent efficiency at all times. A mesh size of 0.3mm was found to represent a compromise between consistent sampling efficiency and retaining as many drifting invertebrates as possible.

Current velocity was measured hourly at the mouth of each net, and mean hourly velocity was calculated from readings taken at the beginning and the end of the sampling hour. Drift densities were calculated as the number of animals collected per volume of water (Allan and Russek, 1985).

Light intensity, reflected from a Kodak Neutral Test card with 18% reflection, which was held just above each sampler, was measured hourly to allow a consistent estimation of sunset and sunrise (= the hour during which light intensity dropped/increased below/above 1 lux).

Both drift and benthic samples were collected on 17 and 18 March, 1983, to compare the relative faunal composition of a specific area of the benthos and the drift originating from it. One of the major problems, when relating benthic samples to drift samples, is that the origin of the drift is unknown; drift could originate from various habitats upstream from the site of benthic sampling. This problem was avoided by restricting the upstream origin of drift with a barrier of the same mesh size as the drift nets. The barrier was placed around an area approximately 3m long and 2m wide, just upstream from 2 drift nets. To minimize the effect of this barrier on current flow, it was pointed into the current and its angle to the current was adjusted so that current velocity was only reduced within 10cm of the barrier (benthic samples were not collected this close to the barrier). Drift sampling began around sunset and finished after sunrise. Then 7 random benthic samples were collected from the area enclosed by the barrier, with a Surber sampler of 0.3mm mesh size.

All samples were preserved in 10% formaldehyde in the field. In the lab arthropods were separated from detritus with a kerosene phase separation technique (Barmuta, 1984) to ensure that all instars were picked from samples consistently, despite variations in the amount of detritus collected (Schreiber, 1994). Chironomids were mounted on slides in Berlese solution (Wiederholm, 1983) for identification.

Due to the number of temporal samples needed to describe diel and seasonal patterns in drift, the number of spatial replicates that can be taken per hour was limited. Yet it was important to establish subjective methods for the description of drift patterns. As samples collected for evaluating diel drift patterns were collected consecutively through time I used a time series analysis (Finch, 1973) to test whether drift increased or decreased between the hours of sunset and sunrise. I have described this procedure in detail elsewhere (Schreiber, in press). I classified a taxon as showing behavioural drift if its densities increased significantly, either at night or during the day, in more than one month in which I sampled.

For the description of seasonal patterns in chironomid drift I used a one-way analysis of variance, followed by a post-hoc comparisons of means test (Ryan's test: Day and Quinn, 1989).

Results

Overall I collected 195 invertebrate taxa from the drift. Most were insects, predominantly of the orders Trichoptera and Chironomidae (Fig.1). Insects also dominated drift numerically, and again the Trichoptera and Chironomidae made up a large proportion of this (Fig.2). Ephemeroptera nymphs contributed relatively more to total abundances than to total number of species (Figs. 1 and 2). This reflects their propensity to drift, which has been documented in many other studies (Brittain and Eikeland, 1988).

Of 41 chironomid taxa identified from the drift only 9 were collected with overall abundances of more than 100 individuals over the whole study period (Table 1).

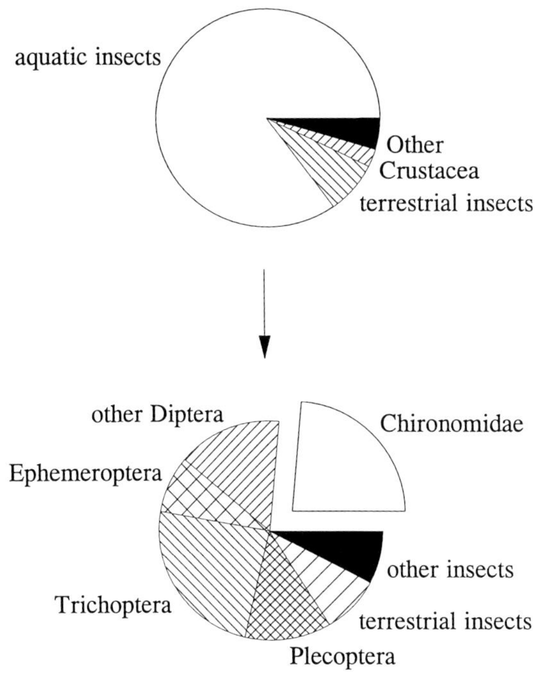

Figure 1. Composition of total drift in terms of species.

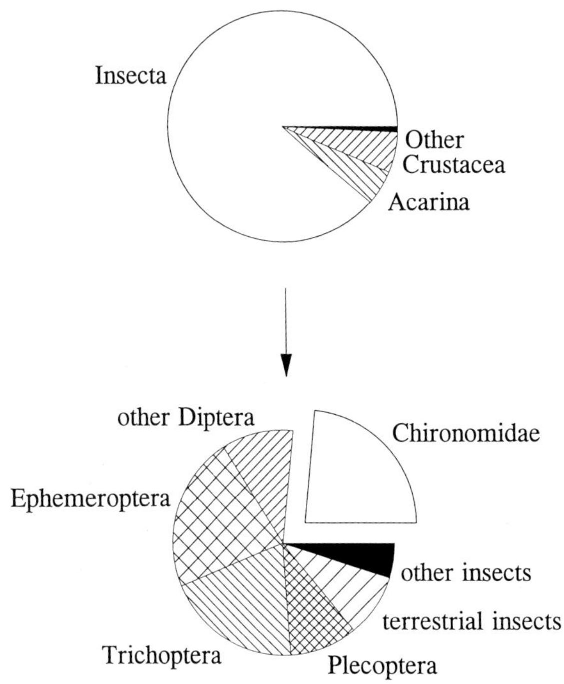

Figure 2. Composition of total drift in terms of abundances.

This pattern of a few individuals from many taxa being present, but only a small number of very common taxa, was also reflected in the drift in general. Overall only 11 taxa, including *Zavreliella* sp. and *Tanytarsus* sp., contributed to nearly half (43.6%) of the total number of invertebrates collected from the drift (Schreiber, 1994). A chironomid, *Zavreliella* sp., dominated the drift and contributed over 7% to total drift abundances (Table 1).

On 17 and 18 March, 1983, the number of taxa collected from nocturnal drift severely underrepresented the number of taxa collected from benthic samples (Fig. 3). This was particularly pronounced in the Chironomidae with only 1/3 of benthic taxa also collected in the drift. In terms of chironomid abundances *Riethia* sp. dominated the benthos, but I collected no *Riethia* in the drift at this time. So, for this particular time and site, drift samples neither represented benthic composition nor densities.

Chironomid drift changed seasonally, with highest densities in summer, 1981/82, and spring, 1982 (Fig.4). This reflected the drift of the common chironomid groups, with *Rheotanytarsus* sp. and *Tanytarsus* sp. showing a peak in drift density in the first summer, and *Podonomopsis* sp. and *Zavreliella* sp. drifting in highest densities in spring (Table 2).

Table 1. Common aquatic taxa in the drift and their relative contributions to total drift.

Taxon	% of total drift
Zavreliella sp.1	7.32
Tanytarsus sp.	2.84
Podonomopsis sp.	1.77
Rheotanytarsus sp.	1.54
Thienemaniella sp.	0.93
Stictocladius sp.	0.73
Orthocladiinae sp. SO1	0.63
Ablabesmyia sp.	0.55
Riethia sp.	0.54

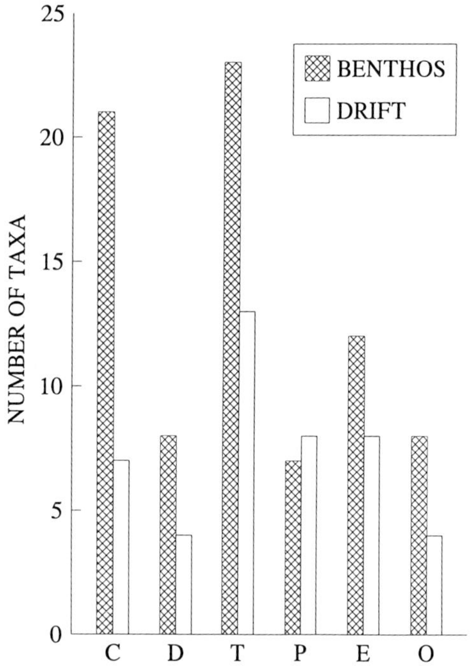

Figure 3. Total number of taxa identified in benthic and drift samples on the 17. and 18. March 1983 (C=Chironomidae, D=non-chironomid Diptera, T=Trichoptera, P=Plecoptera, E= Ephemeroptera, O=Others).

Overall drift was much lower in the second summer than in the first (Fig. 4), which was the case not only for chironomids, but also for total invertebrate drift (Schreiber, in press). This period coincided with a severe drought in Victoria, during which stream discharge was extremely low (Rural Water Commission, Victoria, Australia). However, not all taxa drifted in lower densities in the second summer than the first: drift densities of *Rheotanytarsus* sp. and *Thienemaniella* sp. peaked in summer, and there was no significant difference in densities between the 2 years (Table 2).

Contrary to all other common invertebrate taxa (Schreiber, in press), no chironomid taxa exhibited diel, behavioural drift, as defined in this study. The most abundant invertebrate taxon collected, *Zavreliella* sp., did not show behavioural drift, despite always being present in relatively high densities (Fig. 5, also Schreiber, in press)). Sometimes drift increased after dusk, but then did not decrease after dawn the next day (Fig. 5e). At other times, drift densities increased after dawn, but there was no change in the level before and after sunset on the previous day (Fig. 5f). Overall, the drift of *Zavreliella* sp. did not conform to any previously described drift pattern (Waters, 1972), but was quite erratic.

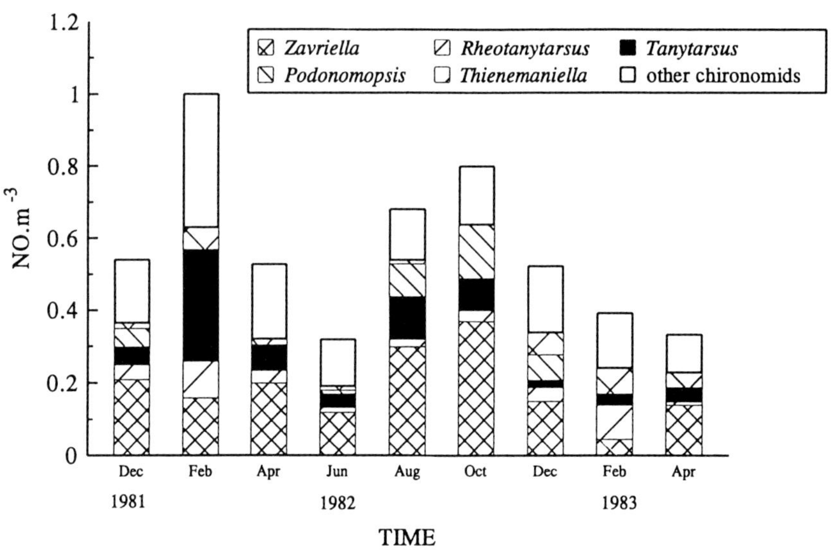

Figure 4. Seasonal changes in chironomid drift.

Table 2. Seasonal patterns in drift: ANOVA and Ryan's test results for the most common chironomid taxa (mean drift densities are listed with sampling month). For all tests: harmonic mean n=2.41; df=10; F=F-ratio; ***=P<0.001, **=P<0.01; *=P<0.05.

total Chironomidae (data ln(x+1) transformed) MS_E=0.041 F=7.308**								
Jun82	Apr83	Feb83	Apr82	Dec82	Dec81	Aug82	Oct82	Feb82
5.769	5.786	5.997	6.201	6.279	6.286	6.530	6.675	6.900
Rheotanytarsus sp. (data not transformed) MS_E=123.918 F=20.856***								
Apr83	Jun82	Aug82	Oct82	Apr82	Dec82	Dec81	Feb83	Feb82
6.185	16.680	20.475	33.020	33.345	46.780	51.843	98.535	110.065
Podonomopsis sp. (data not transformed) MS_E=125.701 F=41.334***								
Apr83	Feb82	Feb83	Apr82	Jun82	Dec81	Dec82	Aug82	Oct82
0.885	2.525	2.870	8.625	11.790	48.803	70.035	92.270	144.960
Thienemanniella sp. (data ln(x+1) transformed) MS_E=0.122 F=14.753***								
Oct82	Aug82	Jun82	Apr82	Dec81	Apr83	Dec82	Feb82	Feb83
1.810	2.138	2.171	2.624	3.042	3.626	4.087	4.147	4.192
Zavreliella sp. (data ln(x+1) transformed) MS_E=0.147 F=4.610*								
Feb83	Jun82	Apr83	Feb82	Dec82	Apr82	Dec81	Aug82	Oct82
3.860	4.831	4.909	4.993	5.048	5.188	5.235	5.704	5.906
Tanytarsus sp. (data ln(x+1) transformed) MS_E=0.104 F=13.883***								
Dec82	Feb83	Jun82	Apr83	Dec81	Apr82	Oct82	Aug82	Feb82
2.935	3.310	3.597	3.632	3.849	4.231	4.484	4.814	5.706

DISCUSSION

Chironomids were an important component of the drift, both in terms of number of taxa and abundances. This contrasts most previous Australian studies of drift, where chironomids were found to contribute between 1% and 5.4% (Cadwallader and Eden, 1977; Bailey, 1981; Arthington *et al.*, 1982; Benson and Pearson, 1987; Suter and Williams, 1977). Only Graesser and Lake (1984), who collected with nets of 0.24mm mesh size, found that chironomids contributed a significant proportion to stream drift. Thus the lack of chironomids in Australian drift studies may be a function of mesh size (Williams, 1989), as at least 3 of the above mentioned studies sampled drift with nets of 0.5mm mesh size (Cadwallader and Eden, 1977; Benson and Pearson, 1987; Suter and Williams, 1977), and in the remainder the mesh size used is not clearly stated (Bailey, 1981; Arthington *et al.*, 1982).

It has been suggested (Waters, 1966; Arthington *et al.*, 1982) that drift sampling could be a potential means of assessing benthic production of a stream, if the composition and relative densities reflected that of the benthos. The proposed advantages of drift sampling over benthic sampling include the relative ease of collecting and sorting drift and the potential for only a few drift samples to represent the benthos (Arthington *et al.*, 1982). Whilst some studies have found a positive relationship between drift and benthic abundances of some species, other studies have not substantiated such a relationship (Brittain and Eikeland, 1988). It can of course be difficult to directly compare benthos and drift, if the origin of the drift is unknown. Thus I compared drift composition and abundances of a spatially restricted area of benthos and found that drift could not be used to assess benthic composition and abundances in this instance. However, as this was an unreplicated experiment, the conclusions that can be drawn from it are spatially and temporally limited. Nevertheless, I suggest that to date no clear relationship has been demonstrated between drift and benthos, and thus drift sampling cannot routinely replace benthic sampling.

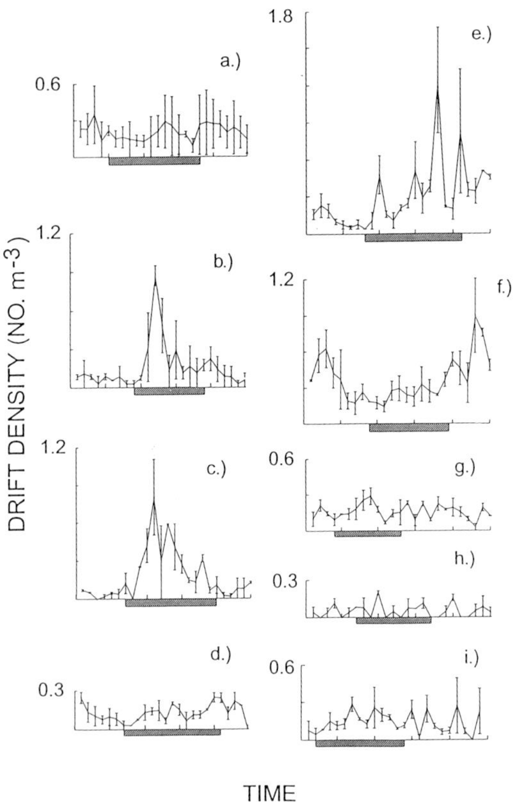

Figure 5. Annual changes in diel drift densities of *Zavreliella* sp.. The vertical bars represent the range of drift denstiies samples. The shaded area underneath the horizontal axis represents night time. a.) 22.& 23. December, 1981; b.) 24.& 25. February, 1982; c.) 26.& 27. April, 1982; d.) 24.& 25. June, 1982; e.) 10.& 11. August, 1982; f.) 19.& 20. October, 1982; g.) 16.& 17. December, 1982; h.) 15.& 16. February, 1983; i.) 16.& 17. April, 1983.

Chironomid drift densities varied between sampling times, with peaks in densities in summer and spring. This was due to the drift of the major taxa rather than an aggregate of all taxa. In the abnormally dry summer in 1982/83 some of the common taxa (eg. *Rheotanytarsus* sp. and *Thienemaniella* sp.) drifted in similar densities to their drift in the previous summer. Other taxa (eg.*Tanytarsus* sp.), drifted in peak densities in the first summer of the study, but not in the second summer. This lack of a repeated drift pattern could have been due to the effects of the drought in Victoria and the associated lower discharge in the Acheron River. However, the suggestion that decreased discharge could be linked with decreased drift contrasts some other studies in which decreases in discharge either did not cause any changes in drift (Elliott, 1967; Allan, 1987), or was accompanied by increases in drift (Minshall and Winger, 1968; Hall and Edwards, 1978; Beckett and Miller, 1982; Corrarino and Brusven, 1983). It is possible that the decrease in discharge in the stream affected different taxa in different ways. Canton *et al.* (1984) found that the ephemeropteran genus *Baetis* drifted in lower densities during low flow than high flow, whilst other taxa were unaffected by different flow conditions. Furthermore the response to a decrease in discharge can depend on the rate of change in flow: different drift responses may be elicited by an abrupt change in flow, such as that associated with changing flow in a regulated river (Minshall and Winger, 1968; Blyth *et al.*, 1984), than by the relatively more gradual reduction in discharge associated with a drought.

I found no evidence of repeated diel drift patterns for any chironomid taxa. All other common invertebrate taxa showed very distinct nocturnal behavioural drift, which was repeated again and again during the 18 months of the study. It has been suggested that for chironomids in general early instars drift in large numbers and that the mesh size used in this study would not have been appropriate for quantitatively collecting these early instars (Williams, 1985). On the other hand, for taxa such as *Baetis*, which have been found to show behavioural drift in many studies, diel rhythms often are particularly strongly exhibited by large instars (Allan, 1978, 1984). This observation has led to the suggestion that nocturnal behavioural drift has evolved in response to predation pressure by visual predators. Large instars, which are relatively more obvious to predators than smaller instars, are more likely to be eaten if they drift during the day than if they drift at night. It is difficult to test a hypothesis for the evolution of a behaviour directly, but predictions from it can be tested. Some workers have compared streams with and without predators and found that in streams with fish many taxa drifted with a very strong behavioural drift pattern. The same taxa did not show diel drift periodicity in streams which historically lacked drift-feeding fish (Flecker, 1992; Malmqvist, 1988). The conclusion was reached that different selection pressures acted on the same taxa in different streams and resulted in different drift behaviour. It is possible that, within a stream, different benthic taxa experience different selection pressures. For example, fish predation could have resulted in the evolution of strong nocturnal behavioural drift in some mayflies in the Acheron River, whilst chironomids could have been under less predation pressure from fish. This could have been the case, in particular, for *Zavreliella* sp., which usually drifted in its case and thus was potentially cryptic to fish. However, this remains speculation as no studies on benthic predation by fish were carried out.

Acknowledgements

This work was part of a Masters of Science thesis in the Department of Zoology at Monash University, Clayton, Australia. I would in particular like to thank Dr. P.S. Lake for his supervision and continuing support of this work. L. Barmuta, R. StClair, O. Hammond, M. Kelly, A. Boulton, C. Bogazz and J. Hammond helped with fieldwork. P.S. Cranston and J. Martin checked the identities of reference specimens. Advice on the analysis of the data was received from M. Cullen, T. MacMahon, P.D. Finch, A. Sokol and L. Barmuta. In the preparation of this manuscript I am particularly indebted to B.J. Downes and P.S. Lake for many helpful discussions.

Temporal and Diel Drift of Chironomidae (Diptera: Insecta) Larvae in a Northern New Zealand Stream

I.K.G. Boothroyd

Dept. of Biological Sciences, University of Waikato, Private Bag, Hamilton, New Zealand

Present address: Hawke's Bay Regional Council, Private Bag, Napier, New Zealand

Introduction

Although downstream drift of aquatic organisms has been the subject of many studies (Brittain and Eikeland, 1988), there have been few studies on the drift of Chironomidae, and they have often been considered only collectively as a family (Williams, 1989). This is particularly so in New Zealand, where there have been few studies on drift in general, and most of the studies on chironomids have been from the lentic environment (Boothroyd, 1987). Lotic studies have been hampered by the poor taxonomic knowledge of the group in New Zealand, although this is improving (Boothroyd, 1990).

Studies on drift in New Zealand have highlighted that the Chironomidae can be a very important component of the drift (Collier and Wakelin, 1992; Sagar and Glova, 1992), and as such are an important in the diet of fish (Sagar and Glova, 1988; Glova *et al.*, 1987), and also in the colonisation dynamics of stream ecosystems (Boothroyd and Dickie, 1991).

This study on chironomid drift formed part of a larger study of the ecology of Chironomidae in a stream in the Waikato region of New Zealand, which also incorporated life history and emergence studies (Boothroyd, 1987, 1988).

STUDY SITE

The Kaniwhaniwha stream is a fast-flowing, stony bottom stream, arising on Mt. Pirongia (37° 95′S, 175°05′E), an extinct volcano west of Hamilton in the North Island of New Zealand. The stream flows through native podocarp-mixed hardwood forest over volcanic deposits, before entering open pastureland, where it runs through sandstone and siltstones.

Samples were taken from a single site (37° 54′S, 175°05′E) from a run approximately 0.3 m deep in normal flows. Like most New Zealand streams, the Kaniwhaniwha stream is subject to unpredictable rainfall and can rise and fall rapidly in times of flood. During normal flows however, velocities ranged from 0.3-0.94 m s^{-1}. Annual maximum and minimum temperature recorded during diel sampling ranged from 7-22° C. Diel changes in water temperature were much less pronounced in winter than summer. Further details are provided in Boothroyd (1987).

METHODS

Drift sampling

Drifting chironomid larvae were collected using a drift sampler modified after Elliott (1970). The sampler consisted of a metal frame, with a 0.06 m^2 inlet and a wider outlet, to which was attached a removable, metre long collecting net with a 250μm mesh. The sampler was raised 50 mm above the stream bed and the top of the inlet was always above the water surface, thereby simultaneously collecting floating pupal exuviae (Boothroyd, 1988). Thus the entire water column was sampled

A single sampler was placed in the fastest section of the run, and just below a riffle. Collections of drifting chironomid larvae were made every 3 h over a 24 h period bi-monthly. Temperature (air and water) and current velocity at the sampler inlet were taken at each collection. Samples were immediately preserved in 4% formalin (w/v), and in the laboratory chironomid larvae were sorted, counted and identified. Identifications were made from existing keys where possible (Freeman, 1959; Stark, 1981), and from notes and drawings made by the author.

Very large numbers of larvae made it necessary to use a subsampler, which comprised a circular bowl modified into a Folsom splitter which halved the sample. Reductions varied but ranged from zero to as low as 3%, but were generally in the range of 25%.

Enclosures

To quantify chironomid larval drift better and to test experimentally the influence of light on larval drift, undisturbed sections of the stream bed were enclosed within collection channels, modified after the design of Wartinbee and Coffman (1976). Six areas, each approximately 0.5 m^2 were enclosed laterally by wooden sideboards of untreated box pine (1.0 m x 0.3 m x 0.02 m), and by removable nets at the upstream and downstream ends. The bottom edges of the sideboards were fitted with strips of foam rubber and strong plastic set approximately 5-20 cm into the stream bed. The channel sides were held firmly in place by bolting them to metal stakes already set into the stream bed. The sideboards thus extended 15-20 cm above the stream bed, and were set parallel to the stream flow.

The end nets (250 μm mesh) were attached to removable frames, to facilitate sample collection. The upstream net comprised a flattened net to prevent the entry of drifting material and was set into the stream bed. The downstream net comprised a 0.5 m long conical collecting net, and the downstream net frame was always 2-3 cm above the substrate to prevent organisms crawling in. Trial studies were carried out to ensure no leakage occurred and that conditions (velocities, temperature etc.) were similar within the channels and external to the channels.

The aim of the enclosures study was to quantify diel drift from the stream bed in winter (July 1985), and summer (February 1986), and under different light regimes on each occasion. To this end, three pairs of adjacent channels were constructed in the stream, with one of each pair covered with thick black polythene to prevent light entering. In addition the upstream and downstream nets of the artificially darkened channels were dyed black to decrease light entry.

Sampling was carried out over a 27 h diel period, beginning at night to prevent any effects of sudden channel darkness. Samples were taken sytematically across the stream at the end of each 3 h diel period, and the sample preserved in 4% formalin (w/v), and later sorted, counted and identified. Temperature (air and water), oxygen levels, and water velocities were recorded at the end of each diel period.

Table 1. Chironomid taxa recorded from the drift in the Kaniwhaniwha stream

Tanypodinae	Orthocladiinae
Podonominae	*Cricotopus zealandicus* Freeman
	Cricotopus planus Boothroyd
Diamesinae	*Cricotopus vincenti* Boothroyd
Maoridiamesa harrisis Pagast	*Cricotopus aucklandensis* Sublette and Wirth
Lobodiamesa campbelli Pagast	*Eukiefferiella brundini* Boothroyd and Cranston
	Paratrichocladius pluriserialis Freeman
Chironominae	*Paratrichocladius* sp. A
Polypedilum opimus Hutton	Orthoclad sp. a
Polypedilum spp.	Orthoclad sp. b
Tanytarsus vespertinus Hutton	Orthoclad sp. c

RESULTS

Taxonomic composition

A total of 17 chironomid larval taxa was recorded from the drift in the Kaniwhaniwha stream (Table 1), which reflected the taxa common in the benthos (Boothroyd, 1987). Podonominae and Tanypodinae were rarely recorded in the drift and *P. pluriserialis* (Freeman), *Paratrichocladius* sp. A, and *Lobodiamesa campbelli* Pagast were recorded only intermittently throughout the year. Orthocladiinae dominated the chironomid taxa in the drift (and benthos), but of the 10 orthoclad taxa, 3 taxa are currently not described or have formal descriptions in press. Identifications in all cases were made by rearing larvae to the adult stage.

Temporal changes in drift

Chironomidae larvae were the dominant group of organisms in the drift at all times of the year. Greatest drift densities of total chironomid numbers occurred in November 1984 (3.89 m^{-3} 24 h^{-1}), and January 1985 (3.87 m^{-3} 24 h^{-1}) (Fig. 1), during periods of high and low flows respectively. Similarly, most species reached peak densities in January 1985, except *Cricotopus planus* Boothroyd, which reached peak densities in November 1984 and only *Cricotopus vincenti* Boothroyd exhibited similar drift densities in both summers (Fig. 1).

With the exception of *Cricotopus zealandicus* Freeman, all taxa showed little tendency to drift in the winter months, but all taxa were present in the drift at all times of the year, although relative abundance varied (Fig. 2).

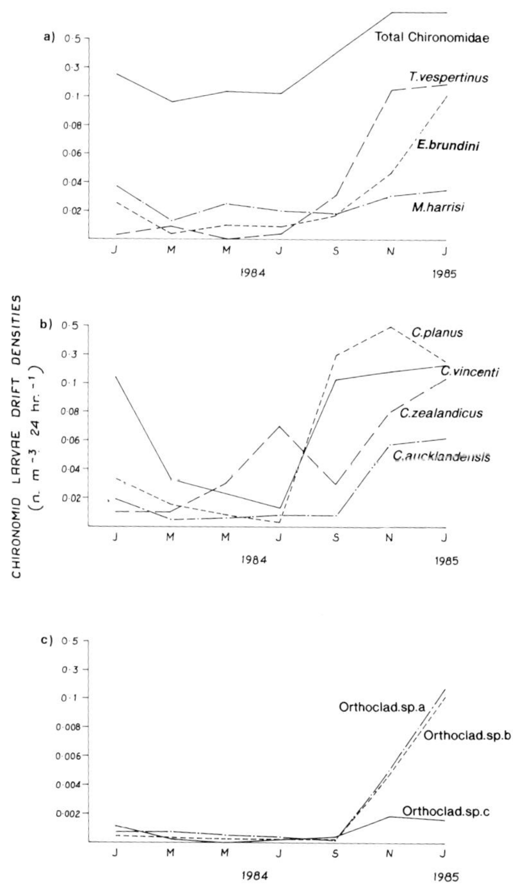

Figure 1. Temporal changes of chironomid larval drift densities (nos. m^{-3} $24h^{-1}$) bimonthly from the Kaniwhaniwha stream. Data log (x+1) transformed.

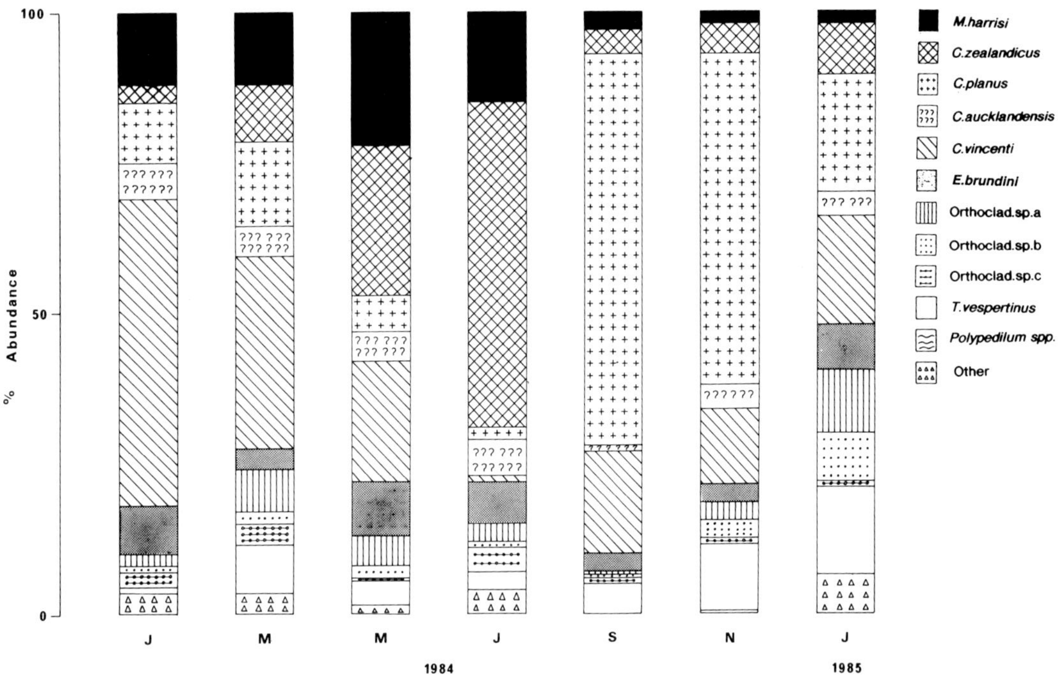

Figure 2. Relative abundance (%) of major chironomid larvae collected bimonthly per 24h in drift samples from the Kaniwhaniwha stream

Table 2. Regression statistics (R^2) between drift (nos. m^{-3} 24h) and benthic (nos. 0.5 m^{-2}) densities of Chironomidae in the Kaniwhaniwha stream; and correlation coefficients (r) between contribution of chironomid taxa to the 24h drift and benthic communities. ns (p>0.05); *(p<0.05); **(p<0.01)

	Drift and benthos regression		Drift and benthos Composition	
Taxon	R^2	p	r	p
M. *harrisis*	0.17	ns	-0.852	*
C. *zealandicus*	0.10	ns	0.656	ns
C. *planus*	0.09	ns	0.151	ns
C. *vincenti*	0.195	ns	0.835	*
C. *aucklandensis*	0.0001	ns	-0.172	ns
E. *brundini*	0.688	**	-0.109	ns
Orthoclad sp. a	0.026	ns	-0.042	ns
Orthoclad sp. b	0.273	ns	0.486	ns
T. *vespertinus*	0.655	*	0.781	*

C. *vincenti* dominated the 24 h drift in January 1984, but C. *planus* was more abundant in January 1985, as well as September and November 1984. C. *zealandicus* dominated in the winter months, with *Maoridiamesa harrisi* Pagast and C. *vincenti* also abundant in May 1984 (Fig. 2).

The relationship between drift and benthos densities of chironomid larvae was investigated by linear regression (Table 2). Only *Eukiefferiella brundini* Boothroyd and Cranston ($p < 0.01$) and T. *vespertinus* Hutton ($p < 0.05$) showed any significant response. On the other hand, M. *harrisi*, C. *vincenti* and *Tanytarsus vespertinus* showed significant correlations ($p < 0.05$) between drift and benthic composition throughout the year (Table 2).

Diel drift

Different patterns of diel drifting activity were evident for individual taxa, which also changed temporally. M. *harrisi* exhibited a similar diel pattern all year, with greatest drift densities occurring during daylight hours, but with increased drift activity during darkness in the summer (Fig. 3). The drift patterns of C. *zealandicus* (Fig. 4) differed over the two summers, with little periodicity in January 1984, but with greater daytime drift in January 1985 peaking at dusk; and which was similar in winter and spring. Drift patterns of the remaining *Cricotopus* species all exhibited little pattern in the winter, but exhibited drift peaks at dusk and/or dawn in spring and summer (Figs. 5, 6, 7).

Daylight drift was prominent for E. *brundini*, and the undescribed orthoclads. E. *brundini* (Fig. 8) and orthoclad sp. a (Fig. 9) exhibited a high daylight drift during the summer, but with peaks in the morning and at dusk. Dusk drift was prominent all year for T. *vespertinus* with increased drift during daylight hours in spring and summer (Fig. 10). However, all chironomid taxa were present in the drift during each 24 h diel period but some taxa were absent from the drift during certain periods of the day.

Enclosures

In general there were fewer chironomid larvae in the enclosures drift during the summer than the winter, but large numbers drifting per enclosure were not recorded on any occasion. Greater numbers of M. *harrisi* drifting

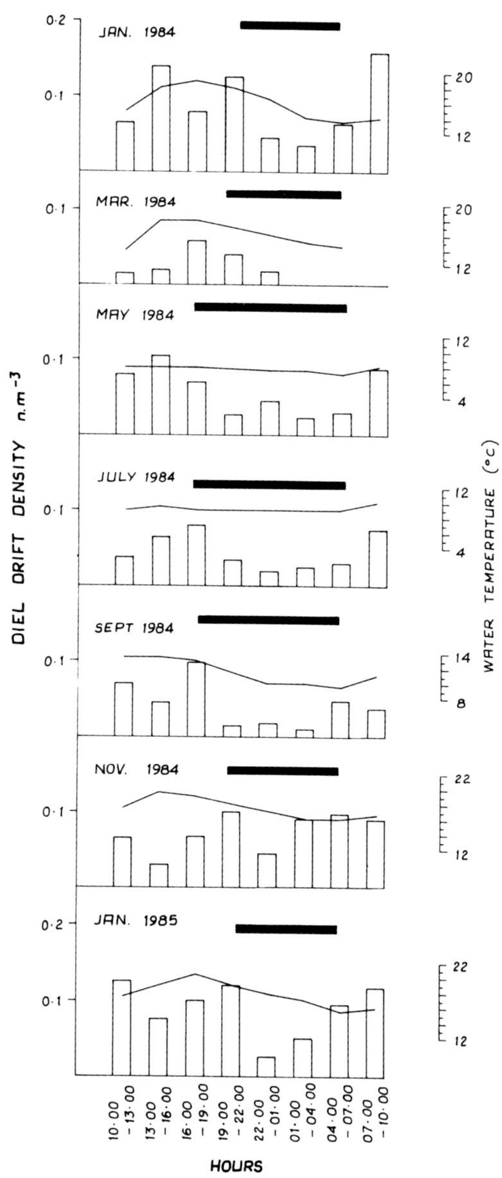

Figure 3. Temporal changes in larval drift densities (nos. m^{-3}) of *Maoridiamesa harrisi* collected from the Kaniwhaniwha stream. Solid bars represent periods of darkness, and solid lines show water temperatures.

during daylight hours were recorded from the artificially darkened channels in both summer and winter, whereas in all other diel periods numbers were similar (Fig. 11).

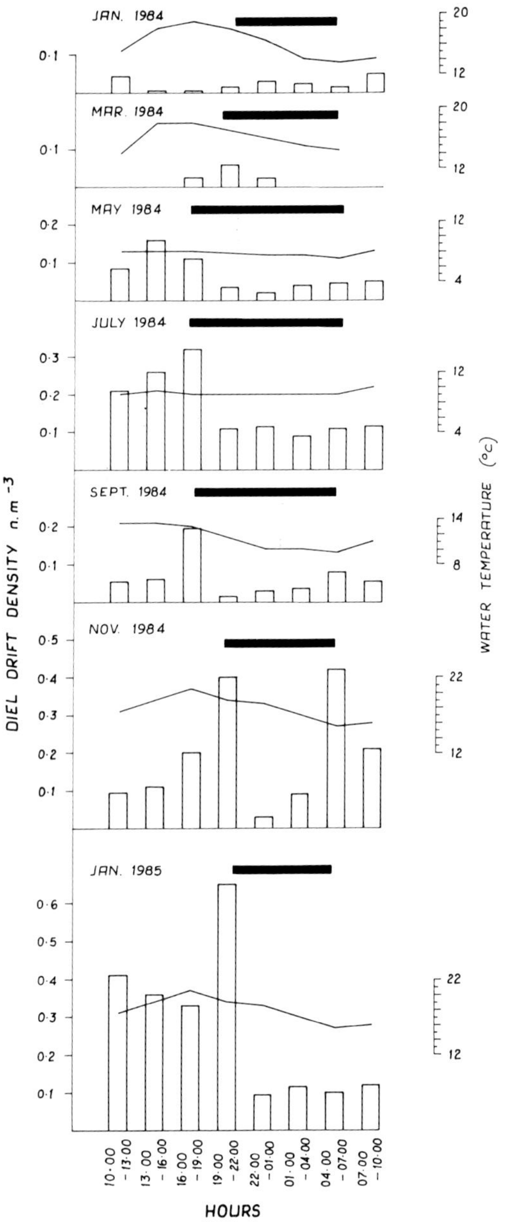

Figure 4. Temporal changes in larval drift densities (nos. m^{-3}) of *Cricotopus zealandicus* collected from the Kaniwhaniwha stream. Solid bars represent periods of darkness, and solid lines show water temperatures.

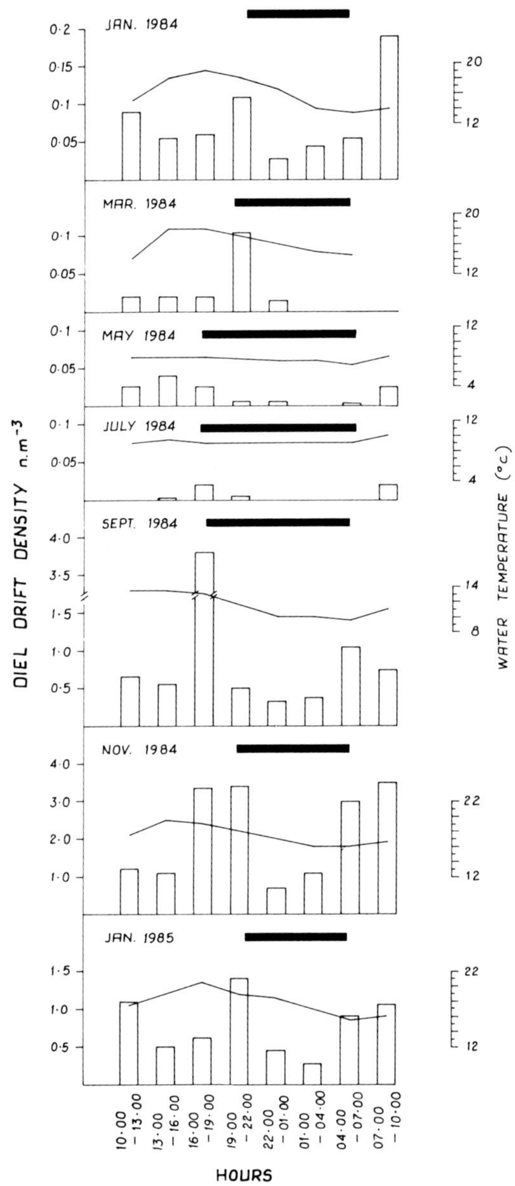

Figure 5. Temporal changes in larval drift densities (nos. m^{-3}) of *Cricotopus planus* collected from the Kaniwhaniwha stream. Solid bars represent periods of darkness, and solid lines show water temperatures.

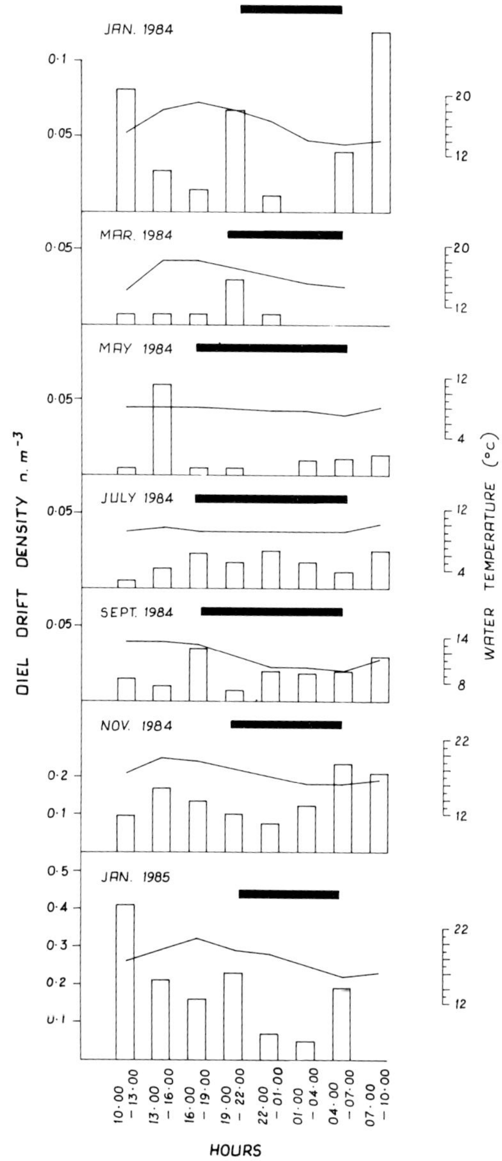

Figure 6. Temporal changes in larval drift densities (nos. m^{-3}) of *Cricotopus aucklandensis* collected from the Kaniwhaniwha stream. Solid bars represent periods of darkness, and solid lines show water temperatures.

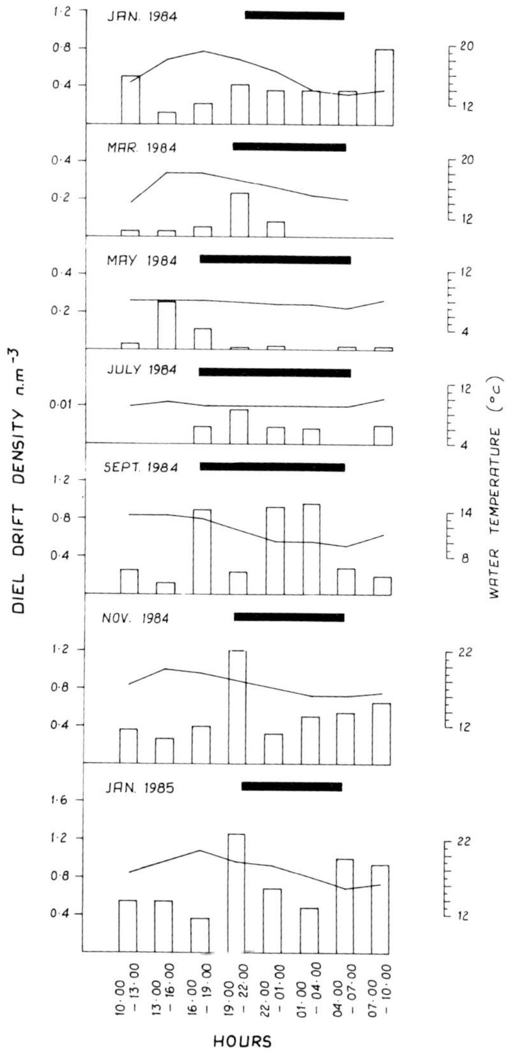

Figure 7. Temporal changes in larval drift densities (nos. m^{-3}) of *Cricotopus vincenti* collected from the Kaniwhaniwha stream. Solid bars represent periods of darkness, and solid lines show water temperatures.

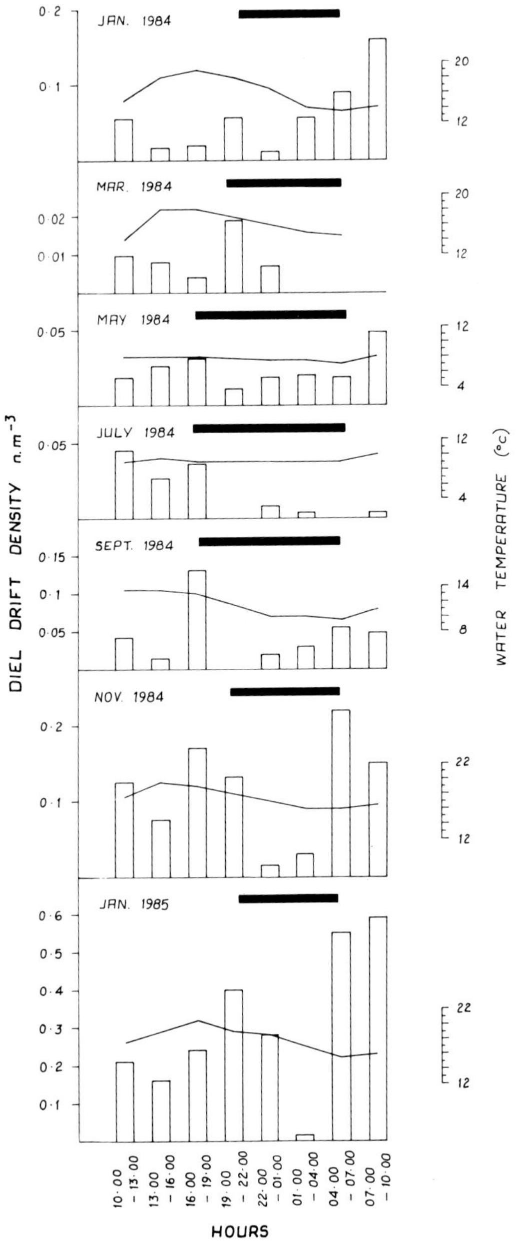

Figure 8. Temporal changes in larval drift densities (nos. m^{-3}) of *Eukiefferiella brundini* collected from the Kaniwhaniwha stream. Solid bars represent periods of darkness, and solid lines show water temperatures.

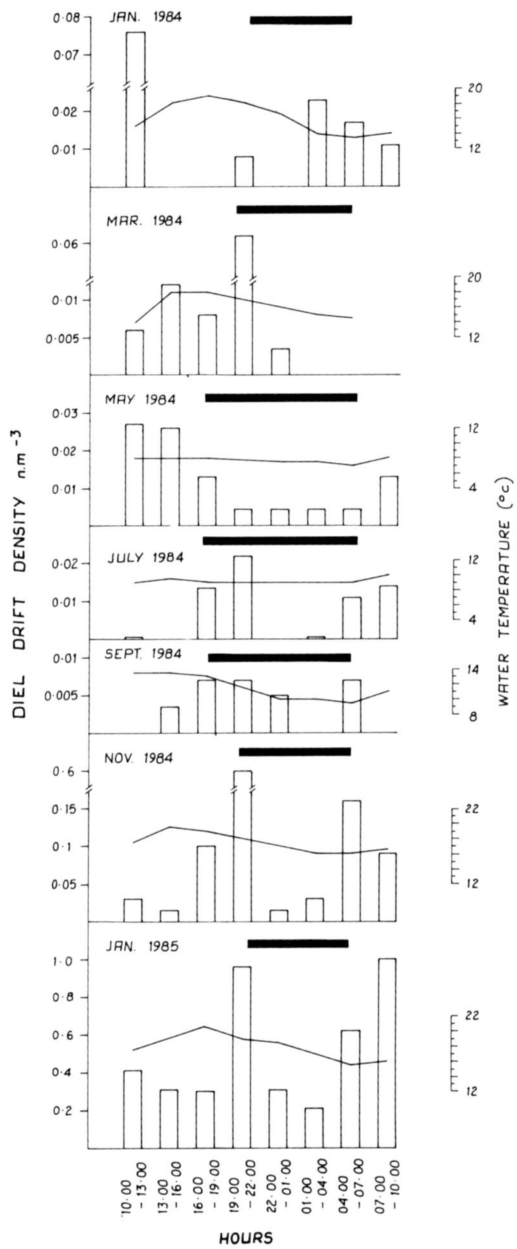

Figure 9. Temporal changes in larval drift densities (nos. m^{-3}) of Orthoclad sp. a collected from the Kaniwhaniwha stream. Solid bars represent periods of darkness, and solid lines show water temperatures.

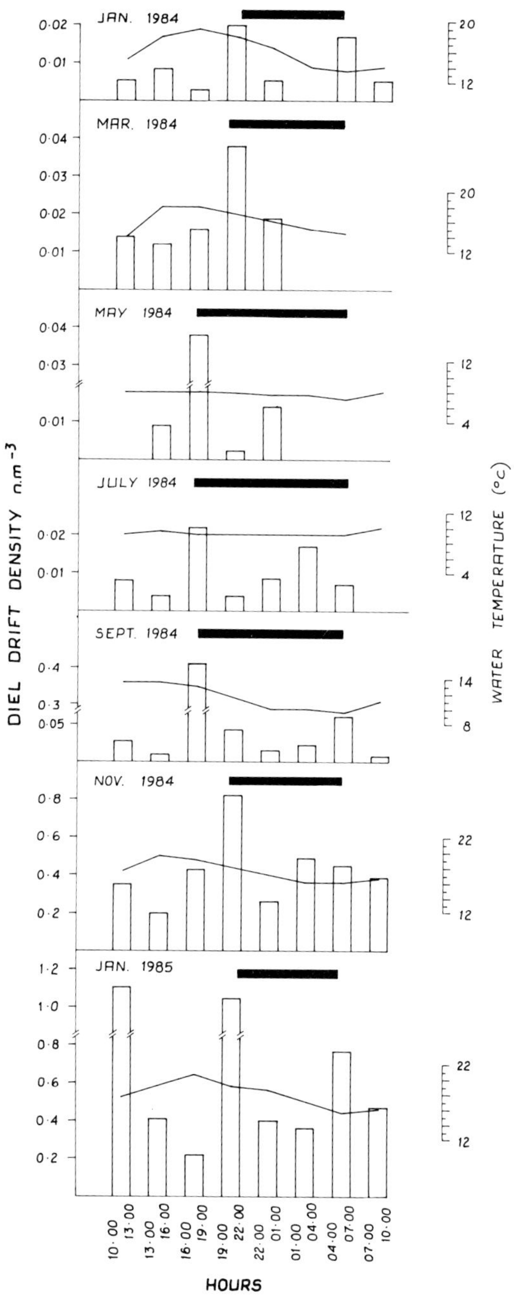

Figure 10. Temporal changes in larval drift densities (nos. m^{-3}) of *Tanytarsus vespertinus* collected from the Kaniwhaniwha stream. Solid bars represent periods of darkness, and solid lines show water temperatures.

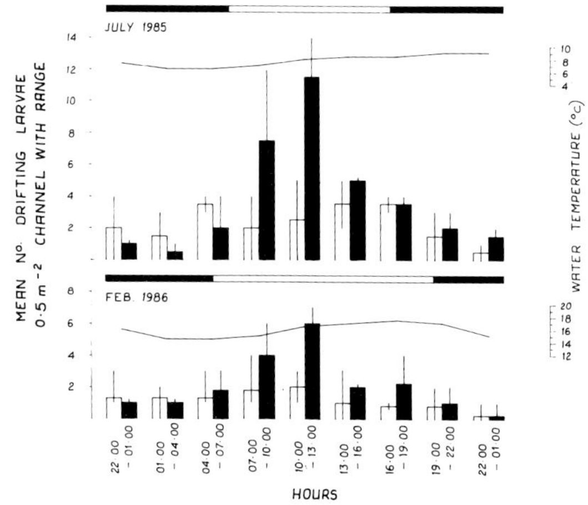

Figure 11. Diel larval drift densities (nos. 0.5 m-2) of *M. harrisi* from natural light (open histograms) and artificially darkened (shaded histograms) enclosures in the Kaniwhaniwha stream. Solid bars represent periods of darkness, and solid lines show water temperatures.

A similar pattern was evident for orthoclad sp. a in winter (Fig. 12). On the other hand, drifting larvae of *E. brundini* (Fig. 13) increased in both the natural and artificially darkened channels during daylight hours in winter. Only *C. vincenti* (Fig. 14) exhibited significantly different ($p < 0.05$) numbers of drifting larvae from the darkened channel and from the natural light channel (Table 3).

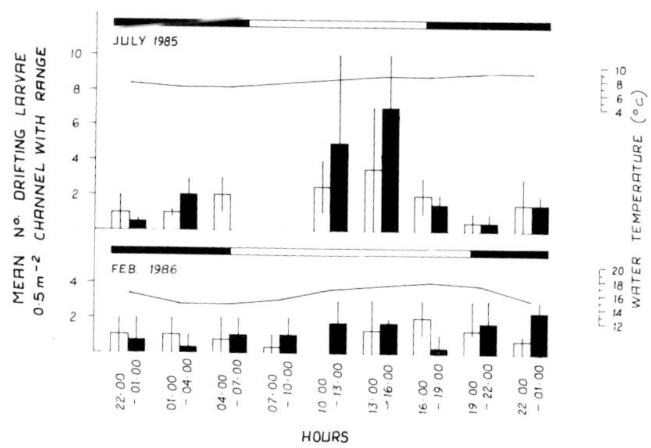

Figure 12. Diel larval drift densities (nos 0.5 m-2) of Orthoclad sp. a from natural light (open histograms) and artificially darkened (shaded histograms) enclosures in the Kaniwhaniwha stream. Solid bars represent periods of darkness, and solid lines show water temperatures.

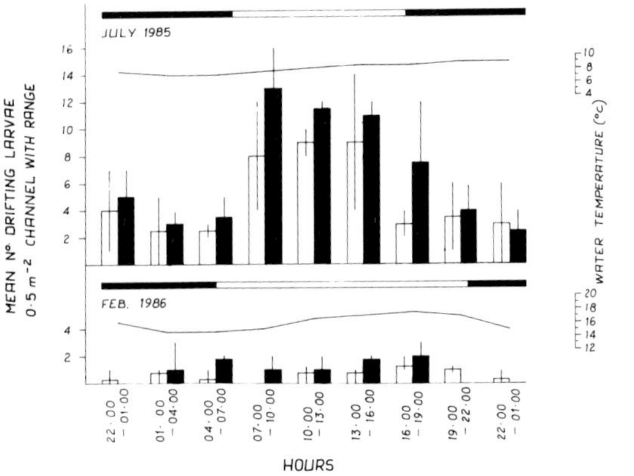

Figure 13. Diel larval drift densities (nos. 0.5 m-2) of *E. brundini* from natural light (open histograms) and artificially darkened (shaded histograms) enclosures in the Kaniwhaniwha stream. Solid bars represent periods of darkness, and solid lines show water temperatures.

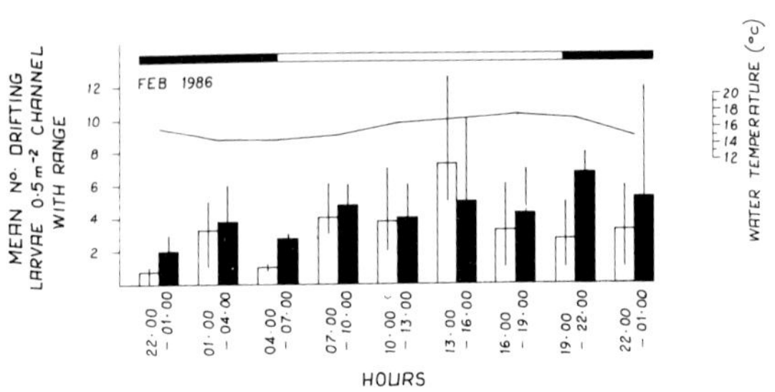

Figure 14. Diel larval drift densities (nos. 0.5 m-2) of *C. vincenti* from natural light (open histograms) and artificially darkened (shaded histograms) enclosures in the Kaniwhaniwha stream. Solid bars represent periods of darkness, and solid lines show water temperatures.

Discussion

Seasonal patterns of drift have been shown to fluctuate with changes in biotic factors such as growth rates of larvae, larval stage, population density, and abiotic factors such as stream temperature and river discharge (Bishop and Hynes, 1969,; Waters, 1972; Hemsworth and Brooker, 1981; Brittain and Eikeland, 1988). Allan (1987) found a positive correlation of chironomid 24 h drift with river discharge, similar to O'Hop and Wallace (1983),

Table 3. Wilcoxin paired sample test of differences between chironomid larval drift from natural light and artificially darkened 0.5 m $^{-2}$ enclosures in the Kaniwhaniwha stream, July 1985 and February 1986. ns (p>0.05); * (p<0.05); n/a = not applicable. [ ] refers to number of pairs where a difference was recorded.

Taxon	July 1985	February 1986
M. harrisi	ns [18]	ns [31]
C. zealandicus	ns [9]	ns [8]
C. planus	ns [17]	ns [9]
C. aucklandensis	ns [5]	ns [17]
C. vincenti	n/a	* [29]
E. brundini	ns [25]	ns [26]
Orthoclad sp. a	ns [17]	ns [22]
Orthoclad sp. b	ns [17]	ns [8]
T. vespertinus	ns [20]	ns [5]

whereas Hemsworth and Brooker (1981) found chironomid drift density correlated with benthic density. In the Kaniwhaniwha stream only *E. brundini* (-ve), and *T. vespertinus* (+ve) showed any relationship between drift and benthic densities; and only *M. harrisi*, *C. vincenti* and *T. vespertinus* correlated with their respective contributions to the drift and benthic composition. In New Zealand, Collier and Wakelin (1992), and Sagar and Glova (1992) have suggested that drift of Chironomidae may be at least partly density dependent, but seasonal factors may be important.

In northern hemisphere temperate streams there is a trend towards low drift densities in the winter months when temperatures are low (Armitage, 1977b; Brittain and Eikeland, 1988), with increasing drift during spring and summer. Similar responses were evident from the Kaniwhaniwha stream for most of the chironomid taxa. Only *C. zealandicus* exhibited any increase in drift during the winter, at a time when benthic densities of this species were also high (Boothroyd, 1987). This same pattern of low winter drift and higher spring and summer drift has also been reported for both chironomids and other aquatic macroinvertebrates from other New Zealand streams (McLay, 1968; Collier and Wakelin, 1992; Sagar and Glova, 1992).

Whereas the life-histories of the New Zealand Chironomidae and the aquatic fauna in general are considered to differ markedly from the northern hemisphere temperate pattern (Towns, 1981; Winterbourn *et al.*, 1981; Winterbourn, 1987; Boothroyd, 1988), being less seasonal and with all life stages present all year round, drift patterns of chironomids at least exhibit more conformity with the northern hemisphere pattern of reduced winter drift. However, in the Kaniwhaniwha at least, the increased drift densities for most species coincided with increased development, shorter generation times (Boothroyd, 1987), and increased emergence (Boothroyd, 1988).

However, Williams (1989) reported that chironomid larvae may drift in large numbers for much of the year; a phenomena unrecorded in many drift studies as a result of inefficient samplers. Many first and second instar larvae pass through the larger mesh (> 200µm) of most studies. Using a finer mesh of 50µm Williams (1989) also discovered that chironomid drift was dominated by first and second instar larvae, but orthocladiinae were represented in all life stages. This was attributed to the mode of life of the chironomid subfamilies, Diamesinae and Orthocladiinae being mobile during their entire life cycle. The results reported in the present study include only fourth instar larvae, and in some cases third instars, but the orthoclad species a, b and c are particularly small larvae, and third instar larvae were more likely to remain unsampled.

Although the Chironomidae are reported to exhibit little pattern in the diel drift (Brittain and Eikeland, 1988), different patterns were evident for different taxa in the Kaniwhaniwha stream, which also differed temporally. Most taxa were more common in the daylight drift, but with peak drift occurring at dusk and/or dawn. The same pattern was less evident for *C. vincenti* and *C. aucklandensis*. Most studies on diel variation in drift in New Zealand streams and rivers have reported greater drift densities of chironomid larvae during the hours of daylight (Winterbourn, 1987).

Chironomidae were found to be dominant in the diet of native fish (Glova and Sagar, 1989) and introduced trout (Fechney, 1988; Sagar and Glova, 1988) during daylight, when Chironomidae were also high in the drift; but with only a small proportion of Chironomidae in the diet at night. Sagar and Glova (1988) suggested that this may be due to selection for dominant prey items rather than for size, but little is known of any fish preferences for individual chironomid taxa. If diel feeding by fish is driven by dominant taxa, then diet is likely to be a reflection of relative abundance in drift. Similarly, prey selection of drifting chironomid taxa by fish may be influenced by the temporal pattern exhibited by chironomid larvae.

The influence of light on chironomid drift was difficult to ascertain, and the enclosure experiments were inconclusive. For most taxa, diel changes of drifting larvae from the enclosures reflected those recorded throughout the year. Evidence from the temporal diel studies suggests that light is an important factor in determining drift patterns, but the influence of benthic densities, stream discharge and temperature cannot be dismissed.

Acknowledgements

I am indebted to Dr. M.A. Chapman and Dr. J.D. Green for the support and supervision of this project. This work was carried out whilst under the tenure of a Hilary Jolly Scholarship at the University of Waikato, Hamilton, New Zealand.

The Chironomidae of Austro-oriental phytotelmata (plant-held waters): *Richea pandanifolia* Hook.f.

Peter Cranston[1] *and Roger Kitching*[2]

[1] CSIRO Division of Entomology, GPO Box 1700, Canberra, ACT 2601, Australia

[2] Arthropod Biodiversity Unit, Australian School of Environmental Studies, Griffith University, Nathan, Qld 4111, Australia

Introduction

Many plants hold small volumes of water that are habitats of a specialized aquatic fauna dominated by species of Diptera (Thienemann, 1932; Kitching, 1971; Frank and Lounibos, 1981). These water bodies occur in specially modified plant parts such as pitchers or "tanks" of bromeliads, internodes of bamboos and other grasses, the leaf or bract axils of many fleshy plants, and watertight holes or hollows in trees. Such habitats occur throughout the world, although they are particularly frequent in moist tropical forests. In recent years these habitats have proved very useful in addressing questions in community ecology (Beaver, 1983, 1985; Kitching and Pimm, 1986).

Plant-held waters were named "phytotelmata" by Varga (1928) although they are often referred to as "container habitats" - a somewhat broader category that includes artificial containers such as discarded cans, tyres, grave containers, and so forth.

Our interests in insects inhabiting phytotelms originated with investigations in the temperate northern hemisphere, and extend over some years (e.g. Kitching, 1971; Cranston and Disney, 1978). Subsequently we have made observations, both sporadic and more structured, in Australia and the Oriental region.

In early 1994, one of us (RK) was commissioned by the Tasmanian National Parks and Wildlife Service to make a preliminary study of the animal fauna associated with Tasmanian phytotelmata.

In Australia, distinctive species assemblages of larval Chironomidae are found in many phytotelmata. One of these, the Tasmanian endemic epacrid ("heath") locally called "pandani" (*Richea pandanifolia* Hook.f. (Epacridaceae)) holds axillary water. Collections from leaf bracts of four pandani, sampled destructively by permission of Tasmanian National Parks, revealed an unexpected and unique chironomid fauna, which is the subject of this paper.

Study sites and methods

Studies on the fauna of *Richea* axils were made in the Hartz Mountains National Park (43°11′S 146°46′E) and at a site along the Anthony Development Road (41°55′S 145°34′E) both in the eastern part of the island. Four medium-sized plants were dissected completely and the contents of axils washed out and sorted for later identification.

Individuals of *Richea pandanifolia* sampled were part of a mixed subalpine vegetation formation with a sparse overstory of *Eucalyptus coccifera* Hook.f. and *E. subcrenulata* Maiden and Blakely. Although eucalypts dominated, the woodland also contained prominent mesic elements, notably low-growing individuals of *Nothofagus cunninghamii* (Hook) Oersted, *Atherosperma moschatum* Labill., *Phyllocladus aspleniifolius* (Labill.) Hook.f. and *Anodopetalum biglandulosum* A.Cunn. ex Hook.f. Kirkpatrick *et al.* (1988) designate the plant alliance as '*E. coccifera-Trochocarpa cunninghamii-Cyathodes glauca* subalpine mixed forest (COC11)'. These authors also present a more extended plant list for the alliance. Ogden and Powell (1979) describe similar vegetation along an altitudinal gradient in nearby Mount Field National Park, Tasmania.

Richea phytotelm Chironomidae

Axils of all four plants examined contained chironomids, comprising *Parochlus bassianus* Brundin (Podonominae), as well as two unidentified species of the subfamily Orthocladiinae, belonging to the genera *Compterosmittia* and near to *Gymnometriocnemus*. All chironomids were present as the larval stage (Fig. 1a-c), although one *Compterosmittia* larvae contained a pharate pupa and some pupae of *P. bassianus* contained pharate adults. Taxonomic, ecological and distributional notes on these species follow.

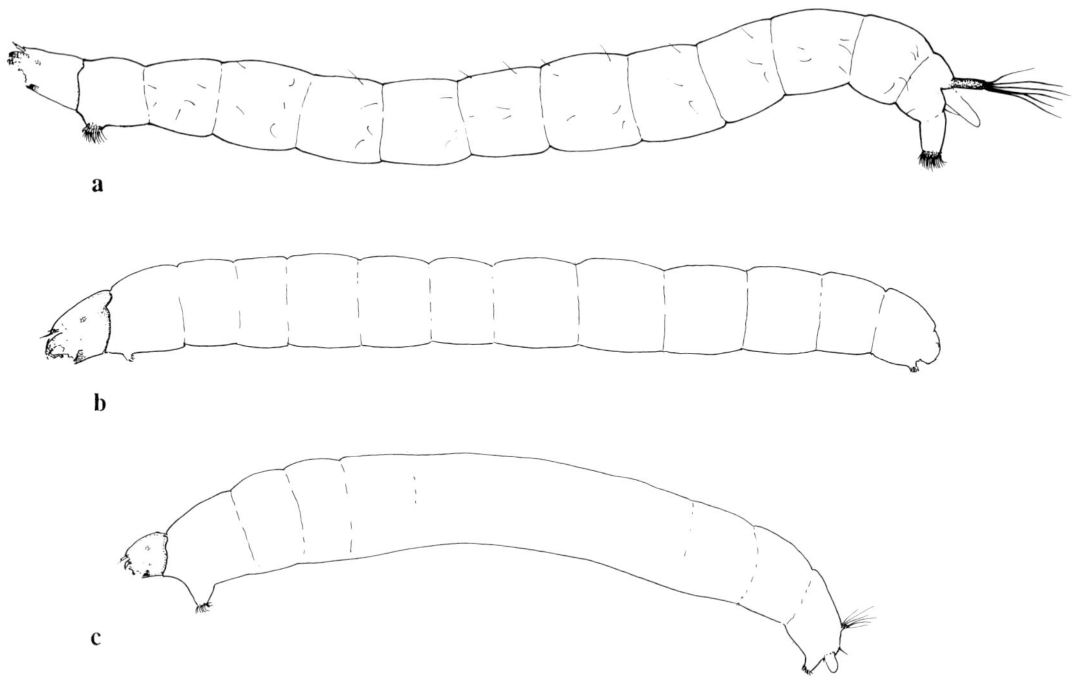

Figure 1 Larvae of (a) *Parochlus bassianus* Brundin, (b) *Gymnometriocnemus* sp., (c), *Compterosmittia* sp.,

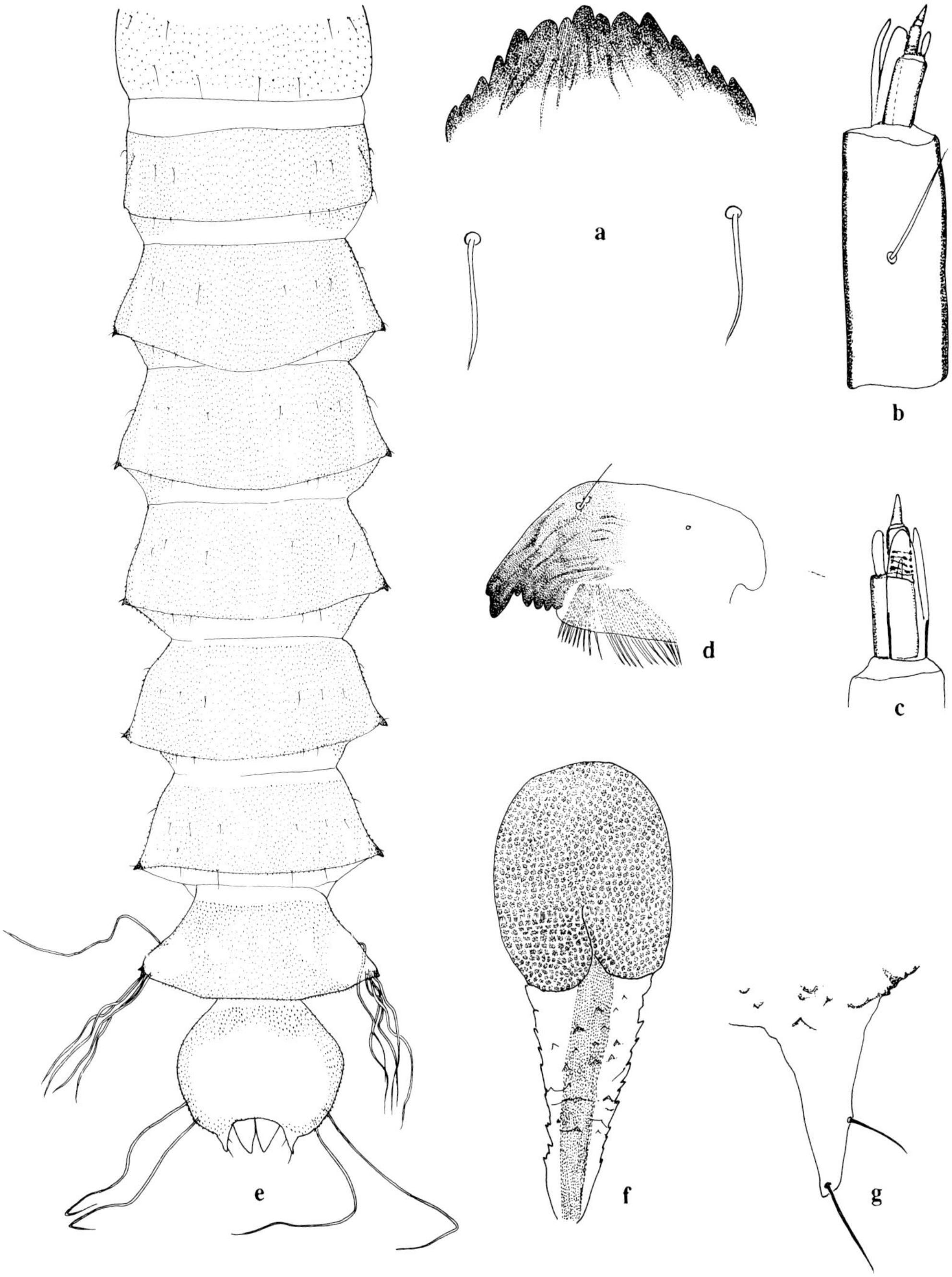

Figure 2 *Parochlus bassianus* Brundin, (a)-(c) larva: (a) mentum; (b) antenna; (c) antennal apex, (d) mandible; (e)-(g) pupa: (e) tergites, (f) thoracic horn, (g) apex of anal lobe.

Parochlus bassianus Brundin (Table 1; Figs. 1a, 2a-g)

Some pupae (Fig. 2e-g) contained pharate adults that were able to be identified to species using Brundin (1966). *Parochlus bassianus* was described by Brundin (1966: 144) from the female pupa and adults of both sexes. This is likely to be the species recorded as an adult from Tasmania by Freeman (1961: 631) as *Podonomus kiefferi*, a taxon referred now to the genus *Parochlus* and restricted to the Holarctic region. This study provides the first description and illustration (Table 1, Fig. 2a-d) of the larva of *P. bassianus*. Unusually within the Podonominae the antenna is apparently four segmented, with the typical annulations restricted to the basal half of the segment. This may indicate that the antenna is five segmented, with the segmental delimitation indicated only by the distribution of annulations.

Parochlus bassianus is the first member of the subfamily Podonominae to be found in phytotelmata. Previous records of the species refer to a female pupa from the Franklin River, and adults from the Derwent River, Nelson River, Hastings Cave and Mount Wellington, all in Tasmania.

Genus nr. _Gymnometriocnemus_ Thienemann
(Table 1; Figs. 1b, 3a-e)

The phytotelm larvae of *Gymnometriocnemus* are unreared, but the species is placed tentatively in the genus based on similarity with the diagnosis for that genus in the northern hemisphere (Cranston *et al.*, 1983). Our original thoughts that this species belonged to the genus *Bryophaenocladius* reflects the great similarity in larval morphology between these genera: however the apparent division of the posterior parapods into anterior clawed and posterior unclawed parts, and the presence of fine setae on the posterior abdominal segment that may represent procercal basal and apical setae, indicates that a placement near *Gymnometriocnemus* is more appropriate at the present time.

One major difference between this larval type and the Holarctic diagnoses for *Gymnometriocnemus* (and *Bryophaenocladius*) lies in the short, four-segmented antenna with an elongate blade (Fig. 3.1b). In this feature the larva resembles *Parasmittia*, but the contruction of the posterior abdomen (Fig. 3.1e) and the simple labral S1 seta (Fig. 3.1d) precludes this placement.

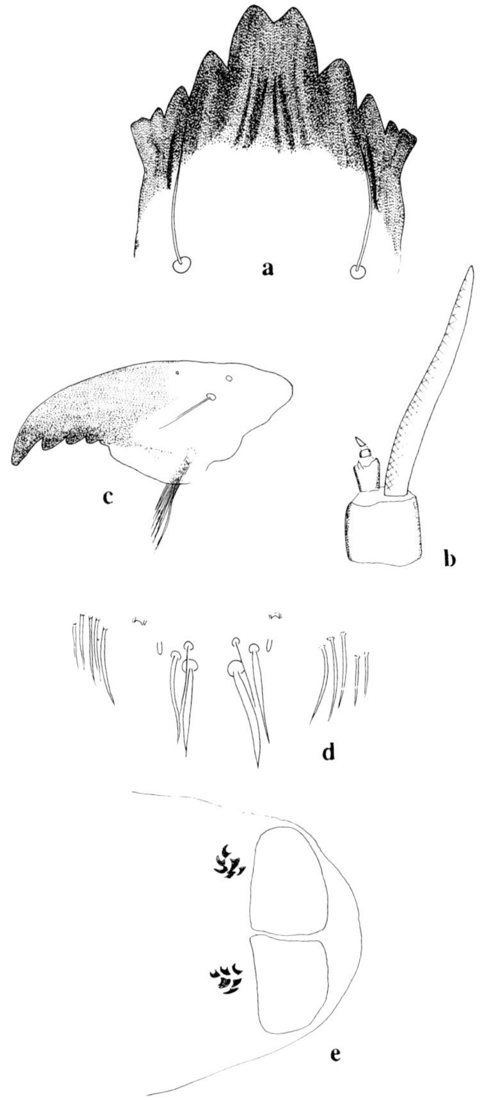

Figure 3 Genus nr. *Gymnometriocnemus* sp., larva: (a) mentum; (b) antenna; (c) mandible, (d) labrum, (e) posterior segment.

The presence of a seta interna of the mandible (Fig. 3.1c) also conflicts with either generic placement, but the diagnostic significance of this feature may be overvalued. Association with the pupa and adult will be necessary to confirm or refute the generic placement made here.

This unreared species of *Gymnometriocnemus* provides the first phytotelm record for the genus. An unreared larva that closely resembles the phytotelm species described here occurs in Carey Brook, Western Australia. Several species of *Gymnometriocnemus* are known as adults from Australia, both described (*G. nitidulus* (Skuse) and *G. wilsoni* Freeman) and undescribed.

Larval *Bryophaenocladius* from Australia that conform exactly with the Holarctic generic diagnosis are known, unreared: from "loamy soil" in Western Australia, rotting fruit in Victoria and from the benthos of the heavy-metal polluted Molonglo River, southern New South Wales.

Compterosmittia Sæther (Table 1; Figs. 1c, 4a-h)

The genus *Compterosmittia* Sæther (1981: 20) was erected on the basis of the adult alone. The immature stages of the genus *Compterosmittia* were known only from a larval rearing from the pitcher plant *Nepenthes mirabilis* in Hong Kong. This latter record (cited by Sæther (1990) as Cranston, 1987) was an abstract of an unpublished presentation and, although not constituting a publication of description of the immature stages, has provided the basis for the subsequent recognition of larvae of putative *Compterosmittia*. The only further associated material is a pupa visible within one of our phytotelm larva. It is appropriate to take advantage of the present discovery to describe the immature stages for the first time, basing the larval description on *Richea* phytotelm material and the pupal illustration on the Hong Kong *Nepenthes* material supplemented by the pharate *Richea* phytotelm specimen.

Characters of the immature stages that differentiate *Compterosmittia* from the genera *Limnophyes* and *Paralimnophyes* have proved elusive. *Compterosmittia* and *Limnophyes* can be separated in the larval stages from *Paralimnophyes* (including the two Australian species) by their appreciably short body setae. All *Compterosmittia* larvae, including the presumptive unreared ones, have the fourth (innermost) mandibular tooth clearly separated from the mola (Fig. 4d), whereas *Limnophyes* larvae have three inner mandibular teeth and the innermost area (the mola) is not tooth-like (Sæther, 1990, *pers. comm.*). Furthermore, the supra-anal seta in *Compterosmittia*, which is about 1/3 the length of the anal setae (Fig. 1c), allows discrimination from *Limnophyes* in which this seta approaches the length of the anal setae (Sæther, 1990). Otherwise all features of immature *Compterosmittia* are as diagnosed for *Limnophyes* (Sæther, 1990), including the extent and depth of branching of

Table 1: Larval Measurements for final instar *Richea* phytotelm chironomid larvae. All measurements in μm except body length in mm.

	Compterosmittia (n=8)	*Parochlus* (n=10)	nr. *Gymnometriocnemus* (n=10)
Body length	2.8-3.3	3.4-4.3	3.5-4.7
Head length	195-250	265-275	350-364
Antenna 1	23-26	39-46	15-20
Antenna 2	10-12	10-11	14-16
Antenna 3	4-5	6-7	7-8
Antenna 4	5-7	3-4	4-5
Antenna 5	2-4	n/a	n/a
Antennal Ratio	0.8-1.1	1.9-2.1	0.6-0.7
Blade	25-30	14-20	50-60
Style	5-6	4-5	3-6
Mandible	69-84	88-92	112-132
Mentum width	63-72	66-79	105-112
Premandible	47-55	n/a	66-81
Procercus height	10-14	105-120	n/a
Procercus width	10-12	35-42	n/a
Anal setal no.	7-8	6	n/a
Anal setal length	80-135	295-350	n/a

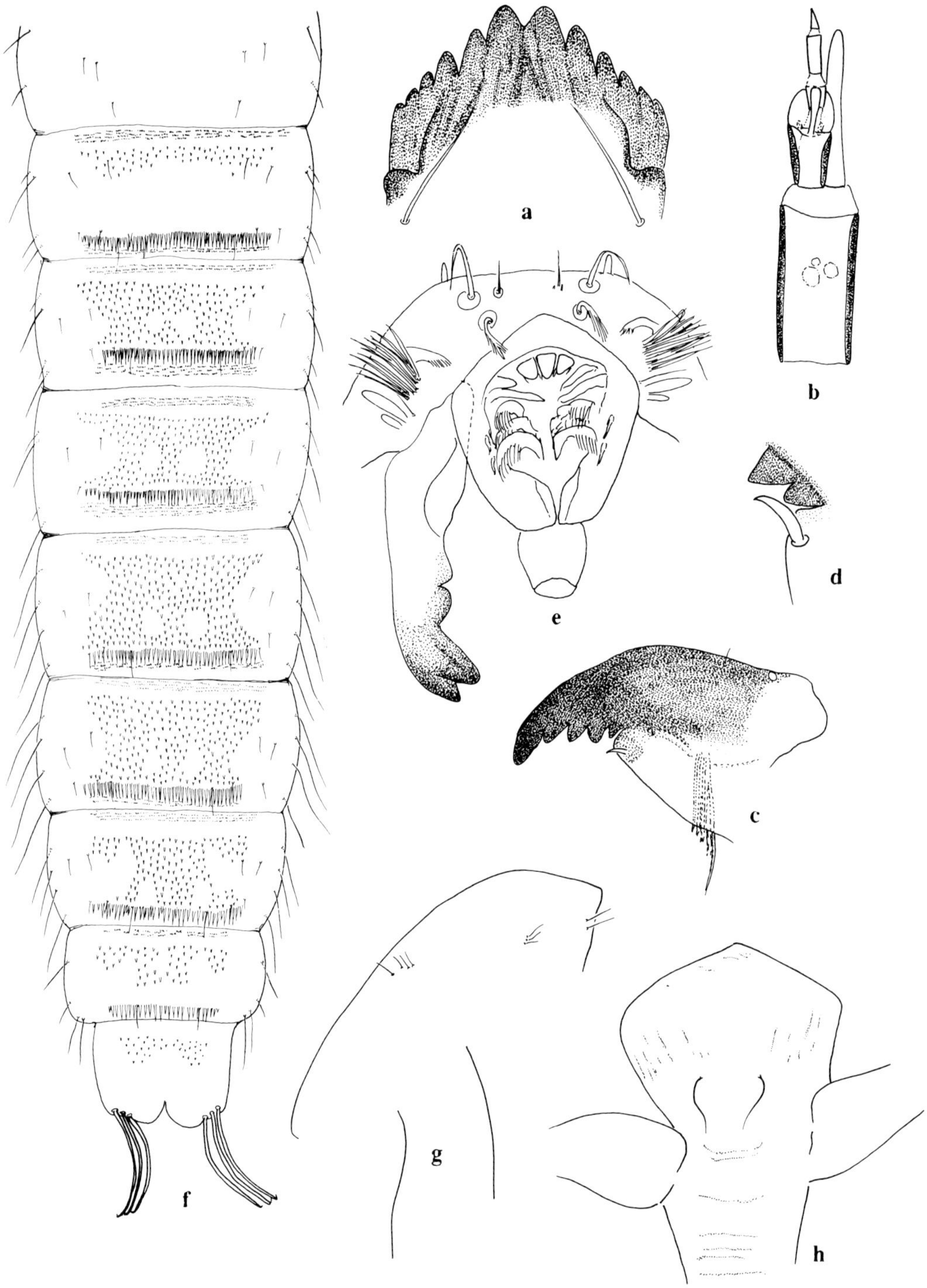

Figure 4 *Compterosmittia* sp., larva of *Richea* phytotelm sp.: (a) mentum, (b) antenna, (c) mandible, (d) inner mandibular teeth, seta subdentalis and mola, (e) labrum; (f)-(h) pupa of *Nepenthes* species: (f) tergites, (g) thorax, (h) cephalic area.

the S1 seta. No pupal features have been found to allow separation of the two genera. The paucity of discriminatory characters in the immature stages does not extend to the adults, males of which are readily separated by features identified in Cranston *et al.* (1989) and discussed in a phylogenetic context by Sæther (1990).

Larvae identifiable as *Compterosmittia* on the basis of the mandible are widespread but never common in a variety of Australian and Oriental phytotelmata (Cranston *pers. obs.*). These range from the internodes of bamboo (*Gigantochloa scortechini*) in Malaysia (leg. D. Kovacs) to rot holes in trees (Daintree, N. Queensland) and axillary liquid in an aroid (Lord Howe Island).

The only species of *Compterosmittia* described from Australia is *C. pectinatus* (Freeman, 1961) which was placed originally in *Gymnometriocnemus*. The relationship (and indeed specific distinction) between *C. pectinatus* and several Micronesian taxa recently recognised as belonging to *Compterosmittia* (Cranston and Martin, 1989) needs further study, but is hindered by the poor quality of Tokunaga's slide-mounted material in the Bishop Museum. Based on the material to hand, and the difficulty in rearing these chironomids, it seems unlikely that variation in the immature stages of *Compterosmittia* will assist in species segregation.

Discussion

The chironomids described here have differing ecological and biogeographic relationships to the total regional phytotelm fauna. *Compterosmittia* is widely distributed (perhaps world-wide) and may be restricted to phytotelmata for its development. *Gymnometriocnemus* has a world-wide distribution, but is previously unrecorded from phytotelmata. The subfamily Podonominae has a more restricted distribution, with strong Gondwanan affinities (Brundin, 1966) and no species has been reported previously from phytotelmata. The presence of three species of Chironomidae in one phytotelm habitat is unusual in a temperate situation. Our observations indicate that only some collections amongst the more complex tropical phytotelms contain more than a single species.

In addition to the chironomid larvae described here, the axils of *Richea pandanifolia* also contain copepods as well as larvae of mosquitoes, psychodids and phorids. The non-insectan fauna, however, is dominated by individuals of a new species of *Anisitsiellides*, a predatory mite belonging to the family Anisitsiellidae (Harvey, 1990) which may be a specialist predator of chironomid larvae (Dr. Mark Harvey, *pers. comm.*).

Acknowledgements

The Tasmanian National Parks and Wildlife Service are thanked for permission to sample in areas under their jurisdiction. We are grateful to Drs P.J. Gullan, Norm Fashing and Ole Sæther for comments on this manuscript, and to Dr. Mark Harvey for his observations on the mite.

Part 5

Chironomidae and Humans

Biology and control of *Chironomus tepperi* Skuse, a pest of rice in New South Wales

M M *Stevens*

Yanco Agricultural Institute, NSW Agriculture,
Private Mail Bag, Yanco, NSW 2703, Australia

Introduction

Larval chironomids cause significant damage to rice in many countries including France, Egypt, Japan, India, and the USA (Way and Wallace, 1989). Although midge larvae may in some instances attack the shoots of young rice plants (Darby, 1962), damage is more often confined to the newly sown seed and developing roots. Aside from impacting on plant growth, damage to root systems renders young plants vulnerable to uprooting by wave action, particularly in larger fields.

Although the potential for chironomids to damage rice in tropical areas has been recognised (Barrion and Litsinger, 1984), crop damage is generally most significant in temperate areas where highly mechanised agricultural practices, particularly aerial sowing, are in use. Aerial sowing leaves pregerminated seed lying on the soil/water interface where it is most vulnerable to both chironomid attack and wind damage. Rice grown as a labour-intensive subsistence crop in tropical areas is often raised in nursery beds and manually transplanted, allowing crop losses during establishment to be compensated for. In addition, the frequent insecticide applications often used for the control of other pests in tropical rice crops presumably provide incidental control of chironomids, leading to an underestimation of their significance as a potential source of crop damage.

Over 120,000 ha of irrigated rice is grown in southern New South Wales (NSW) each year, with approximately 85% of the crop being sown by air. The chironomid fauna of NSW rice fields is diverse, as it is in other temperate countries (Darby, 1962; Kikuchi *et al.*, 1985), with in excess of 30 species being recorded from individual bays during the first 48 days after flooding (Stevens, *pers. obs.*). Despite this high level of diversity two species, *Chironomus tepperi* Skuse and the predatory *Procladius paludicola* Skuse, strongly dominate the fauna during crop establishment. The purpose of this paper is to review available information on the biology and control of *C. tepperi*, the principal species associated with crop damage in NSW.

Chironomid damage to the NSW rice crop was first reported by Hely (1958). Jones (1968) recognised *C. tepperi* as the principal species involved, and demonstrated how relatively low densities of larvae could inflict serious damage to developing plants.

Genetics, development and reproduction

The cytology of *C. tepperi* was investigated by Martin (1974), who found that the banding pattern of salivary gland chromosomes indicated a close relationship between *C. tepperi* and *C.oppositus* Walker. The absence of inversion polymorphisms and the relatively high chiasma frequency found in *C. tepperi* were considered as adaptations to provide genetic variability in a colonising species that does not need to swarm prior to mating.

Atchley and Martin (1971) included *C. tepperi* in a comparative study of final instar chironomid larvae, and found significant morphometric differences between males and females. This variability was also recognised by Stevens (1993a), who showed that all four larval instars could be reliably identified using head capsule morphometrics. Development in *C. tepperi* follows Dyar's Law, with egg-hatch to first adult emergence in small laboratory cultures taking 13 days at 25°C (Stevens, 1993a).

Degree-day (DD) and developmental threshold studies on individually reared *C. tepperi* are currently underway, and preliminary results indicate that females take significantly longer to develop than males (♀, 168 DD; ♂, 143 DD, egg oviposition to adult), due almost entirely to a more persistent fourth instar. The developmental zero for *C. tepperi* is in the range of 9.8 to 10.5°C. Heaviest mortality occurs during the pupal stage. Individually reared larvae appear to develop faster than those reared in low-density cultures, even when larvae raised under both systems are continually supplied with an excess of food.

C. tepperi shows extremely rapid development in comparison to other midge species for which reliable data are available. At 25°C individually reared *C. tepperi* males take c.a. 9.5 days to develop from egg to adult, whilst females take 11.4 days. *C.kiiensis* Tokunaga and *Tanytarsus oyamai* Sasa both take 17 to 18 days to develop at 25°C (Maeda and Yano, 1988; Okazaki and Yano, 1990), whilst *Glyptotendipes tokunagai* Sasa averages 31.2 days (Yano *et al.*, 1991). The relatively short life cycle of *C. tepperi* is undoubtedly advantageous to a species that is a strong coloniser of new environments, but a poor competitor in the longer term.

Reproduction in *C. tepperi* was investigated by Martin and Porter (1977), who found little evidence of courtship behaviour. Unfed females laid one or two egg masses with an average of 351 eggs/egg mass, whilst females fed a dilute treacle solution laid up to six egg masses, with an average of 531 eggs/egg mass. Attempts at hybridisation were unsuccessful.

Ecology

Although *C. tepperi* has been recognised as a rapid coloniser of newly inundated environments (Edward, 1968; Martin, 1974; Martin and Porter, 1977), comparatively little is known about its ecology in the field. Peters (1979) reported *C. tepperi* from sewerage stabilisation ponds, where it formed a minor component of the midge population causing nuisance problems in nearby urban areas. Maher and Carpenter (1984) found *C. tepperi* to be abundant in newly flooded natural wetlands where it, along with other chironomid species, formed an important part of the diet of native waterfowl. Their results imply that *C. tepperi* may have more than one generation in environments with fluctuating water levels, but is nevertheless displaced by other species as the ecosystem develops. White (1970) reported *C. tepperi* in drogue net samples collected up to 300 metres above ground level, highlighting the dispersal ability of the species.

A three year emergence trap study conducted at Yanco Agricultural Institute (34°37'S 146°26'E) showed that *C. tepperi* often has only one generation during crop establishment in small experimental rice bays, with highly synchronised emergence peaks occurring 16 to 17 days after flooding (Stevens, *pers. obs.*). In one of the three seasons there was some evidence of a second, non-overlapping generation, however this second generation comprised less than 0.3% of total *C. tepperi* emergence. Increased predation could have adversely affected *C. tepperi* larvae developing from eggs laid after the emergence of first generation adults, however it cannot account for the almost total absence of

C. tepperi in subsequent trap samples, particularly when the potential fecundity of the first generation is taken into account.

Procladius paludicola Skuse became the dominant species in trap collections after *C. tepperi* emergence declined, and had multiple overlapping generations. *C. tepperi* and *P.paludicola* together comprised between 45.6 and 80.4% (mean 68.1%) of midges emerging during the first 48 days after fields were flooded. The corresponding figures for *C. tepperi* alone were 33.4 and 64.9% (mean 51.6%). Although in excess of thirty chironomid species were recorded during crop establishment, *C. tepperi* was the only phytophagous species recorded in sufficient numbers to warrant individually targeted control.

CHEMICAL CONTROL

Laboratory studies

Extensive laboratory testing of insecticides against *C. tepperi* has been conducted, with a particular emphasis on organophosphorus and pyrethroid compounds. Bioassay results for the species are summarised in table 1. As with other midge species (Ali, 1991), pyrethroids show a higher level of toxicity to *C. tepperi* than organophosphorus compounds, however their efficacy under field conditions has not been evaluated. The benzoylphenylurea growth regulators diflubenzuron and triflumuron both show sufficient ovicidal activity to have potential for preventative control programs, and may have considerable environmental advantages in comparison to broad-spectrum organophosphorus compounds.

Table 1. Toxicity of agricultural chemicals to *C. tepperi* in laboratory bioassays.

Compound	Class[1]	Stage[2]	LC_{50} (ppb)	LC_{90} (ppb)	Slope (±SE)	Reference
chlorpyrifos	OP	4th i	1.3	1.9	6.8 (± 0.5)	Stevens, 1992
chlorpyrifos-methyl	OP	"	2.5	3.6	8.4 (± 1.1)	"
temephos	OP	"	2.8	4.5	6.2 (± 1.1)	"
chlorfenvinphos	OP	"	6.0	11.3	4.7 (± 0.6)	"
malathion	OP	"	8.4	12.9	6.9 (± 0.4)	"
diazinon	OP	"	35.5	64.7	4.9 (± 0.5)	"
trichlorfon	OP	"	67.4	103.0	6.4 (± 0.7)	"
fenthion	OP	"	159.4	447.3	2.9 (± 0.3)	"
deltamethrin	SP	"	0.11	0.25	3.8 (± 0.4)	Stevens, 1993b
cyhalothrin	SP	"	0.14	0.38	3.0 (± 0.5)	"
betacyfluthrin	SP	"	0.23	0.40	5.2 (± 0.5)	"
alphacypermethrin	SP	"	0.28	0.59	4.1 (± 0.2)	"
bifenthrin	SP	"	0.31	0.54	5.3 (± 0.5)	"
cyfluthrin	SP	"	0.32	0.72	3.7 (± 0.4)	"
esfenvalerate	SP	"	0.58	1.37	3.4 (± 0.3)	new data
permethrin	SP	"	0.66	0.87	10.7 (± 1.3)	Stevens, 1993b
tefluthrin	SP	"	2.25	3.72	5.8 (± 0.4)	"
fenvalerate	SP	"	8.44	16.94	4.2 (± 0.5)	"
thiobencarb	H	"	1321	1978	7.3 (± 0.8)	new data
diflubenzuron[3]	IGR	egg	9.2	13.1	8.24 (± 0.29)	Stevens and Warren, 1993
triflumuron[3]	IGR	"	10.8	20.3	4.70 (± 0.17)	"
buprofezin[3]	IGR	"	304	889	2.75 (± 0.24)	"
fenoxycarb[3]	IGR	"	898	1523	5.57 (± 0.27)	"
Bacillus thuringiensis	BI	1st i	-	100-200	-	Treverrow, 1985
Bacillus thuringiensis	BI	4th i	360	790	-	"

[1] OP, organophosphorus compound; SP, synthetic pyrethroid; H, herbicide; IGR, insect growth regulator; BI, bacterial insecticide.
[2] i = instar
[3] see cited reference for cumulative ovicidal/first instar results.

Field studies

Since the withdrawal of organochlorine insecticides control of *C. tepperi* and other rice field chironomids has been based on organophosphorus compounds. At present, pregerminated seed is treated with either malathion or trichlorfon immediately prior to sowing, and either one or two further treatments of malathion, trichlorfon, diazinon or chlorpyrifos are applied at 5 to 7 day intervals. Diazinon and chlorpyrifos are not used on seed for environmental reasons; these compounds have high levels of vertebrate toxicity, and seed that falls beyond the field margins is frequently eaten by birds and other wildlife.

The efficacies of six organophosphorus water treatments were evaluated by Stevens (1991) and Stevens and Warren (1992), who found that at the rates examined malathion on seed and chlorpyrifos post-sowing were the most effective treatments. Stevens (1992) showed that LC_{90}-relative application rates and initial water column toxicity levels were closely related, however persistence in the water column was more strongly influenced by the chemical behaviour of individual compounds.

Water column toxicity levels do not always reflect the level of control achieved, however. Trichlorfon 625 g/l at 700 ml/ha (c.a. 4.6 x LC_{90}) provided high initial water toxicity but only moderate larval control (fig.1); diazinon 800 g/l at 150 ml/ha (c.a. 2.0 x LC_{90}) provided superior initial and residual control, despite providing lower levels of initial water column toxicity (fig. 2) (Stevens, 1991; Stevens and Warren, 1992). Both compounds showed similar levels of residual water column toxicity at the rates examined. Larval population assessments are therefore considered essential if the impact of chemical treatments is to be accurately determined.

More recent studies (Stevens and Warren, *pers. obs.*) on controlled-release (CR) organophosphorus formulations have shown that over 30 days effective control of all chironomids in establishing rice crops can be achieved with a single chemical application containing 200 grams active ingredient per hectare. Pelletised CR formulations avoid the drift problems associated with aerial applications of liquid insecticide, and reduce the risk of toxicant loss to drainage systems since chemical activity in the water column remains low throughout the treatment period. For economic reasons, farmers

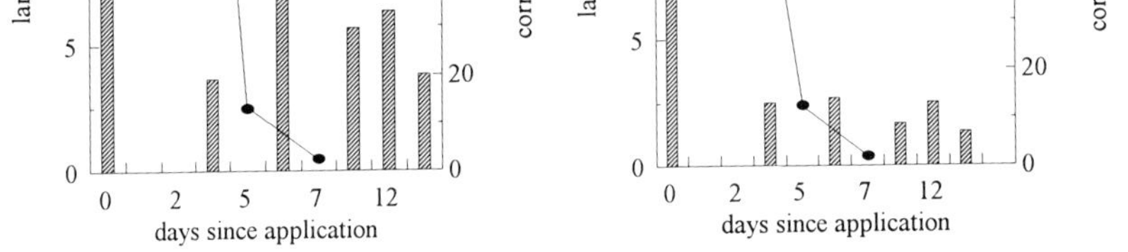

Figures 1 - 2. Influence of trichlorfon and diazinon (at maximum registered rates) on mixed chironomid populations and water column toxicity levels. Bars represent mean larval numbers (predominantly *C. tepperi*-type larvae) in core samples (area 72.4 cm^2), points represent mean mortality (corrected against controls) to fourth instar *C. tepperi* larvae in bioassays with field water samples. Data from Stevens (1991) and Stevens and Warren (1992). Note that notional chemical concentrations (based on water volume rather than area) were c.a. 20% lower in the larval population study (Stevens and Warren, 1992) due to greater water depth. Day zero population assessments were conducted prior to chemical application.

who apply post-sowing treatments generally do so in tank-mixes containing herbicides, frequently leading to the application timing of both materials being compromised. Single CR applications at or before sowing will consequently allow herbicide treatments to be timed more effectively.

Problems with current control procedures

NSW rice farmers frequently report crop damage if chemical treatments for chironomid control are not continued after sowing. This conflicts with the findings of emergence trap studies (Stevens, *pers. obs.*) which indicate that C. *tepperi* generally has only one generation in untreated establishing crops. Theoretically, the malathion applied to the seed immediately before sowing should be eliminating initial C. *tepperi* infestations and further damage should not occur.

There are several possible explanations for this anomaly. Firstly, the malathion seed treatment may not be working. Although malathion is effective as a water treatment (Stevens, 1991; Stevens and Warren, 1992), its efficacy when seed is used as a carrier has not been assessed. Malathion may be protecting the seed from attack in the short term, but failing to disperse sufficiently to eliminate larval populations. Trials are currently being conducted to evaluate this possibility.

A second alternative is that species other than C. *tepperi* may be responsible for crop damage, particularly in the later part of the crop establishment period. Other species with the potential to cause plant damage are generally poorly represented individually in emergence trap captures, however in total these species may be abundant enough to cause some crop injury.

A third possibility is that C. *tepperi* actively recolonises fields after initial infestations are chemically controlled. The phenology of C. *tepperi* in untreated rice bays clearly shows female midges can identify oviposition sites that suit the ecological requirements of the larvae. Although aspects of inorganic water chemistry may be involved in site suitability, the possibility exists that the presence of midge larvae, conspecific or otherwise, may play a role in deterring oviposition. If this is the case then eliminating initial infestations with insecticides may lead to reinfestation by C. *tepperi* when the insecticide and oviposition-deterring compounds decline to ineffective levels. Laboratory studies on factors influencing oviposition site preference are currently underway, and have shown that female C. *tepperi* will strongly avoid ovipositing in rearing solution containing small numbers of homogenised conspecific larvae.

CONCLUSIONS AND FUTURE PROSPECTS

C. *tepperi* is an opportunistic species with several genetic, behavioural and developmental adaptations that allow it to make optimal use of establishing or ephemeral habitats. Although C. *tepperi* has, in practical terms, only one generation in establishing rice crops, it is the most abundant species during early crop growth and constitutes a major problem for the NSW rice industry.

Further research on the role of C. *tepperi* and other chironomid species is required before the midge problem in NSW rice fields can be fully understood. Information is required on the ecology of C. *tepperi* in larger commercial crops, the efficacy of malathion seed treatments, and the role of less abundant species in causing crop damage. Studies on factors affecting oviposition in C. *tepperi* will be of value in understanding processes involved in the formation and colonisation of new freshwater habitats.

In the short term, controlled-release formulations of organophosphorus insecticides hold the greatest potential for improved chironomid control. These formulations can provide thirty days crop protection at environmentally acceptable rates, controlling both C. *tepperi* and other chironomid species which may be associated with crop damage. They do not drift if applied by air, and will allow herbicides to be timed more effectively by removing the need for multiple insecticide applications. The best longer term prospect for entirely eliminating insecticides from chironomid control in rice involves the development of plant varieties resistant to chironomid attack.

A screening program is currently being developed in an effort to locate sources of resistant germplasm.

Infestation of Enclosed Water Supplies by Chironomids (Diptera: Chironomidae): Two Case Studies

Martin B. Berg

Department of Biological Sciences,
University of Notre Dame, Notre Dame,
Indiana 46556, U.S.A.

Present Address: Department of Biology,
Loyola University Chicago, 6525 North Sheridan Road,
Chicago, Illinois 60626, U.S.A

Introduction

Pestiferous emergences of chironomids from recreational lakes and water supply reservoirs occur worldwide and are one of the most common complaints associated with midges (Meillon and Gray, 1937; Freeman, 1950; Bay, 1964; Edwards *et al.*, 1964; Spiller, 1965; Jolivet, 1974; Tabaru, 1975; Tabaru *et al.*, 1978; Ali *et al.*, 1985; also see reviews by Ali, 1980, 1991, 1994). These emergences can interfere with outdoor recreational and work activities and thus, can have a substantial economic impact on affected areas (Tabaru *et al.*, 1987; Mulla *et al.*, 1990, Ali, 1991). It has been suggested that the incidence of future complaints of pestiferous emergences will be increased as anthropogenic impacts on associated aquatic habitats (e.g., nutrient enrichment) also increase (Ali, 1991).

Although problems related to nuisance emergences of adult chironomids have been well-documented, problems associated with larval midges have been reported less frequently (Tokunaga and Kuroda, 1935, 1936; Darby, 1962; Williams, 1974). One such problem involves the infestation of municipal water supplies. Most larval infestations involve water distribution systems with reservoirs that are accessible to ovipositing females. In this situation, modifications of the design of the distribution system and insecticide applications can be used to eliminate or reduce larval populations (Williams, 1974; Mitcham and Shelley, 1980; Bay, 1993). Larval infestations of enclosed water supplies, however, are more difficult to control. Because state and federal laws in the United States preclude the use of insecticides in potable water supplies, other control methods must be used. This paper presents two case studies of the control of chironomids in enclosed water supply systems.

Case studies

Case 1 - Description of the Infestation

The first case study involves the infestation of the water system within a food processing facility. The name and location of the facility have been withheld to ensure confidentiality. Water for the facility is withdrawn from a nearby river and pumped directly into a treatment plant where the water is softened and pH adjusted (Fig. 1). The water is then pumped to a settling basin where particulate matter is removed. Following particulate matter removal, water is directed to an elevated storage tank. Before entering the food processing area, the water passes through a series of carbon filters and is directed to either the hot or cold water systems of the facility. The hot and cold water are then combined into a single pipe to ensure a uniform water temperature for preparation of the final product. Occasionally, scale material and other matter from inside the pipes becomes dislodged and suspended in the water. To prevent this material from entering the food product, fine mesh bags (64µm) are attached to each outflow nozzle. Plant operators first observed the presence of chironomid larvae within these mesh bags.

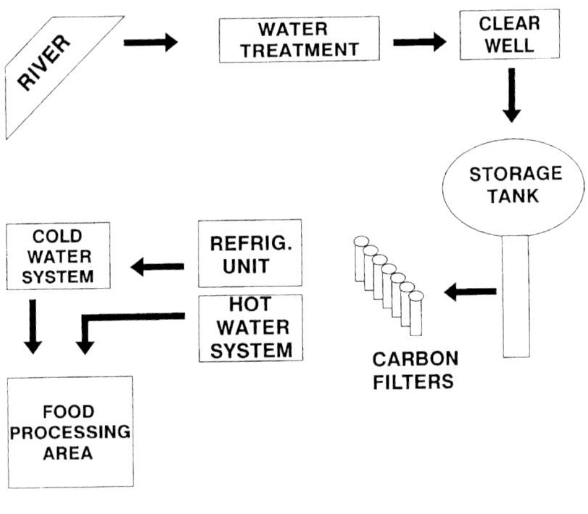

Figure 1. Design and path of water flow in the food processing facility.

Sampling

In an attempt to determine the point of entry of chironomids into the food processing facility, and to assess the extent of infestation, samples of organic matter and water were collected from the following locations: clear well (organic matter), outflow of elevated storage tank (water), carbon filters (carbon), floor drains in the food production area (organic sediment and water), and the water distribution system inside the facility (water) (Fig. 1). Sediment samples from the food processing facility were collected with an Ekman grab and dip nets and were returned to the laboratory for examination under a dissecting microscope at 10X magnification. Water from the elevated storage tank was sampled for larvae from a "downstream" fire hydrant by placing a 64µm plankton net over the open hydrant for three 10 minute periods. Larvae within the distribution system were sampled continuously for seven days at various locations in the facility using 64µm mesh bags. Material from the carbon filters and organic matter from floor drains were placed in white enamel pans and examined for the presence of chironomid larvae. Although adult chironomids were not observed in the production facility, air intake filters were visually examined for the presence of adults. Chironomid larvae from the adjacent river also were collected and identified to assess the "pool" of available midges.

Three chironomid taxa were collected from the water distribution system of the food processing facility. *Dicrotendipes neomodestus* (Malloch) accounted for 98% of all the chironomids collected, and *Phaenopsectra obediens* (Johannsen) and *Eukiefferiella devonica* Edwards sp. gr. comprised the remaining 2%. All three taxa were collected in the benthic samples of the adjacent river with *D. neomodestus* being numerically dominant. Therefore, it appeared that the river may have been either the direct (via the water supply) or indirect (via colonization of the elevated storage tank by aerial adults) source of chironomids in the facility's water system. Because larvae were absent from all sampling locations in the food production facility except the distribution system, the mode and point of entry of midges into the system could not be determined.

Remediation Efforts

In an attempt to directly eliminate chironomids from the water lines, a series of sanitization procedures was initiated. The first sanitization procedure involved backwashing the cold water system with 49°C water for 12h. Laboratory experiments indicated that temperatures in excess of 40°C were lethal to all three species of chironomids infesting the system. This procedure resulted in the absence of larvae from mesh bags for two days following the sanitization. Within five days, however, the number of larvae in the mesh bags returned to pre-sanitization densities. This sanitization effort probably was ineffective because of the extensive network of pipes in the distribution system. It was felt that there were large areas of the system that were not exposed to 49°C water. To determine if all areas were being sanitized, thermistors were placed at several locations along the distribution system, and the system again was flushed with 49°C water. Results from these tests indicated that some areas of the system attained temperatures of only 30°C, a temperature that was sublethal to larvae.

A second sanitization effort using 89°C water was initiated to ensure that water temperatures exceeded the 40°C lethal temperature in all areas of the distribution system. During this sanitization, the thermistors indicated that the minimum water temperature in the system was 55°C, a temperature sufficient to kill all larvae. This treatment reduced larval densities in mesh bags for approximately one week after sanitization. Within two weeks, however, larval densities returned to pre-sanitization levels. Results from this sanitization effort indicated that some area of the distribution system must have been serving as a refuge for chironomids.

Examination of the design of the water distribution system revealed that one area was not exposed to the hot water treatment. This area was the refrigeration unit, which included a large tank that served as a reservoir for the refrigeration system. The third sanitization effort involved backwashing the entire system, including the refrigeration unit, with 89°C water. This treatment resulted in elimination of all chironomid larvae from the water system. Monthly backwashing with 89°C water has maintained a midge-free system for over one year.

Case 2 - Description of the Infestation

The second case study involves the infestation of a municipal drinking-water supply in the town of Lowell, Indiana, U.S.A. The town's water is derived from groundwater sources and is pumped directly into a treatment facility from a series of wells (Fig. 2). Upon entering the treatment facility, the water is aerated, chlorinated, and pumped to an elevated storage tank for eventual distribution to consumers.

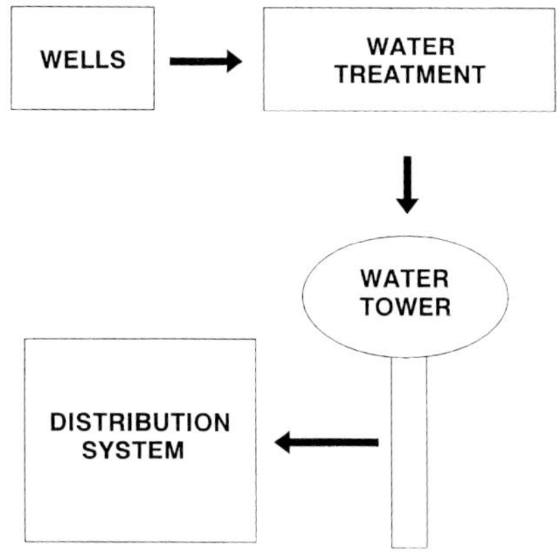

Figure 2. Design and path of water flow in the Lowell, IN, municipal water supply.

Sampling

Operators of the water plant were alerted to the presence of chironomids in the drinking-water supply by a series of consumer complaints. To determine the extent of infestation, groundwater and water from the distribution system and fire hydrants were sampled for larvae by pumping the water through a 64µm plankton net. Larvae were absent from all well-water sites, however, large numbers of larvae were collected from the distribution system. All larvae collected from the distribution system were reared and identified as *Paratanytarsus grimmii* (Schneider).

Remediation Efforts

The first control effort involved an attempt to dislodge as many midges as possible by flushing the distribution system through open fire hydrants. An average of 125 larvae/5 minute flushing period was collected. Flushing the distribution system had no effect on larval abundance, but often increased the number of consumer complaints.

Because no larvae were collected from well-sites, it was concluded that the only point of entry for midges into the system was through a vent located at the top of the elevated storage tank. Therefore, a second control effort was conducted that involved emptying the tower, covering the vent with a fine mesh screen (64µm) to prevent entry of adults, and cleaning the tower with pressurized

50°C water. Before conducting the cleaning procedure, adult and larval chironomids were collected from the inside walls of the elevated storage tank and from sedimented material near a central standpipe inside the tank, respectively. All midges collected in the tower were identified as *P. grimmii* and all adults, pupae, and reared larvae were female.

Two aspects of the life cycle of *P. grimmii* render this chironomid difficult to control in enclosed water systems: 1) mode of reproduction, and 2) oviposition behavior. *P. grimmii* in the Lowell water distribution system reproduces parthenogenetically. Therefore, any attempt at controlling this chironomid by interfering with its mating behavior would be futile. In addition, *P. grimmii* undergoes a modified form of paedogenesis in which the pharate adult oviposits inside the pupal exuviae. As noted by Langton *et al.* (1988), this behavior differs from true paedogenesis in which immatures (larvae) give rise to immatures. Thus, attempts at controlling *P. grimmii* by interfering with emergence behavior also would be unsuccessful because emergence is not necessary to complete the life cycle. Given the inability to control *P. grimmii* mating or emergence behaviors, I attempted to control this chironomid through its food supply.

The diet of *P. grimmii* in the water distribution system was examined using DAPI (4'6-diamidino-2-phenylindole), a fluorochromatic stain with DNA-specific binding properties (Porter and Feig, 1980; Walker *et al.*, 1988). Because different materials stain characteristic colors when treated with DAPI, it was possible to identify food items in the gut contents of *P. grimmii*.

Results from DAPI analyses indicated that *P. grimmii* was feeding on dead bacteria, which appeared as faint fluorescing spots, or "ghosts". These "ghosts" also were observed in water samples collected from the distribution system. Thus, *P. grimmii* probably was filtering dead bacteria from the water column. Although the sources of these bacteria were not determined, they may have been sulfur and iron bacteria that were killed by aeration and pH elevation.

The first attempt at controlling *P. grimmii* via its food supply involved the use of B.t.i. (*Bacillus thuringiensis* var. *israelensis*), a biological agent useful in the control of other aquatic insects such as mosquitoes and blackflies (Ali *et al.*, 1981; Lacey and Undeen, 1987; Merritt *et al.*, 1989; Molloy, 1990). In general, B.t.i. is non-toxic to chironomids when applied at concentrations lethal to blackflies and mosquitoes (Ali *et al.*, 1981; Merritt *et al.*, 1989; Molloy, 1992). When B.t.i. is applied at 5 - 10X the recommended rate for mosquito control, however, filter-feeding chironomids can be negatively affected (Miura *et al.*, 1980; Ali, 1981; Merritt *et al.*, 1989; Molloy, 1992). Laboratory experiments performed in conjunction with this study demonstrated that 5ppm B.t.i. (Teknar HP-D: Potency = 3000 AA units/mg (1200 International Units/mg) resulted in 25% mortality in 48h, however, at 10ppm, mortality was 100% in 48h. Given the successful control of *P. grimmii* at 10ppm, Indiana state officials were approached for approval to add B.t.i. to the Lowell water system. That request was denied because (1) B.t.i. has not been approved for use in potable water supplies and (2) the potential legal implications of adding an insecticide to public drinking water. As a result, I attempted to find a substance that had been approved for addition to drinking water supplies and also negatively affected chironomids.

One such substance that has been used effectively in the elimination of zebra mussels (*Dreissena polymorpha*) in the Laurentian Great Lakes is a food grade polymer, Catfloc® LS (Calgon Corporation) (Hastreiter, 1993). This material interferes with respiration of the mussels and obstructs their filter-feeding mechanism. Thus, this polymer may be useful in controlling *P. grimmii* in the Lowell water system.

Studies of the efficacy of Catfloc® in controlling *P. grimmii* are ongoing. Preliminary results indicate that Catfloc® in concentrations of 12ppm produces very low acute toxicity, however, it appears to produce high mortality in chronic tests conducted at similar concentrations (M. Alexander, Michigan State University, *pers. comm.*).

General Discussion

The most highly publicized instances of problem-causing midges are associated with nuisance emergences. The infestation of potable water supplies by larval midges, however, also is common. To assess the prevalence of chironomids in potable water supplies in the United States, a questionnaire was distributed by the American Water Works Association to state agencies responsible for drinking water. Results from this questionnaire were combined with literature accounts of nuisance emergences in the U.S. to assess the relative prevalence of chironomid infestations of potable water supplies.

Nuisance emergences have been reported from 12 states with six of those states also reporting chironomid infestations of potable water supplies (Fig. 3). Additionally, nine states without nuisance emergences reported infestations of water supplies. Thus, 30% of the states in the U.S. (15 states) reported chironomid infestations of drinking water supplies. Given the relatively high incidence of chironomids in drinking water supplies, it seems surprising that infestations are not more commonly reported. This may be because infestations usually occur in densities too low to elicit consumer complaints. Occasionally, however, larval densities may reach levels sufficient to be noticed by water-users.

Most cases of chironomid infestations in water supplies involve the colonization of readily accessible, and easily treated, open reservoirs. Filters or microstrainers placed at reservoir outflows can remove larval midges before they enter the enclosed distribution system (Resh and Grodhaus, 1983). Although not recommended for potable water supplies, chemical larvicides or other control measures can be effective and easily applied to uncovered reservoirs (Flentje, 1945; Ali, 1980; Creekmur *et al.*, 1982; Ali, 1991). Modifications in the design of distribution systems also have been used in eliminating chironomid infestations. A recent study by Bay (1993) in Tacoma, WA (U.S.A.), demonstrated that replacing reservoirs with enclosed elevated storage tanks could reduce the occurrence of midges in potable water supplies.

A situation more difficult to control, and less commonly reported, is the infestation of persistent chironomid populations in enclosed water distribution systems. The use of heat, as was done in the food processing facility, is one possible alternative. This method, however, generally would be restricted to relatively small and easily controlled systems.

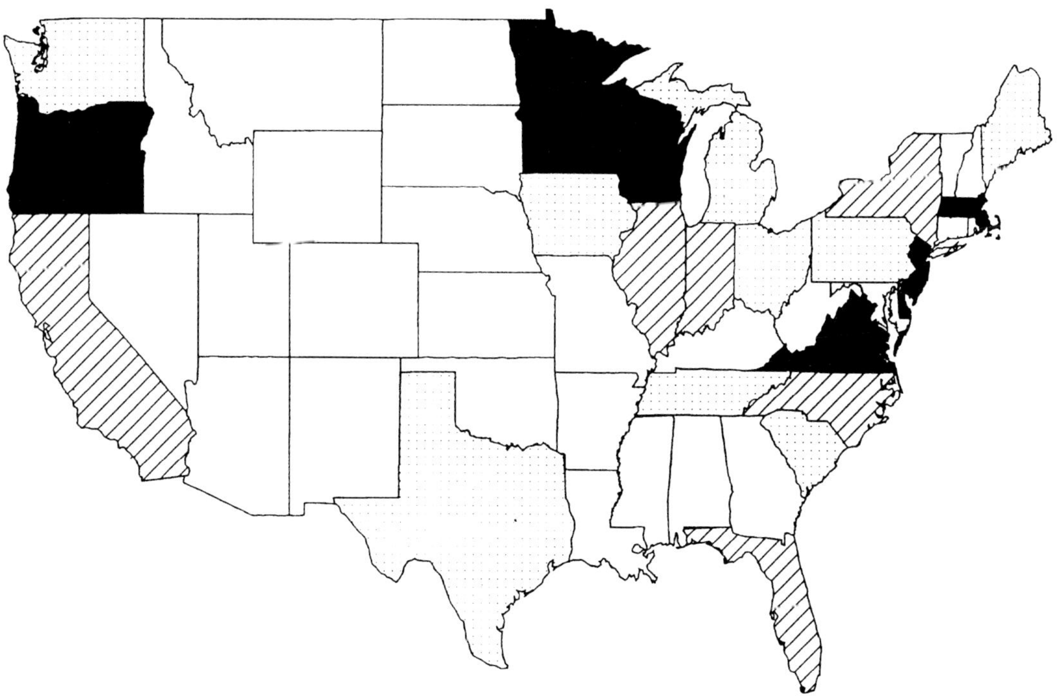

Figure 3. States in the U.S. reporting chironomids as nuisances. Shaded = nuisance emergences, stippled = potable water supply infestations, and hatched = both nuisance emergences and potable water supply infestations.

The most difficult infestation to control occurs in large distribution systems within which chironomids can complete their life cycle, such as parthenogenetic taxa. The infestation of drinking water supplies by parthenogenetic *P. grimmii* is not restricted to the Lowell, IN, water supply. Similar infestations of unidentified *Paratanytarsus* species, presumably *P. grimmii* (Langton *et al.*, 1988), have been reported from distribution systems in California (Resh and Grodhaus, 1983), England (Williams, 1974; Burfield and Williams, 1975), Germany (Kruger, 1941) and Cyprus (as cited in Langton *et al.*, 1988). A variety of control measures have been used in an attempt to control this species. Usual control measures such as pH elevation, chlorination, or the installation of microstrainers have met with little or no success (Williams, 1974). The few cases in which *P. grimmii* has been eliminated involved either the addition of pyrethrins to the water supply (Williams, 1974), or the fortuitous freezing of water mains during winter (Kruger, 1941).

In the United States, the prohibition of adding insecticides to drinking water distribution systems has left few alternatives for the control of chironomid infestations of enclosed water supplies. The approach used in this study of controlling midges through their food supply shows promise for chironomid control in other aquatic habitats. If the food-grade polymer Catfloc®, or other similar material, is found to be effective at controlling midges in the Lowell distribution system, its use could be expanded to other pestiferous situations. This approach would reduce the use of potentially harmful insecticides and other chemicals in eliminating pestiferous chironomid populations from aquatic habitats.

Acknowledgement

I thank P.M. Charlebois for valuable comments on this paper. I also would like to thank the Lowell Water Department, particularly J. Hoshaw (Superintendent), for major logistical support.

Impact of human-induced flow perturbation on the chironomid communities in the first order stream section (Central Poland)

Maria Grzybkowska

Department of Ecology and Vertebrate Zoology, University of Lodz, 190 - 237 Lodz, Banacha 12/16, Poland

Introduction

Flow perturbations that result in the partial drying-up of a streambed have been investigated mainly either in naturally intermittent streams or downstream of hydroelectric power dams. Both types of alternations of flow lead to a break-down of benthic macroinvertebrates (Corrarino and Brusven, 1983; Morgan *et al.*, 1991; Mackay, 1992; Moog, 1993).

Dry sections of the marginal zone of a streambed were monitored in the first-order stream section of the Bzura River flowing in heavily canopied areas in the vicinity of a large city (in Central Poland), downstream of three recreation ponds to observe the effect of refilling of these ponds after their cleaning. The main objectives of this study was to learn the response of a macrobenthic community to this small-scale disturbance. Besides the recolonization rate, community similarity and chironomid secondary production at this station were estimated and these parameters compared both with the period of natural flow pattern at the same station and with the station upstream.

Study area

The Bzura River is a part of the Vistula drainage basin. Study areas were established in the first order section of this river (51°19′N 19°21′W). These two sites were separated by three recreation ponds; the first is 1.8 km from the head of the river, the second 4.8 downstream of the source (Fig. 1).

At the first site (BZA), the Bzura River reached a maximum width of 1.1 m, depth of 0.25 m and current speed of 0.23 m s^{-1}. The riverside was covered by macrophytes such as *Solanum*, *Ranunculus*, *Veronica* and trees (*Padus*, *Quercus* and *Acer*) and the riverbed was mainly sand with scattered pebbles. At the site BZB, the Bzura River reached 2.3 m in width, 0.13 m in depth, and a current speed of 0,20 m s^{-1}. In the mid-channel patches *Potamogeton* rarely occurred. Numerous macrophytes and trees grew at the side. The riverbed was mainly composed of the medium sand and pebbles. At these sites there was much allochtonous organic matter, especially with autumn leaf fall.

Note that the discharge at the both annual cycles was similar at site BZA, while different at BZB. The main reason for this was the release water through the little dam downstream of the recreation ponds due to its cleaning and subsequent slow re-filling (Fig. 1).

A mean annual water temperature was 6.6°C in the first cycle while 6.7° C in the second one.

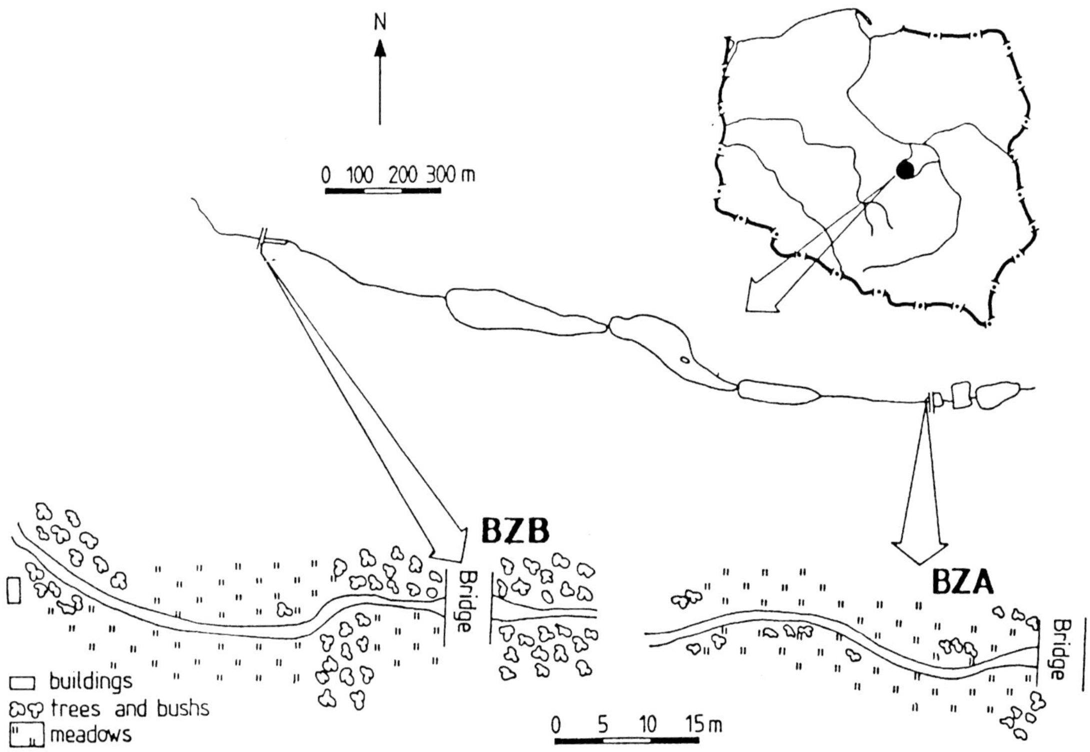

Figure 1. Map of study area showing two stations: BZA and BZB.

MATERIAL AND METHODS

Monthly benthic samples were collected from March 1984 to February 1985 (the first period) and from March 1987 to February 1988 (the second one). At each site (BZA and BZB) 30 samples were taken with a 10 cm^2 tubular sampler. The corer was pushed into sediment to a depth of 15 cm, through vegetation if present.

Additional samples were taken to analyze the composition of particulate inorganic matter according to Cummins (1962). This method divided the dry sediments into size classes that have been shown to be of ecological significance. The analysis of size fraction was made on the basis of weight. To aid the statistical analysis of the substrate data, field size assessments were transfered into a single substrate index (SI) by summing the mid-point value of size classes weighted by their proportional cover (Quinn and Hickey, 1990). These samples were also used to determine the organic content in the sediments. For this purpose, on the basis of 5 1 ml subsamples there the percentage of particulate organic matter was estimated: >1 mm (CPOM) and 1 mm < (FPOM). Next the benthic organic matter was dried at 60°C (for two days), weighted, ashed at 600°C for 24 hours and reweighed. Macroinvertebrates were sorted by hand and preserved in 10% formalin prepared with habitat water. All animals were counted and then weighed. Data obtained were used to calculate density by the transformation of data to $\log_{10}(X-1)$ (Elliott 1979). Three-way analysis of variance (ANOVA) was used to test for significant differences in environmental and biotic factors between sites and cycles (Morrison, 1990).

Chironomid larvae were identified to species (if possible) or from imagines obtained from laboratory rearings. The annual production of dominant taxa of Chironomidae was estimated by the size-frequency method (Hynes and Colemann, 1969) as modified by Hamilton (1969) and Benke (1979). Larval body length was measured with an accuracy of 0.1 mm and individuals were grouped into 0.5 mm length classes relying on the suggestions of Cushman *et al.* (1978). The mean theoretical body weight for each size class was calculated using the equations of Konstantinov (1958), Mackey (1977) and Nolte (1990). The dry weight was established as 15% of the wet weight for the investigated species, enabling determination of the weight loss of larvae stored in formalin (Dermott and Paterson, 1974).

To estimate correct annual production, the obtained P value must be further multiplied by 365/CPI, where CPI is the mean cohort production interval (Benke, 1979). As it was impossible to estimate the CPI's from field data, the Konstantinov (1958a, b) and Mackey (1977a) equations, which predict larval development times from temperature, were used to estimate the CPIs for chironomid species. Since Mackey and Konstantinov equations were based mainly on laboratory data, the CPI calculated for chironomid species were arbitrarily doubled to account for some degree of environmental resistance such as competition, food shortage, etc., according to the suggestion of Benke *et al.* (1984).

RESULTS

Zoobenthic community

The chosen sites of the Bzura River differed in many morphometric parameters such as width, depth, SI (greater at BZB, P>0.000) and POM (higher at BZA, P>0.001).

The macroinvertebrate density did not differ significantly at BZA between investigated cycles. The stream community was typically dominated by Oligochaeta and Chironomidae. At BZA, Oligochaeta varied from 70.2% of the total zoobenthos abundance in the first cycle to 75.6% in the second one, whereas the contribution of Chironomidae to other taxa was respectively 14.9% and 20.3%.

The community (mainly Oligochaeta and Chironomidae) at disturbed site (BZB) varied distinctly between cycles as the effect of reduced stream discharge and then rewatering the streambed. Due to the very low discharge (defined as ten times lower then an average annual flow) at the beginning of the first cycle, a low density of all Chironomidae, similarly to the total benthofauna abundance, occurred. When the discharge returned to its typical level, a strong repopulation begun, firstly by Chironomidae followed by Oligochaeta. Finally the average benthofauna density recorded at BZB was about 20 000 ind. m^{-2} (45.6% of macroinvertebrates belonged to Chironomidae and 49.6% to Oligochaeta). The respective values estimated for the second cycle (without disturbance) were: 7000 ind. m^{-2}, 63.4% of Oligochaeta and 29.3 % of midges.

Spatial and seasonal dynamics of chironomid larvae

The chironomid density at BZA did not differ between cycles. *Prodiamesa olivacea* Meigen was the prevalent species at this site (Fig.2).

At BZB significant differences between cycles were recorded as the influence of flow alternations on the total chronomid abundance (Fig.3). The return of discharge to its normal level caused a very rapid colonization by *Micropsectra notescens* (Walker), one of four co-dominant species in the period without disturbances.

Significant differences between sites showed three of four co-dominant species: *Prodiamesa*, *Micropsectra* and *Polypedilum*. The first of them was significant numerous at BZA (P>0.05) while the other were more abundant at BZB (P>0.00001 for *Micropsectra*, P>0.0003 for *Polypedilum convictum* (Walker). The density of *Micropsectra* was significantly correlated with discharge (P<0.001), *Chironomus thummi* (Kieffer) with abundance of FPOM (P>0.001) while *Prodiamesa* with CPOM (P>0.01).

It is important to notice that a species number was low at these sites: 22 species were recorded over the investigated cycles. Species richness was highest for Chironomini and Orthocladiinae (7 species for each of these taxa) followed by Tanypodinae (5), Prodiamesinae (2) and Tanytarsini (2). As the impact of flow disturbance at BZB the number of species recovered to the "typical" level in late autumn.,

Chironomid production

The differences in the total production values at BZA between cycles were reflected by the differences in the production of *P. olivacea*, which lived in allochthonous coarse particulate organic matter (leaf packs). 98% of the total chironomid production was contributed by this species (Table 1).

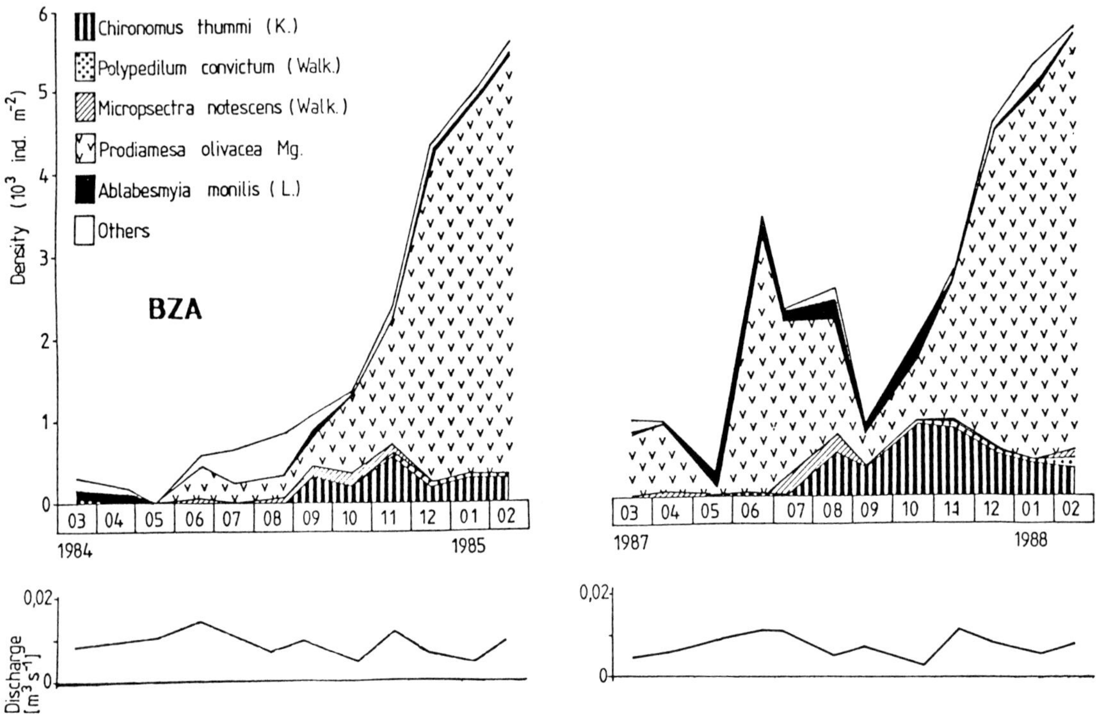

Figure 2. Seasonal fluctuations in chironomid density at the first station (BZA) in two annual cycles.

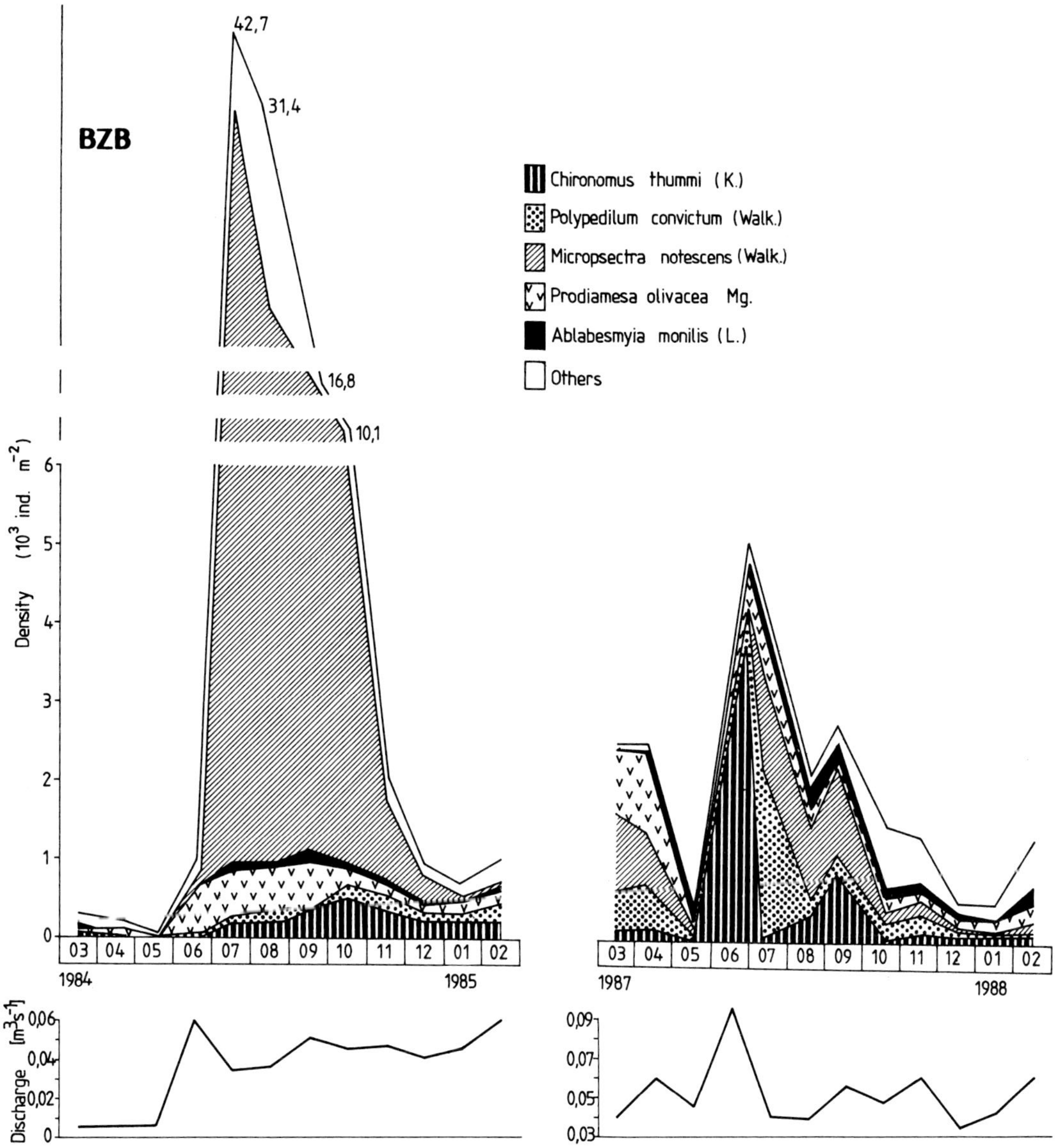

Figure 3. Seasonal fluctuations in chironomid density at the second station (BZB) in two annual cycles.

At the second site (BZB) there were distinct differences between cycles as the effect of reduced stream discharge and then rewatering a majority of the total streambed (the first cycle). *Micropsectra*, representing filtering collectors, made up a major portion of all chironomid production (54.5%). However in the second

Table 1. Biomass (B - g m^{-2} yr^{-1}) and production (P - g m^{-2} yr^{-1}) of larval Chironomidae at two sites of the Bzura River (BZA and BZB) in two annual cycles (I and II). GC - gathering collectors, FC - filtering collectors, P - predators, M - miscellanous,

	Sites and period							
	BZA				BZB			
	I		II		I		II	
Species	B	P	B	P	B	P	B	P
Ch. thummi (GC)	0.095	0.452	0.192	0.867	0.106	0.506	0.313	1.322
P. convictum (GC)	<0.001	0.008	0.006	0.070	0.003	0.041	0.045	0.563
C. defectus (P)	<0.001	0.004	<0.001	0.003	0.003	0.034	0.006	0.093
M. notescens (FC)	0.002	0.036	<0.001	0.013	0.277	5.405	0.007	0.137
P. olivacea (GC)	1.032	7.547	1.601	12.020	0.254	1.773	0.377	2.703
A. monilis (P)	0.009	0.062	0.017	0.108	0.016	0.101	0.022	0.150
Others (M)	0.011	0.113	0.001	0.014	0.200	2.078	0.050	0.551
Total	1.149	8.222	1.817	13.095	0.859	9.938	0.820	5.519

cycle without great flow disturbances, the dispersion of the production between the given species was higher than that at the first cycle (Table 1).

Predatory production made up only a minor part of total chironomid production; both Tanypodinae and Chironomini (*Cryptochironomus*) made up to 2.7% of total chironomid production (at BZB). Annual P:B ratios ranged from 4.7 for large-sized species (*Chironomus*, at BZA) to 20.1 for small-sized species (*Micropsectra*, also BZA). At BZA annual P:B ratios for six dominating of all chironomid species ranged from 7.1 (the first period) to 7.2 (the second one) with an average value for all species at both cycles 7.2. At BZB the turnover ratios reached the values, respectively: 11.9 and 11.6 at the first period while 6.5 to 6.7 at the second one.

Discussion

Ecosystem disturbances produce changes in macrobenthic community structure; the ability of macroinvertebrates to recolonize habitat may vary depending on many factors such as stream size, dispersal ability and the presence of upstream and tributary sources of colonists. According to Minshall and Petersen (1985), the downstream drift was the primary means of redistribution of benthic organisms in streams, especially by specimens that do not move far under their own power such as chironomids. Minshall *et al.* (1983) suggested that midges belong to colonizers which reached the substrate within days of rewatering. Oviposition may also be a means of rapid colonization after drought. The other ways cited by Mackey (1992) in which colonizers may reach a restored habitat, such as crawling or swimming from adjacent substrates seem to be less important for insects like chironomids.

Remarks mentioned above concerned disturbance and recolonization on a large scale. The investigations of the first-order stream section of the Bzura River included rather small scale disturbance (dewatering of near-shore habitat). Taking into account both the season of recovery habitat after disturbance (summer) and rate of recolonization it seems that repopulating at Bzura River was more likely dependent on the influx of adults from a nearby area, although the drift should not be excluded. Independently of the recolonization ways of new available spaces *Micropsectra* reached a very high density (over 30 000 ind. m^{-2}) very quickly. The recovery habitat provided abundant food such as sestonic fine particulate organic matter for these filtering collectors. This species also reached a maximum dominance in a Danish low order stream (Lindegaard and Mortensen, 1988). At BZA, without great flow alternation *P. olivacea* was the prevalent species. It is worth noting that such a high dominance of *P. olivacea* as that at BZA has been reported only from a second-order stream in Belgium (Bisthoven *et al.*, 1992). This seems to be an effect of very homogeneous substrata at BZA - *P. olivacea* is known to be a dweller of soft sediments with organic detritus, especially CPOM (Lindegaard and Mortensen, 1989). Although the precise role of CPOM as substrate or food for *P. olivacea* is still not clear, it was reported in another study (Bisthoven *et al.*, 1992) that CPOM in sediments may be a key factor in controlling this species.

Generally species richness in the Bzura River was low both at BZA and at BZB; it was probably the effect of a small number of Orthocladiinae species in comparison with other low order stream sections.

On the basis of the present investigations, it may be concluded that the Bzura River is typical of many lotic habitats with respect to chironomids dominating insect species richness. However, the distribution of chironomid taxa among subfamilies was only to typical of those low-order stream sections with a large amount of particulate organic matter of allochtonous origin (Lindegaard and Mortensen, 1989; Bisthoven *et al.*, 1992). In lotic ecosystems benthic productivity may be considered in terms of energy produced on the species level, functional feeding groups or total macroinvertebrates in a different kind of microhabitats. The highest production of a single chironomid species, *Orthocladius (E.) calvus* Pinder was recorded in recirculating channels in 37 days at a mean water temperature of 10.5°C. In natural ecosystems such high values as these mentioned have not been recorded, not even for whole benthofaunas, where maximal values varied from 33.5 g m^{-2} yr^{-1} at 19°C (Benke *et al.*, 1984), 22.5 g at 9.5°C (Krueger and Waters, 1983) to 4.1 g at 17.6°C (Smock *et al.*, 1985) and 1.0 - 6.6 g at 12.6°C (Gladden and Smock, 1990). This also refers to chironomid production in natural streams. One of the highest levels of chironomid production was recorded in a temperate stream in the USA, reaching up to 29.7 g (Berg and Hellenthal, 1991), and in Poland (ranged in two rivers from 21.2 to 26.1 g, an average annual water temperature was about 10°C (Grzybkowska, 1989); other values for the same group were: 11.7 g at 7.6°C (Lindegaard and Mortensen, 1988), 4.0 g (Mackey, 1977b) and 3.7 g at 12.9°C (Lugthart *et al.*, 1990).

Annual P:B ratios in the Bzura River were considerably lower than those in the majority of the papers cited above. Such low annual turnover ratios in the present study, especially in comparison with those of Hauer and Benke's data (1991), indicate that moderately high mean annual standing stock rather than rapid biomass turnover may account for the moderately high chironomid production at the Bzura River, similar to that in another north temperate stream (Berg and Hallenthal, 1991). Chironomid P:B ratios varied from stream to stream, although the majority of studies are restricted to the temperate zones and rather to low and mid-order stream sections (Waters, 1977, 1979; Benke *et al.*, 1988); this type of bioenergetic investigations refer rarely to higher-order stream sections (Grzybkowska *et al.*, 1990).

It is important to notice that the high contribution of chironomids to stream insects and the estimation of the chironomid secondary production at the species level instead of by grouping midges at the family level as practiced by many workers (the main reason of underestimation) may lead to the explanation of the Allen paradox that "is still with us" (Waters, 1988).

Acknowledgments

This work was a research grant awarded by the President of the Lodz University. Travel to Australia was funded partially by the University of Lodz, and KBN grant PB 2661/6/9.

Part 6

Ecology and behaviour of Chironomidae

Distribution of Chironomidae (Diptera) in the river continuum

Claus Lindegaard and Klaus P. Brodersen

Freshwater Biological Laboratory, University of Copenhagen

Introduction

Vannote *et al.* (1980) described a framework - the river continuum system - for integrating predictable and observable biological features in lotic systems. They grouped running waters into headwaters (orders 1-3), medium-sized streams (4-6), and large rivers (>6). Many headwaters are characterized by low temperatures and high water velocity resulting in a stony substrate. Further they are often strongly influenced by the riparian vegetation which reduces autotrophic production by shading and provides large amounts of allochthonous material. Accordingly the invertebrate fauna in headwaters is dominated by cold-stenothermous shredders or collectors. As stream size increases, temperature increases, slope and water velocity decrease resulting in a varied substrate of stone, gravel and sand. The reduced importance of terrestrial organic input is replaced by an increased autochthonous production and transport of fine organic matter from upstream increasing the importance of grazers and detrital collectors at the expense of shredders. In large rivers maximum temperatures occur and the substrate often tends to be gravel, sand and silt. The rivers receive quantities of fine particulate organic matter from upstream. A phytoplankton production can occur, but may be limited by depth and turbidity. The invertebrates are dominated by eurythermic detrital collectors.

Thienemann (1954) described a number of chironomid communities in primarily high altitude streams, but had at that time only few data from lowland streams and very little from large rivers. The aim of this paper is to review the faunistic studies on running water chironomids in order to see if the observed chironomid communities fit into the proposed conceptual river continuum framework.

Material and methods

Data were extracted from the studies listed in Table 1, which cover 104 sites in 80 different watercourses. Further data from 11 sites in New Zealand and Australia are listed in Table 2. A criterium for including these studies was that they all had identified chironomids to the same taxonomic level (prevailing to species level), and this had to be consistent for all subfamilies and tribes. In fact, many studies had to be excluded due to inadequate identification especially for Chironomini and Tanytarsini. Also studies published in national reports were in most cases neglected. Because sampling was done in many different ways quantitative aspects were omitted and species lists were based only on presence or absence of a species. In a number of studies altitude was mentioned, otherwise it was determined from topographic maps. Altitude and latitude are important factors determining the temperature. Informations on temperature and substrate types when available were also listed in Table 1. Species composition were extracted from 70 European studies (marked with an (*) in Table 1) and stored in an unpublished database.

Because environmental data and chemical/physical descriptions of sampling sites often are scarce or missing, two indirect ordination techniques were used to summarize and arrange sites and species in an ordination diagram, in the hope that ecological relationships would emerge. A principal components analysis (PCA), based on the percentage composition of the six most common subfamilies/tribes in 104 studies, and a detrended correspondence analysis (DCA) based on presence/absence data (1/0) of 251 species in 67 studies, were carried out through the computer program CANOCO (ter Braak, 1992a). PCA and DCA assume a linear and a unimodal (bell shaped) model for the relationships between the response of each subfamily/species to the ordination (environmental) axes, respectively.

Results

Percentage distribution of chironomid subfamilies/tribes

From each site the percentage of species in each chironomid subfamily and the tribes Chironomini/Tanytarsini was calculated (Table 1). From the original descriptions of the localities, it was decided to delimit eight northern temperate river types: (1) glacial brooks, (2) alpine streams, (3) subalpine streams, (4) mountain streams, (5) lower mountain streams, (6) lowland cold streams, (7) lowland warm streams and (8) lowland warm rivers. Further a few localities in subtropical and tropical streams were established as a separate group. This delimitation was based on a combination of latitude, altitude and temperature. In fact it did not include stream order, because most European studies unfortunately did not state stream order and without very detailed topographic maps this was impossible to determine. However, if we applied the river continuum to this typing, (1), (2) and (3) all would be headwaters, (7) medium-sized streams and (8) large rivers, whereas (4), (5) and (6) included both headwaters and medium-sized streams.

The zoocenoses of chironomids in the eight categories of northern hemisphere temperate running waters (Table 1, nos 1-95) can be described in the following way. In a typical glacial brook, where temperature is low, rarely exceeding 4°C in summer, and substrate is composed mainly of boulders and large stones, there are few species (average 7) characterized by the cold-stenothermous Diamesinae making up 45 to 67% of the taxa. The remaining species are Orthocladiinae. In lower altitudes the temperature increases as does species richness which in the alpine brook on average is 23. The Orthocladiinae takes over the dominance with 60-80% and the Tanytarsini comes in with about 5%.

In subalpine streams, the temperatures in summer may reach 10°C and the substrate becomes more varied with an increase in smaller stones and gravel. Species richness increases to an average of 49 species and most subfamilies/tribes are represented. Orthocladiinae dominates with about 60%, although Diamesinae is still important making up 20%.

The mountain or montane streams are physically very similar to low alpine streams, but always have finer particles in their substrate. Further, water temperature in summer during the day-time may be quite high. This results in a further increase in species richness, with an average of 71 species. This increase occurs in all subfamilies/tribes except the Diamesinae, resulting in lower relative abundances (average 11%).

The lower mountain or lower montane streams are a mixture between the montane and lowland streams with substrate types ranging from boulders to sand and silt; often macrophytes occur. In the preceding water types, primary producers mainly are mosses and periphytes. Average monthly temperatures may still be quite low, but short-termed maximum temperatures may exceed 20°C. Maximum species richness (on average 95 species) is recorded here. The Diamesinae and Orthocladiinae decrease to 6% and 59%, respectively, while Chironomini increases to 10%.

The summer-cold lowland streams are fed mainly by ground water causing similar temperature regimes to lower montane streams. However, the prevailing substrate is more soft and macrophytes are nearly always present. The species richness averages only 60, indicating fewer habitats than in the lower montane streams. The percentage of Diamesinae is unchanged, but the Orthocladiinae averages only 53%, while contribution of the Chironomini increases to 15%.

In the summer-warm lowland streams, the average monthly temperature in summer exceeds 20°C, otherwise they physically resemble summer-cold streams. The species richness is on average somewhat higher, but the average percentage of Orthocladiinae decreases to 46% while Chironomini increases to 19%. The summer-warm lowland rivers correspond the large rivers of the river continuum concept. The temperature in summer is always high, the substrate is mainly sand and silt though areas with stone and gravel occur. Macrophytes, if present, occur only in minor extension along the banks. Species richness is on average 67. Diamesinae and Orthocladiinae decrease to 1% and 36%, respectively, while Chironomini makes up 39%.

The described changes in Diamesinae, Orthocladiinae and Chironomini along the river continuum are - in spite of large fluctuations within the single running water type - distinct and significant (Fig. 1). The changes in the remaining subfamilies/tribes are less distinct. Tanypodinae increases in average from 4-5% in the subalpine and montane streams to 10-15% further downstream the river continuum. In alpine streams Tanytarsini constitutes about 5% and remains stable with 10-15% further downstream. Prodiamesinae is found when water temperature reach 10°C and makes up 1-3% further downstream. They apparently have their maximum in the summer-cold lowland streams and may eventually disappear in the very large rivers. The species poor Podonominae is found scattered in the range from subalpine to lower montane streams, while the single species of Buchonomyiinae is found only from a few lower montane streams to warm lowland rivers.

The chironomid communities in subtropical/tropical streams and rivers are very poorly known (Table 1, nos 96-104, Table 2), but apparently the higher average temperatures benefit Chironomini, which makes up a higher percentage compared to otherwise similar temperate running waters.

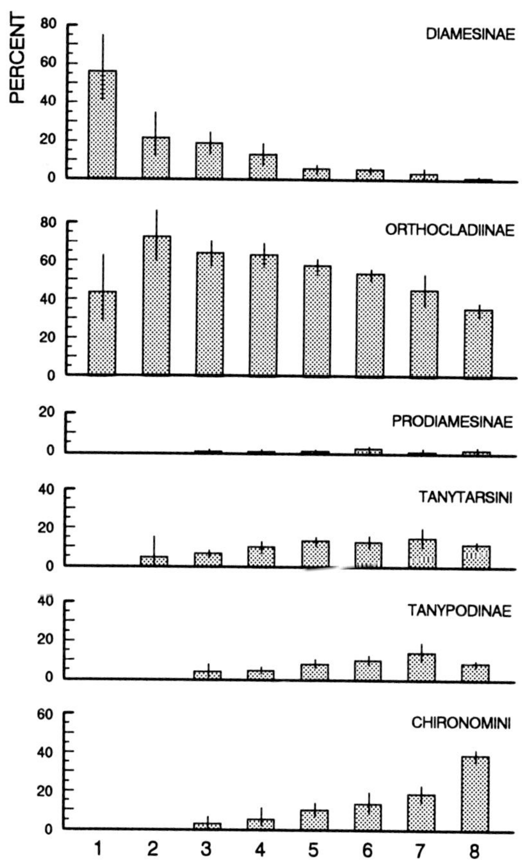

Figure 1. Average percentage composition (± 95% confidence limits) of six subfamilies/tribes in eight running water categories: (1) glacial brooks, (2) alpine streams, (3) subalpine streams, (4) montane streams, (5) lower montane streams, (6) lowland summer-cold streams, (7) lowland summer-warm streams and (8) lowland large rivers.

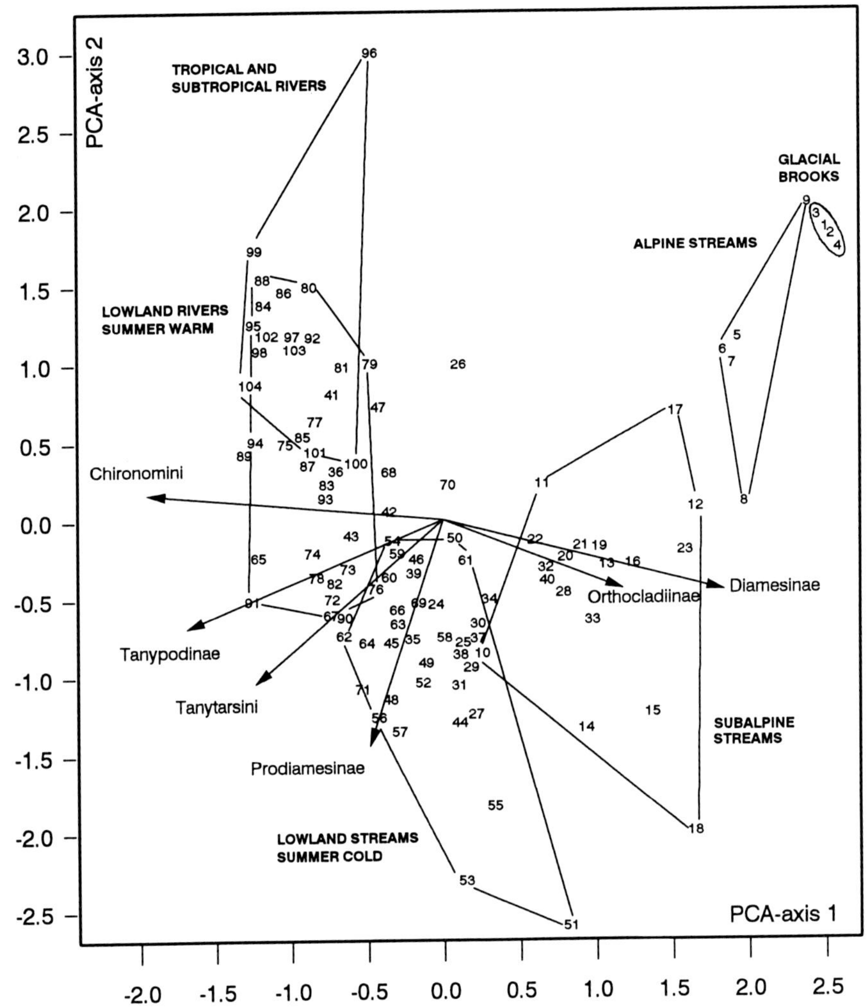

Figure 2. PCA-biplot showing the separation of six running water categories based on the percentage distribution of chironomid subfamilies/tribes in 104 running water sites. Arrows are pointing in the directions of maximum variation in percentage contribution of subfamilies.

Chironomid studies from Australia and New Zealand demonstrate a somewhat similar zonation as in the northern hemisphere, except that other subfamilies/tribes complicate the patterns (Table 2). Aphroteniinae occur in low order streams (Cranston and Edward, 1992); in headwaters and cooler streams Podonominae are abundant; Diamesinae are species-poor although sometimes abundant; Prodiamesinae are absent and Pseudochironomini (*Riethia* spp.) may be numerous (especially in individuals) in middle order streams. These patterns are evident in temperate south-eastern Australia (Metzeling *et al.*, 1984) and Mediterranean south-west Australia even though the headwaters here are of low elevation (Bunn *et al.*, 1986, Storey and Edward, 1989). In tropical, monsoonal northern Australia, where catchment areas are also of modest elevation, subfamily representation is restricted. Diamesinae is absent, Aphroteniinae rare only in the upper reaches and Orthocladiinae frequent only in low order streams. In the middle sections and lowland flood-plains Chironomini dominate; Tanypodinae and Tanytarsini are distributed throughout the continuum (Cranston, 1991).

Wilson (1980) characterized chironomid groups as "rhithral" (Pentaneurini, Diamesinae, Orthocladiinae, Chironomini connectentes) or "potamal" (Tanypodini, Macropelopiini, Prodiamesinae, Chironomini genuini, Tanytarsini) arguing that the first group occurs primarily in rhithron and the latter in potamon according to the zonation suggested by Illies and Botosaneanu (1963). The calculated percentage of rhithral groups declines gradually from 100% in glacial brooks to an average of 58% in lowland warm rivers (Table 1, nos 1-95). Thus, this approach shows a great similarity to the succession described on the basis of the Diamesinae/Orthocladiinae/Chironomini relationships.

In a PCA ordination biplot (Fig. 2) a high separation (69% of variance) was found along axis 1, and six of the named running water categories could be outlined without major overlap. The arrows point in the direction of maximum variation in percentage contribution of subfamilies (ter Braak and Prentice, 1988), and a water temperature and/or altitude gradient could be interpreted along axis 1, mainly due to the distribution of Diamesinae and Chironomini. This was shown by plotting the maximum water temperature and the altitude for each site to the corresponding axis 1 site score from the ordination analyses (Fig. 3). Temperature appears to better explain the separation along axis 1 than does the altitude. However, these two factors are related.

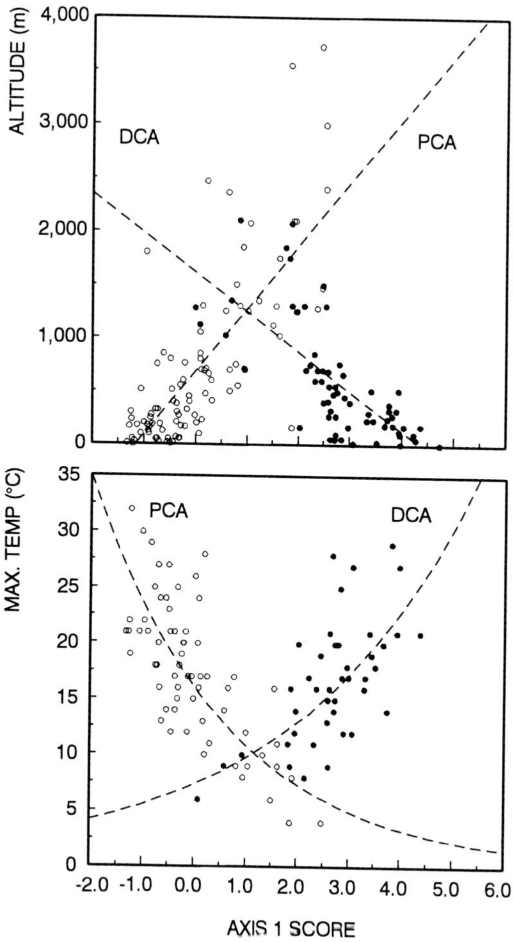

Figure 3. Correlation between ordination-axis 1 site score and the corresponding altitude (m a.s.l.) and maximum temperature (°C) for the site.

Distribution of chironomid species

A total of 514 chironomid species are recorded from 70 European running water sites representing 52 different watercourses. This is about 50% of the total number of Chironomidae known from Europe (Table 3). The species-rich subfamilies are equally represented (44-54%), while the occurrence of Tanypodinae and Diamesinae in running water make up 64% and 71% of total numbers, respectively. About half of the recorded running water chironomids occurred in less than four localities and we eliminated these from the DCA analysis. This did not alter the proportion between the subfamilies/tribes (Table 3). The distribution of the remaining 67 running water sites in a DCA analysis with 251 species is shown in Fig. 4. The alpine, subalpine and montane streams are found to the left, the large rivers are placed to the right, and the lower montane and lowland streams overlap in the centre of the diagram. In fact, the groups which can be distinctly delineated are placed in the same geographically region. Thus, for example, the alpine, subalpine and montane streams in Norway, France and Poland are well separated, and the Mittel and Nieder Rhein are found to be different from other lowland rivers in Germany.

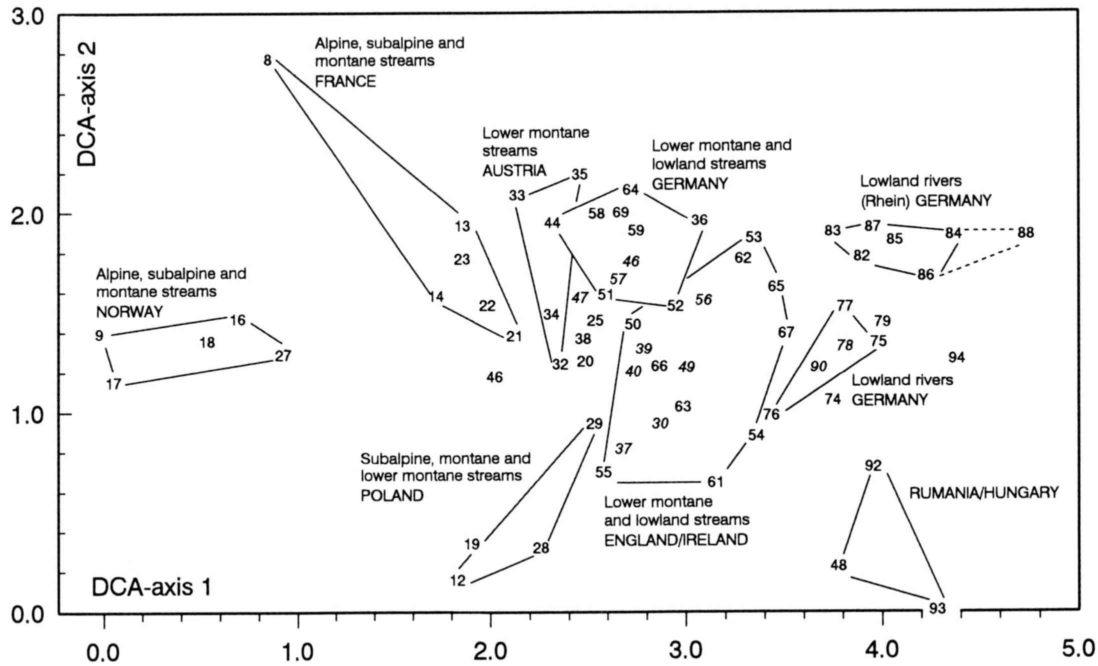

Figure 4. Ordination by DCA showing the scores of 67 European running water sites based on presence/absence data of 251 chironomid species. Sites in *cursive* do not belong to the outlined categories.

The distribution of the 251 species in the same DCA-diagram is shown in Fig. 5. The Diamesinae and Podonominae species are mainly placed to the left with few species in the centre, while the Chironominae concentrates to the right in the diagram. The Orthocladiinae is distinctly placed in the middle section of the diagram with some species to the left. The Tanypodinae is also centred though with a tendency to the right. The Tanytarsini is evenly distributed from left to right although they avoid the extreme margins. The Prodiamesinae is centred but with *Prodiamesa rufovittata* to the right.

The gradient from left to right along axis 1 in Figs 4-5 can be clearly correlated to increasing maximum temperature and less distinctly to decreasing altitude (Fig. 3). The gradient along axis 2 is less obvious, but may to a large degree reflect geographic distribution. The distribution along the temperature gradient reflects again the cold-stenothermic nature of most Diamesinae and of some Orthocladiinae and Tanytarsini, while many Chironomini can be considered warm-stenothermic. Many Orthocladiinae centred in the diagram (Fig. 5) have a wide distribution along the river continuum indicating an eurythermic nature.

Discussion

Although great overlaps occur, especially in the lower mountain and summer-cold lowland stream categories, it was possible to demonstrate distinct shifts in the chironomid community along the river continuum. Also it was possible to correlate these shifts to temperature, which was the only environmental factor stated in nearly all papers. Rossaro (1992) lists optimum mean water temperature (range 4-25°C) for 127 chironomid species from Italian streams and rivers based on their abundance in 991 sites. The cold-stenothermic species were represented by mainly Diamesinae and Orthocladiinae,

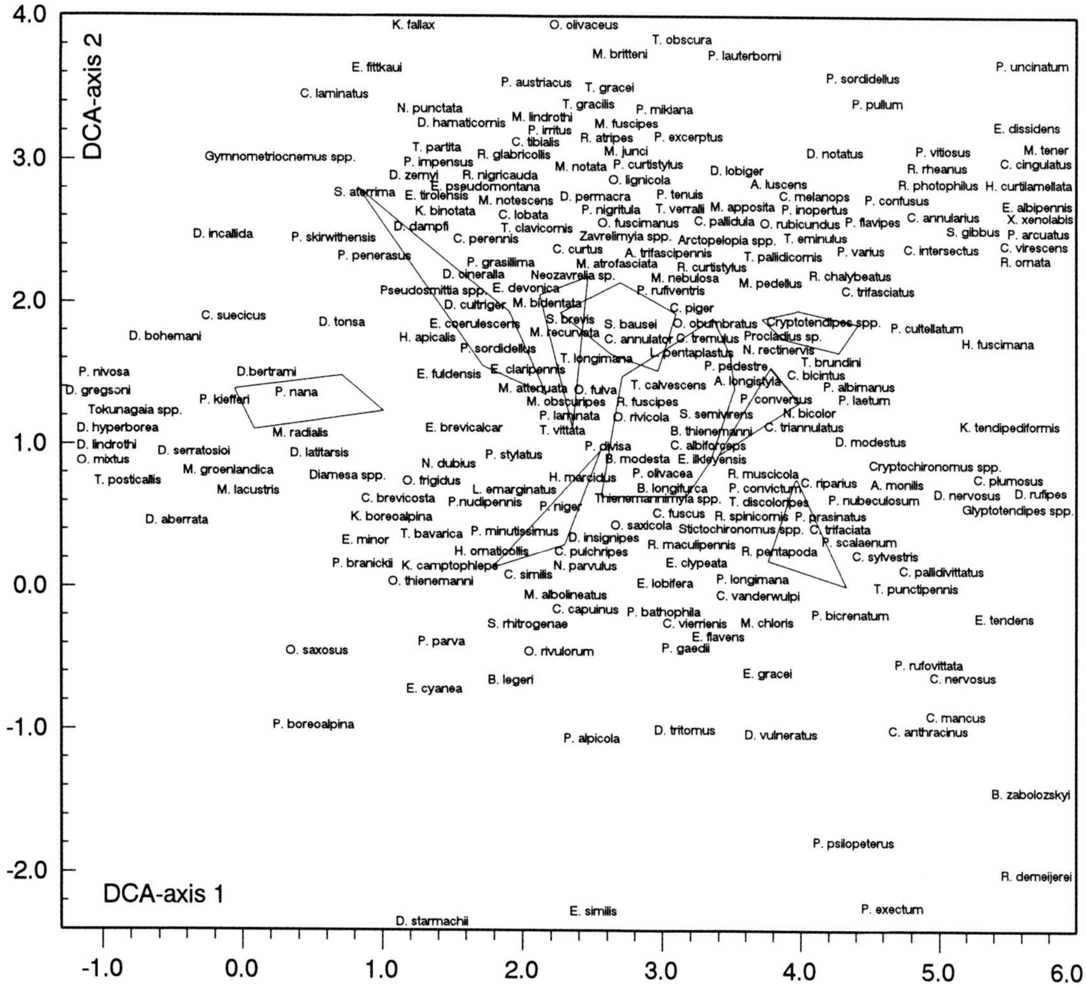

Figure 5. DCA-ordination diagram showing the 251 chironomid species scores of Fig. 4.

while the Chironomini dominated in the range from 17-25°C. A comparison between these optimum mean temperatures of 88 species and the corresponding DCA-axis 1 coordinates (Fig. 5) shows a fairly high correlation (Fig. 6). This correlation makes possible a rough estimate of the remaining species optimum temperature. Rossaro (1992) stated furthermore that most species were cold-stenothermous with an optimum temperature near their minimum tolerance value, whereas the warm water species were more or less eurythermous. This means that temperature may be the key factor for distribution of Diamesinae and cold-stenothermic Orthocladiinae, while other factors may be more important for Chironomini.

Substrate type and available food are supposed to influence chironomid distribution fully as much as temperature. However, these two factors are to some degree linked to temperature as they - according to the river continuum concept - change downstream within the continuum. Functional feeding groups have been linked to the river continuum concept. However, because there is an absence of shredders among Chironomidae, this notion has to be modified if chironomids are considered in isolation. Further, high latitude or altitude headwaters may deviate from the general pattern with regard to autotrophy/heterotrophy by lacking the shading riparian vegetation (Statzner and Higler, 1985). Accordingly algal grazers such as Diamesinae and some Orthocladiinae can dominate glacial brooks, alpine and subalpine streams. At lower altitudes, the shading effect of riparian vegetation may diminish algal grazers in headwaters and increase the importance of scrapers and collectors feeding on a more detritus rich food (many Orthocladiinae and Tanytarsini, some Chironomini). In lowland streams and rivers low water velocity and high amounts of particulate organic matter will benefit some filterers (e.g. *Rheotanytarsus* spp.) and deposit feeders as many Chironomini. Ward and Williams (1986) found a similar zonation of chironomids along the Rouge River in Ontario, Canada, and demonstrated a difference in available food and food uptake along this gradient.

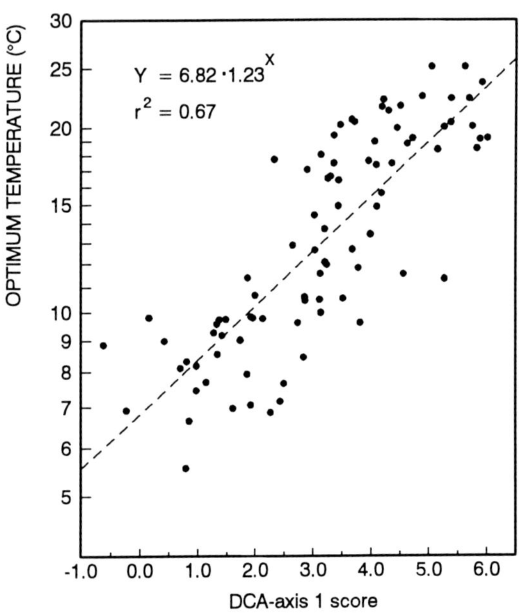

Figure 6. Correlation between DCA-axis 1 species score and weighted average optimum temperature (Rossaro, 1992) for 88 species.

They interpreted this zonation to be influenced by both temperature and substrate. The importance of substrate in determining species composition is demonstrated in Little Boulder Creek, where Tilley (1989) found three species of Chironomini in a canyon site with boulders and stones. Upstream he recorded six species of Chironomini at the same low temperature, but in a meadow site with a gravel and sand dominated substrate. Furthermore, Rae (1985) demonstrated strong differences in habitat preference for deposit-feeding chironomid species when offered a choice of different sediment grain size, sediment heterogeneity and content of organic matter.

The discussed subfamilies/tribes of chironomids all are very rich in species (cf. Table 3) and large differences in temperature optimum, substrate and food preference occur between species within these groups. Therefore, any generalization based on these large systematic groups will be broad. A better delimitation should be achieved by selected groups of species. However, the large number of species (> 500) in European running waters makes a clustering difficult and confused even when the number of species is reduced to about 250 species. Further, a large number of species occur in many of the outlined running water categories, leaving the remaining Diamesinae and Chironomini species to characterize the uppermost and lower parts of the river continuum (cf. Fig. 5). The DCA analysis also grouped running water sites more to geographic position than to running water categories (Fig. 4). A similar separation of running water sites in geographic groups was found by Aagaard (1993) by analyzing 42 sites in 16 different watercourses. This was unexpected, because most European chironomid species seem to be widely distributed throughout Europe (cf. Ashe and Cranston, 1990) and Lindegaard (1995) found 75% of the known European species reported from the Nordic Countries (Denmark, Sweden, Norway, Finland and Iceland). A high separation to geographic areas might therefore be due to either a real difference in geographic distribution or to the fact that identification is not good enough. Some high altitude chironomid species seem to be restricted to single mountain massifs (cf. Laville and Vinçon, 1991; Casas and Vilches-Quero, 1993) or to Scandinavian mountains *versus* the Alps (e.g. Sæther, 1968). Lower mountain and lowland chironomids, however, seem to be very widely distributed.

The steady increase in species richness from glacial brooks to a maximum in lower mountain streams followed by a decrease further downstream agrees with Coffman (1989), who stated a maximum richness of lotic chironomids in 3rd order streams due to primarily a high ecological heterogeneity. These findings are consistent with the river continuum concept.

The observed changes in Chironomidae along the river continuum are - with the implicated modifications - applicable to the river continuum concept. The river continuum concept was proposed by Vannote *et al.* (1980) to be of world-wide applicability. This statement has subsequently been discussed by a number of authors (e.g. Winterbourn *et al.*, 1981, Statzner and Higler, 1985). The described changes in chironomid communities are based almost exclusively on data from northern hemisphere temperate streams. If these changes can be considered valid world-wide, then we must await further progress in southern hemisphere taxonomic and basic faunistic studies on chironomids.

Table 1. Number of chironomid taxa and percentage distribution of chironomid subfamilies/tribes in a wide range of temperate and a few subtropical and tropical running waters representing major sections of the river continuum. Subfamilies/tribes are: Podonominae (1), Tanypodinae (2), Buchonomyiinae (3), Diamesinae (4), Prodiamesinae (5), Orthocladiinae (6), Chironomini (7) and Tanytarsini (8). Rhithral groups are Pentaneurini, Diamesinae, Orthocladiinae, Chironomini connectentes (Wilson 1980). Substrate "types" are delimited as boulder (bo), stone (st), gravel (gr), sand (sa), silt (si), periphyton (pe), macrophytes (ma), mosses (mo), detritus (de), riffles (ri), leave packs (lp), and snag (sn). An (*) indicate localities used in a DCA analyses of species distribution in European running waters.

No	River	Country	Latitude	Altitude m a.s.l.	Tempera. °C ra	mean	% of subfamilies/tribes 1	2	3	4	5	6	7	8	Total no. taxa	% rhit. gr.	Substrate	References
Glacial brooks																		
* 1	Finse Brook	Norway	60°N	1550-1400	1- 4		0	0	0	50	0	50	0	0	4	100	bo, st	Sæther 1968
2	"The Alps"	Europe	47°N	c. 3000		1	0	0	0	63	0	37	0	0	8	100	bo	Thienemann 1954
3	North Boulder Cr.	CO, USA	40°N	3800-3650		1	0	0	0	5	0	55	0	0	11	100	bo	Elgmork and Sæther 1970
* 4	Rio Plima	Italy	46°N	2800-2000			0	0	0	67	0	33	0	0	6	100	st	Kownacki 1991
Average							0	0	0	56	0	44	0	0	7	100		
High mountain brooks, "alpine"																		
5	"The Alps"	Europe	47°N	> 2100	1- 8		0	0	0	16	0	80	0	4	25	96	bo	Thienemann 1954
6	North Boulder Cr	CO, USA	40°N	3650-3450		4	0	0	0	14	0	81	0	5	21	95	bo, mo	Elgmork and Sæther 1970
7	Bathurst Island	NWT, Canada	75°N	163- 0	1- 4		0	0	0	15	0	80	0	5	20	95	st, gr	Hayes and Murray 1987
* 8	Ossau Valley	France	43°N	> 2100			0	0	0	29	0	59	0	12	17	88	bo	Laville and Vinçon 1991
* 9	Atna, Vidjedalsb	Norway	62°N	1280			0	0	0	35	0	65	0	0	31	100		Aagaard *et al.* 1989
Average							0	0	0	22	0	73	0	5	23	95		
High mountain brooks, "subalpine"																		
10	Ll. Boulder Cr. me	Idaho, USA	44°N	2466			1	8	0	10	4	62	12	4	52	85	gr, sa	Tilley 1989
11	Ll. Boulder Cr. ca	Idaho, USA	44°N	2362			1	5	0	15	0	63	7	7	41	90	bo, st, gr	Tilley 1989
* 12	High Tatra streams	Poland	49°N	2000-1500	1-11		0	4	0	20	0	73	0	4	56	96	bo, st, gr, mo	Kownacki 1971
* 13	Estaragne	France	43°N	2300-1850	1- 9		0	5	0	19	0	65	2	9	58	91	bo	Laville and Lavandier 1977
* 14	Ossau Valley	France	43°N	2100-1600			0	6	0	16	2	70	2	9	56	89	bo, mo	Laville and Vinçon 1991
15	"The Alps", übergb.	Europe	47°N		2-10		0	6	0	14	2	73	0	6	51	88	st, mo	Thienemann 1954
* 16	Blesbekken	Norway	62°N	1350	< 10		0	3	0	26	0	59	2	10	54	87	bo, st	Aagaard *et al.*1987
* 17	Atna, Skrangleh.	Norway	62°N	1120	0- 6		5	0	0	32	0	51	3	8	37	89	bo, st	Aagaard *et al.* 1989
* 18	Atna, Dørålsæter	Norway	62°N	1020	0- 9		0	3	0	32	3	53	0	11	38	87		Aagaard *et al.* 1989
Average							1	4	0	20	1	63	3	8	49	89		
Mountain streams, "montane"																		
* 19	High Tatra streams	Poland	49°N	1500-1000	1-12		0	4	0	17	1	70	3	4	69	93	bo, st, gr, mo	Kownacki 1971
* 20	V.- & H. Rhein	Switzerland	47°N	2000-1000			0	4	0	9	0	69	2	16	45	84		Wilson and Wilson 1985
* 21	Ossau Valley	France	43°N	1600-1000			0	2	0	13	1	64	4	9	75	88		Laville and Vinçon 1991
* 22	Neste d'Aure	France	43°N	1500-1000	1-14		0	6	0	12	0	66	5	11	65	88	st, gr, pe	Gazagnes and Laville 1985

* 23	L'Eau d'Olle	France	45°N	1900- 713	3-16		0	3	0	18	0	73	0	8	40	93	bo, st, gr	Gay 1982
24	"Mittelgebirge"	Germany			2-17		0	10	0	5	1	62	10	12	126	82	st, gr, mo	Thienemann 1954
* 25	Sierre Nevada	Spain	37°N	1700- 900			0	8	0	9	1	57	10	14	144	-		Casas and Vilchez-Quero 1989
26	Llobregat a. dam	Spain	42°N	1350- 750	1-24		0	4	0	4	0	73	12	8	26	92	bo, st	Prat *et al.* 1983
* 27	Atna, Vollen	Norway	62°N	710	0-10		2	8	0	14	2	50	12	12	50	74		Aagaard *et al.* 1989
Average							<1	5	0	11	1	65	6	10	71	87		
Lower mountain streams, "lower montane"																		
* 28	High Tatra streams	Poland	49°N	1000- 500	1-17		0	4	0	14	1	59	4	8	101	78	st, gr, sa, mo	Kownacki 1971
* 29	Grajcarek Stream	Poland	50°N	840- 540	0-13		0	7	0	7	2	61	8	12	74	81	bo, st, gr	Kownacki 1982
* 30	Rajcianka Stream	Slovakia	49°N	940- 384	0-17		0	3	0	8	3	63	11	11	71	80	bo, st, gr, sa	Bitusik and Ertlova 1985
31	Windrow Farm St.	Canada	44°N	800	2-16	7	0	8	0	3	2	65	4	18	62	79	bo, st, gr, mo	Singh and Harrison 1984
* 32	Wagrainer Ache	Austria	47°N	700	-16		1	6	0	15	1	66	6	5	164	91	bo, st, sa, pe	Moog and Janecek 1991
* 33	Schreierbach	Austria	48°N	700	4- 8		0	6	0	12	0	64	1	16	81	84	st, mo, pe	Caspers 1983a
* 34	Obere Seebach	Austria	48°N	600	2-11		1	7	0	8	1	67	7	10	102	87	st, gr, sa, mo	Schmid 1986, Bretschko 1991
* 35	Teichbach	Austria	48°N	600	0-19		0	17	0	3	1	57	8	14	109	78	sa, ma	Caspers 1983a
* 36	Krumbach	Germany	51°N		0-27		0	10	0	1	1	50	21	12	68	85	gr	Röser 1980
* 37	Caucasus Mts st.	Grusia?	41°N	950-0	3-28		0	7	0	13	1	51	13	13	107	79	bo, st, sa, ma	Kownacki and Zosidze 1980
* 38	Ossau Valley	France	43°N	1000- 400			0	6	0	6	2	63	9	13	97	81		Laville and Vinçon 1991
* 39	Haute-Lot	France	44°N	1295- 223	4-20		0	6	1	3	1	59	12	19	109	81	st, sa, ma	Laville 1981
* 40	Alpen-Rhein	Switzerland	47°N	600- 400			0	2	0	4	2	78	2	12	49	84		Wilson and Wilson 1985
41	Adirondack st.	NY, USA	44°N	800- 700	2-18		0	13	0	0	0	56	11	16	32	75		Simpson 1983
42	N.Carolina st.	NC, USA	36°N				0	10	0	4	1	54	23	8	168	88		Lenat and Folley 1983
43	Bigoray River	Alb., Canada	53°N	850	1-13		1	18	0	3	0	39	21	19	112	66	sa, si, ma	Boerger 1981
* 44	Fulda, springs	Germany	51°N	850			0	13	0	7	1	53	6	20	86	76	bo	Lehmann 1971
* 45	Fulda, city Fulda	Germany	51°N	300- 300			0	14	0	3	2	59	11	11	100	87	bo, st	Lehmann 1971
* 46	Atna, Solbakken	Norway	62°N	380	0-14		0	9	0	4	0	49	11	26	53	60	st, gr	Aagaard *et al.* 1989
* 47	Skiftesåa	Norway	64°N	170- 160	0-20	10	1	8	0	1	0	63	12	15	147	77	st	Aagaard *et al.* (in press)
Average							<1	9	<1	6	1	59	10	14	95	80		
Lowland streams, summer cold																		
* 48	Bistrita	Rumania	47°N	400			0	9	0	5	4	49	22	12	78	67	bo, st, sa, mo	Candea-Cure 1971
* 49	Bityska Stream	Czech Rep.	49°N	500- 300	1-17	7	0	25	0	9	3	44	16	3	32	78	st, gr, sa, si	Zelinka et al. 1977
* 50	Althahoney	Ireland	54°N	160- 40	5-20		0	14	0	7	0	60	9	10	58	86	bo, st, gr, sa	Fahy and Murray 1972
* 51	Aabach, spring area	Germany	51°N	550	4- 9		0	13	0	8	3	53	0	24	38	66	st, sa, mo	Dittmar 1955
* 52	Aabach, low reach	Germany	51°N	550- 350	3-12		0	7	0	9	2	45	20	16	44	73	st, sa, ma	Dittmar 1955
* 53	River Cynon	England	52°N	350-70	4-17		0	14	0	11	4	46	7	18	28	79	st, pe	Learner *et al.* 1971
* 54	River Wye	England	52°N	600- 0			0	80		4	1	53	22	12	77	79	st, sa, ma	Brooker and Morris 1980
* 55	River Nent	England	55°N	556- 262			0	14	0	11	3	57	5	11	37	84	st, gr	Armitage and Blackburn 1985
* 56	Linding Å	Denmark	56°N	20-10	2-12	7	0	12	0	3	6	55	17	9	67	75	gr, sa, ma	Lindegaard 1972
* 57	Rabis Bæk	Denmark	56°N	55	2-16	8	0	12	0	3	6	62	12	9	34	76	st, gr, sa, ma	Lindegaard and Mortensen 1988
* 58	Breitenbach	Germany	51°N	270	1-15		0	7	0	7	2	58	14	11	81	79	gr, sa, ma	Ringe 1974
* 59	Breitenbach	Germany	51°N	270	1-15		0	11	0	4	1	54	18	10	134	74	gr, sa, ma	Siebert 1980

Table 1 (cont. 2)

No	River	Country	Latitude	Altitude m a.s.l.	Tempera. °C ra	Tempera. °C mean	% of subfamilies/tribes 1	2	3	4	5	6	7	8	Total no. taxa	% rhit. gr.	Substrate	References
60	Juday Creek	IN, USA	42°N	206	2-17	10	0	8	0	4	2	52	23	10	48	79	st, gr, sa	Berg and Hellenthal 1991
* 61	River Frome	England	51°N				0	2	0	7	2	60	16	13	45	78	gr, ma	Pinder 1980
* 62	Tadnell Brook	England	51°N		6-16		0	9	0	1	3	52	16	19	75	68	gr, sa, ma	Pinder 1974
* 63	North Tyne	England	55°N	170	3-18		0	13	0	6	1	49	18	13	71	79	st, sa	Brennan *et al.* 1981
* 64	Fuhlenau	Germany	54°N	c.50	1-14		0	7	0	1	3	54	12	22	67	67	st, gr, sa	Böttger *et al.* 1987
Average							0	11	0	6	3	53	15	13	60	76		
Lowland streams, summer warm																		
* 65	Doulonnes	France	47°N	254 - 209	3-19		0	19	0	0	2	35	27	17	48	65	bo, st, gr, sa	Verneaux 1968
* 66	Caragh River	Eire	52°N	90 - 18	4-25	14	0	21	0	3	1	54	10	10	68	84	bo, gr, sa, si	Dowling *et al.* 1981a, 1981b
* 67	River Chew	England	51°N	25	4-18		0	19	0	2	2	40	21	11	86	60	si, ma	Wilson 1977
68	Linesville Creek	PA, USA	40°N		1-21		0	10	0	3	0	55	17	15	143	68	st, gr, ma	Coffmann 1973
* 69	Rohrweisenbach	Germany	51°N	325	0-21		0	8	0	7	1	51	19	14	85	76	gr, sa, dt	Ringe 1974
70	Llobregat, b. dam	Spain	42°N	400 - 0	6-26		0	9	0	9	0	57	17	9	35	83	st, gr, sa, si	Prat *et al.* 1983
71	Kossau	Germany	54°N	34 - 0	1-24		0	15	0	4	1	34	18	28	74	46	st, si, ma	Nietzke 1938
72	Suså	Denmark	55°N	90 - 6	1-25		0	17	0	2	1	38	23	18	87	52	st, sa, si, ma	Berg 1948
Average							0	15	0	4	1	46	19	15	78	67		
Lowland rivers, summer warm																		
73	Warta River	Poland	52°N	c. 125		11	0	5	0	3	8	30	49	5	37	46	sa	Grzybkowska *et al.* 1990
* 74	Nida River	Poland	50°N	c. 490			0	8	0	1	3	42	33	13	148	57	gr, sa, si, ma	Srokosz 1980
* 75	Oberen Alz	Germany	48°N	517 - 506	2-21		0	18	1	1	0	28	35	18	79	57		Caspers 1983b
* 76	Oberen Alz	Germany	48°N	517 - 506	2-21		0	12	1	5	1	43	26	13	82	66		Kownacki and Margreiter-Kownacka 1993
* 77	Fulda b. city Fulda	Germany	51°N	300 - 250			0	11	0	1	1	44	35	8	80	61		Lehmann 1971
* 78	River Po	Italy	45°N	100 - 70	2-29	18	0	9	0	2	3	37	41	8	148	56	st, sa, si, ma	Rossaro 1984
* 79	River Po	Italy	45°N	7	2-27		0	3	0	3	0	38	38	18	34	59	artificial sub	Battegazzore *et al.* 1992
80	Saskatchewan a. ip.	Canada	54°N		1-21		0	6	0	1	0	32	49	11	79	-		Mason and Lemhkuhl 1983
81	Saskatchewan b. ip.	Canada	54°N		1-20		0	4	0	2	0	33	40	20	84	-		Mason and Lemhkuhl 1983
* 82	Hoch Rhein	Germany	48°N	400 - 250			0	10	0	2	2	39	32	15	59	63		Wilson and Wilson 1985
* 83	Hoch Rhein	Germany	48°N	400 - 250			0	11	0	2	1	43	33	10	107	66		Caspers 1991
* 84	Ober Rhein	Germany	48°N	250 - 100			0	9	0	0	0	40	38	14	58	59		Wilson and Wilson 1985
* 85	Ober Rhein	Germany	48°N	250 - 100			0	10	0	1	1	38	39	11	98	59		Caspers 1991
* 86	M. + N. Rhein	Germany	49-52°N	c. 100			0	13	0	0	0	47	31	9	32	75		Wilson and Wilson 1985
* 87	M. + N. Rhein	Germany	50°N	c. 100			0	9	0	1	1	39	35	14	74	59		Caspers 1991
* 88	Wall	Netherland	52°N	- 0			0	11	0	0	0	33	44	11	18	56		Wilson and Wilson 1985
89	Oder	Germany	53°N	21 - 0	0-21		0	15	0	0	2	28	45	10	60	60	sa, si, ma	Harnish 1922

* 90	Danube	Austria	48°N	187	0-20	0	6	0	2	5	42	33	12	66	56		Schmid 1992
91	Danube	Czech Rep.	48°N	165 - 150		0	7	0	0	7	25	43	18	28	46	st, gr, sa, pe	Ertlová 1963
* 92	Danube	Hungary	48°N	150 - 100		0	3	0	0	3	47	40	7	30	63	st, gr, pe	Berczik 1971
* 93	Danube	Romania	45°N	100 - 0		0	10	0	4	1	24	52	10	83	48	gr, sa, si	Cure 1964
* 94	Thames	England	52°N		0-21	0	8	0	0	4	25	54	8	24	46	st, sa, si, ma	Mackey 1976a,b, 1977
95	Ohio River	Ohio, USA	39°N			0	16	0	0	0	32	40	12	25	52	sa, si	Mason and Sublette 1971
Average						0	9	<1	1	2	36	39	12	67	58		
Subtropical and tropical rivers																	
96	Gt Berg R. spring	South Africa	35°S	1300 - 300	9-23	0	15	0	0	0	70	15	0	20	70	st, mo	Scott 1959
97	Gt Berg R. u. r.	South Africa	35°S	300 - 100	9-30	0	11	0	0	0	50	25	14	64	78	st, sa, ma	Scott 1959
98	Gt Berg R. l. r.	South Africa	35°S	100 - 0	10-32	0	14	0	0	0	34	36	16	56	50	sa, si, ma	Scott 1959
99	Tiga	Nigeria	12°S	300	20-22	0	8	0	0	0	25	55	12	40	50	st, sa, si, pe	Brown 1983
100	Turkey	FA, USA	30°N		12-24	0	19	0	5	0	29	38	10	42	67	sa	Soponis 1980
101	Ridge & Valley st.	GA, USA	33°N	262 - 154		0	23	0	1	0	41	23	13	71	69	ri, lp	Caldwell and Parrish 1987
102	Costal Plain st.	GA, USA	32°N	38 - 20		0	25	0	0	0	38	30	8	53	68	sa, lp, sn	Caldwell and Parrish 1987
103	Kalengo	Zaire	3°S	1800	16-22	0	11	0	0	0	52	22	15	46	80	st	Lehmann 1979
104	Simisi	Zaire	3°S			0	18	0	0	0	29	34	18	38	?		Lehmann 1981
Average						0	16	0	1	0	41	31	12	48	67		

Table 2. Number of chironomid taxa and percentage distribution of chironomid subfamilies/tribes in New Zealand and Australian running waters. Subfamilies/tribes are: Podonominae (1), Tanypodinae (2), Aphroteniinae (3), Diamesinae (4), Orthocladiinae (5), Chironomini (6), Pseudochironomini (7) and Tanytarsini (8). Substrate "types" as in Table 1.

No	River	Country	Lati-tude	Altitude m a.s.l.	Tempera. °C		% of subfamilies/tribes								Total no. taxa	% rhit. gr.	Substrate	References
					ra	mean	1	2	3	4	5	6	7	8				
105	Waitakere River	New Zealand	36°S	110 - 10		11-23	0	25	0	13	38	13	0	13	16		st, sa, si, ma	Towns 1978, 1979
106	Devils Creek	New Zealand	42°S	360 - 280			23	20	3	0	40	11	0	3	35			Cowie 1983
107	North Dandalup	WA Australia	32°S	300 - 0			0	14	0	5	31	36	5	14	42			Storey and Edward 1989
108	Canning	WA Australia	32°S	300 - 0			2	11	0	2	43	34	4	9	56		ri	Storey and Edward 1989
109	La Trobe River	Vic Australia	38°S	930 - 240		2-20	5	11	0	3	30	32	3	16	37		st, gr	Metzeling *et al.* 1984
110	La Trobe River	Vic Australia	38°S	190 - 160		5-20	6	19	0	3	22	34	3	13	32		st, gr, sa	Metzeling *et al.* 1984
111	La Trobe River	Vic Australia	38°S	60 - 50		7-22	7	21	0	3	17	31	3	17	29		gr, sa	Metzeling *et al.* 1984
112	Cooper Creek	NT Australia	13°S				0	13	0	0	7	60	0	20	30			Cranston 1991
113	Radon Spring	NT Australia	13°S				0	21	0	0	9	47	0	24	34			Cranston 1991
114	Gulungal Creek	NT Australia	13°S				0	14	0	0	12	51	0	23	43			Cranston 1991
115	S. Alligator R. (03)	NT Australia	13°S				0	17	0	0	10	48	0	26	42			Cranston 1991

Table 3. Number of chironomid species and percentage distribution of chironomid subfamilies in 70 European running water sites (marked with an (*) in Table 1). Reduced numbers is used in a DCA analyses of species composition (see text). Total number of chironomid species in Europe is enumerated from Ashe and Cranston (1990) and some later descriptions and revisions. (A): Percentage of species found in running water in proportion to species recorded in Europe.

Subfamily	70 running water sites				Total		
	total		reduced		Europe		A
	no.	%	no.	%	no.	%	%
Podonominae	3	1	3	1	8	1	40
Tanypodinae	49	10	22	9	76	8	64
Buchonomyiinae	1	-	1	-	1	-	100
Diamesinae	42	8	27	11	59	6	71
Telmatogetoninae	-	-	-	-	3	-	-
Prodiamesinae	5	1	3	1	10	1	50
Orthocladiinae	230	45	107	43	427	44	54
Chironominae							
Chironomini	103	20	52	21	237	24	44
Pseudochironomini	1	-	1	-	1	-	100
Tanytarsini	80	16	35	14	159	16	50
Total	514	101	251	100	981	100	52

The Chironomids (Diptera, Nematocera) of an Experimental Pond System

Norbert Caspers[1] *and Fred Heimbach*[2]

[1]Bayer AG, Environmental Protection, 51368 Leverkusen, Germany

[2]Bayer AG, Crop Protection, Environmental Biology, 51368 Leverkusen, Germany

Introduction

Artificial ponds have been used increasingly to assess potential hazard effects of chemicals to aquatic ecosystems (Crossland and Wolff, 1988; Crossland and La Point, 1992; Ferrington *et al.*, 1994;Heimbach, 1993; DeNoyelles *et al.*, 1989; Solomon and Liber, 1988). These "microcosms" and "mesocosms" reflect, as far as possible, natural conditions, contain major trophic levels, and, most importantly, are logistically and experimentally manageable when efforts are concentrated on relevant endpoints (Crossland *et al.*, 1993; Hill, *et al.*, 1993; SETAC-Europe, 1992; SETAC-Resolve, 1992).

Depending on the methodological design, the species diversity of these experimental ponds may be somewhat reduced compared to natural habitats of the same size, due to the lack of appropriate niches (Caquet *et al.*, 1992).

At the Crop Protection Research Center of Bayer AG in Monheim (Germany), close to the River Rhine, three artificial ponds are used for studying the fate and effects of pesticides in outdoor multispecies systems. This paper focuses on the chironomids as the most diverse benthic group in the pond system. It presents the species list and comments on the ecological preferences and on the geographic distribution of the species.

Methods and site

The experimental ponds

The ponds (Fig. 1) were constructed in 1988, filled with water and inoculated in 1989 with sediment and macrophytes (*Elodea canadensis* Michx., *Ranunculus aquatilis* s. str. L.) from a mesotrophic pond some 40 kilometers away from the experimental site. Since 1990, the ponds have been used to investigate the biological and ecotoxicological effects of pesticides. After each study period the uppermost sediment layer of about 3 cm was scraped off to eliminate possible residues of test substances. Thus, even in the treated ponds the aquatic biocenosis remained fairly constant throughout the years.

The species list of chironomids is based on macrobenthic samples (Ekman grab), pupal exuviae (hand net) and collections of adult insects (floating emergence traps as in LeSage and Harrison, 1979) in 1992 and 1993.

Species list of chironomids

The species list of chironomids (44 species, Table 1) in the experimental ponds is rather small, as compared to natural ponds of the same size and characteristics (Schleuter, 1986), but it agrees well with the results of other studies in artificial ponds (Caquet *et al.*, 1992 Dewey, 1986). This may be primarily due to the low degree of compartmentalization of the mesocosms which do not provide for a large variety of niches as an important precondition for a higher diversity of the benthic fauna.

There are two main colonization paths contributing at least partly to the chironomid community of the ponds:

(i) the species introduced with the sediments in 1989 (which obviously have been eliminated partly in 1991 and 1992 under the influence of the different ecological conditions in the experimental ponds).

(ii) the species actively colonizing the ponds by the aerial path.

It is far beyond the scope of this paper to assess whether or not this colonization strategy reflects the influence of stochastic factors (e.g. the casual chironomid species composition of other water bodies nearby and the temporal sequence of their flight activities, respectively) or rather the influence of deterministic factors (e.g. the systematic colonization patterns of r-selected ubiquitous pioneer species; see Caspers, 1983).

The common ecological feature to most chironomid species in the experimental ponds is their tolerance of inconstant environmental conditions (e.g. fast and significant fluctuations of water level, temperature, conductivity and hydrochemical factors, respectively) thus favouring these species for a successful colonization of small water bodies.

Apart from the inundation zones of lowland streams natural temporary and permanent pools have become scarce in Central European countries. Therefore our knowledge on species composition and trophic relationships within these water bodies is primarily restricted to few observations in man-made habitats like garden pools, fish ponds, watering-places etc. (Leuchs and Caspers, 1988).

From the biocenotic point of view, the chironomid community of the experimental ponds in Monheim is surprisingly heterogeneous, predacious Tanypodinae species playing an important role with respect to species numbers and biomass. Apparently unbalanced trophic relationships like this have often been reported from insect communities under fluctuating environmental conditions. Some remarks on the ecological preferences of all species involved are presented in Table 1.

From the point of view of regional faunistics, some of the chironomid species are highly remarkable. *Orthocladius (O.) holsatus* Goetghebuer (syn. *Orthocladius (O.) smolandicus* Brundin) is a poorly known holarctic species. Until now only few adult males and some pupae have been found in different Central European countries (Langton and Cranston, 1991), in Sweden (Brundin, 1947), and some places in the Nearctic (Soponis, 1977). The adult female, yet undescribed, can be easily distinguished from all other known females of *Orthocladius sensu strictu* by the diagnostic features summarized in Table 2 and shown in Figure 2.

Few and scattered records of *Labrundinia longipalpis* (Goetghebuer) and *Schineriella schineri* (Strobl) from several mid-European countries imply that the ecological preferences of both species are not yet fully known. According to observations in Southern Germany,

Schineriella schineri is a characteristic species of small, stagnant, and organically enriched water bodies in the inundation area of lowland streams. Pupal exuviae of the rare species *Labrundinia longipalpis* have been collected in a Bavarian wet gravel pit. The collection of Zoologische Staatssammlung München comprises single adult males of the latter species from Corfu (Greece) and Southern Spain, as well as some larvae from the Netherlands (coll. Moller-Pillot). Both species were typical insects of the 1993 emergence of the experimental ponds.

Like the species numbers also the individual numbers of the chironomid emergence of this study were comparatively low: only 4 species made up to about 80% of the total individual numbers of the 1992 emergence rate (Table 1).

Apart from chironomids, other macrobenthic groups like Ephemeroptera (2 species), Trichoptera (3 species), Ceratopogonidae (6 species) also exhibit rather low population densities, indicating their minor role in the aquatic food web of the experimental ponds. This striking reduction of the benthic fauna to a few key species must be attributed to the simple design of the ponds, to the lack of suitable refuges and to the strong foraging pressure of the top consumers (Odonata, Notonectidae, Corixidae, Dytiscidae). Regarding quantitative aspects of energy flow and biological turnover in the test ponds, all mass species of the food web are zooplanktonic. In 1992 and 1993 Cladocera, Copepoda, and Rotatoria were not controlled by any fish predator. As a consequence they developed large population densities utilizing very effectively the ample food source of unicellular algae.

This basic scheme of community development (strong foraging activity of predacious invertebrates, low biomass of chironomids and other benthic organisms, mass development of planktonic organisms) appears to be characteristic for this type of man-made experimental ponds.

Acknowledgements

The authors thank Ms. M. Arnold for her capable technical assistance.

Table 1: Species list of chironomids

Species	Ecological preferences /trophic characteristics *)	1992 emergence rate (%)**)
Ablabesmyia longistyla Fittkau	opportunistic pioneer species in various types of fresh water habitats	27.9
Ablabesmyia monilis (Linnaeus)	same ecological preferences as *A. longistyla*	13.3
Clinotanypus nervosus (Meigen)	ecological preferences poorly known; probably opportunistic pioneer species in various types of fresh water habitats (Leuchs and Caspers, 1988)	3.1
****Labrundinia longipalpis* (Goetghebuer)	according to Fittkau (1962) 'a stenotopic species of European bogs'; this statement does not seem to reflect the true ecological valence of this species	
Monopelopia tenuicalcar (Kieffer)	characteristic species of (small) bogs and swamps	2.0
Paramerina cingulata (Walker)	eurytopic species of the littoral zone of lakes	-
Procladius ? choreus (Meigen)	ubiquitous species	-
****Schineriella schineri* (Strobl)	ecological preferences poorly known; scattered records from England, Belgium, Austria, and Germany	
Tanypus punctipennis (Meigen)	stagnant water bodies; eutrophic and hypertrophic habitats	0.8
Acricotopus lucens (Zetterstedt)	ubiquitous species in all types of small water bodies	-
Chaetocladius sp. (females only)		-
Corynoneura gratias Schlee	inhabits ponds and lakes of different trophic level	-
Cricotopus cylindraceus (Kieffer)	slowly flowing streams and various types of standing water; scattered records from the Western Palaearctic region and Canada	1.3
Cricotopus flavocinctus (Kieffer)	rather few records from different types of flowing and standing waters suggest this species to be ubiquitous	-
Cricotopus obnixus (Walker)	rather few specimens of this species has been recorded from northern lakes, fishponds and ditches of the Western Palaearctic region.	0.5
Cricotopus sylvestris (Fabricius)	ubiquitous species	-
Limnophyes sp. (females only)		-
Nanocladius bicolor (Zetterstedt)	characteristic species of the lower reaches of streams; less abundant in different types of standing water	-
Orthocladius holsatus Goetghebuer	ecological preferences unknown	3.7
Psectrocladius limbatellus (Holmgren)	inhabits primarily lakes, ponds and dam reservoirs	17.0
Psectrocladius obvius (Walker)	ubiquitous species	-
Psectrocladius psilopterus Kieffer	according to Langton (1980) in small, shallow bodies of stagnant water	-
Psectrocladius sp. (n.?)	one adult male only, resembling closely *P. limbatellus* (Holmgren)	-
Chironomus thummi Kieffer	characteristic species of organically enriched sediments with intensive decomposition processes; pioneer species in various types of fresh water habitats	0.5
Chironomus sp. 2		-

Table 1. (cont.)

Species	Ecological preferences /trophic characteristics *)	1992 emergence rate (%)**)
Cladopelma viridula (Fabricius)	characteristic species of littoral mud sediments and aquatic macrophytes in ponds and lakes	20.3
Cryptochironomus albofasciatus (Staeger)	ecological preferences poorly known; according to Reiss (1968) in Lake Constance characteristic species of sandy littoral zones without vegetation	-
Dicrotendipes nervosus (Staeger)	ubiquitous species	-
Endochironomus impar (Walker)	species widely distributed in permanent and temporary pools of the Lake Constance area (Reiss 1968)	-
Endochironomus tendens (Fabricius)	characteristic species of littoral Aufwuchs communities; sometimes mining in aquatic macrophytes	-
Glyptotendipes (s.str.) Pe 2 Langton 1984		-
Microtendipes pedellus De Geer	ubiquitous species	4.9
Parachironomus parilis Walker	ecological preferences poorly known; rather few records from rivers, streams and lakes	0.8
Parachironomus varus Goetghebuer	according to Thienemann (1954) the larvae are ectoparasites of water snails	-
Polypedilum cultellatum Goetghebuer	species characteristic for lower reaches of streams and the littoral zone of ponds and lakes	-
Polypedilum nubeculosum (Meigen)	ubiquitous species	
****Pseudochironomus prasinatus*	inhabits ponds and lakes	
(Staeger)		
Cladotanytarsus sp.		-
Paratanytarsus sp.	pupal exuviae only, keying to-	
P. laccophilus (Edwards)	in Langton (1991), a species inhabiting northern and montain lakes	
Tanytarsus chinyensis Goetghebuer	larvae live in the lower littoral zone of ponds and lakes (primarily oligotrophic)	-
Tanytarsus ejuncidus Walker	inhabits littoral lake sediments	-
Tanytarsus fimbriatus Reiss and Fittkau	opportunistic pioneer species in various types of fresh and brackish water habitats	-
Tanytarsus holochlorus Edwards	inhabits a large variety of eutrophic habitats	0.6
Tanytarsus usmaensis Pagast	key species of sandy littoral sediments in eutrophic ponds and lakes	-

*) referring to observations in Central European habitats

**) The emergence rates of the species refer to a study on an insect growth regulator in 1992. They are indicated as percentage of the total catch (= 2831 specimens); rates <0.5% are not specified. As the control pond and the treated ponds did not differ significantly with respect to species composition and individual numbers of the chironomid emergence, all data were combined as given in this table.

***)These species were not represented in the 1992 emergence.

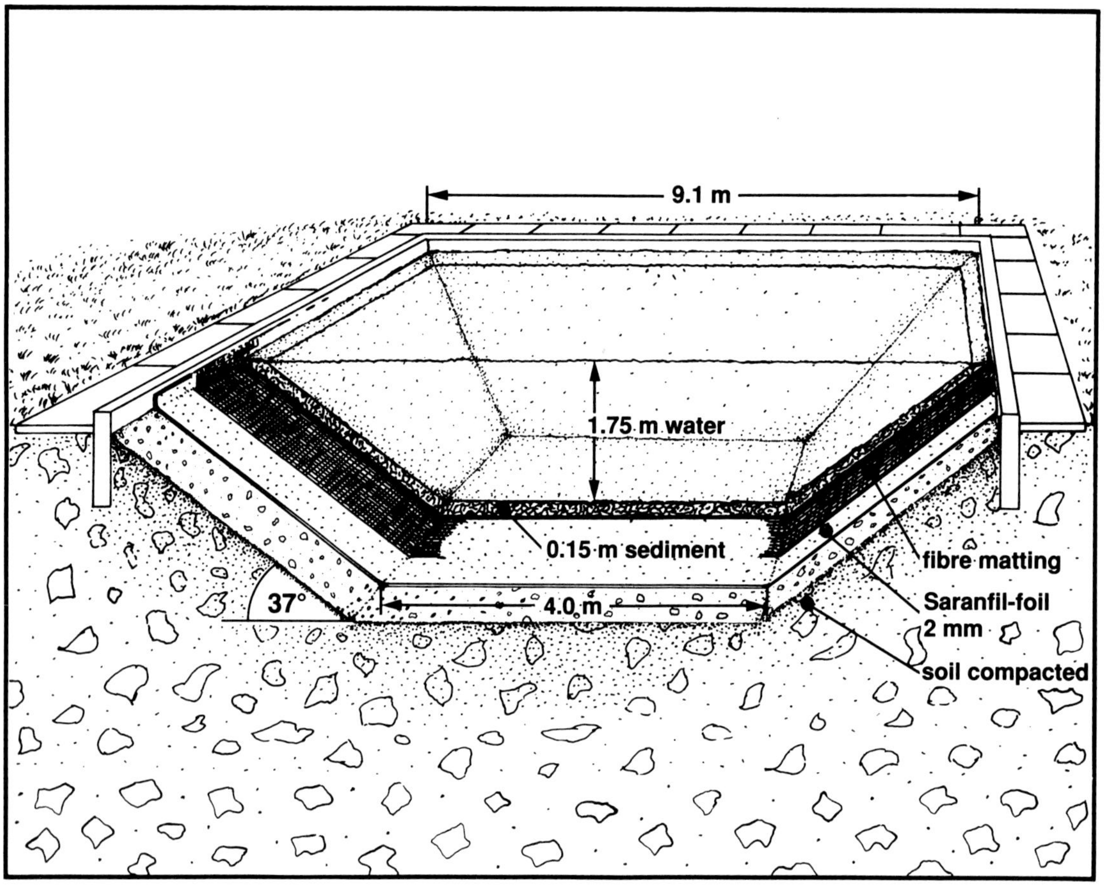

Figure 1. Construction details of the test ponds

Table 2: Some diagnostic characters of the female imago of *Orthocladius (O.)holsatus* Goetghebuer (n = 3)

	Wing length	2.1-2.4 mm
No of	inner verticals	7-8
	outer verticals	4
	postorbitals	3
	setae on clypeus	20-31
	*acrostichals	40-43, biserial
	*dorsocentrals	16-77, bi/triserial
	prealars	5
	scutellars	16-19
	squamal setae	18-19
	setae on R	15-18
	setae on R_1	12
	setae on R_{4+5}	12-14
	Anal lobe of the wing well rounded and produced.	
	*Shape and setation of the genitalia (Fig. 2c, d)	

The combination of the characters indicated by an asterisk seems to be sufficient to distinguish *O. holsatus* from all other known females of the subgenus *Orthocladius* van der Wulp.

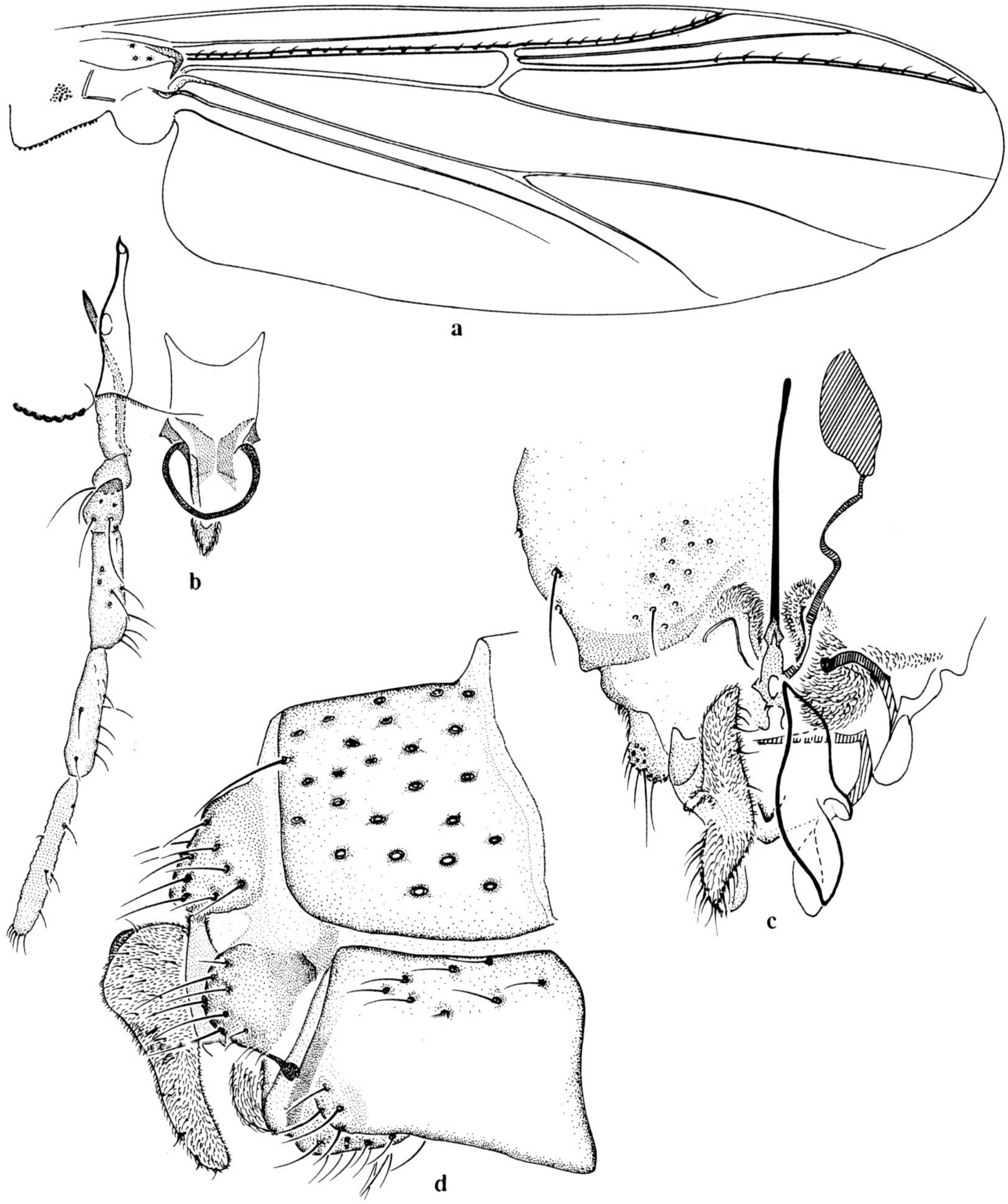

Figure 2. *Orthocladius (O.) holsatus* Goetghebuer, female imago: (a) wing, (b) cibarial pump, tentorium, stipes and palpus, (c) genitalia, ventral view, (d) genitalia, lateral view.

Tanytarsus niger Andersen (Diptera: Chironomidae) and the chironomid community in Gebhartsteich, a carp pond in northern Austria

Berthold F. U. Janecek

Abteilung für Hydrobiologie, Fischereiwirtschaft und Aquakultur, Universität für Bodenkultur, A - 1180 Wien, Austria

Introduction

During the last decades the chironomids in particular from intensively cultivated Eastern European carp ponds have been studied. Such ponds may be highly productive, but their species number is low (Bakhtina, 1978; Borutski *et al.*, 1971; Zieba, 1971). Gebhartsteich, like many other larger carp ponds of Northwestern "Waldviertel" (part of the crystalline Bohemian mass in northern Austria) (Fig. 1), represents a different type of pond, sometimes called the "classical" Central European pond-type. The ponds in the region may be very old, sometimes originating from medieval constructions (Wawrik, 1966). They are often relatively natural, their banks may be intensively interlinked with surrounding wetlands, wet woods or even peat bogs. Because of the extensive cultivation, carp cannot live on artificial feeding alone, but to a large extent must feed on benthic animals.

This study, carried out between 1978 and 1982 for a thesis at the Zoological Institute of the University of Vienna, intended, simply said, to study thoroughly, but nevertheless as a first step, the chironomids of Gebhartsteich qualitatively (e.g. phenology, distribution within the habitat) and quantitatively (larval density, biomass and production). Some novel elements of this thesis are presented here. The first section is a faunistic one: *Tanytarsus niger* is selected as an example of a chironomid species with an interesting distribution and is but one example (of 35 or 36) of chironomid species recorded in Austria for the first time. Quantitative questions of (dem)ecology (larval density, life history, biomass and production) are dealt with at the basis of the single example *Procladius rufovittatus*. The comparable values of the most important species of the pond are briefly presented in a list.

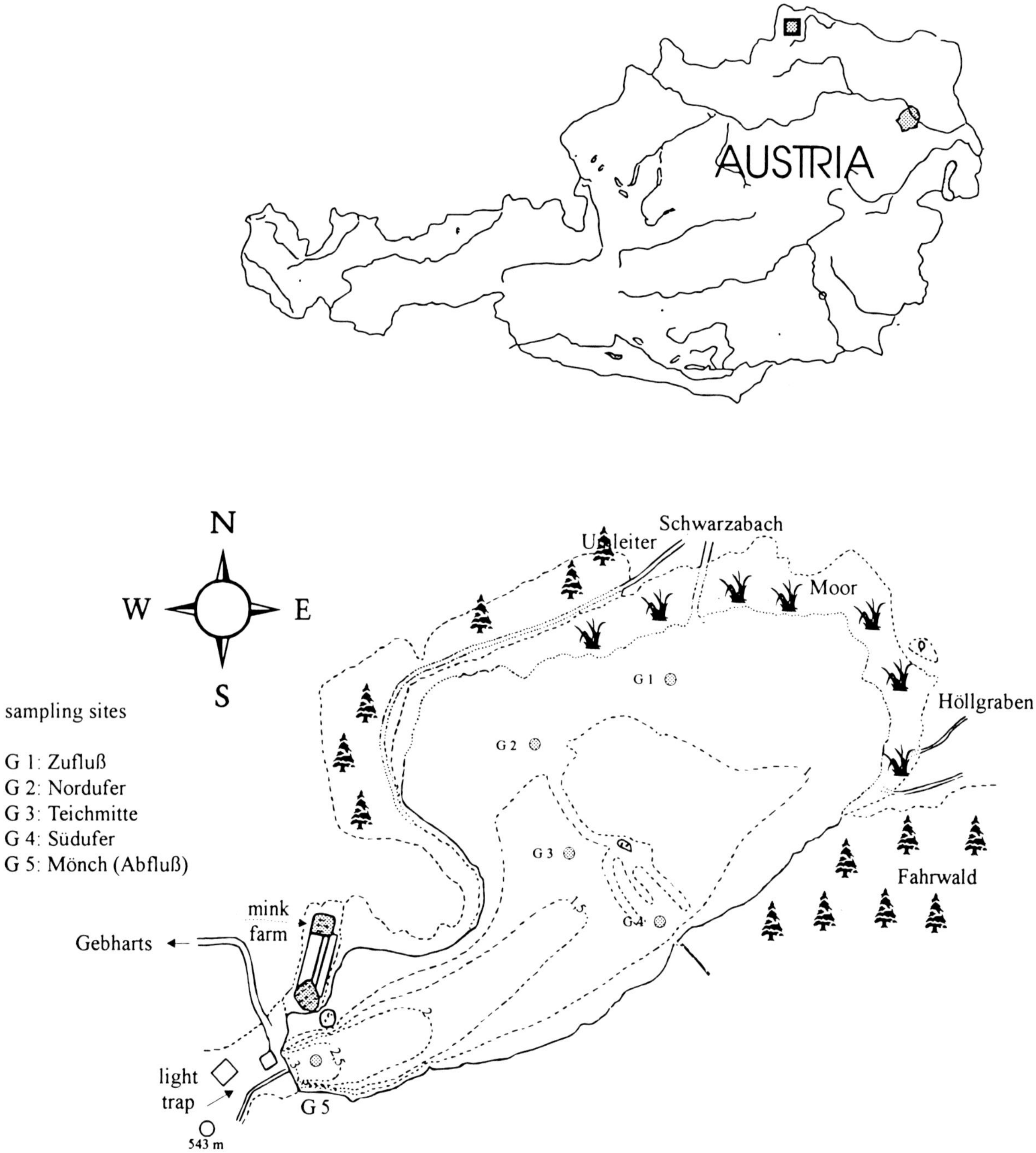

Figure 1. Gebhartsteich

STUDY-SITE

Gebhartsteich has the following morphological parameters: 48°48'N, 15°09'E, 544m a.s.l., surface: 59-62 ha, depth: max.: 320 cm, average: 135 cm, water-volume: about 825,000 m^3. During the period of examination the average water temperature (50 cm depth) amounted to 7,7°C (1978/9), 1979/80 to 11,2°C. 1977/8 the pond was ice-covered for 109 days, 1978/9 for 126 days and 1979/80 for 104 days. Oxygen stratifications with hypersaturations near the surface and depletions near the ground could be measured, but were instable and broke down on windy days. For vegetation and a chemical analysis of the water and ground, see Janecek, 1985. Liming and fertilization (with e.g. Thomas meal and superphosphate) were not carried out annually; the waste and excrements of a mink-farm served as manure. In some years (e.g. 1978) carp growth was too low and, since the pond could not be fished out, water was not released.

MATERIAL AND METHODS

Larval sampling

Between 17.v.1978 and 29.iv.1980 samples were taken monthly (with a gap between 30.x.78 and 21.iii.79) with the help of a core sampler - or in some cases with an Ekman grab. 25 samples were the result of each sampling period, 5 each per sampling site: G1 (depth: 90 cm), G2 (90 cm), G3 (160 cm), G4 (115 cm) and G5 (290 cm). Samples were fixed with formaldehyde, the larvae counted under a binocular microscope and preserved in 80% ethanol. All slide preparations were embedded in Euparal. Many larvae were reared to adults.

Pupal exuviae

Exuviae were taken with hand net on 45 days between 9.iv. and 30.x.1980 at intervals of 2 to 8 days. Between 5.iv. and 9.iv.1982 pupal exuviae were collected in 43 ponds mainly in Northwestern Waldviertel (descriptions of these ponds in Wawrik, 1966) and these are included in the species list (Table 3).

Light trap

Light traps were in permanent service between 5.iv.1979 and 25.i.1980. 13 times the animals were taken out. Additional catches of adults were made with a hand net (also 1981 and 82).

Emergence traps

Five emergence traps (Bretschko, 1974) with an area of 0.25 (or 1) m^2 were exposed near the sample sites (E1-5). Trapped animals were taken 45 times at intervals of 2 to 9 days.

RESULTS

Out of a total of 152 detected chironomid species, 71 certainly develop in Gebhartsteich, but 51 other species most probably may be added to the list. Qualitative results are inserted into the Table 1 as "stenographic" notes.

DISCUSSION

Tanytarsus niger

Like some other species of Gebhartsteich (e.g. *Orthocladius (Pogonocladius) consobrinus*, *Tanytarsus inaequalis* and *T. gracilentus*), *T. niger* has a largely arctic-circumpolar distribution. Recent records from Central Europe on the other hand are very scattered. Reiss (1984b) mentions Stechlinsee, Brandenburg (Mothes, 1966) and Osterseen (Im, Pe from Gartensee, Lustsee, Stechsee, Großer Ostersee) in Germany, in France Lac des Rousses (Jura, 1059 m NN, area 90 ha, maximal depth 18 m, larvae abundant in the whole littoral down to a depth of 6 m) (Verneaux and Vergon, 1974; Verneaux *in litt.*).

In Northwestern Waldviertel *T. niger* has been found not only in Gehartsteich, but also in 12 other carp ponds, situated with 2 exceptions (Gebhartsteich and Großradischenteich) near the villages Pürbach and Hoheneich) (Fig. 2) (see also Wawrik, 1966):

1) **Großradischenteich** (GnG, 48°56'N, 15°07'30"E (same degrees of latitude and longitude for the following ponds), 570 m asl., 6.iv.1982, 16.20h, water temp.(surface): 10.8°C, pH: 6, O_2 (50 cm): 13.8 mg/l; Pe: 0.7% of 203 chironomid)s.
2) **Mitterteich** (HhM, N- 46', E- 03', 510 m asl., 8.iv.1982 (also for the following ponds), 12.30h, temp.: surface: 12.1°C, 300 cm: 9.0°C, pH: surf.: 7.5°C, 300 cm: 7.0, O_2: surf.: 12.6 mg/l, 300 cm: 10.8 mg/l; Pe: 2% of 356).

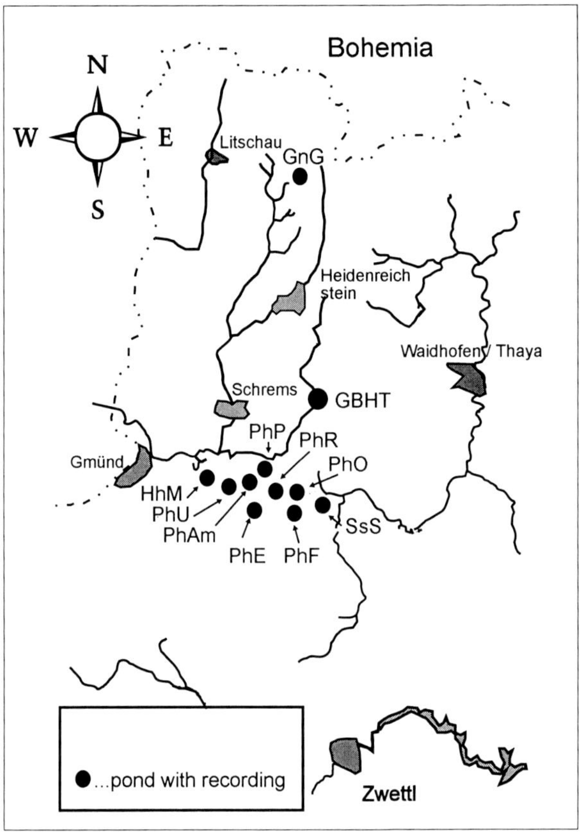

Figure 2. Recordings of *Tanytarsus niger* in Northwestern Waldviertel. Explanations of abbreviations: in the text, rest of ponds see Janecek 1985.

3) **Ullrichsteich** (PhU, N- 46', E- 03' 30", 520 m asl., 13.00h, temp. (50 cm): 10.2°C, pH (50 cm): 7.0, O_2: surf.: 11.9 mg/l, 50 cm: 11.5 mg/l; Pe: 53.4% of 335).
4) **Pürbacher Teich** (PhP, N- 46', E:- 04' 30", 525 m asl., 13.30h, temp.(surf.): 10,7°C, O_2 (surf.).: 12.3; Pe: 7.9% of 341).
5) **Spielberger Teich** (PhS, N- 45' 30", E- 05', 530 m asl., 14.00h, temp. (50 cm): 11.6°C, pH (50 cm): 7.0, O_2 (50 cm): 13.9; Pe: 1,6% of 378).
6) **Althöllteich** (PhA, N- 46', E- 04' 30", 525 m asl., 15.00h, temp.: surf.: 11.8°C, 300 cm: 10.5°C, pH: surf.: 7.5, 300 cm: 7.0, O_2: surf.: 13.8, 300 cm: 10.9; Pe: 89.9% of 367).
7) **Amtwiesteich** (PhAm, N- 46', E- 04', 525 m asl., 15.30h, temp. (surf.): 11.6°C, pH: 9.0; Pe: 9.4% of 308).
8) **Edlauer Teich** (PhE, N- 45', E- 04'; 535 m asl., 16.00h, temp.: surf.: 11.4°C, 250 cm: 11.3°C, pH: surf.: 10.0, 250 cm: 9.5, O_2: surf.: 17.8, 250 cm: 17.7; Pe: 2% of 405).
9) **Frauenteich** (PhF, N- 45', E- 05' 30", 535 m asl., 16.30h, temp. (surf.): 11.1°C, pH (surf.): 7.0; Pe: 5.6% of 178),
10) **Otterteich** (PhO; N- 45', E- 05' 30", 535 m asl., 17.00h, temp. (surf.): 11.9°C, pH: 7.0; Pe: 0.3% of 327).
11) **Stölzlingerteich** (SsS, N- 45', E: 07', 537 m asl., 18.00h, temp. (surf.): 11.2°C, pH: 8.5; Pe: 0.6% of 316).
12) **Kobelteich** (PhKo, N- 46', E- 06' 30", 530 m asl., 18.45h, temp. (surf.): 11.4°C,pH: 6.5; Pe: 0.4% of 240).

In Gebhartstreich, the larvae do not avoid the deeper parts (300 cm depth) near G5 with a 20 cm layer of mud (organic contents: 18.4%, carbonates: 0%, water contents: 18.2%, particle sizes: sand (2.0-0.06 mm): 1%, silt (0.06-0.002 mm): 80%, clay (below 0.002 mm): 9% (Hoheiser, 1981), but were never detected near G4, where the soil is a little peaty and contains dead plants, grass and roots (larva reared from G3: 18.iii.1980, pupae in the samples: 21.iii.1979: G1, G2, G3, G5, 8.xi.79: G3; 23.i.80: G1; 10.iii.80: G1). According to Sæther (1979a), *T. niger* is a characteristic species of the littoral and sublittoral zone of northern ultraoligotrophic and (more rarely) moderately oliogotrophic lakes. In contrast, Reiff *et al.* (1992), investigating Bavarian lakes, found that the species is weakly negatively correlated (r(s)(ϑo) -.296) with P_t. With a location index of 2.74 *T. niger* should tolerate mesotrophic conditions and is a poor trophic indicator ("IG": 1.00). Nevertheless, Althöllteich and Ullrichsteich, where the density of the species seems to be high, are by no means productive carp ponds (Wawrik, *loc. cit.*).

Adults emerge mainly in April (Table 3), in many ponds at the same time as *Microtendipes chloris*, *Tanytarsus miriforceps*, *Tanytarsus sylvaticus* and with the more sparsely detected *Thienemanniola ploenensis*. *T. niger* seems to be a univoltine spring species, but pupae during winter and adults in the light trap at the end of May, the beginning of June and at the beginning of October cannot be easily interpreted, since the larval phase of life-history is poorly known.

Procladius rufovittatus

Life-history

In Gebhartsteich, *P. rufovittatus* is a univoltine species (larval development: Figs. 4, 5; abundance: Fig. 6). A conspicuous peak of emergence lies mV/VI (Fig. 3). Besides this maximum, little emergence takes place till October. Figure 5 suggests that this process is due to unsynchronized individuals, not to a summer generation. The time between hatching and moulting to the second larval instar appears to be very short. Over-wintering takes place especially in the 3rd larval instar. The number of 4th instar larvae increases mainly after ice melting. The curve of average larval weight shows growth running parallel to larval maturation (Fig. 6) and also expresses the development of a univoltine generation.

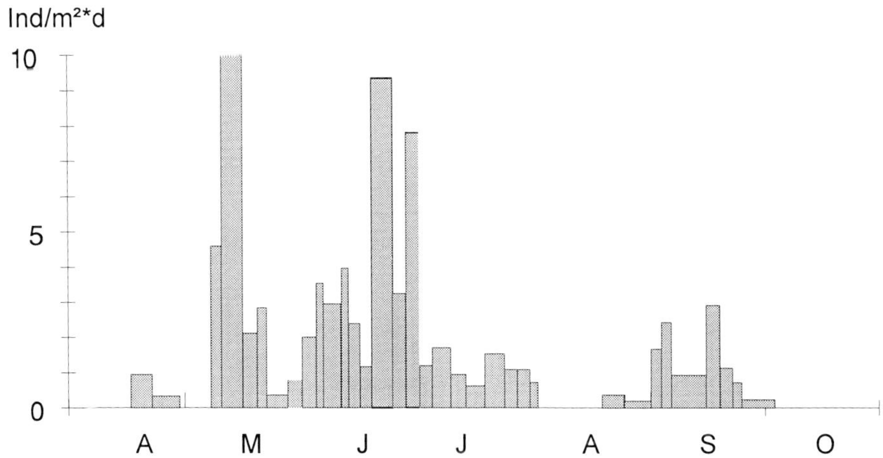

Figure 3. Emergence diagram of *Procladius rufovittatus*

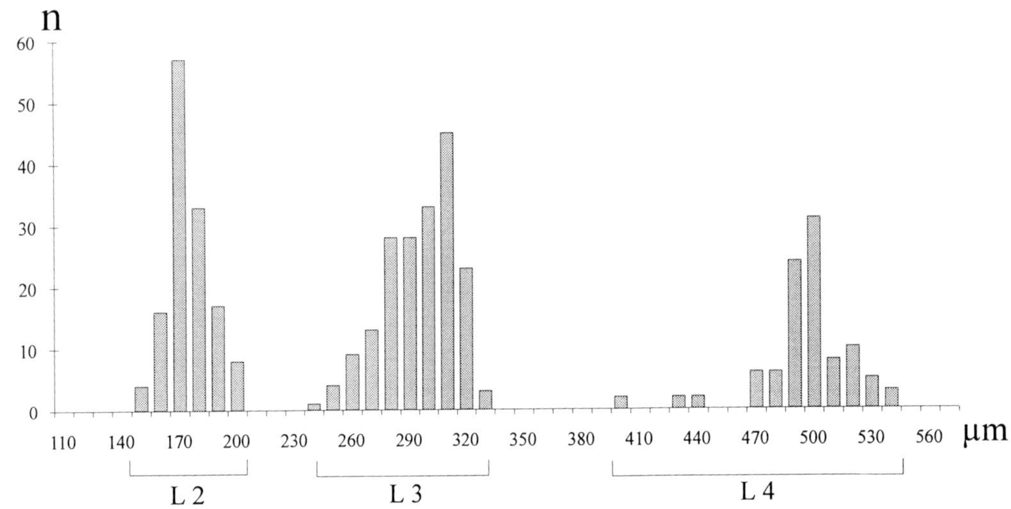

Figure 4. Head-capsule widths of larval instars of *Procladius rufovittatus*

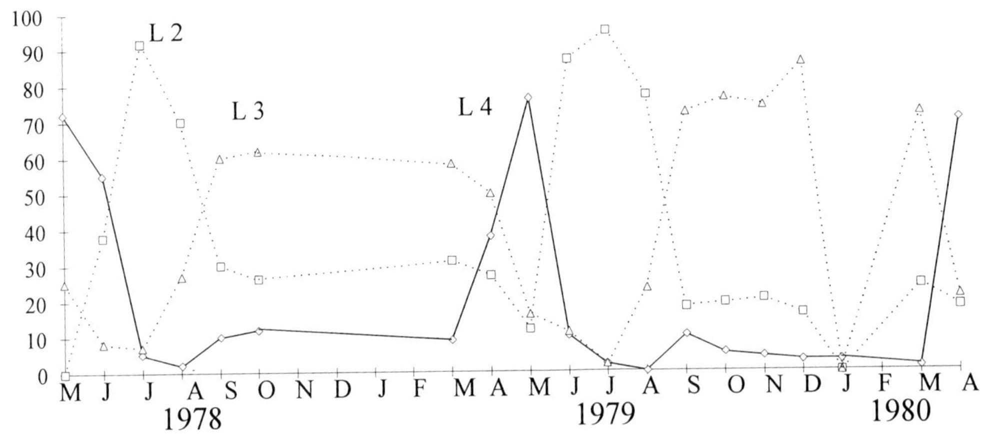

Figure 5. Larval development of *Procladius rufovittatus*

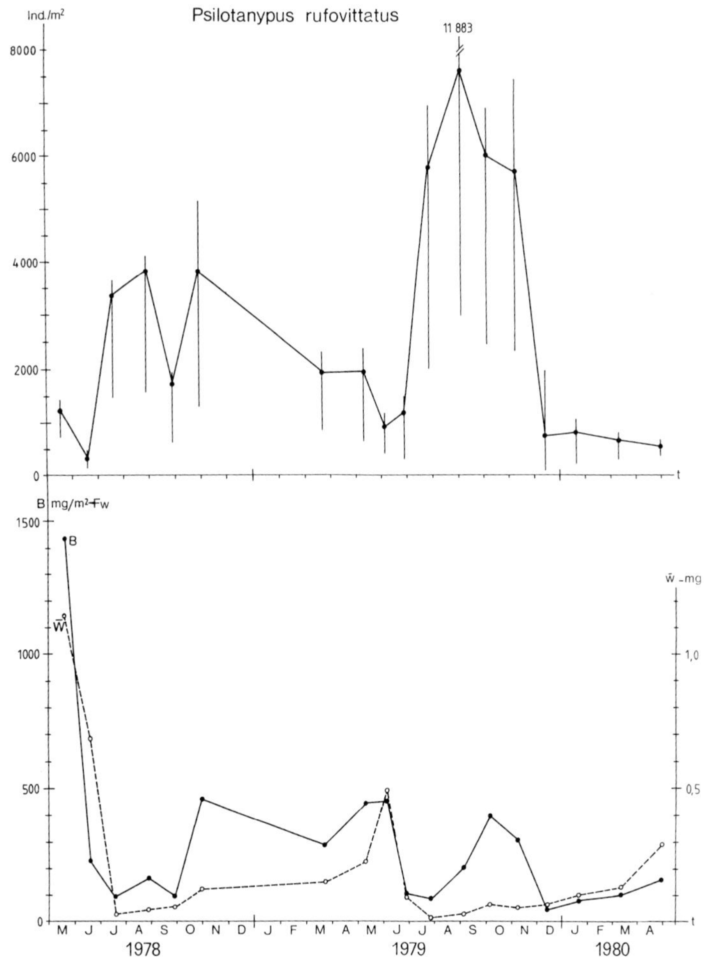

Figure 6. Abundance of *Procladius rufovittatus*
Figure 7. Biomass of *Procladius rufovittatus*

Production

The production of this species (and of *Einfeldia dissidens*: Table 1) is estimated by means of the "classical" Boysen-Jensen" method (Boysen-Jensen, 1919; Schlott, 1976; Janecek, 1985). The production of other species is estimated only roughly on the basis of apparent or minimal production (sum of all increases of biomass during a certain time - a generation, a year) (Winberg, 1971) or with the help of P/B aver. - coefficients (Waters, 1969, Winberg, *loc. cit.*). Figs. 8 and 9 show the production diagrams (and elimination diagrams) for the generations 1978/9 and 1979/80 - on the assumption that the elimination is constant and follows a logarithmic function. The population is considered to be a single cohort and the unsynchronized individuals are neglected. Because of high losses of young larvae in the course of sieving them with a mesh- size of 0.2 mm (1979: 0.1 mm) the values at the beginning of each generation, which show no elimination, have to be corrected.

Table 1: Average abundance (ind./ mΔ), average annual biomass and production (both in mg Fw/mΔ) of species and groups of quantitative importance

Species or group	av. ab. 78	CL 95%	%	av. ab. 79	CL 95%	%	av.B 78/9	%	av.B 79/80	%	P 78/9	%	P 79/80	%
Tanypodinae	5334		17.4	6198		27.3	3013	35	1769	28.8	9126	15.5	3959	10.9
Tanypodinae without Procl. ruf.	2920	2376 - 3377		2975	2485 - 3365									
Procladius rufovittatus	2414	1856 - 2651	7.9	3223	2154 - 4598	14.2	335	3.9	131	2.1	1411	2.4	750	2.1
Chironomini/ Pseudochironomini	3416		11.2	4449		19.6	3062	35.6	3272	53.2	20 378	34.6	23 463	64.8
Chironomini/ Pseud./ without Einf. diss.	2626	1878 - 3167		3931	2871 - 4857							34.6	23463	64.8
Chironomus plumosus	83	27 - 146	0.3	86	40 - 110	0.4	656	7.6	773	12.6	4780	8.1	5770	15.9
Dicrotendipes lobiger	194	58 - 273	0.6	158	80 - 216	0.7	55	0.6	28	0.45	256	0.4	314	0.9
Dicrotendipes modestus	349	213 - 464	1.1	621	354 - 933	2.7	382	4.4	183	3	2106	3.6	1069	3
Einfeldia dissidens	790	463 - 1039	2.6	518	360 - 612	2.3	225	2.6	65	1	1506	2.6	452	1.2
Endochironomus albipennis	46	11 - 53	0.15	550	326 - 721	2.4	43	0.5	330	5.4	400	0.7	2017	5.6
Glyptotendipes paripes	136	52 - 169	0.4	277	132 - 435	1.2	149	1.7	469	7.6	824	1.4	2757	7.6
Microtendipes chloris	338	215 - 420	1.1	163	88 - 217	0.7	500	5.8	48	0.8	2784	4.7	444	1.2
Polypedilum nubeculosum	422	270 - 508	1.4	997	670 - 1404	4.4	365	4.2	752	12.2	2630	4.5	5115	14.1
Polypedilum bicrenatum	354	98 - 783	1.15	116	42 - 230	0.5	20	0.2	3	0.05	92	0.15	25	0.1
Tanytarsini	21 758	17 072 - 25 650	71.1	11 983	8224 - 15 964	52.8	2490	29	1095	17.8	29 203	49.6	8753	24.2
CHIRONOMIDAE	30 596			22697			8595		6146		58 857		36 225	

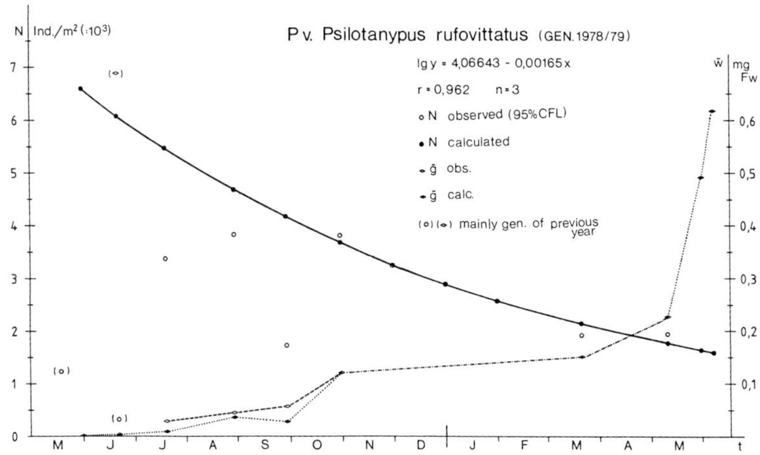

Figure 8. Production diagram of *Procladius rufovittatus* (Gen. 1978/9)

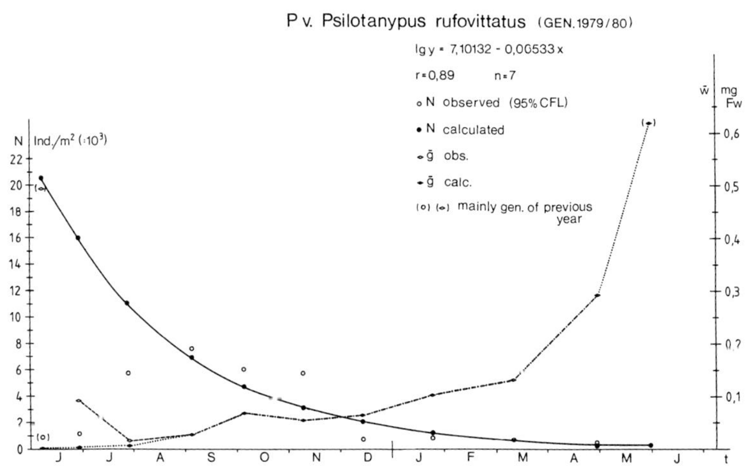

Figure 9. Production diagram of *Procladius rufovittatus* (Gen. 1979/80)

The 1978/9-generation produces 1411 mg fw per m^2 (162 mg dw, 144 mg ash-free dw, 3, 35 kJ/m^2 (McCauley and Tsumura, 1974)). The average (calculated) biomass is 335 mg/m2 (39 mg dw/m^2). P/B aver. is therefore 4.2. The 1979/80- generation produces 750 mg fw per m^2 (86 mg dw, 76.5 mg ash-free dw, 1.78 kJ/m^2). B aver. for this generation is calculated at only 130.74 mg fw/m^2 (15.04 mg dw). P/B aver. is therefore 5.74. 2 periods of higher production, namely autumn (between 28.ix.- and 30.x.1978 11.39 mg fw/m^2 are produced daily) and a short space of time before the emergence confront 2 periods of low production, (mid)summer and the time below the ice.

Table 2: Adults emerging per mΔ and year (1980) and "trophy indices (Orendt 1992, Riss 1992)"
*= also in algal aufwuchs of emergence traps

Taxon	Number	%	ORENDT tLok	ORENDT IG	RISS tLok	RISS IG
1) *Tanytarsus debilis*	1126	20.51	3.70	1.00		
2) *Tanytarsus excavatus*	789	14.36	5.00	3.00		
3)**Glyptotendipes paripes*	375	6.83	4.90	2.00		
4) *Tanytarsus holochlorus*	334	6.08	4.80	2.00		
5) *Procladius signatus*	295	5.38			2.94	2.00
6)**Endochironomus albipennis*	273	4.96	4.50	1.00		
7) *Tanytarsus miriforceps*	271	4.93				
8) *Dicrotendipes modestus*	255	4.64	4.00	3.00		
9) *Polypedilum nubeculosum*	252	4.59	4.30	1.00		
10)**Tanytarsus verralli*	227	4.13	4.00	3.00		
11) *Procladius rufovittatus*	211	3.84			4.13	3.00
12) *Einfeldia dissidens*	206	3.74			4.95	3.00
13) *Tanytarsus occultus*	138	2.51	3.20	1.00		
14) *Tanytarsus lestagei* agg.	124	2.25	4.30	1.00		
15) *Chironomus plumosus*	85	1.54	4.60	1.00		
16) *Cladotanytarsus nigrovittatus*	85	1.54	3.60	1.00		
17)**Ablabesmyia monilis*	78	1.42			2.81	1.00
18)**Glyptotendipes pallens*	70	1.27	4.90	2.00		
19) *Cladotanytarsus atridorsum*	41	0.75	3.80	1.00		
20)**Dicrotendipes lobiger*	39	0.71			2.69	1.00
21) *Cryptochironomus redekei*	37	0.68				
22) **Ablabesmyia phatta*	37	0.67			2.76	2.00
23) *Chironomus ?cingulatus*	35	0.64	5.00	3.00		
24) *Tanytarsus niger*	21	0.38			2.74	1.00
25) *Polypedilum bicrenatum*	18	0.33			2.67	1.00
26) *Cryptochironomus ?obreptans*	17	0.31			3.46	2.00
27) *Psectrocladius ?limbatellus*	17	0.31				
28) *Microtendipes pedellus*	10	0.18	4.40	1.00		
29)**Pseudochironomus prasinatus*	7	0.13	4.00	2.00		
30) *Cladopelma viridula*	7	0.12	3.90	1.00		
31) *Cryptochironomus psittacinus*	6	0.10	4.00	3.00		
32) *Psilotanypus lugens*	4	0.07			4.37	3.00
33) *Microtendipes chloris*	3	0.06	4.80	1.00		
34) *Tanypus vilipennis*	2	0.04			3.89	3.00
Total number	5493	100.00				

Horizontal distribution

Since the calculated densities of larvae (logarithmic transformation of data, 95% CL) at each sample site were statistically compared with each other (t-test), possible differences in density between 2 or more areas could be detected. *P. rufovittatus* is distributed rather evenly.

Depth preferences of Gebhartsteich Chironomidae

Benthic chironomids show a different preference of depth zones even in a shallow pond which appears relatively uniform. Thus:

Group A reaches its largest densities in the deepest (and most polluted) parts of the pond (mainly near

G5): *Procladius (Psilotanypus)* sp. 3 (only G5), *Tanypus kraatzi* (G5), *Tanypus vilipennis* (G3, 5), *Chironomus ?cingulatus*, *Chironomus plumosus* (1979 more likely belonging to Group B).

Group B is represented by larvae, which neither prefer the deeper parts of the pond nor are restricted to more shallow areas (G1, 4): *Procladius ?choreus*, *P. ?signatus*, the above mentioned *P. rufovittatus*, *Cryptochironomus* cf. *obreptans*, *C. psittacinus*, *Einfeldia dissidens*, *Tanytarsus niger*, *T. bathophilus* (G2), *T. excavatus* and *T. inaequalis*. Group C, constituting the largest part of chironomids of the pond, shows preference for the shallow areas (highest densities) or is detected only near G1/G4: *Psectrotanypus varius* (?), *Procladius lugens* (?), *Ablabesmyia monilis*, *Ablabesmyia phatta* (G1, 4, 3), *Corynoneura carriana* (G1, 4), *Psectrocladius ?limbatellus*, *Chironomus tentans*, *Cryptochironomus redekei*, *Demicryptochironomus vulneratus* (G1, 4), *Dicrotendipes lobiger*, *Dicrotendipes modestus*, *Endochironomus albipennis*, *Glyptotendipes pallens*, *Glyptotendipes paripes*, *Microchironomus tener* (G1), *Microtendipes chloris*, *Pagastiella orophila*! (G4), *Parachironomus vitiosus* (G1/4), *Polypedilum bicrenatum*, *Tribelos intextus* (G4), *Zavreliella marmorata*, *Pseudochironomus prasinatus*, *Cladotanytarsus atridorsum*, *C. nigrovittatus*, *Paratanytarsus laccophilus* (?), *Tanytarsus miriforceps*, *T. debilis*, *T. verralli*, *T. holochlorus* and *T. occultus*.

Total Gebhartsteich Chironomidae - trophic status and pollution

The average abundance 1978 is calculated at 30,596 larvae/m^2 (CL 95: 23,690 - 35,988), 1979 at 22,697 lv./m^2. The following B aver. are estimated: 78/9: 8.6 g fw/m^2 (1, 34 g dw/m^2), 79/80: 6, 15 g fw/m^2 (0.99 g dw/m^2). Rough estimations of P amount to: 78/9: 58, 9 g fw/m^2 (9.5 g dw/m^2, 8.4 g af dw/m? - 187 kJ/m^2), P/B aver.: 6.8; 79/80: 36.2 g fw/m? (6.2 g dw/m^2, 5.5 g af dw/m^2 - 123 kJ/m^2), P/B aver.: 5.9. These values are discussed in detail in Janecek, 1985. In a lot of investigated carp ponds especially B and P of chironomids are higher. Russian authors, for example, mention Ps in carp ponds of 335 - 732 kJ/m^2 (Bakhtina, 1978) and 1687 kJ/m^2 (Borutski et al., 1971). Ziebas (1971) values in Polish fingerling ponds are remarkably high (1661-1937 kJ/m^2). Biomass and production of Gebhartsteich can be compared with mesotrophic lakes, for example with P in the free water area of Neusiedlersee (P: 15-148 kJ/m^2) (Schiemer, 1979).

The concentrations of ammonium, nitrate and especially of phosphate are very high at the main measuring point near G5.: NH_4^+: 0.2 (17.v.78) - 1.6 mg/l (13.xii.79), NO_3^- : 0.6 (20.vi., 28.ix.78) - 4.1 mg/l (13.xii.79), P/PO_4^{3-}: 0.1 (10.iii.80) - !4.9 mg/l (28.vi.79). Surely G5 is an extreme place in Gebhartsteich, especially because of the sewage and excrement of a farm once of more than 10 000 mink. But Gebhartsteich is interesting, because it represents a mixture, a transition from a bog water near the Northeastern bank (where Schwarzabach has its mouth) to a highly eutrophicated pond near the outlet, and this fact surely contributes to the species richness of this ponds. Of course many chironomid species develop in Gebhartsteich, which are characterized as "eutraphent" littoral species in the recent investigations of Reiff *et al.* (1992) in Bavarian (German) lakes of different trophic states. In Gebhartsteich not only a lot of semiterrestrical species can be found (forming a reflection of the near-to-nature bank region). There are also some (quite rare) species, which are typical for oligotrophic (e.g. *Pagastiella orophila*) or mesotrophic (*Tanytarsus niger* must be called such a species) conditions. Using the indices (tLok: tropic location, IG: indicatory weight) of *Reiff* et al. *(1992) a trophic index of* Gebhartsteich was calculated, a process, which resembles the calculation of a saprobity index following the Pantle and Buck formula. The list of total chironomid emergence (Table 2) was used particularly, because it is the only (semi)quantitative list, which subdivides the Tanytarsini into species. The value is calculated at 4, 34, indicating less than extreme eutrophic conditions. Probably the index value given to Tanytarsus excavatus in the publications of Reiff et al. (1992) is too high compared to Janecek (1985), and the trophic index of Gebhartsteich might therefore be a little bit "better". The use of the list of larval abundances (percentage - composition of Tanypodinae and Tanytarsini transferred from the emergence list) leads to almost the same results: 1978/9: 4, 20 and 1979/80: 4, 32.

Acknowledgements

I wish to express my gratitude to Drs. E.J. Fittkau, M. Hirvenoja, F. Reiss and L. Säwedal for supporting me in taxonomical questions. Many thanks to Dr. G. Bretschko, who supported me kindly in the field of statistics. For various support, I am grateful the following colleagues at our institute: Mag. Claudia Kriechbaum, Dipl. Ing. Christian Hajny, Dipl. Ing. Gerhard Hutter, Andreas Römer and Thomas Ofenböck.

Table 3: Chironomidae of Gebhartsteich. Abbreviations and codes: species name printed in italics and underlined: species develops in Gebhartsteich; italics, not underlined: not in the pond with certainty, but references can be found in literature; roman type: probably not from the pond, development e.g. in running water; *asterisk in front of a species name: chironomid new to Austrian fauna. G= Gebhartsteich, G1-5: sampling sites for larvae (=Lv); Lt.= light trap, bV= first decade of (for instance) May, mV = second , eV= third decade; E= emergence traps (E1-5); Pe= pupal exuvia(e); T. (or -t.) = pond (from German: Teich).

1) Anatopynia plumipes *(Fries)*
Not G; Pe from Steinfurter T., Herrent., Aut., Steinbrucht. and Neuhaslauer T.

2) Procladius cf. choreus *(Meigen)*
Lv: Neuhaslauer T.(reared.); G: Lt: mV-eV/eVII-bX; E(4): 19.-23.vi.

3) Procladius cf. signatus *(Zetterstedt)*
G: Lv: G1, G2; Lt: mV-eVI/ mIX-bX; E: 15.-29.iv./ 5.v.-2.viii./ 7.viii.-29.ix.; Pe:22.iv.-27.v./ 30.v.-19.vi./ 23.vi.-29.vii./ 2.viii.-12.ix./ 19.ix.-29.ix.

4) Procladius sp.3
G: Lt: eV-bVI/ mVI-mVII; E (1-4): 27.v.-4.vi./ 6.vi.-23.vi.

5) Procladius sp.4
G: Lt: bIX-mIX; E (1, 3-5): 24.vii.-7.viii./ 25.viii.-8.ix./ 16.-19.ix.; Pe: ?24.7-29.vii./ ?2.viii.-7.viii./ 12.viii.-20.viii.

6) Psectrotanypus varius *(Fabricius)*
G: Lt: eV-bX; E: 15.-22.iv.; Pe: 5.-8.ix., also from Langer T., Setzt. near Jägert., Amtwiest and Stölzlingert.

7) *Procladius (Psilotanypus) lugens *Kieffer*
*G: Lv (P. nec rufovittatus): only G5; Lt: bIV-mVI ; E (1, 3) VI-9iv/ 15.-22.iv./9-14v.; Pe (*P. nec rufovittatus*): 15.-22.iv./ 9.-19.v.*

8)*Procladius (Psilotanypus) rufovittatus *(van der Wulp)*
G: Lv: G1-5; Lt: mV-mIX; E: 9.v.-30.vi./ 3.vii.-8.vii./ 16.vii.-22.viii./ 25.viii.-5.ix./ 8.ix.-12.ix./ 29.ix.-10.x.; Pe: 29.iv.-21.vii./ 29.vii.-12.ix./ 19.-29.ix.

9) Procladius (Psilotanypus) sp.3
G: E (E5): 29.iv.-5.v.

10) Ablabesmyia monilis *(Linnaeus)*
G: Lv: G2, G4; Lt: mV-mIX; E (1, 3-5): 5.v.-4.vi./ 10.-13.vi./ 19.vi.-16.vii./ 21.vii.- 12.viii./ 25.viii.-29.ix.; Pe:5.-27.v./ 30.v.-4.vi./ 23.vi.-3.vii./ 8.-24.vii./2.-7.viii./ 12.viii.-19.ix.

11) Ablabesmyia phatta *(Egger)*
G: Lv: G1, G3; Lt: eV-eVI/ bIX-mIX; E (1-4): 19.v.-19.vi./ 3.-8.vii./ 24.-29.vii.; Pe: 9.- 27.v./ 30.v.-6.vi./ 24.-29.vii./ 12.-20.viii./ 25.viii.-2.ix.

12) Conchapelopia melanops *(Wiedemann)*
G: Lt: mV-bX

13) Natarsia punctata *(Fabricius)*
G: Lt: mV-mVI; 1 male pupa: 9.v.81

14) Xenopelopia nigricans *Fittkau 1962*
G: only hand-net: 19., 22.v. 1980; Pe: from Zieringser T.

15) Tanypus kraatzi *(Kieffer)*
G: Lt: eV-mIX; E:(5): 12.vii.-16.vii.; Pe: 8.vii.-12.vii.

16) Tanypus vilipennis *(Kieffer)*
G: only G5 (reared also from Neuhaslauer T.); Lt: eV-mVII; E(3,5): 14.-19.v./ 30.v.- 6.vi./ 13.-16.vi./ 12.-16.vii.; Pe: 9.-30.v./ 10.-23.vi./ 26.vi.-3.vii./2.-7.viii.

17) Prodiamesa olivacea *(Meigen)*
G: Lt: mV-eV

18) *Diamesa hamaticornis Kieffer*
G: Lt: bIV-mV

19) Bryophaenocladius subvernalis *(Edwards)*
G: Lt: mV-bVI

20) *Bryophaenocladius sp.2*
G: Lt: mVII-eVII/ bIX-mIX

21) *Camptocladius stercorarius* (DeGeer)
G: Lt: mV-bVI/ evI-mVII

22) Chaetocladius perennis *(Meigen)*
G: Lt: mV-eV

23) *Corynoneura carriana *Edwards*
G: Lt: mIX-bX; E (1,4): 29.ix.-3.x. (and 16.ix.-4.x. 1979)

24) *Corynoneura fittkaui *Schlee*
G: only hand-net: 22., 27., 30.v.1980

25) *Corynoneura minuscula *Brundin*
G: only hand-net: 27., 30.v.1980/ 24.vii.1980

26) Limnophyes *sp. 1 (cf.* minimus *(Meigen))*
G: Lv: reared; Lt: mV-bIX/ after bIX?; E (3): 7.-12.viii.

27) *Limnophyes pumilio *(Holmgren)*
G: Lt: eV-bVI/ bIX-mXI

28) Limnophyes pentaplastus *(Kieffer)*
G: Lt: biV-eVI/ mVIII-bIX/ mIX-mXI

29) *Limnophyes asquamatus *(Andersen)*
G: only hand-net: 19.,22., 30.v., 4.vi.1980

30) *Limnophyes sp.5*
G: Lt: eV-bVI

31) *Limnophyes sp.6*
G: Lf: ev-bVI

32) Metriocnemus *cf.* atratulus *(Zetterstedt)*
G: Lt: mV-bVI

33) Metriocnemus *cf.* gracei *Edwards*
G: Lt:: mV-eV

34) Metriocnemus hygropetricus *(Kieffer)*
G: Lt: mV-eV

35) Metriocnemus *cf.* picipes *(Meigen)*
G: Lt: bIV-mV/ evI-mVII/aX-mXI

36) Parakiefferiella sp. *(p.)*:
G: only hand-net: 30.v.1980; Pe: Großradischent.

37) Paralimnophyes hydrophilus *(Goetghebuer)*
G: only hand-net: 22.v.1980

38) *Parametriocnemus stylatus (Kieffer)*
G: only hand-net: 24.vii.1980

39) Paraphaenocladius impensus *Walker*
G: *Lt: mV-bVI*
40) *Paratrissocladius excerptus (Walker)*
G: *Lt: eV-eVI*
41) Pseudorthocladius curtistylus *Goetghebuer*
G: *only hand-net: 26.vi., 25.viii., 12.ix. 1980*
42) *Pseudorthocladius filiformis *(Kieffer)*
G: *Lt: mVII-eVII*
43) Pseudosmittia trilobata *(Edwards)*
G: *only hand-net: 27.v.,16.vi.1980*
44) *Smittia cf. aterrima (Meigen)*
G: *Lt: bIV-bVI/ mVI-eVI/ eVII-mVIII/ mIX-mXI*
45) *Smittia superata Goetghebuer*
G: *Lt: mV-bVI*
46) *Smittia sp.3*
Lt: mV-bVI
47) *Thienemannia gracilis Kieffer*
G: *Lt: bVI-mVI*
48) Acricotopus lucens *(Zetterstedt)*
G: *hand-net: 27., 30.v., 4., 16.vi., 24.vii. 1980; Pe: 9.-14.v.; Pe from little pond near Reinberg-Dobersberg*
49) Brillia modesta *(Meigen)*
G: *Lt: mV-eV; Pe: Jägert.*
50) *Cricotopus cylindraceus *(Kieffer)*
G: *Lt: mV-bVI; E (4): 22.-27.v.; Pe: 9.ix.1982*
51) Cricotopus festivellus *(Kieffer)*
G: *Lv: reared (near G4); Lt: mV-bVI; E: (1,3,4,5, probably on "aufwuchs"): 9.-14.v./ 19.-30.v./ 6.-30.vi./ 3.vii.-7.viii./ 25.viii.-2.ix.; Pe: 29.iv.-22.v./ 6.-19.vi./ 26.-30.vi./ 21.vii.-2.viii./ 2.-12.ix.*
52) *Cricotopus flavocinctus *(Kieffer 1924)*
only hand-net: 19.v.1980
53) Cricotopus intersectus *(Staeger)*
G: *Lv: often reared ("aufwuchs", wood, roots, Spongilla); Lt: mV-mXI; E: 13. - 23.vi./ 26.vi.-2o.viii./ 25.viii.-12.ix./ 16.ix.-3.x.; Pe: 14.-22.v./ 6.-10.vi./ 13.-30.vi./ 3.-12.vii./ 16.- 21.vii./ 24.-29.vii./ 2.-7.viii./ 22.viii.-8.ix./ 16.-29.ix.*
54) *Cricotopus obnixus *(Walker)*
G: *Lt: 3V-bV; Pe:29.iv.-5.v.*
55) Cricotopus reversus *Hirvenoja*
G: *Lt: mV-bVI/ eVII-bIX; E (4,5): 21.vii.-2.viii./ 7.-12.viii./ 2.ix.-8.ix.; Pe: 16.-21.vii.*
56) Cricotopus sylvestris *(Fabricius)*
G: *mV-bX; E (1,4,5): 12.vii.-2.viii./ 25.viii.-2.ix./ 19.-29.ix./ 2.-10.x./ 17.-24.x.; Pe: 29.iv.- 5.v./ 22.-27.v./ 19.-23.vi./ 29.vii.-2.viii./ 2.ix.-10.x./ 17.-?24.x.*
57) *Cricotopus tibialis (Meigen)*
Pe: Schönt./ Ullrichs
58) Cricotopus *sp.*
G: *Pe: 8.vi., 16.vii. and 9.ix.1982*
59) *Diplocladius cultriger *Kieffer*
G: *Lt: bIV-mV/ Pe: 7.iv.1982; Pe from a little pond near Reinberg- Dobersberg*
60) *Eukiefferiella brevicalcar (Kieffer)*
not G; Pe: Brandt., little pond near Reinberg-Dobersberg
61) *Tvetenia calvescens (Edwards)*
not from G, Pe: little pond near Reinberg-Dobersberg
62) Eukiefferiella claripennis *(Lundbeck 1898)*
G: *Lt: bIV-bVI/ mVII-eVII/ mVIII-mIX; Pe: GerhartserT.*
63) *Eukiefferiella minor (Edwards) or E. fittkaui Lehmann*
Pe: only from little pond near Reinberg-Dobersberg
64) Tvetenia verralli *(Edwards)*
G: *Lt: mV-eV*
65) *Heterotrissocladius marcidus (Walker)*
Pe: Sagteich near Schandachen
66) **Hydrobaenus pilipes (Malloch 1915)*
Pe only from Großer T. and Setzteich near Jägert.
67) Nanocladius bicolor *(Zetterstedt)*
G: *Lt: mV-mVIi eVII-mVIII/ mIX-bX; E (4): 16.vii.-7.viii.; Pe: 19.-23.vi./ 24.vii.-2.viii./ 2.- 5.ix./ 8.-12.ix./ 16.-19.ix.*
68) *Orthocladius (s.str.) sp.1*
G: *Lt: mV-bVI*
69) Orthocladius *(s.str.) sp. 2*
G: *Lt: mV-bVI/ E (4): 14.-19.v./ 27.-30.v./ 16.-19.vi.; Pe: 9.-14.v./ 19.-30.v./ 23.- 30.vi./ 29.vii.-2.viii.*
70) *Orthocladius (s.str.) sp.3*
G: *Lt: eV-bVI*
71) *Orthocladius (s.str.) cf. rubicundus Meigen*
G: *Lt: eV-bVI*
72) *Orthocladius (Euorthocladius) sp.1*
G: *Lt: eV-bVI*
73) Paratrichocladius rufiventris *(Meigen)*
G: *bIV-bIX/ mIX-mX*
74) Orthocladius (Pogonocladius) consobrinus *(Holmgren 1869)*
G: *Lv: G2-5; E (1,3,4): 27.-30.v./ 12.-16.ix./ 3.-10.x.; Pe: 9.-15.iv./ 22.-30.v./ 23.- 26.vi./ 7.-12.viii./ ?.-8.ix./ 12.-29.ix./ 17.-24.x ; Pe also in Ottert., Mittert. and Fuchst.*
75) Psectrocladius cf. limbatellus *(Holmgren); (P.? edwardsi Brundin*
G: *Lv: G1-4 (not from sapropel); Lt: eV-bVIi E (1,3,4): 13.-23.vi.; Pe: 29.iv.-5.v./ 9.v.- 14.v./ 6.-13.vi.; Pe also in Stölzlingert.*
76) *Psectrocladius schlienzi Wülker*
not from G; Pe: Stölzlingert.
77) Psectrocladius sordidellus *(Zetterstedt)*
G: *Pe: 9.-15.iv./ 22.-29.iv.; Pe also from Hanischt., Mittert. and Schönt.*
78) *Rheocricotopus chalybeatus (Edwards)*
G: *Lt: mV-mXI*
79) *Rheocricotopus ?glabricollis (Meigen)*
G: *only hand-net: 17.x.1980*
80) Chironomus ?cingulatus *(Meigen)*
G: *Lv: reared from mud (G3); Lt: eV-eVII/ mVIII-bIX/ mIX-bX; E: 16.vi.-8.vii./ 12.- 16.vii./ 21.vii.-2.viii./ 12.-20.viii./ 22.viii.-5.ix. (mainly in E5)*

81) *Chironomus ?obtusidens Goetghebuer
G: Lt: bIV-mXI/ E (2,5): 23.-26.vi./ 30.vi.-3.vii.

82) Chironomus plumosus (Linnaeus 1758)
G: Lv: G1-5; also from Neuhaslauer T.; Lt: bIV-mIX/ bX-mX; E: 15.-22.iv./ 29.iv.-6.vi./ 19.-26.vi./ 30.vi.-3.vii./ 12.-16.vii./ 24.-29.vii./ 2.viii.-10.x./ 17.-24.x.; Pe: 1.-9.iv./ 15.iv.- 6.vi./ 10.-13.vi./ 19.-23.vi./ 26.vi.-3.vii./ 8.-16.vii./ 21.-29.vii./ 2.viii.-24.x.; also Pe from: Bruneit., Jägert., Setzt. near Jägert., Stadtt. Waidhofen/Thaya, Spielberger T., Ottert. and Zieringser T.

85) Chironomus (Camptochironomus) tentans (Fabricius 1805)
G: Lv: reeds near G1 and G4 (reared from G1); Lt: eV-mVIII/ bIX-bX; Pe: 19.-27.v./ 7.-12.viii.

86) Cladopelma virescens (Meigen)
G: Lt: eVI-mVII

87) Cladopelma viridula (Fabricius)
G: Lv: G1-G5; Lt: mV-bVI/ mVII-eVII; E (3,4,5): 30.vi.-3.vii./ 12.-16.vii./ 24.vii.-7.viii./ 22.-25.viii./ 2.-8.ix.; Pe: 14.-22.v./ 30.v.-6.vi./ 3.-8.vii./ 21.-24.vii./ 29.vii.-2.viii./ 7.-12.viii.

88) Cryptochironomus cf. albofasciatus (Staeger)
G: eV-bVI/ eVII-mVIII/ bIX-bX

89) Cryptochironomus cf. obreptans (Walker)
Lv not from G, reared from mud of Neuhaslauer T.: 11.i.1980; Lt: eV-bVI/ eVI-mVII/ mVIII-bIX; E: 19.-27.v./ 19.-23.vi./ 3.-8.vii./ 16.-21.vii./ 24.vii.-2.viii./ 2.-5.ix.; Pe: 22.-30.v./ 10.-13.vi./ 16.-23.vi./ 26.vi.-3.vii./ 8.-16.vii./ 21.vii.-2.viii./ 7.-12.viii./ 22.viii.-19.ix.

90) Cryptochironomus psittacinus (Meigen)
Lv: not reared from G, but from sapropel in Neuhaslauer T., near Haslau: 16.v.1979; Lt: eV-bVI, E (2,3,4): 27.-30.v./ 30.vi.-3.vii./ 25.viii.-12.ix.; Pe: 19.-22.v./ 7.-20.viii./ 25.viii.- 2.ix./ 8.-19.ix.

91) Cryptochironomus redekei (Kruseman)
G: Lv: near G4 (reared); Lt: eV-bVI; E: 14.v.-6.vi./ 23.-26.vi. / 3.vii.-2.viii./ 7-12.viii./ 25.viii.- 2.ix. (mainly in E1,4); Pe: 19.-27.v./ 30.v.-6.vi./ 19.-23.vi./ 30.vi.-3.vii./ 12.vii.-12.viii./ 20.viii.-2.ix.

92) Demicryptochironomus vulneratus (Zetterstedt)
G: Lv: only E1,E4; Lt: mV-bVI; E: 3.-8.vii./ 12.-16.vii.; Pe: 9.-22.v./ 30.vi.-3.vii./ 8.-12.vii./ 29.vii.-12.viii./ 2.-5.ix./ 8.-12.ix./ 16.-19.ix.

93) Dicrotendipes lobiger (Kieffer)
G: Lv: G1-5, "aufwuchs", surf-zone, Bryozoa (reared from (e.g.) Plumatella fungosa); Lt:eV-bVI; E (1-5, E5 only 12.-20.viii.): 22.v.-13.vi./ 12.vii.-2.viii./ 12.-20.viii./ 22.viii.-5.ix./ 8.- 16.ix./ 19.-29.ix.; Pe: 19.-22.v./ 30.v.-6.vi./ 12.-16.vii./ 29.vii.-2.viii./ 7.-20.viii./ 22.-25.viii./ 2.- 5.ix.

94) Dicrotendipes nervosus (Staeger)
G: Lv: on flooded wood near G4 (reared); Lt: eV-bVI; Pe: 26.-30.vi.

95) Dicrotendipes modestus (Say)
G: Lv: G1-5; Lt: mV-eVI/ eVII-mVIII; E: 9.-19.v./ 22.-27.v./ 19.vi.-7.viii./ 22.viii.-2.ix. (in G5 only 24.vii.-7.viii.; Pe: 29.iv.-27.v./ 10.-16.vi./ 19.vi.-12.vii./ 21.vii.-25.viii.

96) Einfeldia dissidens (Walker)
G: Lv: G1-G5; Lt: eV-mVI/ mVII-eVII/ mVIII-bIX; E:30.v.-30.vi. (in G5 only 19.- 23.vi.); Pe: 27.-30.v./ 4.-19.vi./ 7.-12.viii./ 22.-25.viii./ 2.-5.ix.

97) Endochironomus albipennis (Meigen)
G: Lv: G1-G5, mainly mud with Potamogeton pusillus, reared also from "aufwuchs"; also from Neuhaslauer T; Lt:.bIV-bX; E: 5.v.-20.viii./ 22.viii.-12.ix.; Pe: 5.v.-16.vii./ 24.-29.vii./ 7.viii.-5.ix./ 16.-19.ix.

98) Endochironomus impar (Walker)
G: only hand-net: 30.v.,4.vi.1980

99) Glyptotendipes gripekoveni (Kieffer)
G: Lv: mainly a mining species, reared from rotten wood and grass; Lt: only 1978: 7.viii.- 15.viii. (leg. Malicky); E (1,3,5 - "aufwuchs"): 7.-20.viii./ 25.viii.-2.ix.; Pe: 4.-6.vi./ 19.vi.-3.vii.

100) Glyptotendipes pallens (Meigen)
G: Lv: G1-5, but mainly on algal covers, on and in wood, in the surf-zone, in Bryozoa (reared (e.g.) from Plumatella fungosa); Lt: eV-bX; E: 22.-30.v./ 10.-23.vi./ 26.-30.vi./ 8.-21.vii./ 29.vii.-16.ix.; Pe: 19.-30.v./ 10.-13.vi./ 8.-12.vii./ 16.-21.vii./ 24.-29.vii./ 7.-25.viii./ 5.-19.ix./ 29.ix.-3.x.

101) Glyptotendipes paripes (Edwards)
G: Lv: G1-5; algal covers on rocks, surf zone; Lt: mV-bX; E: 22.-29.iv./ 9.v.-6.vi./ 10.-13.vi./ 16.-19.vi./ 3.-12.vii./ 21.vii.-3.x.; Pe: 29.iv.-5.v./ 9.v.-3.x./ 24.-30.x.

102) Glyptotendipes signatus (Kieffer)
G: Lt: mVII-eVII/ bIX-mIX/ E (2-5, mainly on Bryozoa): 21.-29.vii./ 20.viii.-12.ix.; Pe: 26.-30.vi./ 7.-12.viii./ 22.viii.-2.ix./ 8.-12.ix.

103) Microchironomus tener Kieffer
G: only 1 lv: 21.v.1979

104) Microtendipes chloris (Meigen)
G: Lv: G1-5; Lt: bIV-mV/ mV-bVI; E (1,2): 1.-9.iv./ 15.-22.iv.; Pe: 1.iv.-9.v. detected in 30 ponds: Edlerseet., Gerhartser T., Großer T., Sagt., Hanischt.,Steinfurter T, Langfurter T., Großradischent., Steinbrucht., Jägert.,Aut., Feldt.,Mittert.,Ullrichst., Pürbacher T., Spielberger T., Althöllt., Amtwiest., Edlauer T., Frauent., Ottert., Schönt., Stölzlinger T., Fuchst., Kobelt., Reitt., Rudmannser T., Zieringser T., Plattent. (underlining: most frequent chironomid-species of the drift)

105) Microtendipes pedellus (de Geer)
G: Lt: mV-eVI/ evII-bIX/ mIX-bX; E (1,3,4,5): 16.-26.vi./ 3.-8.vii./ 24.vii.-2.viii./ 22.viii.-25.viii.; Pe: 13.-19.vi./ 23.vi.-8.vii./ 16.-29.vii./ 2.-12.viii./ 22.viii.-8.ix.

106) Pagastiella orophila (*Edwards*):
G: *only 1 lv: G4 (23.i.1980); Lt: eV-bVI (1 adult male); Pe: 14.-19v.1980 (2 male pe)*

107) Parachironomus arcuatus Goetghebuer
G: *Lv: reared from algae and Plumatella fungosa, also on the ground (probably even near G5); Lt: mV-eV/ mVI-eVI/ bIX-mXI; E: 22.-27.v./ 10.-13.vi./ 12.-16.vii./ 21.-24.vii./ 25.viii.-10.x.; Pe: 19.-22.v./ 7.-12.viii./22.-25.viii./ 2.-5.ix./ 16.-19.ix.*

108) *Parachironomus biannulatus (*Staeger*)
G: *E (3,4,5, "aufwuchs": 12.-20.viii./ 8.-12.ix.?/ 19.-29.ix.*

109) Parachironomus digitalis (*Edwards*)
G: *Lt: evII-mVIII, ?mVIII-bIX*

110) Parachironomus frequens (*Johannsen*)
G: *Lt: eVI-bX*

111) Parachironomus tenuicaudatus (*Malloch*)
G: *Lv: algal "aufwuchs" and Plumatella fungosa (reared); Lt: mV-bVI/ mVI-eVI; E: 14.v.-16.vi./ 19.vi.-29.ix.; Pe: 14.v.-16.vi./ 19.vi.-16.vii./ 24.vii.-20.viii./ 25.viii.-5.ix./ 8.-12.ix.*

112) Parachironomus vitiosus Goetghebuer
G: *Lv: G1,4; Lt: eV-bVI; E: 22.-30.v./ 6.-30.vi./ 8.-12.vii./ 12.-20.viii./ 22.viii.-29.ix.; Pe: 14.-19.v./ 6.-13.vi./ 16.-?23.vi./ 30.vi.-3.vii./ 7.-12.viii./ 8.-19.ix.*

113) Parachironomus sp. 1 (*cf. ?*Parachironomus *Pe. "a" sensu Langton 1991*)
G: *E: 8.vii.-20.viii.; Pe: 30.vi.-3.vii./ 22.-25.viii./ 2.-5.ix./ 16.-19.ix.*

114) Paratendipes albimanus (*Meigen*) (*incl.*P. plebejus (*Meigen*))
G: *Lt: eV-eVI/ bIX-mIX*

115) Polypedilum (Pentapedilum) sordens (*van der Wulp*)
G: *Lv: "aufwuchs", down to a depth of 2m; Lt: eV-mXI; E: 30.v.-23.vi./ 3.vii.-8.ix./ 12.-29.ix./ 2.-10.x.; Pe: 30.v.-16.vi./ 19.-30.vi./ 12.-29.vii./ 2.-20.viii./ 22.viii.-19.ix.*

116) Polypedilum (Pentapedilum) tritum (*Walker*)
G: *only hand-net: 27.,30.v., 4.vi.1980*

117) Polypedilum nubeculosum (*Meigen*)
G: *Lv: G1-5 (most frequent P. - species; reared (e.g.) from G3); Lt: bIV-bX; E: 22.iv.-27.v./ 13.vi.-7.viii./ 25.viii.-5.ix.; Pe: 22.iv.-27.v./ 10.vi.-20.viii./ 22.viii.-5.ix./ 8.-12.ix./ 16.-19.ix.*

118) Polypedilum (Tripodura) bicrenatum Kieffer
Lv: G1-5, in the mud of G5 only on 8-11-79 (pond empty); reared from Neuhaslauer T. (6-2-80); Lt: eV-bVI/ mVII-eVII; E (1): 6.-16.vi.; Pe: 4.-6.vi./ 10.-23.vi./ 16.vii.1981

119) Stenochironomus sp.
G: *Pe: 2.-7.viii.*

120) Tribelos intextus (*Walker*)
G: *Lv: only G4, probably in the reed zone; Lt: mV-bVI/ Pe: 9.-22.v.*

121) Zareliella marmorata (*van der Wulp*)
G: *Lv: G4, 5 ("aufwuchs"?); hand-net (26.vi., 3.vii.1980); E (1): 23.-26.vi./ 3.-8.vii./ 16.-21.vii.; Pe: 10.-13.vi.*

122) Pseudochironomus prasinatus (*Staeger*)
G: *Lv: G1-5, also reared from roots and rotten wood; mainly in winter; Lt: eV-eVI/ eVII-mVIII; E (3,4): 22.-30.v./ 4.-6.vi./ 23.-26.vi./ 16.-21.vii.; Pe: 14.-30.v./10.-23.vi./ 2.-20.viii.*

123) Cladotanytarsus atridorsum Kieffer
G: *Lt: mV-bVI/ mVI-eVI; E (1-4): 14.-19.v./ 22.v.-4.vi./ 16.-19.vi./ 23.-26.vi./ 30.vi.-16.vii./ 21.-29.vii.; Pe: 5.-22.v./ 6.vi.-24.vii./ 29.vii.-7.viii./ 25.viii.-2.ix./ 8.-12.ix.*

124) Cladotanytarsus cf. nigrovittatus (*Goetghebuer*)
G: *Lv: reared from G1; Lt: eV-mVI/ eVII-mVIII; E (1-4, 2 only 16.-24.vii.); Pe: 14.v.- 26.vi./ 3.-8.vii./ 12.-16.vii./ 21.vii.-12.ix.*

125) *Micropsectra atrofasciata agg.*
G: *Lt: mV-bVI/ mVI-eVII; Pe: Gerhartser T., Jägert. (Pu)*

126) *Micropsectra junci (Meigen)*
G: *Lt: bIX-mIX*

127) **Micropsectra recurvata (Goetghebuer)*
G: *hand-net: 29.iv.1980: 1 adult male; Lt: mIX-bX; adult males from little pond near Reinberg-Dobersberg*

128) Micropsectra *Pe sp. 1*
G: *Pe: 9.-15.iv./ 22.iv.-9.v./ 7-4-1982*

129) Micropsectra *Pe sp. 2*
G: *Pe: 22.-29.iv.*

130) Micropsectra *Pe sp. 3*
G: *Pe: 22.-29.iv.*

131) Paratanytarsus austriacus (*Kieffer*)
G: *Lt: mV-bVI*

132) Paratanytarsus dissimilis *Johannsen*
G: *Lt: bIV-bIX/ mIX-bX*

133) Paratanytarsus laccophilus (*Edwards*)
G: *E (1,2,4,5): 3.-8.vii./ 25.viii.-8.ix./ 12.-16.ix.; Pe: 19.-22.v./ 23.-30.vi./ 21.-29.vii./ 22.viii.-8.ix.*

134) Paratanytarsus lauterborni (*Kieffer*)
G: *Lt: mV-eV*

135) *Paratanytarsus tenellulus (*Goetghebuer*)
G: *only hand-net: 4-6-1980; lv probably live in the "kinal" (Fittkau - in Säwedal 1982)*

136) *Rheotanytarsus photophilus* Goetghebuer
G: *Lt: mV-eVII/ mVIII-bIX*

137) *Rheotanytarsus ringei *Lehmann*
G: *Lt: mV-mXI; E (3): 22.-27.v.*

138) *Tanytarsus gracilentus *Holmgren*
G: *only hand-net: 4.vi.1980*

139) Tanytarsus miriforceps (*Kieffer*)
G: *Lv: G1-5, reared (G1,3): 21.iii.1979, 18.iii.80; Lt: bIV-mIV (82,1% of Chironomidae), mV-eV (only 3 adults); E: 1.iv.-5.v.; Pe. 1.iv.-5.v.; detected in 29 ponds: Edlerseet., Gerhartser T., Müllert., Großer T., Sagt., Hanischt., Steinfurter T., Bruneit., Großradischent., Steinbrucht., Neuhaslauer T., Aut., Mittert., Ullrichst., Pürbacher T., Spielberger T., Althöllt., Amtwiest., Edlauer T., Karfreitagsteich, Frauent., Ottert., Stölzlingert., Fuchst., Kobelt., Reitt., Rudmannser T., Zieringser T. (underlining: most frequent chironomid-species of the drift)*

140) *Tanytarsus niger *Andersen*
G: *Lv: G (1,2,3,5); reared (e.g.) 18.iii.1980; Lf: bIV-mV (10,2%), mV-eV (1 adult male), eV-bVI (2 adult males)/ mIX-bX (1 adult male); E (1,2,3): 1.-22.iv.; Pe: 1.-29.iv.; detected in 13 ponds: Großradischent., Mittert., Ullrichst., Pürbacher T., Spielberger T., Althöllt., Amtwiest., EdlauerT., Frauent., Ottert., Stölzlingert., Kobelt.*

141) Tanytarsus sylvaticus *(van der Wulp)*
G: *Pe: 7-4-1982; detected in 28 other ponds: Edlerseet., Gerhartser T., Großer T., Sagt., Hanischt., Bruneit., Großradischent., Neuhaslauer T., Setzt. near Jägert. and Stadtt.Waidhofen/ Thaya, Aut., Mittert., Ullrichst., Pürbacher T., Spielberger T., Althöllt., Amtwiest., Edlauer T., Karfreitagst., Frauent., Ottert., Schönt., Stölzlingert., Fuchst., Kobelt., Rudmannser T., Zieringser T.*

142) Tanytarsus debilis *(Meigen 1830)*
G: *Lv: G 2,3,4, reared (e.g.)5-5-1980, probably the most abundant species of G; Lt: mV-eV (77%), eV-bVI (62%), bVI-mVIII; E: E1-4: 29.iv.-7.v./ 30.v.-6.vi./ 10.vi.-2.viii./ 25.viii.-16.ix.; E5: 14.-19.v./ 22.viii.-16.ix.; Pe: 29.iv.-10.vi./ 13.vi.-12.ix.*

143) Tanytarsus verralli *Goetghebuer*
G: *Lv: G2,3,4; also reared from "aufwuchs" (E4); Lt: eV-bX; E: E1-4: 19.-30.v./ 23.-26.vi./ 30.vi.-7.viii./ 25.viii.-5.ix.; E5: 22.-27.v./ 2.-7.viii./ 25.viii.-5.ix., Pe: 9.-?14.-30.v./ 16.-?19.vi./ 23.vi.-19.ix.*

144) Tanytarsus holochlorus *Edwards*
G: *Lv: reared G1 (17-7-1980), G4 (12-6-1981), G2; Lt: eV-mVII/ eVII-bIX; E: 30.v.-16.vi./ 19.vi.-5.ix./ 8.-12.ix./ 16.-19.ix./ 29.ix.-3.x.; Pe. 30.v.-8.vii./ 12.-16.vii./ 21.vii.-12.ix./ 16.ix.-10.x./ 17.-24.x.*

145) Tanytarsus occultus *Brundin*
G: *lv. G2,3,4; reared (G1: 22.vii.1980); Lt: bV-eVI/ eVII-mVIII/ bIX-bX; E: E4: 3.-8.vii./ 24.vii.-12.viii./ E1: 19.-26.vi./ 30.vi.-3.vii./ 2.-12.viii./ 20.-25.viii./ E3: 26.-30.vi./ 3.-8.vii./ 7.-20.viii./ 22.-25.viii./ 2.-5.ix./ E2: 30.vi.-8.vii./ E5: 16.-19.vi./ 22.viii.-8.ix.7 12.-16.ix.; Pe: 4.-30.vi./ 3.-8.vii./ 12.-?16.vii./ 21.vii.-12.ix./ 16.-29.ix./ 3.-?10.x.*

146) Tanytarsus bathophilus *(Kieffer)*
G: *E (2): 23.-26.vi. (1 male); Pe: 10.-13.vi.*

147) Tanytarsus excavatus *Edwards*
G: *Lv: G1-4; also reared; Lt: mV-bIX/ mIX-bX; E: E1-4: 9.-19.v./ 22.-30.v./ 4.-6.vi./ 19.vi.-8.ix.; E5: 12.-16.vii./ 22.viii.-5.ix.; Pe: 29.iv.-27.v./ 10.vi.-5.ix.*

148) Tanytarsus inaequalis *Goetghebuer*
G: *Lt: mV-eVI; E (2,3): 23.-26.vi./ 3.-8.vii./ 29.viii.-5.ix.; Pe: 10.-13.vi.; 8.vi.1981*

149) (+150?) Tanytarsus lestagei agg.
G: *Lv: reared (G4), Pu: G1-G4; Lt: mV-mVI; E: E1-4: 9.-19.v./ 19.vi.-16.vii./ 21.vii.-2.viii./ 22.-25.viii./ 2.-16.ix./ 19.-29.ix.; E5: 29.vii.-2.viii./ 2.-5.ix./ 16.-29.ix./ 2.-10.x.; Pe: 29.iv.-27.v./ 6.vi.-29.ix./ 3.-10.x.*

151) *Thienemanniola ploenensis *Kieffer*
Pe in 5 ponds: Großradischent., Ullrichst., Pürbacher T., Edlauer T., Fuchst.

152) Zavrelia pentatoma *Kieffer*
G: *only hand-net: 30.v.1980 (3 adult males), 4.vi.1980 (1 adult male).*

POPULATION DYNAMICS OF CHIRONOMIDAE ASSOCIATED WITH *NYMPHAEA ALBA* LEAVES IN SPRINGS IN THE EBRO DELTA (NE SPAIN)

Narcís Prat[1] and Alberto Rodrigues-Capitulo[2]

[1] Dept. Ecologia, Univ. Barcelona, Diagonal 645, 08028 Barcelona, Spain.

[2] Instituto de Limnología, cc 712, 1900 La Plata, Argentina.

INTRODUCTION

During 1992 we studied the ecology of springs in the Ebro Delta in N.E. Spain (Rodrigues-Capitulo *et al.*, 1994). In one of the springs nearly 50% of the surface was occupied by emergent leaves of the white water lily *Nymphaea alba.* This species is known for its abundance in ponds and resistance to adverse environmental conditions (e.g. Smits *et al.*, 1988; Arts *et al.*, 1990; 1990; Kok *et al.*, 1990; Papastergiadou and Babalonas, 1992).

Chironomids associated with different species of water lilies (especially the mining species living inside the leaves and stems) are relatively well known (e.g. Mackey, 1977b; Kondo *et al.* 1989; Kondo and Hamashima, 1993; Van der Velde and Hiddink, 1987) and this provided us with the possibility of comparison of our community with those studied in other countries.

Study Site and Methods

The study site was a spring in the Ebro Delta in N.E. Spain (Fig. 1). Springs are present in the inner border of the delta plain. To date more than 30 springs have been identified and studied but many others have disappeared in the past due to the agricultural activities. The study spring ("Ullal del Teix") is the largest found, and is nearly round, with a maximum diameter of 56.6 m and surface area of 1,682 m^2. Its mean depth is 2.45 m and the maximum depth 5.90 m, with a volume of 4,199 m^3. Water inflow is of subterranean and surface origin, the latter with a mean rate of 11.89 $l\ s^{-1}$. As the mean outflow is 51.04 $l\ s^{-1}$ the difference should be accounted for by underground water rising in the spring. The mean renewal time can be calculated, therefore, as 21.09 hours.

The area covered by the white water-lily roughly corresponds with the depths of less than 3.5 m. Samples for chironomids were taken each month by wiping the undersurface of leaves with a soft brush and collecting all the animals in a 150 µm net laid over a plate. The material collected from each leaf was preserved in ethanol and counted and identified in the laboratory. Seven leaves were brushed in each sampling period. In February 1992, we took samples from 15 leaves taken from three different points of the spring in the search for varying chironomid densities. No significant differences were found at this time and we decide to collect successive samples at random from different parts of the spring. Larval instars were deduced from the length and width of the larval head.

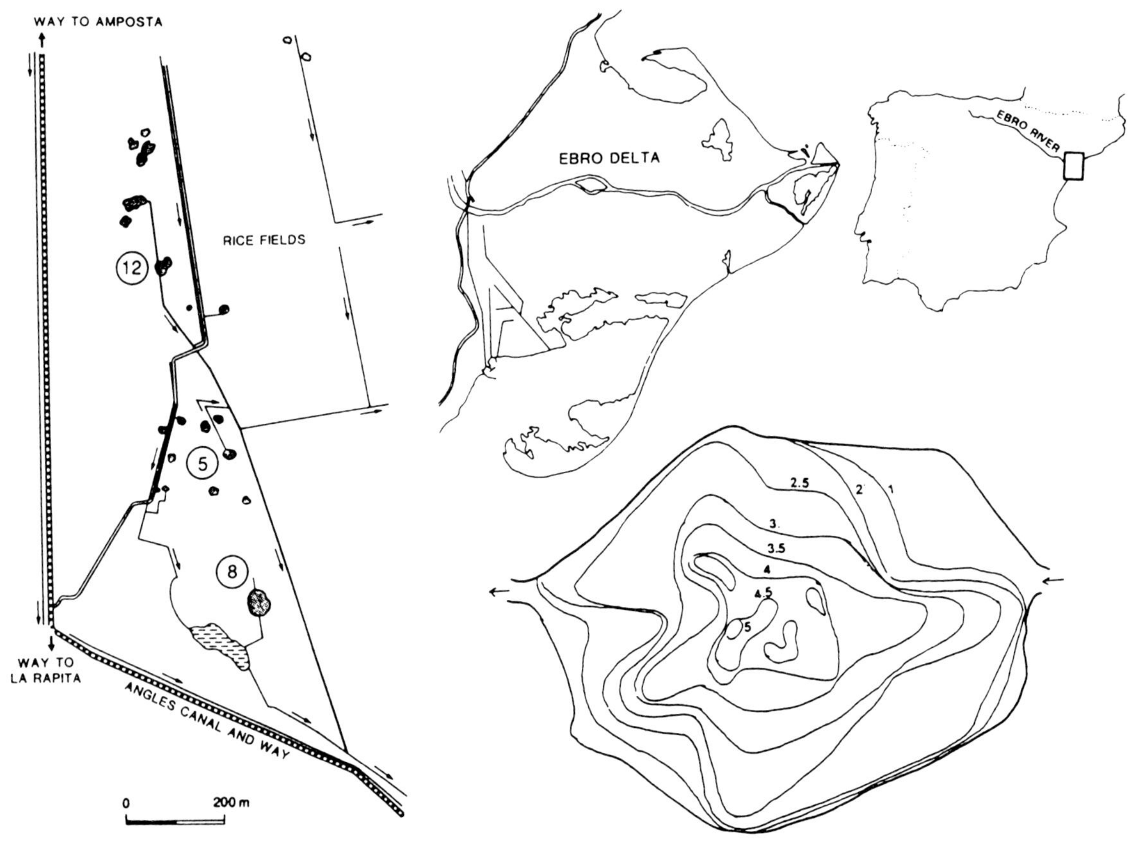

Figure 1. Map showing the location of the "Ullal del Teix" Spring (number 8) in the Ebro Delta and its batimetric map.

RESULTS

Characteristics of the spring

The springs of the Ebro Delta are fed by a mixture of fresh and salt-water; the conductivity is 1740 µS with sodium and calcium as main ions (269 and 113 ppm respectively). The water is alkaline (4 to 5 meq/l) and oligotrophic. Phosphate is not detectable, although nitrates are relatively high (up to 253 µmol/l) probably due to the agricultural fields that surround the spring. The water is relatively transparent (Secchi disc depth between 2 and 4.7 m); the suspended solids are less than 4.7 ppm, mostly organic, and the bottom sediments are highly organic (peat). Due to the ionic composition, pH is nearly constant at 7.3 to 7.5. Water temperature ranges from 15.4°C in March to 22.4°C in August. There is no vertical stratification due to the high water turnover, although small thermoclines can develop daily in summer. Oxygen saturation is between 64 to 97%. Due to the lack of –phosphoros, chlorophyll *a* was always low at the surface (less than 3.4 µg/l), although in summer an important population of *Bothryococcus* develops in the deep waters (85.86 µg/l of chlorophyll *a* at 4 m in August).

Chironomids collected on the undersurface of water lilies

Most of the material collected were larvae but mature pupae, pupal exuviae and adults were also found. In some cases the presence of larval skins attached to pupae allows identification of the larvae. Three species were most abundant, two *Corynoneura* (*C. coronata* Edwards and *C. scutellata* Winnertz) and *Cricotopus sylvestris* (Fabricius). These species accounted for 86% of the total individuals collected (Table 1). Other species were found frequently but with low abundances (larvae of *Cricotopus* (*Cricotopus*) and *Paratanytarsus*.) We found adults of *P. mediterraneus* Reiss and Sàwedal and exuviae of three species of *Cricotopus* (*C. annulator* Goetghebuer, *C. festivellus* (Kieffer) and *C. triannulatus* Macquart). Other exuviae or pupae found were those of *Psectrocladius sordidellus* (Zetterstedt), *P. schlienzi* Wülker, *Phaenopsectra flavipes* (Meigen) and *Thienemaniella vittata* (Edwards). A male of *Endochironomus tendens* (Fabricius) was also collected close to the spring and one larva of *Ablabesmyia phatta* (Egger) was also living on the leaves in April.

Table 1. Mean annual density and relative abundance of the macroinvertebrates found under the leaves of *Nymphaea alba*.

	ind/m2	%
Crustacea	5.6	1.2
Mollusca	11.9	2.6
Other insecta	8.9	1.9
Cricotopus sylvestris	260.5	56.0
Other *Cricotopus*	10.8	2.3
Corynoneura spp.	143.7	30.9
Other chironomids	23.9	5.1
TOTAL	465.2	100.0

Population dynamics

The population dynamics of the community is shown in Table 2. One of the most striking features is the sudden change in chironomid density between June and the beginning of August. The *Corynoneura* disappeared and the *Cricotopus* were reduced to few individuals. After this, the community never reached its initial densities. The decline in chironomid densities are coincident with an increase in the densities of other macroinvertebrates, especially when *Acroloxus* was present (Table 2 and Fig. 2).

Life cycles

The density changes of the larval stages of the two most abundant chironomid genera were studied (Figs. 3 and 4). In *Cricotopus sylvestris* no clear trend emerged. Pupae were found from February to June and pupal skins and egg masses throughout the year. After June the larvae are very scarce (Table 2), but different larval instars were present (Fig. 3). In addition, for each leaf, differences in instar composition can be found at each sampling date, suggesting a continuous recruitment of larvae and the emergence of adults during the year. The water temperature was never below 14°C, and at this temperature development of larvae took place on only 28 days (in agreement with laboratory experiments, e.g. Menzie 1981), therefore continuous development of larvae and emergence of adults is possible. In summer with temperatures of 22°C the larvae can develop until imago in only 10 days (Menzie, *loc. cit.*).

The *Corynoneura* species offers an additional complication because two species were found under the leaves of water lily (*C. coronata* and *C. scutellata*). At each sam-

Table 2. Mean monthly densities (ind/leaf) of the most abundant species or groups of macroinvertebrates under the surface of the leaves of water-lily (*Nymphaea alba*) in the "Ullal del Teix" well. The mean and the standard deviation is indicated. In January all the leaves were mixed and no sd can be provided. The number of leaves are seven for each month except in February, when 15 leaves were collected.

DATE (1992)	Crustacea	Mollusca	Other Insects	*Cricotopus sylvestris*	*Corynoneura*	Other Chironomidae	TOTAL
19.i.	0	0.2	1	11.6	4.75	5	21'5
21.ii.	0.2±0.56	0	0.07±0.26	50.4±34.11	11.33±10.73	3.07±2.28	65.07±40.74
4.iii.	0	0	0	51.86±34.08	10.43±8.56	1.43±1.51	63.71±37.03
20.iv.	0	0	0	18.86±13.73	26.71±20.52	0	45.57±32.43
20.v.	0.14±0.38	0	0.57±1.51	30.57±39.25	30.43±19.92	11.86±7.01	73.57±56.03
8.vi.	3.71±3.95	0.71±1.11	1.57±2.57	15.71±16.05	32.43±20.67	2.43±2.51	56.57±37.70
6.viii.	0.29±0.76	1.43±3.78	0.71±1.5	2.14±1.77	0	0	4.57±4.39
31.viii.	0	6.14±3.78	0	1.57±2.88	0	0	7.71±7.61
20.ix.	0	3.43±5.16	0.57±1.13	6.71±2.88	0.29±0.49	0	11.00±8.79
4.xi.	0.57±0.98	1±1.41	0.71±1.5	1.57±1.27	1.71±1.50	1.29±1.38	6.86±4.78
14.xii.	1.29±1.11	0	0.43±0.79	4.57±3.95	0.57±0.53	0	6.86±5.5 8

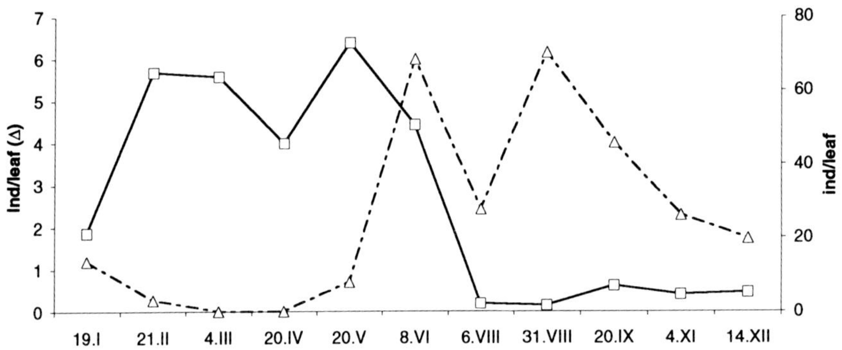

Figure 2 Densities (ind/leaf) of Chironomidae (squares) and all the other macroinvertebrates (triangles) in the different sampling dates under the leaves of *Nymphaea alba* from the "Ullal del Teix" well.

pling date, when available, we examined 20 larvae of *Corynoneura* in order to determine the proportion of each species at each sampling date (Table 3). During the first three months all the larvae were from *C. coronata*. *C. scutellata* appeared at beginning of April and at the end of this month it was the dominant species in the larval population. At this time all the pupae examined were from this species, while in February, beginning of April and June the pupae found were those of *C. coronata*. In November and December although the densities were very low, both species were found.

These data suggest different generations though the year as in *Cricotopus* and, as in the species of this genus, many differences exist between individual leaves. Data show alternative dominance of the two species in April, May and June (Table 3) and we believe that there is continuous recruitment in these species,

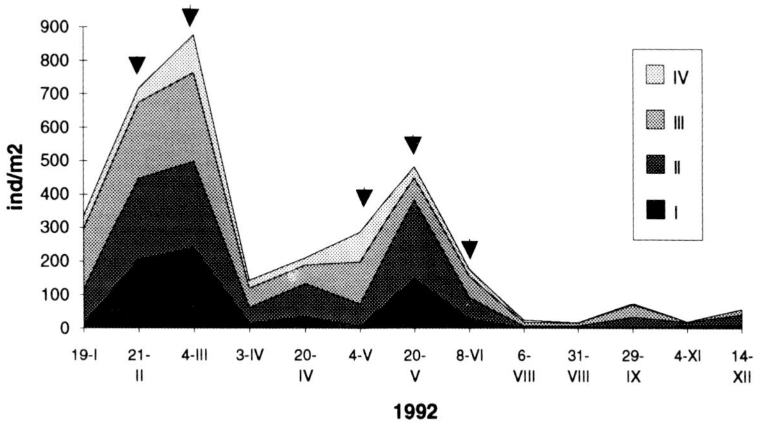

Figure 3. Population dynamics of each larval instar of *Cricotopus sylvestris* during the sampling period. Arrows mean presence of pupae.

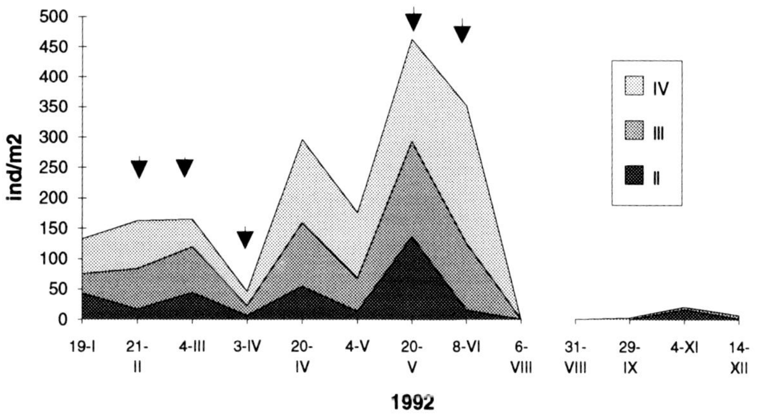

Figure 4. Population dynamics of each larval instar of *Corynoneura* spp. during the sampling period. Arrows mean presence of pupae.

as Mackey (1977b) suggested for *C. coronata* living under the leaves of yellow water-lilies from the river Thames. He did not deduce the life-cycle of the species with certainty. No data on the development of these species in the laboratory are available.

Interference of other species with the Chironomidae

If we plot the abundance of the Chironomidae with respect to other species (Fig. 2) we find that the abundance of chironomids decreases when abundance of other macroinvertebrates increases, especially when more than 20 ind/m2 (approx. 3 ind/leaf) of Mollusca, Coleoptera, Isopoda and/or Decapoda are present. These large forms seem to deplete the chironomid density by interference competition for space (Mollusca, moths), exploitation of resources (Mollusca, Crustacea), or predation (Odonata). Mollusca seem to exert the highest interference pressure on chironomids, because when its densities increased in June and August the chironomid densities fell considerably (Table 2). On the other hand there is no relationship between the mean surface area of leaves collected and the abundance of macroinvertebrates. Large leaves appear in summer and as the chironomid densities are low at this time, there is no relationship between densities and area (Fig. 5).

Table 3. Percentage of abundance of the two species of *Corynoneura* under the surfaces of leaves of water-lily. Percentages were deduced from the examination of several larvae from all collected on each date.

1992	19.i.	21.ii	4.iii.	3.iv.	20.iv.	4.v.	20.v.	8.vi.	29.ix.	4.xi.	14.xii.
C. coronata	100	100	100	87,5	10	66,6	87,5	5	100	88	33,3
C. scutellata				12,5	90	33,3	12,5	95		12	66

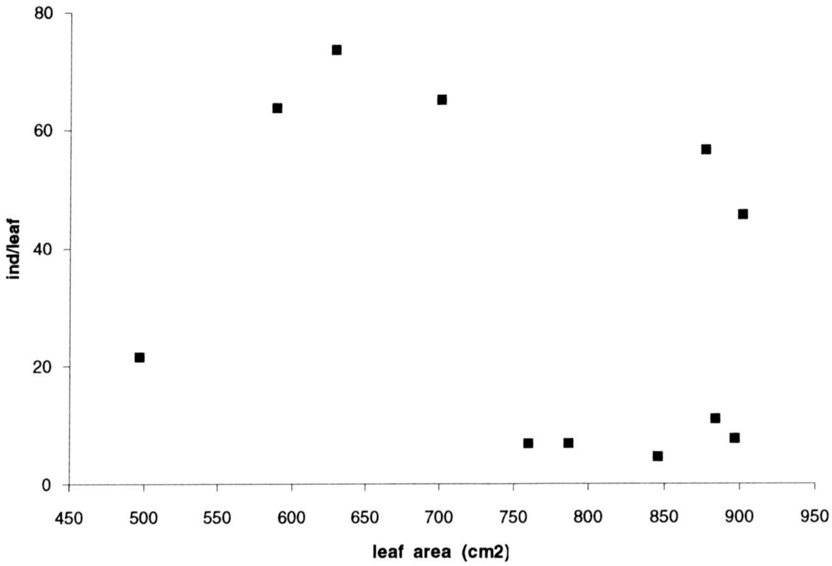

Figure 5. Relationship between the mean leaf area in each sampling period with the macroinvertebrate density.

Discussion

The community of chironomids found under the leaves of white water-lilies in one spring in the Ebro Delta are similar to those collected in water-floating plants from ponds (Kondo and Hamashima, 1993). One of the main differences is the absence of leaf-miner species like *Pentapedilum*, or *Endochironomus*, as has been described in yellow water-lilies (Van der Velde and Hiddink, 1987). Mines or galls from chironomids were not observed in the white water-lilies collected during our study. This is coincident with the pupal skin collections in our spring and other studies conducted on floating leaves, where the species use the leaves only as a substrate (Learner *et al.*, 1989) and animals do not eat the leaves (Friedrich and Meier-Drook, 1987). The dominant species seem to be the opportunists, like *Cricotopus sylvestris*. Larvae of *Corynoneura*, which are scarcely found in other studies, seem to be abundant under plants with floating leaves (Learner *et al.*, 1989; Kondo and Hamashima, 1993; Mackey, 1977b) as in our study. In rivers, associated with yellow water lilies growing in marginal areas subjected to a certain degree of flow, the chironomid community was richer in species. Although the larvae of *Cricotopus sylvestris* were abundant other species were also abundant, such as *C. bicinctus* (Mackey, 1977b). In this study, the larvae of *Corynoneura* were abundant only in summer just after a flow, but disappeared afterwards.

The life cycle of the most abundant chironomids is difficult to ascertain in our study. We believe that as the water temperature is always above 14°C, both species can develop to adult in less than a month throughout the year. We suggest that there is a continuous recruitment of larvae. Water temperature is regulated due to its subterranean origin and does not follow the atmospheric temperature. Thus, if the mean monthly temperature a few kilometers from the spring is compared with the measured water temperature, the temperature is higher than air temperature in winter (Fig. 6). The water temperature is buffered by its karstic origin. With these water temperatures we can expect fast turnover of generations (low CPI in the sense of Benke, 1992) in the spring, and the life cycle will be more dependent on other factors (like food availability, predation or competition) than temperature. For other macroinvertebrates living beneath floating leaves of the yellow water lily, short generation times (between 13 and 42 days in a beetle) have been described in Georgia (Otto and Wallace, 1989). For this reason, in chironomids (smaller than beetles) and with the temperature of the spring above 14°C even shorter generation times might be predicted.

One of the main factors influencing the density and longevity of each generation seems to be the abundance of other macroinvertebrates, especially gastropods. The density of chironomids is negatively correlated with those of other invertebrates (Fig. 2). This fact has not been studied previously in plants with floating leaves. Disruptive interactions have been noted in macroinvertebrates mostly in rivers, especially intra and interspecific competition for space in filter-feeding macroinvertebrates (Wiley and Kolher, 1984; Hershey and Hiltner, 1988). In our case, the foraging behaviour of *Acroloxus* (Mollusca) moving under the surface of the leaf may displace the tube building larvae of *Cricotopus* and the small *Corynoneura*. This behavioural interaction could be one of the main factors

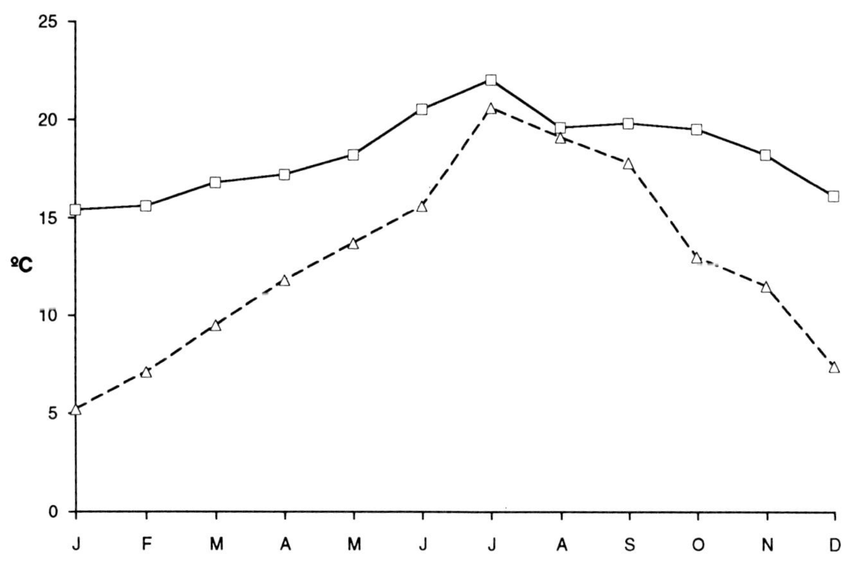

Figure 6. Mean monthly air temperature in the Natural Park Meteorological Station (triangles) compared with the surface temperature measured in the "Ullal del Teix" well each month (squares).

explaining the sudden fall in chironomid density. The reason why Mollusca are absent during the first six months of 1992 under the leaves of *Nymphaea alba* are unknown, but may be related to the life cycle of these species or the renewal time of leaves,which can be as high as 21 days in *Nuphar lutea* (Kouki, 1991). Kouki (*loc. cit.*) notes that the leaf longevity in the water-lily is shorter than the generation time of the Coleoptera, which must migrate in order to complete their life-cycle. We did not measure the leaf-longevity of *Nymphaea alba* in our pond, but with the high spring temperature this may be short and we may hypothesize that the molluscs or other macroinvertebrates can die by drowning. However, due to the dense stand of leaves covering the spring, is not difficult for any species to change from one leaf to another and for these reason we do not believe that this is an important factor in explaining the absence of mollusca from underneath of leaves during the first months of our study. The difficulty of completing the entire life cycle in beetles on leaves of floating water plants has also been reported by Otto and Wallace (1989), who note that one leaf of *Nuphar lutea*, if consumed by a beetle, is replaced by a new one in only 28 days and many larvae of the beetle die by drowning but this does not seem to be the case in our spring.

The biological interactions of larval populations of chironomids with other macroinvertebrates, especially the gastropods, by interference and resource competition, seems to be the main factor determining the abundance of midges on the undersurface of the leaves of *Nymphaea alba* in our study although some experimental research must be done in the future before definitive conclusions are derived.

Acknowledgments

We thank Carles Ibañez and Toni España for field assistance. Peter Langton kindly identified the *Corynoneura* pupae. Robin Rycroft has improved the English. "La Caixa" has provided financial assistance for this research and the Ebro Delta Natural Park provided facilities to study the spring, which is included in the nature reserve.

Utilization of Anterior Headcapsule Structures in Locomotion by Larvae of *Constempellina* Sp. (Diptera: Chironomidae)

Leonard C. Ferrington Jr.

Department of Entomology, Foley Hall, West Campus, University of Kansas, Lawrence, Kansas 66047 USA

Introduction

Construction of portable sand-grained cases by chironomid larvae is limited in the Orthocladiinae to the genus *Abiskomyia*, and in the Tanytarsini to a group of poorly known genera of the subtribe Zavreliina (= Tanytarsariae connectentes, sensu Brundin, 1948), including *Constempellina*, *Stempellina*, *Stempellinella*, *Zavrelia* and *Thienemanniola*. Larvae of these genera also share several morphological adaptations of the anterior and dorsal regions of the head capsule that have been used for taxonomic differentiation, but the functional significance of the structures remains unknown. Larvae of two Chironomini genera that construct portable cases, *Lauterborniella* and *Zavreliella*, lack the structural modifications of the head capsule, and their case composition and architecture differs.

Larvae of a *Constempellina* species that appears to be conspecific with the larva photographed in Oliver *et al.* (1978) were commonly encountered in a study of Chironomidae of a small forested watershed in southwestern Pennsylvania. Live larvae were collected for rearing purposes and their behaviors observed in the lab. Observations were also made at various times of *Lauterborniella* and *Zavreliella* larvae, and their locomotory sequences also recorded. In this paper I describe the locomotory behaviors of *Constempellina* larvae and discuss the functional significance of the anterior head capsule morphology in the locomotory sequence. The locomotory behaviors are contrasted with those of *Lauterborniella* and *Zavreliella* larvae. In addition, I propose that similar morphological adaptations of the antennal pedicel of *Abiskomyia* larvae may be used to assist in transporting the portable sand-grained case, and thus represent convergent adaptations which facilitate the case-building behaviors.

Methods

Live larvae of *Constempellina* were collected from first order streams of the Bear Run watershed located in Fayette County, southwestern PA during late winter and early spring 1977, and again in spring 1978 and 1980. Larvae were collected from sand substrates in areas of stream with reduced, but detectable, current. Larvae were transferred in the lab to small petri dishes and observed under dissection microscopes at various magnifications. Locomotory behavior was observed and recorded after larvae were given at least 24 hours to habituate. Some larvae were then transferred to depression slides or modified sedgewick rafter cells without cover slips for more detailed observations under higher magnification. Observations of this type were limited because larvae avoided intense light or died as the water in the cell warmed. Contact of the anterior head capsule structures with the surface of the glass in the cell was confirmed by placing a thin film of mud along the surface of the cell. Movement of the mud as larvae exhibited the locomotory sequence confirmed the contact of head capsule structures with the glass surface. Close inspection of the mud film at higher magnification after larvae were removed revealed small lines scraped into the film where the apico-mesal spur contacted the glass surface.

Larvae of *Lauterborniella* were collected and observed during summers of 1975 and 1979 from a small fish hatchery pond near the University of Pittsburgh's Pymatuning Laboratory of Ecology in Linesville PA. The pond is known locally as "Rogers Pond" and larvae were consistently collected from fine organic sediments within beds of cattails (*Typha* spp.). Larvae of *Zavreliella* were collected during summer of 1984 from a small, organically enriched slough habitat near the South Fork of the Republican River in Cheyenne County, Kansas, near the town of Saint Francis. These larvae were also collected from fine organic sediments among beds of cattails. In both instances, larvae were transported live to the lab and observed in white plastic trays, then in petri dishes under low magnification with a dissecting microscope.

As indicated earlier, larvae of *Constempellina* were collected for rearing purposes when the locomotory behaviors were first observed. Although most larvae survived well in the laboratory and several closed their cases for pupation, no male specimens successfully emerged as adults, despite several rearing attempts over three years. Thus the species identity remains unknown. Comparison of larvae with the illustration in Oliver *et al.* (1978) leads me to believe that the two taxa are conspecific. However, since no formal description of the larvae has been published, it is necessary to provide some additional description of the head capsule morphology so that identification can be confirmed if larvae are reared and the species formally described in the future.

The antennal pedicel of the larva is well developed and is armed with a single strong apico-mesal spur (Fig. 1). In lateral view the spur is slightly angled dorsal to the plane of the head capsule, and the apex of the spur projects anterior to the labrum when in reflexed position. There is a well developed membraneous region between the antennal socle on the pedicel and the basal segment of the antenna. The first segment of the antenna is slightly curved in the basal one-half, and is usually held so that it points antero-laterally or laterally, rather than anteriorly, during locomotory activities. Preserved larvae often display similar orientation of the antennae. Lauterborn organs originate on opposite sides at the apex of the second antennal segment and are large, as is characteristic for the genus. The anterior frontoclypeal setae originate medially near the bases of the pedicels, and are strongly sclerotized. Basally the setae originate from setal sockets that have extra thickening of the sclerite, and are larger in diameter than sockets of other frontoclypeal setae. The setae are larger in diameter basally than other frontoclypeal setae, and undergo numerous bifircations along the main axis. The setae extend anteriorly to approximately the apex of the apico-mesal spur of the antennal pedicel, and also extend beyond the labrum when it is reflexed.

Description of locomotory behaviors

For purposes of description the locomotory behaviors are described as a sequence of four relatively distinctive steps (Fig. 2), however in the laboratory larvae often completed the entire sequence as a continuous uninterupted event. Larvae were also observed to feed during steps 1 and 2 of the locomotory sequence, and as a consequence, these two steps were more variable than the last 2 steps, depending on the intensity of feeding activity.

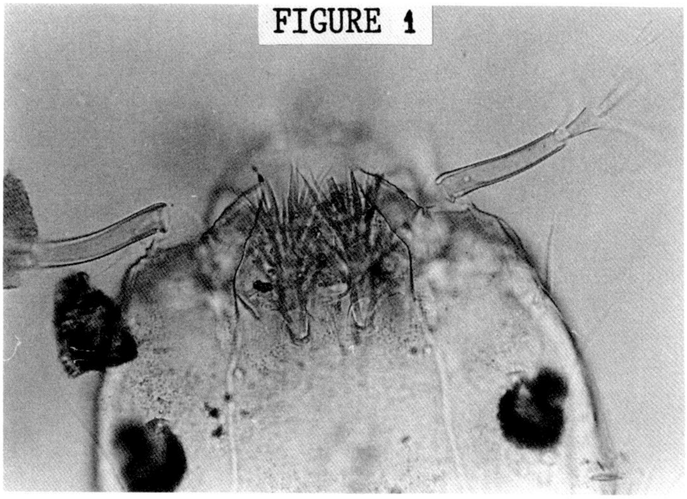

Figure 1. Anterior half of head capsule of *Constempellina* sp. (dorsal view).

Step 1

Larvae extend out of the case, which is resting upon, and in full contact with the substratum (Fig. 2A). It appears that the extension of the larvae out of the case is facilitated by stretching the thorax and abdomen so that the first and second, and occasionally third thoracic segments extend out of and beyond the anterior edge of the portable case. There was no indication that the larvae release their grasp of the case by the claws of the posterior prolegs, and it is assumed that continuous contact with the inner wall of the case by the posterior prolegs is maintained through out the locomotory sequence. After extending out of the case the larvae move the head and thorax variously to the left and right, contacting the substratum and brushing it with their anterior prolegs and occasionally the labrum, antennae and/or ventral mouthpart surfaces. Periphyton and/or fine detrital materials are often taken in at this point in the sequence, but the left-right movement of the head and contacting of the sediments with the antennae most likely enable the larvae to assess the texture of the substratum. After one or more lateral passes over the substratum the larvae select a portion of the substrate and firmly grasp it with the claws of the anterior prolegs.

Step2

While grasping the substratum with the anterior prolegs, the larvae begin to pull the case forward by contracting their thoracic and abdominal muscles, thereby reducing the length of the body (Fig. 2B). The posterior prolegs maintain contact with the interior of the case, thus facilitating forward movement of the case. At the same time, the larvae bend their head ventrally and scrape the substratum with the antero-ventral portion of the labrum and palatum. This movement of the labrum also basically appears to be a feeding movement, with the labral setae and other palatal structures brushing periphyton and fine detritus into the pharyngeal region. The claws of the anterior prolegs retain contact with and grasp the substratum through the end of this step.

Step 3

Larvae continue the ventral flexion of the head and the apico-mesal spur of the antennal pedicel contacts the substratum (Fig. 2C). At this point the labrum is mostly or fully retracted and the mandibles begin to close, pushing periphyton and detrital materials further into the pharyngeal region. Continued downward movement of the head capsule is accompanied by

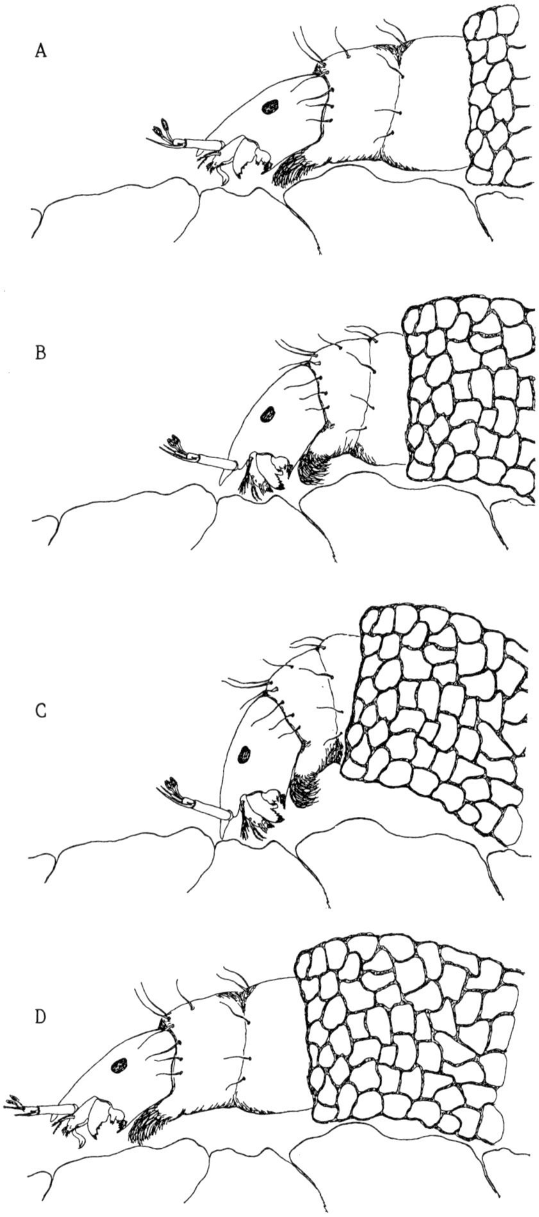

Figure 2A-D. Steps in the locomotory behavior of *Constempellina* sp.

release of the substratum by the claws of the anterior prolegs. The larvae then rapidly and forcefully bend the head capsule and often the first thoracic segment ventrally such that the anterior edge of the larval case is forced upward off of the substratum and forward, with the apico-mesal spur of the antennal pedicel acting as a fulcrum. As the anterior edge of the case is forced upward larvae release their grasp of the substratum with the claws of the anterior prolegs.

Step 4

After moving the case forward and upward, larvae contract the abdominal muscles, pulling the body backward into the case (Fig. 2D). As the body is contracting the thorax and head capsule are raised antero-dorsally, and the case falls back to, and rests in full contact with the substrate. Larvae do not contact the substrates with either the head or the anterior prolegs while pulling the body back into the case. After retracting into the case, larvae then begin to repeat the locomotory sequence by extending out of the case and moving their head left and right, contacting and brushing the substratum in front of the case.

Discussion

Building of portable cases by chironomid larvae is not common. Although many taxa build tubes and filtering devices with secretions from the salivary glands, construction of portable cases is limited to *Abiskomyia* among Orthocladiinae larvae, *Zavreliella* and *Lauterborniella* in the Chironomini, and *Constempellina*, *Stempellina*, *Stempellinella*, *Zavrelia* and *Thienemaniolla* in the Tanytarsini. This behavior provides obvious benefits to the larva in the form of freedom of movement during foraging, and protection of the membranous abdomen from attacks by both insect and fish predators. However behavioral modifications of locomotion are necessary in order to transport the case. The locomotory behaviors that I have observed for *Constempellina* larvae provide a strong contrast with those of *Lauterborniella* and *Zavreliella* larvae and allow for developing hypotheses regarding portable case-building behaviors in chironomids, and also for understanding the morphological modifications of the head capsule that are characteristic of Tanytarsini taxa that build the portable sand-grained cases.

Zavreliella and *Lauterborniella* larvae construct a case that is open at both ends and is composed primarily of silken secretions. Although small amounts of detrital material may stick to the case or be incorporated into it by larvae, no sand grains are used in case construction. Locomotory behaviors that I observed by these larvae differ markedly from *Constempellina* larvae, and the Chironomini larvae lack the modifications of anterior head capsule structures characteristic of the Tanytarsini "Connectentes" larvae, all of which build some form of portable cases with sand grains incorporated into them.

Locomotion by larvae of *Zavreliella* and *Lauterborniella* is accomplished by extending the body out of the case and grasping sediment with the anterior prolegs. After securing a grip on the sediments, the larvae begin to contract the abdominal muscles, thereby shortening the body. At the same time the larvae grasp the inside of the case with the posterior prolegs and arch the eighth and ninth abdominal segments so that the dorsal hump on segment eight contacts the opposing inner edge of the case. After completing contraction of the body the larvae release the grasp of substrates with the anterior prolegs and begin to repeat the locomotory sequence by extending out of the case. At times, the larvae will retract into the case, and turn 180 degrees so that the head is extended out of the opening at the other end of the case. This behavior is common when an obstacle is encountered in the direction of movement, but is not limited to this, since larvae were observed to retract into cases and reverse movement numerous times without any obvious obstacles in their path.

Larvae also withdrew into the case and reversed their orientation when detrital materials accumulated at the forward opening of the case or just inside of it. Apparently this accumulation of detritus blocked water flow into the case, and larvae reorientated before undulating the body to set up a current in the case opposite of the previous direction. The behavior is similar to that observed by Walsh (1947, 1951) for *Chironomus* larvae when their filtering tubes became clogged with detrital material, and it can be assumed to have a similar function in *Lauterborniella* and *Zavreliella*. Resting larvae retract fully into the case and readily undulate their bodies while not in the process of locomotion.

Occasionally larvae were observed to supplement their forward movement by releasing their grip on the interior of the case by retracting the claws of the posterior prolegs, then extending the posterior prolegs further backward into the case while maintaining contact with the top of the case by the dorsal hump of the eighth segment. Larvae then reattached the posterior prolegs to the inside of the case and re-established the arch in the abdomen, thereby pulling the thorax and anterior abdominal segments further back into the case. This behavior allowed the larvae to move the case forward an additional distance without releasing the grip of the anterior prolegs on the substratum, however it was not a common supplement to the locomotory sequence.

The evolution of case building by larvae of *Lauterborniella* and *Zavreliella* larvae can be readily conceptualized as derived from modification of tube-building behaviors of other Chironomini. Characteristics that the cases share with tubes are (1) they have openings at either end that larvae can extend to forage from, (2) they are constructed primarily out of secretions of the salivary glands, with only small amounts of detrital materials incorporated into the case, (3) they are flexible and large enough in diameter to allow the larvae to reorientate within the tube and (4) they can be used in ventilation by undulating larvae.

The portable case can be visualized as simply a shortened tube that is dragged along with the larvae rather than being firmly attached to a substrate. I have observed tube construction behavior by other Chironomini larvae and it is obvious that several elements of their tube construction behavior are homologous to the locomotory behaviors of *Lauterborniella* and *Zavreliella* larvae. For instance, among tube-building taxa it is common to see larvae extend out of the tubes and select small pieces of detrital materials that are pulled to the entrance of the case as the larva contracts the abdominal muscles. The detrital material is often incorporated into the case near the opening, as the case is lengthened. It is also common to see larvae extend out of the case, select a large piece of detrital material and attempt to pull it to the tube by arching the posterior end of the abdomen while gripping the interior of the tube with the posterior prolegs. Occasionally larvae will release their grip and extend the prolegs posteriorly to secure a new grip on the tube and continue the tugging movement on the detritus by contracting the abdominal muscles or re-arching the posterior segments of the abdomen. It seems logical that these behaviors could be readily adapted into locomotory sequences if one hypothesizes that the evolution of portable case-building habits in Chironomini larvae evolved from tube-building taxa by shortening of the "tube" and not attaching it firmly to a piece of substrate.

Larvae of tube-building species readily reverse the direction of the body by reorientating within the case. Reorientation allows the larvae to extend the tube construction in either direction, and to forage from either end. It also allows the larvae to redirect current flow within the tube to remove fecal material from the tube, or to push out detrital particles that accumulate near the tube entrances. All of these behavioral elements are retained among the larvae that build portable cases, and it thus seems to me that the portable case-building habits of *Lauterborniella* and *Zavreliella* are derived from tube-building habits, and required only minor degrees of morphological and behavioral modification.

The cases that *Abiskomyia* and the Tanytarsini species construct differ in several respects from the cases of *Zavreliella* and *Lauterborniella*. They are composed of sand grains held together by salivary secretions and are presumably proportionally heavier. The cases are linear (*Stempellinella* and *Zavrelia*) or cornucopia shaped (*Constempellina* and *Stempellina*) or gently tapering from front to back (*Thienemanniola*). They are rigid, and have distinct anterior and posterior ends to the cases. It is not possible for larvae to turn within the case and reorientate so that the head can be pushed out of the posterior end. Consequently larvae can only drag the case in one direction.

The differences in case architecture in *Constempellina* require more elaborate behavioral and morphological changes to facilitate locomotion. Incorporation of sand grains into the case provides increased structural rigidity, however at the expense of increased case weight and probably increased drag. Consequently, the posterior prolegs are used to anchor the larvae within the case as it is pulled forward. The posterior prolegs were never observed to assist in forward movement of the case by releasing and extending backward into the case to regrasp. It is presumed that the taper of the case assists in providing a more solid grasp of the interior wall of the case by the claws of the prolegs as as they are wedged posteriorly into the decreasing diameter of the case, but this limits the ability of the larva to regrasp the case to assist in forward movement. The posterior prolegs were never observed to be extended from the posterior opening of the case, and it seems that their principal function is to anchor the larvae to the case and are never used in locomotion.

The anterior prolegs firmly grasp the substrate in front of the case as the abdomen is contracted and are therefore used in forward movement of the case. As the case is pulled forward the head is flexed ventrally until the antero-mesal spur of the antennal pedicel contacts the substrate, then the head is forcefully reflexed and the spur acts as a fulcrum to lift the case from the sediment and continue the forward movement of the case. Next, the head is quickly lifted and the case drops to the sediment surface. Although ventral head flexing movements obviously contribute to the forward movement of the case and larva, I propose that it has a second function associated with minimizing the build up of fine detrital material and periphyton along the anterior edge of the case. As the case is dragged forward along the sediment it is likely that detritus and periphyton material accumulates along the anterior edge of the case, and if allowed to remain along the anterior margin could block water flow into the case and around the larval abdomen, thus interfering with cuticular absorption of oxygen for respiration. By lifting the case and allowing it to drop back to the substrate the accumulation of material along the anterior edge of the case is disturbed and does not accumulate in front of the case and block the flow of oxygenated water through it. This is particularly important since larvae cannot reorientate with in the case to establish a counter current if materials accumulate within the case and block the flow of water around the larvae. There are no obvious homologous behaviors among larvae of *Lauterborniella* and *Zavreliella* larvae, thus the ventral head flexing and use of the apico-mesal projection of the antennal pedicel are novel elements in the locomotory sequence that arise with the incorporation of sand into the case and the inability of larvae to re-orientate within the case.

Although I have only observed the locomotory behavior in *Constempellina* larvae, as a consequence of the similarities in case configuration it is presumed that the locomotory behaviors of the other genera with portable sand-grained cases will all be very similar, and they all are likely to differ in the same or similar significant respects from those of *Lauterborniella* and *Zavreliella*.

Modifications of the anterior region of the head capsule observed among larvae of the other Tanytarsini genera (see Figs. 3-7) presumably are used in a similar manner in locomotion. By close examination of the morphology of several undescribed species in my museum collection it is possible to propose a set of characteristics,

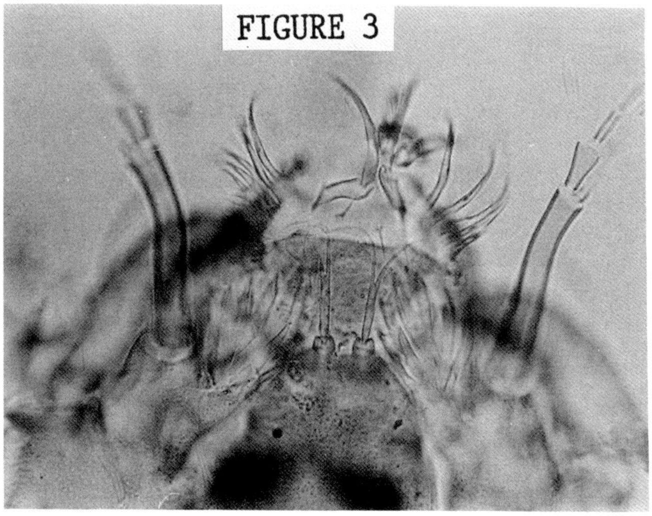

Figure 3. Anterior half of *Stempellina* sp. larval head capsule showing antennal pedicel and spur with numerous spines, and strong, bifurcating frontoclypeal setae with well developed sockets.

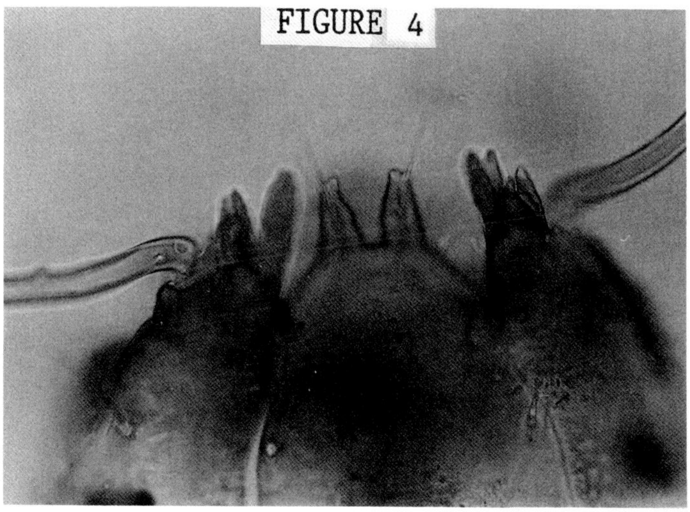

Figure 4. Anterior half of *Stempellina* sp. larval head capsule showing antennal pedicel and spur with 3-4 strong, blunt spines, and simple frontoclypeal setae originating subapically from enlarged sockets.

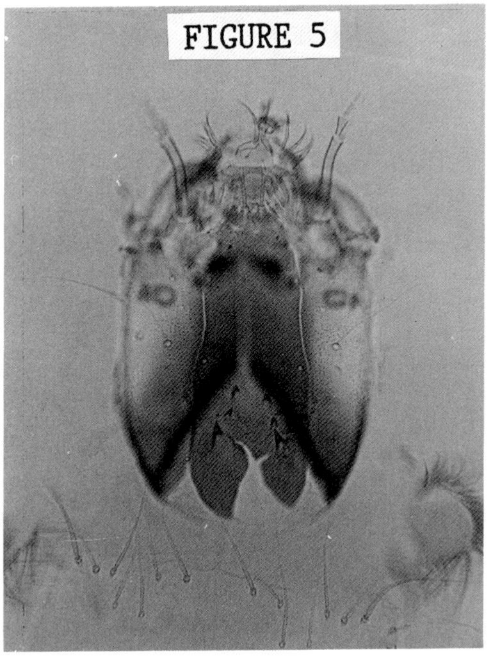

Figure 5. Head capsule and anterior portion of prothorax of *Stempellina* larva showing dorsal tubercles along posterior margin of frontoclypeal apotome, and lamellar setae of prothorax. (Same specimen as shown in Figure 2).

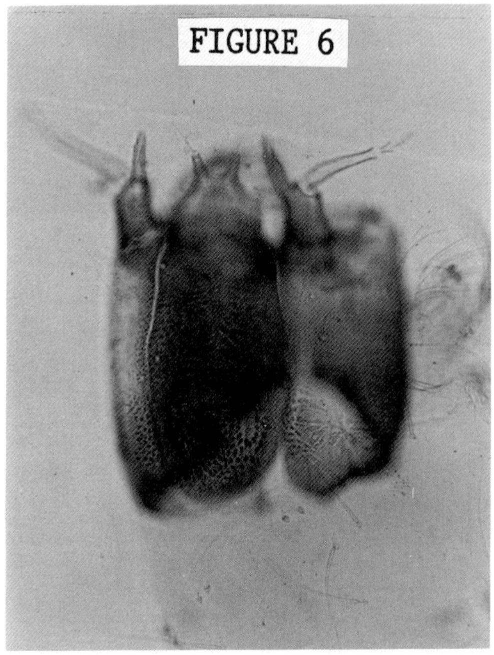

Figure 6. Head capsule and anterior portion of prothorax of *Constempellina* larva showing extensive surficial rugulosity on genae and frontoclypeal apotome. Large, single spur of antennal tubercle and well developed sockets of frontoclypeal setae also visible.

in addition to the apico-mesal spur, that appear to be morphological correlates of sand-case building behaviors. The morphological correlates function to facilitate locomotion among larvae that have evolved the habit of building portable sand-cases.

Morphological correlates evident among sand-case building larvae consist of modifications of the genae and frontoclypeal sclerites of the head capsule, elaboration of the thoracic setation and, in at least two species, elaboration of frontoclypeal setae or their sockets for use in locomotion, in addition to the modifications of the antennae and antennal pedicel already discussed.

Modifications of the head capsule include increased sclerotization of the frontal apotome and genae, along with increased rugulosity and occasionally a few to numerous tubercles among some species (Figs. 5-7). The increased sclerotization should impart increased strength to the head capsule, as would the increased rugulosity. Additionally the rugulosity and tubercles provide increased surface area on the inner wall of the sclerite for attachment of muscles that assist in reflexing the head. I presume that there are increased mechanical stresses on the head capsule when the apico-mesal spur is acting as a fulcrum, and the rugulosity and tubercles represent morphological mechanisms to deal with them.

In some species the anterior labral and frontoclypeal setae or their sockets are modified (see Figs. 1, 3 and 4 for close up views). The modifications are of basically two types. In *Constempellina* the S3 setae are strongly sclerotized and ramified and extend anteriorly to approximately the apex of the apico-mesal spur of the antennal pedicel (Fig. 1). The setae assist in forward movement of the case as the head is flexed ventrally.

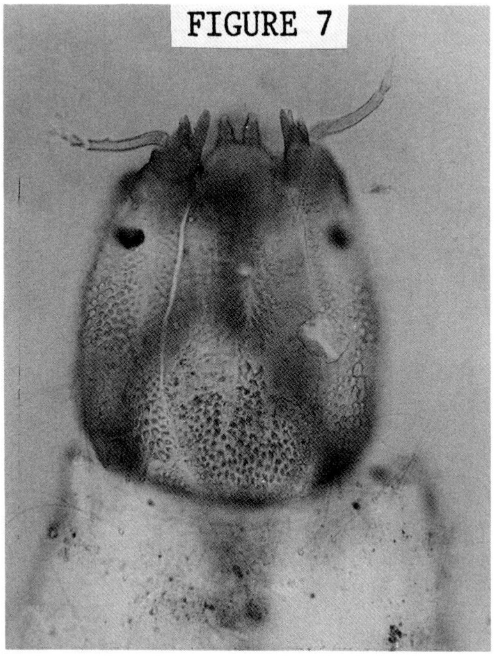

Figure 7. Head capsule and anterior portion of prothorax of *Stempellina* larva showing extensive surficial rugulosity on genae and frontoclypeal apotome. Large antennal tubercle with 3-4 spines on spur and well developed sockets of frontoclypeal setae also visible. (Same specimen as shown in Fig. 4).

In other species the same setae are not modified and presumably do not function in moving the case forward. However, as can be seen in Figures 3 and 4, the setal socket is stongly sclerotized and raised from the surface of the frontoclypeal apotome. In the species shown in Figure 4, the point of origin of the seta has migrated from the apex of the socket to the dorsal edge of it. It seems logical to propose that the modified setal socket functions in forward movement of the case as the head is reflexed ventrally, and may even be used as a fulcrum in addition to the apico-mesal spur. Migration of the point of origin for the setae to the dorsal surface can be interpreted as a mechanism to protect the base of the setae from abrasion or breakage when the socket contacts the substrate.

Close inspection of the anterior edges of the first thoracic segment of several taxa that construct portable sand-cases revealed a series of elongated lamellar setae are present (visible in Figs. 5-7). The setae most likely have two functions that relate to the case. The first function is one of thigmotaxis, allowing the larvae to determine the position of the case relative to the thorax as the larvae are completing the contraction of the abdomen. As the case is pulled forward during contraction of the abdominal muscles the setae would be contacted by the anterior edge of the case, thereby signifying to the larvae that the ventral flexion behavior should be initiated.

The second function of the elongated lamellar setae is in rheotaxis. With the case of the larvae restricting the direction of current movement around the abdomen of the larvae it would be beneficial to have well developed setae on the portions of the body that extend out of the case in order to detect current direction and/or velocity. The elongated length of the setae would also ensure that the tips of the setae extend far enough over the edge of the case in order to be moved by the current flowing over or around the case.

As can be seen from the previous discussion, several morphological adaptations are correlated with portable sand-case building habits of Tanytarsini larvae. By using the observations of locomotory behaviors of *Constempellina* as a model, it must be concluded that behavioral modification are also necessary in order for larvae to pull the sand-case during locomotion. From both an energetic and evolutionary sense it must be argued that the morphological and behavioral modifications are costly to the larvae, at least relative to the alternative of making a stationary tube or constructing a case of salivary secretions. In order to more fully understand the case-building behavior of the Tanytarsini larvae one must consider hypotheses relating to the advantage of building the sand-grained case. Two of the most obvious explanations, of course, would be that the sand grains provide better protection from predators than the portable cases of *Lauterborniella* and *Zavreliella* larvae, and the portability of the case provides for more efficient foraging or microhabitat selection than the stationary tube. While I think that there is certainly some degree of validity to each of these hypotheses, I don't think they are sufficient to account for the differences in case construction since the same selective forces would have resulted in *Lauterborniella* and *Zavreliella* also incorporating sand-grains into the case architecture. It seems to me that there are other alternative hypotheses for the differences in case-building behaviors that can account for the incorporation of sand grains into the cases by Tanytarsini larvae.

Lauterborniella and *Zavreliella* larvae are most commonly collected from beds of vascular hydrophytes in lentic habitats or from similar stands of vascular hydrophytes in pools or backwater areas of larger streams and rivers (Pinder and Reiss 1983, Ferrington, *pers. obs.*). They are are not common in lotic erosional habitats, nor in wave-swept littoral zones of lentic habitats. By contrast, the larvae of sand-case building Tanytarsini are often found in lotic erosional habitats or littoral zones of lakes. This habitat segregation among the larvae that build the two types of portable cases suggests that current velocity conditions in which the larvae occur are salient features in the adaptive significance of the alternative case-building behaviors.

It has been hypothesized by Wiggins (1977) that rigid portable cases of Trichoptera larvae may function as balast for larvae in lotic habitats, providing orientation into the current so that larvae can efficiently use their legs to grasp the substratum or forage in a counter current manner. It seems logical that the sand-grained case of the Tanytarsini larvae would function in a similar manner, and would provide more structural rigidity in the current than a case made of salivary secretion such as made by *Lauterborniella* and *Zavreliella* larvae. The sand grains would also resist wear and abrasion as larvae drag the cases over the rock and gravel substrates that characterize erosional zones in lotic habitats and wave-swept littoral zones. In contrast, however, it would seem that wear and abrasion of the case for larvae that select lentic habitats with soft organic sediments be minimal, and the energetic costs of repairing the soft case would be less than the cost associated with dragging the heavier sand-grained case in the same habitat. Also in the lentic habitats chosen by *Lauterborniella* and *Zavreliella* larvae, where current velocities are negligible or reduced, the ballasting function of the heavier sand-grained case would not be a consideration.

It is also possible that dissolved oxygen tensions of the two types of microhabitats are related to the two types of cases that are constructed. It is likely that the average dissolved oxygen concentrations at the sediment-substratum interface is lower in the microhabitats that *Lauterborniella* and *Zavreliella* larvae select. Their case architecture, with two large openings, provides greater diameters for water to pass through the case when larvae are irrigating. Due to diel and seasonal fluctuations in dissolved oxygen concentrations it seems logical to assume larvae may experience cyclic or even extended periods of reduced oxygen tension, especially in highly organic sediments among beds of emergent hydrophytes. Thus the case architecture, which which readily facilitates irrigation by abdominal undulation, would be favorable.

In contrast, the lotic and littoral habitats of the Tanytarsini larvae would likely be saturated with oxygen, or nearly so, through most of the diel and annual cycles. Larvae presumable would be less likely to experience reduced oxygen tension to the point where irrigation of the case would be critical for survival. This then could account for the change in case architecture seen in *Constempellina* and *Stempellina*, in which the posterior opening of the case is reduced due to the cornucopia-case shape. At the same time, it would also seem logical to assume that the cornucopia-case architecture would provide reduced drag and may allow larvae to exist in areas of greater current velocity. This would, in turn, further reduce the possibility of larvae experiencing reduced oxygen tensions, and would allow for the reduced posterior opening of the case since irrigation would be less critical to the larvae.

Abiskomyia larvae are the only Orthocladiinae that build portable cases with sand-grains incorporated into them. They occur in oligotrophic trophic lakes, where they are common in the littoral zones, and in moss zones of small bodies of flowing water (Cranston *et al.*, 1983). Nothing is recorded with regard to their locomotory behavior, however the case architecture is similar to that of Tanytarsini larvae that construct the sand-grained cases, and it must be presumed that locomotory behaviors are similar. The larval head capsules possess well developed antennal tubercles with strong apico-mesal projections, and the thoracic setation includes elongate and what appear to be slightly flattened setae. These morphological attributes strongly suggest that the head is used to assist in transporting the case and that larvae extend out of it to grasp the substratum. The case is tapered posteriorly and it seems unlikely that larvae re-orientate within the case, therefore the posterior prolegs must only function to anchor the larvae within the case. Since phylogenetic analyses show that *Abiskomyia* is not closely related to the case-building Tanytarsini, it must be concluded that the morphological similarities are the result of convergences, and that morphological correlates of sand-grain case-building behaviors in chironomids are functional adaptations.

In conclusion, it seems likely to me that although the evolution of portable case-building behavior by Chironomidae can be explained by the benefits provided to the larvae in terms of reduced predation and increased foraging efficiency relative to stationary tube-building behaviors, the differences in case construction and architecture among Chironomini larvae and Tanytarsini larvae cannot be accounted for in these terms. One must propose that the construction and architecture differences are a function of the microhabitats chosen by larvae, and the concommittant differences in current velocity and possibly dissolved oxygen concentrations to which the larvae are exposed. It is also necessary to propose that the habit of building sand-grained portable cases for exploiting lotic erosional and littoral zone habitats requires more elaborate locomotory behaviors and morphological correlates in order to accomodate physical and energetic constraints imposed by the case. The presence of similar but non-homologous morphological adaptations in *Abiskomyia* larvae which live in microhabitats similar to those of the Tanytarsini larvae provide additional support to my hypotheses.

Acknowledgements

I would like to thank Bill Coffman for initially stimulating my studies of *Constempellina* by suggesting that the antennal pedicels and spurs may be associated with larval feeding. Although the initial hypothesis was not supported by my observations, it was his enthusiasm and curiosity that provided the spark to ignite my observations of these larvae. I would also like to acknowledge the following faculty of the Department of Entomology at the University of Kansas for their insight and stimulating discussions of behavior, homology and evolution: Charles Michener, George Byers and Rudolph Jander. Ms. Rosella Cox assisted in making editorial changes in the final draft of this paper. This research is published under the auspices of the Water Quality and Freshwater Ecology Program of the Kansas Biological Survey at the University of Kansas.

Eurycnemus sp. (Diptera: Chironomidae) larvae ectoparasitic on pupae of *Goera japonica* (Trichoptera: Limnephilidae)

Tadashi Kobayashi

Caritas Senior High School, 4-6-1, Nakanoshima, Tama-ku, Kawasaki-shi, Kanagawa-ken, 214 Japan

Introduction

Many pupal cases of *Goera japonica* (Trichoptera: Limnophilidae) collected in Yamaguchi-ken, West Japan, contained larvae and pupae of a species of *Eurycnemus* (Diptera: Chironomidae). Positive proof has been obtained that *Eurycnemus* sp. feed on trichopteran pupae within the cases.

In a comprehensive review of ectosymbiosis in aquatic insects, Steffan (1967) reported that pupae of the trichopteran family Limnephilidae are subject to parasitism by various chironomid larvae. Hudson (1987) mentioned two chironomids - *Cardiocladius* and *Dratnalia* - symbiotic on Trichoptera. Parker and Voshell (1979) found larvae of a *Cardiocladius* species (now known to be *albiplumus* Sæther) inside the pupal cases of Hydropsychidae (Trichoptera). They also reported the *Cardiocladius* larva to be predaceous on trichopteran pupae. Their observation and those reported here have much in common although there are no previous reports of ectosymbiosis involving *Eurycnemus* sp. with Limnephilidae.

Materials and Methods

Takao Nozaki (*pers. comm.*), a specialist of caddis flies, informed the author that chironomid larvae may feed on the pupal and prepupal trichoptera in their cases. Subsequently, a large number of pupal cases of *Goera japonica* collected from the Nishiki-River, Yamaguchi-Prefecture, West Japan and preserved in 70% alcohol were sent to the author's laboratory for identification. Sixty-four pupae, prepupae and larvae of chironomid in total were found in the trichopteran cases.

In May 1992, the author collected 2,473 pupal cases of *Goera japonica* in the Nishiki-River, and succeeded in collecting the adults of *Eurycnemus* sp. More than 10 pupae swam out of the cases; one male and two female adults emerged. These were slide mounted in Balsam. Thirty-nine pupae and sixty larvae were collected from within the trichopteran cases.

Results

The pupal and larval morphological characteristics almost agreed with the description on *Eurycnemus crassipes* (Panzer) by Murray and Ashe (1981b) except for the lack of a seta interna on the larval mandible and the larger central tooth of larval mentum. Only one larva or pupa was found per trichopteran pupal case. Close examination of the trichopteran pupae revealed signs of feeding by *Eurycnemus* sp. In some pupae, one deep hole was found in the trichopteran pupal abdomen (Fig. 1a). The size of these holes corresponded to the size of a *Eurycnemus* larva. Often the caddis fly was completely devoured and the case contained only a late-instar chironomid larva or pupa which was encapsulated within trichopteran body fragments (Fig. 1b). Besides, trichopteran chitin fragments were observed in the gut contents of the *Eurycnemus* larva. The larva feed on trichopteran fragments (Fig. 2) leading to suggestions that *Eurycnemus* larva invaded and stayed in the case and fed on trichopteran prepupa or pupa.

The morphological characteristics of the adults agreed with the description of the genus *Eurycnemus* by Cranston *et al.* (1989). According to Steffan (1967), the association reported here represents inquilinism and ectoparasitism. When and how the *Eurycnemus* larva invades the trichopteran case is not altogether clear so far, but this may take place at an early stage by a small-sized (c.4mm) larva.

Morphology

Larva *(Fig. 3a-g).*

Eurycnemus sp. larvae ranged from 4.8mm to 11.5mm total body length. Antenna four segmented, less than half mandible length, segments consecutively smaller. Lauterborn organs absent. Antennal blade shorter than second antennal segment. AR 1.0. SI, SII and SIII setae of labrum simple. Labral lamellae small. Pecten epipharyngis consists of three broad triangular spines. Ungula broadly U-shaped; basal sclerite apparently absent. Premandible broad, with two apical teeth; brush absent. Apical tooth of mandible shorter than combined width of four inner teeth. Seta subdentalis long, apically pointed. Seta interna absent. Mentum with one median and six pairs of lateral teeth; median tooth and first lateral teeth larger than remaining teeth. Frontal parapods very long, separate, retractile, each one bearing many scales and apical crown of claws. Anal setae present.

Pupa *(Fig. 3h-j).*

7.5mm long. Coloration brownish. Thorax with pair of blunt blackish projections extending from anterodorsal margin of scutum. Thoracic horn with apical 2/3 gradually tapering to a point. Tergite I without shagreen; II-VIII with coarse shagreen on median area consisting of large spinules to small spines. Tergite II-VI with long posterior spines. Anal lobe rounded, with 5-7 long broad anal macrosetae. Fringe absent. Male genitalia sac subequal in length to anal lobe.

Adult male *(Fig. 4a-d).*

Body length 6.2mm. Antenna with 13 flagellomeres. A.R. 2.4. Eyes bare with long, almost parallel-sided dorsomedial extension. Temporal setae multiserial. Strong, dorsally-produced anterior margin of the scutum. Wing membrane with long setae; some setae dark, giving wing spotted appearance. Costal extension present. RM long, oblique. Squama completely fringed. Hind tibia without comb, but with two spurs. Pulvilli well developed. Gonocoxite parallel-sided and somewhat elongate, well developed; long and narrow superior volsella with setae on ventral surface. Distomedial lobe present. Gonostylus bifurcate with subapical lobe shorter than apical lobe; apical lobe with four stout apical lamelliform setae.

Adult female (Fig. 4e-f).

Body length 7.0mm. Eyes bare with dorsal extention. Scutum conically produced. Pulvilli well developed. Antenna six segmented. Palp five segmented. Hind tibial comb absent. Wing membrane densely covered with macrotrichiae. Dorsal part of cerci elongate.

Acknowledgments

The author wishes to express his gratitude to Dr. M. Sasa, International University of Toyama, and Dr. K. Moriya, Kanagawa Prefectural Museum for their many helpful suggestions throughout this research. Many thanks are also expressed to Dr. K. Yasuraoka, University of Tsukuba, for his useful advice and for reading the manuscript; to Mr. T. Nozaki, Kanagawa Prefectural Research Center for Environmental Studies and Mr. H. Shimomura, University of Hiroshima, and Mr. Yokota for their assistance in collecting specimens; and to Mr. K. Watanabe, Kawasaki City Research Center for Education, for his generosity in allowing the author to use the scanning electronic microscope.

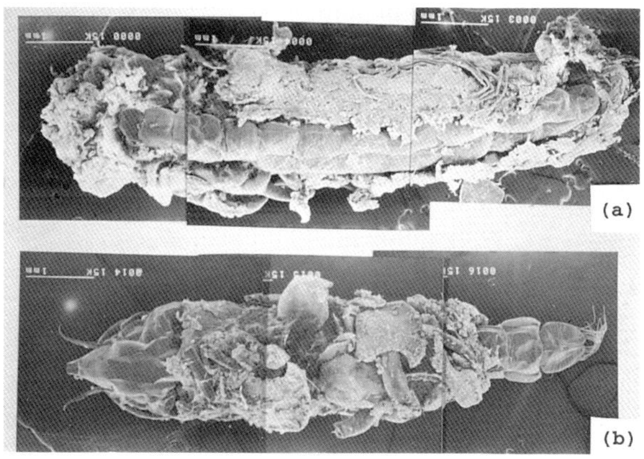

Figure 1. *Eurycnemus* sp. (a) larva within pupal trichoptera; (b) pupa encapsulated within pupal trichoptera.

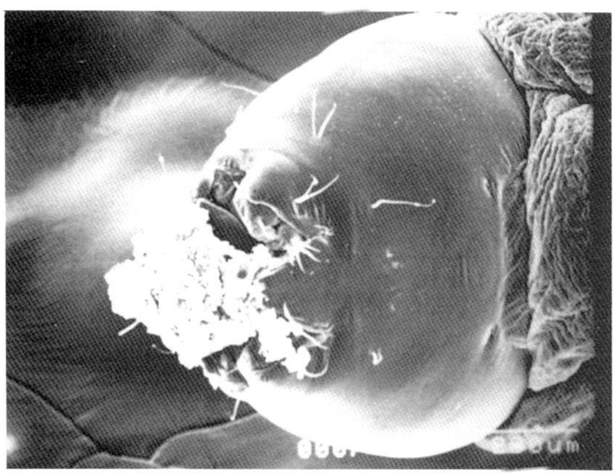

Figure 2. *Eurycnemus* sp. Larva feeding on trichopteran fragments.

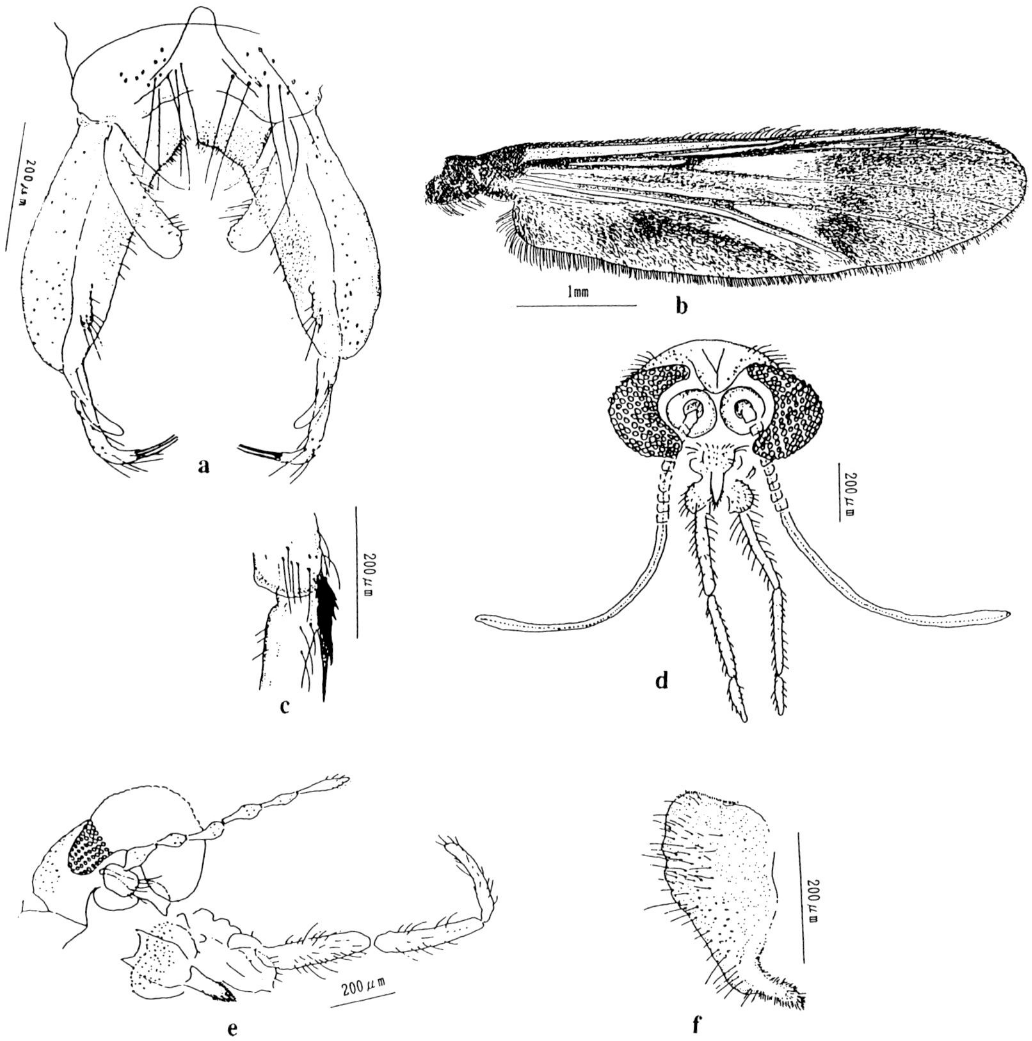

Figure 3. *Eurycnemus* sp. (a)-(d) male, (a) hypopygium, (b) wing, (c) apex of hind tibia, (d) head; (e)-(f) female, (e) head, (f) cercus.

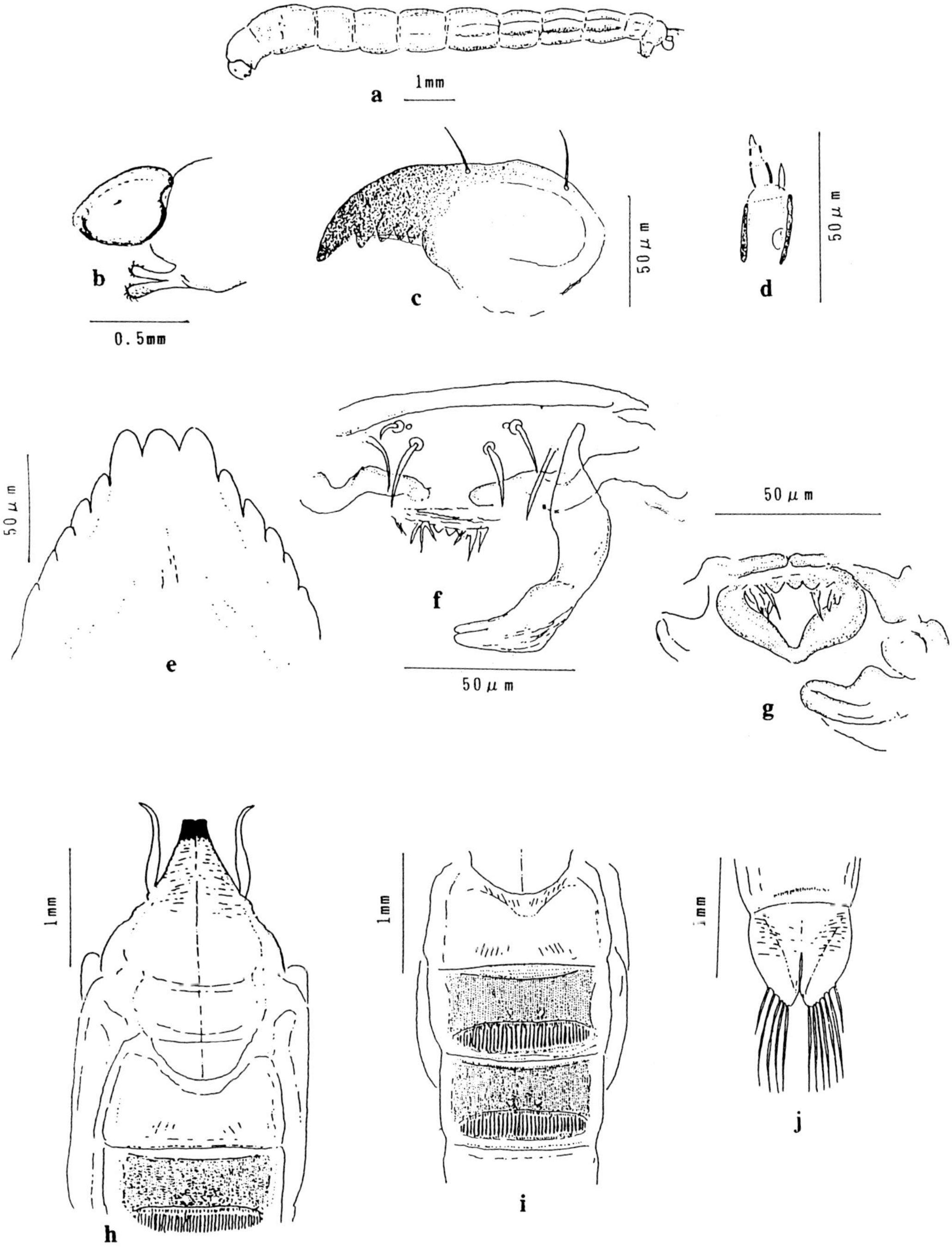

Figure 4. *Eurycnemus* sp. (a)-(g) larva, (a) whole larva, (b) head capsule, (c) mandible, (d) antenna, (e) mentum, (f) labrum, (g) ungula; (h)-(j) pupa, (h) cephalothorax, (i) abdominal tergites I-III, (j) anal lobe.

On the relationship between the oxygen microstratification in a pond and the spatial distribution of the benthic chironomid fauna

L. Int Panis[1,3], *B. Goddeeris*[2] *and R.F. Verheyen*[1]

[1]Universitaire Instelling Antwerpen, Departement Biologie, Universiteitsplein 1, B-2610 Wilrijk/Antwerpen. Belgium

[2]Koninklijk Belgisch Instituut voor Natuurwetenschappen, Afd. Zoetwaterbiologie, Vautierstraat 29, B-1040 Brussel. Belgium

[3]Research assistant of the Belgian National Fund for Scientific Research (N.F.W.O.).

Introduction

Our objective is to elucidate the influence of oxygen on the spatial distribution of the benthic fauna in shallow waters. In this study, our attention was focused on the Chironomidae that are abundant in our study site - a small eutrophic pond in Belgium. It is well known that oxygen controls the specific occurrence and macrodistribution of chironomid larvae in standing waters (Heinis, 1993). However, it is unclear how the "respiratory environment" (Brundin, 1951) of these organisms is defined. Our study of the oxygen microstratification is intended to elucidate its significance for the vertical distribution of chironomid larvae. Therefore, we have measured the oxygen concentrations at several locations above and in the sediment as well as in the water column.

Materials and Methods

Sampling procedure

The investigated pond is situated in the nature reserve "Waelenhoek" at Niel (prov. Antwerp, Belgium). It is an old clay pit of about 1 ha with a maximum depth of approximately 4 m in its centre. The water is highly eutrophic. A summer stratification has been observed with an oxycline at a depth of approximately 2.5 m. At least twenty chironomid species have been reported from this pond, some of which appeared to be new for the Belgian fauna (Int Panis *et al.*, 1992; Int Panis *et al.*, 1993).

In July 1992, samples were taken at 14 stations widely spread along the shore and in the central part of the pond. A hand corer with an area of 40 cm^2 was used for quantitative sediment sampling. At each station 4 replicates were taken, one of which was used for the measurement of the sediment oxygen microprofile. The three remaining cores of each station were vertically sliced at depths of 1,2,3,4,5,7 and 10 cm and subsequently fixed with neutralised formalin. All slices were sieved (mesh size = 300 μm) and sorted separately. Chironomid larvae were identified using the keys of Moller Pillot (1984).

Oxygen measurements

Oxygen concentrations near the bottom were obtained in two ways. Measurement of the bathymetric distribution of oxygen in the water column was done by lowering a Clark Au-Ag oxygen electrode (WTW EO 196) from a boat, in steps of 10 cm, at the deepest point of the pond, i.e. the centre of the pond. The oxygen concentration above the bottom at each station was considered to be equal to that observed at the same depth in the water column. This crude estimate was made in order to find out how much oxygen could approximately be available at each station regardless of the influence of different substrate characteristics. Secondly, oxygen microprofiles in the sediment were measured in the fourth replicate core sample for each station immediately after sampling. A Clark-style microelectrode (Diamond 737GC) was lowered into the sediment with a micromanipulator in steps of 0.1 mm. In this way oxygen concentration profiles have been measured from 1.2 mm above the sediment-water interface down to the sediment oxygen penetration (SOP) depth, i.e. the zero concentration depth in the sediment.

Statistical Analysis

There are many possible methods of analysing the structure of a multivariate dataset and finding the most likely abiotic factor that can explain the observed structure (e.g. Int Panis & Verheyen, 1994). We chose the Mantel-test (Mantel, 1967) as described by Manly (1991) because it is an objective method that makes no assumptions about the statistical distribution of the data. Two square matrices are needed for each test. Matrix A contains the Bray-Curtis dissimilarities (Bray and Curtis, 1957) of the total chironomid fauna, between all pairs of sampling units. Matrix B was constructed for each of the different oxygen measurements. A simple index was needed to express the dissimilarity of oxygen concentrations between sites. This was done by taking the absolute difference between the oxygen readings for all pairs of sampling units. Then a correlation r, between the corresponding off-diagonal elements of the lower triangle of each matrix, was calculated with the help of the formula given by Manly (1991). The significance of this correlation statistic was estimated by means of a Monte Carlo procedure. The rows and columns of matrix B were randomly permutated and a new correlation was calculated. This procedure was repeated 10000 times. Finally, was compared to the randomization distribution to find the level of significance (Manly, 1991).

Results

At most sites we could observe a clear microstratification. The sediment microprofiles demonstrate a sharp dropping of the oxygen concentration to zero in the upper sediment film (e.g. Fig. 1). The average reduction of the oxygen concentration in the water overlying the sediment was about 100μMol/l. SOP values were very small, ranging from 0 mm in the "profundal" mud to a maximum of 1.8 mm in sediments of the littoral zone.

The correlations between corresponding elements of matrix A, containing the community dissimilarities between the sampling units, and matrix B, containing the differences between a certain O_2 measurement for the same pairs of sampling units, are given in Table 1. Chironomid larvae were found at 11 stations and thus calculations are based on 11 samples.

We find a correlation between chironomid fauna and depth that is highly significant ($r = 0.66$, $p<0.001$).

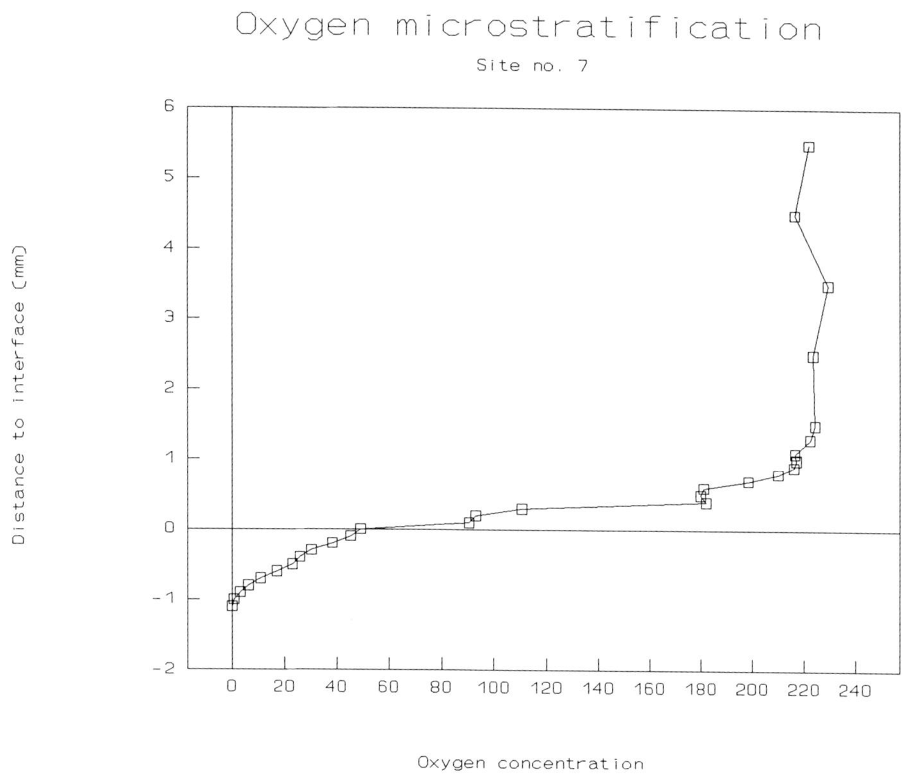

Fig. 1. Oxygen microprofile near the sediment-water interface. O_2 concentrations are given in µMol/l.

The correlation with the oxygen measurements from the water column is much higher: the r value of 0.82 was only exceeded in two out of 10000 randomizations (p<0.001).

Table 1. Correlations of various oxygen measurements with the Bray-Curtis dissimilarity between pairs of sampling units (Mantel test).
(*:p< 0.05; ** : p<0.01; ***: p < 0.001).

Depth	r = 0.66 ***
O_2-profile	r = 0.82 ***
O_2 at 1.2 mm	r = 0.32 *
O_2 at 0.6 mm	r = 0.27
O_2 at 0.2 mm	r = 0.12
O_2 at interface	r = 0.24
SOP depth	r = 0.14

The correlations between the chironomid fauna and the accurate oxygen measurements near and in the sediment-water interface were much lower. Only the correlation with the oxygen measurement at 1.2 mm above the sediment-water interface is significant. All other correlations are not significant.

DISCUSSION

The first part of our data analysis confirmed our expectations. The results suggest that the chironomid fauna changes with depth. However, this correlation is believed to be indirect; depth should not cause these changes. Our hypothesis is that oxygen stratification causes the faunal differences that we observe. The oxygen profile is not linear but changes suddenly at a depth of 2.5 m (Fig. 2). Consequently, the correlation between the chironomid fauna and the oxygen profile measurements is higher. Significant correlation means that the larger the difference in oxygen concentration between two sites, the higher their faunistical dissimilarity. Two sites, one with high and the other with low oxygen concentrations, have few species in common. It is amazing that our first rough estimate of the bottom oxygen concentration (by taking the oxygen concentration in the water column at the same depth) was so strongly correlated with the dissimilarity-matrix.

The correlations between the fauna and the microelectrode oxygen measurements near the sediment-water interface (i.e. 1.2 mm above the sediment) were much lower. We had two reasons to expect that these correlations would be higher than the one based on the oxygen profile.

1) Taking the oxygen values for stations, situated relatively far apart, from a single oxygen profile that was measured at the centre of the pond might be inaccurate. We have not tested how stable the stratification is and we do not know if spatial factors influence the stratification depth.

2) Measuring the oxygen directly with a microelectrode is much more accurate. It takes into account all possible influences of the bottom. The most important effect is the existence of a microstratification near the sediment. Therefore the oxygen concentration of the water surrounding the benthos is lower.

These observations may be explained by assuming that the oxygen concentration at the sediment-water interface is less important to chironomids than the concentration

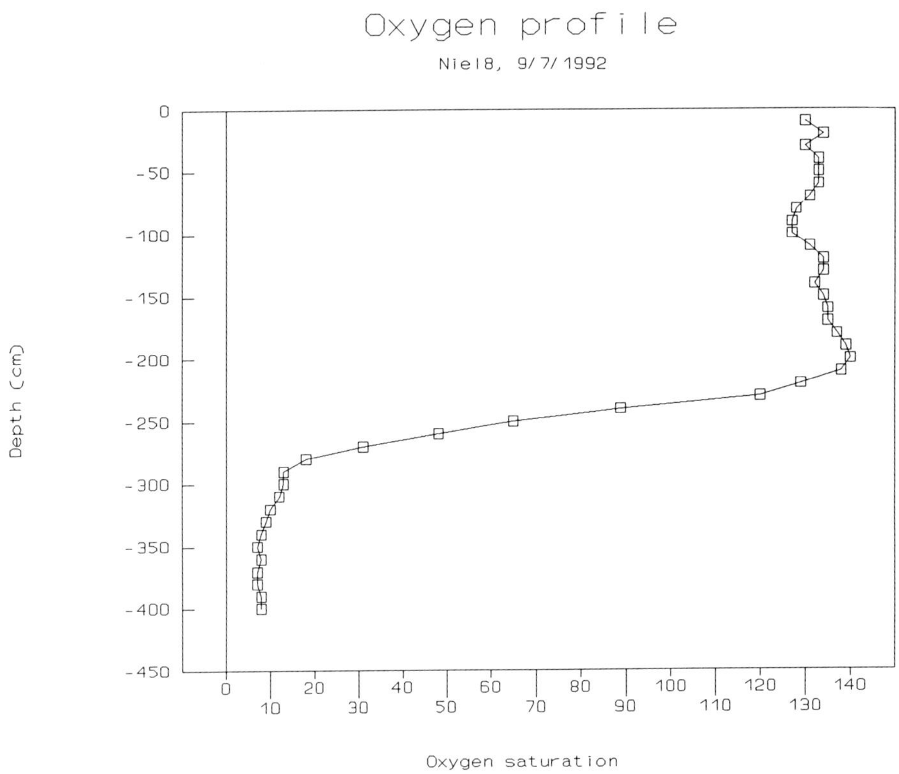

Fig. 2. Oxygen saturation profile measured on July 9th 1992 at the centre of the pond. There is a sharp discontinuity at a depth of 2.5 metres.

in the water column. Several behavioural adaptations could lead to this observation. It is possible that chironomids are able to ventilate sufficiently to break the microstratification just above the sediment. Some chironomid larvae are known to build a tube, which extends the sediment surface several millimetres, sometimes even centimetres (Thienemann, 1954). In this way they can use the oxygen that is available at one or more millimetres above the sediment. This is indicated by the weak correlation that we have found with the measurements taken at 1.2 mm above the interface. It is not too difficult to imagine a sediment surface dwelling chironomid that can ventilate oxygen-rich water down over a distance of several millimetres. In our pond, however, some species were mostly found several centimetres below the sediment surface (Table 2). It is hard to see how they could pump oxygenated water deep into their tubes. Heinis *et al.* 1994 has shown that the oxygen consumption of the surface of the burrow can be important. Some species have to compete for the available oxygen with the surrounding sediment. Because oxygen only penetrates about 1 mm into the sediment film, the larvae would be surrounded by anoxic sediment if they did not ventilate (Frenzel, 1990). Therefore we have abandoned our community approach and we have calculated correlation coefficients between the numbers of all chironomid species separately and the oxygen concentration in the water and at the interface. Although the overall result for the chironomid community suggests that the sediment oxygen concentrations are less important, we expected this not to be the case for the burrowing species. The results are given in Table 2 for those species that were significantly correlated with at least one of both measurements.

These results suggest species-specific behaviour. We find that there are two groups of species. A median depth threshold of 1.5 cm in the sediment was chosen arbitrarily to distinguish between surface-dwelling and burrowing species. A depth of 1.5 cm is representative for the second slice (1-2 cm) in a core. The densities of the species of the first group, i.e. surface-dwelling species, are correlated with the oxygen concentration in the water column. The correlation with the oxygen concentration at the sediment-water interface, however is lower or not significant. Testing of the significance of the differences between both correlation coefficients was hampered by the low number of sample units. We think that these species can ventilate oxygen-rich water from above the sediment-water interface down to the sediment surface.

Table 2. Correlations between the density and two of the oxygen measurements. The median depth of occurrence is given in centimetres. Group 1 are surface-dwelling species, Group 2 are burrowing species. (NS : not significant; * : $p < 0.05$; ** : $p < 0.01$; *** : $p < 0.001$).

Group 1 : Species	Median depth	Sediment	Water
Ablabesmyia	0.7	NS	0.49 *
Tanytarsus	1.1	NS	0.67 **
Procladius cf. *choreus*	1.3	0.63 *	0.77 ***
Cryptochironomus	1.3	NS	0.69 **
Microtendipes gr. *chloris*	1.4	0.56 *	0.69 **
Group 2 : Species	**Median depth**	**Sediment**	**Water**
Polypedilum nubeculosum agg.	1.5	0.57 *	NS
Glyptotendipes cf. pallens	2.5	0.64 **	NS
Stictochironomus cf. *maculipennis*	5.7	0.58 *	0.58 *

Group 2 consists of three sediment-dwelling species. The results for *Polypedilum nubeculosum* agg. and *Glyptotendipes* cf. *pallens* were as expected. Their densities are positively correlated with the sediment oxygen concentration. The more oxygen there is at the sediment-water interface, the higher their density. The oxygen that is present above this interface does not seem to be important for these species.

Stictochironomus cf. *maculipennis* is an exception. Its density is weakly correlated with both oxygen measurements. Therefore the habits of this species do not appear clearly from the oxygen measurements made in its microhabitat. Fourth instar larvae of *Stictochironomus* generally occur in deeper sediment layers. The following hypothesis could explain our observations. It is possible that fourth instar larvae of *Stictochironomus* migrate between the sediment surface and the deeper sediment layers where most animals were observed. In this way they could benefit from the higher oxygen concentrations at the sediment surface to breathe and store oxygen by their haemoglobin. The advantages of migrating to deeper sediment layers might include other food supply and reduced risk of predation. Such a migratory behaviour might explain why both the oxygen in the water column and the oxygen at the sediment-water interface are important to this species. However, there are no data to support this hypothesis and it should be tested experimentally.

Conclusions

We think that the fact that the densities of some burrowing species are correlated with oxygen concentrations near the sediment-water interface could indicate that their capacity to ventilate oxygen-rich water down is limited. If this is true, we must assume that some adaptive mechanism exists among these species that enables them to cope with diminished oxygen concentrations (Heinis, 1993). One obvious possibility is that they could have higher hemoglobin concentrations than the surface-dwelling Chironomidae. However, Heinis (1993) did not discover a clear relationship between the hemoglobin concentrations and the depth at which species occur in the sediment. Several authors have observed that larger species and late instars generally occur deeper in the sediment than the younger smaller animals and early instars. Brundin (1951) has suggested that body size is related to the ability of larvae to manage under unfavourable respiratory conditions.

Adaptations of chironomid larvae to low oxygen concentrations could be of a physiological, morphological or behavioural nature (Heinis, 1993). It is likely that all three types of adaptation occur together and determine the microhabitat of these species.

Acknowledgements

Hilde Sluyts, Bert Int Panis and all the members of the research group Ethologie en Natuurbehoud of the U.I.A. are thanked for their help with the field work. P. Dumont is thanked for his help with the sediment oxygen measurements and P. Swalus for technical help. The authors also acknowledge the N.F.W.O. (Belgian National Fund for Scientific Research) for financial assistance.

A STUDY ON PHOTOTAXIS OF ADULT CHIRONOMIDAE BY ARTIFICIAL LIGHT IN LAKE SUWA, JAPAN

Kimio Hirabayashi[1], *Ryoji Nakazato*[2], *Akio Ohara*[3] *and Tokio Okino*[2]

[1] Liberal Arts, Yamanashi Women's College, Kofu, Yamanashi Prefecture, 400, Japan

[2] Faculty of Science, Suwa Hydrobiological Station, Shinshu University, Suwa, 392, Japan

[3] Toshiba Lighting and Technology Co., Yokosuka, 237, Japan

INTRODUCTION

In many eutrophic lakes of Japan, there are large populations of larval chironomids, especially *Chironomus plumosus* Linnaeus and *Tokunagayusurika akamusi* Tokunaga (Yasuno *et al.*, 1983; Iwakuma *et al.*, 1988). Lake Suwa is surrounded by several cities with total populations of 200,000; many resort hotels and business establishments in this area are severely affected by the dense swarms of adult chironomid midges that emerge from the lake. In the Lake Suwa area, massive accumulations of living and dead chironomid adult midges have been observed (Hirabayashi, 1991b). Aspects of midge adult behaviour, including dispersal, patterns of abundance, diel eclosion periodicity and resting behaviour have been reported (Hirabayashi, 1991a, b, c; Hirabayashi *et al.*, 1992). In this paper, we attempt to clarify the attractive effect of colours, light intensity and wavelength on the adult midges of two species, *C. plumosus* and *T. akamusi*, in the field.

STUDY SITE

Lake Suwa is a tectonic lake located on the central highland of Honshu at an altitude of 759m above sea level (36°03'N, 138°05'E at the center of the lake). The surface area of the lake is 13.3 km^2, maximum depth is 6.5 m. A single stream, Tenryu River, drains the lake water into the Pacific Ocean (Fig. 1). The lake has been filled with sediments transported by rivers from its relatively wide drainage basin to form a typical eutrophic lake with an average depth of about 5m. The drainage basin is 80% forested, whilst the greater part of the population is concentrated in a few cities near the lake shore. The spectacular growth of industrial activity is indicated by heavy blooms of *Microcystis* (blue-green algae) that take place every summer (Anon, 1984).

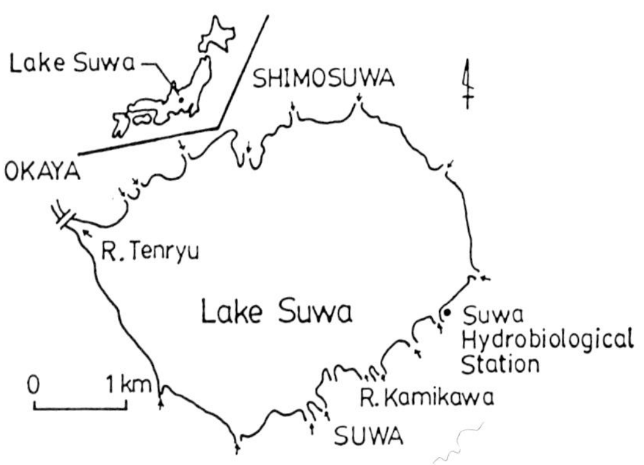

Figure 1. Map of the study area of Lake Suwa

MATERIALS AND METHODS

These studies were conducted during the September to November periods of 1989 and 1990 on the rooftop of the Suwa Hydrobiological Station which is 18 m above ground level, and located on the east shore of Lake Suwa (Fig. 1). Five light traps were used. Each trap was hung from a pole 1.5 m above floor level.

Number of adult midges attracted by visible lights

Different colour lamps with the same wattage

The 100W lamps used were white (100-WL), yellow (100-YL), blue (100-BL), green (100-GL) and red (100-RL) in colour. The qualitative characteristics (wavelengths) of the lamps used are previously been reported. All lamps emitted broad band wavelengths in the visible spectrum. The relative spectral distribution of the lamps used for the experiments is shown in Figure 2.

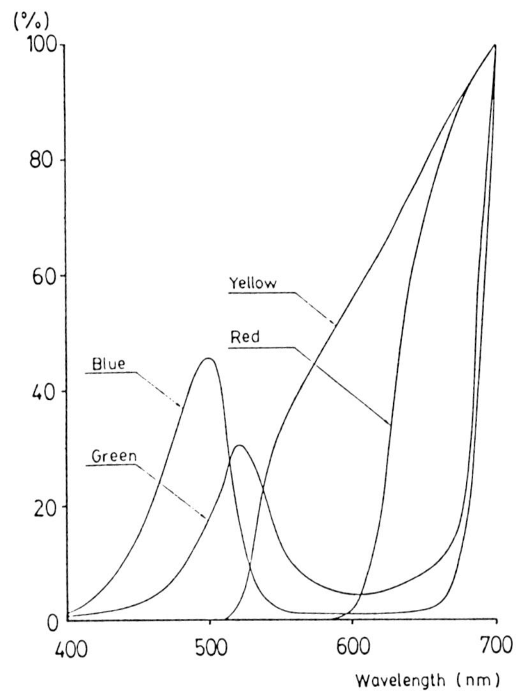

Figure 2. Relative spectral distribution of the lamps used for light attraction experiments. The energy of a white lamp of 40 W (100 V) was calculated as 100%. Source: Technical data from Toshiba Lighting and Technology Co.

The energy of the white 40 W lamp (100 V) was calculated as 100%. To estimate the relative light intensity of the lamps, the illuminance of each lamp was measured at night. The intensities of illumination for 100-WL, 100-YL, 100-BL, 100-GL and 100-RL were 474, 299, 14.8, 18.3 and 43.7 lx, respectively.

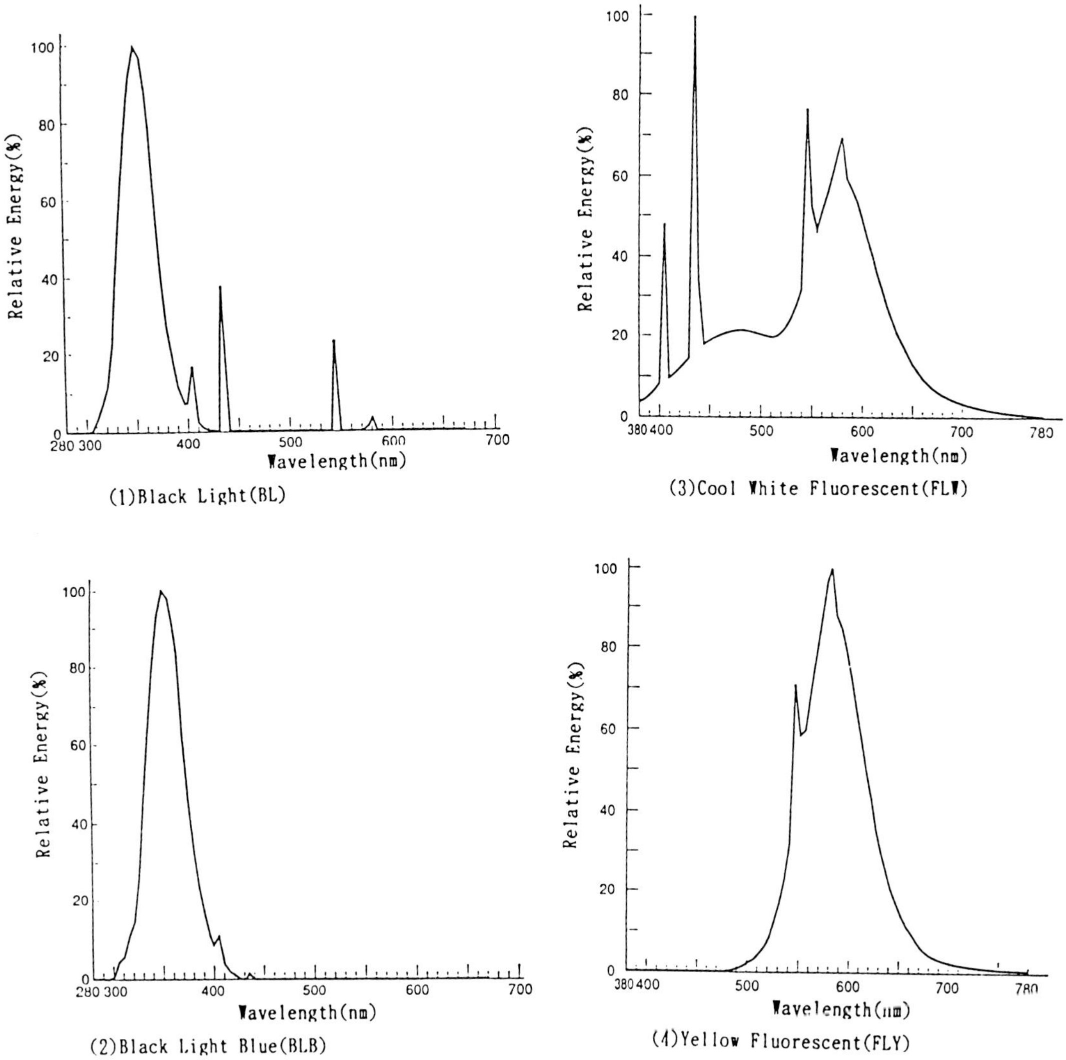

Figure 3. Relative spectral distribution of the lamps used for light attraction experiments. Source: Technical data from Toshiba Lighting and Technology Co.

Different wattage lamps with the same colour

White lamps of 100 W (100-WL), 60 W (60-WL), 40 W (40-WL), and 20 W (20-WL) were used. The intensities of illumination for 100-WL, 60-WL, 40-WL and 20-WL were 474, 251, 171 and 52.5 lx, respectively.

Number of adult midges attracted by near-ultraviolet lights

Five lamps of Black Light Blue lamp (BLB; completely reduced visible lights), Black Light lamp (BL), Cool White Fluorescent lamp (FLW), Non-attractive Yellow Fluorescent lamp (FLY) and 100 W white lamp

(100-WL) were used. Except for the 100-WL which was a 100 W lamp, all lamps used were 40 W lamp. The qualitative characteristics or wavelength of the lamps used are available. Except for the 100-WL, all lamps emitted broad band wavelengths in the ultraviolet spectrum. The wavelengths emitted by BLB were between 290 and 410 nm, those of BL were between 290 and 500 nm, those of FLW were between 350 and 705 nm, and those of FLY were between 500 and 705 nm. Both BLB and BL light peaks at ca. 345-355 nm, which is shorter than the FLW and FLY peak wavelengths (Fig. 3).

To attract adult midges emerged from Lake Suwa, the traps in each test were activated about half an hour before sunset and the midges trapped in the field were left until the next morning about two hours after sunrise. The adults trapped during each period were identified and counted in the laboratory. No visible sources of artificial light were noticed in this area during the study periods at night.

Results

Number of adult midges attracted by visible lights

Comparison between different colour lamps of same wattage

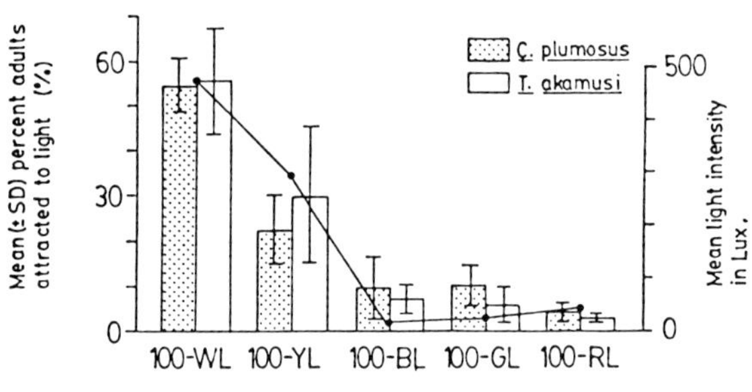

Figure 4. Attraction of adult midges (*C. plumosus* and *T. akamusi*) by five different colour lamps of the same wattage (100W) tested in the field. Bars represent the percentages of adults attracted by light; the line graphs indicate the light intensity: 100-WL, 100-YL, 100-BL, 100-GL and 100-RL indicate white, yellow, blue, green and red incandescent lamps respectively. Data on *T. akamusi* are from Hirabayashi (1991c).

The attraction of adult midges by five different colour lamps of the same wattage tested in the field is shown in Figure 4. Bars represent the percentages of adults attracted by light, while the line graphs indicatethe light intensity. In case of *C. plumosus*, 100-WL attracted 56.2%, 100-YL 22.1%, 100-BL9.0%, 100-GL 9.5%, and 100-RL 3.2% of the total midges trapped. *T. akamusi* responded the most with 57.0% to 100-WL and the least with only 2.2% to 100-RL. The response of *T. akamusi* to the traps was generally similar to that of *C. plumosus*. All lamps in this test were 100 W, but due to the different colours the intensity differed and ranged from 474 lx for 100-WL to 14.8 lx for 100-BL. Thus, 100-WL with the highest intensity invariably attracted significantly higher numbers of adults than other lamps. The 100-YL was the second best preference for both species. The midge responses to the other three lamps, 100-BL, 100-GL, and 100-RL, were statistically insignificant. We also ran a preliminary test which revealed that both species were attracted in almost equal numbers by the different-coloured lights of the same intensity.

Comparison between different wattage lamps of same colour

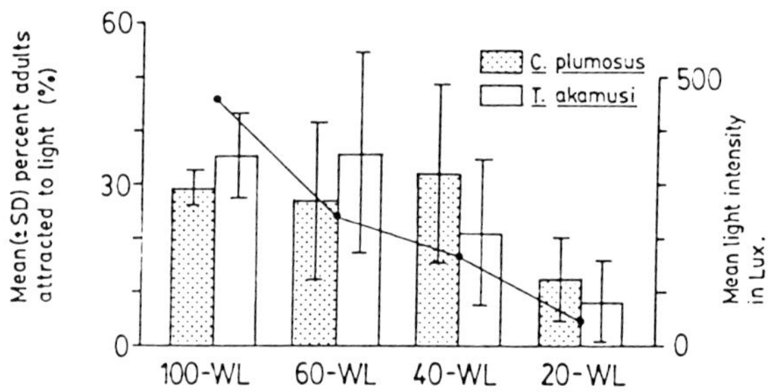

Figure 5. Attraction of adult midges (*C. plumosus* and *T. akamusi*) by four different lights of the same colours but different wattages in the field. Bars represent the percentage of adults attracted to white light; the line graphs indicate the light intensity: 100-WL, 60-WL, 40-WL and 20-WL indicate white incandescent lamps, 100, 60, 40 and 20 W, respectively. Data on *T. akamusi* are from Hirabayashi (1991c).

The attraction of adult midges by four different lights of the same colours but of different wattages is shown in Figure 5. The white lamp of 100W and an illuminance of 474 lx attracted 28.8% of the total C. *plumosus* midges, while 20-WL and an illuminance of 52.5 lx induced only 12.0%. The 100-WL with the highest intensity invariably attracted significantly higher numbers of adults than the 20-WL. The midge responses to the three lamps, 100-WL, 60-WL and 40-WL, were statistically insignificant. On the one hand, the response of *T. akamusi* was similar to that of *C. plumosus*, but the midge responses to the two lamps, the 100-WL with 35.2% and the 60-WL with 35.6%, were statistically insignificant. The 100-WL and the 60-WL lamps attracted significantly higher numbers of adults than the 40-WL with 20.8% or 20-WL with 8.4%.

Number of adult midges attracted by near-ultraviolet lights

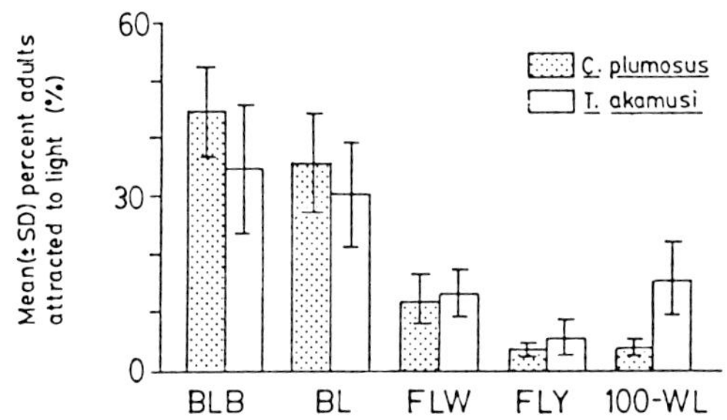

Figure 6. Attraction of adult midges (*C. plumosus* and *T. akamusi*) by five different wavelengths of ultraviolet light tests in the field. Bars represent the percentage of adults attracted to light: BLB, BL, FLW, FLY and 100-WL (control) indicate black light blue, black light, cool white fluorescent and non-attractive yellow fluorescent lamps (40W) respectively.

The attraction of adult midges to five different wavelengths of near-ultraviolet light is shown in Figure 6. With *C. plumosus*, BLB with 44.8% invariably attracted significantly higher numbers of adults than BL with 36.0%. Moreover, BLB and BL lamps attracted more than 80% of the total adult midge captured in the traps used in this test. After BLB and BL, the adult midge responded in significantly higher numbers to the FLW with 11.8% as compared with the FLY with 3.7% or 100-WL with 3.7%. Number of adult midges attracted by FLY and 100-WL lamps were statistically insignificant. FLY attracted only one - eleventh the number of adult midges as did BLB. The response of *T. akamusi* was similar to that of *C. plumosus*, although *T. akamusi* responded in significantly higher numbers to the 100-WL with 15.6% than the FLY with 6.3%. FLY attracted as one - sixth the number of adult midges as did BLB.

DISCUSSION

Many studies on the attraction to light of a variety of nocturnal insects have shown that insect species exhibit a specific response to certain ranges of wavelengths in the electromagnetic spectrum (Hienton, 1974). The studies of Ali *et al.* (1984, 1986) in the laboratory and field in Lake Monroe on the attraction of the most predominant species, *Glyptotendipes paripes*, involving the combination of white, yellow, orange, blue, green and red lamps of the same wattage, showed that these species were generally most attracted by white light (highest intensity) and least by blue, green or red light (low intensity). Thus, these species responded more to the quantity (power or intensity) than to the quality (colour or wavelength) of the visible area of the electromagnetic spectrum. The present findings are in agreement with those of Ali *et al.* (1984, 1986) (Fig. 4 and 5). However, in the near-ultraviolet region the behaviour of *C. plumosus* and *T. akamusi* is in contrast to that of *G. paripes*; the former respond to a certain range of wavelengths, particularly near the ultraviolet region as shown in Fig. 6. In addition, the phototactic response of *C. plumosus* in these tests were not similar to *T. akamusi*; in other words, radiant energy of a specific wavelength does not equally attracted all species of chironomid midges. Such information is of practical significance for the development of an integrated control program against pestiferous chironomid midges.

This investigation on attraction of chironomid midges by light demonstrated that more work using different approaches is necessary. For example, adult midges might be eliminated by electrocutor traps. In any case, it is clear that by appropriate use of the proper lights on and around a lake adults of these species can be drawn to uninhabited or relatively less urbanized areas for suitable midge control.

Acknowledgements

We wish to express our thanks to Dr. Y. Nasu of the Department of Public Health and Prof. N. Murayama of the Department of Hygiene, School of Medicine, Shinshu University, for their considerable assistance and valuable suggestions. We are pleased also to acknowledge the full support of the members of the Hydrobiological Station, Department of Biology, Faculty of Science, Shinshu University, and Mr. Shinada (Toshiba Lighting and Technology Co.).

Chironomids in the diet of fish in Lake Banyoles (Catalonia, Spain)

Maria Rieradevall[1], *Emili García-Berthou*[2] *and Narcís Prat*[1]

[1] - Dept. Ecologia, Univ. Barcelona, Diagonal, 645, 08028 Barcelona, Spain.

[2] - Institut d'Ecologia Aquàtica, Pl. Hospital, 6. 17071 Girona, Spain.

Introduction

The effects of fish predation on freshwater benthic communities is well studied (Schlosser and Ebel, 1989; Winkel *et al.*, 1987). In the simplest cases, an increase in predation intensity generally produces a decrease in abundance and diversity of benthos, including chironomids. The population structure also is often affected. Laboratory experiments have shown that chironomid larvae reduce their activity and growth in the presence of predatory fish (Macchiusi and Baker, 1991, 1992).

Previous studies have provided detailed information on composition, abundance, population dynamics, and production of the chironomids from the sublittoral and profundal zones of Lake Banyoles (NE Spain) (Rieradevall, 1991). One of us is currently studying the habitat and resource use of the fish assemblage in the lake. Since chironomids are frequent in the fish gut contents, we now attempt to assess their importance in fish diet and to improve understanding of the relationship between the benthic and fish communities in the lake.

Study Site and Methods

Fish were caught in Lake Banyoles, situated at 42°7′N, 2°45′E and at 175 m above sea level. The maximum length of the lake is 2.13 km and surface area is 1.18 km^2. The lake is the largest water mass of a karstic system, which includes several accompanying ponds. Mean depth is 15 m and the maximum depths are around 40 m. The littoral zone (0-3 m) represents a thin strip, since the lake rapidly reaches a depth of 20 m. Macrophytic vegetation (*Phragmites*, *Schoenoplectus* etc.) and several *Myriophyllum verticillatum* meadows as well as travertinic rocks are abundant in the littoral. The lake is oligotrophic with cosmopolitan phyto- and zooplankton. Detailed descriptions of the lake and the limnology of the zone are given elsewhere (Garcia-Gil, 1990; Planas, 1991; Rieradevall, 1991).

The littoral fish assemblage was sampled using boat electrofishing and trammel nets. Mosquitofish and some alevins of several species were obtained with dip-nets. Sublittoral and profundal fish were sampled with trammel nets at 5, 10, 15 and 20 m. To prevent digestion, captured fish were immediately held on ice until they were frozen in the laboratory.

A total of 1321 gut contents were examined. In the fish with a well-defined stomach, only the stomach contents was examined. In the other species (i.e. cyprinids) the whole gut was examined. Prey were sorted by taxon, counted and weighed to the nearest 0.1 mg. For the smallest fish (i.e., mosquitofish and alevins) the volume was measured instead of weight. To describe fish diet, numbers and amount (biomass or biovolume) of all prey categories were summed over all fish guts. The importance of a food category was then calculated as a percentage of total number or amount in the gut contents.

Chironomids were sorted under stereoscope, rinsed carefully, cleared and mounted individually on an slide using euparal. Large forms (*Procladius* and *Chironomus* fourth instars) were not prepared since they are easy to identify. The larval instar was deduced from the head capsule length and/or width according to the measurements taken previously in the study of benthos by Rieradevall (1991). Biomass of each instar is also calculated using the values provided in the studies of this author in the same paper.

Results

Eleven species of fish were captured in the lake. Several species, e.g., eel, perch, and barbel, are now very rare in the lake. The most frequent and abundant species is roach in the limnetic zone, and pumpkinseed sunfish and largemouth bass in the littoral zone. Most abundant species are introduced fish and the native fish are now scarce.

Chironomids were frequent in the guts of many fishes (as much as 66% of presence in some species, Table I), especially in those of more benthic habits like carp or freshwater blenny. Cephalic capsules were also found with some frequency in roach and pumpkinseed sunfish (18 and 23 % of the guts examined respectively). However, their importance in the fish diet was low (Table 1), especially if we consider the biomass of the midges compared with other components of the diet. Chironomid biomass was most important in the sunfish and the blenny, although it never reached 5% of the total mean biomass present in the fish guts examined. Other benthic animals (especially in the blenny, sunfish, and mosquitofish) together with unidentifiable digested matter and detritus (in roach, rudd, and carp) were the major components of the diet of Banyoles fishes (Fig 1).

Table 1. Chironomid importance (frequency of occurrence %FO, percent number %N, and percent biomass %B) in the fish diet of Lake Banyoles.

fish species	%FO	%N	%B
Anguilla anguilla	0.00	0.00	0.00
Blennius fluviatilis	37.50	14.43	1.45
Barbus meridionalis	66.67	2.13	0.01
Cyprinus carpio	64.00	6.49	0.18
Gambusia holbrooki	9.84	0.53	0.51
Leuciscus cephalus	0.00	0.00	0.00
Lepomis gibbosus	22.98	10.54	4.31
Micropterus salmoides	2.30	0.52	0.00
Perca fluviatilis	0.00	0.00	0.00
Rutilus rutilus	17.67	1.04	0.25
Scardinius erythrophthalmus	5.66	0.74	0.00

28 chironomid species were identified in the fish guts, including 2 Tanypodinae, 9 Orthocladiinae, 14 Chironomini, and 3 Tanytarsini (Table 2). The most abundant prey species were *Procladius* and *Chironomus* but relative abundance changed among species.

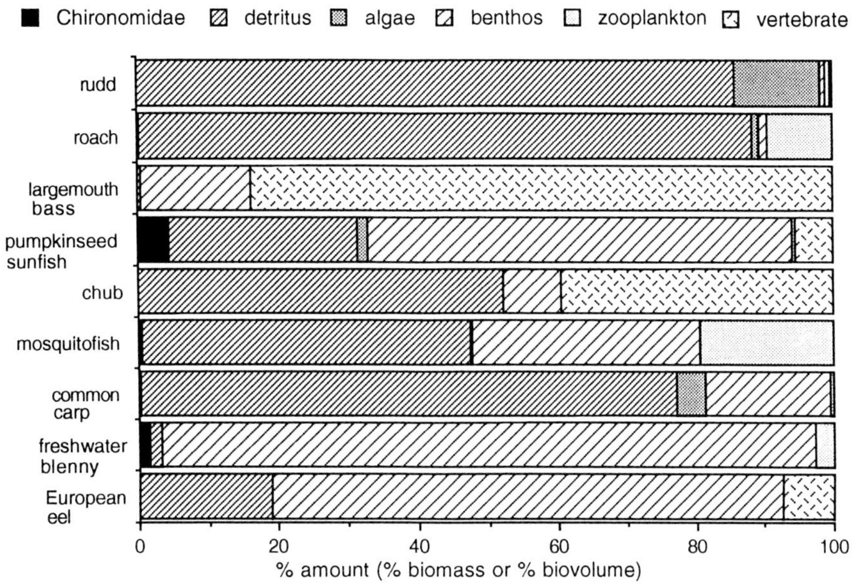

Figure 1. Percentage of importance (by biomass) of each food type in the guts of the differents species of fish studied in Lake Banyoles. Note the small importance of chironomids.

However, for some fishes the number of guts examined was low (e.g. blenny) or the number of chironomid remains was low (e.g. mosquitofish, largemouth bass) and for this reason the figures in Table 2 should be treated with caution. Only for three species (carp, sunfish and roach) can the number of chironomid head capsules be considered adequate to establish some conclusions.

Carp feed especially on the large benthic forms present in the benthos (third and four instars of *Procladius* and *Chironomus*). These two species are also the most abundant chironomids at all depths of the lake (Rieradevall and Prat, 1991, see also Table 2). The pumpkinseed sunfish preys on a diverse array of chironomids but half the head capsules found in the guts were from *Microtendipes*, a species scarcely found in the benthos of the lake and more common in the sublittoral than in the profundal zones (Rieradevall, 1991). The presence of *Glyptotendipes* and *Polypedilum* also suggests that this species is mostly littoral in its feeding habits (0-5 m depth in the lake). The same can be deduced (from specific composition) for the mosquitofish and the freshwater blenny (Table 2). These fishes and the pumpkinseed sunfish have been caught exclusively or preferentially in the littoral zone of the lake (0-5 m, Table 3).

In the roach, the abundance of *Procladius* and *Chironomus* and the scarce importance of the other species also show the benthic habits of this fish, very similar to those of carp, with an abundance of third and fourth larval instars.

The importance of each chironomid species at different depths of the lakes is shown in Table 2. From this table it can be seen that the importance of *Procladius* and *Chironomus* in roach and carp diet is similar to the abundance of the two chironomids in the benthos at 12 meters. In fact many of the individuals of carp (60%) and up to 56% of the roach were caught at similar depths (Table 3), and thus the importance of the preys reflects their abundance in the lake benthos as well as the preferred distribution of the fishes in the lake.

Table 2. Relative percentage of chironomid larvae in the fish gut contents and in the benthos of Lake Banyoles at three different depths. For each fish species the number of guts examined and the pooled number estimate of chironomid larvae is also given.

	fish gut contents							benthos		
	freshwater blenny (*B. fluviatilis*)	common carp (*C. carpio*)	mosquitofish (*G. holbrooki*)	pumpkinseed sunfish (*L. gibbosus*)	largemouth bass (*M. salmoides*)	roach (*R. rutilus*)	rudd (*S. erythrophthalmus*)	20 m	12-13 m	5-7 m
undetermined Chironomidae	3.6	1.0	12.5	0.8		0.1				
undetermined Tanypodiinae		0.6								
Procladius sp.		43.8		14.4		67.4		84.1	56.5	33.6
Tanypus sp.		2.0		1.5				+	4.5	1.0
Corynoneura cf. *scutellata*	3.6		9.4							
Cricotopus (*I.*) *sylvestris* gp.	3.6		3.1					+		
Cricotopus (*C.*) *?triannulatus*	3.6			0.2		0.3				
Cricotopus (*C.*) *albiforceps*				0.2						
Cricotopus (*C.*) *annulator*	25.0	0.0						+		
Cricotopus (*C.*) *bicinctus*				0.2				+		
Cricotopus (*C.*) *flavocinctus*		1.0		3.1	22.2	0.2	78.5			
Parametriocnemus sp.	17.9		12.5		11.1					
Psectrocladius (*A.*) sp.				0.3						
undetermined Chironomini		0.0	12.5	0.5	11.1	0.3				
Cladopelma virescens		0.6		0.3		1.5		2.0	13.7	18.5
Cryptochironomus sp.		0.0						+	+	+
Dicrotendipes sp.	3.6									
Chironomus sp.	3.6	27.6	6.2	15.9		18.6		5.2	14.1	6.5
Glyptotendipes sp.		0.5		9.4	11.1	0.0				
Harnischia sp.						0.0		+	+	
Kiefferulus tendipediformis		0.4		0.2						
Microchironomus tener		5.8				0.0		4.1	5.6	19.0
Microtendipes sp.	17.9	5.9	6.2	50.0	22.2	0.5		+		
Parachironomus sp.					11.1	0.3				
Paratendipes sp.		0.7				2.0		+	+	
Phaenopsectra sp.	3.6	0.0	3.1							
Polypedilum gr. *nubeculosum* sp.	14.3	0.6	31.2	2.0				+		
Stictochironomus maculipennis	6.2					0.1		1.7	+	+
undetermined Tanytarsini			3.1			0.4		+	+	
Cladotanytarsus sp.		2.9		0.5		1.1		+	3.3	13.3
Paratanytarsus sp.				0.3	11.1	0.0	21.5			
Tanytarsus sp.		0.3		0.2		7.1		1.5	1.4	2
number of examined guts	16	50	254	384	261	261	56			
total number of chironomids	28	5537	32	604	9	3869	270			

The most striking feature is the under-representation in the guts of these two fish of some abundant species such as *Cladopelma virescens*, *Microchironomus tener* and *Cladotanytarsus atridorsum*, the latter well represented in the sublittoral zone, especially at 5 metres. Conversely it is interesting to note the higher relative abundance of *Tanytarsus* in the gut of roach respect to its presence in the benthos (Table 2).

Table 3. Relative abundance (as % of the total fish catch) of the more important fish of Lake Banyoles.

fish species	0 m	1-5 m	6-15 m	> 15 m
Blennius fluviatilis	100	0	0	0
Gambusia holbrooki	100	0	0	0
Micropterus salmoides	94	6	0	0
Scardinius erythrophthalmus	8	88	4	0
Lepomis gibbosus	42	55	3	0
Rutilus rutilus	30	11	56	3
Cyprinus carpio	13	23	60	4

The larger fishes also preyed on larger chironomids than small fishes as can be seen clearly in Table 4. Thus roach and carp eat mostly chironomids larger than 500 µm of head capsule length, which are mainly III and IV instars of *Procladius Microtendipes* and *Chironomus*. In contrast, the mosquitofish and the blenny eat smaller larvae, mostly species associated with vegetation, as noted above (Table 2). Even the pumpkinseed sunfish, which prey mostly on *Microtendipes*, eat larger forms whose head capsules were larger than 500 µm (Table 4).

Table 4. Percentage of each size class of larval head capsules of chironomids present in the fish guts of Lake Banyoles and mean size of each fish species.

fish species	< 250 µm	250-500 µm	>500 µm	Fish size mm.
Blennius fluviatilis	52	32	16	32
Gambusia holbrooki	41.4	44.8	13.8	22
Lepomis gibbosus	0.2	5.2	93.6	115
Rutilus rutilus	1.7	8.7	89.6	192
Cyprinus carpio	0.3	9.7	90	475

DISCUSSION

Although chironomids are important elements in the fish diets in lakes with taxonomic composition similar to Banyoles (e.g. Bazzanti *et al.*, 1989), few studies identify the different species present in the guts. In these cases, and for cyprinids, there are direct qualitative and quantitative relationships with the benthic fauna present (Winkel, 1987). There does not appear to be any specific selection.

From the Table 2 it seems clear that the more abundant fish in the open area of the Lake Banyoles (roach and carp), eat the most abundant chironomid in the benthos at a percentage similar to their presence in the mud except for the small species. The fish seem to prefer the large mobile forms (as *Procladius*) or those that can be easily seen in the bottom (the large red *Chironomus*). These larger forms are easily found in the fish mouth, while small species are not taken by the fish due to their size and even rejected. This feeding behaviour (size selection) has been described in detail in several papers (e.g. Winkel *et al.*, 1987). On the other hand, small species like *Cladopelma virescens*, *Microchironomus tener* and *Cladotanytarsus* spp. that contribute more than 50% of the chironomid densities in benthos at 5 meters and 20% at 12 meters are not taken to the same extent by the fishes (8% in carp and less than 3% in roach, Table 2). The same situation has been described from a Dutch lake where the abundant, but small *Cladotanytarsus* is not taken according its benthic abundance, while the larger *Stictochironomus* is found in higher proportions in the gut of breams than in the benthos (Winkel *et al.*, 1987). Our data support the way of feeding of cyprinids with selection on large preys as in other studies, e.g. Richeux *et al.* (1993) who found that roach species feed in an opportunistic way and the gut content reflects the available prey in the reservoir they studied. In that study, and similarly to that found in Banyoles (Fig. 1), detritus and zooplankton are the most abundant food in the gut of the roach. The same authors found a seasonal change in the prey capture according with the abundance of preys. The changing hierarchy of preys with the season in Banyoles fishes has still to be examined, although we know that there are important changes in density and relative abundance in benthic chironomids (Rieradevall, 1991; Prat and Rieradevall, in press), therefore changes in the importance of each species during the year can be expected.

Another general conclusion of the studies of fish-chironomid interactions in lakes is that high densities of fishes, especially cyprinids, depleted the density of benthic chironomids. This has been demonstrated by experiments in enclosure cages, which show how chironomid density, in cages where fishes are excluded, increased 5 to 10 fold (Kajak, 1977). The reduction of fishes in ponds is currently used to diminish eutrophication in lakes that sometimes increases chironomid densities (Moss and Turner, 1989) but in some cases does not affect their density (e.g. Riemann *et al.*, 1990). The mean annual densities of chironomids in lake Banyoles ranged from less than 400 to more than 9000 ind m^{-2} (Rieradevall and Prat, 1991), the highest densities were found in the sublittoral zone and they are mostly small individuals. The mean biomass is between 40 and 500 mg m^{-2}, far from the current values of eutrophic lakes and ponds (Bazzanti *et al.*, 1989). The predominance of small individuals in Banyoles and the low biomass and production of the lake (Prat and Rieradevall, in press) are in accordance with the oligotrophic state of the lake, but could also be related to the predation pressure of fishes. Although cyprinids are known to feed preferentially on macrophytes rather than on chironomids (Prejs, 1984), in absence of plants they are more detritivorous and use the food available in the benthos. Lake Banyoles was known in the past by the presence of a dense stands of *Chara*, which have now disappeared and many subfossil remains of these plants are found in the sediments. Disappearance of plants could be related to the introduction of roach and carp in the lake, which in turn, after the destruction of the submerged meadows, have shifted their feeding habits to a more detritivorous diet, which includes the chironomids present in the benthos.

In summary, the study of chironomid diet in the fishes of lake Banyoles has shown that the importance of each midge species is similar to the benthic relative abundance at the preferred depth distribution of the two more abundant fish species and that a size selectivity existed and large fishes eat large prey.

Part 7

Chironomid morphology and systematics

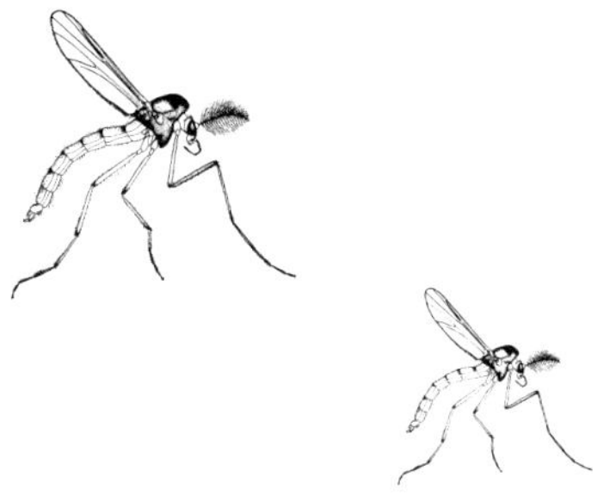

Morphological variation in secondary sexual characteristics of gynandromorphic *Chironomus* (*Camptochironomus*) *tentans* Fabricius (Diptera, Chironomidae)

James E. Sublette[1] *and Jon Martin*[2]

[1] Department of Biology, University of Southern Colorado, Pueblo, CO, 81001, USA

[2] Genetics Department, University of Melbourne, Parkville, Victoria 3052, Australia

Introduction

In a laboratory culture of *Chironomus (Camptochironomus) tentans* Fabricius originally derived from a wild population collected near Winnipeg, Manitoba, Canada, a gynandromorphic expression of a presumed single mutation has occurred (Martin, 1995). There are two axes of variable expression: anteroposterior and lateral, with the most conspicuous secondary sexual characteristics, the antennae and genitalia, demonstrating slight to pronounced sexual dimorphism.

Methods

Seventeen reared specimens ranging from normal males and females to those showing various degrees of sexual intermediacy were examined with an ISI SR-50 scanning electron microscope. Specimens were fixed, dehydrated and coated with gold-palladium following conventional procedures. Some abdominal terminalia (genitalia) were removed and mounted in a vertical position on examination pedestals in order to provide 360° examination of all surfaces.

Details of the laboratory crosses which provided the specimens examined are presented in Martin (1995). Specimens examined are retained in the laboratory of the second author. Terminology follows Sæther (1977a, 1980).

Results

The antennae of most chironomids demonstrate a pronounced sexual dimorphism with the female exhibiting a small pedicel (Fig. 1) (about half the size of the male, cf. Fig. 35) and a short flagellum with reduced flagellomeres and sparse macrotrichia (setae). The male antennal pedicel is enlarged (Fig. 2) and houses the specialized Johnston's organ, a sound receptor, which presumably aids the individual in recognizing conspecific individuals with which it forms an aerial swarm. Ground mating species may demonstrate a range of pedicel size ranging from an enlarged pedicel characteristic of those species which form aerial swarms to species with pedicels reduced to the appearance of the female.

The male flagellum is elongate and has more numerous flagellomeres than the female. The flagellum is densely clothed with very long macrotrichia (Fig. 2) except for the tip (Fig. 3). The long plume-like macrotrichia are postulated to assist in sound detection through transmission of vibrations to the Johnson's organ in the pedicel.

The male genitalia, which have been illustrated by Townes (1945, Fig. 151), are exerted and conspicuous (Figs. 4, 5, 6) while the female structures, which have been illustrated by Sæther (1977a, fig. 82), are more cryptic. The cerci (Ce) are enlarged, exerted leaf-like appendages (Figs. 7, 8) while the genital orifice is surrounded by the dorsomesial lobes of gonapophysis VIII (DmL) (Fig. 8) and the membranous labia (L) (Fig. 9). Projecting from beneath the DmL is the ventrolateral lobe of gonapophysis IX (VlL) (Figs. 8, 9). In those species in which the male has an anal point there is usually a well defined sulcus (Su) for reception of the anal point during copulation (Figs. 8; cf. Sublette, 1979, fig. 59). The internal apodemes which are of significance in female taxonomy (Sæther, 1977a) cannot be seen with the SEM without dissection.

The male and female antennae arise from the same anlagen. Thus, the antennal structure can be expressed as either male or female or some intermediate. Expressivity can be bilateral with left being male and right female, or *vice versa*, with varying degrees of penetration.

The genitalia, however, arise from separate anlagen; thus, expressivity can include a complete (or almost complete) female genitalia but with additional male components. The reverse condition of complete (or almost complete) male genitalia with additional female components was not observed, although one specimen showed a lateral amorphous tissue mass which probably represented the female component.

The following describes some of the variants of expression observed.

Antennae gynandromorphic; genitalia normal

Complete anteroposterior gynandromorphism was observed in six specimens with complete male antennae but with female genitalia occurring in four specimens, while two specimens had complete female antennae but with the specimens having complete male genitalia.

An additional three specimens showed varying degrees of penetration on the lateral axis. These are described as follows:

Specimen 9

Head - Complete bilateral sexual dimorphism (Reverse of that seen in Fig. 35); the right antenna is normal male (Figs. 10,11); the left antenna is normal female (Figs. 12,13).

Genitalia - Normal female (Fig. 14); the DmL, VlL and L appear normal (Fig. 15).

Specimen 15

Head - Antennae largely female with only a slight indication of male expression (Fig. 16). The right antennal apex (Fig. 17) has a few male-like macrotrichia mostly on the lateral face. The left antennal flagellum was unfortunately lost during preparation but the pedicel (Fig. 16) is slightly larger indicating that the left antenna is more masculinized.

Genitalia - The female terminalia appear completely normal (Fig. 18).

Specimen 16

Head - Left antenna almost completely feminized but with the right showing only slight feminine characteristics (Figs. 19, 20). While the left antenna is clearly of the female pattern, the pedicel is larger than a normal female and the lateral surface of the flagellomeres have some of the elongate macrotrichia (plumes) characteristic of the male (Fig. 20). The right antenna is superficially masculinized in appearance with large pedicel and long macrotrichia; however, the number of flagellomeres is reduced from that of a normal male (Fig. 21). The macrotrichia appear to be more densely packed than normal.

Genitalia - Typical male terminalia (Figs. 22, 23). The specimens were not ultra-sonically cleaned for fear of destroying salient features.

Antennae normal; genitalia gynandromorphic

Specimen 14

Head - Normal male.

Genitalia - Female components ventral; male, dorsal (Fig. 24). Both Ce normally shaped but other external female components distorted and misshapen. Of the male, only the ninth tergum and anal point (Fig. 25) are developed; both are essentially normal in proportions.

Both antennae and genitalia gynandromorphic

Specimen 10

Head - Antennae foreshortened and superficially female-like (Fig. 26). However, the basal flagellomeres, although reduced in number, are disc-like (Fig. 26) which is similar to that of the male antenna. The apical flagellomere is female-like on the lateral surface (Figs. 27, 28) but the medial surface bears a dense brush of male-like macrotrichia (Figs. 26, 27) although these are much more densely packed (Fig. 28) than in the normal male.

Genitalia - Left side (Fig. 29), essentially normal male, including anal point and ninth tergum (T IX). Right side, female components represented by a twisted, amorphous mass (Fig. 30).

Specimen 8

Head - Not examined under SEM but slide mounted for light microscopy. The head was normal male.

Genitalia - Right male component incomplete (Fig. 31), but with essentially normal gonocoxite (Gc), gonostylus (Gs) and inferior volsella (I Vo); the anal point is completely lacking and the T IX is essentially normal female (Fig. 32). The female genitalia are complete but with the Ce compressed and somewhat distorted (Fig. 33); the DmL, VlL, L and Su are normal (Figs. 33, 34).

Specimen 17

Head - Complete differentiation of antenna; right antennas apparently normal female, left, apparently normal male (Fig. 35).

Genitalia - Incomplete differentiation; left, almost complete male genitalia (Fig. 36) ; right, male component with ninth tergum and anal point slightly abnormal (Fig. 36). Female largely vestigial (Fig. 37) with Ce and DmL represented by an unorganized tissue mass (Fig. 38), but with VlL and L readily recognizable (Fig. 39).

Specimen 5

Head - Right antenna masculinized, left, feminized as indicated by a large (male) and a small pedicel (female).

Genitalia - Left male component normal (Fig. 40); anal point and ninth tergum well developed, but distorted (Fig. 41); right male component represented by a vestigial but discernable gonocoxite (Gc) and gonostylus (Gs) (Fig. 40); of interest is a clearly defined SVo associated with the vestigial Gc/Gs (Fig. 42); lateral to the vestigial Gc and Gs is a distorted mass of tissue which may represent female components although no recognizable morphological feature can be seen.

Discussion

Although referred to here as gynandromorphs, the features of a number of the specimens, taken in conjunction with the conclusion of Martin (1995) from other specimens, suggest that many of them should be more correctly called mosaics. However since no measurements of wings or legs, or examination of the internal reproductive organs was possible on these specimens mounted for SEM-examination, no conclusions as to the extent of mosaicism is possible.

The results reported here indicate that the male and female antennae arise from the same anlagen. However, the male and female genital structures arise from a different anlagen, or at least from different components within the same basic anlagen, permitting the simultaneous production of both male and female structures on the same side of the animal. The observation of a virtually complete female genitalia with additional male components, but not the reverse, may be a consequence of the nature of these gynandromorphs. All individuals are presumed to have begun as genetic males, with female tissue resulting from mitotic recombination in the nuclear divisions of the embryo (Martin, 1995). The timing of the mitotic recombination event may preclude the retention of full male structures within the genitalia when this anlagen is involved.

Gynandromorphs in C. *tentans* have previously been reported by Sæther and Galloway (1980) from Lake Winnipeg in Manitoba. The majority of these were due to nematode parastism and were Type 1 intersexes. They did report one individual which was a Type 2 intersex, with a partially feminized antenna. This specimen is more similar to the individuals described here. Whether it arose from a similar cause can only be speculated upon.

Acknowledgements

We are grateful to Mary Sublette for typing most of the manuscript, to Karen McCoy and Darren Martinus who helped prepare specimens for SEM examination, and to Quentin Lang who printed the photographs.

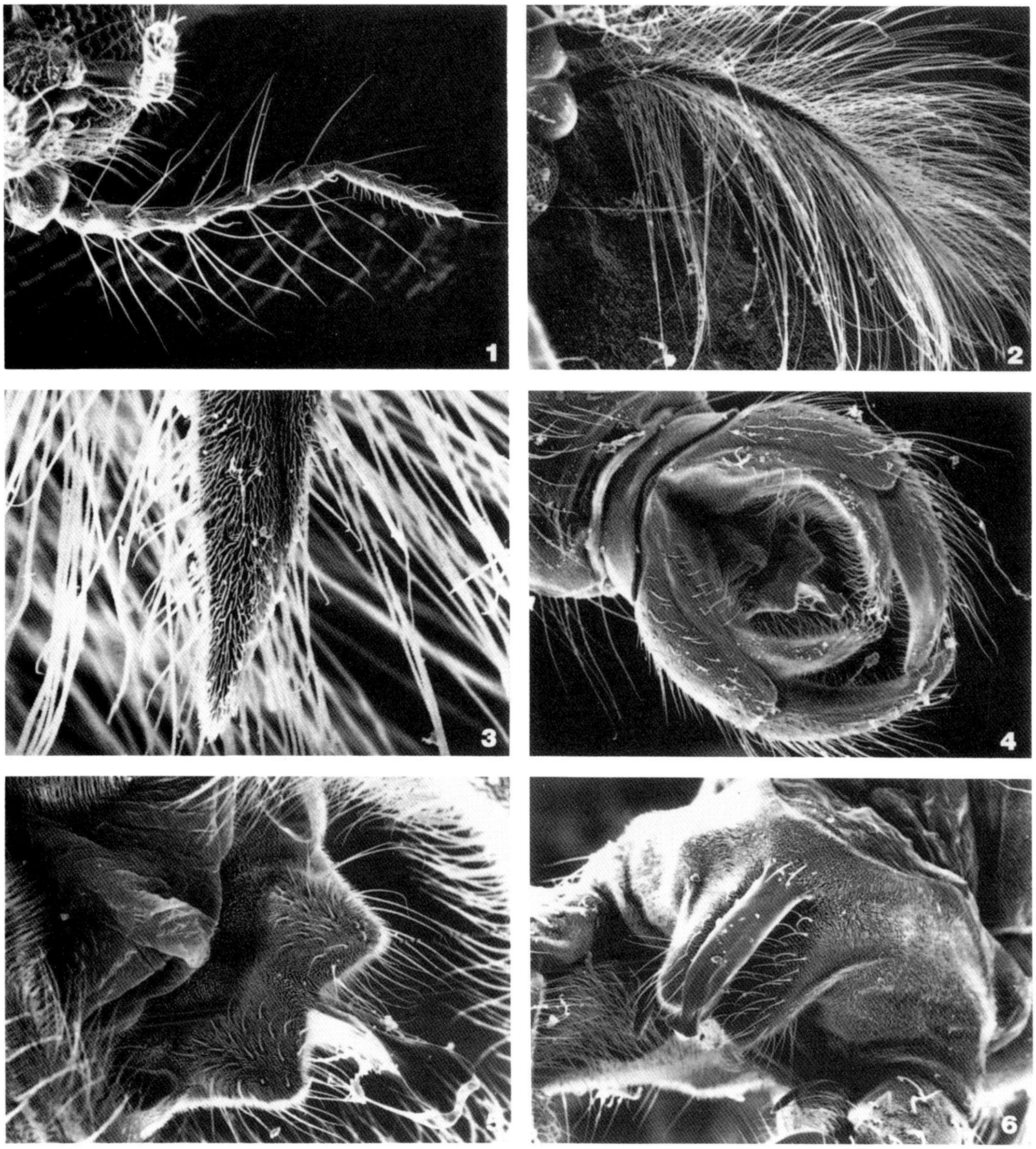

Plate I.

1. Normal female antennae (0.12 kx)
2. Normal male antennae (0.07 kx)
3. Apex of normal male antennae (0.70 kx)
4. Normal male genitalia, ventral view (0.12 kx)
5. Anal point and ninth tergum, normal male, ventral view (0.35 kx)
6. Anal point and ninth tergum, dorsal view, slightly tilted (0.20)

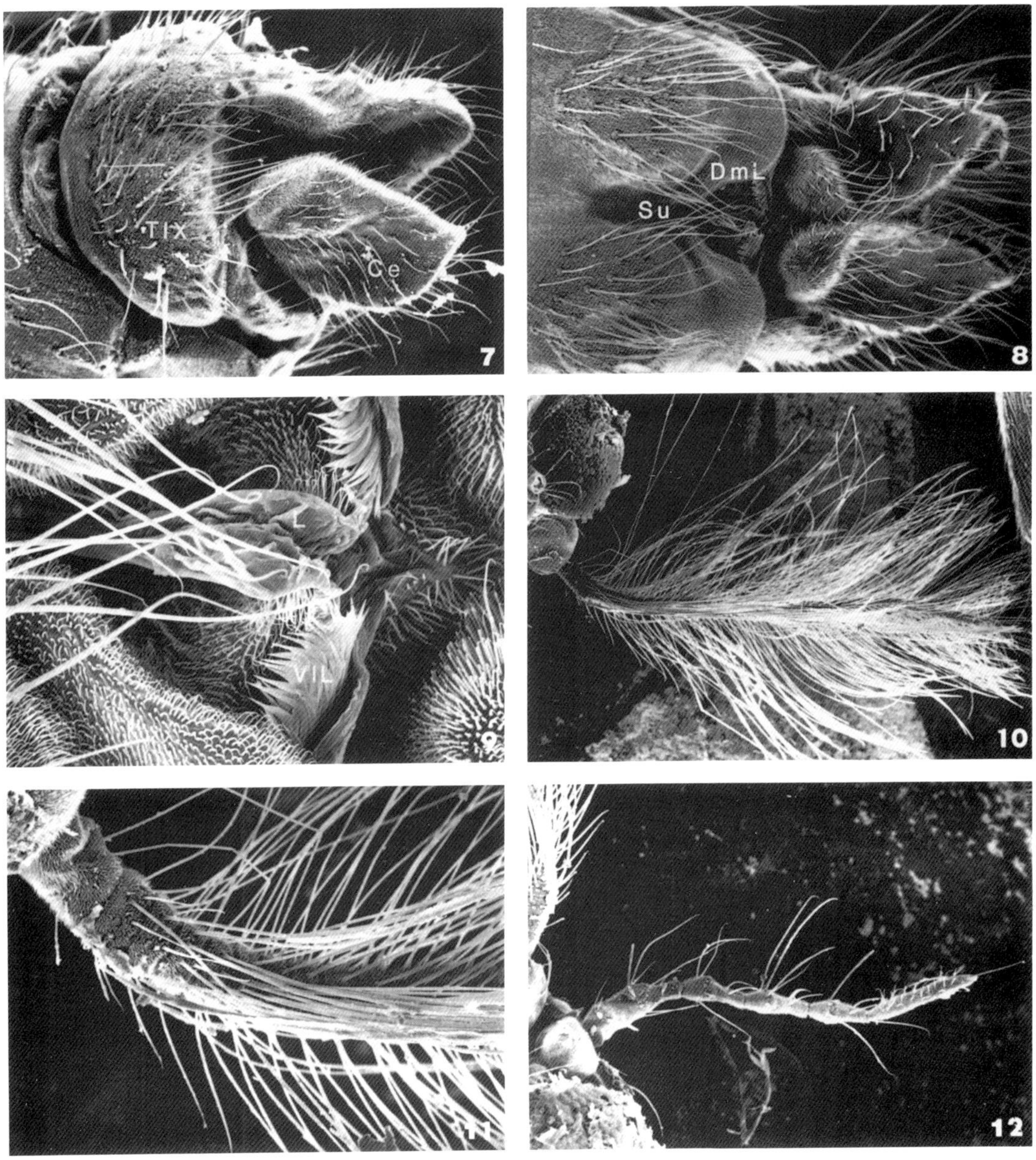

Plate II.

7. Normal female genitalia, dorsal view. (TIX, Tergum IX; Ce, cerci) (0.16 kx)
8. Normal female genitalia, ventral view Note sulcus (Su) and dorsomesal lobe of gonopophysis VIII (DmL) (0.16 kx)
9. Labia (L) and ventrolateral lobe of gonapophysis IX (VIL) from Fig. 8 (0.71 kx)
10. Normal male antenna, gynandromorph (0.06 kx)
11. Basal flagellomeres, normal male antenna, gynandromorph from Fig. 10 (0.26 kx)
12. Normal female antenna, gynandromorph from same specimen as Fig. 10 (0.11 kx)

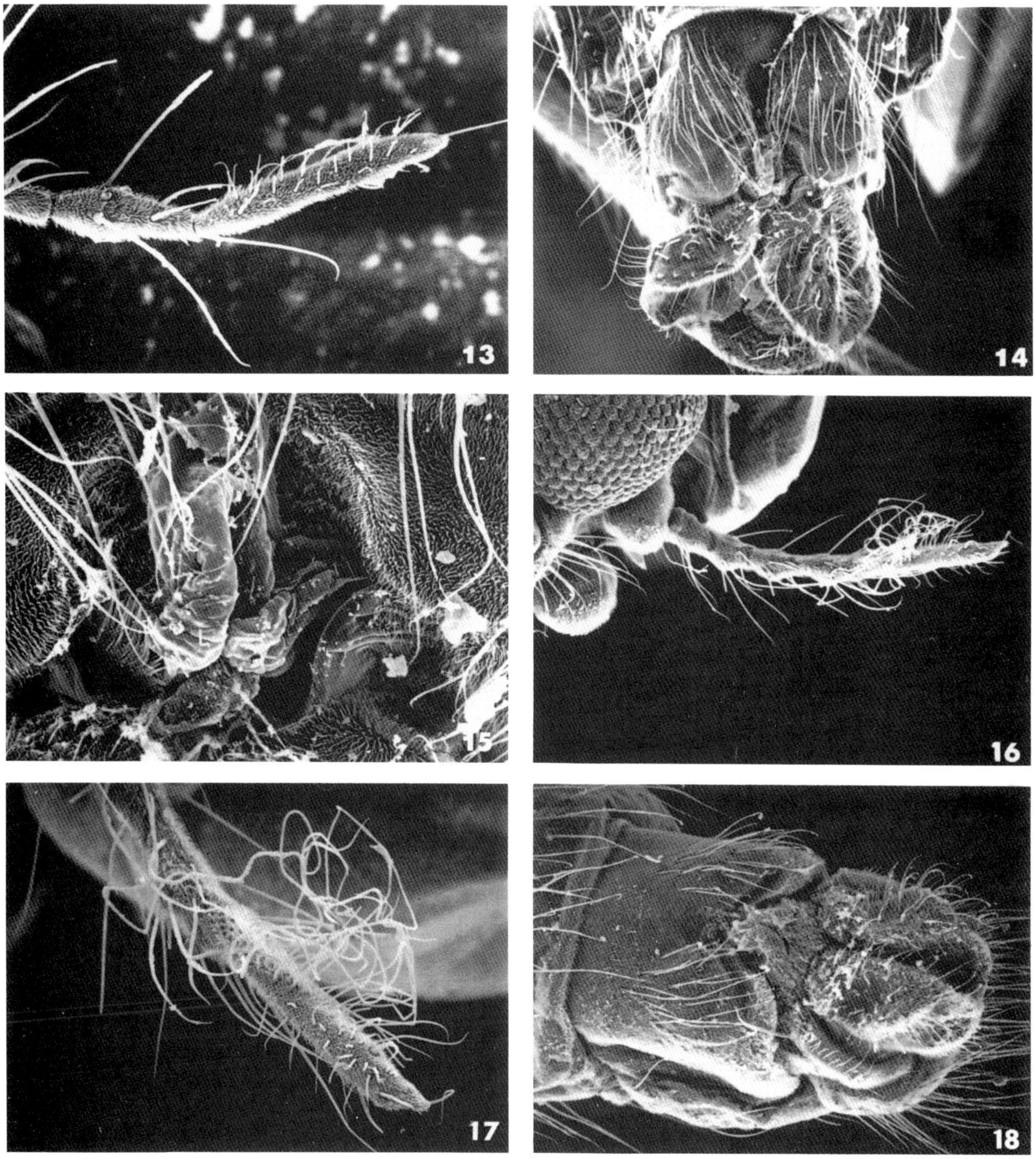

Plate III.

13. Apical flagellomeres, normal female antenna, gynandromorph from Fig. 12 (0.26 kx)
14. Normal female genitalia, gynandromorph, ventral view (0.12 kx) (slight distortion is attributed to preparation)
15. External genital structures, normal female, gynandromorph from Fig. 17 (0.55 kx)
16. Gynandromorphic female-type antenna (0.13 kx)
17. Apex of antenna shown in Fig. 19 with apex bearing a few male-like macrotrichia (0.28 kx)
18. Normal female genitalia, gynandromorph (0.14 kx)

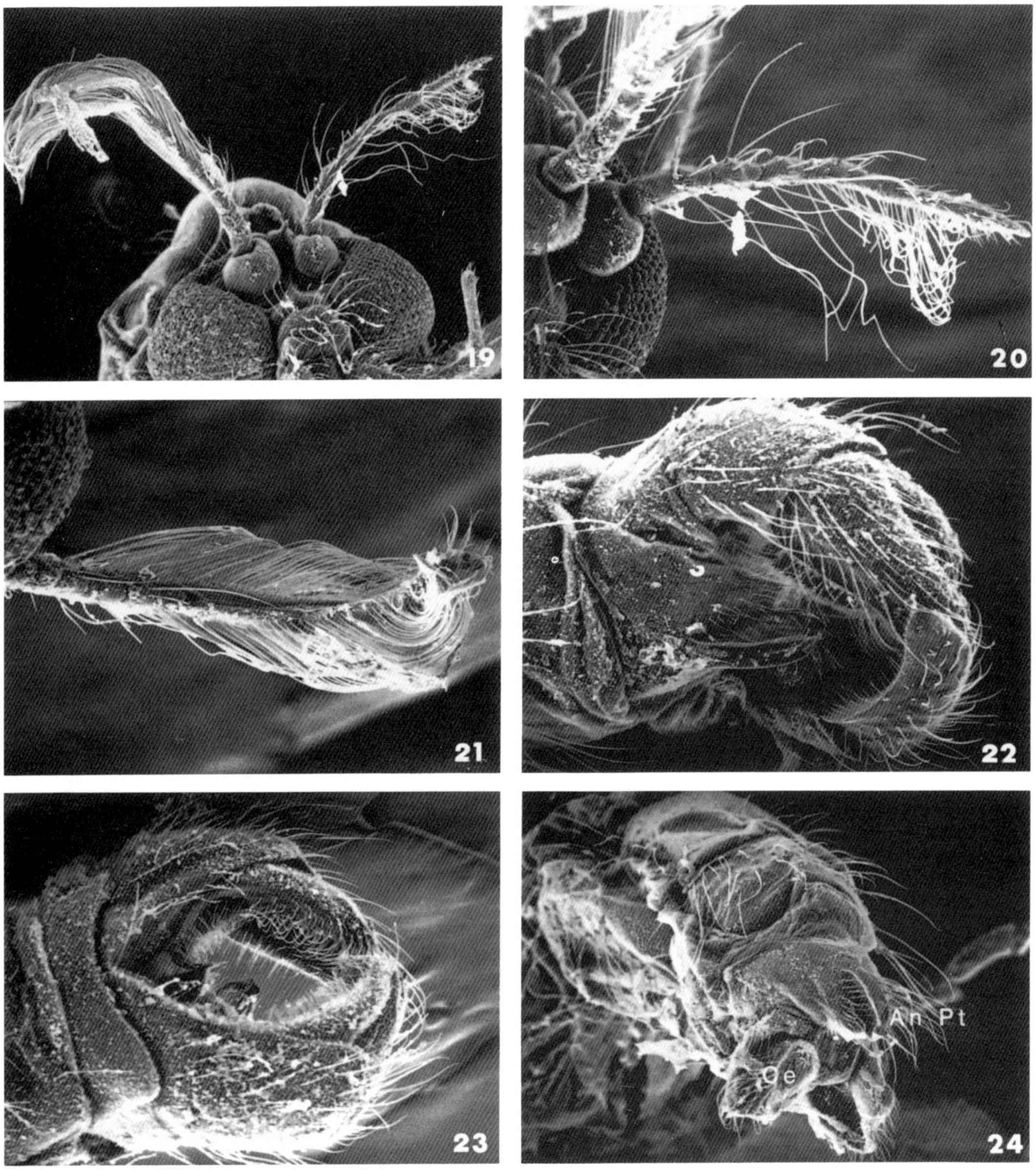

Plate IV.

19. Gynandromorphic antennae; the right is abnormal male, the left, abnormal female (0.07 kx)

20. Details of abnormal female antennae shown in Fig. 19; note male-like macrotrichia on lateral surface (0.12 kx)

21. Details of abnormal male antenna from Fig. 19; note reduced number of flagellomeres and the more densely packed macrotrichia (0.12 kx)

22. Normal male genitalia, gynandromorph, dorsal view, slightly tilted (0.14 kx)

23. Ventral view of male genitalia shown in Fig. 22 (0.13 kx)

24. Gynandromorphic genitalia; male component dorsal, female, ventral (0.09 kx) (An Pt, anal point, ♂; Ce, cercus, ♀)

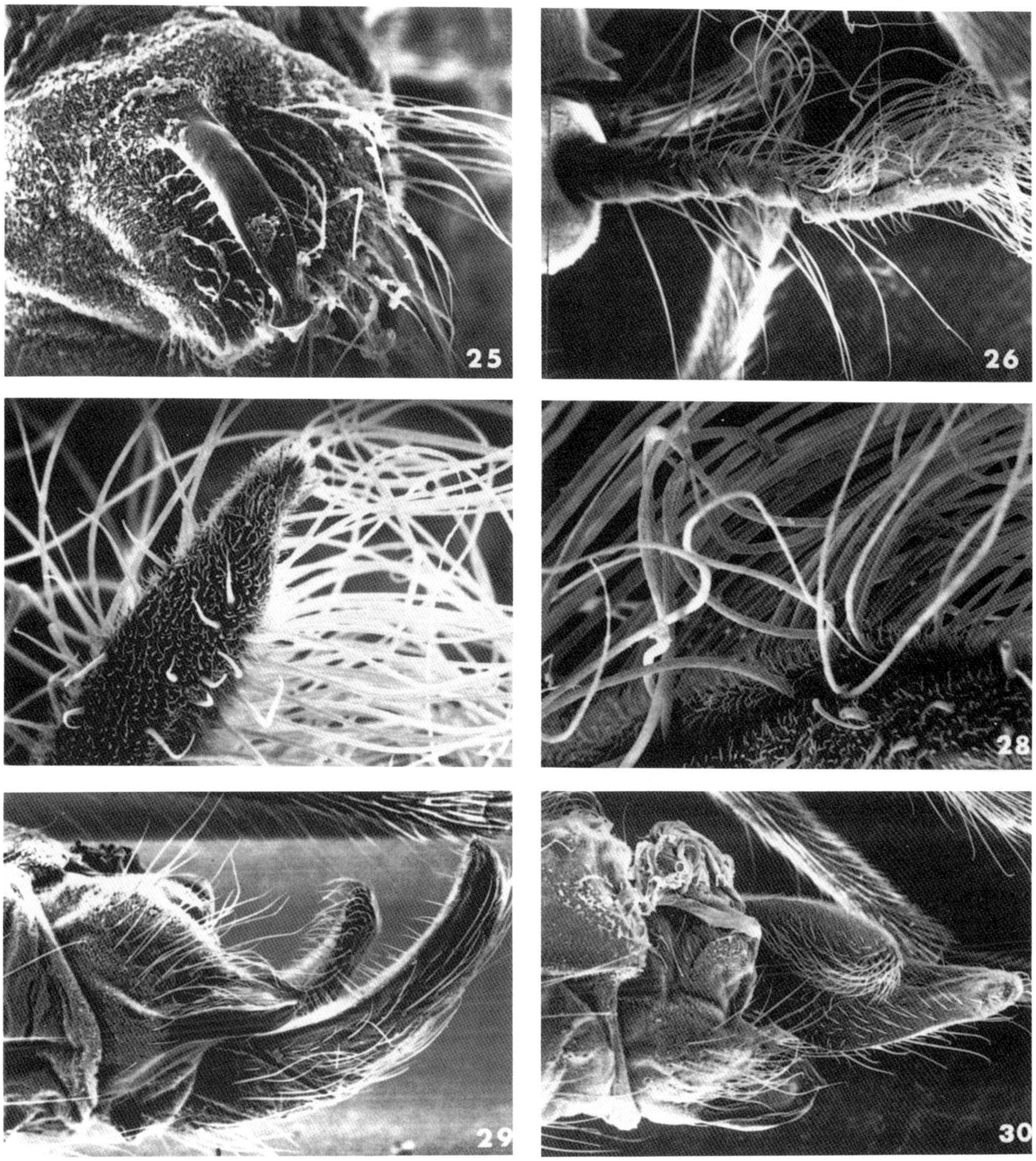

Plate V.
25. Male anal point and ninth tergum from specimen in Fig. 24 (0.29 kx)
26. Gynandromorphic antenna (0.18 kx)
27. Apex of gynondromorphic antenna (0.87 kx)
28. Terminal flagellomere, gynandromorphic, showing densely packed macrotrichia (0.97 kx)
29. Gynandromorphic genitalia, dorsal view (0.15 kx)
30. Genitalia from Fig. 29, lateral view (♂ An, presumptive unorganized female structures) (0.13 kx)

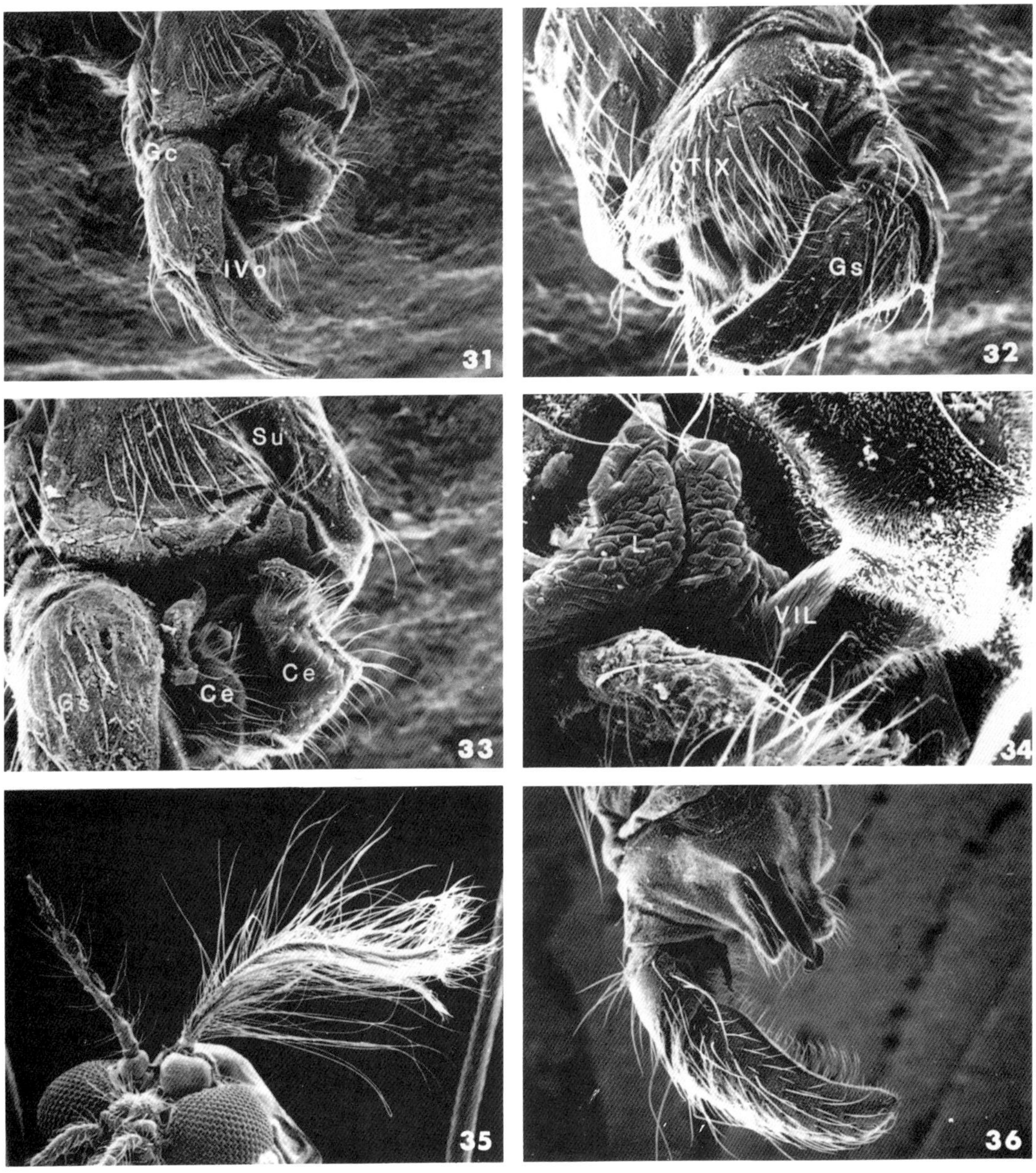

Plate VI.

31. Gynandromorphic genitalia, ventral view (Gc, gonocoxite; Ivo, inferior volsella) (0.10 kx)
32. Gynandromorphic genitalia from Fig. 31, dorsolateral view Note the absence of an anal point as the IX tergite is essentially female (♂TIX; Gs, gonocoxite) (0.15 kx)
33. Ventral view of genitalia from Fig. 32. Note complete, but spatially distorted, female components (Ce & Su as previously defined). (0.15 kx)
34. Details of female genitalia from Fig. 33 (0.56 kx)
35. Gynandromorphic antennae (0.06 kx)
36. Gynandromorphic genitalia, dorsal view; note slight distortion of anal point tip and right side of IX tergite (0.13 kx)

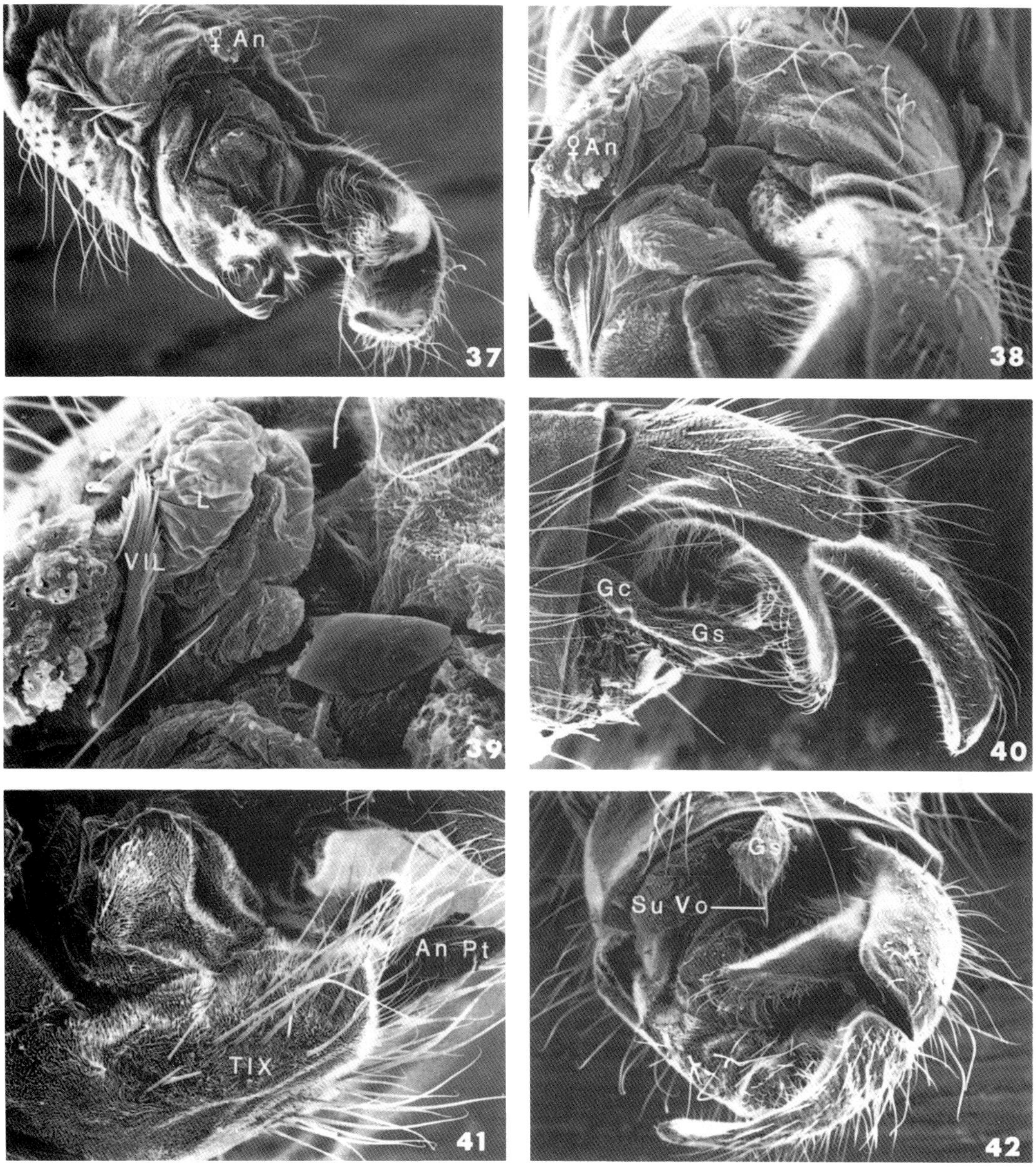

Plate VII.

37. Lateral view of genitalia from Fig. 36 (0.11 kx)
38. Ventrolateral view of genitalia from Fig. 36 (0.21 kx)
39. Details of female components from Fig. 38 (0.51 kx)
40. Gynandromorphic genitalia, ventral view (0.15 kx)
41. Anal point of ninth tergum from Fig. 40, ventrolateral view (0.26 kx)
42. Ventroapical view of genitalia from Fig. 45 (♂ An, presumptive unorganized ♂ structures; Su Vo, superior volsella) (0.15 kx)

Change in the size of midges (Diptera: Chironomidae) emerging from temporary ponds -the probable effect of larval interactions

R.S. Wotton[1] and P.D. Armitage[2]

[1]Department of Biology, Darwin Building, University College London, Gower Street, London WC1E 6BT, U.K.

[2]Institute of Freshwater Ecology, River Laboratory, East Stoke, nr. Wareham, Dorset BH20 6BB, U.K.

Introduction

Midges (Diptera: Chironomidae) are successful as early colonists of aquatic habitats and their larvae often occur in high densities over the substratum of newly-filled ponds (Street and Titmus, 1979; McLachlan, 1985; Williams, 1987). Some midges require the development of fine-grained, organically-rich sediment before they can achieve high population density (McLachlan and Cantrell, 1976), whereas others readily colonise sand free of sediment (Ali and Mulla, 1976).

In an earlier study of the midges inhabiting the temporary ponds formed in slow sand filter beds we found that *Cricotopus sylvestris* (Fabricius) was the dominant midge colonising the clean sand of newly-filled ponds (Wotton *et al.*, 1993). Adults of this species are among the smallest of those colonising these habitats and the life cycle of the first-emerging midges is completed in less than three weeks (Wotton *et al.*, 1993). In the present study we monitored the size of midges emerging from ponds over time, knowing the time of commencement of colonisation by *C. sylvestris*. We expected that the midges emerging first would be the smallest, with an increase in size through time as the substratum became enriched (cf. Welch *et al.*, 1988). Previous studies have shown that intraspecific variation in size of midges is caused by environmental variables and by the effects of intraspecific competition (Mackey, 1977a; Delettre, 1988; McLachlan, 1989). However, it has been shown that the size of a species may remain relatively constant across seasons (Caspers, 1983).

We give two sets of data from ponds of relatively long duration and two from ponds of short duration (the filter beds being drained at different intervals). In addition to the comparison of size in *C. sylvestris* across the four ponds, we compared this species with two other small species of midges (*Psectrocladius limbatellus* [Holmgren] and *Tanytarsus fimbriatus* Reiss and Fittkau) that co-exist with, and have the same life cycle as, *C. sylvestris* (Wotton *et al.*, 1993). In all collections, males and females were measured separately to record differences in the size of the two sexes, Delettre (1988) having shown that females can increase in wing length relative to males.

Methods and materials

Each slow sand filter consists of a homogeneous bed of sand overlying cobbles in a large (100 m x 30 m) tank. Reservoir water is pumped to the beds to form a pond of near-constant depth (1.2 - 1.6 m) with water draining through the filter to the domestic supply. Particles and dissolved organic matter are removed by filtration and/or uptake by the macrobiota, and microbiota, as water traverses the filter. At intervals, the sand filter loses efficiency, water is drained away and the surface of the sand layer removed. The pond is subsequently re-filled, and the time when colonisation by midges begins is therefore known.

An emergence trap was located on the water surface of four filter beds immediately after the ponds had been re-filled so that emergence of adult midges could be monitored. The traps were of the box type described by Morgan (1971) and covered 0.5 m^2 of the water surface. Each was anchored to the sides of the bed by a continuous loop of string which ran through a pulley and thus allowed the return of the trap to the same location on the pond surface after midges had been collected (Wotton *et al.*, 1993).

Each trap was brought to the side of the bed every two days so that estimates of midge numbers could be made. Adults were collected at the start of the first mass emergence, when thousands of midges suddenly appeared in the traps. This occurred 15-18 days after re-filling of the pond. An aspirator was used to collect adults but so many midges were usually present that subsamples had to be taken (Wotton *et al.*, 1993). After collection, midges were retained in the aspirator tube and placed in a deep freeze to chill the flies. This served to prevent loss, or damage, when flies were transferred to labelled bottles of 70% alcohol.

The wing length (from arculus to apex) of a sample of flies of each sex was measured using an eyepiece micrometer mounted in a binocular microscope. Wing length measurements were taken on a total of 1848 midges, ten flies of each sex usually being measured on each sampling occasion. The depth of the thorax (from the most dorsal part of the post-scutum to the most ventral part of the sternum) was also measured in individuals of a range of thorax depths to produce a second correlate of adult size.

Results

The constant supply of food, by a process that can be described as "active deposition", is an unusual feature of filter beds compared with most natural ponds, and allows colonisation by large numbers of midge larvae. Physico-chemical factors during the bed runs (the duration of the ponds) have been summarised in Table 1.

Table 1. Physico-chemical variables in the filter beds studied (with permission of Thames Water plc). Week no. refers to the weeks of the year: week 30 = 23rd July to 29th July: week 36 = 3rd to 9th September, etc.. Bed 2 was filled from week 32 to week 40, Bed 9 was running from week 30 to week 34, etc. Water temperature was measured each day, chlorophyll *a* and POC were measured once a week.

Week no.	Beds running	t°C range	Chlorophyll *a* (μg l^{-1})	POC (mg l^{-1})
30	9	20.2 - 22.5	4.8	368
31	9	18.0 - 21.0	8.4	597
32	2,9,19,26	18.0 - 19.1	21.5	974
33	2,9,19,26	18.0	7.8	490
34	2,9,19,26	18.5 - 20.0	4.4	417
35	2,19	19.0 - 20.0	2.7	237
36	2,19	18.7 - 19.0	5.0	463
37	2,19	17.8 - 18.0	5.2	620
38	2,19	17.9 - 18.0	2.6	390
39	2,19	17.0 - 18.0	4.8	389
40	2,19	16.0 - 17.0	4.1	348

Table 2. Comparison of mean wing length (W, in mm from arculus to apex ± SD) and mean thorax depth (T, in mm from the most dorsal part of the post-scutum to the most ventral part of the sternum ± SD) for each sex and species. n = number of individuals measured; r = the value of the product-moment correlation coefficient; *** = P < 0.001.

Species	Sex	n	W	T	r
C. sylvestris	M	35	1.94 ± 0.18	0.96 ± 0.10	0.956***
	F	32	2.03 ± 0.19	0.95 ± 0.09	0.961***
P. limbatellus	M	32	1.89 ± 0.14	0.99 ± 0.08	0.920***
	F	24	1.87 ± 0.12	0.95 ± 0.07	0.866***
T. fimbriatus	M	29	1.82 ± 0.14	0.76 ± 0.05	0.901***
	F	33	1.56 ± 0.14	0.63 ± 0.05	0.878***

Comparison of wing length and thorax depth shows a high correlation (Table 2), so wing length provides a good estimate of body size in the midges investigated. In all four beds the largest adults of *C. sylvestris* were those that emerged first (Fig. 1) and the size of adults then diminished markedly (on average by 18% for male *C. sylvestris* in Bed 19 and 22% for females of this species in the same bed). In the longer bed runs, this was followed by an increase in size (respectively 16% and 17% in the example used above) followed by a further decrease (15% and 20% in the example used above). This pattern was arrested in the short bed runs when the ponds were drained, thus interrupting the cycle of change of size. Comparison of the wing lengths of the first adults of *C. sylvestris* emerging from Beds 19 and 26 show that they had a different starting size compared to those from Beds 2 and 9 where midges were smaller overall. This difference was highly significant for males and females first emerging from Beds 19 and 26 compared to those emerging initially from Beds 2 and 9 (t values 3.84 - 10.66, P < 0.01 in all cases in t-tests).

All differences between peaks and troughs of size were significant (Table 3) and nearly identical patterns were observed in the dominant *C. sylvestris* and in *P. limbatellus* and *T. fimbriatus* (midges having the same life cycle as *C. sylvestris*). Comparison of Figs. 1 and 2 emphasises this similarity, but the amplitude of the cycle of change in size of *C. sylvestris* is largest.

DISCUSSION

Several explanations can be put forward for the change in wing length over time shown in this study: (i) that it was the result of different genotypes, large individuals being the progeny of the previous generation of large adults; (ii) that it was a result of changes in water temperature, food availability, or substratum quality; and (iii) that it resulted from larval interactions.

Explanation (i) appears, at first sight, to be attractive as peaks of large, and of small, adults are separated by approximately one generation cycle (c. 20 days). However, the size of adults varies in different beds at the same time of year (the ponds formed in beds 2, 19 and 26 were re-filled within four days of each other) and this would mean that adults must be selecting for certain ponds. As cues from the water surface will be similar (the water being from the same source) this seems very unlikely and this explanation can be discounted.

Turning to explanation (ii), water temperature ranged from 18.0 - 21.0°C during the first cycle of midge size and temperature differences can be ruled out as a factor promoting change in size because of its effect on growth rate. There were differences in the amount of particulate organic carbon (POC) in the ponds through time with a peak of chlorophyll *a* at the time

Table 3. Table of t-test results for comparison of wing lengths of midges in the peaks and troughs of size shown in Figs 1 and 2. n = the number of adults in each comparison, * = $P < 0.05$, ** = $P < 0.01$, *** = $P < 0.001$. (data from Bed 26 have not been included because one "half cycle" was not completed).

Species	Sex	Bed	Days for comparison	n	t
C. sylvestris	M	2	18-28	29	9.43***
	M	2	28-38	25	6.26***
	M	2	38-50	20	5.56***
C. sylvestris	F	2	18-28	27	9.95***
	F	2	28-38	21	2.92**
	F	2	38-50	21	4.51***
C. sylvestris	M	9	16-30	20	4.47***
C. sylvestris	F	9	16-30	20	7.95***
C. sylvestris	M	19	18-28	21	14.96***
	M	19	28-36	19	10.81***
	M	19	36-50	17	10.48***
C. sylvestris	F	19	18-28	19	8.04***
	F	19	28-36	23	4.67***
	F	19	36-50	23	8.20***
P. limbatellus	M	19	18-30	16	8.24***
	M	19	30-40	18	5.12***
	M	19	40-48	20	6.15***
P. limbatellus	F	19	18-30	17	6.52***
	F	19	30-40	17	2.92*
	F	19	40-48	16	2.72*
T. fimbriatus	M	19	18-26	20	7.51***
	M	19	26-42	20	5.82***
	M	19	42-52	20	6.26***
T. fimbriatus	F	19	18-26	16	10.47***
	F	19	26-42	16	5.07***
	F	19	42-52	20	6.15***

Beds 2, 19 and 26 were filled. This good quality source of food explains the large numbers of midges emerging (Welch *et al.*, 1988) but it does not explain why first-emerging adults from some beds were larger than those from others. Neither does it explain the second peak of emerging large midges, because the levels of chlorophyll *a* and POC remained relatively constant after the initial high levels. However, the small size of adults emerging from Bed 9 may be the result of changes in substratum quality as this bed was re-laid with new sand immediately before the bed run and will therefore have been poorly colonised with biofilm (Lock, 1994), a potentially valuable food source for larvae.

For operational reasons, it was not possible to sample the surface of the sand during the running of beds so we cannot provide data on the change of larval numbers needed as evidence for explanation (iii).

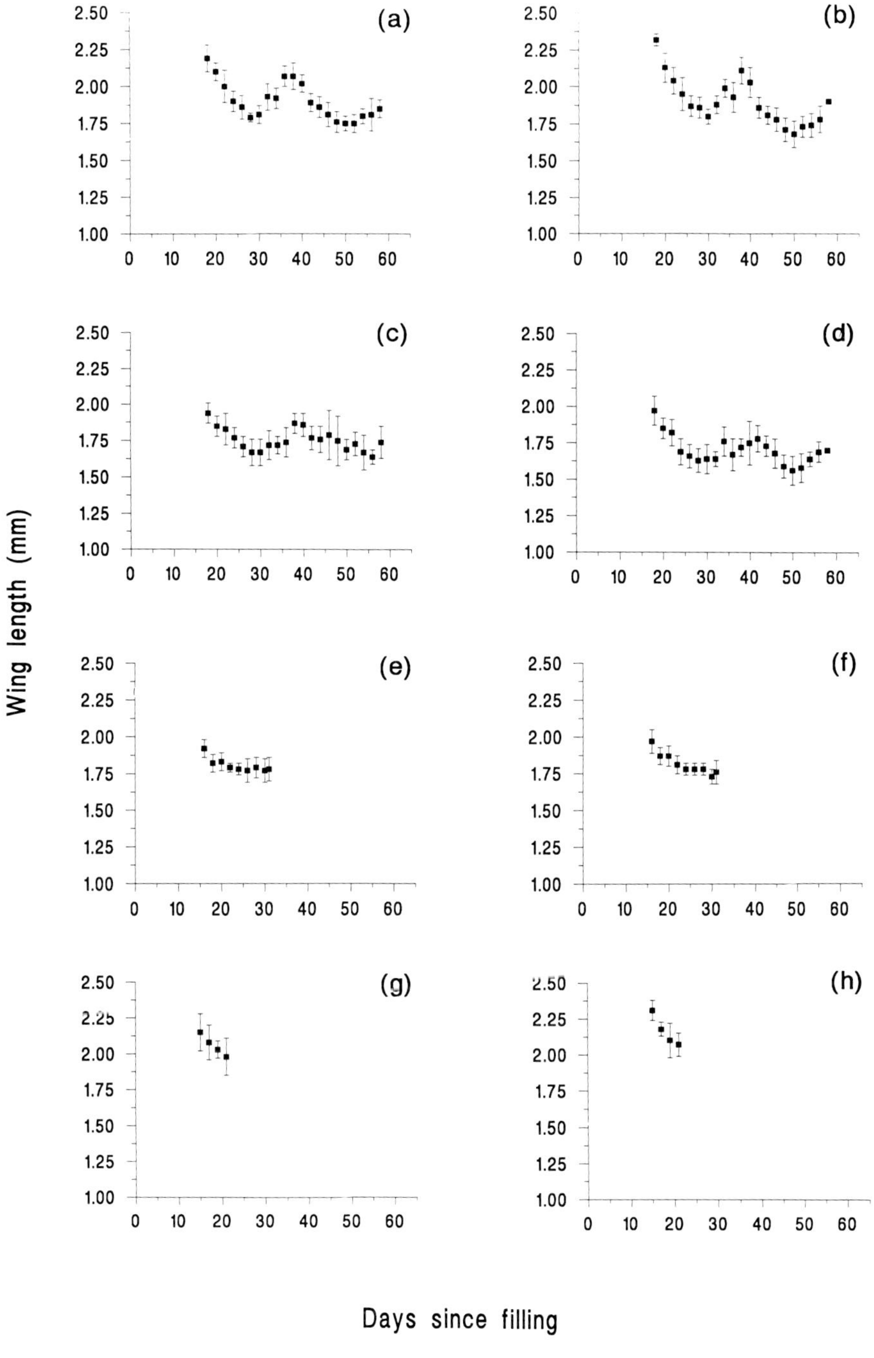

Figure 1. The size (mean ± SD) of *Cricotopus sylvestris* male and female adults emerging from Beds 2, 9, 19 and 26. In Beds 9 and 26, the ponds were drained after the last collection in each case. (a) = males, Bed, 19; (b) = females, Bed, 19; (c) = males, Bed 2; (d) = females, Bed 2; (e) males, Bed 9; (f) females, Bed 9; (g) = males, Bed 26; (h) females, Bed 26.

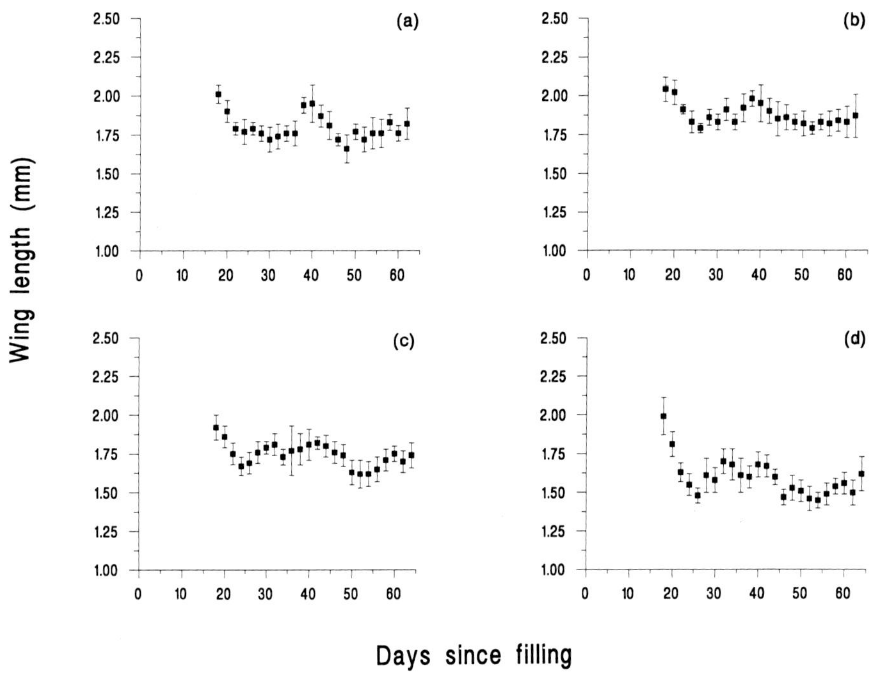

Figure 2. The size (mean ± SD) of *Psectrocladius limbatellus* and *Tanytarsus fimbriatus* males and females emerging from Bed, 19. (a) = *P. limbatellus* males; (b) = *P. limbatellus* females; (c) = *T. fimbriatus* males; (d) *T. fimbriatus* females.

Extrapolation from the numbers of adults in emergence traps indicates that larval density was consistently high during the long bed runs (Wotton *et al.*, 1993). Larvae of the first peak of emerging adults will grow in the ponds unhindered by competition with larger larvae and this would be expected to reduce size-related interference competition for the available space (food not being considered limiting). Larvae hatching from eggs laid later will have to compete with conspecifics that are already large and this may explain why they are smaller as last instar larvae and therefore as adults. Larvae of the next batch of emerging large midges will have coexisted with relatively small last instar larvae. This may have allowed them to reduce intraspecific competition and thus grow into larger fourth instars and, subsequently, to produce larger adults. This also applies to interspecific interactions since *P. limbatellus* and *T. fimbriatus* have strikingly similar life cycles and patterns of adult size to those of *C. sylvestris*. We recognise, however, that this explanation must await tests before it becomes anything other than speculation. We also have, as yet, no explanation for the coexistence of the three species, though we would expect there to be partitioning of limiting resources as shown by Rasmussen (1985) for two other species of midges.

Some comment is required on why *C. sylvestris* dominates in these habitats and why it shows the largest variation in size through time. Overall, *C. sylvestris* is the largest of the three species studied and is the most numerous midge species in this habitat (Wotton *et al.*, 1993), two factors that are very likely to be related. It has been reported as a dominant species in wet gravel pits (Titmus, 1979) where organic-poor mineral substrata develop a covering of detritus over time, conditions similar to those in the ponds we studied. It is well pre-adapted to the conditions pertaining in these man-made habitats and is known to colonise other man-made habitats such as newly-filled rice paddy fields (Clement *et al.*, 1977). *C. sylvestris* shows considerable phenotypic plasticity, occurring in conditions of high salinity (Menzie, 1981), and in both lentic and lotic freshwater systems over a wide geographical area (LeSage and Harrison, 1980; Ali and Mulla, 1977; Drake, 1985; Jonsson *et al.*, 1986; Menzie, 1981). We believe it is able to take advantage of the favourable conditions in filter beds to effectively promote growth.

Acknowledgements

We are very grateful to Karen Aston for collecting midges and to Miriam Hamburger, Daniel Ladle, Karen Aston, and Tony Woodward for technical help and for discussion. Athol McLachlan, Dominic Chaloner and Kari Vepsäläinen read earlier versions of the manuscript and we are grateful for the many suggestions they made for improvements. The work was carried out during a contract from Thames Water plc and we thank them for allowing us to work at one of their filter works, and for allowing us to use data collected by Phil Renton and the Operations staff at Wraysbury. Baden McCann of Thames Water plc provided much helpful support for our study.

The first record of *Buchonomyia* Fittkau and the subfamily Buchonomyiinae from the New World (Diptera: Chironomidae)

T. Andersen and O. A. Sæther,

Museum of Zoology, University of Bergen, Muséplass 3, N-5007 Bergen, Norway

Introduction

The enigmatic subfamily Buchonomyiinae includes the single genus *Buchonomyia* Fittkau. This genus was described by Fittkau (1955) on the base of a single male of *B. thienemanni* Fittkau from a mountain spring in Middle Germany. The species has since been found in England, France, Germany, Ireland, Luxembourg, and possibly Iran (Murray and Ashe, 1981a; Reiss, 1984a; Sæther, 1983, 1989a). As mentioned by Brundin (1966) an apparent male has also been found in Baltic amber (Upper Eocene). Brundin and Sæther (1978) described *B. burmanica* from Northern Myanmar (Burma) and erected the subfamily Buchonomyiinae. They, as well as Sæther (1979b, 1983, 1989b), regarded the subfamily as the plesiomorphic sistergroup of the Diamesinae, Prodiamesinae, Orthocladiinae, and Chironominae combined, i.e. a member of the semifamily Chironomoinae, whereas Murray and Ashe (1981a, 1985) regarded the subfamily as an integral member of the semifamily Tanypodoinae. Although pupal exuviae have been obtained from rivers (Murray and Ashe, 1981a; Sæther, 1986) and eggs, first instar and late instar larva are known (Ashe, 1985; Ashe and Murray, 1983; Ashe, 1995), the controversy about the phylogenetic position of the subfamily, due to contradictory evidence, remains unresolved.

During a course in tropical entomology, six males of a third species were collected in Malaise traps at the Rio San Josecito, Cerro Cacao, Guanacaste Province, Costa Rica. This is the first record of the subfamily Buchonomyiinae from the New World.

Methods and terminology

The specimens were mounted on slides following the procedure outlined by Sæther (1969). The general terminology follows Sæther (1980). The measurements are given as ranges, followed by a mean, followed by the number of specimens measured in parentheses (n).

Genus *Buchonomyia* Fittkau

The generic diagnoses given by Fittkau (1955) and emended by Brundin and Sæther (1978) and Sæther (1989a) can be emended as follows:

Temporals not or only slightly divided into groups. AR 0.3-1.3. VR 0.6-0.9. Pseudospurs present or absent.

Buchonomyia brundini sp. n.

(Figs. 1-2)

Type locality: Costa Rica: Guanacaste Province, Guanacaste Conservation Area, Cerro Cacao, Rio San Josecito.

Material examined

Holotype m, Costa Rica, Guanacaste Prov., Guanacaste Conservation Area, Cerro Cacao, Rio San Josecito, 1000 m a.s.l., 3.-7.v.1993, Malaise trap, T. Andersen (ZMB No. 178).

Paratypes: 5 m, as holotype.

Etymology: - The species is named in memory of the late Prof. Lars Brundin, the dean of chironomidology, and a dear friend and colleague.

Diagnostic characters: the male imago can be separated from the other species of the genus by having lower chaetotaxy, a wing length of 2.3-2.6 mm, an AR of 0.36-0.46, no pseudospurs, longer and more slender main lobe of gonostylus, and apical lobes of aedeagus curved.

Male imago (n=6, except when otherwise stated)

Total length 3.84-4.21, 4.00 mm. Wing length 2.28-2.59, 2.44 mm. Total length/ wing length 1.60-1.71, 1.64. Wing length/ length of profemur 2.04-2.10, 2.06. Thorax blackish brown with darker vittae and more pale humeral, anepisternal and dorsal prepisternal areas; legs yellowish brown; abdomen brown.

Head. AR 0.36-0.46, 0.43. Ultimate flagellomere 33-57, 44µm long; penultimate flagellomere 165-217, 196µm long. Temporal setae 34-43, 38; including 11-22, 14 frontals; 13-17, 15 inner verticals; 4-7, 5 outer verticals; and 4-5, 5 postorbitals. Clypeus with 19-25, 22 setae. Tentorium (Fig. 1a) 221-240, 227µm long; 30-34, 33µm wide; sieve pore apparently absent. Stipes (Fig. 1a) 128-154, 143µm long; 45-53, 49µm wide; without distinctly sclerotized outer margin. Cibarial pump as in Fig. 1a. Palpomeres lengths (in µm): 45-53, 50; 53-71, 63; 158-173, 168; 236-278, 253; 218-281, 249. Third palpomere with apical sensillum capitatum with about 15 sensilla clavata.

Thorax (Fig. 1c). Antepronotum with 7-12, 9 setae. Dorsocentrals 63-73, 68 in 3-4 rows; acrostichals 66-76, 72; prealars 20-29, 23 including 4-8, 6 anterior and 15-21, 17 posterior, well separated prealars. Scutellum with 36-53, 45 (5) setae. Postnotum with 8-13, 10 setae.

Wing (Fig. 1d). VR 0.63-0.74, 0.70. Anal lobe very weak. C extension 4-15, 9 mm long. R_{2+3} absent. Setae of wing veins indistinct on most veins and thus counts only approximate. First axillary sclerite with 5-8, 6 setae; brachiolum with 5-7, 6; Sc with about 60-100, 70; C extension with 13-18, 16 non-marginal; R with 121-166, 103; R_1 with about 165-290, 195; R_{4+5} with about 155-185, 170; M with 34-52, 42 setae; RM and MCu bare; M_{1+2} with about 80-110, 100 setae; M_{3+4} with about 60-90, 75; Cu with 35-46, 41; Cu with about 50-70, 60; PCu with about 50-75, 55; An with about 30-49, 36 setae. Wing membrane with numerous setae and microtrichiae in all cells except sc and r, cell m basally of MCu with 36-66, 51 setae. Sensilla campaniformia about 10 at base of brachiolum, 3 below setae, and about 12 at apex of brachiolum; 4 on Sc; one at base of R_1; and 2 at base of R_{4+5}. Squama with 20-31, 23 setae.

Legs. Spur of front tibia 69-90, 84µm long, of middle tibia 68-75, 73µm long, of hind tibia 83-96, 89µm long. Width at apex of front tibia 71-86, 80µm, of middle tibia 68-75, 73µm, of hind tibia 79-96, 87µm. Pseudospurs absent. All legs rubbed, but bristle ratios apparently lower than 2.5 on all legs, and scale-like setae shorter, wider and denser than in the other described species (Fig. 1e,f). Lengths (in µm) and proportions of legs (n=5 on ta_1-ta_5 on front leg):

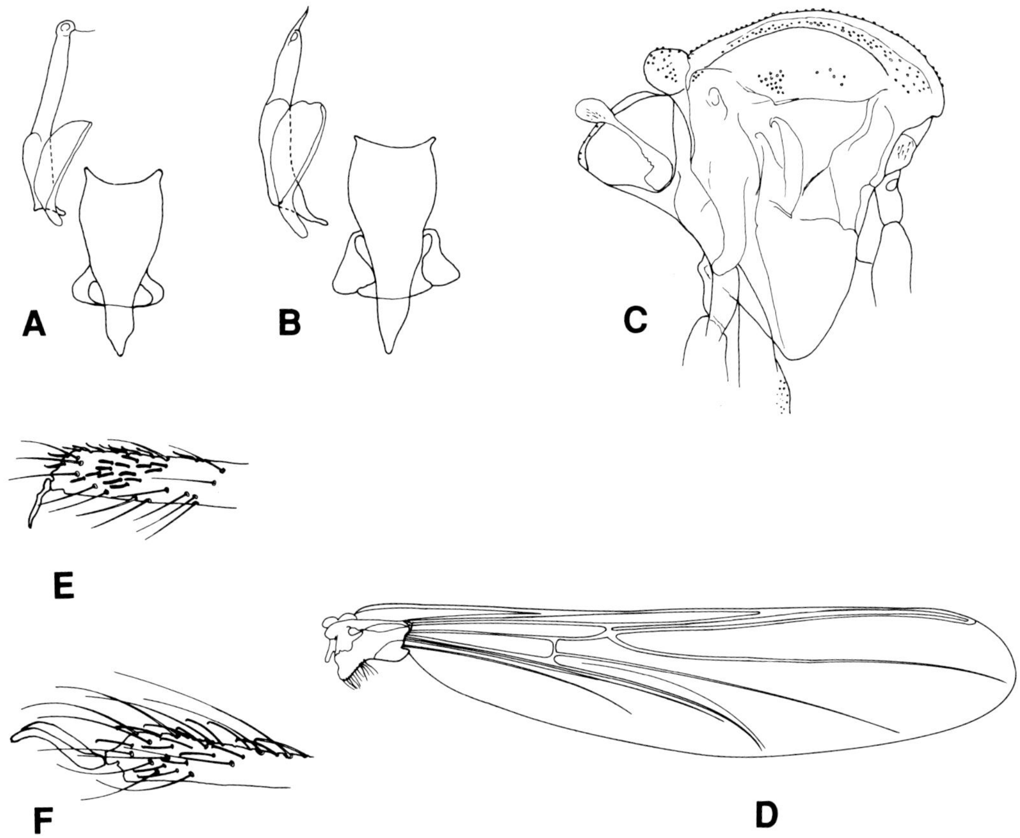

Figure 1. Male imagines of *Buchonomyia brundini* sp. n and *B. thienemanni* Fittkau. (a)-(b) tentorium, stipes and cibarial pump of (a) *B. brundini*, (b) *B. thienemanni* ; (c) thorax of *B. brundini*; (d) wing of *B. brundini*; (e)-(f) tarsomere 5 of (e) *B. brundini*, (f) *B. thienemanni*.

	fe	ti	ta_1	ta_2	ta_3
p_1	1087-1266, 1178	1625-1899, 1742	992-1115, 1058	635-728, 686	435-501, 471
p_2	945-1086, 1016	1550-1786, 1666	945-1148, 1039	657-803, 729	473-539, 498
p_3	983-1134, 1065	1956-2259, 2059	1219-1418, 1324	737-898, 808	482-576, 521
	ta_4	ta_5	LR	BV	SV
p_1	293-340, 321	180-217, 197	0.60-0.63, 0.62	2.31-2.40, 2.35	2.64-2.80, 2.59
p_2	331-397, 366	198-265, 235	0.60-0.64, 0.62	2.02-2.07, 2.04	2.50-2.64, 2.58
p_3	378-435, 407	227-265, 246	0.62-0.65, 0.63	2.21-2.30, 2.26	2.33-2.44, 2.39

Hypopygium (Fig. 2a-c, e). Tergite IX with 83-90, 88 setae. Sternite IX weak, bare; laterosternites barely distinguishable, with 0-2, 0 setae. Dorsomedian and dorsolateral rods of aedeagus each with pair of curved apicolateral appendices. Ventral portion of aedeagus plate-like with apical notch and lateral connection to volsella and to shorter dorsal lobe of gonostylus, and with articulation against longer ventral lobe.

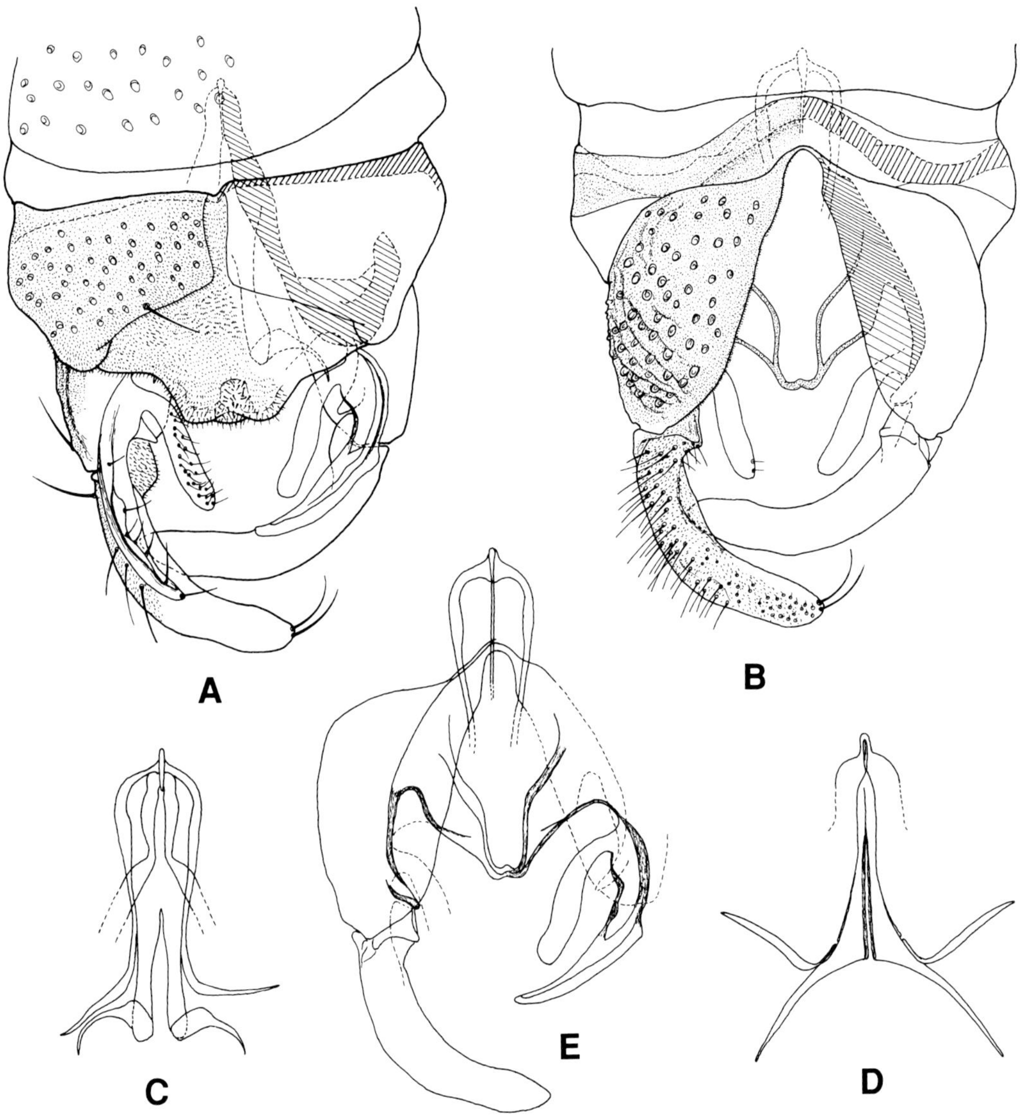

Figure 2. Male imagines of *Buchonomyia brundini* sp. n and *B. thienemanni* Fittkau. (a)-(b) hypopygium (a) dorsal, (b) ventral, view of *B. brundini*. (c)-(d) aedaegus, dorsal view of (c) *B. brundini*, (d) *B. thienemanni*; (e) aedeagus, ventral view of *B. brundini*.

Gonocoxite 255-266, 261µm long. Volsella slender, cylindrical; 120-135, 130µm long. Gonostylus divided, main lobe 199-233, 220µm long, slender; accessory lobe 124-139, 133µm long, pointed. HR 1.13-1.30, 1.17. HV 1.71-1.93, 1.82.

Distribution and ecology

The species is known only from a single station at an elevation of 1000 m at Rio San Josecito, a tributary to Rio Tempisque on the western slope of the Volcano

Cacao in the Guanacaste Conservation Area in Costa Rica, 1 km south of Estacion Biologica Cacao. The stream is relatively cool and flow through areas of Mesoamerican cloud forest. At the trapping site it is 1.5-2 m wide, 10-20 cm deep, and rather slow flowing. The bottom substratum consists of mud, gravel and stones and there are much rotting leaves, twigs and tree trunks. Although this area has been extensively collected (Coffman *et al.*, 1993) no adults or pupal exuviae of this easily recognizable genus has been encountered previously. However, as stated by Coffman *et al.*, site differences in species composition between different stream sites in this area are great.

COMPARISON WITH *B. THIENEMANNI* AND *B. BURMANICA*.

B. brundini has a much lower AR (0.3-0.5) than both *B. thienemanni* (about 1.3) and *B. burmanica* (0.6-0.7), lower chaetotaxy except on some wing veins, and lower bristle ratios. It also differs from both in being smaller with a wing length of 2.3-2.6 mm as opposed to 2.8-3.4 mm, lacking pseudospurs, having more dense and more clearly scale-like setae on the legs, and a more slender ventral lobe of gonostylus. It differ from *B. burmanica* in lacking any vestige of pulvilli or R_{2+3} and in having distinct ventral apicolateral appendages of the aedeagus. These appendages are present in *B. thienemanni* (Fig. 2d, Fittkau, 1955 figs. 12, 13), but were wrongly interpreted by Fittkau as consisting of one fork-shaped appendage ("gabelförmige Anhang"). In *B. thienemanni*, however, the appendages are straight, not curved, and better sclerotized. In *B. burmanica* there apparently are no such appendages although some broader lateral plates may be homologous. It is interesting to note that the volsella is directly connected to the ventral part of the aedeagus indicating that the volsella is mainly of gonapophyseal origin. The lower VR ratios in *B. brundini* and *B. thienemanni*, lack of vestige of R_{2+3} or pulvilli, and the presence of similar ventral apicolateral appendages indicate that *B. brundini* is the sister species of *B. thienemanni* with *B. burmanica* forming a more plesiomorphic sister group of both combined.

Segment IX apparently was incorrectly interpreted in Brundin and Sæther (1978). As segment I appeared to consist of partly fused segments I and II, segment VIII was regarded as part of IX. However, there is a sternite IX (although lacking setae) and the vestiges of laterosternites form an apparent gonosternite (Sæther, 1989b)

Acknowledgements

The expedition to Costa Rica was funded by The Norwegian Research Council for Science and the Humanities (NAVF), and the University of Bergen. Thanks are also due to Dr. Dan Janzen, University of Pennsylvania, and the staff at INBIO, Costa Rica, for their cooperation and support. Gladys Ramirez made the slide preparations, Maria Diaz the illustrations.

Afrotropical Species of *Parakiefferiella* Thienemann, with a Review of Species with Palpal Projections (Diptera: Chironomidae)

Leonard C. Ferrington Jr.[1] *and Ole A. Sæther*[2]

[1] Kansas Biological Survey and Department of Entomology, University of Kansas, Lawrence, Kansas USA 66047-2906

[2] Zoological Museum, University of Bergen, Bergen, Norway

Introduction

As part of a revision of the genus *Pseudosmittia* Goetghebuer, we examined the types placed in that genus by Freeman and Cranston (1980). One of the types actually belongs in the genus *Parakiefferiella* Thienemann. This species shows several interesting features, most of which are present in a previously undescribed species of *Parakiefferiella* collected in the West Usumbara Mountains of Tanzania by an expedition by the Museum of Zoology, the University of Bergen in 1990.

The two species share some unusual features such as apical projections of the third or third and fourth palpomeres carrying 1-2 spiniform setae, a keel on the front trochanter, and a knob on the ventral apex of the preepisternum. A survey shows that in addition to *Parakiefferiella*, palpal projections with or without spiniform setae occur among species of *Odontomesa*, *Sympotthastia*, *Syndiamesa*, *Pseudochironomus*, *Bryophaenocladius*, *Pseudosmittia*, *Thienemanniella*, *Kuschelius*, *Smittia* and *Lipurometriocnemus*. Spiniform setae on the third palpomere are found in *Trichochilus*, *Metriocnemus*, *Bryophaenocladius* and *Pseudosmittia*.

Methods and terminology

Male imagines were mounted on slides following the procedure outlined in Sæther (1969). The general terminology follows Sæther (1980). The measurements are given as ranges followed by a mean when four or more measurements were made, followed by the number measured (n) in parentheses.

The type species of *P. minax* sp. n. is deposited in the Museum of Zoology, the University of Bergen, Norway (ZMB). The type of *P. harrisoni* (Freeman) **comb. n.** has been returned to the Natural History Museum, British Museum (Natural History), London (BMNH).

Parakiefferiella minax sp. n.

(Fig. 1, 3b)

Type locality: **Tanzania**, Tanga region, West Usumbara Mts., Mazumbai.

Type material: Holotype male, **Tanzania**, Tanga region, West Usumbara Mts., Mazumbai, Kaputu Stream, Malaise trap, Nov. 1990, ZMB's Tanzania Expedition (ZMB No. 177).

Diagnostic characters

The characteristic coloration, yellow with dark markings on thorax, legs and abdomen; the projection of third palpomere carrying 1-2 apical spiniform setae; and the presence of a knob at the ventral apex of the preepisternum will separate the species from other orthoclads except *P. harrisoni* (Freeman) comb. n. described below. It differs from *P. harrisoni* by having a projection only on the third palpomere (Fig. 3b), its larger size and fully straight gonostylus.

Description

Male imago (n=6-7 except when otherwise indicated)

Total length 2.70-3.02, 2.84 mm (6). Wing length 1.52-162, 1.58 mm. Total length/wing length 1.72-1.90, 1.80. Wing length/length of profemur 2.47-2.67, 2.61. Thorax yellow with three dark vittae, posterior margin of scutum, anterior margin of scutellum and postnotum brown. Legs yellow with front trochanter, apex of fe, especially fe_2 and fe_3, base and apex of ti, and base of ta_1 dark brown; ti yellowish brown. Abdomen with tergite I mostly dark with small yellow anteromedian area; II-III dark with larger anteromedian area, IV pale in anterior 1/2; V pale in anterior 1/3; VI pale in anterior 2/3; VII mostly pale, darker along posterior margin; VIII-IX mostly dark.

Head (Fig. 1a, b). Ultimate flagellomere 378-432, 418µm (5) long; AR 0.74-0.88, 0.82 (5). Temporal setae consisting of 3-4, 4 outer verticals; and 0-1, 0 postorbitals. Clypeus with 8-14, 9 setae. Tentorium 126-142, 129µm long;16-23, 19µm wide. Stipes 114-139, 126µm long; 41-48, 45µm (5) wide. Palpomere lengths (in µm): 25-27, 26; 37-41, 38; 144-160, 151; 96-119, 109; 133-158, 146. Third palpomere with 25-41, 33µm long apical projection carrying an apical spiniform seta 16-18, 17µm long and usually a preapical spiniform seta 25-30, 27µm long (Fig. 3b)(seta bristle-like in one specimen).

	fe	ti	ta_1	ta_2
P1	572-630, 605	774-846, 808	621 (1)	—
P2	648-720, 677	693-783, 719	329-338 (2)	180 (2)
P3	612-684, 657	770-819, 788	318-414, 394	207-234, 216

	ta_3	ta_4	ta_5	LR
P1	—	—	—	0.73 (1)
P2	140-144 (2)	81 (2)	45 (2)	0.43-0.46 (2)
P3	171-189, 180	86-99, 94	54-63, 58	0.49-0.51, 0.50

	BV	SV	BR
P1	—	—	—
P2	3.85-3.93 (2)	4.22-4.24 (2)	3.1 (1)
P3	3.33-3.45, 338	3.52-3.70, 3.63	4.0-4.4, 4.2 (4)

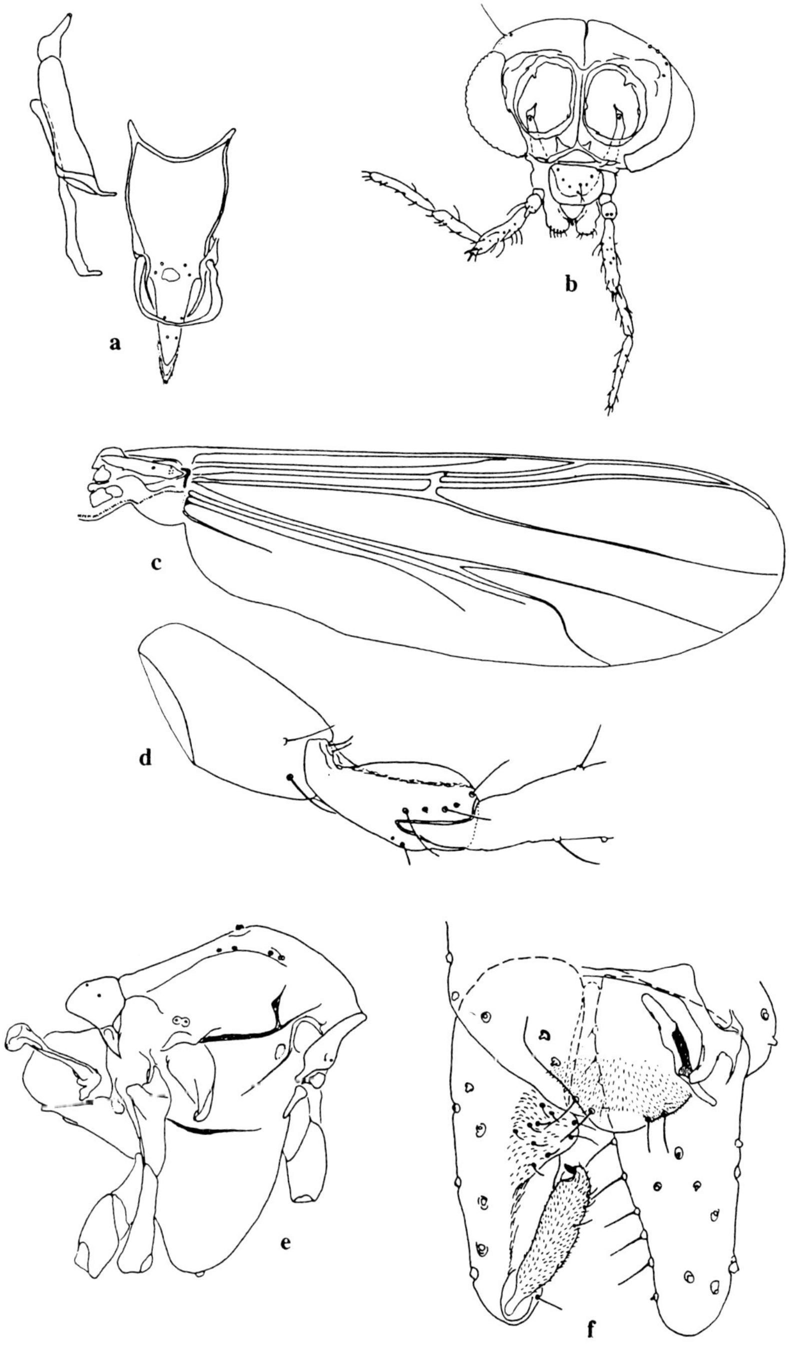

Figure 1. *Parakiefferiella minax* n. sp.: (a) cibarial pump, tentorium and stipes. (b) head, (c) wing, (d) coxa, trochanter and basal portion of femur of fore leg, (e) thorax, (f) hypopygium.

Thorax (Fig. 1e). Antepronotum with 1-2, 2 lateral setae. Dorsocentrals 3-4, 3; microtrichial tuft prominent; prealars 2; Scutellum with 4-6, 6 setae. Ventral apex of preepisternum with small knob.

Wing (Fig 1c). VR 1.11-1.16, 1.15. C extension 117-169, 149µm long. Anal lobe slightly projecting. R_{2+3} fused with R_{4+5}, R_{4+5} ends above apex of M_{3+4}; Cu_1 strongly sinuate. Brachiolum with 1 seta, R_1 with 1 seta in one specimen, other veins bare.

Legs. Front trochanter with indication of keel (Fig. 1d). Spur of front tibia 39-43, 41µm long; spurs of middle tibia 25-30, 28µm and 21-27, 25µm long; of hind tibia 41-48, 44µm and 27-30, 28µm long. Width at apex of front tibia 31-34, 32µm; of middle tibia 25-39, 36µm; of hind tibia 41-46, 44µm. Comb of 10-12, 11 setae; shortest setae 23-27, 25µm; of longest setae 23-27, 25µm. ta_1-ta_3 each with an apical sinuous seta. Lengths (in µm) and proportions of legs (n=5 for hind tarsomeres):

Hypopygium (Fig. 1e). Hyaline portion of anal point 11-16, 14µm (5) long; 23-37, 30µm (5) wide at base; tergite IX with 14-20, 16 setae; laterosternite IX with 2-3, 2 setae. Phallapodeme 64-69, 67µm long. Transverse sternapodeme 46-64, 49µm long. Virga absent. Gonocoxite 183-203, 187µm long; inferior volsella reaching 0.54-0.57, 0.55 gonocoxite length. Gonostylus straight, 85-91, 88µm long; megaseta 7-9, 8µm long. HR 2.05-2.23, 2.11; HV 3.03-3.34, 3.21.

Parakiefferiella harrisoni (Freeman) comb. n.

(Fig. 2, 3c)

Smittia harrisoni Freeman, 1956: 355

Pseudosmittia harrisoni (Freeman), Freeman and Cranston, 1980: 185, generic placement.

Material examined: **South Africa**: Transvaal, Mica-Gravelotte Road, paratypes male, female, 6.v.1955, A.D. Harrison, B.M. 1955-800 (BMNH)

Diagnostic characters.

The species differs from *P. minax* sp. n. by having projections with spiniform setae on both palpomere 3 and 4 (Fig. 3c), smaller size, and slightly curved gonostylus.

Description

Male imago (n=1)

Total length 1.77 mm. Wing length 1.14 mm. Total length/wing length 1.56. Wing length/length of profemur 2.84. Thorax yellow with dark vittae, scutellum and postnotum. Legs yellow with apex of fe, base and apex of ti, and base of ta_1 dark. Abdomen with tergites I-III dark except anteromedian half, IV pale, V pale in anterior 1/2; VI and VII pale, VIII pale on anterior 1/3.

Head (Fig. 2a, e). Antenna lost, AR 0.6 according to Freeman (1956). Temporal setae consisting of 3 outer verticals. Clypeus with 7 setae. Tentorium 103µm long, 14µm wide. Stipes 98µm long, 27µm wide. Palpomere lengths (in µm): 23, 27, 101, 85, 107. Third palpomere with 27µm long apical projection carrying an apical spiniform seta 14µm long, fourth palpomere with 9µm long projection carrying 15µm long apical spine (Fig. 3c).

Thorax (Fig. 2b). Antepronotum without lateral setae. Dorsocentrals 3, microtrichial tuft prominent, prealars 1. Scutellum with 2 setae. Ventral apex of preepisternum with knob.

Wing (Fig. 2c). VR 1.16. C extension 117µm long. Anal lobe reduced. R_{2+3} running close to R_{4+5} from base to apex, R_{4+5} ends proximal to apex of M_{3+4}; Cu_1 strongly sinuate. Brachiolum with 1 seta, other veins bare.

Legs. Front trochanter with indication of keel (Fig. 2d). Spur of front tibia 30µm long, spurs of middle tibia 23µm and 16µm long, of hind tibia 34µm and 18 µm long. Width at apex of front tibia 22µm, of middle tibia 25µm, of hind tibia 30µm. Comb of 9 setae, 21-34µm long. Setae at apex of ta_1-ta_3 curved, occasionally more sinuous. Lengths (in µm) and proportions of legs:

	fe	ti	ta_1	ta_2	ta_3	ta_4
P1	401	531	351	257	153	99
P2	450	446	316	113	90	45
P3	459	522	—	—	—	—

	ta_5	LR	BV	SV	BR
P1	45	0.66	2.32	2.65	4.5
P2	36	0.48	3.92	4.15	6.0
P3	—	—	—	—	—

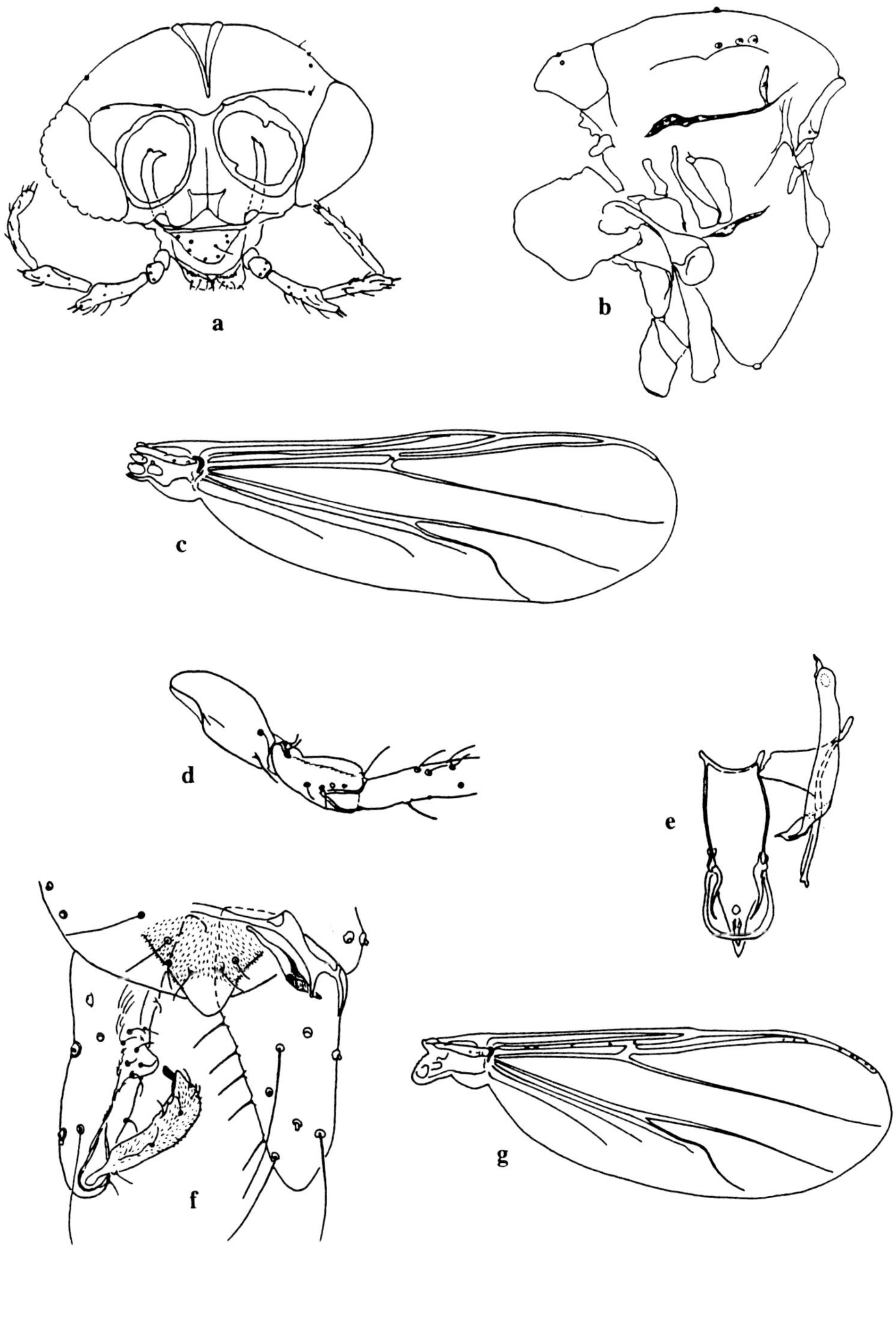

Figure 2. *Parakiefferiella harrisoni* n. comb.: (a) head, (b) thorax, (c) wing of male, (d) coxa, trochanter and basal portion of femur of fore leg, (e) cibarial pump, tentorium and stipes, (f) hypopygium, (g) wing of female.

Hypopygium (Fig. 2f). Hyaline portion of anal point 23µm long, 34µm wide at base; tergite IX with 5 setae; laterosternite IX with 3 setae. Phallopodeme 53µm long. Transverse sternapodeme 37µm long. Virga absent. Gonocoxite 123µm long; inferior volsella ending at 0.54 gonocoxite length. Gonostylus slightly curved, 66µm long; megaseta 9µm long. HR 1.86; HV 2.69.

Female imago (n=1)

Total length 0.98 mm. Wing length/length of profemur 3.11. Coloration slightly paler than male on thorax and legs, with posterior half of scutellum; abdomen lost but, darker than male according to Freeman.

Head. Flagellomere lengths (in µm): 57, 37, 43, 46, 75; AR 0.41. Temporal setae consisting of 3 outer verticals. Clypeus with 10 setae. Tentorium 78µm long, 7µm wide. Stipes 91µm long, 23µm wide. Palpomere lengths (in µm): 21, 23, 78, 82, fifth segment lost. Third palpomere with 27µm long apical projection carrying an apical spiniform seta 15µm long, fourth palpomere with 11µm long projection carrying 14µm long apical spiniform seta.

Thorax. Antepronotum with 2 lateral setae. Dorsocentrals 3, microtrichial tuft prominent, prealars 1. Scutellum with 2 setae.

Wing (Fig. 2g). VR 1.19. C extension 153µm long. Brachiolum with 1 seta, costal extension with 5 nonmarginal setae, R with 4 setae, R_{4+5} with 3 setae.

Legs. Spur of front tibia 18µm long, spurs of middle tibia 21µm and 14µm long, of hind tibia lost. Width at apex of front tibia 21µm, of middle tibia 25µm. Hind femur 369µm long. Lengths (in µm) and proportions of front and middle legs:

	fe	ti	ta_1	ta_2	ta_3	ta_4
P1	315	422	270	180	113	63
P2	387	378	180	86	72	36

	ta_5	LR	BV	SV	BR
P1	41	0.64	2.55	2.73	3.1
P2	27	0.48	4.29	4.25	3.3

Abdomen lost

Systematic remarks

The two species are obviously closely related as shown by the peculiar palps, the knob on the preepisternum, the wing venation, coloration and absence of virga. In some respects the two species are intermediate between *Parakiefferiella* and *Epoicocladius*. However, *P. harrisoni* has a bend in the gonostylus as occurs in other *Parakiefferiella* Thienemann and although *P. minax* lacks the bend in the gonostylus it should be placed in *Parakiefferiella* on the basis of the above mentioned synapomorphies. Other characters distinguishing *Epoicocladius* from *Parakiefferiella* are the elongated palpomeres in *Epoicocladius*, with palpomere 5 about 10X as long as wide (as opposed to less than 7X as long in *Parakiefferiella*) and vein R_{2+3} ending at 2/3 the distance from R_1 to R_{4+5} (Cranston, Oliver and Sæther, 1989). In the two species described here, R_{2+3} is fused or nearly fused with R_{4+5} as in *Parakiefferiella*. The ultimate palpomere is also clearly longer than 7X as long as wide, but not as much as 10X longer than wide. Both species lack a virga, which is present and strong in both *Epoicocladius* and *Parakiefferiella*.

Palpal projections and spiniform setae in Chironomidae.

The elaboration of a palpomere into an distal projection (or tooth) with or without apical spiniform setae may appear to be a valuable synapomorphy. However, such ornamentations occur scattered throughout the Chironomidae (see Figs. 3, 4 for some examples). Palpal projections have been shown to occur in all *Odontomesa* Pagast (Sæther, 1985a)(Fig. 3d), in *Sympotthastia takatensis* (Tokunaga)(Sasa and Kawai, 1985, fig 1B as *Syndiamesa*), *Sympotthastia huldeni* Tuiskunen (1986), *Pseudochironomus fulviventris* (Johannsen) (Sæther, 1977a, fig. 28), *Lipurometriocnemus vixlobatus* Sæther (1982, but not mentioned in the description) and another undescribed species of this genus (Fig. 3e), *Bryophaenocladius niger* Albu, *B. pectinatus* Albu (Albu, 1974, as subgenus *Odontocladius*), *B. brincki* (Freeman)(Sæther, 1973, fig. 2F), *B. flavoscutellatus* (Malloch) (Cranston and Oliver, 1988, fig. 6), *Kuschelius dentifer* Sublette and Wirth (1980, fig. 20c), *Smittia retracta* Freeman (1961) (Fig. 3a), *Thienemanniella similis* (Malloch) (Sublette, 1970, indicated in description, but considered as highly variable by Boesel and Winner, 1980) and *Pseudosmittia digitata* Sæther (1981, fig. 11a).

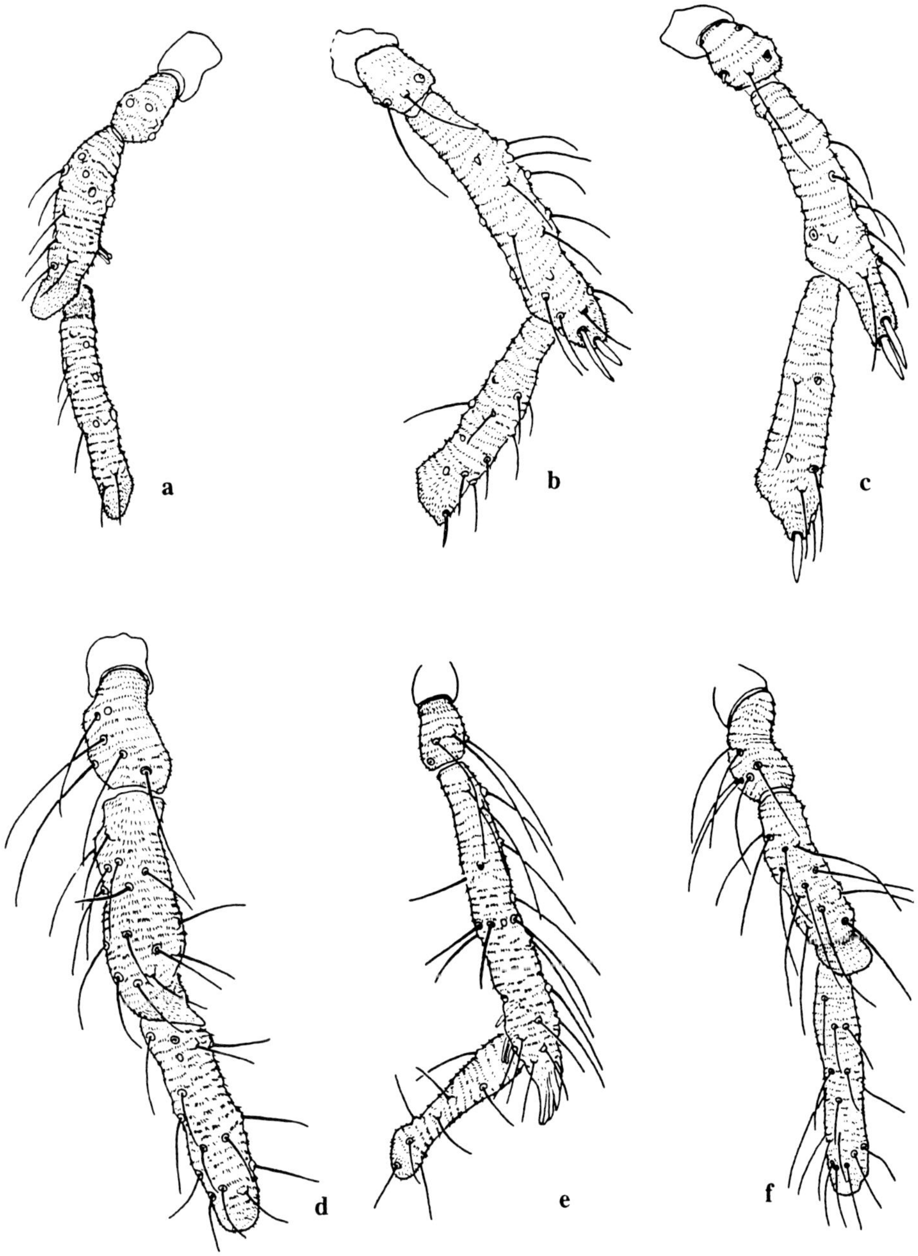

Figure 3. Palp segments 1-4 of selected taxa with varying degrees of development of ornamentation at apex of segment 3: (a) *Smittia retracta* Freeman, holotype, (b) *Parakiefferiella minax* n. sp., holotype, (c) *Parakiefferiella harrisoni* (Freeman), paratype 2, (d) *Odontomesa ferringtoni* Sæther, from Blue River, north of Silverthorne, Colorado, USA, (e) *Lipurometriocnemus* sp., from window of lab at Pymatuning Laboratory of Ecology, University of Pittsburgh, Linesville, PA, USA, (f) *Sympotthastia* sp., from Shoal Creek at Schermerhorn Park, south of Galena, KS, USA.

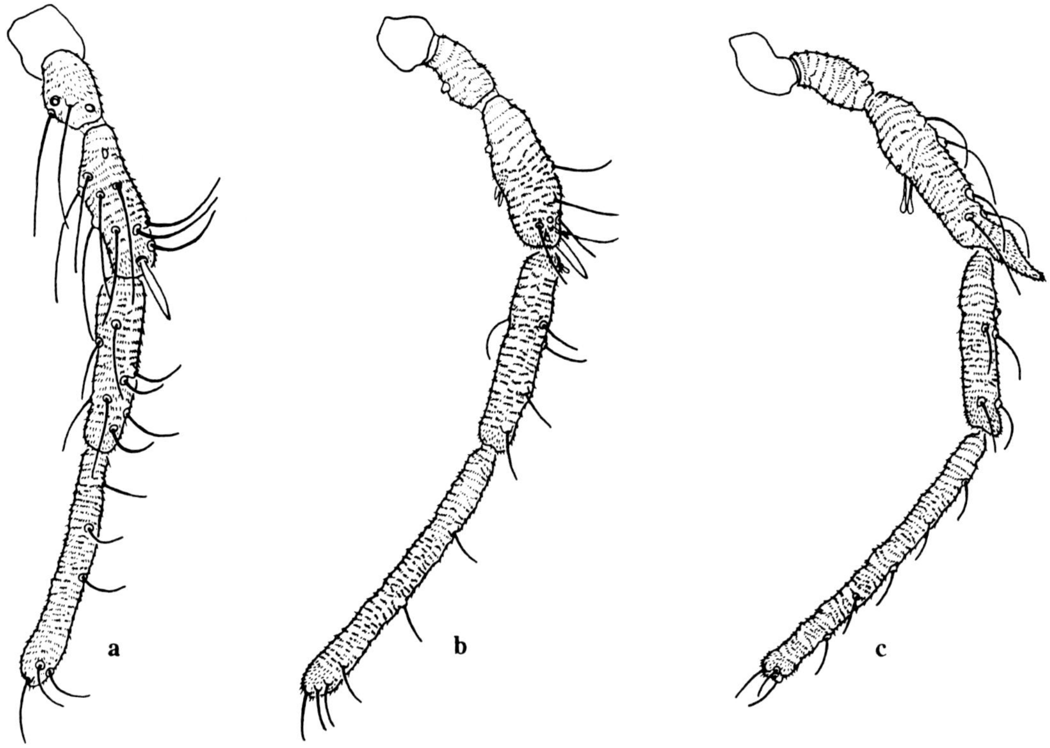

Figure 4. Palp segments 1-5 of selected taxa with differing degrees of ornamentation at apex of palp segment 3: (a) *Pseudosmittia insulsa* (Johannsen), from Suspe Lake, Saipan, (b) *Pseudosmittia* sp., from Seychelles, (c) *Pseudosmittia* sp., from Rio de Janeiro, Brazil.

(The type series of the last species was mixed, however, and the specimen that was illustrated is actually a new species which will be described elsewhere, but see Fig 4c.). A palp similar to the two species described here is found in a new species of *Pseudosmittia* from the Seychelles (Fig. 4b). Spiniform setae at the apex of the third palpomere apparently are rare, but they are noticed in *Metriocnemus lautus* Sublette and Sasa (1994, fig. 103), *Bryophaenocladius flavoscutellatus* (Cranston and Oliver, 1988), *Trichochilus lacteipennis* (Johannsen) (Sæther, 1985b), and *Pseudosmittia insulsa* (Johannsen) (Fig. 4a). Sensilla clavata are often lacking when the apex of the palpomere is elongated and/or when there is a spiniform seta. In *Pseudosmittia*, however, sensilla clavata are always present.

The widespread occurrence of palpal elongations among chironomids shows that it cannot be considered a valuable synapomorphy in higher level phylogenetic analysis of the family. It certainly, however, appears to be valuable on in a more limited context. For instance, it appears to be an objective synapomorphy for *Odontomesa*, and a valuable autapomorphy within several species. The monotypic genus *Kuschelius* Sublette and Wirth and the subgenus *Odontocladius* Albu of *Bryophaenocladius* were erected for taxa with palpal projections. *Kuschelius* is a close relative and possible synonom of *Bryophaenocladius* as presently defined. Ironically, in this grouping the palpal projection does not appear to be restricted to most closely related species, and therefore can be considered to be an underlying synapomorphy within a broadened concept of the genus *Bryophaenocladius*. In *Pseudosmittia*, our preliminary phylogenetic analysis suggests that species with the projection may be closely related, yet do not form a monophyletic group within the genus.

If the palpal elongation with pronounced apical spine or spines is not a good synapomorphy, then the sporadic appearance among numerous groups suggests that they are the result of similar selective pressures, most likely associated with feeding habit. It has been shown that several genera of chironomids feed on honeydew (Downes, 1974), and one of us has observed adults congregating on leaves of riparian vegetation that have active extra-floral nectaries. It is possible that the spines or sensilla clavata are used in detecting honeydew, however this needs to be tested. It would be interesting to determine if the taxa with the most pronounced palpal projections with modified setae have similar feeding habits.

Acknowledgements

We are grateful to Dr. Brian Pitkin, The Natural History Museum, London, for supplying us with the type material and to David Goldhammer, University of Kansas, for making some of the slide preparations. Ms Rosella Cox assisted in typing and revising earlier versions of the manuscript. *Parakiefferiella minax* was collected during an expedition to the West Usumbara Mountains of Tanzania which was funded by the Museum of Zoology, the University of Bergen in 1990. This paper was completed while the second author was on a sabbatical leave at the University of Kansas supported by the Norwegian National Research Council. The paper is published under the auspices of the Water Quality and Freshwater Ecology Program of the Kansas Biological Survey at the University of Kansas.

CHIRONOMIDAE IN FRESHWATER HABITATS IN TENERIFE, CANARY ISLANDS

P.D. Armitage[1], *J.H. Blackburn*, *A.N. Nilsson* [2], *and B. Malmqvist*

[1]Institute of Freshwater Ecology, River Laboratory, East Stoke Wareham, Dorset BH20 6BB, England.

[2]Department of Animal Ecology, University of Umeå, Sweden

INTRODUCTION

Some groups of the freshwater fauna of the Canary Islands have been well studied. These include beetles, blackflies, mosquitoes, odonates and water bugs (see Malmqvist *et al.*, 1993 for a review of studies). The Chironomidae of the Canary Islands in contrast have been the subject of only occasional investigations (Becker, 1908; Santos Abreu, 1918; Storå, 1936). Subsequent collections of adult material have been made (Armitage, 1987; Armitage and Tuiskunen, 1988) and Cranston and Armitage (1988) redescribed material from the earlier collections of Becker and Santos Abreu. More recently 7 permanent freshwater streams on the island of Tenerife have been examined by Malmqvist *et al.* (1993). Chironomid data are presented in that paper and this present study analyses those records together with new information from a number of additional sites, based largely on larval material.

Study Area and Methods

Tenerife is the largest (2057km^2) and highest (3718m) of the Canary Islands and is situated at latitude 28°15"N, 300km off the coast of Morocco in the trade wind belt. The island has a volcanic origin and has never been connected to the African mainland (Schmincke, 1976). This isolation together with the arid nature of the southern part of the island which lies within a rain shadow result in a paucity of freshwater fauna and natural freshwater habitats. This situation is exacerbated by the continuing pressure on existing water sources.

The habitats sampled included stream riffles (7 localities on 'permanent' streams), stream pools (11 localities), streams general, that is not specifically pool or riffle samples (17 localities), residual pools in dried-up stream beds (12 localities), 4 ponds, 4 reservoirs, 14 artificial pipes, channels ('canals') and 15 seepages/trickles (wet walls). The location of the sites is indicated in Fig. 1 and the altitudinal distribution is illustrated in Fig. 2. Some localities were sampled on more than one occasion in 'spring' and 'autumn/winter' in 1991 and 1993 and details are presented in the appendix. Details on sampling methodology and information on the physical and chemical features of the main stream sites are presented in Malmqvist *et al.* (1993).

The majority of material was larval hence most records are at genus level but the inclusion of some pupae and adults facilitated identification to species in some cases. Light traps located by three streams also provided some adult material. Semi-quantitative data on spatial and temporal changes in the chironomid communities of the 7 'permanent' streams are presented in

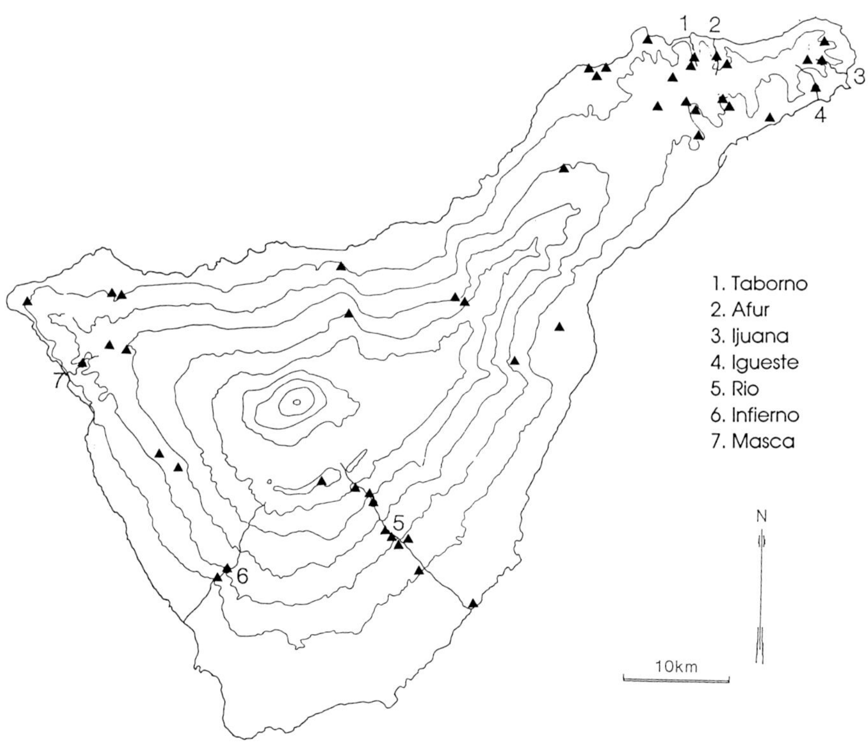

Figure 1. Tenerife; location of sampling locations in relation to the 7 main stream sites.

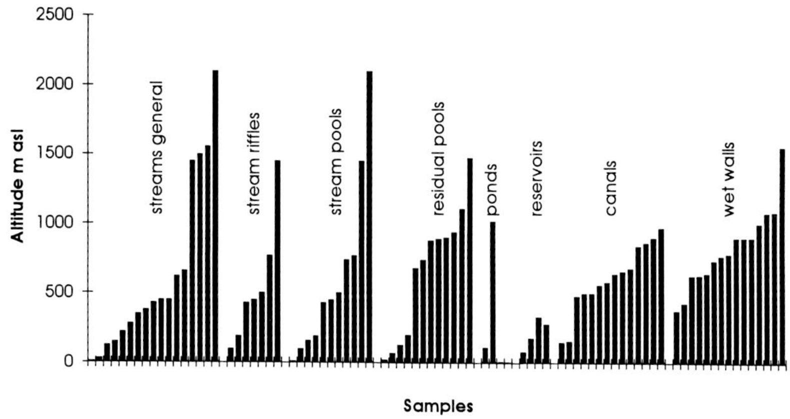

Figure 2. The altitudinal distribution of sample sites per habitat type.

Malmqvist *et al.* (1993) but in this present work most samples were not quantitative and to ensure compatibility all are treated qualitatively.

The data from the collections were analysed to examine the relationship between the chironomid community as a whole to the particular habitat types using ordination techniques. In addition the richness of habitat types in relation to sample effort is examined briefly.

RESULTS

The distribution of taxa in the habitat types is presented in Table 1. Forty-three separate taxa were recorded but the list includes some taxa groups and larvae unidentifiable to species. Of the species, most are cosmopolitan or European in distribution and the African element is weak.

The most commonly occurring taxa in stream pools were *Ablabesmyia longistyla* Fittkau, *Zavrelimyia ?nubila* (Meigen) and *Chironomus* spp.. In riffles *Rheotanytarsus* spp. and *Thienemaniella clavicornis* (Kieffer) were the common forms and in residual pools, ponds and reservoirs *Chironomus* spp. dominated the records. In 'canals' *Paratrichocladius* spp. (probably *rufiventris* (Meigen)) with *Rheotanytarsus* spp. were the most commonly occurring taxa. Chironomidae were seldom found in 'wet wall' samples and here the most common taxa were *Rheocricotopus* sp, *Paratrichocladius* sp. and *O. (Eudactylocladius)* sp.

The association of assemblages of chironomid taxa with habitat type was tested using Detrended Correspondence Analysis [DCA] (DECORANA; Hill, 1979) in the program CANOCO (ter Braak, 1988). The data matrix consisted of a reduced list of 40 taxa (see Fig. 3) at 118 sites. Ordination resulting from DCA, groups samples on the basis of their faunal composition and the relationship between axis scores and environmental variables allows the detection of general trends governing the distribution of samples. A preliminary run showed that the taxon *Halocladius*, which occurred at only one site not associated with any other species, was exerting a distorting effect on the ordination and it was omitted from the list of taxa. Fig. 3a presents the results of the site ordination. Clearly there is much variation and considerable overlap between habitat types. Nevertheless, lentic and lotic components are recognizable. The ordination of taxa is shown in Fig. 3b. and reflects the same lentic-lotic trend from right to left along axis 1, but no well-defined assemblages of taxa could be detected in response to underlying environmental gradients, a feature which may be linked to the opportunistic response of chironomid species to habitat availability (Rossaro, 1993). Even the very large altitudinal variation in sites was not evident in the ordination.

Table 1 List of taxa found in 8 habitat categories with number of records per habitat

Habitats	streams general	stream pools	stream riffles	residual pools	'canals'	wet walls	ponds	reservoirs
Sampling occasions	25	23	14	21	20	15	7	8
Ablabesmyia longistyla Fitt.	4	12	7	1	2	-	-	-
Ablabesmyia sp.	-	1	-	2	2	-	-	-
Bryophaenocladius sp.	-	-	-	-	1	-	-	-
Cardiocladius capucinus? (Zett.)	-	-	-	-	1	-	-	-
Cardiocladius sp.	-	-	-	-	1	-	-	-
Chaetocladius sp.	-	-	-	1	-	-	-	-
Chironomus sp.	2	9	2	10	4	-	4	5
Cladotanytarsus sp.	-	2	1	-	-	1	-	-
Corynoneura sp. A.	-	-	4	-	-	1	-	-
Cricotopus (C.) vierriensis Goetgh.	2	-	-	-	6	1	-	-
Cricotopus (I.) ornatus Mg.	-	-	-	-	-	-	-	1
Cricotopus (I.) sylvestris Fabr.	-	1	-	-	-	-	1	-
Cricotopus sp.	1	7	5	-	3	-	-	-
Dicrotendipes sp.	-	2	-	1	-	-	2	-
Eukiefferiella ? ilkleyensis (Edw.)	1	1	4	-	1	-	-	-
Eukiefferiella ? minor (Edw.)	5	2	3	-	-	1	-	-
Halocladius sp.	-	1	-	-	-	-	-	-
Limnophyes ? minimus (Mg.)	1	-	-	-	-	-	-	-
Limnophyes sp.	1	2	1	-	-	1	-	-
Macropelopia nebulosa (Mg.)	1	5	1	-	-	-	-	-
Macropelopia sp.	3	-	-	-	-	-	-	-
Metriocnemus ? fuscipes (Mg.)	-	1	-	-	-	-	-	-
Metriocnemus obscuripes (Holmgr.)	-	1	-	-	-	-	-	-
Metriocnemus sp.	1	-	-	-	-	-	-	-
Micropsectra ? notescens (Walk.)	1	3	1	-	-	-	-	-
Micropsectra sp.	1	2	1	-	-	1	-	-
Orthocladius ? (Eudactylocladius) sp.	2	-	-	-	1	2	-	-
Orthocladius (Euorthocladius) rivicola Kieff.	2	4	5	-	-	-	-	-
Orthocladius (Orthocladius) sp.	-	4	5	1	1	-	-	-
Orthocladius sp.	1	-	-	-	1	-	-	-
Orthocladius/Cricotopus sp.	5	-	-	2	2	-	-	-
Parakiefferiella sp.	1	-	-	-	-	-	-	-
Paramerina vaillanti (Fitt.)	-	1	1	-	-	-	-	-
Paramerina mauritanica (Fitt.)	1	-	-	-	-	-	-	-
Paramerina sp. Pe1 (sensu Langton 1991)	-	5	3	-	-	-	-	-
Paramerina sp. 1	-	-	1	-	-	-	-	-
Parametriocnemus stylatus (K.)	4	5	7	1	1	1	-	-
Paratrichocladius rufiventris (Mg.)	3	4	7	1	6	2	-	-
Paratrichocladius sp.	7	1	-	-	12	-	-	-
Phaenopsectra sp.	3	6	1	3	2	-	-	-
Polypedilum sp.	5	6	1	3	2	-	2	4
Procladius (Holotanypus) choreus (Mg.)	1	4	-	1	-	-	1	-
Procladius (Psilotanypus) sp.	-	-	-	-	1	-	-	-
Procladius sp.	-	-	-	1	-	-	-	1
Psectrocladius limbatellus Holmgr.	1	2	2	2	1	-	1	-
Psectrocladius ? octomaculatus Wülk.	-	-	-	-	-	-	-	1
Psectrocladius sp.	-	-	-	-	-	-	2	2
Rheocricotopus atripes K.	4	4	4	1	-	-	-	-
Rheocricotopus sp.	2	3	4	2	1	3	-	-

Table 1. cont.

Habitats	streams general	stream pools	stream riffles	residual pools	'canals'	wet walls	ponds	reservoirs
Sampling occasions	25	23	14	21	20	15	7	8
Rheotanytarsus sp.	3	4	11	1	10	-	-	-
Stictochironomus sp.	1	1	-	3	-	-1	-	-
Tanytarsus sp.	-	7	2	1	-	-	-	-
Thienemanniella ? clavicornis K.	-	5	8	-	-	-	-	-
Thienemanniella sp.	2	-	-	1	2	1	-	-
Thienemannimyia gp.	1	-	-	-	-	-	-	-
Trissopelopia ? flavida (K.)	1	1	1	-	-	-	-	-
Virgatanytarsus albisutus (Santos Abreu)	1	4	6	-	-	-	-	-
Zavrelimyia ? nubila (Mg.)	4	11	7	5	1	-	-	-
Zavrelimyia sp.	4	1	-	5	1	-	-	-

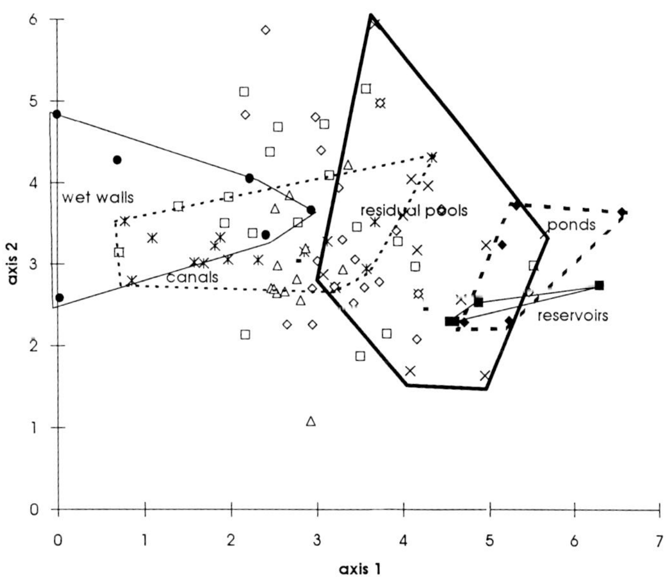

Figure 3a. Plots of ordination sample scores derived from DECORANA analysis of the entire sample site matrix. The polygons surround the data points in particular habitat types; (open square=streams general, open diamond=stream pools, open triangle=stream riffles, X=residual pools, closed circle=wet walls and seepages, closed square=reservoirs, closed diamond=ponds, ='canals' (artificial channels), =light trap)

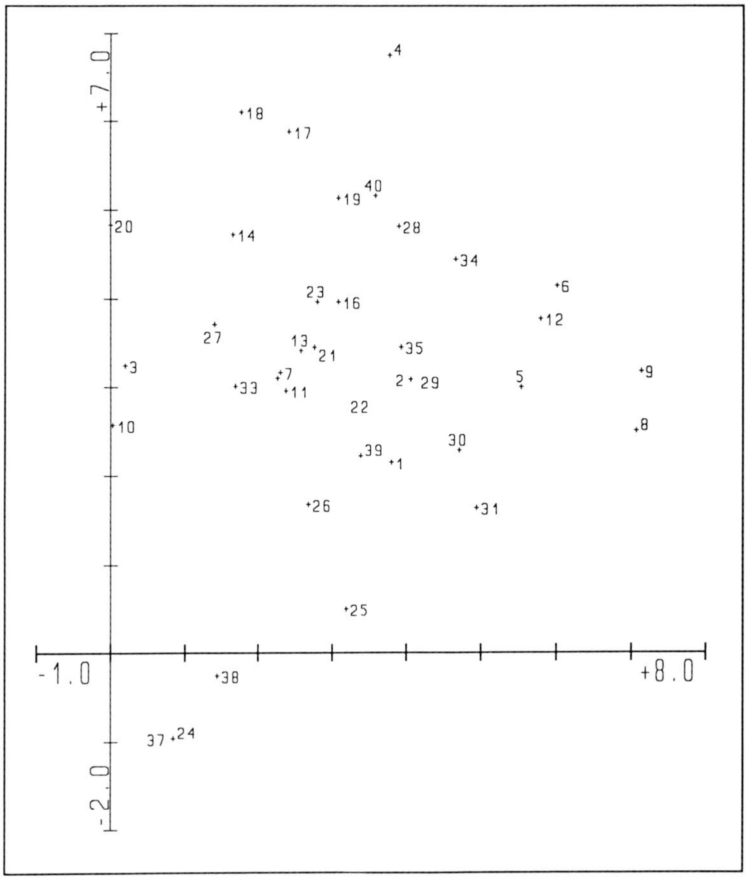

Figure 3b. Plots of ordination species scores. (axis1=x axis, axis2=y axis). 1 *Ablabesmyia* sp., 2 *Bryophaenocladius* sp., 3 *Cardiocladius* sp., 4 *Chaetocladius* sp., 5 *Chironomus* spp., 6 *Cladotanytarsus* sp., 7 *Corynoneura* sp., 8 *Cricotopus ornatus*, 9 *C. sylvestris*, 10 *C. vierriensis*, 11 *Cricotopus* sp., 12 *Dicrotendipes* sp., 13 *Eukiefferiella ?ilkleyensis*, 14 *E. ?minor*, 16 *Limnophyes* sp., 17 *Macropelopia* sp., 18 *Metriocnemus* spp., 19 *Micropsectra* sp., 20 *Orthocladius (Eudactylocladius)* sp., 21 *O.(Euorthocladius)* sp., 22 *O.(Orthocladius)* sp., 23 *Orthocladius/Cricotopus* spp., 24 *Parakiefferiella* sp., 25 *Paramerina* spp., 26 *Parametriocnemus stylatus*, 27 *Paratrichocladius rufiventris?*, 28 *Phaenopsectra* sp. 29 *Polypedilum* sp., 30 *Procladius* spp. 31 *Psectrocladius* spp., 32 *Rheocricotopus* sp., 33 *Rheotanytarsus* spp., 34 *Stictochironomus* sp., 35 *Tanytarsus* sp., 36 *Thienemanniella* sp., 37 *Thienemannimyia* group, 38 *Trissopelopia* sp. 39 *Virgatanytarsus* sp. 40 *Zavrelimyia* sp.

The relative richness of the habitat types is illustrated in Fig. 4. In the upper part of the figure it is clear that there is a positive relationship between total number of taxa and the number of sampling occasions, with the exception that stream riffles have slightly higher number of taxa than expected and wet walls a lower number than expected. This point is borne out when the mean number of taxa per sample is considered (lower

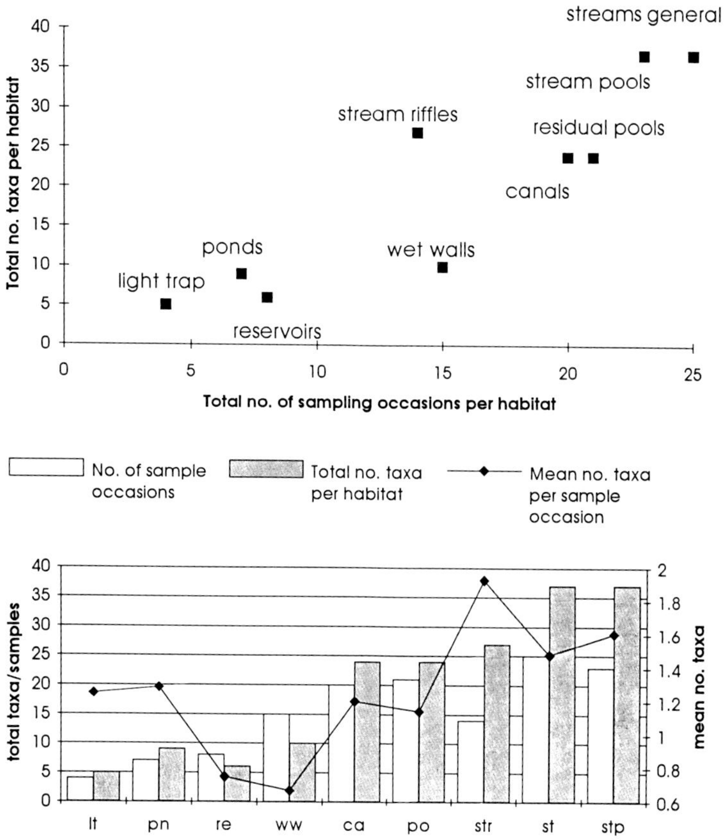

Figure 4. Taxon richness per sample effort (upper) and per habitat type (lower). (lt- light trap, pn - ponds, re - reservoirs, ww - wet walls, ca - canals, po - residual pools, and streams - general, st, riffles str and pools stp).

section of Fig. 4). The stream habitats contain more taxa than standing waters and within the latter ponds are richer than reservoirs.

DISCUSSION

The study of Chironomidae of Tenerife is still at an early stage. Although the work of Malmqvist *et al.* (1993) was systematic and detailed, collections of larval material do not usually provide the information necessary for species identifications. Adult material in good condition is required.

Some taxa that were represented by adult, pupal and larval material were still difficult to identify to species. The genus *Paratrichocladius* definitely contained *P. rufiventris* but also included some forms which did not readily fit that species, with pupal exuviae near *rufiventris* but with a point patch on tergite VII and adult features (gonocoxite lobe) which did not fit *P. rufiventris*. Within the *Rheotanytarsus* complex there are probably three species.

One of these is new (*Rheotanytarsus* Pe 2, sensu Langton (1991) and is currently being described from all life-history stages (collected by PDA in 1983 and 1985). No pupal or adult material is available for the other *Rheotanytarsus* species. All queried species require more adult material to be certain of their identity.

The paucity of taxa found in this present survey may reflect not only the isolated and arid nature of much of the island but also the absence of suitable material which would have permitted further identification of the genera recorded. In neighbouring Morocco, Azzouzi and Laville (1987) recorded 65 species based on data from the literature between 1955 and 1986; with more collections this total rose to 134 species. Subsequent studies in lotic habitats (Azzouzi *et al.*, 1992) have now raised this number to 223 species. In Tenerife a large increase in the recorded species can also be expected with more systematic study, despite the limited availability of freshwater habitat. Standing water habitats in the form of small reservoirs are increasing and it is likely that taxa lists from these habitats will increase over the next few years. Omitted from this study are terrestrial and marine habitats which can be expected to contribute a number of species (Armitage, 1987; Armitage and Tuiskunen, 1988).

It is too early for biogeographic speculation and this should await the availability of comprehensive species lists based largely on adult material.

Acknowledgements

We are most grateful to Dr. Marcos Baez (Dept of Zoology, University of La Laguna, Tenerife, Spain) for his help in locating the freshwater habitats.

APPPENDIX

List of sampled localities. Abbreviations: lt- light trap, pn - ponds, re - reservoirs, ww - wet walls, ca - canals, po - residual pools, and streams - general, st, riffles str and pools stp.

site name	type	alt	date
Afur	st	350	10.iv.91
Afur	st	280	10.iv.91
Afur	st	220	10.iv.91
Afur	stp	190	15.iv.91
Afur	str	190	15.iv.91
Afur	stp	190	31.x.91
Afur	str	190	31.x.91
Afur	lt	180	6.iii.93
Afur	lt	180	6.iii.93
Afur	ww	375	1.iii.93
Agua Garcia	po	880	3.xi.91
Agua Garcia	po	880	25.ii.93
Aguamansa	ww	1080	2.xi.91
Aguamansa	ww	1080	28.ii.93
Bajamar	pn	110	26.ii.93
Bco del Anesma	st	620	7.iii.93
Bco del Ruiz	po	890	4.iii.93
Bco del Ruiz	ca	680	10.xi.91
Bco del Ruiz	po	890	10.xi.91
Bco del Ruiz	ca	680	4.iii.93
Bufadero	re	80	6.xi.91
Bufadero	po	70	6.xi.91
Bufadero	po	70	2.iii.93
Bufadero	re	80	2.iii.93
Chamorga	st	380	6.xi.91
Chamorga	st	380	7.iii.93
Chiguerque	ca	820	9.xi.91
Chiguergue	ca	820	4.iii.93
Chinobre	ww	770	6.iii.93
Chirche	ca	955	9.xi.91
Chirche	ca	955	4.iii.93
1 km E Bailadero	ww	735	6.iii.93
El Bailadero	ww	625	5.iii.93
Erjos	pn	1035	14.iv.91
Erjos	pn	1035	1.xi.91
Erjos	pn	1035	1.xi.91
Erjos	pn	1035	3.iii.93
Fuente Paloma	st	30	3.xi.91
Fuente Paloma	st	30	3.xi.91
Fuente Paloma	st	30	26.ii.93
Guimar	ca	640	5.xi.91
Guimar	ca	640	26.ii.93
Ifonche	sp	1025	5.iii.93
Igueste	stp	100	14.iv.91
Igueste	str	100	14.iv.91
Igueste	stp	100	4.xi.91
Igueste	str	100	4.xi.91
Ijuana	stp	770	16.iv.91
Ijuana	str	770	16.iv.91
Ijuana	stp	770	6.xi.91
Ijuana	str	770	6.xi.91
Ijuana	st	770	5.iii.93
Ijuana	lt	770	5.iii.93
Ijuana at road	st	660	7.iii.93
Infierno	st	450	15.v.91
Infierno	ca	500	
1.xi.91Infierno	stp	500	14.iv.91
Infierno	str	500	14.iv.91
Infierno	stp	500	1.xi.91
Infierno	str	500	1.xi.91
Infierno	ca	500	4.iii.93
Infierno	ww	1085	5.iii.93
La Hidalga	re	180	5.xi.91
La Hidalga	re	180	26.ii.93
La Jardina	ww	630	6.iii.93
La Jardina	ww	645	6.iii.93
Las Aguas	po	1110	2.xi.91
Las Aguas	po	1110	5.iii.93
Las Cañadas	stp	2100	1.xi.91
Las Cañadas	st	2100	9.xi.91
Las Cañadas	st	2100	9.xi.91
Las Cañadas 2	st	2100	9.xi.91
Las Cañadas	st	2100	5.iii.93
Bco Las Canteras	stp	740	9.iv.91
Bco Las Canteras	stp	740	30.x.91
Bco Las Canteras	po	740	1.iii.93
Llanos los Viejos	ww	785	1.iii.93
Los Campitos	re	330	4.xi.91
Los Campitos	re	330	2.iii.93
Los Silos	ca	160	10.xi.91
Los Silos	st	125	10.xi.91
Los Silos	ca	160	3.iii.93
Los Silos	st	125	3.iii.93
Los Silos	st	150	9.iii.93
Masca	st	450	8.xi.91
Masca	stp	450	12.iv.91
Masca	str	450	12.iv.91
Masca	stp	450	8.xi.91
Masca	str	450	8.xi.91
Mte del Agua	ww	900	3.iii.93
Pta Hidalgo	po	20	30.i.91
Pta Hidalgo	po	20	30.x.91
Pta Hidalgo	po	20	30.x.91
Pta Hidalgo	po	20	26.ii.93
Pta Teno	po	130	10.xi.91
Pta Teno	po	130	3.iii.93
El Rio	ca	500	27.ii.93
El Rio above dam	po	680	27.ii.93
El Rio above dam	po	680	5.xi.91
El Rio Galleria	ca	660	5.xi.91
El Rio levada	ca	480	4.xi.91

Playa del Rio	pn	10	5.xi.91	San Andrés	st	10	2.iii.93
Playa del Rio	pn	5	27.ii.93	San Cristobal	ca	560	9.iv.91
Rio canal	ca	900	5.xi.91	San Cristobal	ca	560	31.x.91
Rio	stp	1450	11.iv.91	San Cristobal	ca	560	2.iii.93
Rio	str	1450	11.iv.91	Taborno	st	430	17.iv.91
Rio	stp	1450	2.xi.91	Taborno	stp	430	17.iv.91
Rio	str	1450	2.xi.91	Taborno	str	430	17.iv.91
Rio	lt	1450	2.xi.91	Taborno	stp	430	7.xi.91
Rio	st	1450	11.iv.91	Taborno	str	430	7.xi.91
Rio	po	1475	28.ii.93	Taborno	ww	570	1.iii.93
Rio	st	1500	28.ii.93	Tahodio	po	200	4.xi.91
Rio	ww	1555	28.ii.93	Tahodio	po	200	4.xi.91
Rio	st	1555	28.ii.93	Tahodio	re	280	2.iii.93
Rio	ca	900	27.ii.93	Tahodio	po	200	2.iii.93
Rio	po	900	27.ii.93	Tejina	stp	160	8.xi.91
Rio	ww	900	27.ii.93	Tejina	stp	160	30.x.91
Rio	ww	900	27.ii.93	Tejina	stp	160	26.ii.93
San Andrés	stp	10	1.xi.91				

Two Orthocladiinae (Chironomidae) genera common to New Zealand and Australia: *Pirara* n. gen. and *Eukiefferiella* Thienemann

I.K.G. Boothroyd[1] *and Peter Cranston*[2]

[1]Department of Biological Sciences, University of Waikato, Private Bag, Hamilton, New Zealand,

[2] CSIRO Division of Entomology, GPO Box 1700, ACT 2601, Australia

[1]Present address: Hawke's Bay Regional Council, Private Bag 6006, Napier, New Zealand

Introduction

The described chironomid fauna of Australia and New Zealand poorly reflects the actual fauna. The taxonomy, distribution and ecology of many species are either unknown, or unpublished and/or hidden in "grey literature". This is especially true in the case of the subfamily Orthocladiinae. In New Zealand, only 27 species of this subfamily have been recorded, and of these, eleven are known only from the subantarctic islands (Freeman, 1959; Sublette and Wirth, 1980). Freeman (1961) documented 41 species of orthoclad from Australia that were placed, along with all Australasian and Oceanian regional species, into currently recognised genera in the catalogue of Cranston and Martin (1989). Although the catalogue included some previously unrecognised genera for Australia, no additional species of Australian Orthocladiinae have been described since Freeman's work.

Most regional species are described on the basis of adults alone, and few publications concern the ecologically significant immature stages. This adult-based perspective has led to the belief that the chironomid fauna of New Zealand is somewhat impoverished (Boothroyd, 1987), and similar inferences have been drawn regarding Australia. However, our studies on the immature stages suggest otherwise, and many new, but as yet undescribed species await formal description.

This paper derives from a study of the ecology of chironomids in a stream in the North Island of New Zealand (Boothroyd, 1987). Distinctive features of one New Zealand species that indicated potential new generic status to IKGB were found subsequently amongst undescribed orthoclads reared by PSC in Australia. A new genus, *Pirara* Boothroyd and Cranston, is described from the North Island of New Zealand and from south-eastern Australia. *Pirara* is distinguished in the adult by the combination of bare eyes, absence of pulvilli and acrostichals, and fringed squama; in the pupa by the absence of thoracic horn, presence of elongate fine spinules on the abdominal conjunctives, and by the reduced size of the pupal anal macrosetae; in the larva by the palmate setae on the submentum and maxilla, absence of seta interna and the deeply bifid SI seta. The New Zealand species, *Pirara matakiri*, is designated as the type of the genus.

Additionally, New Zealand species belonging to the cosmopolitan genus *Eukiefferiella* Thienemann warranted comparison with the sole Australian representative of the genus, *Eukiefferiella insolida* (Skuse). Distinctions, particularly in the pupa, indicate separate status for one of the New Zealand species, which is here described in all stages as *Eukiefferiella brundini* Boothroyd and Cranston, and the immature stages of *Eukiefferiella insolida* are described for the first time from Australia and New Zealand.

Methods and Morphology

Imagines and larvae were prepared by clearing in 10% KOH, either cold overnight or by gentle heating for a few minutes. Specimens were mounted in Euparal either after several ethanol washes (IKGB) or neutralisation in glacial acetic acid and transfer to isopropanol. Morphological nomenclature follows Sæther (1980). Most measurements have been tabulated (individual measurements are available on request) with all measurements in µm, except for body length in mm. Spurious precision has been eliminated by rounding, e.g. to nearest 5 or 10µm as appropriate.

All Australian specimens have been entered in the Australian National Insect Collection (ANIC) database - the collector is P.S. Cranston and the depository is ANIC, unless stated otherwise.

Abbreviations: morphology

Adult features, unless stated otherwise.

ac	acrostichals.
Anl	Anal lobe length (pupa).
Ant 1-5	Length of antennal segments 1st to 5th (larva).
aps	antepronotals.
Apt	Anal point length.
A.R.	Antennal ratio = Length of ultimate flagellomere : combined lengths of flagellomeres 1 - penultimate (adult) or Length of basal segment : combined lengths of segments 2 - apex (larva).
A.t.l.	Anal tubule length.
A.t.w.	Anal tubule maximum width.
B.l.	Body length (thorax + abdomen of adult, otherwise total length of larva, pupa) in mm.
B	No. of setae on brachiolum.
BR	Beard ratio (max. setal length on ta: width of ta).
BV	Combined length fe+ta+ti_1 : Combined lengths ta_{2-5}.
cb	number of spines in comb of posterior ti.
cbl	length of longest comb spine.
Cerc	Cercus length.
Cext	Costal extension (beyond R_{4+5}).
dc	dorsocentral setae (adult).
$dc_{1,2,3,4}$	dorsocentral setae (pupa).
dc_{1-2} etc	distance from dc_1-dc_2 etc.
fe	femur.
$fl_{(1-13)}$	antennal flagellomeres (1-13, proximal to distal).
fs	frontal setae (pupa).
gcx	gonocoxite length.
gc.s	gonocoxite setal number.
gst	gonostylus length.
hcl	head capsule length (larva).
LR	Leg Ratio - ta_1:ti.
$maps_{1,2}$	median antepronotals (1-2 -dorsal, dorso-lateral) (pupa).
Md	Mandible length (larva).
M.w.	Mentum width (compressed larval head capsule).
M.m.w.	Width of median tooth of mentum (larva).
ms (1-3)	Anal macrosetae, 1-3 = outer to inner (pupa).
Ms	Megaseta length.
P_{1-3}	Legs (1-Fore, 2-Mid, 3-Hind).
pa	prealar setae. pcs_{1-3} Precorneal setae, anterior to posterior (pupa).
Pc.h.	Procercus height (larva).
Pc.s.	Procercal seta, maximum length (larva).
Pmd	Premandible length
pp.l	posterior parapod length.
pp.w	posterior parapod, maximum width.
R, R_1, R_{4+5}	No. of setae on wing veins R, R1 and R4+5 respectively.
RO	Distance of ring organ from base of first antennal segment (larva).
Sc.l.	Length of largest seminal capsule.
Sc.w.	Width of largest seminal capsule.

scts	No. of setae on scutellum.
sq	No. of setae on squama.
SV	Ratio of fe + ti : ta1.
$ta_{1\text{-}5}$	tarsomere 1-5, proximal to distal.
temp	temporal setae.
th.l	thoracic horn, length (pupa).
th.w	thoracic horn, maximum width (pupa).
ti	tibia
$tisp_{1\text{-}2}$	length of tibial spur (1=longest, 2=shortest)
VR	Ratio of length of Cu1 : length of M.
W.l.	Wing length (arculus to apex) in mm.

ANIC	Australian National Insect Collection, CSIRO, Canberra, Australia
BMNH	The Natural History Museum, London, England
Ck	Creek
CM	Canterbury Museum, Christchurch, New Zealand
L(e)	Larva (exuviae)
NP	National Park
NZAC	New Zealand Arthropod Collection, Auckland, New Zealand
P(e)	Pupa (exuviae)
R.	River
SF	State Forest
SPHTM	Sydney School of Public Health and Tropical Medicine (collection now incorporated into ANIC)
ZMBN	Zoological Museum, University of Bergen, Norway

PIRARA BOOTHROYD AND CRANSTON

Type-species: *Pirara matakiri* Boothroyd and Cranston, n.sp. by present designation.

Etymology: From the Maori *pirara* (= separated) referring to the branched and separating submental setae of the larva.

Generic diagnosis

The bare eyes, fringed squama, vestigial to absent pulvilli, and absence of acrostichals and close approximation of R_{2+3} to R_{4+5} separates adult *Pirara* from all other described orthoclads. Pupae are distinguished by elongate fine spines on the conjunctives of many tergal and some sternal abdominal segments, absence of thoracic horn, reduced anal macrosetae, lack of median anal setae and setose prealar projection. The elongate branched setae, notably on the submentum, maxilla and mandible readily distinguish the larvae of *Pirara*.

Generic description

Imago (Figs. 1,2)

Small to medium sized species. Male antenna with 13 flagellomeres, groove in male reaching flagellomere 3, plume normally developed (Fig. 1c), A.R. less than 0.7; female antenna (Fig. 1d) with 5 flagellomeres, A.R. 0.5-0.6.

Head. Eyes naked, without dorsal elongation. Temporal setae restricted to few linear outer vertical / postorbitals. Palp 5-segmented with basal segment indistinctly delimited, sensilla clavata at apical sensillum coeloconicum of third segment not evident.

Thoracic setation (Fig. 1e): acrostichals absent, dorsocentrals few, uniserial, prealars few, scutellars in single or irregular double row.

Wing (Fig. 1a). Membrane without setae, with very fine punctation of microtrichia. Anal lobe normal. Costa slightly extended; R_{4+5} ends distal to M_{3+4}; R_{2+3} closely approximated to R_{4+5} with unclear ending in Costa; An ends distal to RM; FCu lies distal to RM. Cu_1 slightly curved. Squama with fringe of few setae.

Legs. Pulvilli vestigial or absent. Tarsal pseudospurs present or absent at apex of tarsomere 1 and/or tarsomere 2 and 3 of the mid and hind legs, occasionally absent on tarsomere 3 of hind leg. Hind tibial comb and tibial spur normal.

Abdomen. Setae of tergites in more or less distinct rows of basal setae, marginal setae absent.

Male genitalia (Fig. 1f,g). Anal point small and bare, or absent. Anterior margin of transverse sternapodeme convex, phallopodeme and aedeagal lobe normal. Virga small. Gonocoxite with small to modestly developed inferior volsella.

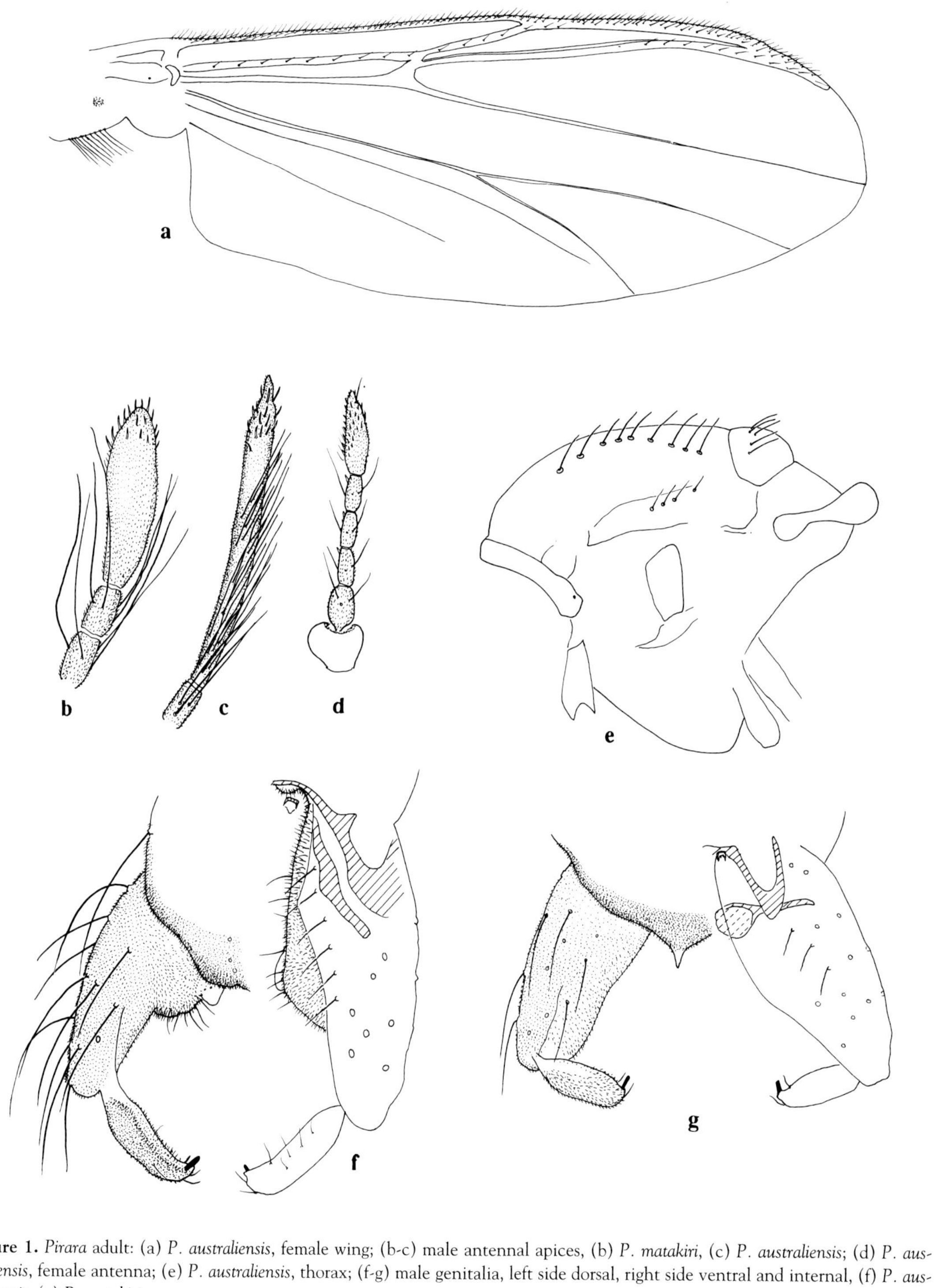

Figure 1. *Pirara* adult: (a) *P. australiensis*, female wing; (b-c) male antennal apices, (b) *P. matakiri*, (c) *P. australiensis*; (d) *P. australiensis*, female antenna; (e) *P. australiensis*, thorax; (f-g) male genitalia, left side dorsal, right side ventral and internal, (f) *P. australiensis*, (g) *P. matakiri*.

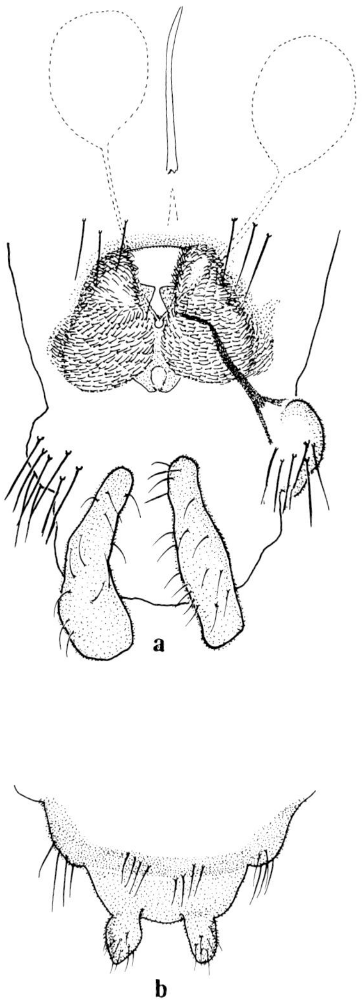

Figure 2. *Pirara* female genitalia: (a) ventral, (b) dorsal.

Female genitalia (Fig. 2a,b). Gonapophysis VIII comprising large ventrolateral lobe, extensively in contact medially with opposite lobe, and with barely delimited hyaline separation from microtrichiose dorsomesal lobe that forms inner contoured surface of vagina; apodeme lobe not visible. Tergite IX simple, bearing two groups of 4-5 strong setae. Gonocoxite IX large, with several strong setae. Postgenital plate undeveloped. Seminal capsules round to ovoid, weakly pigmented, without necks. Spermathecal ducts relatively straight, non-microtrichiose, bulbous at separate terminations. Cerci large.

Pupa (Fig. 3)

Small to medium sized, up to 2.5mm long.

Cephalothorax. Frontal setae weak (Fig. 3a); frontal apotome without warts. Thorax (Fig. 3b) smooth or nearly so; thoracic horn absent; thoracic setae long; three precorneal setae, medium to long, in triangle; 4 dorsocentrals, forming single group in mid-cephalothorax. Wing sheath smooth, without pearl row.

Abdomen (Fig. 3c). Tergites I-IV without shagreen, more posterior tergites with sparse anterior shagreen; sternite I without shagreen, sternites II-III with or without, IV-VIII with anterior weak shagreen. Tergites with posterior transverse bands of spines of two sizes on TII-IX; tergal conjunctives II/III - V/VI with transverse band of long pale spinules, many directed away from the body. Sternite II with transverse band of pale spines. Sternal conjunctives II/III-V/VI VI/VII with long, pale anteriorly-directed spinules. Pedes spurii A and B absent. Apophyses absent. Setation: 5 or more pairs of D setae, up to 10 pairs of V setae, 4-5 L setae, O-setae weak or absent. Anal lobe (Fig. 3d,e) with short macrosetae, either 2 unequal spiniform or three subequal fine, without median seta, fringe, or apical spines. Genital sac of male overreaching anal lobe.

Fourth-instar larva (Fig. 4)

Small to medium sized larvae. Head capsule dark brown with pale area around ocelli and dark occipital region.

Head. Antenna (Fig. 4c) 5-segmented; Lauterborn organs distinct, subequal to segment 3; blade distinct, may extend to apex of flagellum; ring organ in basal half of segment 1. Labrum (Fig. 4f) with SI strong and bifid, other S setae weak and simple. All chaetae simple. Pecten epipharyngis consisting of 3 scales; chaetulae laterales simple or apparently weakly serrated. Premandible with single, broad apical tooth, without brush. Maxilla (Fig. 4d) with short sensilla and seta, numerous fine chaetulae, lacking any pecten, with plumose SM_{1-4}, appendix and basal lacinial chaeta. Mandible (Fig. 4e) with apical tooth shorter than combined width of 3 inner teeth; seta subdentalis lanceolate;

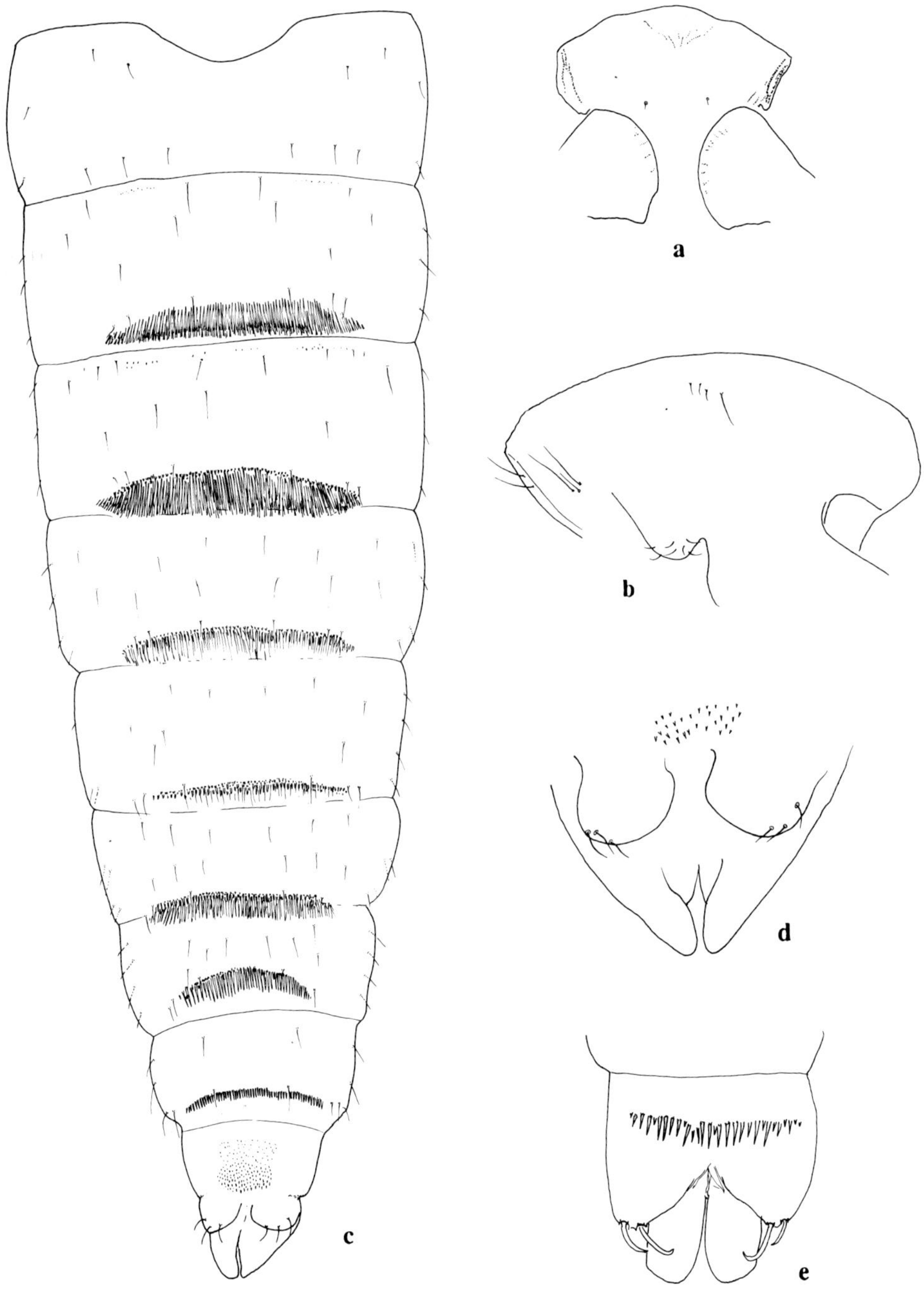

Figure 3. *Pirara* pupa: (a) *P. australiensis* cephalic area; (b) *P. australiensis* thorax; (c) *P. australiensis* tergites; (d-e) anal lobe, (d) *P. australiensis*, (e) *P. matakiri*.

seta interna absent, but long muscle filaments very evident; mola smooth. Mentum (Fig. 4a,b) with double median tooth and 3 or 5 lateral teeth; with pale, less sclerotised longitudinal grooves appearing as striations. Ventromental plate without beard, elongate, terminating posteriorly in bulge. Submental setae with 7-8 branches.

Abdomen. Parapods well developed. Claws of anterior parapods all with multiple fine serrations. Procercus (Fig. 4g) small, basal seta apparently displaced subapically, half length of 3-4 anal setae. Anal tubules shorter than posterior parapods.

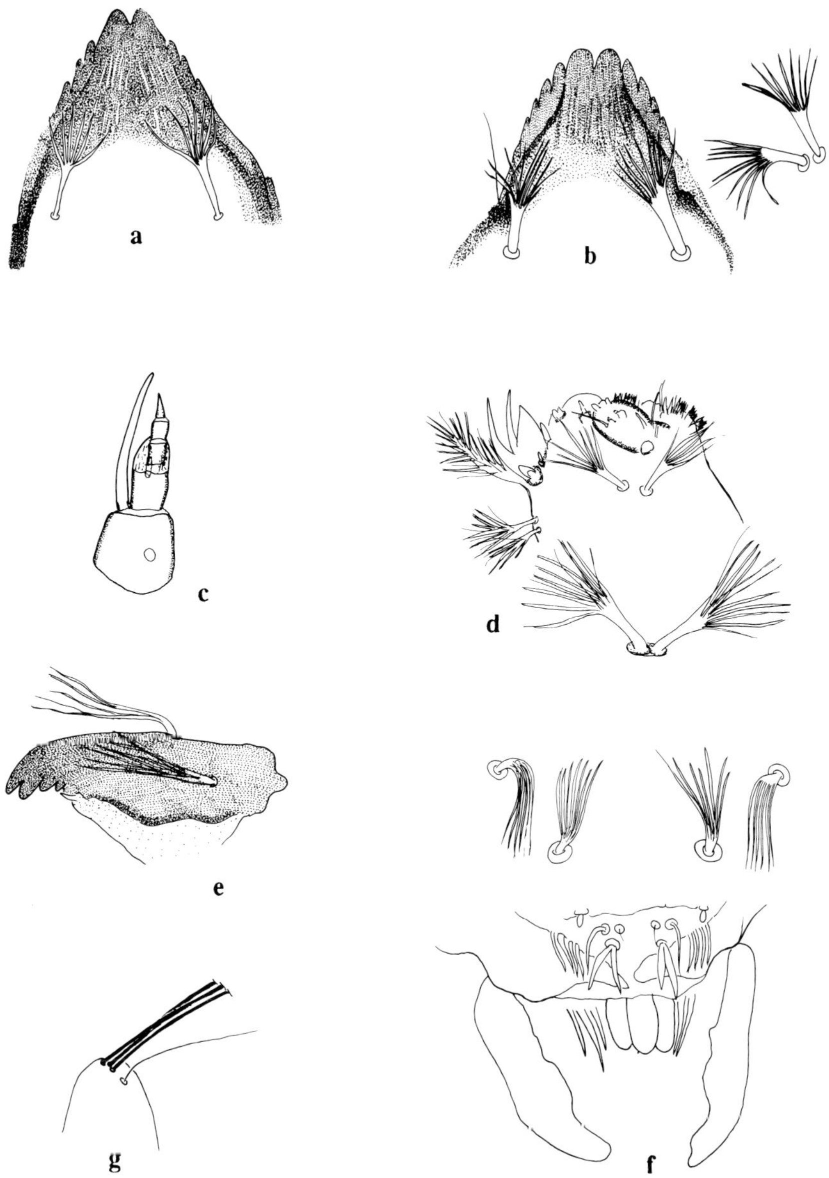

Figure 4. *Pirara* larva: (a-b) Menta of (a) *P. matakiri*, (b) *P. australiensis*; (c-g) *P. australiensis*, (c) antenna, (d) maxilla, (e) mandible, (f) labrum and anterior dorsal head setae, (g) procercus.

Remarks

Features of the two species described here indicate substantial morphological disparity, notably differences in the larval mental teeth, posterior pupal abdomen and male hypopygium. These differences invalidate *a priori* expectations that any particular character or life-history stage should provide unambiguous or "better" evidence of relationship than others: in this case the autapomorphic features of the larval setation and pupal abdomen confirm that the two species are related in a manner that is not remotely indicated by the male genitalia. This morphological disparity is addressed further in the Biogeography section.

According to the Holarctic key of Cranston *et al.* (1989), the adult male of *Pirara* keys to *Eukiefferiella* if the anal point is absent (*P. australiensis*), whereas it keys to *Parorthocladius* Thienemann if the anal point is present (*P. matakiri*). However, any close relationship to *Eukiefferiella* is denied by the immature stages, and since the female keys to *Parorthocladius* in Sæther (1977b), a relationship to the latter genus appears better supported. In both sexes, adults of *Pirara* differ from *Parorthocladius* by the reduction of vein R_{2+3} and its apparent fusion to R_{4+5}, and the organisation of the abdominal setae into anterior rows of tergal setae.

Using the pupal key of Coffman *et al.* (1986), the anal lobe macrosetae, small frontal setae, lack of thoracic horn, and presence of elongate fine spines on tergites IV-VIII apparently associates *Pirara* with the *Limnophyes* grouping. However, although the long fine spines on the conjunctives do not occur in any genus of Holarctic Orthocladiinae, this feature and the multiple prealar lobe setae do occur in some undescribed and generically unplaced Australian taxa.

The palmate setae on the submentum and maxilla which serve to distinguish larvae of *Pirara* from all others appear to be uniquely derived. *Pirara* larvae key close to *Mesocricotopus* Brundin in Cranston *et al.* (1983), yet bear a closer resemblance to the larvae of *Parorthocladius* and *Synorthocladius*, in the shape of the mentum, lack of seta interna and large Lauterborn organs. *Pirara* differs from both in lacking a ventromental beard and having the typical bifid SI of the *Cricotopus* / *Orthocladius* grouping.

Without a rigorous phylogenetic analysis involving a much wider grouping, firm conclusions on the relationships of *Pirara* are speculative; however, the larval features in particular support a sister group relationship to *Parorthocladius* and *Synorthocladius* combined, with the mentum shape and loss of seta interna being larval synapomorphies for the three genera combined.

Key to species of Pirara *gen. nov.*

Adult male:

1. Antennal apex clubbed (Fig. 1b). Inner volsella of gonostylus barely developed; anal point present (Fig. 1g) .... *Pirara matakiri* sp. nov.

\- Antennal apex an elongate nipple (Fig. 1c). Inner volsella of gonostylus strongly developed; anal point absent (Fig. 1f) ....*Pirara australiensis* sp. nov.

Pupa:

1 Anal macrosetae comprising two spines (Fig. 3e) . ...*Pirara matakiri* sp. nov.

\- Anal macrosetae comprising three fine setae (Fig. 3d) ....*Pirara australiensis* sp. nov.

Larva:

1 Mentum with 3 pairs of lateral teeth (Fig. 4e). Dorsal setae of head simple ....*Pirara matakiri* sp. nov.

\- Mentum with 5 pairs of lateral teeth (Fig. 4b). Dorsal setae of head plumose (Fig. 4f) ....*Pirara australiensis* sp. nov.

Species descriptions

Pirara matakiri sp. nov., Boothroyd and Cranston

Etymology: From the Maori *matakiri*, meaning minute, tiny.

Diagnostic characters: The characters used in the keys distinguish *P. matakiri* from *P. australiensis*.

Male Imago

Colour: head dark, thorax dark brown with slightly paler shoulders, abdomen dark. Mensural characters as in Table 1; leg lengths (μm) and proportions of legs as in Table 2. Figures: antennal apex (Fig. 1b), hypopygium (Fig. 1g).

Female imago

Colour: as male. Mensural characters as in Table 3; leg lengths (μm) and proportions of legs as in Table 2.

Pupa

Colour: exuviae dark with dark spines. Mensural characters as in Table 5. Figures: apex of anal lobe (Fig. 3e).

Fourth-instar larva

Mensural characters as in Table 6. Figures: mentum (Fig. 4a).

Material examined

Holotype, Le/Pe/m, 37°23'S 175°52'E, **New Zealand, North Island**, Waihi, Ohinemuri River, 100m asl, 22.ix.1987 (Boothroyd) (NZAC).

Paratypes: 4m, same data as holotype; 5L, many Pe, 37°54'S 175°05'E, **New Zealand, North Island**, Mt. Pirongia, Kaniwhaniwha Stream, 60m asl, various dates 1984-5. 4 paratypes, 3Pe, 6L deposited in NZAC; 1 paratype, 1Pe, 3L in CM; 1 paratype m, 17Pe, 1L, deposited in ANIC.

Distribution and ecology

P. matakiri is known from the northern North Island, and is commonly found throughout the Waikato region. Larvae are found in a range of habitats, but are more common on the surface of cobbles and small boulders in small, fast-flowing streams where they can be locally abundant. In studies of life cycles and emergence, *P. matakiri* larvae were found to occur all year but with distinct population increases during the summer months, with a similar pattern of emergence (Boothroyd, 1988). Larval drift occurs during daylight hours at all times of the year (Boothroyd, 1995).

Pirara australiensis sp. nov., Boothroyd and Cranston

Etymology: denoting the Australian provenance of this species.

Diagnostic characters: the characters used in the keys distinguish *P. australiensis* from *P. matakiri*.

Male imago

Colour: head dark, thorax dark brown with slightly paler shoulders; abdomen dark. Mensural characters as in Table 1; leg lengths (μm) and proportions of legs as in Table 2. Figures: antennal apex (Fig. 1c), thorax (Fig. 1e), wing (Fig. 1a), hypopygium (Fig. 1f).

Female imago

Colour: as male. Mensural characters as in Table 3; leg lengths (μm) and proportions of legs as in Table 2. Genitalia (Fig. 2a,b).

Pupa

Colour: exuviae dark with dark spines. Mensural characters as in Table 5. Figures: cephalic area (Fig. 3a), thorax (Fig. 3b), tergites (Fig. 3c), anal lobe (Fig. 3d).

Fourth-instar larva

Colour: head capsule mid-brown, occipital margin, mandible and mentum dark brown, anterior parapods mid-brown, finely serrate, posterior parapods dark brown to black, simple. Mensural characters as in Table 6. Figures: mentum and maxillary setae $MS_{3,4}$ (Fig. 4b), antenna (Fig. 4c), maxilla (Fig. 4d), mandible (Fig. 4e), labrum and anterior dorsal setae (Fig. 4f), procercus (Fig. 4g).

Material examined

Holotype, Le/Pe/m, 36°36'S 149°47'E, **Australia, New South Wales**, Brown Mt., Rutherford Creek, 2.x.1989 (Cranston).

Paratypes: **Australia: New South Wales**: 36Pe, 36°36'S 149°47'E, Brown Mt., Rutherford Ck, 16.x.1990 (Cranston and Edward); 3L, 35°53'S 148°29'E, Kosciusko NP, nr Kiandra, Permanent Creek, 14.xi.1993; 1Pe, 35°37'S 148°39'E, Cave Ck, 13-14.xi.1993; 1Le, 1Pe, 1Le/Pe/f, 35°34'S 150°02'E, Currowan SF, Cabbage Tree Ck, 30.iii.1994.

Australian Capital Territory: 1Pe/m, 4Pe, 2Pm, 1Pf, 35°20'S 148°56'E, Pierces Ck, at Cotter, 23.ix.1991 (Drayson); 10Pe, Corin, Gibraltar Falls, 11.viii.1989; same data except 3Pf, 3Pe, 25.viii.1989; same data except 1Pe, 24.x.1991 (Drayson).

Victoria: 1L, Tambo R., Bindi, 31.vii.1990 (Hortle).

2 paratype Pe deposited in NZAC, 2 paratype Pe deposited in ZMBN.

Distribution and Ecology

Pirara australiensis occurs along the montane eastern coast of mainland Australia. A single larval record from Yuccabine Creek, N. Queensland (S. Schreiber *pers. comm.*) is a northern outlier from an otherwise typically "cool, Gondwanan-type" distribution that lacks only Tasmanian representation. *P. australiensis* in south-east Australia is predominantly a winter/spring developing and spring emerging species, most often found in clean, fast-flowing, cool, relatively low-order streams at moderate elevations (escarpment edge and southern tableland). There is one larval record from a larger (4th order) cool river, the Victorian Steavenson River (37°28'S 145°45'E) (S. Schreiber *pers. comm.*)

Larval association with hygropetric (thin water-film) zones such as waterfalls or weirs within streams is suspected, but records are too scarce to confirm this habit.

The species is known from single or few specimens from any one locality, with the exception of an early spring collection from Rutherford Creek, a second order Dividing Range stream that still resembles Brundin's photograph (1966: Plate 3), with the exception of a possibly more recent weir construction. Numerous pupal exuviae were found in daytime drift at this site.

EUKIEFFERIELLA THIENEMANN

Eukiefferiella Thienemann, 1926: 32

Type species: *Spaniotoma (Orthocladius) gracei* Edwards (see Ashe, 1983: 23).

Generic diagnosis

Diagnoses of *Eukiefferiella* are as follows: adult males (Sæther and Halvorsen, 1981; Cranston *et al.*, 1989); adult females (Sæther, 1977b); pupae (Coffman *et al.*, 1986); larvae (Sæther and Halvorsen, 1981; Cranston *et al.*, 1983).

Additional features of the two species described here include: adult - acrostichals very small and difficult to see, eyes weakly pubescent, squama setose, virga of clustered fine spines, female genitalia with 5-6 posterior setae on large undivided tergite IX; pupa - with medially-divided transverse row of hooklets (3-6 each side) on tergal conjunctiva of III/IV - V/VI, well developed, stout to spine-like L_2, L_3 setae on segment VIII; larva - antenna with 4 segments, mentum with 5 pairs of lateral and 2 median teeth, all S setae simple.

Key to Australian/New Zealand species of Eukiefferiella

Adult male:

1. Volsella of gonostylus approximating right angled (Fig. 5c). . . . . *Eukiefferiella brundini*

1. Volsella of gonostylus narrower (Fig. 5a,b) . . . . *Eukiefferiella insolida*

Pupa:

1 Lateral setae L_2, L_3 fine, flexible, L_4 shorter than either L_2 or L_3 (Fig. 6e). Apex of anal lobe with spines between bases of anal macrosetae (Fig. 6f) . . . . *Eukiefferiella brundini*

\- Lateral setae L_2, L_3 stiff, inflexible, L_4 longer than either L_2 or L_3 (Fig. 6d). Apex of anal lobe without apical spines (Fig. 6g) . . . . *Eukiefferiella insolida*

Larva:

1 Mentum with broader, somewhat bulbous median teeth (Fig. 7c) . . . . *Eukiefferiella brundini*

\- Mentum with narrower, more tapering, median teeth (Fig. 7b) . . . . *Eukiefferiella insolida*

Species descriptions

Eukiefferiella brundini **sp. nov., Boothroyd and Cranston**

Etymology: named in gratitude for and recognition of the influence of the late Professor Lars Brundin in austral chironomid studies.

Diagnostic characters: the characters used in the keys above distinguish *E. brundini* from *E. insolida*.

Male Imago

Colour: head dark, thorax yellowish brown, abdomen pale brown. Mensural characters as in Table 1; leg lengths (μm) and proportions of legs as in Table 4. Figure: hypopyium (Fig. 5c).

Female Imago

Colour similar to male. Mensural characters as in Table 3; leg lengths (μm) and proportions of legs as in Table 4.

Pupa

Colour: exuviae dark with dark spines. Mensural characters as in Table 5. Figures: tergite VIII (Fig. 6e), anal lobe (Fig. 6f).

Fourth-instar larva

Colour: head capsule mid-brown, with darker occipital margin, labrum, mandibles and mentum. Mensural characters as in Table 6. Figure: median mentum (Fig. 7c).

Material examined

Holotype: m, **New Zealand, North Island,** Waihi, Ohinemuri River, 37° 23'S 175° 52'E, 100m asl. September 1987, light trap (Boothroyd) (NZAC).

Paratypes; 1f, 2m, as for holotype; 1f reared from L, 1m reared from L, Mt. Pirongia, Kaniwhaniwha Stream, 37° 54'S 175° 05'E, 60m asl., 1986; 4m, light trap, as for holotype; 1Pe reared from L, 5Pe, Kaniwhaniwha Stream; 5L, Kaniwhaniwha Stream; 2L, Ohinemuri River. 2Pe, 2L in NZAC, 1m, 1f, 1P, 1L in CM, 9L, 24Pe, 1m in ANIC.

Ecology and distribution

E. brundini is widely distributed in the North Island of New Zealand, with fewer records from the less well sampled South Island. Larvae are commonly encountered in a large range of habitats from small, fast-flowing streams to larger rivers. Studies on habitat preferences suggest that *E. brundini* favours intermediate velocities (0.7-1.5m s^{-1}), and larger, stable habitats (Quinn and Hickey, 1990; Collier, 1993). *E. brundini* tolerates a wide range of temperatures and moderate organic pollution. The larvae are found year-round, with no dominant season of emergence (Boothroyd, 1987, 1988), whereas diel emergence generally occurs at dusk in the summer, and during daylight hours in winter (Boothroyd, 1988). In contrast, larval drift is common during daylight hours in summer, with peaks in the morning and at dusk (Boothroyd, 1995).

Eukiefferiella insolida (Skuse)

Orthocladius insolida Skuse, 1889: 258

Eukiefferiella insolida (Skuse), Freeman, 1961: 644

Diagnostic characters: The characters used in the keys above distinguish *E. brundini* from *E. insolida*.

Male imago

Colour: Mid-brown, with darker brown vittae and paler brown scutellum, palps and legs. Mensural characters as in Table 1; leg lengths (μm) and proportions of legs as in Table 4. Figures: hypopygium (Fig. 5a, b).

Female imago

Colour: as male. Mensural characters as in Table 3; leg lengths (μm) and proportions of legs as in Table 4. Figures: hypopygium (Fig. 5d, e).

Pupa

Colour: exuviae mid-brown, apophyses, spines and macrosetae darker brown, tergal conjunctival hooklets nearly black. Mensural characters as in Table 5. Figures: tergites (Fig. 6a), sternites (Fig. 6b), thoracic horn (Fig. 6c), tergite VIII (Fig. 6d), anal lobe (Fig. 6g).

Fourth-instar larva

Colour: head capsule mid- to dark brown, body green to blue-tinged, anterior parapods golden brown, posterior parapods dark brown, procerci mid-brown. Mensural characters as in Table 6. Figures: mentum (Fig. 7a), median mentum (Fig. 7b), mandible (Fig. 7d), antenna (Fig. 7e), procercus (Fig. 7f).

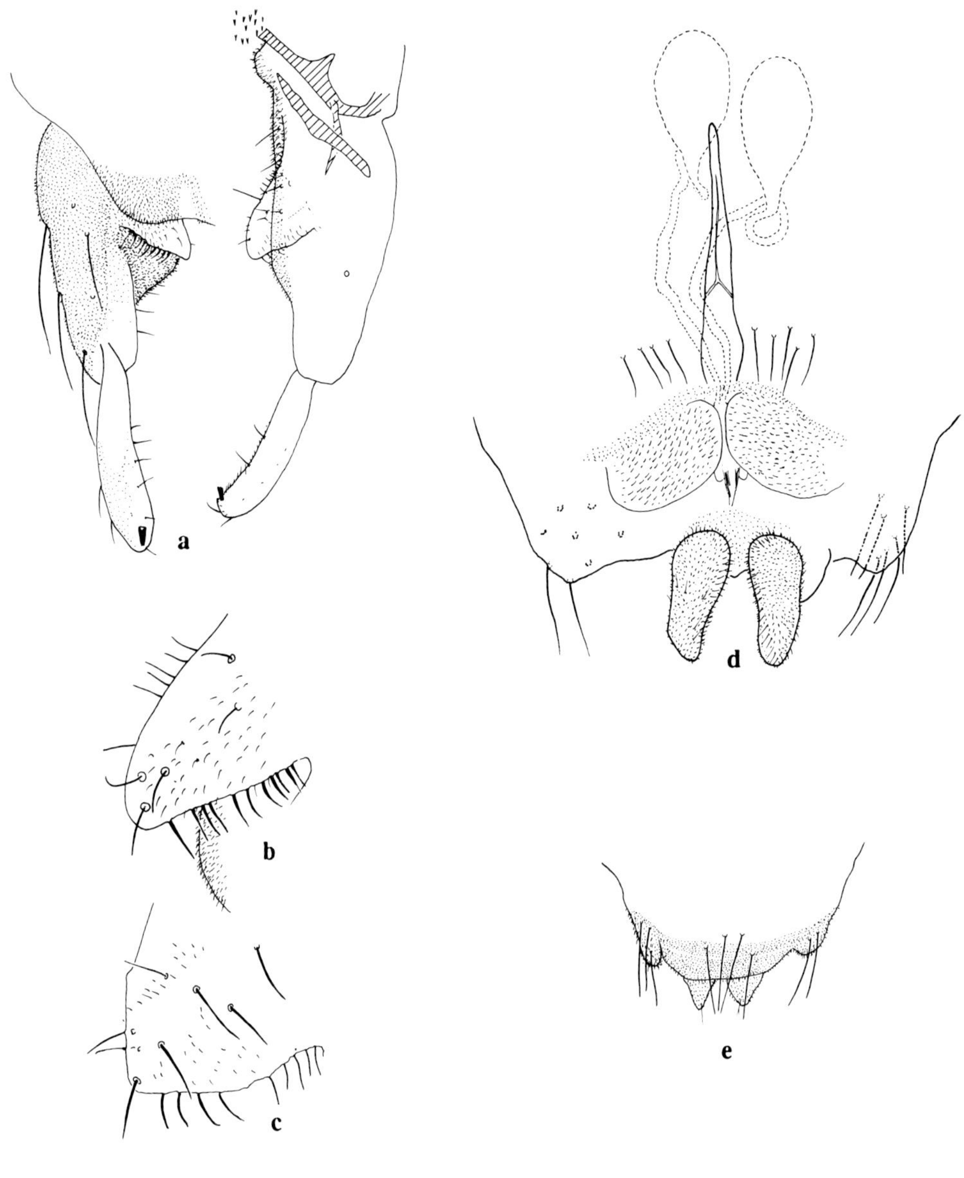

Figure 5. *Eukiefferiella* adults: (a) *E. insolida* male genitalia, left side dorsal, right side ventral and internal; (b-c) volsellae, (b) *E. insolida*, (c) *E. brundini*; (d-e) *E. insolida* female genitalia, (d) ventral, (e) dorsal.

Material examined

Holotype m, glued to card (hypopygium missing), **Australia,** Sydney, Middle Harbour (Skuse) (ANIC) (Freeman, 1961:644 states that he removed and remounted hypopygium - but missing from SPHTM when received by ANIC).

Australia:

Queensland: Pe, 25°04'S 148°14'E, Carnarvon Ck, W. of camp, 4-5.vii.1991 (Black); 5L, 3Pe, Pm, 26°42'S 152°36'E, Kenilworth SF, Bundaroo Ck, 27.ix.1989 (Cranston & Edward).

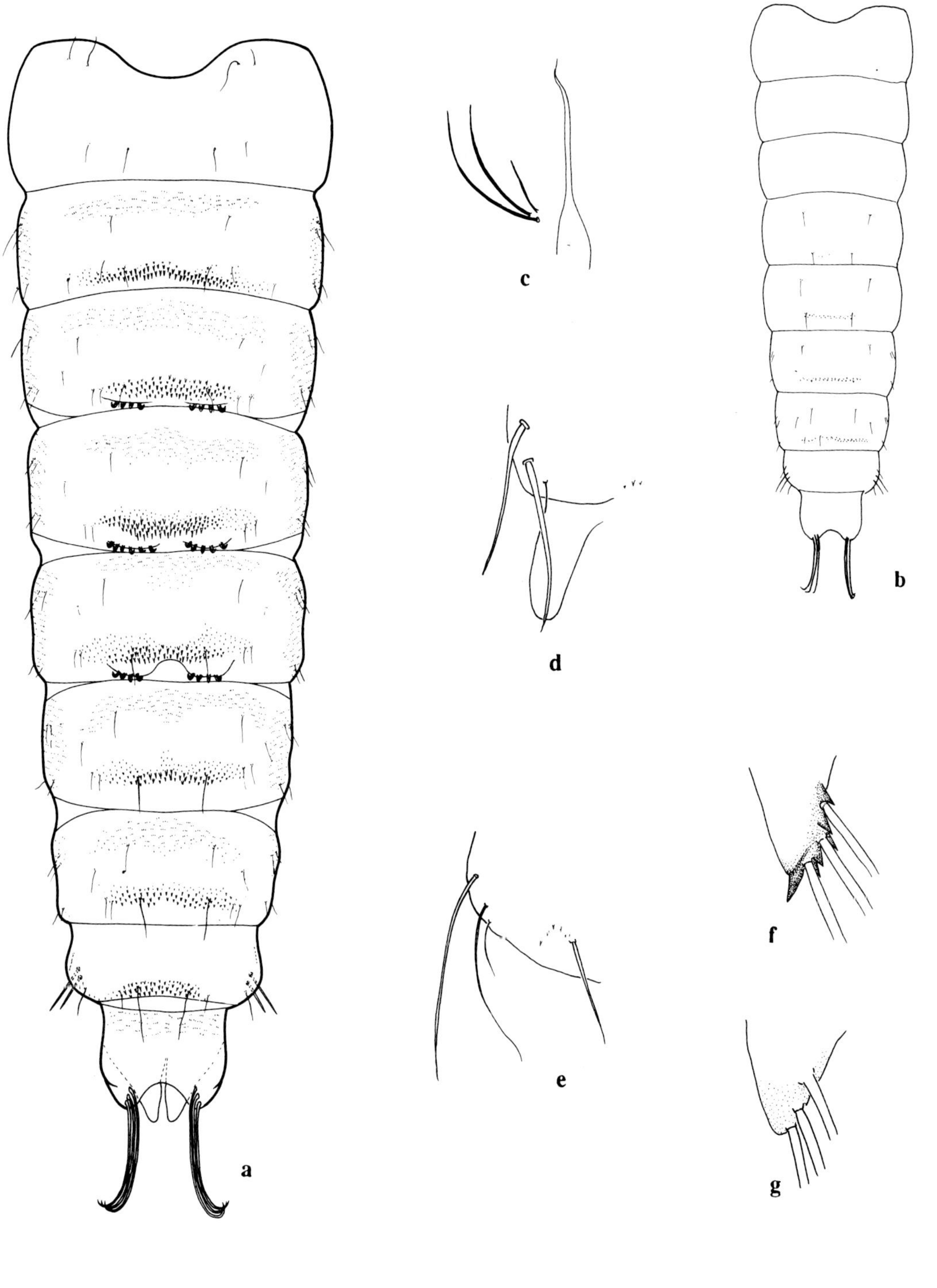

Figure 6. *Eukiefferiella* pupa. (a-c) *E. insolida*, (a) tergites, (b) sternites, (c) thoracic horn and precorneal setae; (d-e) posterolateral corner of tergite VIII of (d) *E. insolida*, (e) *E. brundini*; (f-g) apex of anal lobe of (f) *E. brundini*, (g) *E. insolida*.

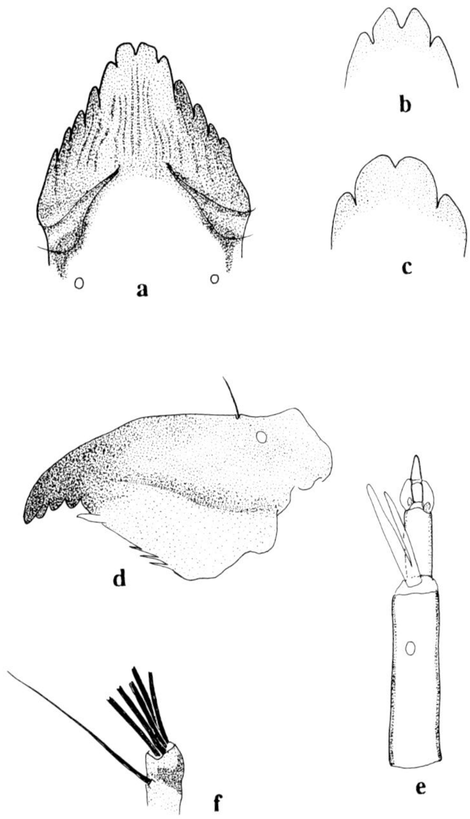

Figure 7. *Eukiefferiella* larva: (a) Mentum of *E. insolida*; (b-c) median mental teeth of (b) *E. insolida*, (c) *E. brundini*; (d-f) *E. insolida*, (d) mandible, (e) antenna, (f) procercus.

New South Wales: Le/P, 12Pe, Pm, 31°16'S 149°9'E, Warrumbungle Range, Timor Rock, Shawn's Ck., 15.ix.1989; Pe, 31°54'S 151°34'E, Barrington Tops, Dilgry R., 14.iv.1990; Pe, 34°35'S 150°05'E, Paddy's R., 16.x.1993; 2Pe, 34°44'S 150°05'E, Tallong, Barbers Ck, 16.x.1993; L, 34°47'S 150°0'E, Bungonia, Bungonia Falls, 11.xi.1988; 2Pe 35°5'S 150°8'E, Endrick R., 6km N.E. of Nerriga, 1.xi.1988; 7L, 14Pe, 35°35'S 149°28'E, upstream of Captains Flat, Molonglo R., 7.iii.1989, 3L, Pe/f, Pe/m,, 20.ii.1989, 3Pe, 6.iii.1993; 2L, 35°37'S 148°39'E, Kosciusko NP, Cave Ck, 13-14.xi.1993; Pe, 35° 39'S 148°28'E, Yarrango-billy, Yarrangobilly R., 14-15.i.1991; 2L, 35°53'S 148°31'E, Snowy Ck, 30.i.1988; L, 35°53'S 148°31'E, Black Walters Ck, 30.i.1988; 4Pe, 36°29'S 148°48'E, Kosciusko NP, Upper Snowy R., 12.i.1967 (Brundin); 6Pe, 36°36'S 149°47'E, Brown Mt., Rutherford Ck, 16.x.1990 (Cranston & Edward); 1Pe, 37°05'S 149°35'E, Tantawangalo SF, Wog R., 13.iii.1993.

Lord Howe Island (NSW): Pe, 31°33.15'S 159°04.50'E, Soldier's Ck, 3/4.xi.1993; 2Pe, 31°33.20'S 159°05.35'E, upper Rocky Run Ck, 4.xi.1993; 10L, 16Pe, P, 31°33.25'S 159°05.35'E, Rocky Run Ck, track crossing, 30.xi.1993; 2L, 1L/P, 5Pe, 1P, 31°33.25'S 159°05.35'E, Rocky Run Ck, at shore, 30.xi.1993; 3Pe, 31°33.35'S 159°05.52'E, Boat Harbour stream, 4.xi.1993.

Australian Capital Territory: 3L, 35°10'S 150°45'E, Jervis Bay, Stony Ck, 19.xii.1991; L, Le/P, 35°19'S 148°51'E, Brindabella Range, Condor Ck, 27.x.1991; L, Le/P, 23.iii.1992 (Cranston & Gullan), Pe, 17.v.1993; Pe/m, 35°20'S 148°57'E, Paddy's R. at Cotter, 15.ix.1988; 7L, 2Le/P, 2Pe, P, Pe/f, 35°21'S 148°52'E, Brindabella Range, Warks Road, Warks Ck, 9-13.iv.1988, L, 20.iv.1998; L, Le/Pm, 35°21'S 148°52'E, Lees Ck, 21.ix.1988; L, 35°22'S 148°50'E, Blundells Ck, 14.i.1988, Pe, 13.iv.1988, Pe, 24.viii.1988, Pe, 27.viii.1988; 3L, 2Pe, 35°27'S 148°57'E, Tidbinbilla, Tidbinbilla R., 19.ii.1989, L, Pe, 8.iv.1991; 5Pe, 35°28'S 148°55'E, Gibraltar Falls, 11.viii.1989, 2L, Pe, 25.viii.1989, Le/Pm, Le/Pf, Le/Pe/f, 23.xii.1992, L, 5.xi.1989, L, 5.xi.1993; 3L, Pe, P, 35°41'S 149°0'E, Namadgi NP, Gudgenby R., 14.ii.1988.

Victoria: 2Pe, 36°48'S 146°51'E, Buckland R., 6.v.1991 (Cook, Cranston, Nielsen); Le/Pf, 36°9'S 146°57'E, Albury-Wodonga, Middle Ck, Whytes Road, 31.vii.1990 (Cook); L, 37°5'S 147°55'E, Tambo R., Bindi #6, 24.viii.1989 (Hortle); 4Pf, 37°28'S 145°45'E, Steavenson R., 7.vi.1993 (Downes).

Tasmania: 1Pe, 41°04'S 147°15'E, Ansons R., "The Bottleneck", 23.iii.1994; 2Pe, 41°10'S 147°08'E, nr. Weldborough, Weld R., 22/3.v.1993; 2Pe, 41°50'S 146°2'E, Cradle Mt-Lake St Clair NP, Douglas Ck, Ranger Hut, 25.i.1990; 4Pe, 41°50'S 146°1'E, Cradle Mt-Lake St Clair NP, Douglas Ck, Old Pelion Hut, 25.i.1990.

South Australia: 3Pe, f, 35°27'S 138°35'E, Hindmarsh Falls, 9.v.1972 (Hergstrom); m, Eric Bonython Res., 23.ix.1970 (Hergstrom).

Western Australia: 9L, 2L/P, 2Pe, 34°22'S 118°14'E, Stirling Range NP, Bluff Knoll, cascade trail, 19-20.xi.1990.

New Zealand:

North Island, 4Pe, Mokau River, 6.ix.1967 (Brundin) (ZMBN); **South Island,** 4Pe, Canterbury, Porter R.,

"2337 ft", 22.x.1961 (Brundin) (ZMBN); Lake Paringa, 14.ix.1972 (Brundin) (ZMBN).

Table 1. *Pirara* and *Eukiefferiella*, adult males.

	P.matakiri (n=5)	*P.australiensis* (n=1-2)	*E.brundini* (n=3-5)	*E.insolida* (n=2)
B.l.	1.23-1.58	2.9	1.89-2.02	2.18-2.48
W.l.	1.08-1.23	1.5	1.58-1.80	1.65
A.R.	0.27-0.28	0.67-0.70	0.77-0.83	0.62-0.75
temp	3-4	5	2-4	1-2
Palp2	25-40	38	45-48	28-38
Palp3	52-58	70	60-73	42-70
Palp4	58-68	88	88-98	63-93
Palp5	78-90	108	133-145	91-112
aps	0	2	1	1
dc	6-7	6-8	7-9	6-8
pa	2-3	3	3-4	3
scts	8-12	8-11	8-12	4-6
VR	1.32-1.47	1.4	1.25-1.34	1.34
B	0-1	1	0-1	1
R	2-3	3	3	0
sq	?	10-15	6-8	5
Cext	45-65	35	55-65	60
Apt	15-17	n/a	n/a	n/a
gcx	120-130	157	130-155	115-154
gst	50-55	75	75-85	70-80
Ms	6-8	9	8-10	7-8

Remarks

The systematic placement of *Eukiefferiella insolida* (Skuse), then known only as an adult, was questioned by Sæther and Halvorsen (1981), who suggested that it might belong to *Tokunagaia* Sæther. As they acknowledged, the unknown immature stages would provide substantive evidence of the generic placement. The larva and pupa of both *E. insolida* and *E. brundini* indicate clearly that belong to *Eukiefferiella*, and to the *E. claripennis* species-group as keyed by Cranston *et al.* (1983) and Coffman *et al.* (1986). Although no species-groups were recognised by Cranston *et al.* (1989) for adult males, features of *E. claripennis* (Lundbeck) reasonably might be compared with the present material. According to Halvorsen (*pers. comm.* and unpubl. ms), his and our observations are consistent with this placement. Furthermore, female adult genitalia features, which are likely to be phylogenetically valuable, in all features of *E. insolida* and *E. brundini* are consistent with those of *E. claripennis*.

Distribution and Ecology

Eukiefferiella insolida is widely distributed in southern Australian streams. The only records from north of 30°S are from two cool, shaded streams in which specimens were collected in winter - both sites are northern outliers for other cooler "Gondwanan" taxa (P.S. Cranston *pers. obs.*). Further south, records (of the immature stages) predominantly are from autumn/winter/spring (May to November), with exceptions being some summer samples from elevations of above 1000m asl in the southern Highlands, including Kosciusko N.P., and on the central Tasmanian plateau.

E. insolida immature stages are found predominantly in relatively low order, well-oxygenated, permanent streams - the species appears to be sensitive to elevated temperatures (through latitude, loss of shade, or summer reduction in flow) but is very tolerant of organic enrichment (e.g. Hardwick *et al.*, 1995). In favourable conditions, *E. insolida* is one of the most numerous aquatic macroinvertebrates. These suitable conditions vary from the cool, 4th order, riffle-dominated sections of the Steavenson River, Victoria (Downes *et al.*, 1993) to the immediate downstream of effluent discharge from local sewage treatment in low order streams in the Blue Mountains, New South Wales (Hardwick *et al.*, 1995).

BIOGEOGRAPHY OF *PIRARA* AND AUSTRAL *EUKIEFFERIELLA*

Brundin (1966) documented many trans-Tasman (New Zealand/ Australian) distributions at generic level amongst the Podonominae. Examination of the regional catalogue (Cranston and Martin, 1989) shows several other examples amongst the Orthocladiinae and Chironominae, but these generally concern cosmopolitan genera (e.g. *Cricotopus*, *Tanytarsus*, *Chironomus*, and *Polypedilum*). At the species level, only the cosmopolitan *Paratanytarsus grimmii* (Schneider) is reported previously as occurring in both Australia and New Zealand (Langton *et al.*, 1988). *Eukiefferiella insolida*, which occurs on Lord Howe Island (isolated in the Pacific several hundred kilometres from Australia and even more distant from New Zealand) and is widely distributed in southern Australia, and less widely in New Zealand, is thus unusual amongst the Chironomidae. Whether this pattern represents the historic remnant of a contiguous distribution across land masses whose contiguity has been fragmented (see below) or results from recent dispersal (and if so, from where), can only be addressed with a reliable phylogeny of the worldwide *Eukiefferiella claripennis* grouping.

The case of the trans-Tasman, but non-cosmopolitan, genus *Pirara* may be less problematic. Once again, a consideration of the distribution allows either a dispersal or a vicariance explanation. However, the clear monophyly with substantial morphological disparity between the two constituent species, tends to indicate ancient divergence and rules unlikely any recent dispersal from one of the two countries to the other.

If a vicariance explanation is favoured, the geological framework can be examined for plausibility. The following derives principally from Veevers (1991), supplemented by Coleman (1980) and Walley and Ross (1991).

In the late Jurassic (160 million years before present, mybp), Australia and New Zealand were in contact (via the Lord Howe Rise to the east of the future Australian continent, and via eastern Antarctica to the south of the continent) and, although Tasmania was at more than 80°S it was relatively temperate. Throughout the Cretaceous, Australia separated from the Antarctic through a west-to-east opening of the future Southern Ocean. On the eastern side, the Lord Howe Rise (running north-westwards from New Zealand to the mid-east of Australia) was spreading from Australia. Australia was covered with shallow seas, although the eastern cordillera was rising to give elevation and commence the asymmetrical (dominantly westward) drainage pattern still seen today. The precise timing of the loss of contiguous, above-sea, connections between Australia and New Zealand (via Tasmania, or the Lord Howe Rise) are unclear because of variations in marine regressions in the late Cretaceous (100-65 mybp) - however a dating of about 80 mybp is generally agreed for complete continental separation and new sea floor creation. This minimum date for trans-Tasman vicariance provides adequate time for the observed morphological divergence within *Pirara*. The cool stenothermic lotic conditions currently favoured by both *Eukiefferiella* and *Pirara* were present in the past, even as the continental fragments moved northwards, due to (i) the cooling of Antarctica, (ii) the development of the cool Circum-Antarctic current, and (iii) orogeny in New Zealand and eastern Australia.

The dating of the genus *Pirara* by vicariance as being older than 80 million years may seem surprising. However, the dates implied by Brundin's (1966) studies on Podonominae, extended and supplemented by those of Cranston *et al.* (1987) and Cranston and Edward (1992), allow a much earlier origin and diversification of the Chironomidae than was previously suspected.

The possibility that the ancestral distribution of *Pirara* was greater than the combined current-day distribution (plus any included parts of the east Antarctic) would be confirmed by a wider distribution, namely in the Neotropics. We are unaware of any records of a taxon that resembles *Pirara* from the southern part of South America, but any such discovery would substantiate the early dating and allow more substantial interpretation of the relationship between chironomid taxa and geography.

Acknowledgements

IKGB is indebted to M.A. Chapman and J.D. Green for assistance and constructive comments during his study of the Chironomidae. The study was made possible through a Hilary Jolly Scholarship at the University of Waikato, Hamilton, New Zealand; additional funds were provided by the Entomological Society of New Zealand, and the New Zealand Lottery Science Research Grants Board.

We are grateful to Anker Halvorsen (Bergen) for sharing his knowledge and unpublished observations concerning the worldwide *Cardiocladius* group of genera including *Eukiefferiella*, and for loaning us Brundin's slides. Sabine Schreiber (Melbourne) willingly revealed much about the ecology and distribution of *E. insolida* and *P. australiensis*, for which, many thanks. Professor O.A.Sæther kindly provided advice and comments on both an early draft and the final version of this manuscript and Patrick Ashe loaned us undescribed *Eukiefferiella* material from Sulawesi.

Much of the Australian material described here has depended upon permits to collect in National Parks, notably in Western Australia and New South Wales (Kosciusko National Park and Lord Howe Island). PSC is grateful to the authorities for issuing licences to collect in these areas. The Australian Biological Resources Survey is acknowledged for assistance with some illustrations.

The junior author thanks Renate Sadler, for making many of the microscope slides, Wendy Lee for checking the distributional data and databasing the Australian specimens, and Drs Penny Gullan and Don Edward for refereeing the manuscript.

Table 2. *Pirara*, legs.

	P.matakiri(m) (n=5)	*P matakiri*(f) (n-2)	*P.australiensis*(m) (n=1)	*P.australiensis*(f) (n=4)
P1				
fe	310-320	400-410	490	445-470
ti	320-370	370-420	525	500-560
ta_1	155-195	280-300	310	295-310
ta_2	100-145	150-165	220	205-220
ta_3	80-85	120-125	160	135
ta_4	50-55	65-70	100	80-90
ta_5	50-55	60	80	70
LR	0.48-0.59	0.67-0.81	0.59	0.54-0.59
BV	2.58-2.77	2.68-2.72	2.37	2.52-2.62
SV	3.33-4.63	2.60-2.93	3.27	3.11-3.41
BR	2.2-2.5	2.3-2.6	2.4	2.0-2.2
tisp	24-29	27	28	24-31
P2				
fe	330-370	370-390	545	525-570
ti	330-380	400-450	580	545-595
ta_1	140-170	170-210	265	260-265
ta_2	90-95	100-130	170	150-170
ta_3	75-90	80-105	135	115-125
ta_4	50-65	60	90	70-80
ta_5	50-55	55	95	60-70
LR	0.40-0.45	0.42-0.46	0.46	0.44-0.47
BV	2.93-3.25	3.06-3.23	2.84	3.25-3.31
SV	4.17-5.91	4.05-4.60	4.25	4.13-4.52
BR	2.3-3.3	1.68-1.77	2.6	2.4-2.6
tisp1	17-21	20	28	32-35
tisp2	11-17	13-14	-	
P3				
fe	330-340	410-420	550	535-580
ti	390	450-460	605	585-640
ta_1	160-180	250-300	300	295-320
ta_2	120-140	140-145	200	195-215
ta_3	105-110	105-120	180	150-175
ta_4	55-60	55-60	105	80-90
ta_5	50-55	55-65	85	70-85
LR	0.40-0.46	0.56-0.65	0.50	0.50-0.53
BV	2.53-2.71	2.92-3.28	2.55	2.61-2.83
SV	4.06-5.11	2.93-3.44	3.85	3.6-3.8
BR	2.6-3.4	1.6-1.8	2.4	2.6
tisp1	28-30	32-37	42	35-38
tisp2	16-19	20-22	12	10
cb	9-10	10-12	12	12-14
cbl	9-12	19-32	15	24

Table 3. *Pirara* and *Eukiefferiella*, adult females.

	P. matakiri (n=2)	*P.australiensis* (n=3-4)	*E.brundini* (n=4-5)	*E.insolida* (n=2)
B.l.	1.3-1.5	2.2-2.6	1.8-2.0	1.8-2.0
W.l.	1.4-1.5	1.5-1.6	1.7-2.1	1.1-1.36
fl1	60	35-54	50-68	38-52
fl2	50-58	28-35	38-40	24-35
fl3	53-55	31-36	38-45	28-38
fl4	45-55	30-35	40-48	31-46
fl5	80-88	73-91	90-110	70
A.R.	0.4	0.5-0.6	0.5-0.6	0.4-0.5
temp	5-6	4-5	4-5	1-2
Palp 2	53	28-35	40-53	28-35
Palp 3	55-58	49-66	60-70	45-60
Palp 4	73-78	66-98	80-100	59-74
palp 5	85-98	94-126	149-145	105-108
aps	1	2-3	1	0-1
dc	12-14	8-11	8-10	5-9
pa	3	3	3	3
scts	12	5-9	8-10	4-6
VR	1.24-1.29	1.30-1.36	1.36-1.40	1.37-1.44
B	1	1	1	0-1
R	8-9	9-12	9-11	3-9
R_1	9-11	7-14	7-11	2-3
R_{4+5}	19-21	8-13	6-7	4-5
sq	6	8-11	8-9	4-5
Cext	145-163	-	75-113	70-80
gc.s	7-8	8-10	10-11	7-11
Cerc	80-88	?	77-94	76-85
Sc.l	72-80	72-80	60-74	40-52
Sc.w	49-50	42-45	47-54	30-35

Table 4. *Eukiefferiella*, legs.

	brundini(m) (n=3-5)	*brundini*(f) (n=4-5)	*insolida*(m) (n=2)	*insolida*(f) (n=2)
P1				
fe	450-480	450-490	355-580	340-400
ti	460-500	480-530	400-575	340-470
ta_1	350-410	300-390	260-445	240-280
ta_2	220-270	259-260	180-290	160-195
ta_3	150-190	150-180	140-215	115-135
ta_4	100-110	90-100	90-115	60-90
ta_5	80-90	70-80	65	55-70
LR	0.70-0.84	0.62-0.78	0.65-0.76	0.60-0.71
BV	2.08-2.67	2.12-2.37	2.15-2.33	2.34-2.36
SV	2.32-2.77	2.51-3.10	2.60-2.93	2.80-3.04
BR	1.70-1.90	2.10-2.50	na	na
tisp	17-21	25-40	27-45	17-20
P2				
fe	400-450	450-510	375-500	310-400
ti	420-470	460-570	355-460	320-460
ta_1	200	190-230	170-265	150-205
ta_2	140-180	140-190	90-140	90-105
ta_3	90-100	90-110	70-115	60-90
ta_4	60-70	60-70	45-60	35-45
ta_5	60-70	70	55-70	45-60
LR	0.44-0.48	0.35-0.49	0.45-0.57	0.44-0.47
BV	2.69-3.11	2.69-3.11	3.13-3.48	3.38-3.53
SV	4.25-4.60	4.26-5.25	3.60-4.30	4.17-4.22
BR	1.90-3.40	1.53-2.00	na	na
tisp1	37-44	18-26	15-17	10
tisp2	16-21	12-15	9-10	14-15
P3				
fe	450-530	520-600	380-550	355-490
ti	530-590	560-630	410-605	375-550
ta_1	290-340	280-350	205-365	195-275
ta_2	190-220	170-220	135-185	105-175
ta_3	100-150	110-190	90-170	90-140
ta_4	70-100	70-80	55-80	35-60
ta_5	80-90	70	55-80	45-70
LR	0.54-0.59	0.48-0.56	0.50-0.59	0.50-0.52
BV	2.46-3.09	2.71-3.28	2.93-3.02	2.90-3.35
SV	3.06-3.69	3.51-4.04	3.25-3.87	3.72-3.77
BR	2.10-3.00	1.57-2.90	na	na
tisp1	42-52	42-52	29-45	28-42
tisp2	18-25	18-25	13-14	10-21
cb	13	11-13	13-14	9-11
cbl	20-33	22-42	31-35	22-35

Table 5. *Pirara* and *Eukiefferiella*, pupae.

	P.matakiri (n=4)	*P.australiensis* (n=5)	*E.brundini* (n=6)	*E.insolida* (n=10)
B.l.	1.7-2.0	2.0-2.4	2.4-2.8	2.0-2.7
fs	2-5	3-5	-	-
maps 1	30-32	35-60	60-85	130-190
maps 2	48-50	20-25	90-110	65-110
pcs1	29-31	60-70	90-120	120-190
pcs2	34-40	55-60	60-115	95-122
pcs3	31-38	25-35	105-120	30-55
th.l	-	-	250-280	180-230
th.w	-	-	25-40	30-40
dc_1	30-32	10-155	45-65	40-65
dc_2	24-33	10	35-55	20-30
dc_3	20-22	10-20	30-50	20-30
dc_4	42-59	40-55	30-55	35-60
dc_{1-2}	11	5-15	10-25	5-10
dc_{2-3}	6-10	10-20	75-85	125-190
dc_{3-4}	10-13	20-35	10-20	5-15
Anl	100-110	110-130	150-170	145-170
ms1	15-20	10-15	100-130	155-175
ms_2	25-30	10-15	120-135	145-165
ms_3	-	10-15	110-130	135-155

Table 6. *Pirara* and *Eukiefferiella*, larvae.

	P.matakiri (n=9)	*P.australiensis* (n=1-3)	*E.brundini* (n=6-7)	*E.insolida* (n=12)
B.l.	2.5-3.0	3.3	3.0-3.2	2.3-3.4
hcl	0.23-0.26	0.28-0.31	310-350	275-330
Ant 1	19-23	14-15	39-46	40-45
Ant 2	9-11	7-8	11-13	12-14
Ant 3	2-4	4	2-3	3-4
Ant 4	2-4	3	2-3	5-6
Ant 5	5-7	3	n/a	n/a
A.R.	0.8-1.1	0.8-0.9	2.1-2.5	1.8-2.1
RO	9-11	5-6	10-12	-
Pmd	35-45	48-50	?	?
Md	62-68	75-87	72-87	70-76
M.w.	45-57	50-63	10-13	-
M.m.w.	19-20	12-14	50-63	45-51
Pc.h	11	10-11	15-21	16-19
Pc.s	68-79	48-87	215-310	290-365
A.t.l	46-52	?	75-100	50-90
A.t.w	17-24	?	35-50	25-40
pp.l	90-140	315	225-310	175-295
pp.w	18-26	110	110-130	60-120

A new genus and new species of Neotropical Orthocladiinae (Diptera, Chironomidae)

Sebastião José de Oliveira, Maria Conceição Messias and Andréa Ribeiro dos Santos

Coleção Entomológica, Departamento de Entomologia, Instituto Oswaldo Cruz, Av. Brasil, 4365, Manguinhos CEP 21045-900, Rio de Janeiro, RJ, Brazil

Introduction

Amongst a collection made in Curitiba, Paraná State, Brazil for the Oswaldo Cruz Institute Collection were 3 unusual male Orthocladiinae. These midges possess a unique antepronotum and are described here as a new genus.

In this paper the terminology follows Sæther (1980).

Corytibacladius Oliveira, Messias and Santos, gen. n.

Type species: *Corytibacladius gercinoi* sp. n. by present designation.

Generic diagnosis

The unique antepronotal structure and some characteristics of the hypopygium diagnose *Corytibacladius* gen. n. and are sufficient to distinguish it from all other Orthocladiinae genera.

Etymology: From Curitiba city and Greek *klados* (branch) with - *cladius* a common suffix of many Orthocladiinae genera.

Generic description

Eyes naked, without trace of pubescence between ommatidia. Antennae with 13 flagellomeres, without sensilla chaetica, plumes long, fine, silky; antennal groove reaching flagellomere 4. Temporals not visible. Palpi 4-segmented, ultimate segment about 2x as long as third.

Antepronotum well developed, median lobes enlarged, strongly projecting over head. Some small dorsocentrals and acrostichals, 4 very small scutellars.

Wing membrane very translucent, void of setae, without microtrichia. Anal lobe absent. Costa extended; R_{4+5} ends distad to M_{1+2}; M_{1+2} vestigial; vannal fold ends basad to FCu; An ends distad to FCu; Cu1 straight ending distad of wing border. Brachiolum, R and R1 without setae. Squama without setae.

Pulvilli present. Pseudospurs absent on all legs. Comb and tibial spur normal. Setae of tergites scattered, without any obvious pattern. Anal point short, semitriangular, void of lateral setae, with macrotrichia. Phallapodeme well developed. Virga apparently absent. Gonocoxite without distinct inferior volsella. Gonostylus with well developed subapical megaseta, row of about 10 very short spurs and apical seta.

Species description

Corytibacladius gercinoi Oliveira, Messias and Santos, sp. n.

Figs. 1a-f; 2a-c.

Etymology: Named in honour of Dr. Gercino F. de Carvalho, Sanilar Ltd, Curitiba, for his assistance in collecting this material.

Diagnostic characters: cf diagnosis of the genus.

Male imago (n=2, except when otherwise stated)

Total length 1.36 - 1.48, 1.42 mm. Wings length 0.96 mm (1). Total length/wing length 1.53 (1).

Antennae, pedicellum light brown, flagellomeres yellowish. Thorax brown. Legs yellowish. Abdomen pale brown.

Head: Antenna (Fig. 1f) last flagellomere length 0.08-0.09, 0.08 mm. AR = 0.21 Cibarial pump, tentorium and clypeus as in Fig. 1d. Palp (Fig. 1e) length 0.25 mm. (1)

Thorax: (Fig. 1a-c) Antepronotum with 2-3 setae on median lobe.

Wing: (Fig. 2a).

Legs: Yellowish. Comb and tibial spurs normal. Pseudospurs absent in all legs.

Lengths (in µm) and proportions of legs (n= 1)

	Fe	Ti	Ta1	Ta2	Ta3	Ta4	Ta5	LR
P1	384	480	252	155	107	61	55	0.52
P2	372	373	176	84	63	36	39	0.47
P3	314	427	220	111	103	39	33	0.51

Hypopygium (Fig. 2b-c).

Female, pupa and *larva* unknown.

Type locality: **Brazil**, Paraná State, Curitiba

Material examined

Holotype, male, Brazil, Paraná State, Curitiba, Parque João Paulo II, 620 m a.s.l., 4.ii.1988, caught at the ground with an entomological net, (S. J. de Oliveira and M. C. Messias), Inst. Oswaldo Cruz Coll.

Paratypes: 2 males, same data as holotype.

Acknowledgments

We are grateful to Dr. Gercino F. de Carvalho for his assistance in collecting this material. The study was finacially supported by the Conselho Nacional de Desenvolvimento Científico e Tecnológico (CNPq).

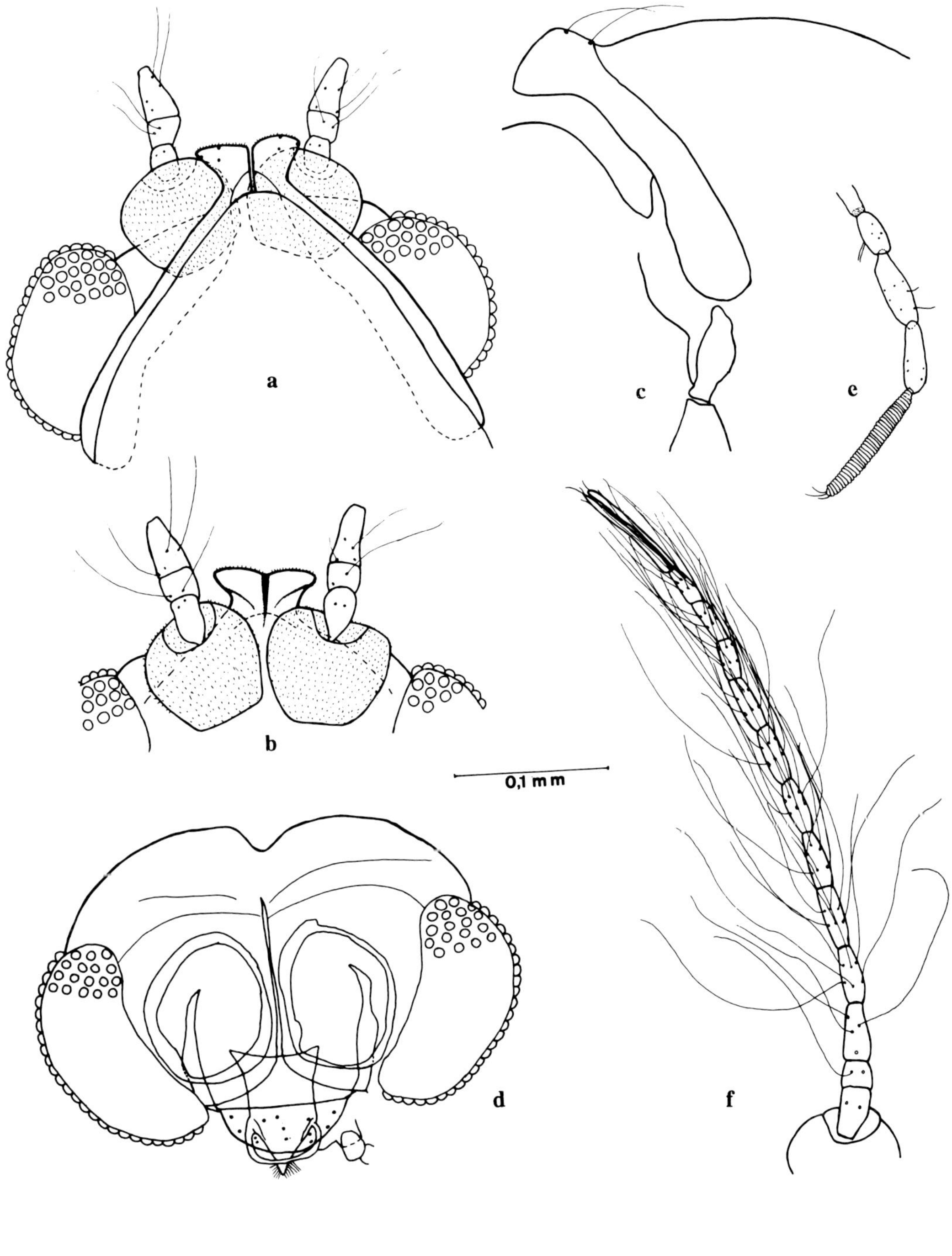

Figure 1. *Corytibacladius gercinoi* gen. n., sp. n.: (a) antepronotum, dorsal view (b) same, ventral view, (c) same, lateral view; (d) head: tentorium and cibarial pump, (e) palp, (f) antenna.

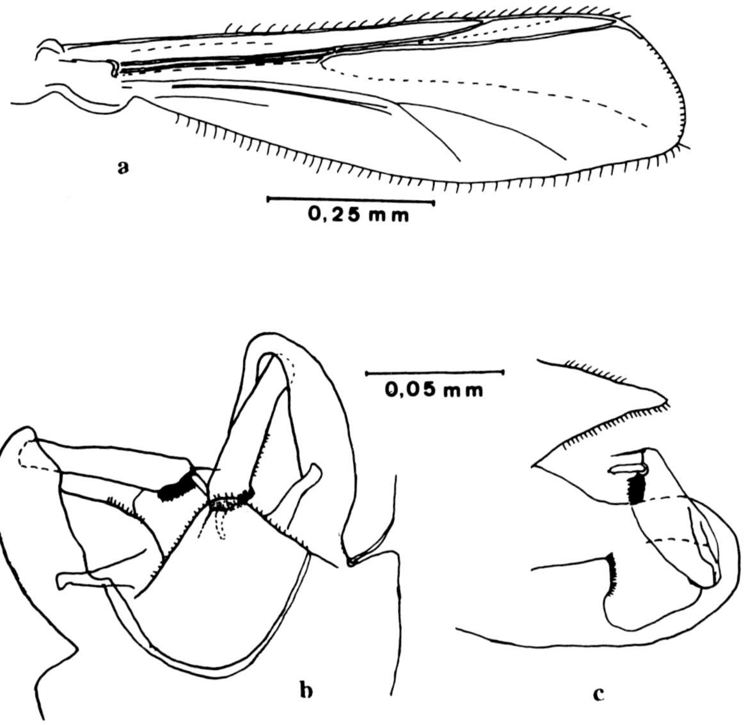

Figure 2. *Corytibacladius gercinoi* gen. n., sp. n.: (a) wing, (b) hypopygium, dorsal view, (c) same, lateral view.

Contribution to the knowledge of the females of *Demeijerea rufipes* (Linnaeus) (Diptera; Nematocera, Chironomidae)

Ruth Contreras-Lichtenberg

Naturhistorisches Museum Wien,
2. Zoologische Abteilung, Burgring 7, P.O. Box 417,
A-1014 Wien, Austria

Introduction

There are very few records from Austria of *Demeijerea rufipes* (Linnaeus), a species with larvae that mine in freshwater sponges. In the collection of the Natural History Museum of Vienna there exists one female from Austria (Freistadt), of *Glyptotendipes glaucus* (Meigen) - misidentified by Goetghebuer as *Demeijerea rufipes*. In 1989 and 1993 some males and females were collected at a fishpond about 27km south of Vienna.

Sublette and Sublette (1965) placed *Demeijerea* Kruseman in *Endochironomus* Kieffer. Rodova (1969) was the first to describe the female of *Demeijerea rufipes*. Sæther (1977b) dealt with the morphology of the female of the nearctic species *Demeijerea brachialis* (Coquillett). Following Lenz (1954-1962) and Hamilton *et al.* (1969), Sæther treated *Demeijerea* as a subgenus of *Glyptotendipes* Kieffer.

The morphology of larvae, pupae and male adults differs clearly from that of *Endochironomus* and *Glyptotendipes* (Pinder and Reiss, 1983, 1986; Cranston *et al.*, 1989; Langton, 1984, 1991).

The purpose of this paper is to redescribe the female, based mainly on characters that allow separation of *Demeijerea* Kruseman from *Glyptotendipes* Kieffer.

Material

Demeijerea rufipes: 1 female: Kiel, labelled as "D. rufipes FB." ex Coll. Wiedemann; 1 female: labelled: Holland v.d.W.; rufipes det. Schiner; Schiner 1869; 2 females from Berlin (1 female labelled: Berlin-Grunewald 6-98 Oldenberg); 1 female: Schönau, NÖ, 12.ix.1989 (Contreras-Lichtenberg).

Glyptotendipes barbipes (Staeger):

1 female (reared), England, Romney March, "New Sewer" 8.v.82 (Langton);

Glyptotendipes glaucus (Meigen):

2 females Lab. stock, 20.i.1987 (Michailova); 2 females (reared), England, GWS-Fishpond, 10.v.1976 (Langton).

Glyptotendipes gripekoveni (Kieffer):

1 female (reared), England, Wicken Main Pond, 1.vii.1977 (Langton); 1 female (reared), Wicken Main Pond, 30.vii.1977 (Langton); 2 females, Austria: Schönau, 26.vii.1993 (Contreras-Lichtenberg).

Glyptotendipes paripes (Edwards):

1 female (reared), Scotland, Loch of Stenen, outlet, 23.viii.1982 (Langton).

Glyptotendipes salinus Michailova:

3 females, Lab. stock, 1.iii.1988 (Michailova)

Description of female *Demeijerea rufipes*

(n=4 unless stated otherwise):

Length: 7 mm (n=5).

Head: dark brown, frons with fine hair; pedicel dark brown, remaining antennomeres yellow; antenna with 7 segments; length of pedicel and antennomeres (in µm): 89 (80-98); 133; 102 (98-111); 122 (116-125); 129 (n=3); 271 (n=1) AR: 0.44 (n=1)

Frontal tubercles present in male but reduced to small tuberosities in female.

Temporal setae: 19 (16-22); clypeus with 44 (41-49) very strong setae; clypeus with 5 palpomeres; length of palpomeres in micrometers: 66; 78 (66-89); 212 (200-231); 203 (178-245); 336 (267-378);

Thorax: pronotum well developed, tapering dorsally, with fine hair, widely separated at V-shaped notch, mesonotum, metanotum, mesosternum and scutellum brown, mesonotal vittae nearly black; humeral pit clear and with granulate structure;

Chaetotaxy: 30 (21-40) acrostichals, reaching far beyond the middle of the mesonotum; 21 (16-30) uni-biserial dorsocentrals; humerals absent; 8 (6-10) pre-alars, 1 postalar seta; 48 scutellar setae (n=1);

Legs: fore tarsus without beard, pulvilli large, lobe-shaped; mid and hind tibiae with two combs with spurs; femora and tibiae of middle and hind leg very broad; leg colour: femur and tibia orange; forelegs with proximal part of Ta_1 yellow, distal half of Ta_1 and Ta_3 to Ta_5 dark brown; mid- and hindlegs with Ta_1, distal part of Ta_2 and Ta_3 to Ta_5 dark brown.

Wing (Fig. 1a, b): brown, with a dark band reaching from costa to Cu; from Ar to end of An yellow; Ar distally pointed (Fig. 1b); squama fringed; length: (n=3) 4.3mm (4.3-5.0), width: 1.5mm (1.4-1.7), Ar-rm: 1.9mm (1.8-2.0), Ar-FCu: (n=3) 2.0 (1.9-2.1).

Abdomen: colour dark brown, segment 1 yellow, hind margin of segment 3 and 4 yellow.

Lengths (in micrometers) and ratios of legs:

	Fe	Ti	Ta_1	Ta_2
PI	1579-1668, 1615	1612-1779, 1705	2090-2224, 2184	867-889, 870
PII	1857-2112, 1946	1723-2023, 1828	856-945, 900	500-556, 522
PIII	2068-2335, 2173	2010-2279, 2157	1112-1223, 1172	667-778, 697

	Ta_3	Ta_4	Ta_5	LR
PI	800-845, 826	611-656, 626	300-311, 307	1.22-1.3, 1.26
PII	411-444, 436	266-278, 276	222-244, 227	0.49-0.86, 0.54
PIII	533-611, 560	300-322, 309	244-266, 251	0.5-0.6, 0.54

Figure 1. *Demeijerea rufipes* (L.): (a), wing, (b), wing base, (c)-(e) genital segments: (c), ventral view, (d) lateral view, (e) Gonapophysis VIII.
Scale bar length: (a) 1mm, (b)-(e) 100μm.

Genital segments (Fig. 1c-e) (terminology and abbreviations follow Sæther 1977b, 1980): TIX big, somewhat overlapping the ventral side, distally rounded, strongly sclerotized, with fine hair and strong yellow setae dispersed all over TIX; segment X with 6-8 setae; postgenital plate triangular, short, caudally pointed and bent dorsally, strongly sclerotized, covered densely with microtrichia (length/width ratio of postgenital plate: 0.33-0.56); Gc IX characteristically ovoid in shape, not separated from TIX, with 6-8 setae; Gp VIII with VlL of the same size as DmL; DmL big, medially curved, with fine shagreen, ApL strongly sclerotized, with microtrichia from the base to end; Csa black broadly curved, medially divided into two short sclerotized ends; GCa caudally rounded, caudally and medially sclerotized, medial ends straight, converging; two short and straight SDu ending separately; two ovoid SCa.

Diagnostic characters for the separation of Demeijerea and Glyptotendipes

Demeijerea rufipes (L.):

Head: sexual dimorphism in the frontal tubercles (females only with very small tuberosities;

Thorax: pronotum big, tapering dorsally;

Chaetotaxy: humerals missing, dorsoventrals uni- or biserial, prealars uniserial;

Wing: Ar distally pointed;

Genital morphology: TIX big, somewhat overlapping the ventral side, distally rounded, strongly sclerotized; postgenital plate triangular, strongly sclerotized; Gc IX ovoid, not separated from TIX; ApL distinct, strongly sclerotized; medial ends of Gca straight, converging; SDu short, straight.

Glyptotendipes

Head: generally with small frontal tubercles;

Thorax: Pronotum small, dorsally rounded;

Chaetotaxy: humerals present; dorsocentrals always multiserial; Wing: Ar distally never pointed; TIX generally small, distally straight; postgenital plate triangular, not as sclerotized as in *Demeijerea*, generally long, distally rounded; Gc IX triangular, separated from TIX; only proximal end of ApL sclerotized; medial ends of Gca parallel; SDu short, curved.

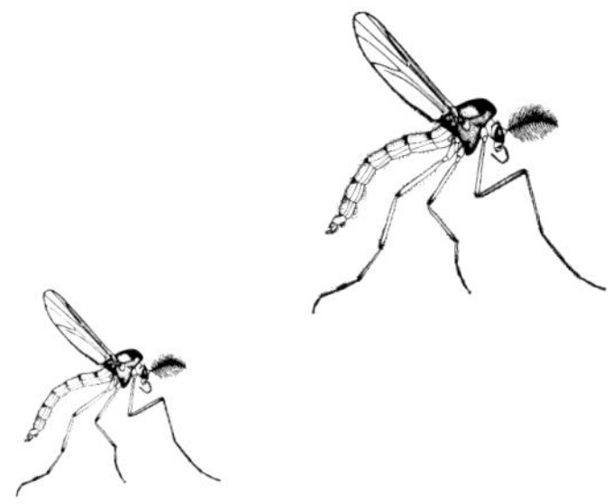

Conchapelopia insolens, n.sp., a new species of Tanypodinae (Diptera: Chironomidae) from Sulawesi

D.A. Murray

Department of Zoology, University College Dublin, Belfield, Dublin 4, Ireland.

Introduction

Ashe (1990) reported on the diversity of chironomids obtained on the 1985 Royal Entomological Society "*Project Wallace*" expedition to Sulawesi, Indonesia where he undertook an extensive chironomid sampling and larval rearing regime from habitats in the vicinity of the River Toraut, Dumoga-Bone National Park, Sulawesi. Part of this reared material included all life-history stages of a new species of Tanypodinae belonging to the *Thienemannimyia* series (Tribe Pentaneurini) collected in kick samples. Dr. Ashe has made this material available for study and the present paper gives descriptions of this new species, herein described as *Conchapelopia insolens* n.sp. The terminology used follows Sæther (1980).

CONCHAPELOPIA INSOLENS N. SP

Material examined

Holotype: adult male, associated pupal and larval exuviae, slide mounted in Euparal, **Indonesia,** Dumoga-Bone National Park, Utura, River Toraut, 0° 33'N, 123° 54'E., 8.x.1985 (P. Ashe), in the collection of the Natural History Museum, Dublin. NMI 70:1993.

Paratype: Adult male with associated pupal and larval exuviae slide mounted in Euparal, same locality data as for holotype, in Museum Zoologicum Bogoriense, Indonesia. Additional paratype material will be deposited in the Australian National Insect Collection, CSIRO, Canberra, Australia and Zoologische Staatssammlung, Munich, Germany.

Etymology: from the Latin meaning "strange" or "unusual", referring to the combination of characters hitherto not observed in *Conchapelopia.*

Description

Adult male (Fig. 1)

Total length 3.5mm, wing length 2.5mm.

Colour. Head pale; antennal pedicel dark brown, plume on middle of penultimate flagellomere distinctly brownish; clypeus and palps pale. Ground colour of thorax pale yellow. Light brown pigmentation lateral of anterior ends of both median vittae. Median vittae pale, lateral vittae with brown seam, region between lateral vittae brown, this area extending anteriorly to surround scutal tubercle and posteriorly to pre-scutellar region between lateral vittae. Basal $^{2}/3$ of pre-episternum, most of median anepisternum, epimeron II and postnotum light brown. Legs pale with narrow brown ring at apices of femora and bases of tibiae. Wings with two transverse bands, one before crossveins RM and MCu and broader band between ends of R_1 and R_{4+5}. Abdomen yellowish with brown pigmentation; tergite I all pale; tergite 2 brownish on apical $^{4}/5$, with anterior median pale mark; tergite 3 brownish in anterior $^{1}/3$ and with median brown "strap" extending to beginning of distal $^{1}/3$; tergite 4 mostly brownish, paler anteriorly and posteriorly and with pale median strap extending to end on anterior $^{1}/3$; tergite 5 with narrow basal band occupying $^{1}/4$ tergite length and separated in middle; tergites 6 and 7 uniformly pale brown; tergite 8 mostly pale with transverse median brown band; gonocoxite and gonostylus pale.

Head. A.R. about 1.7, terminal flagellomere 3x as long as broad, penultimate flagellomere 6.5x terminal flagellomere. Eyes with apically expanded dorsomedial extension, with 3-4 ommatidia at minimum width. Temporal setae 13, comprised of 4 postorbitals, 6 outer verticals and 3 inner verticals; post orbitals and outer verticals forming continuous row. Clypeal setae 32. Pedicel with 2 ventral and 1 lateral setae.

Thorax. Scutal tubercle distinct and medial scar present. Antepronotal setae 7. Acrostichals 22, biserial between vittae, diverging in front of scutal tubercle and ending in prescutellar field. Dorsocentrals 18-20, biserial between median and lateral vittae, irregular in the prescutellar field. Prescutellars 2, supraalars 2, prealars 13-14, humerals 4, scutellars 25. Preepisternum with 4 setae.

Wing. Costa slightly produced beyond R_{4+5}, ends before M_{1+2}. R_{2+3} well developed, R_3 joins the costa midway between R_1 and R_{4+5}. MCu almost at FCu. Membrane with setae slightly more dense in mid region of cells r_{4+5}, m_{1+2}, m_{3+4} and at apex of cell an. Anal lobe rounded. Squamal setae 28-30.

Legs. Tibial spurs with main tooth and 7-8 side teeth. Tarsomere 3 of mid legs with tarsal brush of 9-11 setae. Tibia of P_3 with comb of 7 setae of which middle seta is shortest with remaining setae becoming progressively longer towards outsides of comb. Claws slender, slightly curved, distally pointed and devoid of basal spines. Pulvilli absent. Measurements and ratios as in Table 1.

Hypopygium. Tergite IX with an irregular concave row of 8-10 setae, arranged as two median and 3-4 lateral setae on each side. Anal point broad and conical. Gonocoxite 1.5x as long as broad, shaped like a mussel-shell in the basal $^{2}/3$ and somewhat flattened dorsally. Dorso median surface with 4 large, erect, pre-apical setae. Median volsella well developed, reaching $^{2}/3$ gonocoxite length, with lateral row of 20 filaments and baso-lateral setose appendage. Gonostyle large, broad basally and abruptly bent through 90° at middle, narrow and tapering in apical $^{1}/3$, terminal spur long. Phallapodeme reaching under anal point and terminally expanded.

Pupa (Fig. 2)

Length 5.0-5.5mm, pale in colour.

Cephalothorax. Thoracic horn flattened, expanded apically and about 2.5 x as long as maximum preapical width, external membrane with basally interconnected spines.

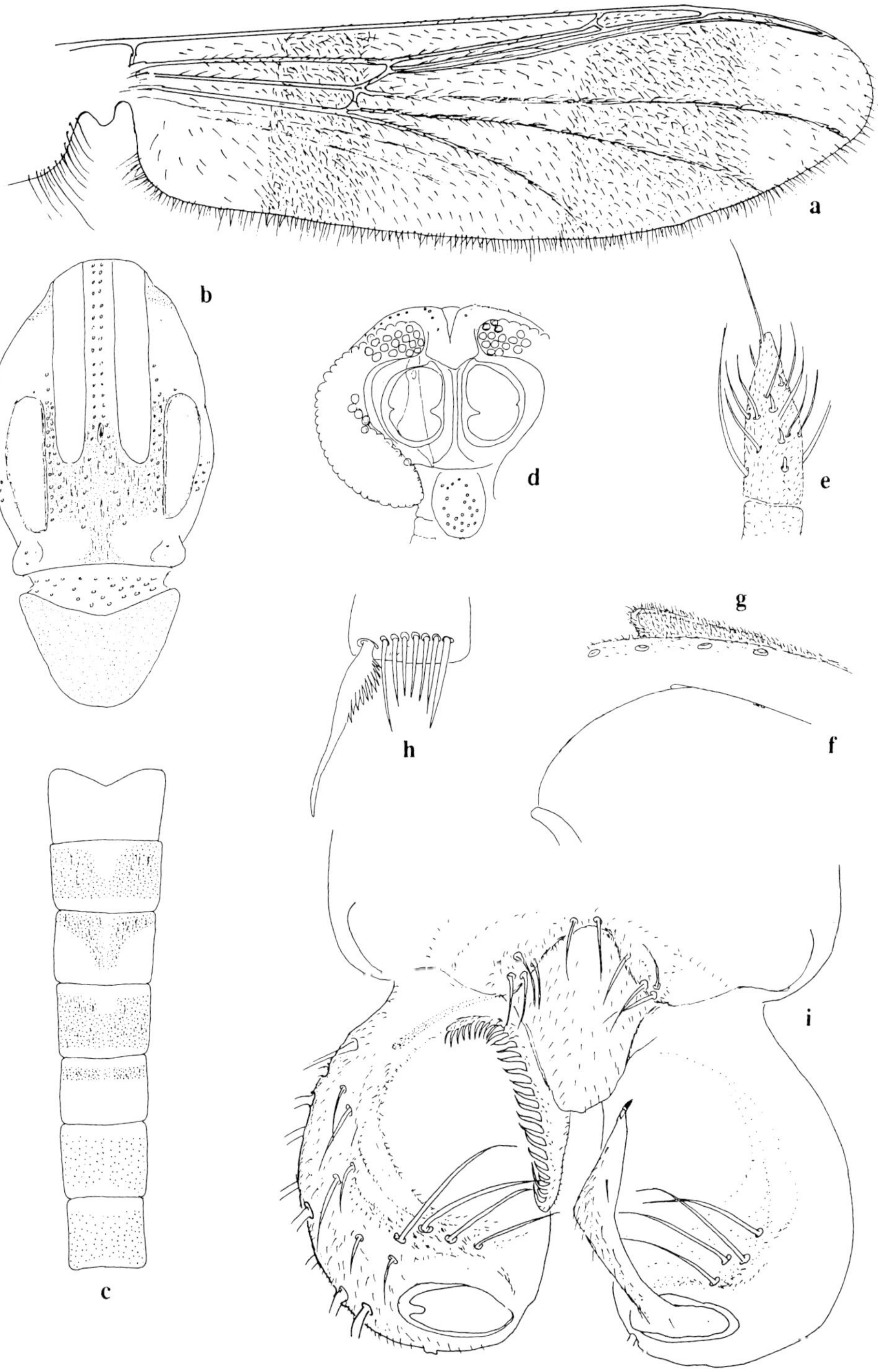

Figure 1. *Conchapelopia insolens* n.sp. Adult male. (a) wing, (b) thorax, dorsal, c) abdomen, (d) head, (e), apex of antenna, (f), outline of thorax, lateral, (g) scutal tubercle, (h) comb on tibia of hind leg (note - only one spur shown), (i) hypopygium.

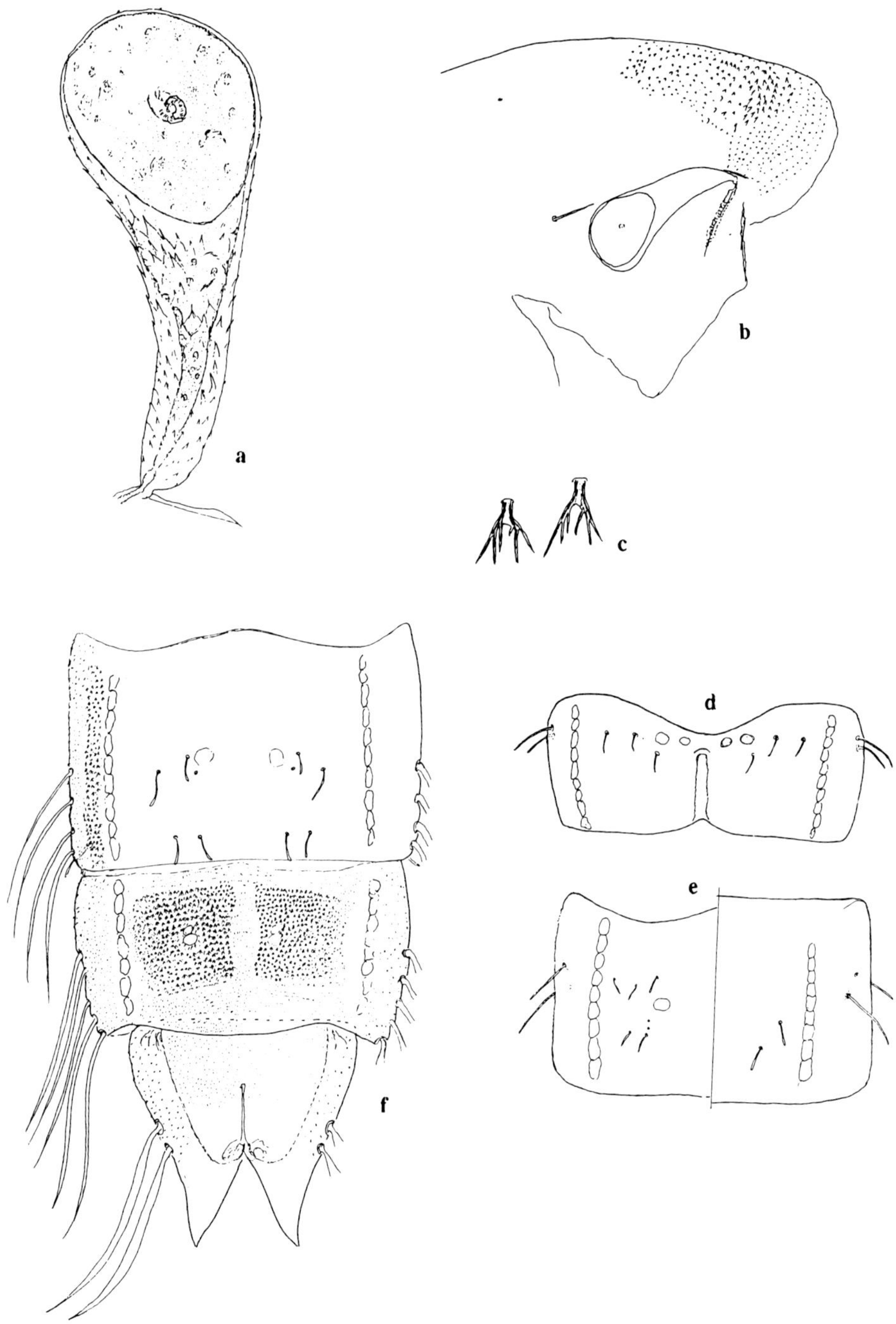

Figure 2. *Conchapelopia insolens* n.sp. Pupa. (a) thoracic horn, (b) thorax lateral, (c) multibranched shagreen, (d) tergite I, (e) segment IV, left dorsal, right, ventral, (f) segments VII-VIII and anal lobe.

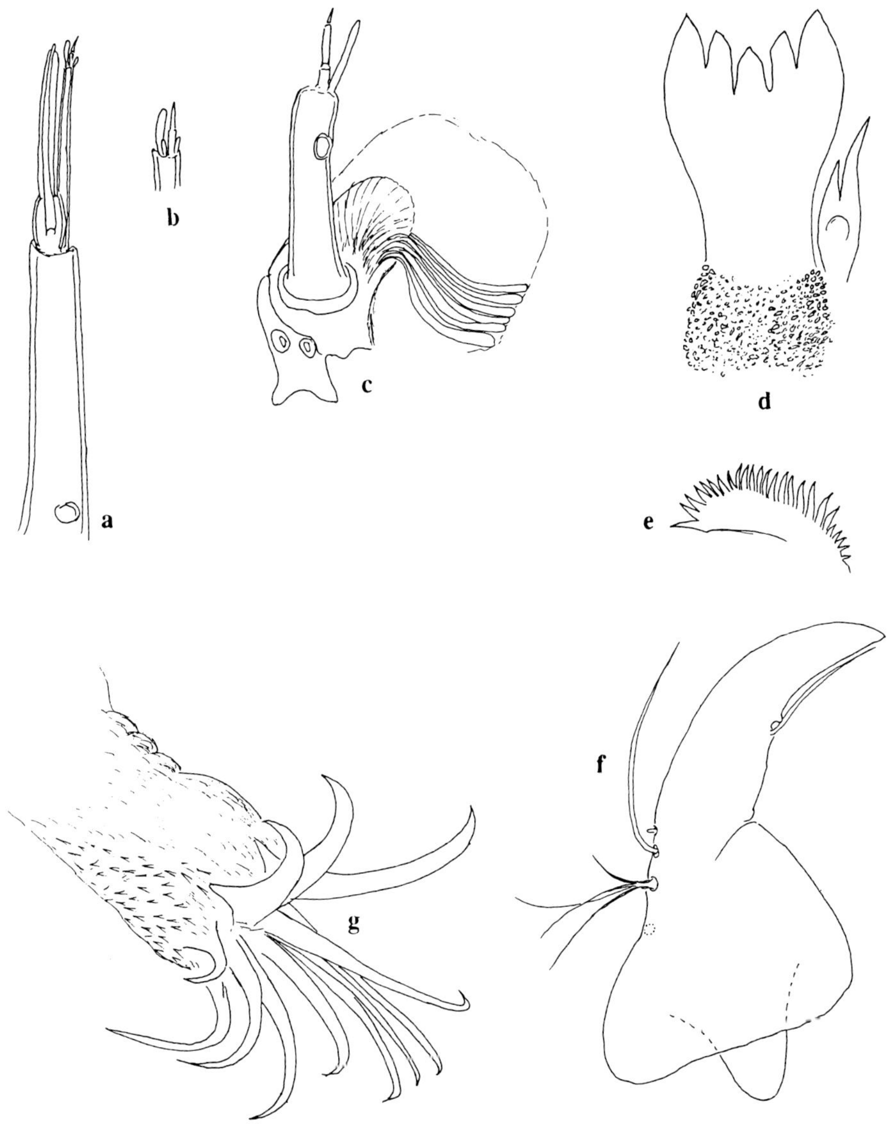

Figure 3. *Conchapelopia insolens* n.sp. Larva. (a) antenna, apex of segment 1, (b) detail apex of segment 2, (c) maxillary palp and b seta, (d) Ligula and paraligula, (e) pecten hypopharyngis, (f) mandible, (g) posterior parapod.

Table 1 . Leg measurements (μm) and ratios for *Conchapelopia insolens* n.sp.

	Fe	Ti	Ta_1	Ta_2	Ta_3	Ta_4	Ta_5	LR	BR
P1	806	1103	930	459	310	223	99	0.84	3.51
P2	868	979	595	273	198	164	87	0.60	3.69
P3	781	1179	917	508	372	248	99	0.77	4.64

Horn sac expanded apically and terminating in short neck connected to aeropyle. Plastron plate absent, corona present, about 0.48x horn length. Small elongate basal lobe present, about 0.25x horn length. Thoracic comb weak. Surface of thorax with apically pointed teeth dorsal to horn, tuberculate anteriorly, otherwise sparsely granulate. Scutal tubercle and postnotal tubercle absent. Median antepronotal seta 1, precorneal seta 1. Dorsocentral setae 1 and 2 present, simple; supraalar seta lamelliform and distally round.

Abdomen. Tergite I with scar, lateral muscle marks obvious on tergites I-VII. Abdominal shagreen predominantly of multibranched spinules with small spines anteriorly on tergites and intersegmental membranes. Shagreen less extensive on tergite VIII where the multibranched spinules are restricted to two median roughly elongate rectangular fields, remainder of tergite with simple spine shagreen only. Tergite IX with sparse covering of spine shagreen only. Multibranched spinules less extensive on sternites I-VIII. 2 L setae on segments I-VI, 4 LS and 1 L seta on segment VII, 5 LS on segment VIII. LS_1 on segment VII at 0.6x segment length. LS_1 on seg. VIII at 0.5x segment length. D and V setae filamentous, apically round. 3 D-setae on segment I, 5 on segs. II-VI, 4 on seg VII. Anal lobe about 2x as long as broad, dorsal surface with simple spine shagreen, inner and outer borders with spinules, inner border convex, outer border somewhat concave posteriorly. Anal macrosetae at 0.5x segment length. Gonopodial sheath ending slightly beyond distal anal macroseta, somewhat compressed apically in cast exuviae.

Larva (Fig. 3)

Medium sized, up to 8.0 mm long.

Colour. Head pale with distinct dark posterior margin. Body pale.

Head. Capsule longish-oval, cephalic index 0.6.

Antenna. Almost 0.5x as long as head and nearly 2x as long as mandible. Antennal ratio 4.0. Basal segment 10 as long as basal width, with ring-organ at base of apical 1/3. Distal margin of basal segment straight, not produced into a point on outer side. Segment 2 approximately 10 as long as basally wide. Segment 3 approximately 4/5 as long as segment 4. Style slightly broadened apically and extending to apical 1/3 of segment 4. Lauterborn organs minute. Blade slightly shorter than flagellum, extending beyond middle of segment 3, 5.6x as long as height of basal ring; basal ring 2x as high as wide. Accessory blade longer than blade, reaches middle of segment 4.

Mandible. Uniformly curved, distal 1/2 strongly narrowed and dark; apical tooth 2x as long as basal width, distal 1/3 dark brown, apex black; basal tooth inconspicuous, apically directed and not protruding beyond inner margin of mandible, with long seta subdentalis arising from inner side; minute apically round accessory tooth present on ventral inner margin of apical tooth immediately distal to basal tooth. Ventrolateral setae arising close together on basal 1/2 of outer margin, seta 1 is a peg sensillum, seta 2 is simple and reaches 2/3 length of apical tooth, seta 3 is divided in 4 branches, distance between seta 2 and 3 twice that between seta 1 and 2; sensillum minusculum displaced ventrally, proximal to seta 3.

Maxilla and mentum. Basal segment of maxillary palp about 2.7x as long as wide, with ring organ at base of apical 1/3; b seta 3 segmented, middle segment 2 as long as segments 1 and 3. Dorsomentum without teeth; sclerites on each side of base of M appendage ending in two, superimposed and apically round, points; M appendage triangular. Pseudoradula gradually tapering to 1/2 its basal width at the apex, granulation coarser in basal 1/2, consisting of about 8 parallel rows.

Ligula. With 5 teeth, approximately 2 as long as apical width, apical 1/3 brown, noticably broad. Points of inner teeth curved towards outer teeth, points of outer teeth curved slightly inwards. Paraligula bifid, 1/2 as long as ligula, inner tooth 1/3 as long as outer. Pecten hypopharyngis with about 25 teeth, those on outer 1/4 0.5x smaller than remainder.

Abdomen. body without a fringe of swim setae. Anal tubules slender, tapering apically, about 5x as long as basally wide. Procercus about 4x as long as wide, with 7 anal setae. Subbasal seta of posterior parapod simple. Posterior parapods with an area of fine points on anterior margin of apical 1/4; with 12 to 14 large yellow/brown claws, some with some fine points on inner margins, and one smaller pale claw.

Discussion

The suite of character states present in the three life stages of this species clearly indicates that it belongs to the tribe Pentaneurini, and to the *Thienemannimyia* series, currently considered to be composed of the genera, *Arctopelopia*, *Conchapelopia*, *Hayesomyia*, *Helopelopia*,

Table 2. Comparison of some larval, pupal and adult features in C. *insolens* n.sp with genera of the *Thienemannimyia* series. + present, - absent. (Abbreviations: Conc= *Conchapleopia*, Helo=*Helopelopia*, Rheo= *Rheopelopia*, Arct= *Arctopelopia*, Haye= *Hayesomyia*, Mero= *Meropelopia*, Telo= *Telopelopia*, Thie= *Thienemannimyia*)

	n.sp	Conc	Helo	Rheo	Arct	Haye	Mero	Telo	Thie
Larva									
Segments of b seta	3	3	3	3	2	2	2	2	2
No. of pecten teeth	25	20	20	18-20	25	22	22	22	20
Pseudoradula rows	8	8	12	10	10	10	11	coarse	10
AR	4.0	3.3-4.4	4.5	3.5	5	4	5	3.8	6
Margin of head	Dark	Pale	Dark	Dark	Dark	Dark	Dark	Dark	Dark
Adult									
Brush on Tm_3, P_2	+	+	+	+	-	-	-	-	-
Comb on Ti, P_3	+	+	+	+	-	+	-	-	+
Scutal tubercle	+	+	+	-	-	+	-	+	-
T VIII setae	-	-	+	-	-	-	-	-	-
Pupa									
Thoracic comb	+/-	+	-	-	-	-	-	+	+
Scutal tubercle	-	-	-	+	+	+	-	-	+
Postnotal tubercle	-	-	+	-	+	+		-	+
Plastron plate	-	+	+	-	-	-	+	+	+/-?
Aeropyle	+	-	-	-	-	+	-	-	+/-

Meropelopia, *Rheopelopia*, *Telopelopia* and *Thienemannimyia* (Murray and Fittkau, 1989). A comparison of selected relevant characters of these genera with the new species from Sulawesi is given in Table 2. Features of the larva and adult of the new species indicate that it belongs in *Conchapelopia*. A 3-segmented b seta of the larval maxillary palp (the apomorphic alternative to the 2-segmented condition) within the Pentaneurini is, up to now, only known to occur in larvae of *Helopelopia*, *Rheopelopia* and *Conchapelopia*. Other genera in the series have a 2-segmented b-seta (Fittkau and Roback, 1983). All members of the three above-mentioned genera have a brush of setae on tarsomere 3 of the middle leg in the adult (Murray and Fittkau, 1989) in common with the new species, but its volsella more closely resembles that of *Helopelopia* and *Conchapelopia*. The absence of the distinct tuft of setae on tergite VIII clearly distinguishes this new Sulawesi species from members of the genus *Helopelopia* and suggests a closer affinity with *Conchapelopia*. However, the absence of a distinct plastron plate on the pupal respiratory thoracic horn, present in the pupa of all known species of *Conchapelopia* (and *Helopelopia*, *Meropelopia* and *Telopelopia*) is a trait at variance with the current concept of the pupal stage of this genus (Fittkau and Murray, 1986). The thoracic horn of the Sulawesi species resembles that seen in pupae of the genus *Hayesomyia* Murray and Fittkau, but other features of the pupa, such as the lack of both scutal and postnotal tubercles, are not consistent with the concept of *Hayesomyia* and again suggest a closer association with *Conchapelopia*. Such plasticity of generic character expression has also been observed in the Nearctic species *Thienemannimyia barberi* (Coquillett) where the adult clearly belongs to *Thienemannimyia*, whose larva also has the typical 2-segmented b-seta, present in all members of the genus *Thienemannimyia* but whose pupa, atypically, has a plastron plate (Roback and Ferrington, 1983) - a feature not seen in other *Thienemannimyia* species.

The combination of characters in the different life-stages of the new species suggests an affinity closer to *Conchapelopia* than other *Thienemannimyia* series genera and it is thus assigned to this genus as *Conchapelopia insolens*. This assignment may be provisional until such time as a world-wide review of *Conchapelopia* and related genera is undertaken.

Acknowledgement

I am grateful to Dr. Patrick Ashe for providing the material on which this paper is based.

Description of a late-instar larva of *Buchonomyia thienemanni* Fittkau and further data on its ecology with diagnoses for the subfamily Buchonomyiinae and the genus *Buchonomyia* (Diptera: Chironomidae)

Patrick Ashe

Research Associate, Department of Zoology, University College, Belfield, Dublin 4, Ireland

Introduction

The larva of the phylogenetically significant genus *Buchonomyia* has been elusive, despite the search of a very wide range of habitats. An extensive and intensive rearing programme was undertaken in the hope that *Buchonomyia* would be successfully reared. Simultaneously, a mated adult female was captured in order to obtain an egg mass (Ashe and Murray, 1983), from which first-instar larvae were reared and described (Ashe, 1985).

The discovery of the first-instar larvae of *Buchonomyia* (Ashe, 1985), together with discussions on the adult male, female and larval features (Murray and Ashe, 1985), resolved many of the phylogenetic questions posed by Brundin (1966), Brundin and Sæther (1978) and Murray and Ashe (1981a). None-the-less, the habitat, ecology and morphology of subsequent instars remained unknown. The intensive rearing and collecting between 1976 and 1981 in search of later instars was believed initially to have been unsuccessful. However, one particular larva, discovered in a case attached to a rock in the River Flesk, S.W.Ireland on the 19 July, 1978, and slide mounted ventral side upwards, without clearing the head musculature, was tentatively identified as "? *Symbiocladius* sp." It was obvious it could not be a true *Symbiocladius*, but the reduction or absence of several features (i.e. reduced anterior parapods with few claws; reduced posterior parapods; small, triangular shaped head; procercus absent; body white in colour) indicatated a similar parasitic mode of life.

An important feature of this larva was the absence of premandibles. In the original description of the larva of *Phaenocladius rhithrogenae* Zavřel, the type species of *Symbiocladius* Kieffer, Sulc and Zavřel (1924) stated that the premandibles are absent. This error persisted for over 50 years in the taxonomic literature and was only corrected with the redescription and illustration of *Symbiocladius* by Cranston *et al.* (1983). Prior to this correction, Sæther (1977b) had stated that the premandible is occasionally absent in Orthocladiinae; a view that could have been derived only from Zavřel's error. Indeed, there are no described larvae of the monophyletic group Diamesinae + Prodiamesinae + Orthocladiinae + Chironominae that lack a premandible.

In assessing the phylogenetic relationships of *Buchonomyia*, Murray and Ashe's (1981a) placement of the subfamily Buchonomyiinae in the monophyletic group Tanypodinae + Aphroteniinae + Podonominae predicted the absence of the larval premandible. When the revised information on *Symbiocladius* appeared (Wiederholm, 1983), the "? *Symbiocladius* " identification needed review. In a comparison with re-examination of the first-instar material of *Buchonomyia*, striking differences were evident, but examination of head features (the mandible and labral area in particular) demonstrated that the "? *Symbiocladius*" larva was indeed the long-sought late-instar larva of *B. thienemanni*. The first-instar larva was crucial in confirming that the the third-instar described here was indeed the larva of *Buchonomyia*. The marked differences between the first and subsequent instars of *Buchonomyia* support the idea that larvula (Dorier, 1933) may be a more appropriate term for first-instar larvae of Chironomidae and other Diptera. A knowledge of the morphology of the first-instar larvae for all chironomid subfamilies and other nematocerous Diptera would be of immense value in resolving many phylogenetic questions, because the first-instar larva appears to be more conservative in retaining features which are modified in later instars.

Materials and Methods

The single third-instar larva described here was collected on the 19 July, 1978 in the River Flesk, S. W. Ireland. Originally the head of the larva was uncleared and mounted ventral side uppermost under a single cover slip. In order to provide a complete description of the larva it was necessary to dissolve the mountant in Euraral essence, hydrate to distilled water and clear the muscle in 10% KOH. Before this procedure was undertaken the whole head was drawn (Fig. 1) in order to indicate the main features and shape. After clearing, the head was dissected and mounted on the original slide, in Euparal, under several cover slips.

Terminology and abbreviations used are in accordance with the glossary of Sæther (1980).

Genus *Buchonomyia*

Larval diagnosis (3rd instar)

Larvae medium-sized, body white and head yellow in colour. Eye spots single. Head triangular in shape, tapering anteriorly.

Antenna. Very small and short, less than half (0.42) mandible length; 5 segmented; basal segment about as high as wide; remaining segments smaller, each becoming consecutively smaller; segment 3 not annulate; blade simple, club-shaped, reaching basal third of segment 3; accessory blade short, reaching basal half of segment 4; Lauterborn organs not discernable; Ring organ apparently absent; basal segment with 2 setae.

Labrum. Labral area reduced; all S setae of labrum simple; SII well developed, arising from relatively large pedestals; labral rod apparently present. Pecten epipharyngis consists of five well developed lamellae, outer lamellae slightly smaller than three median lamellae; chaetulae laterales and chaetulae basales not visible, probably present. Premandible absent.

Mandible. With 7 teeth, consisting of 1 dorsal subapical outer tooth, 1 large apical tooth and 5 inner teeth; seta subdentalis present; seta interna present, consisting of 4 simple branches.

Mentum. Median part worn, but consists of 13 teeth (1 broad median tooth and 6 pairs of lateral teeth). Ventromental plate very weakly developed. Setae submenti present.

Maxilla. Relatively large, with very few chaetae or sensillae, maxillary palp well developed.

Body. With very few, mostly inconspicuous setae. Anterior parapods strongly reduced, with few claws. Posterior parapods short, projecting laterally from the body, each with about 22 claws and 5 setae around parapod margin. Procercus absent. With 6 weakly developed, short anal setae. Three pairs of large, bulbous anal tubules.

Description of *Buchonomyia thienemanni*

Material examined: Ireland, Co. Kerry, River Flesk (grid ref.: V988900), one 3rd instar larva, 19.vii.1978, leg. D.A. Murray and P. Ashe. Slide preparation, mounted in Euparal, deposited in the collections of the National Museum of Ireland, NMI 69 :1993.

3rd instar larva (n = 1)

Total length 4.0 mm long (Fig. 2g). Head capsule 0.46 mm long, yellow in colour, triangular in shape (Fig. 1 exaggerates the triangular shape due to compression after mounting). Body white in colour with gut contents brown in colour.

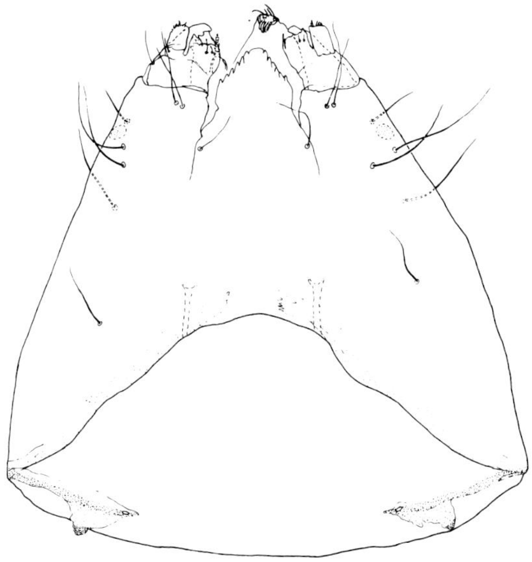

Figure 1. *Buchonomyia thienemanni* larva. Ventral view, complete head.

Head. Antenna (Fig. 2a), with antennal segment lengths (in µm): 14.5; 8.3; 6.2; 4.1; 2.7 (only one antenna measurable); Antennal ratio: 0.64; basal segment apparently without Ring organ but with two setae; blade about 4µm long; accessory blade about 3µm long; Lauterborn organs not visible (either very small as in the first-instar larva or absent). Labrum and epipharynx (Fig. 2b); SI simple, 11-12µm long; SII simple and well developed, 22-23µm long, arising from relatively large pedestals; SIII simple, 12-13µm long; SIV A short and peglike, about 3µm long; SIV B present; pecten epipharyngis of 5 scales, outer scales slightly smaller than three median scales; chaetulae laterales and chaetulae basales not visible in slide preparation but probably present; premandible absent; presumed labral rod very small and difficult to see (apparently occupying the same position as in first-instar larvae); Mandible (Fig. 2c), 87-91µm long, with 7 teeth (1 subapical outer tooth, 1 large apical tooth and 5 inner teeth - one mandible deformed and the two most basal inner teeth missing); seta subdentalis simple, 18µm long; seta interna with 4 simple branches. Mentum (Fig. 2d) about 80µm wide, worn apically but with 13 teeth (1 median tooth and 6 pairs of lateral teeth); with weak ventromental plate. Maxilla (Fig. 2e) with relatively few chaetae or sensillae, maxillary palp well developed.

Abdomen. Anterior parapods separate but poorly developed, each with relatively few (probably less than 50), very small (6-11µm long), comblike claws, mostly with 4 to 5 teeth (Fig. 2f). Anal end (Fig. 3) lacking procercus; with 6 fine, pale anal setae, 62-100µm long; posterior parapods reduced, projecting laterally from the body, each with about 22 simple, yellow claws, 20-25µm long; around the margin of each parapod are 5 setae, 76-114µm long; with three pairs of well developed anal tubules, 83-126µm wide, 228-311µm long.

ECOLOGY OF *BUCHONOMYIA THIENEMANNI*

The single third-instar larva of *B. thienemanni* described here was discovered when a large stone (approx. 2 kg. in weight) was removed by hand from a shallow part of the bottom of the River Flesk. On this rock an elongated case constructed of small detritus particles and silk was found, opened in the hand and the white chironomid larva described here was discovered inside. The case had a diameter of about 1-1.5 mm and was about 2 cm long which is about 4 times longer than the third-instar larva of *B. thienemanni*. The lack of well developed anterior parapods (used to manipulate silk) and the peculiarly developed posterior parapods indicates that the larva would not be able to cling to a rock in flowing water and seems to be quite incapable of constructing the case in which it was found. The laterally projecting posterior parapods, pressed alternately against the inner walls of the case for grip, may enable the larva to move backwards or forwards within the case. The combination of the triangular shape of the head, the case being about four times longer than apparently necessary and the probable incapacity of constructing the case itself, suggests that *B. thienemanni* may live in a commensal, parasitic or predatory association with another invertebrate animal.

The identity of the postulated host is uncertain but a trichopteran host would be very likely. From information on *B. thienemanni* and discussions with Dr. J. P. O'Connor, the most likely trichopteran hosts are hydropsychid caddis. Four species of Hydropsychidae are recorded in the larval stage from the stretch of the River Flesk in which *B. thienemanni* is very common. These are *Hydropsyche pellucidula* (Curtis), *H. contubernalis*

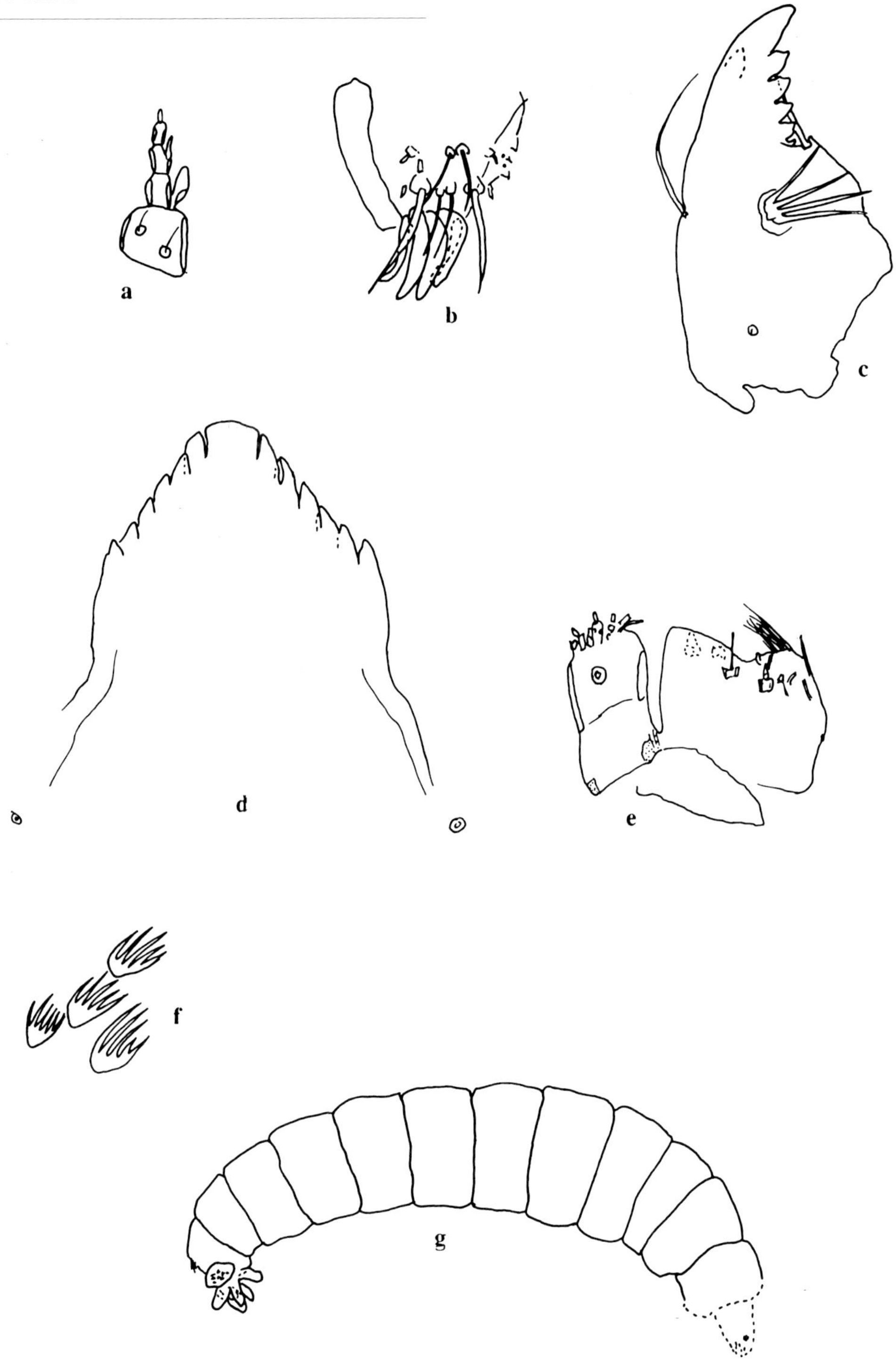

Figure 2. *Buchonomyia thienemanni* larva. (a) antenna, (b) labrum, (c) mandible, (d) mentum, (e) maxilla, (f) anterior parapod claws, (g) whole larva.

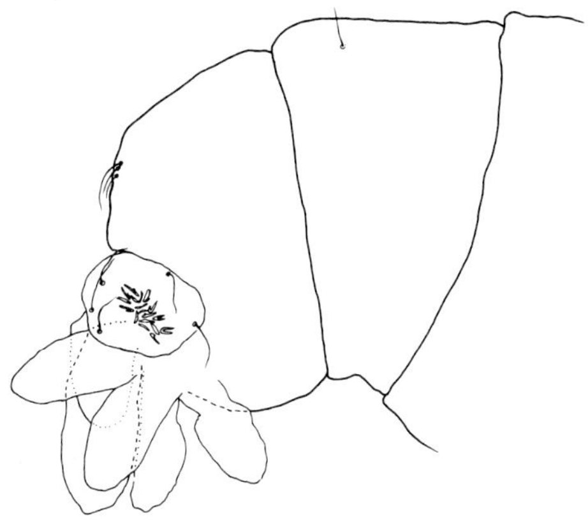

Figure 3. *Buchonomyia thienemanni* larva. Lateral view, posterior abdomen.

McLachlan, *H. siltalai* Döhler and *Cheumatopsyche lepida* (Pictet), with *H. siltalai* being very common, *H. pellucidula* and *C. lepida* are quite common and *H. contubernalis* is uncommon (J.P. O'Connor *pers. comm.*).

Further evidence of an association with Trichoptera can be postulated from both pupal and larval evidence in *Buchonomyia* and from two unrelated chironomid genera, *Eurycnemus* and *Collartomyia*, which are associated in some way with Trichoptera. The development of a heavily chitinised structure on the pupal cephalothorax occurs in *Buchonomyia* (base of antennal sheaths, Murray and Ashe, 1981a, Fig. 1A), *Eurycnemus* (the scutal process - SP, Murray and Ashe, 1981b, Fig. 1A) and *Collartomyia* (thoracic projection, Amakye and Sæther, 1992, Fig. 5A). These independently-evolved chitinised structures may enable the pupa to force its way out of a trichopteran case. The presence of similar structures on other chironomid pupal exuviae may be used to predict an association with Trichoptera.

There are also some parallel developments in the larvae of all three genera which may also indicate an association with Trichoptera; anal tubules well developed; anterior and posterior parapods reduced; procercus absent or reduced in size; antennae very small. Other indicative features (not shared by all three genera) include: body of larva white (*Buchonomyia* and *Eurycnemus*); posterior parapods project laterally from the body (*Buchonomyia* and *Collartomyia*). The association between *Eurycnemus* and a trichopteran, first suggested in Murray and Ashe (1981b), has been confirmed by Kobayashi (1995), who states that a Japanese species of *Eurycnemus* is ectoparasitic on trichopteran pupae.

Modified subfamily key to Holarctic larval Chironomidae

The key to larvae of Holarctic chironomid subfamilies, given in Cranston and Reiss (1983), is modified below to include the Buchonomyiinae.

2. Premandibles absent 2a
- Premandibles present 3

2a. Procercus well developed, 8-10x as high as wide . . . . .Podonominae (p. 23)
- Procercus absent . . . . .Buchonomyiinae

Acknowledgements

I wish to thank Dr. D. A. Murray, Zoology Department, University College, Dublin 4, Ireland for his constructive criticism of the paper; Dr. J. P. O'Connor, National Museum of Ireland, Kildare Street, Dublin 2, Ireland for some very useful discussion and data on the Trichoptera of the River Flesk; and an award from the Systematics Association (towards the cost of purchasing microscopical equipment) is gratefully acknowledged.

A NEW SPECIES OF *SUBLETTEA* ROBACK FROM SULAWESI, INDONESIA (DIPTERA: CHIRONOMIDAE)

Patrick Ashe[1] *and James P. O'Connor*[2]

[1]Research Associate, Department of Zoology, University College, Belfield, Dublin 4, Ireland

[2]National Museum of Ireland, Kildare Street, Dublin 2, Ireland.

INTRODUCTION

Adult males and pupal exuviae of a new species of *Sublettea* Roback are described and figured from northern Sulawesi, Indonesia. The material was collected during the "Project Wallace" expedition in 1985. The new species, *Sublettea wilesi* n. sp., is only the second species of the genus to be described. However, undescribed species are known in the pupal stage from S.W. China (Pinder and Reiss, 1986) and in the adult stage from the Neotropical region (Cranston *et al.*, 1989).

Materials and Methods

Although the material on which the present paper is based was collected and preserved in ethanol in 1985 already it is quite difficult to clear the musculature using the traditional method of treatment in hot 10% KOH. An additional problem is that chitin may be weakened during KOH treatment and various poorly chitinised structures may collapse resulting in very unsatisfactory slide preparations. A new method of enzymatic clearing of musculature was tested on specimens of this new species, with encouraging results.

The terminology used is in accordance with Sæther (1980) and adult males were mounted using a slightly modified method to that described by Pinder (1989).

Sublettea wilesi Ashe and O'Connor, sp. nov.

Type-locality: **Indonesia**, Sulawesi Utara, River Toraut, Dumoga-Bone National Park.

Material examined

Holotype male mounted in Euparal on a single slide, labelled "*Sublettea wilesi* n. sp., male, 9-14.x.1985, River Toraut, Dumoga-Bone National Park, Sulawesi Utara, Indonesia, 0°33'N, 123°54'E, 212 m. a.s.l., P. Ashe, Malaise trap, S269". Deposited in the National Museum of Ireland, Dublin, Ireland, NMI 58: 1993.

Paratypes: 24 males, same data as holotype, on slide and in alcohol; 3 males and their associated pupal exuviae, same data as holotype except, 20.x.1985, reared from drift net, sample S267; 37 pupal exuviae, same data as holotype except, 20.x.1985, collected by drift net, sample S267. Paratype material deposited in the following institutions: 2 males, 2 pupal exuviae, in Museum Zoologicum Bogoriense, Bogor, Indonesia; 1 male and its associated pupal exuviae, 2 pupal exuviae, in Australian National Collection, Canberra, Australia; 1 male, 2 pupal exuviae, in each of the following: Zoologische Staatsammlung, München, Germany; The Natural History Museum, London, England; Museum of Zoology, University of Bergen, Norway; the remaining paratype material deposited in the National Museum of Ireland, Dublin, Ireland, NMI 58: 1993.

Etymology: Named in honour of Dr. P. R. Wiles, University of Buckingham, England, who greatly assisted the senior author in collecting chironomids in Sulawesi and Malaysia in 1985.

Description

Male imago (n = 8)

Wing length 0.805-0.894, 0.866 mm.

Colouration (in alcohol): ground colour of whole body (antennae, head, thorax, abdomen and legs) pale yellow, with pale brownish markings on thoracic vittae; wings hyaline, without contrasting spots or markings; legs without distinct markings.

Antenna: With 13 flagellomeres, last flagellomere 374-472, 420 μm. Setal plume poorly developed. AR 0.19-0.27, 0.21.

Head: Eyes mostly bare but hairy in ventromedial area, with dorsomedial extension. Frontal tubercule indistinct, 4-5μm long. Palps with 5 segments of lengths (in μm): 81-98; 65-89; 211-244; 260-309; 398-496. Clypeus with 11-15, 13 setae.

Thorax: Antepronotal lobes widely separated, scutum overreaching antepronotum; tubercle absent; 9-10 irregularly biserial acrostichals; 5-8 uniserial dorsocentrals; 1 prealar; 4 scutellar setae, one distinct pair located close together posteriomedially.

Table 1. Lengths (in μm) and ratios of legs for *Sublettea wilesi* m.sp.

	fe	ti	ta_1	ta_2	ta_3
PI	390-439	195-228	479-553	243-276	178-207
PII	422-455	276-313	154-179	81-93	56-61
PIII	471-488	341-390	211-244	138-155	138-146

	ta_4	ta_5	LR	BV	SV
PI	130-175	73-89	2.43-2.46	1.63-1.71	1.21-1.22
PII	36-40	32-37	0.56-0.57	4.10-4.16	4.29-4.53
PIII	81-89	48-57	0.62-0.63	2.51-2.53	3.60-3.85

Wing: VR 1.48-1.58. R_{2+3} indistinct; R_{4+5} running close to costa, ending about halfway between apices of M_{1+2} and M_{3+4}; FCu well distal to RM. R, R_1, R_{4+5}, M_{1+2}, M_{3+4}, Cu, Cu_1, Vf and An with setae, other veins bare. Costa not extended. Wing membrane bare basally, hairy apically. Anal lobe not developed. Squama bare.

Legs: Fore tibia with pale spur, 12-20µm long; mid and hind tibia each bearing two separated combs, only one comb with spur, 22-26µm long and 24-29µm long respectively. Lengths (in µm) and ratios of legs (Table 1)

Genitalia (Fig. 1a): Anal tergite bands separate. Anal point tapering to point, with paired crests which extend onto anal tergite. Superior volsella well developed. Digitus sinuous, very elongate (apices of digiti often touching or overlapping one another), of about equal width except near tip which tapers to blunt point. Inferior volsella large and wide, extending to half length of relatively short gonostylus. Median volsella short, with lamellae distally, hidden beneath superior volsella.

Female imago

Not known, but probably present in the collected material.

Pupa (n = 10)

Total length: 1.76-1.93, 1.85 mm. Exuvial colouration: cephalothorax and segments VIII and IX brown, remainder weakly infuscated or clear.

Cephalothorax (Fig. 1b): Frontal setae 25-41µm long, arising from frontal apotome. Frontal apotome rugulose, without tubercules or warts. Pedicel of antennal sheaths with pearl row. Thoracic horn (Fig. 1b) 58-70µm long, with apical tuft of chaetae, chaetal tuft 66-93µm long. Anterior precorneal seta 37-60µm long, median seta 116-135µm long and posterior seta 86-114µm long. Thorax smooth along median suture. Wing sheath with nose, without pearl row.

Abdomen (Fig. 2): Shagreen, spine patterns and chaetotaxy of tergites I-IX as in Fig. 2. Tergite I bare; II and III with paired patches of spines anteriorly; IV and

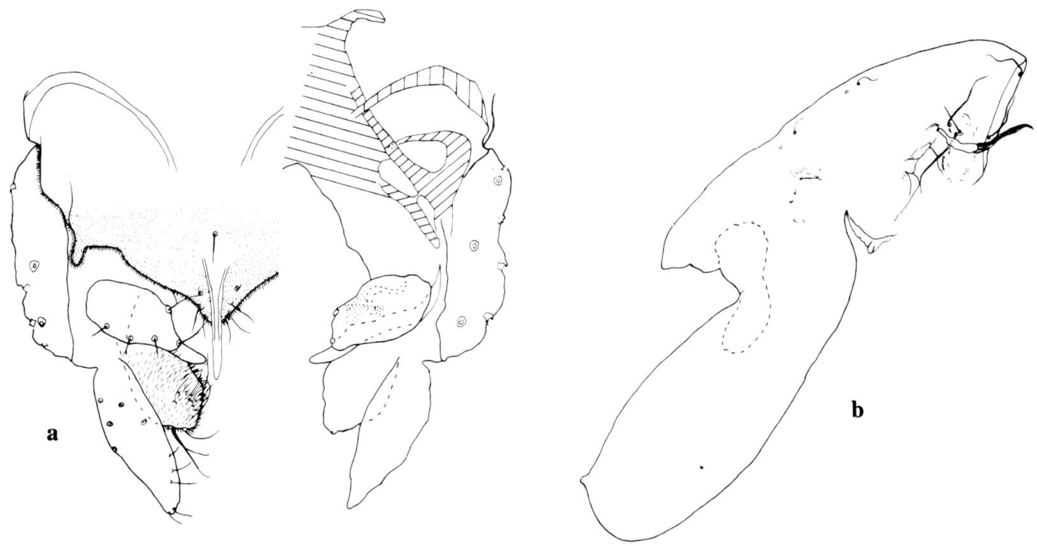

Fig. 1. *Sublettea wilesi* n. sp., (a) holotype male hypopygium, (b) pupal cephalothorax.

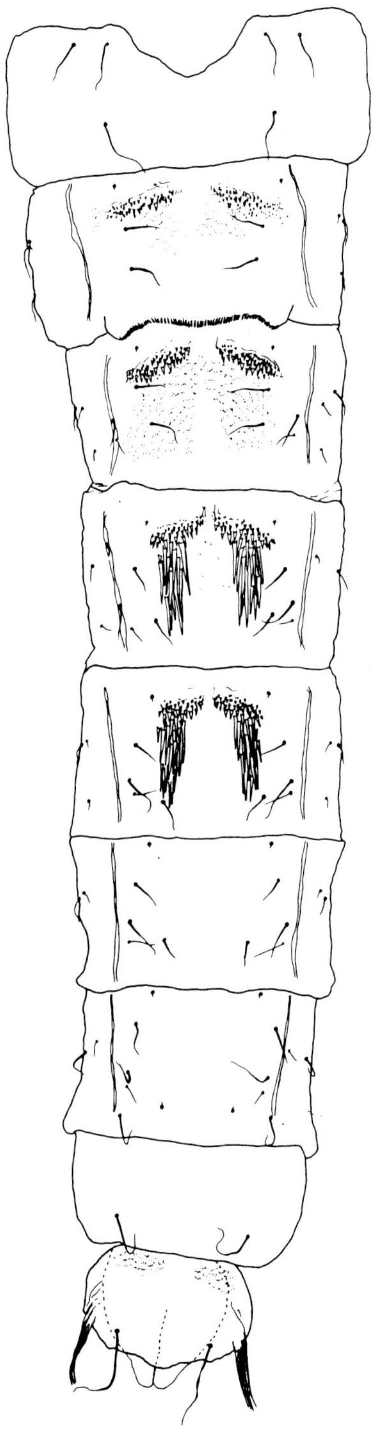

Fig. 2. *Sublettea wilesi* n. sp., pupal tergites.

V also with paired patches of spines anteriorly as well as long spines extending posteriorly from these patches; VI-VII bare. Hook row 165-176 μm wide, not interrupted medially. Lateral setation of abdominal segments: segment I with 0 setae, II-VII with 3 setae, VIII with 0 setae. Posterior margin of sternite VIII medially with pair of combs (16-19μm wide) of 7-9 small spines. Anal lobe weakly developed; with 1 lamelliform dorsal seta 78-93μm long, and 2-5 basally fused lamelliform setae 54-68μm long. Sexual dimorphism evident on sternite IX and posterior part of sternite VIII.

Larva

The larva of *S. wilesi* is unknown. The larval generic diagnosis for *Sublettea* (Pinder and Reiss, 1983), based on *S. coffmani* Roback, should enable recognition of other members of the genus.

Ecology and Distribution

The only other named species of *Sublettea* is *S. coffmani* which was described from the Nearctic region in the larval, pupal and adult stages by Roback (1975). The genus *Sublettea* is known from the Nearctic, Neotropical and Oriental regions (Ashe *et al.*, 1987; Cranston *et al.*, 1989) and appears to be confined to flowing waters including cool, temperate streams and warm, tropical rivers and streams.

The diel emergence pattern of *S. wilesi* was given in Ashe (1990, Fig. 1e, as *Sublettea*) and two emergence peaks were evident, the major peak in the hour before sunset and the secondary peak in the hour before sunrise. Diel emergence data for *S. coffmani* was given in Roback (1975).

Taxonomic Remarks

Adult males of *S. wilesi* are easily separated from those of *S. coffmani* by comparing the wing length, antennal ratio, figures of the genitalia as well as various other diagnostic characters in the respective descriptions. A comparison of the figures of the pupal exuviae of *S. wilesi* (Fig. 1b, 2) and those of *S. coffmani* and *Sublettea* "sp. (China)" (Pinder and Reiss, 1983: Fig. 10.78) indicates that *S. wilesi* more closely resembles *Sublettea* "sp.

(China)" which may not be unexpected since they are geographically closer. With the description of *S. wilesi* the generic diagnoses of the pupa and adult male of *Sublettea* as given in Pinder and Reiss (1983) and Cranston *et al.* (1989) requires modification but it would be best to wait until the Neotropical and Chinese material has been examined in detail so that it can be included in any modified diagnoses.

Acknowledgements

This paper is based on material collected whilst the senior author was a participant on Project Wallace, sponsored by the Royal Entomological Society of London and the Indonesian Institute of Sciences. An award to the senior author from the Systematics Association, towards the cost of purchasing microscopical equipment, is gratefully acknowledged.

Mapucheptagyia brundini n.gen., n.sp. - A new Diamesinae (Diptera: Chironomidae) from Chile

Endre Willassen

Museum of Zoology, University of Bergen, Norway

Introduction

The subfamily Diamesinae was an integral part of Lars Brundin's (1966) famous study of chironomid midges of the southern hemisphere. A few years ago, Professor Brundin left me some undescribed Diamesinae collected on his travels in South America. Unfortunately, the material was in an extremely poor state after long time storage in alcohol. The male of the species to be described below was strongly bleached and also deformed from intermediate desiccation and shrinking. Fortunately, Brundin had already made beautiful drawings of the hypopygium and details of the leg and because of the poor state of the specimen, I was allowed to use his illustrations in a future description of the species. In spite of lacking certain morphological detail, I trust that this species can be recognized from the following description.

METHODS

In an attempt to restore the shape of the adult specimen, it was treated with cold KOH for about 12 hours and mounted on slides in Canada Balsam according to normal procedure (Pinder 1989).

MAPUCHEPTAGYIA WILLASSEN, N. GEN.

Type species: Mapucheptagyia brundini Willassen, n. sp., by monotypy.

Etymology: The generic name is composed from *Mapuche* (indigenous people of Araucania and W. Patagonia) and *Heptagyia* (a genus of Diamesinae). The species is named in memory of its discoverer, Lars Brundin.

Generic diagnosis

Male antenna plumose with 13 flagellomeres, innermost flagellomeres somewhat globular to barrel-shaped, ultimate flagellomere club-shaped. Eyes hairy. Crossvein MCu inserted well proximally of FCu. Tarsomere 3 slightly expanded distally, tarsomere 4 distinctly cordiform with strong dorsomedial excavation. Hypopygium with well developed, distally acute anal point; gonostylus stout and simple, without crista dorsalis, posterior heel or megasetae. Pupa with extremely weak cephalothoracic setae, thoracic horn lacking; abdominal tergites II-IV with week shagreen composed of an anterior, transverse band of small spinules near the apophysis and 1-2 rows of small, anteriorly directed spinules near posterior margin; posterolateral corners of tergites II-VIII with progressively larger spurs. L-setae all inserted ventrally and generally not reaching lateral margin; anal lobe with conspicuous corniform projection and 2 short macrosetae shaped as rigid hooks.

***Mapucheptagyia brundini* Willassen, n.sp.**

Figs. 1a-d, 2a-g

Material examined

Holotype male and paratypes, 2 pupal exuviae (on one slide), **South Chile**, Llanquihue, Peulla, Rio Canteras, 19.i.1959 (Brundin). To be deposited in Naturhistoriska Riksmuseet, Stockholm, Sweden.

Male imago

Small species. Total length not measurable. Wing length ca. 1.2 mm.

Head (Figs.1a-b): Antenna plumose with 13 flagellomeres, proximal flagellomeres swollen to more or less globular, ultimate flagellomere clubshaped, AR about 0.5. Coronal suture not visible. Temporal setae either extremely weak or absent. Eyes distinctly hairy, without dorsomedial extension. Clypeus broad, densely covered with very weak setae. Palp segments 2-4 broad, ultimate segment thin and shorter than penultimate; palp segment 3 without sensory pit. Tentorium with weak anterolateral protrusion.

Thorax: Antepronotum broad with widely separated dorsal lobes, ventrolateral part with distinct setae. Anterodorsal surface of scutum with circular to polygonous patches of microtrichia. Chaetotaxy poorly visible, except 2 distinct seta bases in front of scutellum. Scutellum with scattered seta bases.

Wing: membrane greyish, apparently without setae; microtrichia visible under 400x magnification. MCu ending distinctly proximal to FCu (L. Brundin, *in litt.*), patterns of veins otherwise difficult to discern (specimen caught during emergence). Anal lobe apparently moderately developed. Squama with well developed fringe of setae.

Legs: ta3 slightly expanded apically; ta4 (Fig. 1c) distinctly cordiform, with strong dorsal excavation and well developed membraneous sole. Claws serrated. Remaining parts of legs missing or damaged.

Genitalia (Fig. 1d): Tergite IX with about 12 weak setae on each side, posterior margin convex and ending in a well developed anal point, bearing a delicate setae on the pointed apex. Laterosternite well delineated from tergite IX, with a few fine setae. Sternapodeme apparently narrow and gently rounded.

Phallapodeme moderately long and somewhat furcate. Aedeagal lobe complex, apparently with two sclerotized pointed projections. Gonocoxites with hyaline, rounded basal plates on each side and with interconnecting strongly sclerotized transverse lamella which is medially incised. Gonostylus simple, stout and distally rounded; with long setae on outer surface, but without megasetae.

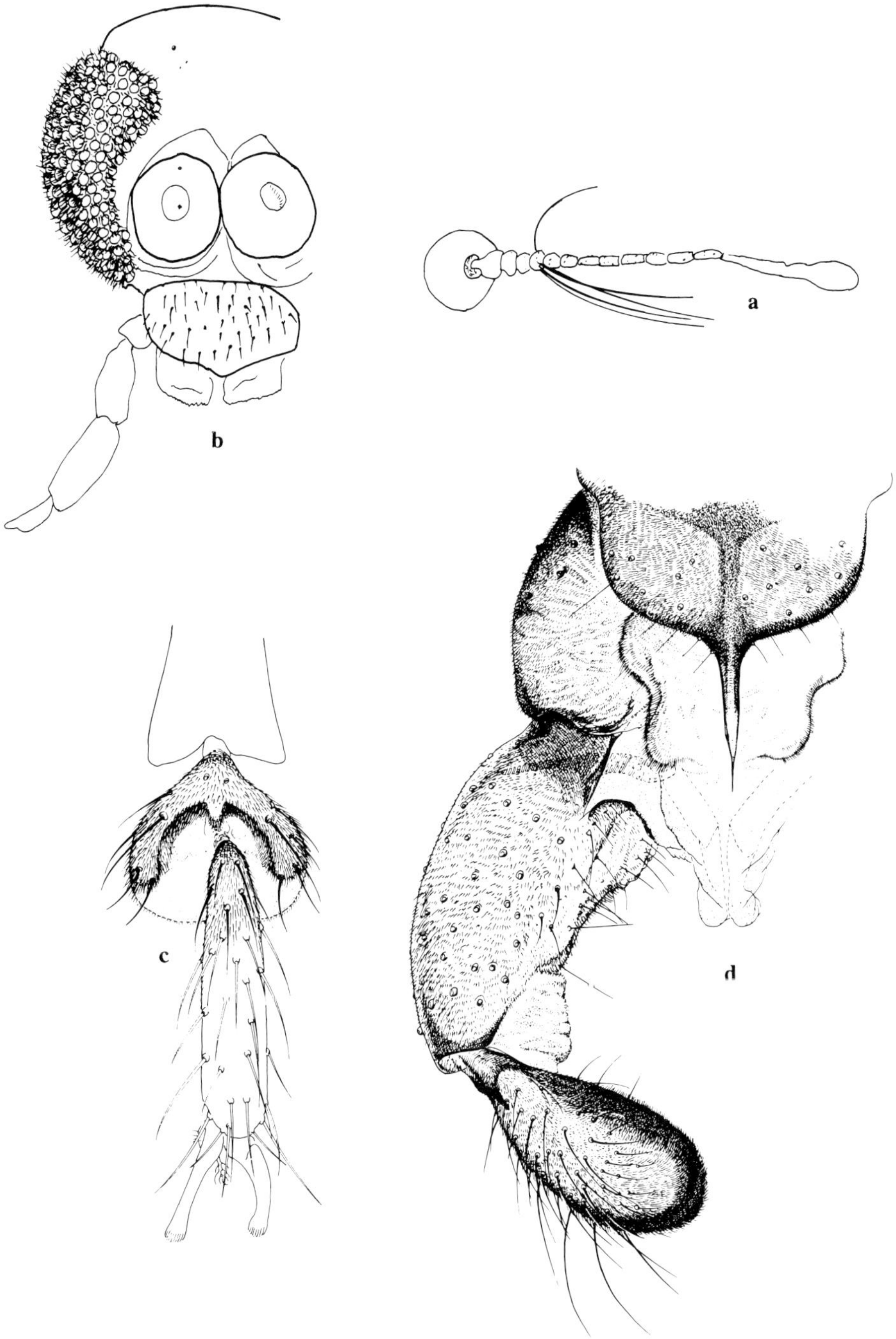

Figures. 1a-d. *Mapucheptagyia brundini* n.sp., adult male. (a) antenna (mostly denuded), (b) head, (c) ultimate part of mid leg, (d) genitalia. Figs. 1c-d drawn by Lars Brundin.

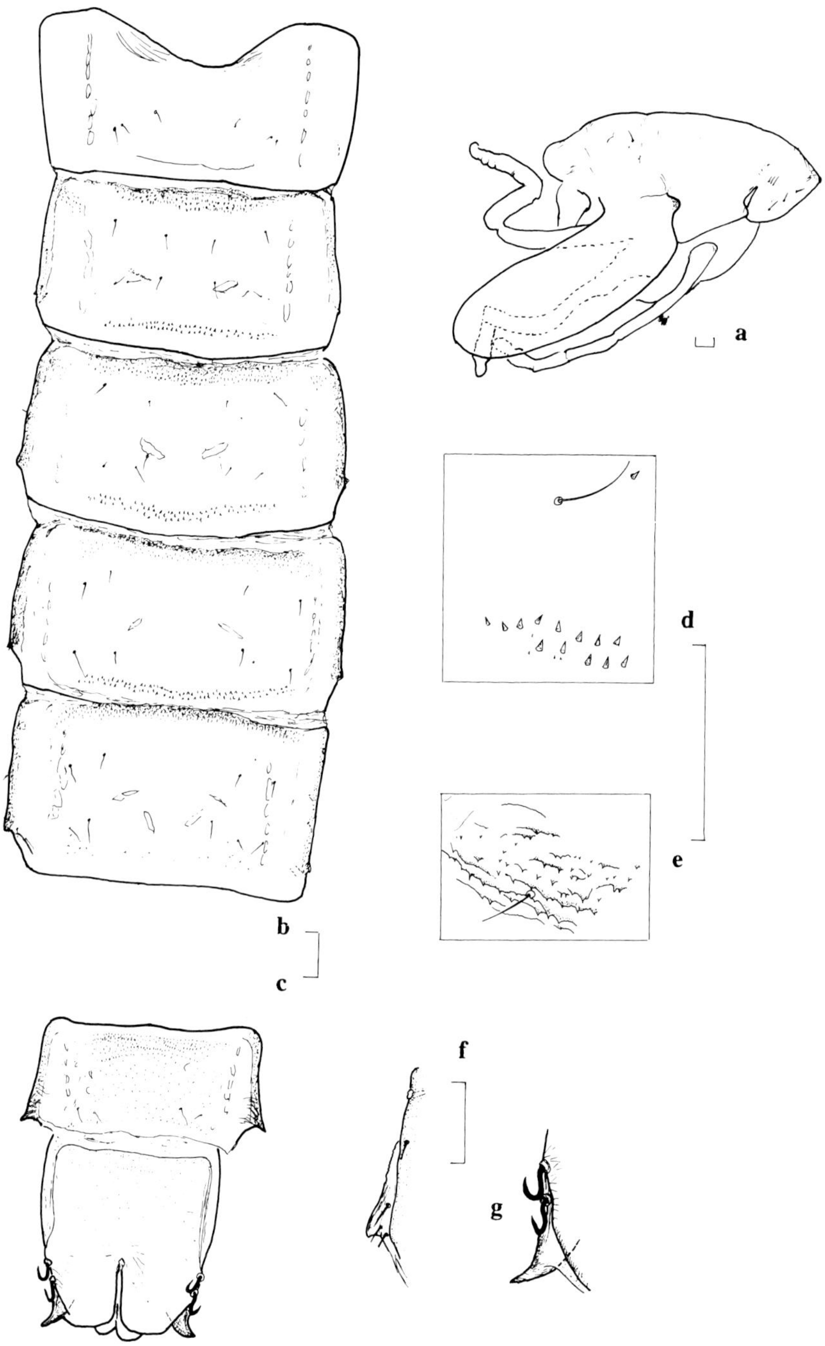

Figures 2a-g. *Mapucheptagyia brundini* n.sp., pupal exuviae; all scales = 0.1 mm.

(a) thorax, (b) abdominal tergites I-V, (c) tergites VIII-IX, (d) shagreen and D-seta on TIII magnified, (e) shagreen and V-seta on SVIII, (f) posterolateral spur and L-setae on segment VI in ventral view, (g) anal macrosetae and posterior spur in dorsal view.

Pupa

Exuviae yellowish, abdomen with brown apophyses and lateral margins. Total length about 5.0 mm.

Cephalothorax: Antennal sheath without bead-like ring near base. Frontal apotome very slightly rugulose, without tubercles or warts, with 2 weak frontal setae. Antepronotum (Fig. 2a) with 2 median and 1 lateral weak setae. Precorneal setae very weak, all 3 closely inserted. One or possibly 2 weak dorsocentrals above wing base. Other setae missing on thorax. Wing sheath smooth, without pearl row or nose. Front leg sheath slightly curved in tarsal area and apparently reaching wing sheath tip, mid leg sheath similarly curved (the 90° bend in Fig. 2a is an artefact) but apparently reaching slightly beyond wing tip, hind leg with S-bend under distal part of wing sheath.

Abdomen (Figs. 2b-g): Pedes spurii absent. Lateral margins sclerotized and posterolateral corners projecting into progressively larger spurs from segment II to VIII. Tergite I without shagreen; TII-VI with transverse raws of weak shagreen over and slightly posterior of apophyses, also with weak lateral patches near spiracles; TII-IV with additionally 1-2 rows of anteriorly directed, cone-shaped spinules (Figs. 2b.d) near posterior margin; TVII-VIII with weak shagreen more or less all over the area delimited by lateral muscle scars. TIX with weak anteriomedial shagreen. Sternites with much more shagreen than tergites, reaching to spiracles anteriorly, but confined to median area posteriorly; shagreen pattern is composed in the anterior area by transverse punctuated rows of minute spinules, this pattern becomes progressively more irregular towards the caudal end of each tergite and the spinules increase to about the same size as those in the posterior rows on tergite III.

TI-VIII with following number of D-setae on each side: 3, 5, 5, 5, 5, 6, 5, 2. SI-VIII with following number of V-setae on each side: 3, 4, 5, 5, 5, 5, 5, 1. Segments I-VIII with following number of L setae: 1?, 1, 4, 4, 4, 4, 4, 2. L-setae very weak, inserted ventrally and generally not reaching beyond lateral margin; L_1 situated slightly behind spiracle, L_2 about midway on anterior tergites but displaced towards L_3 on posterior tergites; L_3 and L_4 closely set on spurlike corner of lateral margin, L_4 slightly shorter and in a more posteromedian position. TIX with 2 pairs of closely set rigid and hookshaped macrosetae; other setae lacking; anal lobe without rounded lateral flanges, but with strongly sclerotized, dorsolaterally curved apical spur which is smooth and without setae. Genital sac extending approximately to margin of anal lobes.

HABITAT

The material was collected by drift net in the Rio Canteras near Peulla in S.Chile. Rio Canteras is a tributary to the Rio Tronadór. It drains meltwater from the glaciated Mt. Tronadór, but runs through *Nothofagus* forest in its lower reaches (Brundin 1966:pl.13.) and had fairly clear water at the time of Brundin's visit. Brundin (1966) described the stream bed as consisting of stones and large blocks, with water temperatures of 6-8.5°C (9.30-16.00h) on 19th January 1959.

DISCUSSION

The male hypopygium of *Mapucheptagyia* superficially resembles those of typical Diamesini, particularly in the configuration of tergite IX and the well developed anal point. However, although the interpretation of the internal structures is difficult due to the damage of the single available specimen, an affiliation with the "Heptagyiae" is clearly apparent. The hyaline basal plates of the gonocoxite are discernable and there is also a structure described by Brundin (1966: 396, under *Heptagyia*) as a strongly sclerotized transverse lamella. The shape of this structure in *Mapucheptagyia* seems to conform with that found, for instance, in *Reissia antiqua* (Brundin, 1966: fig. 538).

Most Diamesinae with plumose male antennae have more or less cylindrical flagellomeres. The somewhat globular inner flagellomeres in *Mapucheptagyia* resemble those found only in *Heptagyia* (Brundin, 1966: 396) and *Boreoheptagyia* (Serra-Tosio, 1989: fig. 3). The very narrow penultimate flagellomere and the distally expanded ultimate flagellomere give the antenna a clubbed form which to my knowledge occurs distinctively only in *Heptagyia*. In *Boreoheptagyia*, the ultimate flagellomere is shortened to an extent that the club-shape is less conspicuous, but still apparent. This character probably points to a close relationship between these three genera.

The pupal exuviae are highly likely associations with the male. The extremely short ta_4 is clearly apparent in the pupal leg sheath. *Mapucheptagyia* pupae differ from previously known "Heptagyae" in many features, including the fact that three precorneal setae are retained. I have failed to identify thoracic horns on the *Mapucheptagyia* exuviae, but I should mention that there is a grainy substance covering the area where the tracheae, according to my interpretation, end blindly. Brundin described the thoracic horns of *Heptagyia* as smooth and thin-walled with a lumen filled with a grainy substance. A general resemblance to *Heptagyia* is otherwise apparent in the sparse shagreen of the abdominal tergites, the reduced size of the setae and the sharply marked posterior corners of the abdominal segments. In this latter feature, *Mapucheptagyia* is actually more similar to *Syndiamesa*, *Diamesa*, and *Pseudokiefferiella*, because the posterolateral corners are strongly sclerotized and progressively more spurlike towards the caudal end of the abdomen. The anal macrosetae resemble those of *Boreoheptagyia* in shape, although only two pairs are present in *Mapucheptagyia*. It is presently unclear whether the well developed anal lobe spur is strictly homologous to the spur in many Diamesini (see Oliver, 1986), where three macrosetae are also present. Smooth anal lobe spurs similar to the ones in *Mapucheptagyia* are found in certain orthoclad species and in the latter case, they are obviously derived from the macrosetae or their bases (G. A. Halvorsen, *pers. comm.*).

Brundin (1966) found evidence for phylogenetic relationships within the Diamesinae from the arrangement of the pupal leg sheaths. Unfortunately, his classification of pupal types can sometimes be difficult to apply in the study of exuviae because the position of the leg sheaths may obviously change in slide mounted pupal skins. In the material available of *Mapucheptagyia*, both of the anterior leg pairs seem curved and precludes *Mapucheptagyia* from being characterised as a *Heptagyia*-type of pupae. At least the sheath of the mid leg appears to extend slightly beyond the tip of the wing sheath, thereby approaching the *Boreoheptagyia* type (Brundin, 1966: fig. 624) of leg configuration.

More evidence is certainly needed in order to speculate much further on the phylogenetic relationships of *Mapucheptagyia*. A position near *Heptagyia* or Boreoheptagyia is indicated by some of the characters mentioned above. This and the similarities with the Diamesini places *Mapucheptagyia* in a key position with respect to the further understanding of the biogeographical relationships between the Diamesinae of the northern and southern continents (see Willassen and Cranston, 1986). More complete morphological data from the male and pupa, and certainly from the unknown larvae and the female adult are therefore urgently needed.

Acknowledgement

This paper is dedicated to the memory of Lars Z. Brundin in humble admiration for his scientific achievements and in gratitude for the inspiring meetings with him when I was a young student.

REFERENCES

Aagaard, K. (1986) The chironomid fauna of North Norwegian lakes, with a discussion on methods of community classification. *Holarctic Ecology*, **9**, 1-12.

Aagaard, K. (1993) Ordination or typology - The search for a stable classification of running water communities. *Netherlands Journal of Aquatic Ecology*, **26**, 441-5.

Aagaard, K., Olsen, A. and Solem, J. O. (1987) Chironomids of Blesbekken, an alpine tundra stream at Dovrefjell National Park, Norway. *Entomologica Scandinavica Supplement*, **29**, 349-54.

Aagaard, K., Solem, J. O., Lillehammer, A., Hanssen, O., Nost, T. and Dalen, T. (1989) Forsknings - og referansevassdrag ATNA. Utbredelse, sonering og årsvariasjoner hos bunndyr i Atna og Atnasjoen. *MVU-rapport (Trondheim, Norway)*, **B57**, 1-50.

Aagaard, K., Solem, J. O., Nost, T. and Hanssen, O. (in press) The macrobenthos of the pristine stream, Skiftesåa, Hoylandet, Norway.

Acton, A.B. (1957) Chromosome inversions in natural populations of *Chironomus tentans*. *Journal of Genetics*, **55**, 61-94.

Acton, A.B. (1959) A study of the differences between widely separated populations of *Chironomus* (=*Tendipes*) *tentans* (Diptera). *Proceedings of the Royal Society of London*, **151**, 277-97.

Acton, A.B. (1962) Incipient taxonomic divergence in *Chironomus* (Diptera)? *Evolution*, **16**, 330-7.

Acton, A.B. and Scudder, G.G.E. (1971) The zoogeography and races of *Chironomus* (=*Tendipes*) *tentans* Fab. *Limnologica*, **8**, 83-92.

Adams, M.S., and Niswander, J.D. (1967) Developmental noise and a congenital malformation. *Genetical Research*, **10**, 313-7.

AFNOR. (1985) Essais des eaux. Détermination de l'indice biologique (IBG). *ANFOR T*, 90-350.

Albu, P. (1974) A new subgenus of the genus *Bryophaenocladius* and two new species (Diptera: Chironomidae). *Entomologisk Tidskrift Supplement*, **95**, 9-12.

Ali, A. (1980) Nuisance chironomids and their control: a review. *Bulletin of the Entomological Society of America*, **26**, 3-16.

Ali, A. (1981) *Bacillus thuringiensis* serovar. *israelensis* (ABG-6108) against chironomids and some nontarget aquatic invertebrates. *Journal of Invertebrate Pathology*, **38**, 264-72.

Ali, A. (1991) Perspectives on management of pestiferous Chironomidae (Diptera), an emerging global problem. *Journal of the American Mosquito Control Association*, **7**, 260-81.

Ali, A. and Mulla, M.S. (1976) Substrate type as a factor influencing spatial distribution of chironomid midges in an urban flood control channel system. *Environmental Entomology*, **5**, 631-6.

Ali, A. and Mulla, M.S. (1977) The [use of] IGR diflubenzuron and organophosphorus insecticides against nuisance midges in man-made residential-recreational lakes. *Journal of Economic Entomology*, **70**, 571-7.

Ali, A., R.D. Baggs, and J.P. Stewart. (1981) Susceptibility of some Florida chironomids and mosquitoes to various formulation of *Bacillus thuringiensis* serovar. *israelensis*. *Journal of Economic Entomology*, **74**, 672-7.

Ali, A., Stafford, S.R., Fowler, R.C. and Stanley, B.H. (1984) Attraction of adult Chironomidae (Diptera) to incandescent light under laboratory conditions. *Environmental Entomology*, **13** (4), 1004-9.

Ali, A., G. Majori, G. Ceretti, F. D'Andrea, M. Scattolin, and U. Ferrarese. (1985) A chironomid (Diptera: Chironomidae) midge population study and laboratory evaluation of larvicides against midges inhabiting the lagoon of Venice, Italy. *Journal of the American Mosquito Control Association*, **1**, 63-8.

Ali, A., Stanley, B.H. and Chaudhuri, P.K. (1986) Attraction of some adult midges (Diptera: Chironomidae) of Florida to artificial light in the field. *The Florida Entomologist*, **69** (4), 644-50.

Allan, J.D. (1978) Trout predation and the size composition of stream drift. *Limnology and Oceanography*, **23**, 1231-7.

Allan, J.D. (1984) The size composition of invertebrate drift in a Rocky Mountain stream. *Oikos*, **43**, 68-76.

Allan, J.D. (1987) Macroinvertebrate drift in a Rocky Mountain Stream. *Hydrobiologia*, **144**, 261-8.

Allan, J.D. and Russek, E. (1985) The quantification of stream drift. *Canadian Journal of Fisheries and Aquatic Sciences*, **42**, 210-5.

Amakye, J.S. and Sæther, O.A. (1992) The immature stages and imagines of the Afrotropical species *Microtendipes lentiginosus* Freeman and *Collartomyia hirsuta* Goetghebuer (Diptera: Chironomidae). *Entomologica Scandinavica*, **23**, 429-42.

Anon (1984) *Data Book of World Lakes*, LECS '84, Kyoto.

Anon (1985) *Standard Methods for the Examination of Water and Wastewater*. 16th Edition. American Public Health Association Inc., New York.

Anon (1990) *Proposed Regional Sewerage Scheme for the Upper Blue Mountains: Environmental Impact Statement*. Prepared by Camp Scott Furphy, Mitchell McCotter and Associates. Water Board, Sydney.

ANZECC. (1992) *Australian Water Quality Guidelines for Fresh and Marine Waters*. Australian and New Zealand Environment and Conservation Council.

Appelberg, M. and Aldén U. (1992) Integrerad uppföljning av kalkningen effekter på sjöar och vattendrag - en treårsrapport. *Information från Sötvattenslaboratoriet Drottningholm*, **4,** 1-60.

Armitage, P.D. and Blackburn, J.H. (1985) Chironomidae in a Pennine stream system receiving mine drainage and organic enrichment. *Hydrobiologia*, **121,** 165-72.

Armitage, P.D. and Tuiskunen, J. (1988) *Thalassosmittia atlantica* (Stora) comb. nov. Description of female and immature stages from Tenerife, Canary Islands. *Spixiana Supplement*, **14,** 25-8.

ARRRI (1989) Alligator Rivers Region Research Institute. Annual Research Summary 1988-89. Supervising Scientist for the Alligator Rivers Region.

Arthington, A.H., Conrick, D.L., Connell, D.W. and Outridge, P.M. (1982) The Ecology of a Polluted Urban Creek. Technical Paper No.68, Australian Water Resources Council, Department of National Development and Energy. Australian Government Publishing Service, Canberra.

Arts, G.H.P., Van der Velde, G., Roelofs, J.G.M., and Van Scraay, C.A.M. (1990) Sucessional changes in the soft-water macrophyte vegetation of (sub)atlantic sandy low-lands regions during this century. *Freshwater Biology*, **24(2),** 287-94.

Ashe, P. (1983) A catalogue of chironomid genera and sub-genera of the world including synonyms (Diptera: Chironomidae). *Entomologica Scandinavica Supplement*, **17,** 1-68.

Ashe, P. (1985) A larval diagnosis for the subfamily Buchonomyiinae and the genus *Buchonomyia* with a description of the 1st instar larva of *Buchonomyia thienemanni* Fitt. (Diptera: Chironomidae). *Spixiana Supplement*, **11,** 143-8.

Ashe, P. (1990) Ecology, zoogeography and diversity of Chironomidae (Diptera) in Sulawesi with some observations relevant to other insects, in *Insects and the Rain Forests of South East Asia (Wallacea)* (eds W.J. Knight and J.D. Holloway), The Royal Entomological Society, London, pp.261-8

Ashe, P. (1995) Description of a late-instar larva of *Buchonomyia thienemanni* Fittkau and further data on its ecology with diagnoses for the subfamily Buchonomyiinae and the genus *Buchonomyia* (Diptera: Chironomidae), in *Chironomids: from Genes to Ecosystems* (ed. P.S. Cranston), CSIRO, Melbourne, pp.425-9.

Ashe, P. and Cranston, P.S. (1990) Family Chironomidae, in *Catalogue of Palaearctic Diptera*, volume 2, (eds A. Soós and L. Papp), Akadémiai Kiadó, Budapest, pp.113-441.

Ashe, P. and Murray, D.A. (1983) Observations on and descriptions of the egg-mass and eggs of *Buchonomyia thienemanni* Fitt. (Diptera: Chironomidae). *Memoirs of the American Entomological Society*, **34,** 3-13.

Ashe, P., Murray, D.A. and Reiss, F. (1987) The zoogeographical distribution of Chironomidae (Diptera). *Annales de Limnologie*, **23,** 27-60.

Atchley, W. R. and Martin, J. (1971) A morphometric analysis of differential sexual dimorphism in larvae of *Chironomus* (Diptera). *Canadian Entomologist*, **103,** 319-27.

Azzouzi, A. and Laville, H. (1987) Premier inventaire faunistique de chironomidés (Diptera, Chironomidae) du Maroc. *Annales de Limnologie*, **23,** 217-24.

Azzouzi, A., Laville, H. and Reiss, F.(1992) Nouvelles récoltes de chironomidés (Diptera) du Maroc. *Annales de Limnologie*, **28,** 225-32.

Bailey, P.C.E. (1981) Insect drift in Condor Creek, Australian Capital Territory. *Australian Journal of Marine and Freshwater Research*, **32,** 111-20.

Bakhtina, V. (1978) Life cycles and production of mass species of chironomids in fattening ponds. *Acta Universitatis Carolinae - Biologica*, 13-20.

Barclay, M.H. (1966) An ecological study of a temporary pond near Auckland, New Zealand. *Australian Journal of Marine and Freshwater Research*, **17,** 239-58.

Barmuta, L.A. (1984) A method for separating benthic arthropods from detritus. *Hydrobiologia*, **112,** 105-7.

Barrion, A.T. and Litsinger, J.A. (1984) Chironomid, corixid, and ostracod pests of irrigated rice seedling roots. *International Rice Research Newsletter*, **9,** 19.

Battegazzore, M., Petersen Jr., R.C., Morettti, G. and Rossaro, B. (1992) An evaluation of the environmental quality of the River Po using benthic macroinvertebrates. *Archiv für Hydrobiologie*, **125,** 175-206.

Bay, E.C. (1964) An analysis of the "Sayule" (Diptera: Chironomidae) nuisance at San Carlos, Nicaragua, and recommendations for its alleviation. *World Health Organization mimeographed document, WHO/EBL/20.* pp.18.

Bay, E.C. (1993) Chironomid (Diptera: Chironomidae) larval occurrence and transport in a municipal water system. *Journal of the American Mosquito Control Association*, **9,** 275-84.

Bayley, P.B. (1991) The flood pulse advantage and the restoration of river-floodplain systems. *Regulated Rivers: Restoration and Management*, **6,** 75-86.

Bazzanti, M., Ferrara, O. and Mastrantuono, L. (1989) Composition and structure of planktonic and benthic communities as a basic information in fishpond culture. *Centro*, **1,** 28-40.

Beaver, R.A. (1983) The communities living in *Nepenthes* pitcher plants: fauna and foodwebs, in *Phytotelmata: Terrestrial Plants as Hosts for Aquatic Insect Communities* (eds J.H. Frank and L.P. Lounibos), Plexus Publishing, Marlton, New Jersey, pp.129-59.

Beaver, R.A. (1985) Geographical variation in food web structure in *Nepenthes* pitcher plants. *Ecological Entomology*, **10,** 241-8.

Becker, T. (1908) Dipteren der Kanarischen Inseln. *Mitteilungen aus dem Zoologischen Museum in Berlin*, **4,** 1-180.

Beckett, D.C. and Miller, M.C. (1982) Macroinvertebrate colonization of multiplate samplers in the Ohio River:

the effect of dams. *Canadian Journal of Fisheries and Aquatic Sciences*, **39**, 1622-7.

Beerman, W. (1955) Cytologische Analyse eines *Camptochironomus*- Artbastards. I. Kreuzungsergebnisse und die Evolution des Karyotypes. *Chromosoma*, **7**, 198-259.

Belbin, L. (1991) The analysis of pattern in bio-survey data, in *Nature Conservation: Cost Effective Biological Surveys and Data Analysis* (eds C.R. Margules and M.P. Austin), CSIRO, Melbourne, pp.176-90.

Belyanina, S.I., Loginova, N.V., Sigareva, L.E. (1992) Morphology and karyopool of *Chironomus entis* (Chironomidae, Diptera) from different reservoirs. *Zoological Zhurnal* **71**, 32-8 (in Russian with English summary).

Benke, A.C. (1979) A modification of the Hynes method for estimating secondary production with particular significance for multivoltine populations. *Limnology and Oceanography*, **24**, 168-71.

Benke, A.C. (1993) Concepts and patterns of invertebrate production in running waters. *Verhandlungen der Internationalen Vereinigung für Theoretische und Angewandte Limnologie*, **25**, 15-35.

Benke, A.C., Hall, C.A.S., Hawkins, C.P., Lowe-McConell, R.H., Stanford, J.A., Suberkropp, K. and Ward, J.V. (1988) Bioenergetic considerations in the analysis of stream ecosystems. *Journal of the North American Benthological Society*, **7**, 480-502.

Benke, A.C., Van Arsdall, T.C., Gillespie, D.M. and Parrish, F.K. (1984) Invertebrate productivity in a subtropical blackwater river: the importance of habitat and life history. *Ecological Monographs*, **54**, 25-63.

Benson, L.J. and Pearson, R.G. (1987) Drift and upstream movement in Yuccabine Creek, an Australian tropical stream. *Hydrobiologia*, **153**, 225-39.

Berczik, A. (1971) Die Chironomiden und ihre Lebenstätten auf dem ungarischen Donauabschnitt. *Limnologica*, **8**, 61-71.

Berg, K. (1948) Biological studies on the River Susaa. *Folia Limnologica Scandinavica*, **4**, 1-318.

Berg, M.B. and Hellenthal, R.A. (1991) Secondary production of Chironomidae (Diptera) in a north temperate stream. *Freshwater Biology*, **25**, 497-505.

Bishop, J.E. and Hynes, H.B.N. (1969) Upstream movements of the benthic invertebrates in the Speed River, Ontario. *Journal of the Fisheries Research Board of Canada*, **26**, 279-98.

Bishop J.A. (1974) The fauna of temporary rain pools in eastern New South Wales. *Hydrobiologia*, **44**, 319-23

Bitusik, P. and Ertlova, E. (1985) Chironomid communities (Diptera, Chironomidae) of the River Rajcianka (North-Western Slovakia). *Biologia (Bratislava)*, **40**, 595-608.

Blyth, J.D., Doeg, T.J. and St Clair, R.M. (1984) Response of the macroinvertebrate fauna of the Mitta Mitta River, Victoria, to the construction and operation of Dartmouth Dam. I. Construction and initial filling period. *Occasional Papers from the Museum of Victoria* **1**: 83-100.

Bodmer, W. and Ashburner, M. (1984) Conservation and change in the DNA sequences coding for alcohol dehydrogenase in sibling species of *Drosphila*. *Nature*, **309**, 425-30.

Boerger, H. (1981) Species composition, abundance and emergence phenology of midges (Diptera: Chironomidae) in a brown-water stream of West-Central Alberta, Canada. *Hydrobiologia*, **80**, 7-30.

Boesel, M.W. and Winner, R.W. (1980) Corynoneurini of Northeastern United States, with a key to adults and observations on their occurrence in Ohio (Diptera: Chironomidae). *Journal of the Kansas Entomological Society*, **53**(3), 501-8.

Boothroyd, I.K.G. (1987) Taxonomic composition and life cycles of Chironomidae (Diptera) in a Northern New Zealand stream. *Entomologica Scandinavica Supplement*, **29**, 15-22.

Boothroyd, I.K.G. (1988) Temporal and diel emergence of Chironomidae (Diptera: Insecta) from a New Zealand stream. *Verhandlungen der Internationale Vereinigung für Theoretische und Angewandte Limnologie*, **23**, 1399-404.

Boothroyd, I.K.G. (1990) Additions to the Chironomidae (Diptera: Insecta) of New Zealand: *Cricotopus* van der Wulp species from a North Island stream. *New Zealand Journal of Zoology*, **17**, 419-36.

Boothroyd, I.K.G. (1995) Temporal and diel drift of Chironomidae (Diptera: Insecta) larvae in a northern New Zealand stream, in *Chironomids - from Genes to Ecosystems* (ed. P.S. Cranston), CSIRO, Melbourne, pp.213-23.

Boothroyd, I.K.G. and Dickie, B.N. (1991) Macroinvertebrate drift and community colonisation on perspex artificial substrates in the Ohinemuri River, New Zealand. *New Zealand Journal of Marine and Freshwater Research*, **25**, 167-76.

Borutski, E.V., Sokolova, N.Yu. and Yablonskaya, E.A. (1971) A review of Soviet studies into production estimates of chironomids. *Limnologica*, **8**, 183-91.

Boysen Jensen, P. (1919) Valuation of the Limfjord. I. Studies on the fish-food in the Limfjord 1909-1917, its quantity, variation and annual production. *Report of the Danish Biological Station to the Board of Agriculture*, 1-44.

Böttger, K., Holm, U. and Mikowski, K. (1987) Vergleichende Emergenzstudien an einem naturnahen und einem naturferner Abschnitt des Fliessge-wässersystems der Fühlenau in Schleswig-Holstien. *Internationale Revue der Gesamten Hydrobiologie*, **72**, 339-68.

Bray, J. and Curtis, J. (1957) An ordination of the upland forest communities of southern Wisconsin. *Ecological Monographs*, **27**, 325-49.

Brennan, A., Walentowicz, A.T. and McLachlan, A.J. (1981) Midges (Diptera: Chironomidae) from the upper reaches of a spate river. *Hydrobiologia*, **78**, 147-51.

Bretschko, G. (1974) The chironomid fauna of a high-mountain lake (Vorderer Finstertaler See, Tirol, Austria, 2237m asl.) *Entomologisk Tidskiftr, Supplement*, **95**, 22-33.

Bretschko, G. (1991) The limnology of a low order alpine gravel stream (Ritrodat-Lunz study area, Austria. *Verhandlungen der Internationale Vereinigung für Theoretische und Angewandte Limnologie*, **24,** 1908-12.

Briggs, S.V. and Maher, M.T. (1985) Limnological studies of waterfowl habitat in South-western New South Wales. II. Aquatic macrophyte productivity. *Australian Journal of Marine and Freshwater Research*, **36,** 707-15.

Brinkhurst, R.O. (1974) *The Benthos of Lakes*. The Macmillan Press Ltd, London, England.

Brittain, J.E. and Eikeland, T.J. (1988) Invertebrate drift - a review. *Hydrobiologia*, **166,** 77-93.

Brock, T.C.M., Crum, S.J.H., van Wijngaarden, R., Budde, B.J., Tijink, J., Zuppelli, A. and Leeuwangh, P. (1992) Fate and effects of the insecticide Dursban 4E in indoor *Elodea* - dominated and macrophyte-free freshwater model ecosystems: I. Fate and primary effects of the active ingredient chlorpyrifos. *Archives of Environmental Contamination and Toxicology*, **23,** 69-84.

Brooker, M.P. and Morris, D.L. (1980) A survey of the macroinvertebrate riffle fauna of the River Wye. *Freshwater Biolology*, **10,** 437-58.

Brown, A.E. (1983) A preliminary study of the Chironomidae (Diptera) from a stream in Northern Nigeria. *Memoirs of the American Entomological Society*, **34,** 35-46.

Brundin, L. (1947) Zur Kenntnis der schwedischen Chironomiden. *Arkiv för Zoologie*, **39A,** 1-95.

Brundin, L. (1948) Über die Metamorphose der Sectio Tanytarsariae Connectentes (Diptera, Chironomidae). *Arkiv für Zoologie*, **41A,** 1-22.

Brundin, L. (1949) Chironomiden und andere Bodentiere der südschwedischen Urgebirgsseen. *Report of the Institute of Freshwater Research Drottningholm*, **30,** 1-914.

Brundin, L. (1951) The relation of O_2-microstratification at the mud surface to the ecology of the profundal bottom fauna. *Report of the Institute of Freshwater Research Drottningholm*, **32,** 32-43.

Brundin, L. (1956) Zur Systematik der Orthocladiinae. *Report of the Institute of Freshwater Research, Drottningholm*, **37,** 5-185.

Brundin, L. (1966) Transantarctic relationships and their significance, as evidenced by chironomid midges with a monograph of the subfamilies Podonominae and Aphroteniinae and the austral Heptagyiae. *Kunglica Svenska Vetenskapsakademiens Handlingar*, **11,** 1-472.

Brundin, L. and Sæther, O.A. (1978) *Buchonomyia burmanica* sp. n. and Buchonomyiinae, a new subfamily among the Chironomidae (Diptera). *Zoologica Scripta*, **7,** 269-75.

Brundin, L. (1956) Die bodenfaunistischen Seetypen und ihre Anwendbarkheit auf die Südhalkugel. Zugleich ein Theorie der produktionbiologischen Bedeutung der glazialen Erosion. *Report of the Institute of Freshwater Research Drottningholm*, **37,** 186-235.

Bunn, S.E., Edward, D.H. and Loneragan, N.R. (1986) Spatial and temporal variation in the macroinvertebrate fauna of streams of the northern jarrah forest, Western Australia: community structure. *Freshwater Biology*, **16,** 67-91.

Burfield, I. and Williams, D.N. (1975) Control of parthenogenetic chironomids with pyrethrins. *Water Treatment and Examination*, **24,** 57-67.

Butler, M.G. and McMillan, A.M. (1990) Cohort structure and voltinism in two profundal *Chironomus* populations. *Verhandlungen der Internationale Vereinigung für Theoretische und Angewandte Limnologie*. **24,** 438-44.

Cadwallader, P.L. and Eden, A.K. (1977) Effect of a total solar eclipse on invertebrate drift in Snobs Creek, Victoria. *Australian Journal of Marine and Freshwater Research*, **28,** 799-805.

Cairns, J. Jr. and Pratt, J.R. (1993) A history of biological monitoring using benthic macroinvertebrates, in *Freshwater Biomonitoring and Benthic Macroinvertebrates*, (eds D.M. Rosenberg and V.H. Resh), Chapman and Hall, New York, pp.10-27.

Cairns, J. Jr., Crossman, J.S., Dickson, K.L. and Herricks, E.E. (1971) The recovery of damaged streams. *Association of Southeastern Biologists*, **18,** 79-106.

Cairns, J. Jr. (1974) Indicator species versus the concept of community structure as an index of water pollution. *Water Resources Bulletin*, **10,** 338-47.

Caldwell, B.A. and Parrish, F.K. (1987) Chironomidae (Diptera) of small Georgia streams. *Entomologica Scandinavica Supplement*, **29,** 375-9.

Candea-Cure, V. (1971) Contributions à la connaissance de la fauna de chironomidés (larves) du bassin de la Rivière Bistrita dans la zone du Lac de Barrage Bicaz avant, au début et six ans après la mise en eu. *The Canadian Entomologist*, **103,** 424-31.

Canton, S.P., Cline, L.D., Short, R.A. and Ward, J.V. (1984) The macroinvertebrates and fish of a Colorado U.S.A. stream during a period of fluctuating discharge. *Freshwater Biology*, **14,** 311-6.

Cantrell, M.A. and McLachlan, A.J. (1977) Competition and chironomid distribution patterns in a newly flooded lake. *Oikos*, **29,** 429-33.

Caquet, T., Thybaud, E., Le Bras, S., Jonot, O. and Ramade, F. (1992) Fate and biological effects of lindane and deltamethrin in freshwater mesocosms. *Aquatic Toxicology* **23,** 261-78.

Casas, J.J. and Vilchez-Quero, A. (1989) A faunistic study of the lotic chironomids (Diptera) of the Sierra Nevada (S.E. of Spain): changes in the structure and composition of the populations between spring and summer. *Acta Biologica Debrecina Supplementum Oecologica Hungarica*, **3,** 83-94.

Casas, J.J. and Vilchez-Quero, A. (1993) Altitudinal distribution of lotic chironomid (Diptera) communities in the Sierra Nevada mountains (Southern Spain). *Annales de Limnologie*, **29,** 175-87.

Caspers, N. (1983a) Chironomiden-Emergenz zweier Lunzer Bäche, 1972. *Archiv für Hydrobiologie Supplement*, **65,** 484-549.

Caspers, N. (1983b) Die Chironomiden der Oberen Alz. *Nachrichtenblatt der Bayerischen Entomologen*, **32,** 97-108.

Caspers, N. (1983) Chironomiden-Emergenz zweier Lunzer Bäche, 1972. *Archiv für Hydrobiologie Supplementbande*, **65,** 484-549.

Caspers, N. (1983) Sukzessionsanalyse des Makrozoobenthos eines neu angelegten stehenden Gewässers. *Archiv für Hydrobiologie Supplementbande*, **65,** 300-70.

Caspers, N. (1991) The actual biocoenotic zonation of the River Rhine exemplified by the chironomid midges (Insecta, Diptera). *Verhandlungen der Internationale Vereinigung für Theoretische und Angewandte Limnologie*, **24,** 1829-34.

Chandler, J.R. (1970) A biological approach to water quality management. *Water Pollution Control*, **69,** 415-21.

Cheong, M.F. (1984) *A sex-influenced protein in* Chironomus *- detection and characterization*. M.Sc. Thesis, University of Melbourne.

Chessman, B.C. (in review) Rapid river assessment using macroinvertebrates: a procedure based on habitat-specific sampling, family level identification, and a biotic index.

Clarke, G.M. (1993a) Fluctuating asymmetry of invertebrate populations as a biological monitor of environmental quality. *Environmental Pollution*, **82,** 207-11.

Clarke, G.M. (1993b) Patterns of developmental stability in *Chrysopa perla* L. (Neuroptera: Chrysopidae) in response to environmental pollution. *Environmental Entomology*, **22,** 1362-6.

Clarke, G.M. (1994) Developmental stability analysis: an early-warning system for biological monitoring of water quality. *Australian Biologist*, **7,** 94-104.

Clarke, G.M. (1995a) Relationships between developmental stability and fitness: application for conservation biology. *Conservation Biology*, (in press).

Clarke, G.M. (1995b) The genetic basis of developmental stability. III. A character-based model of stability and some consequences. *Philosophical Transactions of the Royal Society of London* B (in press).

Clarke, G.M., Brand, G.W., and Whitten, M.J. (1986) Fluctuating asymmetry: a technique for measuring development stress caused by inbreeding. *Australian Journal of Biological Science*, **39,** 145-53.

Clement, S.L., Grigarick G.G. and Way, H.O. (1977) The colonisation of Californian rice paddies by chironomid midges. *Journal of Applied Ecology*, **14,** 379-89.

Coffman, W.P. (1973) Energy flow in a woodland stream ecosystem. II. The taxonomic composition and phenology of the Chironomidae as determined by the collection of pupal exuviae. *Archiv für Hydrobiologie*, **71,** 281-322.

Coffman, W.P. (1989) Factors that determine the species richness of lotic communities of Chironomidae. *Acta Biologica Debrecina Supplementum Oecologica Hungarica*, **3,** 95-100.

Coffman, W.P., de la Rosa, C., Cummins, K.W. and Wilzbach, M.A. (1993) Species richness in some Neotropical (Costa Rica) and Afrotropical (West Africa) lotic communities of Chironomidae (Diptera). *Netherlands Journal of Aquatic Ecology*, **26,** 229-37.

Coffman, W.P., P.S. Cranston, D.R. Oliver and Sæther, O.A. (1986) The pupae of Orthocladiinae (Diptera: Chironomidae) of the Holarctic region - Keys and diagnoses. *Entomologica Scandinavica Supplement*, **28,** 147-296.

Cohen, J. (1988) *Statistical Power Analysis for the Behavioral Sciences* (2nd edition). Hillsdale, N.J. Lawrence Erlbaum.

Coleman, P.J. (1980) Plate tectonics background to biogeographic development in the southwest Pacific over the last 100 million years. *Palaeography, Palaeoclimatology, Palaeoecology*, **31,** 105-21.

Collier, K.J. (1993) Flow preferences of larval Chironomidae (Diptera) in Tongariro River, New Zealand. *New Zealand Journal of Marine and Freshwater Ecology*, **27,** 219-26.

Collier, K.J. and Wakelin, M.D. (1992) Drift of aquatic macroinvertebrate larvae in Manganuiateao River, Central North Island, New Zealand. *New Zealand Natural Sciences*, **19,** 15-26.

Corrarino, C. A. and Brusven, M. A. (1983) The effects of reduced stream discharge on insect drift and stranding of near shore insects. *Freshwater Invertebrate Biology*, **2,** 88-98.

Courtemanch, D. (1987) Trophic classification of Maine lakes using benthic chironomid fauna. Presented at 7th International Symposium, North American Lake Management Society, Orlando Florida, Nov. 1987. pp.20

Cowie, B. (1983) Macroinvertebrate taxa from a southern New Zealand montane stream continuum. *New Zealand Entomologist*, **7,** 439-47.

Cranston, P.S. (1982) A key to the larvae of the British Orthocladiinae (Chironomidae). *Scientific Publications of the Freshwater Biological Association*, **45,** 1-152.

Cranston, P.S. (1987) *Compterosmittia* Sæther (Diptera: Chironomidae) - the immature stages and revisionary notes. Abstract in *Bulletin of the North American Benthological Society*, **4** (1), 116.

Cranston, P.S. (1990) Biomonitoring and invertebrate taxonomy. *Environmental Monitoring and Assessment*, **14,** 265-73.

Cranston, P.S. (1991) *Immature Chironomidae of the Alligator Rivers Region*. Open File Report 82. Supervising Scientist for the Alligator Rivers Region, pp.269

Cranston, P.S. and Armitage, P. (1988) The types of Canary Isles Chironomidae described by Becker and Santos Abreu. *Deutsche Zeitschrift für Entomologie*, **35,** 341-54.

Cranston, P.S. and Disney, R.L. (1978) [Major habitats -] Standing waters, in *A Dipterist's Handbook* (eds A. Stubbs and P. Chandler), *The Amateur Naturalist*, **15,** 111-5.

Cranston, P.S. and Edward, D.H.D. (1992) A systematic reappraisal of the Australian Aphroteniinae (Diptera: Chironomidae) with dating from variance biogeography. *Systematic Entomology*, **17,** 41-54.

Cranston, P.S. and Martin, J. (1989) 'Family Chironomidae' in *Catalog of the Diptera of the Australasian and Oceanian*

Regions (ed. N.L. Evenhuis), Bishop Museum Press, Honolulu and E.J. Brill, Leiden. pp.252-74.

Cranston, P.S. and Oliver, D.R. (1988) Additions and corrections to the Nearctic Orthocladiinae (Diptera: Chironomidae). *Canadian Entomologist*, **120**, 425-62.

Cranston, P.S. and Reiss, F. (1983) The larvae of Chironomidae (Diptera) of the Holarctic region - key to subfamilies. *Entomologica Scandinavica Supplement*, **19**, 11-5.

Cranston, P.S., Edward, D.H.D. and Colless, D.H. (1987) *Archaeochlus* Brundin: a midge out of time (Diptera: Chironomidae). *Systematic Entomology*, **12**, 313-34.

Cranston, P.S., Oliver, D.R. and Sæther, O.A. (1983) The larvae of Orthocladiinae (Diptera: Chironomidae) of the Holarctic region - keys and diagnoses. *Entomologica Scandinavica Supplement*, **19**, 149-291.

Cranston, P.S., Oliver, D.R. and Sæther, O.A. (1989) The adult males of Orthocladiinae (Diptera: Chironomidae) of the Holarctic region - Keys and diagnoses. *Entomologica Scandinavica Supplement*, **34**, 165-352.

Cranston, P.S., Dillon, M.E., Pinder, L.C.V. and Reiss, F. (1989) The adult males of Chironominae (Diptera: Chironomidae) of the Holarctic Region - keys and diagnoses. *Entomologica Scandinavica Supplement*, **34**, 353-502.

Creekmur, G.D., M.P. Russell, M.P. and J.E. Hazelrigg. (1982) Field evaluation of the effects of slow-release wettable powder formulation of Altosidr on nontarget organisms. *Proceedings and Papers of the Annual Conference of the California Mosquito and Vector Control Association*, 95-7.

Crossland, N.O. and Wolff, C.J.M. (1988) Outdoor ponds: Their construction, management, and use in experimental ecotoxicology. *The Handbook of Environmental Chemistry*, **2**, 51-69.

Crossland, N.O. and La Point, T.W. (1992) The design of mesocosm experiments. *Environmental Toxicology and Chemistry*, **11**, 1-4.

Crossland, N.O., Heimbach, F., Hill, I.R., Boudou, A., Leeuwangh, P., Matthiessen, P. and Persoone, G. (1993) Summary and recommendations of the European Workshop on Freshwater Field Tests, Potsdam (Germany), June 25-26, 1992, in *Freshwater Field Tests for Hazard Assessment of Chemicals* (eds I.R. Hill, F. Heimbach, P. Leerwangh, and P. Matthiessen), Lewis Publishers, Michigan (in press).

Culp, J.M. and Davies, R.W. (1985) Responses of benthic macroinvertebrate species to manipulation of interstitial detritus in Carnation Creek, British Columbia. *Canadian Journal of Fisheries and Aquatic Science*, **42**, 139-46.

Cummins, K.W. (1962) An evaluation of some techniques for the collection and analysis of benthic samples with special emphasis on lotic waters. *American Midland Naturalist*, **67**, 477-504.

Cure, V. (1964) Beiträge zur Kenntniss der Tendipeddiden (Larven) im rumänishen Donaugebiet. *Archiv für Hydrobiologie Supplement*, **27**, 418-41.

Currey, J. (1993) *Water Quality of Blue Mountains Streams 1992*. Australian Water Technologies, Science and Environment, Sydney. Report No. 93/14.

Cushman, R.M. (1984) Chironomid deformities as indicators of pollution from a synthetic, coal-derived oil. *Freshwater Biology*, **14**, 179-82.

Darby, R.E. (1962) Midges associated with California rice fields, with special reference to their ecology (Diptera: Chironomidae). *Hilgardia*, **32**, 1-206.

Davies, L.J. and Hawkes, H.A. (1981) Some effects of organic pollution on the distribution and seasonal incidence of Chironomidae in riffles in the River Cole. *Freshwater Biology*, **11**, 549-59.

Day, R.W. and Quinn, G.P. (1989) Comparisons of treatments after an analysis of variance in ecology. *Ecological Monographs*, **59**, 433-63.

De Pauw, N. and Vanhooren, G. (1983) Method for biological quality assessment of watercourses in Belgium. *Hydrobiologia*, **100**, 153-68.

De Pauw, N., Ghetti, P.F., Manzini, D.P. and Spaggiari, D.R. (1993) Biological assessment methods for running water, in *River Water Quality, Ecological Assessment and Control* (eds P.J. Newman, M.A. Piavaux, and R.A. Sweeting), Commission of the European Communities, EUR 14606 En-Fr, pp.217-48.

Delettre, Y.R. (1988) Chironomid wing length, dispersal ability and habitat predictability. *Holarctic Ecology*, **11**, 166-70.

DeNoyelles, F., Kettle, W.D., Fromm, C.H., Moffett, M.F. and Dewey, S.L. (1989) Use of experimental ponds to assess the effects of a pesticide on the aquatic environment, in *Using Mesocosms to Assess the Aquatic Ecological Risk of Pesticides: Theory and Practice* (ed. Voshell, J. R.), *Entomological Society of America, Miscellaneous Publications*, **75**, 41-56.

Dermott, R. M. and Paterson, C. G. (1974) Determining dry weight and percentage dry matter of chironomid larvae. *Canadian Journal of Zoology*, **52**, 1243-50.

Dermott, R.M. (1991) Deformities in larval *Procladius* spp. and dominant Chironomini from the St Clair River. *Hydrobiologia* **219**, 171-85.

Devai, Gy., Miscolczi, M. and Wülker, W. (1989) Standardization of arms B, C and D in *Chironomus*. *Acta Biologica Debrecina Supplementum Oecologica Hungarica*, **2**, 79-92.

Devai, Gy., Wülker, W, and Scholl, A. (1983) Revision der Gattung *Chironomus* Meigen (Diptera). IX: *C. balatonicus* sp. n. aus dem Flachsee Balaton (Ungarn). *Acta Zoologica Academiae Scientiarum Hungaricae* **29**, 357-74.

Dewey, S.L. (1986) Effects of the herbicide atrazine on aquatic insect community structure and emergence. *Ecology*, **67**, 148-162.

Dickman, M., Brindle, I. and Benson, M. (1992) Evidence of teratogens in sediments of the Niagara River watershed as reflected by chironomid (Diptera: Chironomidae) deformities. *Journal of Great Lakes Research*, **18**, 467-80.

Dickman, M., Lan, Q and Mathews, B. (1989) Teratogens in the Niagara River watershed as reflected by chironomid (Diptera: Chironomidae) labial plate deformities. *Water Pollution Research Journal of Canada*, **24,** 47-79.

Dittmar, H. (1955) Ein Sauerlandbach. Untersuchungen an einem Wiesen-Mittelgebirgsbach. *Archiv für Hydrobiologie*, **50,** 305-552.

Dodson S.I. (1987) Animal assemblages in temporary desert rock pools: aspects of the ecology of *Dasyhelea subletti* (Ceratopogonidae). *Journal of the North American Benthological Society*, **6,** 65-71

Dorier, A. (1933) Sur la biologie et les metamorphoses de *Psectrocladius obvius* Walk. *Travaux du laboratoire d'hydrobiologie et de pisciculture de l'Université de Grenoble*, **25,** 135-45.

Dostine, P.L., Humphrey, C.L. and Spiers, A.G. (1992) Benthic Macroinvertebrate Communities in Rockhole Mine Creek: Review of 1991 data. *Office of the Supervising Scientist, Sydney, Internal Report 72.*

Dostine, P.L., Humphrey, C.L. and Spiers, A.G. (1993) *Benthic Macroinvertebrate Communities in Rockhole Mine Creek: Review of 1992 data.* Internal Report 116.

Dowling, C., O'Connor, O'Grady, M.F. and Clynes, E. (1981a) A baseline survey of the Caragh, an polluted river in southwest Ireland: Topography and water chemistry. *Journal of Life Science Royal Dublin Society*, **2,** 137-45.

Dowling, C., O'Connor and O'Grady, M.F. (1981b) A baseline survey of the Caragh, a polluted river in southwest Ireland: Observations on the macroinvertebrates. *Journal of Life Science Royal Dublin Society*, **2,** 147-59.

Downes, J.A. (1974) The feeding habits of adult Chironomidae. *Entomologisk Tidskrift Supplement*, **95,** 84-90.

Downes, B.J., Lake, P.S. and Schreiber, E.S.G. (1993) Spatial variation in the distribution of stream invertebrates: implications of patchiness for models of community organization. *Freshwater Biology*, **30,** 119-1323.

Drake, C.M. (1985), Emergence patterns of Diptera in a chalk stream. *Aquatic Insects*, **2,** 97-110.

Edward, D.H.D. (1968) Chironomidae in temporary freshwaters. *Australian Society for Limnology Newsletter*, **6,** 3-5.

Edward, D.H.D. (1986) Chironomidae (Diptera) of Australia, in *Limnology in Australia* (eds W.D. Williams and P. De Dekker), C.S.I.R.O., Melbourne. pp.153-73.

Edwards, F.W. (1929) British non-biting midges (Diptera, Chironomidae). *Transactions of the Entomological Society of London*, **77,** 279-439.

Edwards, R.W., Egan, H., Learner, M.A. and Maris, P.J. (1964) The control of chironomid larvae in ponds, using TDE(DDD). *Journal of Applied Ecology*, **1,** 97-117.

Einarsson, A. (1982) The palaeolimnology of Lake Myvatn, northern Iceland: plant and animal microfossils in the sediment. *Freshwater Biology*, **12,** 63-82.

Einarsson, A. and Haflidason, H. (1988) Predictive palaeolimnology: effects of sediment dredging in Lake Myvatn, Iceland. *Verhandlungen der Internationalen Vereinigung für Theoretische und Angewandte Limnologie*, **23,** 860-9.

Elgmork, K. and Sæther, O. A. (1970) Distribution of invertebrates in a high mountain brook in the Colorado Rocky Mountains. *University of Colorado Stududies Series in Biolology* **31,** 1-55.

Elliott, J.M. (1967) Invertebrate drift in a Dartmoor stream. *Archiv für Hydrobiologie*, **63,** 202-37.

Elliott, J.M. (1977) Some methods for the statistical analysis of samples of benthic invertebrates. *Scientific Publications of the Freshwater Biological Association*, **25,** 1-156.

Epler, J.H. (1988) Biosystematics of the genus *Dicrotendipes* Kieffer, 1913 (Diptera: Chironomidae) of the world. *Memoirs of the American Entomological Society*, **36,** 1-214.

Epler, J.H. (1988) A reconsideration of the genus *Apedilum* Townes, 1945 (Diptera: Chironomidae). *Spixiana Supplement*, **14,** 105-16

Ertlová, E. (1963) Zur Kenntnis der Zuckmücken (Chironomidae) der Donau. *Biologia, Bratislava*, **18,** 612-20.

Fabb, S.A., Cheong, M.F., Lee, B.T.O. and Martin, J. (1990) A sex-influenced protein in *Chironomus* (Diptera). *Insect Biochemistry*, **20,** 135-40.

Fahy, E. and Murray, D.A. (1972) Chironomidae from a small stream system in Western Ireland with a discussion of species composition of the group (Dipt.). *Entomologisk Tidskrift*, **93,** 148-55.

Faith, D.P. (1990) *Submission to the Kakadu Enquiry (90/147)*, Public Enquiry into Mining in Kakadu National Park, AGPS, Canberra.

Faith, D.P., Humphrey, C.L. and Dostine, P.L. (1991) Statistical power and BACI design in biological monitoring: comparative evaluation of measures of community dissimilarity based on benthic macroinvertebrate communities in Rockhole Mine Creek, Northern Territory, Australia. *Australian Journal of Marine and Freshwater Research*, **42,** 589-602.

Faith, D.P., Dostine, P.L. and Humphrey, C.L. (in press). Detection of mining impacts on aquatic macroinvertebrate communities: results of a disturbance experiment and the design of a multivariate BACIP monitoring program at Coronation Hill, N.T. Australian Journal of Ecology

Fechney, L.R. (1988) The summer diet of brook trout (*Salvelinus fontinalis*) in a South Island high-country stream. *New Zealand Journal of Marine and Freshwater Research*, **22,** 163-8.

Ferrington, C.L.J. (1984) Drift dynamics of Chironomidae larvae: I.Preliminary results and discussion of importance of mesh size and level of taxonomic identification in resolving Chironomidae diel drift patterns. *Hydrobiologia*, **114,** 215-27.

Ferrington, L.C. Jr., Blackwood, M.A., Wright, C.A., Anderson, T.M. and Goldhammer, D.S. (1994) Sediment transfers and representativeness of mesocosm test fauna, in *Aquatic Mesocosm Studies in Ecological Risk Assessment* (eds R.L. Graney, J.H. Kennedy and J.H. Rogers, Jr), Lewis Publishers, Boca Raton, pp.179-200.

Finch, P.D. (1973) Change in small time series. *Scima,* **2,** 53-62.

Fittkau, E.J. (1955) *Buchonomyia thienemanni* n. gen., n. sp. Chironomidenstudien IV (Diptera: Chironomidae). *Beiträge zur Entomologie,* **5,** 403-14.

Fittkau, E.J. (1962) Die Tanypodinae (Diptera, Chironomidae). Die Tribus Anatopyniini, Macropelopiini und Pentaneurini. *Abhandlung der Larvalsystematik Insekten* **6,** 1-453.

Fittkau, E.J. and Reiss, F. (1979) Die Zoogeographische Sonderstellung der neotropischen Chironomiden (Diptera). *Spixiana,* **2,** 273-80.

Fittkau, E.J. and Murray, D.A. (1986) The pupae of Tanypodinae (Diptera: Chironomidae) of the Holarctic Region - Keys and diagnoses. *Entomologica Scandinavica Supplement,* **28,** 31-113.

Fittkau, E.J. and Roback, S.S. (1983) The larvae of Tanypodinae (Diptera: Chironomidae) of the Holarctic Region - Keys and diagnoses. *Entomologica Scandinavica Supplement,* **19,** 34-110.

Flecker, A.S. (1992) Fish predation and the evolution of invertebrate drift periodicity: evidence from Neotropical streams. *Ecology,* **73,** 438-48.

Flentje, M. (1945) Elimination of midge fly larvae with DDT. *Journal American Water Works Association,* **37,** 1053.

Frank, J. and Lounibos, L.P. (1983) *Phytotelmata: Terrestrial Plants as Hosts for Aquatic Insect Communities* (eds J. Frank and L.P. Lounibos), Plexus Publishing, Marlton, New Jersey,

Freeman, P. (1950) A species of chironomid (Diptera) from the Sudan suspected of causing asthma. *Proceedings of the Royal Entomological Society of London Series B Taxonomy,* **19,** 55-9.

Freeman, P. (1956) A study of the Chironomidae (Diptera) of Africa south of the Sahara. Part II. *Bulletin of the British Museum of Natural History (Entomology),* **4,** 287-368.

Freeman, P. (1959) A study of the New Zealand Chironomidae (Diptera, Nematocera). *Bulletin of the British Museum (Natural History), Entomology,* **7,** 393-437.

Freeman, P. (1961) The Chironomidae (Diptera) of Australia. *Australian Journal of Zoology,* **9,** 611-737.

Freeman, P. and Cranston, P.S. (1980) Family Chironomidae, in *Catalogue of the Diptera of the Afrotropical Region',* (ed. R.W. Crosskey), British Museum (Natural History), London, pp.175-202.

Frenzel, (1990) The influence of chironomid larvae on sediment oxygen microprofiles. *Archiv für Hydrobiologie,* **119(4),** 427-37.

Friedrich, G and Meier-Drook, C. (1987) The macrofauna of floating and submerged macrophytes in pools near Tuebingen (SW Germany). *Carolinea,* **45,** 99-106.

Friedrich, V.G. (1990) Eine Revision des Saprobiensystems. *Zeitshrift für Wasser-Abwasser-Forschung,* **23,** 141-52.

Garcia-Gil, J. (1990) *Bacteris Fototròfics i Cicle del Ferro a l'Estany de Banyoles.* Ph.D. Thesis, Autonomous University of Barcelona, pp.295

Gardarsson, A. and Einarsson, A. (eds) (1991) *Nattura Myvatns.* Hid islenska natturufraedifelag, Reykjavik.

Gardarsson, A. and Einarsson, A. (1994) Responses of breeding duck populations to changes in food supply. *Hydrobiologia,* **279/280,** 15-27.

Gardarsson, A. and Snorrason, S.S. (1993) Sediment characteristics and density of benthos in Lake Myvatn, Iceland. *Verhandlungen der Internationalen Vereinigung für Theoretische und Angewandte Limnologie,* **25,** 452-7.

Gardarsson, A., Gislason, G.M. and Einarsson, A. (1989) Long term changes in the Lake Myvatn ecosystem. *Aqua Fennica,* **18,** 125-35.

Gardarsson, A., Olafsson, J.S., Hrafnsdottir, Th., Gislason, G.M. and Einarsson, A. (1995) Monitoring chironomid numbers at Myvatn, Iceland: the first sixteen years. *Chironomids: from Genes to Ecosystems,* (ed. P.S. Cranston), CSIRO, Melbourne, 141-54.

Gauch, H.G. Jr. (1982) *Multivariate Analysis In Community Ecology.* Cambridge University Press, London.

Gauss, J.D., Woods, P.E., Winner, R.W. and Skillings, J.H. (1985) Acute toxicity of copper to three life stages of *Chironomus tentans* as affected by water hardness - alkalinity. *Environmental Pollution Series* **37,** 149-57.

Gay, C. (1982) Les communautés benthiques d'un torrent des Alpes françaises: l'Eau d'Olle (Isère). *Travaux du Laboratoire d'Hydrobiologie et de Pisciculture de Grenoble,* **71-73,** 7-31.

Gazagnes, G. and Laville, H. (1985) Etude faunistique des chironomidés (Diptera) de la Haute Neste d'Aure (Pyrénées Centrale): impact des aménagements hydroélectriques. *Annales de Limnologie,* **21,** 149-59.

Gerstmeier, R. (1989) Lake typology and indicator organisms in application to the profundal chironomid fauna of Starnberger See (Diptera, Chironomidae). *Archiv für Hydrobiologie,* **116,** 227-34.

Gislason, G.M. (1985) The life cycle and production of *Simulium vittatum* Zett. in the River Laxa, North-East Iceland. *Verhandlungen der Internationalen Vereinigung für Theoretische und Angewandte Limnologie,* **22,** 3281-7.

Gislason, G.M. (1994) River management in cold regions: A case study of the River Laxa, North Iceland, in *The Rivers Handbook* (eds P. Calow and G.E. Petts), Blackwell Scientific Publications, Oxford, pp.464-83.

Gislason, G.M. and Johannsson, V. (1991) Effects of food and temperature on the life cycle of Simulium vittatum Zett. (Diptera: Simuliidae) in the River Laxa, N-Iceland. *Verhandlungen der Internationalen Vereinigung für Theoretische und Angewandte Limnologie,* **24,** 2912-6.

Gislason, G.M. and Gardarsson, A. (1988) Long term studies on *Simulium vittatum* Zett. (Diptera: Simulidae) in the River Laxa, North Iceland, with particular reference to different methods used in assessing population changes. *Verhandlungen der Internationalen Vereinigung für Theoretische und Angewandte Limnologie,* **23,** 2179-88.

Gislason, G.M., Hrafnsdottir, Th. and Gardarsson, A. (1994) Long term monitoring of numbers of Chironomidae and Simuliidae in the River Laxa, North Iceland. *Verhandlungen der Internationalen Vereinigung für*

Theoretische Angewandte Limnologie, **25**, (1492-5).

Gladden, J.E. and Smock, L.A. (1990) Macroinvertebrate distribution and production on the floodplains of two lowland headwater streams. *Freshwater Biology*, **24**, 533-45.

Glova, G.J. and Sagar, P.M. (1989) Prey selection by *Galaxias vulgaris* in the Hawkins River, New Zealand. *New Zealand Journal of Marine and Freshwater Research*, **23**, 153-61.

Glova, G.J., Sagar, P.M. and Docherty, C.R. (1987) Diel feeding periodicity of torrentfish (*Cheimarrichthys fosteri*) in two braided rivers of Canterbury, New Zealand. *New Zealand Journal of Marine and Freshwater Research*, **21**, 555-61.

Glover, H.G. (1973) Acidic and ferruginous mine drainages, in *The Ecology of Resource Degradation and Renewal* (eds M.J. Chadwick and G.T. Goodman), Blackwell Scientific Publications, London, pp.173-95.

Goodman, M., Pedwaydon, J., Czelusniak, J., Susuki, T., Gotoh, T., Moens, L. Shishikura, F., Walz, D and Vinogradov, S. (1988) An evolutionary tree for invertebrate globin sequences. *Journal of Molecular Evolution*, **27**, 236-49.

Goonan, P.M., Beer, J.A., Thompson, T.B. and Suter, P.J. (1992) Wetlands of the River Murray flood plain, South Australia. 1. Preliminary survey of the biota and physicochemistry of ten wetlands from Chowilla to Mannum. *Transactions of the Royal Society of South Australia*, **116**, 81-94.

Graesser, A. and Lake, P.S. (1984) Diel changes in the benthos of stones and of drift in a southern Australian upland stream. *Hydrobiologia*, **111**, 153-60.

Growns, J.E., Chessman, B.C., McEvoy, P.K. and Wright, I. A. (1995) Rapid river assessment using macroinvertebrates: case studies in the Nepean River and Blue Mountains, N.S.W. (In review.)

Grzybkowska, M. (1989) Production estimates of the dominant taxa of Chironomidae (Diptera) in the modified, River Widawka, and the natural, River Grabia, Central Poland. *Hydrobiologia*, **179**, 245-59.

Grzybkowska, M., Hejduk, J. and Zielinski, P. (1990) Seasonal dynamics and production of Chironomidae in a large lowland river upstream and downstream from a new reservoir in Central Poland. *Archiv für Hydrobiologie*, **119**, 439-55.

Guchte, C. van de and Van Leeuwen, C.J. (1988) Sediment pollution, in *Manual on Aquatic Ecotoxicology* (eds H.A. de Kruijf, de Zwart, P.N. Viswanathan, and P.K. Ray), Proceedings of the Indo-Dutch training course on aquatic ecotoxicology, Lucknow, India, pp.180-91.

Haines, T.A. (1981) Acid precipitation and its consequences for aquatic systems: a review. *Transactions of the American Fisheries Society*, **110**, 669-707.

Hall, J.C. (1978) Behavioral analysis in *Drosophila* mosaics. *Results and Problems in Cell Differentiation* (ed. W. Gehring), Springer Verlag **9**, 259-305.

Hall, K.J. and Bindra, K.S. (1979) Geochemistry of selected metals in sediments and factors affecting organism concentration, in *Management and Control of Heavy Metals in the Environment*, C.E.P. consultants, Edinburgh, pp.337-40.

Hall, R.O. and Edwards, A.J. (1978) Observations on the settling of *Simulium damnosum* larvae on artificial substrates in the Ivory Coast. *Hydrobiologia*, **57**, 81-4.

Hamilton, A.L. (1969) On estimating annual production. *Limnology and Oceanography*, **14**, 771-82.

Hamilton, A.L. and Sæther, O.A. (1971) The occurrence of characteristic deformities in the chironomidae larvae of several Canadian lakes. *Canadian Entomologist*, **103**, 363-8.

Hamilton, A.L., Sæther, O.A. and Oliver, D.R. (1969) A classification of the Nearctic Chironomidae. *Technical Report of the Fisheries Research Board of Canada*, **124**, 1-42.

Hankeln, T., Keyl, H.-G., Ross, R. and Schmidt, E.R. (1993) Evolution of histone gene loci in chironomid midges. *Genome*, **36**, 852-62.

Hanson, M.A. and Butler, M.G. (1994) Responses of plankton, turbidity, and macrophytes to biomanipulation in a shallow prairie lake. *Canadian Journal of Fisheries and Aquatic Sciences* **51** (in press).

Hardwick, R.A. (1993) *Tests on the Utility of Cast Pupal Exuviae of Chironomidae (Diptera) in Biological Monitoring of the Alligator Rivers Region*, Open file record No. OFR 106, ARRRI: Office of the Supervising Scientist, Sydney.

Hardwick, R.A., Wright, I.A., Chessman, B.C. and Holleley, D.E. (1995) Rapid Biological Assessment of streams in the Blue Mountains, Australia: Characteristics of the Chironomidae (Diptera) fauna, in *Chironomids: from Genes to Ecosystems* (ed. P.S. Cranston), CSIRO, Melbourne, pp.155-60.

Hare, L. and Carter, J.C.H. (1976) The distribution of *Chironomus* (s.s.)? *cucini* (*salinarius* group) larvae (Diptera: Chironomidae) in Parry Sound, Georgian Bay, with particular reference to structural deformities. *Canadian Journal of Zoology*, **54**, 2129-34.

Harnish, O. (1922) Zur Kenntnis der Chironomidenfauna der Brassenregion schlesischer Flüsse. *Archiv für Hydrobiologie*, **14**, 125-43

Harvey, M.S. (1990) A review of the water mite family Anisitsiellidae in Australia (Acarina). *Invertebrate Taxonomy*, **3** (6), 629-46.

Hastreiter, J. (1993) Zebra mussels subdued; flow boosted 300 percent. *Journal, American Water Works Association*, **85**, 121-2.

Hatakeyama, S. (1987) Chronic effects of Cd on reproduction of *Polypedilum nubifer* (Chironomidae) through water and food. *Environmental Pollution*, **48**, 249-61.

Hatakeyama, S. (1988) Chronic effects of Cu on reproduction of *Polypedilum nubifer* (Chironomidae) through water and food. *Ecotoxicology and Environmental Safety*, **16**, 1-10.

Hatakeyama, S. and Shiraishi, H. (1991) Chronic effects of waterborne or dietary exposure to a herbicide, chlornitrofen (2,4,6 - trichlorophenyl - 4' nitrophenyl ether) on reproduction of *Polypedilum nubifer*. *Water Research*, **25** (8), 945-51.

Hauer, F.R. and Benke, A.C. (1991) Rapid growth of snag-dwelling chironomids in a blackwater river: the influence of temperature and discharge. *Journal of the North American Benthological Society*, **10,** 154-64.

Hawking, J.H. (1993) *A Preliminary Guide to the Identification of Dragonfly Larvae (Odonata) from the Alligator Rivers Region of the Northern Territory*, Open File Record No. OFR 102, ARRRI: Office of the Supervising Scientist, Sydney.

Hayes, B. P. and Murray, D. A. (1987) Species composition and emergence of Chironomidae (Diptera) from three high arctic streams on Bathurst Island, Northwest Territories, Canada. *Entomologica Scandinavica Supplement*, **29,** 355-60.

Heimbach, F. (1993) Methodologies of aquatic field tests: system design for field tests in still waters, in *Freshwater Field Tests for Hazard Assessment of Chemicals* (eds I.R. Hill, F. Heimbach, P. Leerwangh, and P. Matthiessen), Lewis Publishers, Michigan (in press).

Heinis, F. (1993) *Oxygen as a Factor Controlling Occurrence and Distribution of Chironomid Larvae*. University of Amsterdam, PhD dissertation, pp.155

Heinis, F., Sweerts, J-P., and Loopik, E. (1994) Micro-environment of chironomid larvae in the littoral and profundal zones of Lake Maarsseven I, The Netherlands. *Archiv fur Hydrobiologie*, 130: 52-67.

Hellawell, J.M. (1989) *Biological Indicators of Freshwater Pollution and Environmental Management*, Elsevier Applied Science Publishers, London.

Hely, P.C. (1958) Insect pests of the rice crop. *Agricultural Gazette of New South Wales*, **69,** 29-32.

Hemsworth, R.J. and Brooker, M.P. (1981) Macroinvertebrate drift in the upper Wye catchment, Wales. *Hydrobiologia*, **85,** 145-55.

Hershey, A.E. and Hiltner A.L. (1988) Effect of caddisfly on black fly density: interspecific interactions limit blackflies in an arctic river. *Journal of the North American Benthological Society*, **7**(3), 188-96.

Hienton, T.E. (1974) Summary of investigation of electric insect traps. *Agricultural Research Services and Technical Bulletin*, 1498. U.S. Dept. of Agriculture, Washington, D.C.

Hilfiker-Kleiner, D., Dübendorfer, A., Hilfiker, A. and Nöthiger, R. (1993) Developmental analysis of two sex-determining genes, M and F, in the housefly, *Musca domestica*. *Genetics*, **134,** 1187-94.

Hill, I.R., Heimbach, P., Leeuwangh, P. and Matthiessen, P. (eds.) (1993) *Freshwater Field Tests for Hazard Assessment of Chemicals*, Lewis Publishers, Michigan (in press).

Hill, M.O. (1979) *DECORANA - A FORTRAN Program for Detrended Correspondence Analysis and Reciprocal Averaging*. Cornell University Ithaca, New York.

Hill, M.O. (1979) *TWINSPAN - A FORTRAN program for arranging multivariate data in an ordered two-way table by classification of the individuals and attributes*. Cornell University, Ithaca, NY. pp.49

Hinton H.E. (1951) A new chironomid from Africa, the larva of which can be dehydrated without injury. *Proceedings of the Zoological Society of London*, **121,** 371-80

Hirabayashi, K., Nakazato, R., Nasu, Y., Okino, T. and Murayama, N. (1992) Ecological studies on adult midges of Chironomidae (Diptera)., 1. Resting habit of adult midges of *Tokunagayusurika akamusi*. *Japanese Journal of Environmental Entomology and Zoology*, **4** (2), 71-7 (in Japanese).

Hirabayashi, K., Nakazato, R., Ohara, A. and Okino, T. (1993) A study on phototaxis for adult Chironomidae by artificial light in Lake Suwa, Japan. - Response of adult chironomid midges to near ultraviolet and visible light. *Japanese Journal of Sanitary Zoology*, **44** (1), 33-9.

Hirabayashi, K. (1991a) Studies on the massive flights of chironomid midges (Diptera:Chironomidae) as nuisance insects and plans for their control in the Lake Suwa Area, central Japan. 1. Occurrence of massive flights of *Tokunagayusurika akamusi*. *Japanese Journal of Hygiene*, **46,** 652-61 (in Japanese).

Hirabayashi, K. (1991b) Studies on the massive flights of chironomid midges (Diptera: Chironomidae) as nuisance insects and plans for their control in the Lake Suwa Area, central Japan. 2. Quantitative evaluations of the nuisance of chironomid midges. *Japanese Journal of Hygiene*, **46,** 662-75 (in Japanese).

Hirabayashi, K. (1991c) Studies on the massive flights of chironomid midges (Diptera: Chironomidae) as nuisance insects and plans for their control in the Lake Suwa Area, central Japan. 3. Some experimental trials for control of nuisance midges and proposed counterplans. *Japanese Journal of Hygiene*, **46,** 676-87 (in Japanese).

Hoheiser, H. (1981) *Ausgewählte Kapitel zur Ökologie einiger Oligochaeten species zweier Fischteiche des Waldviertels (N.Ö), des Neuhaslauer Teiches und des Gebhartsteiches*. Dissertation am Zoologisches Institut de Üniversität Wien, 145 pp.

Hudson, P.L. (1978) Unusual larval habitats and life history of Chironomid (Diptera) genera. *Entomologica Scandinavica Supplement*, **29,** 369-73.

Humphrey, C.L., Bishop, K.A. and Brown, V.M. (1990) Use of biological monitoring in the assessment of effects of mining wastes on aquatic ecosystems of the Alligator Rivers Region, Tropical Northern Australia. *Environmental Monitoring and Assessment*, **14,** 139-81.

Hynes, H.B.N. (1960) *The Biology of Polluted Waters*. Liverpool University Press, Liverpool, England. pp.202

Hynes, H.B.N. and Coleman, M.J. (1968) A simple method of assessing the annual production of stream benthos. *Limnology and Oceanography*, **13,** 569-73.

Illies, J. and Botosaneanu, L. (1963) Problèmes et méthodes de la classification et de la zonation écologique des eaux courantes, considerées surtout du point de vue faunistique. *Mitteilungen. Internationalen Vereinigung für Theoretische und Angewandte Limnologie*, **12,** 1-57.

Int Panis, L., Bervoets, L. and Verheyen, R. (1993) The Chironomidae (Diptera) of the "Waelenhoek" nature reserve. *Bulletin et Annales de la Société Royale Belge d'Entomologie*, **129,** 139-43.

Int Panis, L., Goddeeris, B., Bervoets, L. and Verheyen, R. (1992) *Ablabesmyia longistyla* Fittkau, 1962 (Diptera: Chironomidae), new for the Belgian fauna. *Bulletin et Annales de la Société Royale Belge d'Entomologie*, **128,** 316-8.

Int Panis, L. and R. Verheyen (1994) On the use of split moving window analysis for boundary detection in ordered dataseries from benthic communities. *Netherlands Journal of Aquatic Ecology*, in press.

ISO (International Organization for Standardization), (1979) *Assessment of the Biological Quality of Rivers by a Macroinvertebrate "Score"*. ISO/TC 147/SC 5/WG 6 N 5, Draft Proposal pp.18

Iwakuma, T., Yasuno, M., Sugaya, Y. and Sasa M. (1988) Three large species of Chironomidae (Diptera) as biological indicators of lake eutrophication, in *Biological Monitoring of Environmental Pollution* (eds M. Yasuna and B.A. Whitton), Tokai University Press, Tokyo. pp. 101-13.

Janecek, B. (1985) Beiträge zur Faunistik und Ökologie der Chironomidae (Diptera) des Gebhartsteiches, eines vom Menschen mäßig beeinflußten Karpfengewässers im nordwestlichen Waldviertel (Niederösterreich). *Dissertation an der Universität Wien, formal- und naturwissenschaftliche Fakultät*, pp.370

Janssens de Bisthoven, L. and Ollevier, F. (1989) Some experimental aspects of sediment stress on *Chironomus* gr. *thummi* larvae (Diptera: Chironomidae). *Acta Biologia Debrecen Oecologia Hungarica*, **3,** 147-55.

Janssens de Bisthoven, L., Timmermans, K.R. and Ollevier, F. (1992a) The concentration of cadmium, lead, copper and zinc in *Chironomus* gr. *thummi* larvae (Diptera: Chironomidae) with deformed versus normal menta. *Hydrobiologia*, **239,** 141-9.

Janssens de Bisthoven, L., Postma, J., Vermeulen, A. and Ollevier, F. (1992b) Assessment of morphological deformities in *Chironomus riparius* larvae (Diptera, Nematocera) in a chronic exposure test with cadmium. *Belgian Journal of Zoology*, **122,** 240.

Johannesson, H. and Birkisson, S.F. (1991) Setflutningar i Myvatni. Svifaursmaelingar og setgildrur. Vegagerd rikisins, Reykjavik. Mimeographed report, in Icelandic [Sediment transport in Myvatn. Seston and sediment traps].

Johnson, R.K., Wiederholm, T. and Eriksson, L. (1990) The influence of season on the classification and ordination of profundal communities of nutrient poor, oligo-mesohumic Swedish lakes using environmental data. *Internationale Vereinigung für Theoretische und Angewandte Limnologie Verhandlungen*, **24,** 646-52.

Johnson, R.K., Wiederholm, T. and Rosenberg, D.M. (1993) Freshwater biomonitoring using individual organisms, populations, and species assemblages of benthic macroinvertebrates, in *Freshwater Biomonitoring and Benthic Macroinvertebrates* (eds D.M. Rosenberg and V.H. Resh), Chapman and Hall, New York, pp.40-158.

Johnson, R.K. (1989) Classification of profundal chironomid communities in oligotrophic/humic lakes of Sweden using environmental data. *Acta Biologica Debrecina Oecologica Hungarica*, **3,** 167-75.

Jolivet, P. (1974) Le problème des Chironomides (Diptères: Nematocères) dans le Languedoc-Rousillon (Note preliminaire). *Vie et Milieu Série C Biologie Terrestre*, **23,** 269-90.

Jonasson, P.M. (ed.) (1979a) Ecology of eutrophic, subarctic Lake Myvatn and the River Laxa. *Oikos*, **32,** 1-308.

Jonasson, P.M. (1979b) The Lake Myvatn ecosystem. *Oikos*, **32,** 289-305.

Jones, C.W., Dalton, M.W. and Townley, L.H. (1991) Interspecific comparisons of the structure and regulation of the *Drosophila* ecdysone-inducible gene *E74*. *Genetics*, **127,** 535-43.

Jones, E.L. (1968) *Chironomus tepperi* Skuse (Diptera: Chironomidae) as a pest of rice in New South Wales. *Australian Journal of Science*, **31,** 89.

Jones R.E. (1975) Dehydration in an Australian rockpool chironomid larva (*Paraborniella tonnoiri*). *Journal of Entomology*, **49,** 111-9

Jongman, R.H., ter Braak, C.J.F. and van Tongeren, O.F.R. (eds.) (1987) *Data Analysis in Community and Landscape Ecology*. Pudoc, Wageningen.

Jonsson, E., Gardarsson, A. and Gislason, G. M. (1986) A new window trap used in the assessment of the flight periods of Chironomidae and Simuliidae (Diptera). *Freshwater Biology*, **16,** 711-9.

Junk, W.L., Bayley, P.B. and Sparks R.E. (1989) The flood pulse concept in river-floodplain systems. Proceedings of the International Large River Symposium. *Canadian* (ed. D.P. Dodge) *Special Publications in Fisheries and Aquatic Science*, **106,** 110-27.

Kainz, E. and Schwarz, K. (1986) Chemische, biologische und fischereiliche Untersuchungen an mehreren Waldviertler Karpfenteichen. *Wissenschaftliche Mitteilungen aus dem Niederösterreichischen Landesmuseum* **4,** 43-210.

Kajak, Z. (1977) Factors influencing benthos biomass in shallow lake environments. *Ekologia Polska*, **25,** 421-9.

Kansanen, P.H., Aho, J. and Paasivirta, L. (1984) Testing the benthic lake type concept based on chironomid associations in some Finnish lakes using multivariate statistical methods. *Annales Zoologici Fennici*, **21,** 55-76.

Kerkis, I., Kiknadze, I., Filippova, M. and Gunderina, L. (1989) Cytogenetic differentiation of the *Chironomus* species of the *plumosus* group. *Acta Biologica Debrecina Supplementum Oecologica Hungarica*, **2,** 103-14.

Keyl, H.G. (1962) Chromosomenevolution bei *Chironomus* II. Chromosomenumbauten und phylogenetische Beziehungen der Arten. *Chromosoma (Berlin)* **13,** 464-514.

Keyl, H.G. and Keyl, I. (1959) Die cytologische Diagnostik der Chironomiden. *Archiv für Hydrobiologie*, **56,** 43-57.

Kiknadze, I.I. (1987) Chromosomal polymorphism in natural populations of the *plumosus* species-group of West Siberia. *Entomologica Scandinavica Supplement*, **92,** 113-121.

Kiknadze, I.I., Shilova, A.I., Kerkis, I.E., Shobanov, N.A., Zelentzov, N.I., Grebenjuk, L.P., Istomina, A.G., and Prasolov, W.A. (1991) Karyotypes and morphology of larvae in the Tribe Chironomini. *Nauka, Novosibirsk*, pp.114 (in Russian with English summary).

Kiknadze, I., Butler, M.G., Aimanova, K.G., Gunderrina, L.I. and Cooper, J.K. (in prep) Geographic variation in polytene chromosome banding pattern of the Holarctic midge *Camptochironomus tentans* Fabr.

Kikuchi, M., Kikuchi, T., Okubo, S., and Sasa, M. (1985) Observations on the seasonal prevalence of chironomid midges and mosquitoes by light traps set in a rice paddy area in Toksuhima. *Japanese Journal of Sanitary Zoology* **36,** 333-42.

Kirkpatrick, J.B., Peacock, R.J., Cullen, P.J. and Neyland, M.G. (1988) *The Wet Eucalypt Forests of Tasmania*. Tasmanian Conservation Trust, Hobart.

Kitching, R.L. (1971) An ecological study of water-filled treeholes and their position in the woodland ecosystem. *Journal of Animal Ecology*, **40,** 281-302.

Kitching, R.L. and Pimm, S.L. (1985) The length of food chains: phytotelmata in Australia and elsewhere. *Proceedings of the Ecological Society of Australia*, **14,** 123-40.

Kobayashi, T. (1995) *Eurycnemus* sp. (Diptera: Chironomidae) larvae ectoparasitic on pupae of *Goera japonica* (Trichoptera: Limnephilidae), in *Chironomids - from Genes to Ecosystems* (ed. P. S. Cranston), CSIRO, Melbourne, pp.317-21.

Kok, C.J., Van der Velde, G. and Landsbergen, K.M. (1990) Production, nutrient dynamics and initial decomposition of floating leaves of *Nymphaea alba* (L) and *Nuphar lutea* (L) Sm (Nymphaceae) in alkaline and acid waters. *Biogeochemistry*, **11**(3), 235-50.

Kolasawat, P. and Knight, A.W. (1987) Chronic toxicity of copper to a partial life cycle of the midge *Chironomus decorus*. *Archives of Environmental Contamination and Toxicology*, **16,** 283-90.

Kolkwitz, R. and Marsson, M. (1909) Ökologie der tierischen Saprobien. Beiträgezur Lehre var de biologischen Gewasserbeurteilung. *Internationale Revue der Gesamten Hydrobiologie und Hydrographie*, **2,** 126-52.

Kondo, S. and Hamashima, S. (1985) Chironomid midges emerged from aquatic macrophytes in reservoirs. *Japanese Journal of Limnology*, **46,** 50-5.

Kondo, S. and Hamashima, S. (1993) Habitat preferences of four chironomid species associated with aquatic macrophytes in an irrigation reservoir. *Netherlands Journal of Aquatic Ecology*, **26**(2-4), 371-7.

Kondo, S., Hamashima, S. and Hashimoto, H. (1989) Life history and seasonal occurrence of *Pentapedilum tigrinum* Hashimoto associated with *Nymphoides indica* O. Kuntze in an irrigation reservoir. *Acta Biologica Debrecina, Oecologica Hungarica*, **2,** 237-45.

Konstantinov, A.S. (1958a) Biologija chironomid i ich razwedene. *Trudy Saratovskogo Otdelenja Kaspijskogo Filiala*, **5,** 1-363.

Konstantinov, A.S. (1958b) Wlijanie temperatury na skorost' rosta i razwitija licinok chironomid. *Doklady Akademii Nauk SSSR*, **120,** 1362-5.

Kosalwat, P. and Knight, A.W. (1987) Chronic toxicity of copper to a partial life cycle of the midge *Chironomus decorus*. *Archives of Environmental Contamination and Toxicology*. **16,** 283-90.

Kouki, J. (1991) The effects of water-lily beetle, *Galerucella nymphaeae*, on leaf production and leaf longevity of the yellow water-lily *Nuphar lutea*. *Freshwater Biology*, **26**(3), 347-54.

Kownacki, A. (1971) Taxocens of Chironomidae in streams of the Polish High Tatra Mts. *Acta Hydrobiologia*, **13,** 439-64.

Kownacki, A. (1982) Stream ecosystems in mountain grassland (West Carpathians). 8. Benthic invertebrates. *Acta Hydrobiologia*, **24,** 375-90.

Kownacki, A. (1991) Zonal distribution and classification of the invertebrate communities in high mountain streams in South Tirol (Italy). *Verhandlungen der Internationale Vereinigung für Theoretische und Angewandte Limnologie*, **24,** 2010-4.

Kownacki, A. and Margreiter-Kownacka, M. (1993) 4.4. Diptera-Chironomidae, in *Die Makrozoobenthos-Gemeinschaften der Obern Alz-Bestandaufnahme* (ed. M. Margreiter-Kownacka), Krakòw, pp.22-8.

Kownacki, A. and Zosidze, R. S. (1980) Taxocens of Chironomidae (Diptera) in some rivers and streams of the Adzhar ASSR (Little Caucasus Mts). *Acta Hydrobiologia*, **22,** 67-87.

Köhn, T. and Frank, C. (1980) Effect of thermal pollution on the chironomid fauna in an urban channel, in *Chironomidae. Ecology, Systematics, Cytology and Physiology* (ed. D.A. Murray), Pergamon Press, Oxford, pp.187-94.

Krueger, C.C. and Waters, T.F. (1983) Annual production of macroinvertebrates in three streams of different water quality. *Ecology*, **64,** 840-50.

Kruger, R. (1941) Parthenogenetische *Stylotanytarsus*-Larven als Bewohnereiner Trinkwasser-leitung (*Tanytarsus*-Studien III. Die Gattung *Stylotanytarsus*). *Archiv für Hydrobiologie*, **38,** 214-53.

Lacey, L.A. and A.H. Undeen. (1987) The biological control potential of pathogens and parasites of black flies, in *Black flies: Ecology, Population Management, and Annotated World List* (eds K.C. Kim and R.W. Merritt), Pennsylvania State University Press, University Park, Pennsylvania, USA, pp.327-40

Ladle, M., Cooling, D.A., Welton, J.S. and Bass, J.A. (1985) Studies on Chironomidae in experimental recirculating stream systems. II. The growth, development and production of a spring generation of *Orthocladius (Euorthocladius) calvus* Pinder. *Freshwater Biology*, **15,** 243-55.

Lake, P.S, Bayly, I.A.E. and Morton, D.W. (1989) The phenology of a temporary pond in western Victoria,

Australia, with special reference to invertebrate succession. *Archiv für Hydrobiologie*, **15**, 171-202.

Langton, P.H. (1980) The genus *Psectrocladius* Kieffer (Diptera: Chironomidae) in Britain. *Entomologist's Gazette*, **31**, 75-88.

Langton, P.H. (1984) *A key to Pupal Exuviae of British Chironomidae*. P.H.Langton, March, Cambridgeshire, pp.324.

Langton, P.H. (1991) *A Key to Pupal Exuviae of West Palaearctic Chironomidae*. P. H. Langton, Huntingdon, pp.386.

Langton, P.H. and Cranston, P.S. (1991) Pupae in nomenclature and identification: West Palaearctic *Orthocladius* s.str. (Diptera: Chironomidae) revised. *Systematic Entomology* **16**, 239-52.

Langton, P.H., Cranston, P.S. and Armitage, P. (1988) The parthenogenetic midge of water supply systems, *Paratanytarsus grimmii* Schneider (Diptera: Chironomidae). *Bulletin of Entomological Research*, **78**, 317-28.

Laville, H. (1981) Récoltes d'exuvies nymphales de chironomides (Diptera) dans le Haut-Lot, de la source (1 295 m), au confluent de la Truère (223 m). *Annales de Limnologie*, **17**, 255-89.

Laville, H. and Lavandier, P. (1977) Les chironomides (Diptera) d'un torrent de haute montagne, l'Estaragne. *Annales de Limnologie*, **13**, 57-81.

Laville, H. and Vinçon, G. (1991) A typological study of Pyrenean streams: Comparative analysis of the Chironomidae (Diptera) communities in the Ossau and Aure Valleys. *Verhandlungen der Internationale Vereinigung für Theoretische und Angewandte Limnologie*, **24**, 1775-84.

Learner, M.A., Williams, R., Harcup, M. and Hughes, B.D. (1971) A survey of the macro-fauna of the River Cynon, a polluted tributary of the River Taff (South Wales). *Freshwater Biology*, **1**, 339-67.

Learner, M.A., Wiles, P.R. and Pickering, J.G. (1989) The influence of aquatic macrophyte identity on the composition of the chironomid fauna in a former gravel pit in Berkshire, England. *Aquatic Insects*, **11**, 183-91.

Lehmann, J. (1971) Die Chironomiden der Fulda. (Systematische, ökologische und faunistische Untersuchungen). *Archiv für Hydrobiologie Supplement*, **37**, 466-555.

Lehmann, J. (1979) Chironomidae (Diptera) aus Fliessgewässern Zentralafrikas (Systematik, Ökologie, Verbreitung und Produktionsbiologie) Teil I: Kivu-gebiet, Ostzaire. *Spixiana Supplement*, **3**, 1-144.

Lehmann, J. (1981) Chironomidae (Diptera) aus Fliessgewässern Zentralafrikas Teil II. Die Region um Kisangani, Zentralzaire. *Spixiana Supplement*, **5**, 1-85.

Lenat, D.R. (1983) Chironomid taxa richness: natural variation and use in pollution assessment. *Freshwater Invertebrate Biology*, **2**, 192-98.

Lenat, D.R. (1988) Water quality assessment of streams using a qualitative collection method for benthic macroinvertebrates. *Journal of the North American Benthological Society*, **7**, 222-33.

Lenat, D.R. and Folley, D.R. (1983) Lotic chironomids of the North Carolina mountains. *Memoirs of the American Entomological Society*, **34**, 145-64.

Lenz, F. (1954-1962) Tendipedidae (Chironomidae). b) Subfamilie Tendipedinae (Chironominae). B. Die Metamorphose der Tendipediinae, in *Die Fliegen der palaearktischen Region 13c* Schweizerbart'sche Verlagsbuchhandlung, Stuttgart, pp.139-260.

Lerner, I.M. (1954) *Genetic Homeostasis*. Oliver and Boyd, Edinburgh.

LeSage, L. and Harrison, A.D. (1979) Improved traps and techniques for the study of emerging aquatic insects. *Entomology News* **90**, 65-78.

Lesage, L. and Harrison, A.D. (1980) The biology of *Cricotopus* (Chironomidae: Orthocladiinae) in an algal-enriched stream: Part I. Normal Biology. *Archiv für Hydrobiologie Supplementbande*, **4**, 375-418.

Leuchs, H. and Caspers, N. (1988) Chironomiden-Gesellschaften in stehenden Gewässern der Niederrheinauen bei Rekdes (Diptera: Chironomidae). *Decheniana (Bonn)*, **141**, 288-95.

Lindeberg, B. and Wiederholm, T. (1979) Notes on the taxonomy of European species of *Chironomus* (Diptera: Chironomidae). *Entomologica Scandinavica Supplement* **10**, 99-116.

Lindegaard, C. (1972) An ecological investigation of the Chironomidae (Diptera) from a Danish lowland stream (Linding Å). *Archiv für Hydrobiologie*, **69**, 465-507.

Lindegaard, C. (1979) The invertebrate fauna of Lake Myvatn, Iceland. *Oikos*, **32**, 151-61.

Lindegaard, C. (1994) Classification of water bodies and pollution, in *Chironomidae: Biology and Ecology of Non-biting Midges* (eds P. Armitage, P.S. Cranston and L.C.V. Pinder), Chapman and Hall, London, pp.385-404.

Lindegaard, C. (1995). Diptera Chironomidae, non-biting midges, in *Aquatic Insects of N.W. Europe - A Taxonomic Handbook* (ed. A. Nilsson), Apollo Books, Strenstrup, Denmark.

Lindegaard, C. and Brodersen, K.P. (1995) Distribution of Chironomidae (Diptera) in the river continuum, in *Chironomids - From Genes to Ecosystems* (ed. P.S. Cranston), CSIRO, Melbourne, pp.257-71.

Lindegaard, C. and Mortensen, E. (1988) Abundance, life history and production of Chironomidae (Diptera) in a Danish lowland stream. *Archiv für Hydrobiologie, Supplement*, **81**, 563-87.

Lindegaard, C. and Jonasson, P.M. (1979) Abundance, population dynamics and production of zoobenthos in Lake Myvatn, Iceland. *Oikos*, **32**, 202-27.

Lipsey, M.W. (1992) *Design Sensitivity: Statistical Power for Experimental Research* Sage Publications, London.

Lock, M.A. (1994) Dynamics of particulate and dissolved organic matter over the substratum of water bodies, in

Particulate and Dissolved Matter in Aquatic Systems (ed. R.S. Wotton). Lewis Publishers. Chelsea, Michigan, U.S.A. In press.

Lugthart, G. J., Wallace, J.B. and Huryn, A.D. (1990) Secondary production of chironomid communities in a insecticide-treated and untreated headwater streams. *Freshwater Biology*, **24,** 417-27.

Luoma, S.N. (1989) Can we determine the biological availability of sediment-bound trace elements? *Hydrobiologia*, **176/177,** 379-96.

Luoma, S.N. and Bryan, G.W. (1978) Trace metal bioavailability: modeling chemical and biological interaction of sediment-bound zinc, in *Interfacing with Biological Models, Chemical Modelling in Aqueous Systems* pp.577-609.

Macchiusi F. and Baker R.L. (1991) Prey behaviour and size-selective predation by fish. *Freshwater Biology*, **25,** 533-8.

Macchiusi F. and Baker R.L. (1992) Effects of predators and food availability on activity and growth of *Chironomus tentans* (Chironomidae: Diptera). *Freshwater Biology*, **28,** 207-16.

Mackey, A.P. (1976a) Quantitative studies on the Chironomidae (Diptera) of the Rivers Thames and Kennet. I. The *Acorus* zone. *Archiv für Hydrobiologie*, **78,** 240-67.

Mackey, A.P. (1976b) Quantitative studies on the Chironomidae (Diptera) of the Rivers Thames and Kennet. II. The Thames Flint zone. *Archiv für Hydrobiologie*, **78,** 310-8. Mackey, A.P. (1977a) Growth and development of larval Chironomidae. *Oikos*, **28,** 270-5.

Mackey, A.P. (1977b) Quantitative studies on the Chironomidae (Diptera) of the rivers Thames and Kennet. III The *Nuphar* zone. *Archiv für Hydrobiologie*, **79,** 62-102.

Mackey, A.P. (1977c) Quantitative studies on the Chironomidae (Diptera) of Rivers Thames and Kennet. IV. Production. *Archiv für Hydrobiologie*, **80,** 327-48.

Madden, C.P., Austin, A.D. and Suter, P.J. (1995) Pollution monitoring using chironomid larvae: what designates a deformity? in *Chironomids: From Genes to Ecosystems* (ed. P.S. Cranston), CSIRO, Melbourne. pp. 89-100.

Madden, C.P., Suter, P.J., Nicholson, B.C. and Austin, A.D. (1993) Deformities in chironomid larvae as indicators of pollution (pesticide) stress. *Netherlands Journal of Aquatic Ecology*, **26,** 551-7.

Madden, C.P. (1990) *Deformities in Chironomid Larvae (Insecta: Diptera) Associated with Exposure to Pesticides*. Master of Environmental Studies Thesis, The University of Adelaide.

Maeda, M. and Yano, K. (1988) Biology of *Chironomus kiiensis* Tokunaga (Diptera: Chironomidae). *Bulletin of the Faculty of Agriculture, Yamaguchi University*, **36,** 37-47.

Maher, M.T. and Carpenter, S.M. (1984) Benthic studies of waterfowl breeding habitat on south-western New South Wales. II. Chironomid Populations. *Australian Journal of Marine and Freshwater Research*, **35,** 97-110.

Malmqvist, B. (1988) Downstream drift in Madeiran Levadas: test of hypotheses relating to the influence of predators on the drift of insects. *Aquatic Insects*, **10,** 141-52.

Malmqvist, B., Nilsson, A.N., Baez, M., Armitage, P.D. and Blackburn, J.H. (1993) Stream macroinvertebrate communities in the Island of Tenerife. *Archiv für Hydrobiologie*, **128,** 209-35.

Manly, B. (1991) *Randomization and Monte Carlo Methods in Biology*. Chapman and Hall, pp.281

Mantel, N. (1967) The detection of disease clustering and a generalized regression approach. *Cancer Research*, **27,** 209-20.

Marchant, R. (1988) A subsampler for samples of benthic invertebrates. *Bulletin of the Australian Society for Limnology*, **12,** 49-52.

Marchant, R., Metzeling, L., Graesser, A. and Suter, P. (1985) The organisation of macroinvertebrate communities in the major tributaries of the La Trobe River, Victoria, Australia. *Freshwater Biology*, **15,** 315-31.

Marshall, W.K., and Roberts, J.R. (1978) *Ecotoxicology of Chlorpyrifos*. National Research Council, Canada.

Martin, J. (1969) The salivary gland chromosomes of *Chironomus oppositus* Walker (Diptera: Nematocera). *Australian Journal of Zoology*, **17,** 473-86.

Martin, J. (1971a) A review of the genus *Chironomus* (Diptera: Chironomidae) IV. The Karyosystematics of the australis group in Australia. *Chromosoma*, **35,** 418-30.

Martin, J. (1971b) A review of the genus *Chironomus* (Diptera: Chironomidae) II. Added descriptions of *Chironomus cloacalis* Atchley and Martin from Australia. *Studies in Natural Sciences (Portales, New Mexico)* **1**(2), 1-21.

Martin, J. (1974) A review of the genus *Chironomus* (Diptera, Chironomidae). IX. The cytology of Chironomus tepperi Skuse. *Chromosoma (Berlin)*, **45,** 91-8.

Martin, J. (1979) Chromosomes as tools in taxonomy and phylogeny of Chironomidae (Diptera). *Entomologica Scandinavica Supplement* **10,** 67-74.

Martin, J. (1995) A mutation in *Chironomus (Camptochironomus) tentans* Fabricius that leads to gynandromorph and mosaic formation, in *Chironomids: from Genes to Ecosystems*, (ed. P.S. Cranston), CSIRO Publications, Melbourne, pp.55-60.

Martin, J. and Lee, B.T.O. (1984a) Are there female heterogametic strains of *Chironomus tentans* Fabricius? *Canadian Journal of Genetics and Cytology* **26,** 743-7.

Martin, J. and Lee, B.T.O. (1984b) A phylogenetic study of sex determiner location in a group of Australasian *Chironomus* species (Diptera, Chironomidae). *Chromosoma*, **90,** 190-7.

Martin, J. and Lee, B.T.O. (1988) The chromosomal location of the malate dehydrogenase and the phosphoglucomutase loci in *Chironomus* and their relationship with a sex determining region. *Genetics (Life Science Advances)*, **7,** 57-63.

Martin, J. and Porter, D.L. (1977) Laboratory biology of the rice midge, *Chironomus tepperi* Skuse (Diptera:

Nematocera): Mating behaviour, productivity and attempts at hybridization. *Journal of the Australian Entomological Society*, **16**, 411-6.

Martin, J. and Sublette, J.E. (1972) A review of the genus *Chironomus* (Diptera: Chironomidae) III. *Chironomus yoshimatsui* new species from Japan. *Studies in Natural Sciences (Portales, New Mexico)*, **1** (3), 1-59.

Martin, J. and Wülker, W. (1971) Inversion polymorphism in *Chironomus staegeri* Lundbeck. *Canadian Journal of Genetics and Cytology* **13**, 306-21.

Martin, J. and Wülker, W. (1974) A review of the genus *Chironomus* (Diptera, Chironomidae) VIII. Cytology of *Chironomus stigmaterus* Say. *Studies in Natural Science (Portales, New Mexico)* **1(11)**, 1-17.

Martin, J., Kuvangkadilok, C., Peart, D.H. and Lee, B.T.O. (1980) Multiple sex determining regions in a group of related *Chironomus* species (Diptera: Chironomidae). *Heredity*,**44**, 367-82.

Martin, J., Lee, B.T.O. and Connor, E. (1978) Apparent incipient speciation in the midge *Chironomus oppositus* Walker (Diptera: Chironomidae). *Australian Journal of Zoology*, **26**, 323-9.

Martin, J., Wülker, W. and Sublette, J.E. (1974) Evolutionary cytology in the genus *Chironomus* Meigen. *Studies in Natural Sciences, (Portales, New Mexico)* **1**(12), 1-12.

Mason, P. G. and Lemhkuhl, D. M. (1983) Effects of the Squaw Rapids hydroelectric development on Saskatchewan River Chironomidae (Diptera). *Memoirs of the American Entomological Society*, **34**, 187-210.

Mason, W. T. and Sublette, J. E. (1971) Collecting Ohio River basin Chironomidae (Diptera) with a floating sticky trap. *The Canadian Entomologist*, **103**, 397-404.

McCahon, C.P. and Pascoe, D. (1991) Brief-exposure of first and fourth instar *Chironomus riparius* larvae to equivalent assumed doses of cadmium: effects on adult emergence. *Water, Air, and Soil Pollution*, **60**, 395-403.

McCauley, V. and Tsumura, K. (1974) Calorific values of Chironomidae (Diptera). *Canadian Journal of Zoology*, **52**, 581-6.

McGill, J.D. (1980) *The distribution of Chironomidae throughout the River Chew drainage system, Avon, England*. PhD Thesis, University of Bristol.

McKnight, D.M. and Feder, G.L. (1984) The ecological effect of acid conditions and precipitation of hydrous metal oxides in a Rocky Mountain stream. *Hydrobiologia*, **119**, 129-38.

McLachlan, A.J. (1985) What determines the species found in a rain-pool? *Oikos*, **45**, 1-7.

McLachlan, A.J. (1989) Animal populations at extreme densities:size dimorphism by frequency dependent selection in ephemeral habitats. *Functional Ecology*, **3**, 633-43.

McLachlan, A.J. and Cantrell, M.A. (1976) Sediment development and its influence on the distribution and tube structure of *Chironomus plumosus* L. (Chironomidae, Diptera) in a new impoundment. *Freshwater Biology*, **6**, 437-43.

McLachlan, A.J. and Cantrell M.A. (1980) Survival strategies in tropical rain pools. *Oecologia*, **47**, 344-51.

McLay, C.L. (1968) A study of drift in the Kakanui River, New Zealand. *Australian Journal of Marine and Freshwater Research*, **19**, 139-49.

Meillon, B. de and Gray, F.C. (1937) The control of a species of *Chironomus* Meigen (Diptera: Chironomidae) in an artificial lake by increasing the salinity. *South African Medical Journal*, **11**, 658-60.

Menzie, C.A. (1981) Production ecology of *Cricotopus sylvestris* (Fabricius) (Diptera: Chironomidae) in a shallow estuarine cove. *Limnology and Oceanography*, **26**, 467-81.

Meriläinen, J.J. and Hamina V. (1993) Recent environmental history of a large, originally oligotrophic lake in Finland: a paleaeolimnological study of chironomid remains. *Journal of Paleolimnology*, **9**, 129-40.

Meriläinen, J.J. and Hynynen, J. (1990) Benthic invertebrates in relation to acidity in Finnish forest lakes, in *Acidification in Finland* (eds P. Kauppi, P. Anttila, and K. Kenttämies), Springer-Verlag, Berlin F.R.G. pp.1029-49.

Merritt, R.W., Walker, E.D. Wilzbach, M.A., Cummins, K.W. and Morgan, W.T. (1989) A broad evaluation of *B.t.i.* for black fly (Diptera: Simuliidae) control in a Michigan river: efficacy, carry and nontarget effects on invertebrates and fish. *Journal of the American Mosquito Control Association*, **5**, 397-415.

Metcalfe, J.L. (1989) Biological water quality assessment of running waters based on macroinvertebrate communities: history and present status in Europe. *Environmental Pollution*, **60**, 101-39.

Metzeling, L., Graesser, A., Suter, P. and Marchant, R. (1984) The distribution of aquatic macroinvertebrates in the upper catchment of the La Tobe River, Victoria. *Occasional Papers from the Museum of Victoria*, **1**, 1-79.

Michailova, P. (1989) The polytene chromosomes and their significance to the systematics and phylogeny of the family Chironomidae, Diptera. *Acta Zoologica Fennica* **186**, 1-107.

Michailova, P. and Belcheva, R. (1990) Different effect of lead on external morphology and polytene chromosomes of *Glyptotendipes barbipes* (Staeger) (Diptera: Chironomidae). *Folia Biologica (Krakow)*, **38**, 1-4.

Michailova, P. and Fischer, J. (1986a) *Chironomus vancouveri* sp. n. from Canada (Diptera, Chironomidae). *Reichenbachia* **23**, 99-106.

Michailova, P. and Fischer, J. (1986b) Speciation within the plumosus group of the genus *Chironomus* Meigen (Diptera, Chironomidae). *Zeitschrift für zoologische Systematik und Evolutionsforschung* **24**, 207-22.

Minshall, G.W. and Petersen, R.C. (1985) Towards a theory of macroinvertebrate community structure in stream ecosystems. *Archiv für Hydrobiologie*, **104**, 49-76.

Minshall, G.W. and Winger, P.V. (1968) The effect of the reduction in stream flow on invertebrate drift. *Ecology*, **49**, 580-5.

Minshall, G.W., Andrews, D.A. and Manuel-Faler, C.Y. (1983) Application of island biogeographic theory to streams: macroinvertebrate recolonization of the Teton River, Idaho, in *Stream Ecology: Application and Testing of General Ecological Theory* (eds J.R. Barnes and G.W. Minshall), Plenum, New York, pp 279-97.

Mitcham, R.P. and Shelley, M.W. (1980) The control of animals in water mains using permethrin, a synthetic pyrethroid. *Journal Institution of Water Engineers and Scientists*, **34**, 474-83.

Miura, T., Takahashi, R.M. and Mulligan, F.S. III. (1980) Effects of the bacterial mosquito larvicide, *Bacillus thuringiensis* serotype H-14 on selected aquatic organisms. *Mosquito News*, **40**, 619-22.

Moller Pillot, H.K.M. (1984) De larven der Nederlandse Chironomidae (Diptera) Deel 1A. *Nederlandse Faunistische Mededelingen*, (ed. European Invertebrate Survey Nederland), pp.277

Molloy, D. (1990) Progress in the biological control of black flies with *Bacillus thuringiensis* var. *israelensis* with emphasis on temperate climates, in *Bacterial Control of Mosquitoes and Black flies: Biochemistry, Genetics and Applications of* Bacillus thuringiensis israelensis *and* Bacillus sphaericus (eds H. de Barjac and D. Sutherland), Rutgers University Press, New Brunswick, New Jersey, USA, pp.161-86.

Molloy, D.P. (1992) Impact of the black fly (Diptera: Simuliidae) control agent *Bacillus thuringiensis* var. *israelensis* on chironomids (Diptera: Chironomidae) and other nontarget insects: results of ten field trials. *Journal of the American Mosquito Control Association*, **8**, 24-31.

Moog, O. (1993) Quantification of daily peak hydropower effects on aquatic fauna managment to minimize environmental impacts. *Regulated Rivers: Research and Managment*, **8**, 5-14.

Moog, O. and Janecek, B.F.U. (1991) River flow, substrate type and *Hydrurus* density as major determinants of macroinvertebrate abundance, composition and distribution. *Verhandlungen der Internationale Vereinigung für Theoretische und Angewandte Limnologie*, **24**, 1888-96.

Morgan, N.C. (1971) Factors in the design and selection of insect emergence traps, in *A Manual for Assessment of Secondary Productivity in Freshwaters* (eds W.T. Edmondson and G.G. Winberg), Blackwell, Oxford, U.K. pp.93-108.

Morgan, R.P., Jacobsen, R.E., Weisberg, S.B., McDowell, L.A. and Wilson, H.T.(1991) Effects of flow alternation on benthic macroinvertebrate communities below the Brighton Hydroelectric Dam. *Journal of Freshwater Ecology*, **6**, 419-29.

Morrison, D.F. 1990. *Wielowymiarowa Analiza Statystyczna*, PWN, Warszawa, pp.589

Moss, B. and Timms, M. (1989) Predation, sediment stability and food availability as determinants of the benthic invertebrate fauna in two shallow lakes. *Hydrobiologia*, **185**, 249-57.

Moss, D., Furse, M.T., Wright, J.F. and Armitage, P.D. (1987) The prediction of the macro-invertebrate fauna of unpolluted running-water sites in Great Britain using environmental data. *Freshwater Biology*, **17**, 41-52.

Mosteller, F and Tukey, J.W. (1977) *Data Analysis and Regression: a Second Course in Statistics*. Addison - Wesley Publishing Co. Inc., Philippines.

Mothes, G. (1966) Die makroscopische Bodenfauna des oligotrophen Stechlinsees im Vergleich zu eutrophen Nachbarseen. *Verhandlungen der Internationale Vereinigung für Theoretische und Angewandte Limnologie*, **16**, 258-62.

Muir, D.C.G., Grift, N.P., Townsend, B.E., Metner, D.A. and Lockhart, W.L. (1982) Comparison of the uptake and bioconcentration of fluridone and terbutryn by rainbow trout and *Chironomus tentans* in sediment and water systems. *Archives of Environmental Contamination and Toxicology*, **11**, 595-602.

Mulla, M.S., Chaney, J.D. and Rodcharoen, J. (1990) Control of nuisance aquatic midges (Diptera: Chironomidae) with the microbial larvicide *Bacillus thuringiensis* var. *israelensis* in a man-made lake in southern California. *Bulletin of the Society for Vector Ecology*, **15**, 176-84.

Murkin, H.R. and Kadlec, J.A. (1986) Responses by benthic macroinvertebrates to prolonged flooding and marsh habitat. *Canadian Journal of Zoology*, **64**, 65-72.

Murray, D.A. and Ashe, P. (1981a) A description of the pupa of *Buchonomyia thienemanni* Fittkau, with notes on its ecology and on the phylogenetic position of the subfamily Buchonomyiinae (Diptera, Chironomidae). *Spixiana*, **4**, 55-68.

Murray, D.A. and Ashe, P. (1981b) A description of the larva and pupa of *Eurycnemus crassipes* (Panzer) (Diptera: Chironomidae). *Entomologica Scandinavica*, **12**, 357-61.

Murray, D.A. and Ashe, P. (1985) A description of the adult female of *Buchonomyia thienemanni* Fittkau, with a reassessment of the phylogenetic position of the subfamily Buchonomyiinae (Diptera, Chironomidae). *Spixiana Supplement*, **11**, 149-60.

Murray, D.A and Fittkau, E.J. (1989) The adult males of Tanypodinae (Diptera: Chironomidae) of the Holarctic Region - Keys and diagnoses. *Entomologica Scandinavica Supplement*, **34**, 37-123.

Neter, J., Wasserman, W. and Kutner, M.H. (1985) *Applied Linear Statistical Models*. 2nd Edn. Irwin, Homewood, Illinois.

Nietzke, G. (1938) Die Kossau. Hydrobiologisch-faunistische Untersuchungen an schleswig-holsteinischen Fliessgewässern. *Archiv für Hydrobiologie*, **32**, 1-74.

Nolte, U. (1989) Observations on neotropical rainpools (Bolivia) with emphasis on Chironomidae. *Studies in Neotropical Fauna*, **24**, 105-12.

Nolte, U. (1990) Chironomid biomass determination from larval shape. *Freshwatater Biology*, **24**, 443-51.

Nolte, U. (1993) Egg masses of Chironomidae (Diptera). *Entomologica Scandinavica Supplement*, **43**, 1-75.

Norris, R.H. and Georges, A. (1993) Analysis and interpretation of benthic macroinvertebrate surveys, in *Freshwater Biomonitoring and Benthic Macroinvertebrates* (eds D.M. Rosenberg and V.H. Resh), Chapman and Hall, New York, pp.234-86.

Nowak, B., Clarke, G., Pettigrove, V., Hughes, J., and Madden, C. (1993) Biological indicators and biomonitoring of water quality in agricultural areas, in *Proceedings of the Ecotoxicology Specialist Workshop, Wee Waa, NSW, Australia*. Published by the Land and Water Resources Research and Development Corporation. Occasional Paper No. 07/93 pp.83-91.

Ogden, J. and Powell, J.A. (1979) A quantitative description of the forest vegetation on an altitudinal gradient in Mount Field National Park, Tasmania, and a discussion of its history and dynamics. *Australian Journal of Ecology*, **4,** 293-325.

Ohlmeyer, R.G. (1991) Report on Implications of Manipulating Water Levels in the River Murray Between Lock 1 and 10. *Engineering and Water Supply Departmental Report* 91/11.

O'Hop, J. and Wallace, J.B. (1983) Invertebrate drift, discharge and sediment relations in a southern Appalachian headwater stream *Hydrobiologia*, **98,** 71-84.

Okazaki, A. and Yano, K. (1990) Biology of *Tanytarsus oyamai* Sasa (Diptera: Chironomidae). *Transactions of the Shikoku Entomological Society*, **19,** 89-99.

Okland, R.H. (1990) *Vegetation Ecology: Theory, Methods and Applications with Reference to Fennoscandia.* Sommerfeltia Supplement 1.

Olafsson, J. (1979a) Physical characteristics of Lake Myvatn and River Laxa. *Oikos*, **32,** 38-66.

Olafsson, J. (1979b) The chemistry of Lake Myvatn and River Laxa. *Oikos*, **32,** 82-112.

Olafsson, J.S. (1987) *Faeda mylirfa (Diptera: Chironomidae) i Myvatni.* Thesis. Department of Biology, University of Iceland. pp.62

Oliver D.R. (1971) Life history of the Chironomidae. *Annual Reviews of Entomology*, **16,** 211-30

Oliver, D.R. (1985) Review of *Xylotopus* Oliver and description of *Irisobrillia* n.gen. (Diptera: Chironomidae). *Canadian Entomologist*, **117,** 1093-110.

Oliver, D.R. (1986) The pupae of the Diamesinae (Diptera: Chironomidae) of the Holarctic region - keys and diagnoses. *Entomologica Scandinavica Supplement*, **28,** 119-37.

Oliver, D.R. and Roussel, M.E. (1983) Redescription of *Brillia* Kieffer (Diptera: Chironomidae) with descriptions of Nearctic species. *Canadian Entomologist*, **115,** 277-8.

Oliver D.R., Dillon M.E. and Cranston P.S. (1990) A catalog of Nearctic Chironomidae. *Publications of the Research Branch, Agriculture Canada* **1857/B,** pp.89

Oliver, D.R., McClymont, D. and Roussel, M.E. (1978) A Key to Some Larvae of Chironomidae (Diptera) from the Mackenzie and Porcupine River Watersheds. *Fisheries and Environment Canada, Fisheries and Marine Services Technical Report*, **791,** 1-73.

Otto, C. and Wallace, J.B. (1989) Life cycle variation and habitat heterogeneity in water-lily leaf beetles. *Holarctic Ecology*, **12**(2), 144-51.

Paijmans, K., Galloway, R.W., Faith, D.P., Fleming, P.M., Haantjens, H.A., Heyligers, P.C., Kalma, J.D., & E. Loffler (1985) *Aspects of Australian Wetlands.* CSIRO Division of Water and Land Resources, Technical Paper **44,** 1-71.

Palmer, A.R. and Strobeck, C. (1986) Fluctuating asymmetry: measurement, analysis, patterns. *Annual Review of Ecology and Systematics*, **17,** 391-421.

Palmer, A.R. and Strobeck, C. (1992) Fluctuating asymmetry as a measure of developmental stability: implications of non-normal distributions and power of statistical tests. *Acta Zoologica Fennica*, **191,** 57-72.

Palmer, A.R., Strobeck, C., and Chippindale, A.K. (1993) Bilateral variation and the origin of macroscopic asymmetries. *Genetica*, **89,** 201-18.

Pankratova, V. Ya. (1970) Lichinki i kukolki komarov podsemeistva Orthocladiinae fauny SSSR (Diptera, Chironomidae - Tendipedidae). *Opred Faune SSSR*, **102,** 1-343.

Pantle, R. and Buck, H. (1955) Die biologische Überwachung der Gewässer und die Darstellung der Ergebnisse. *Gas Wasserfach* **96,** 604.

Papastergiadou, E. and Babalonas, D. (1992) Ecological studies on aquatic macrophytes of a dam lake: Lake Keskini, Greece. *Archiv für Hydrobiologie, SupplementBande*, **90**(2), 187-206.

Parker, C.R. and Voshell, J.R. (1979) *Cardiocladius* (Diptera: Chironomidae) larvae ectoparasitic on pupae of Hydropsychidae. *Environmental Entomology*, **8,** 357-61.

Parsons, P.A. (1990) Fluctuating asymmetry: an epigenetic measure of stress. *Biological Reviews*, **65,** 131-45.

Pascoe, D., Williams, K.A. and Green, D.W.J. (1989) Chronic toxicity of cadmium to *Chironomus riparius* Meigen - effects upon larval development and adult emergence. *Hydrobiologia*, **175,**109-15.

Paterson, C.G. and Fernando, C.H. (1970) The macroinvertebrate colonization of a new reservoir with particular reference to the Chironomidae. *Journal of the Canadian Fisheries Research Board, Canada*, **27,** 213-32.

Percy, J., Kuhn, K.L. and Kalthoff, K. (1986) Scanning electron microscopic analysis of spontaneous and UV-induced abnormal segment patterns in *Chironomus samoensis* (Diptera, Chironomidae). *Roux's Archives of Developmental Biology*, **195,** 92-102.

Peters, B. (1979) Control of midge flies in sewage stabilization ponds. *Entomological Society of Queensland News Bulletin*, **7,** 91-7.

Pettigrove, V. (1989) Larval mouthpart deformities in *Procladius paludicola* Skuse (Diptera: Chironomidae) from the Murray and Darling Rivers, Australia. *Hydrobiologia*, **175,** 111-7.

Phillips, A.M., Martin, J. and Bedo, D.G. (1994) *In situ* hybridization to polytene chromosomes of *Drosophila melanogaster* and other dipteran species, in *In situ*

Hybridization Protocols (ed. K.H.A.Choo), *Methods in Molecular Biology*, **10,** (In Press).

Pinder, L.C.V. (1974) The Chironomidae of a small chalk-stream in southern England. *Entomologisk Tidskrift, Supplement*, **95** 195-202.

Pinder, L.V.C. (1978) A key to the adult males of British Chironomidae. *Scientific Publications of the Freshwater Biological Association*, **37** 1-189.

Pinder, L.C.V. (1980) Spatial distribution of Chironomidae in an English chalk stream, in *Chironomidae - Ecology, Systematics, Cytology and Physiology* (ed. D.A. Murray), Pergamon Press, Oxford, pp.153-61.

Pinder, L.C.V. (1986) Biology of freshwater Chironomidae. *Annual Review of Entomology*, **31,** 1-23.

Pinder, L.C.V. (1989) The adult males of Chironominae (Diptera: Chironomidae) of the Holarctic Region - Introduction. *Entomologica Scandinavica Supplement*, **34,** 5-9.

Pinder, L.C.V. and Reiss, F. (1983) The larvae of Chironominae (Diptera: Chironomidae) of the Holarctic Region keys and diagnoses. *Entomologica Scandinavica Supplement*, **19,** 293-435.

Pinder, L.C.V. and Reiss, F. (1986) The pupae of Chironominae (Diptera: Chironomidae) of the Holarctic Region - keys and diagnoses. *Entomologica Scandinavica Supplement*, **28,** 299-456.

Planas M. D. (1991) Factors controlling phytoplankton community structure in an alkaline versus a softwater lake, in *Homage to Ramón Margalef; or, Why there is such pleasure in studying nature* (eds J.D. Ros and N. Prat). *Oecologia Aquatica*, **10,** 95-111.

Porter, K.G. and Y.S. Feig. (1980) The use of DAPI for identifying and counting aquatic microflora. *Limnology and Oceanography*, **25,** 943-8.

Prat, N. and Rieradevall M. (in press) Life cycle and production of Chironomidae (Diptera) from the Lake Banyoles (NE Spain). *Freshwater Biology*.

Prat, N., Puig, M.-A., Gonzales, G. and Millet, X. (1983) Chironomid longitudinal distribution and macroinvertebrate diversity along the Llobregat River (NE Spain). *Memoirs of the American Entomological Society*, **34,** 267-78.

Prejs, A. (1984) Herbivory by temperate freswater fishes and its consequences. *Environmental Biology of Fishes*, **10,** 281-96.

Pressey, R.L. (1986) *Wetlands of the River Murray below Lake Hume*. River Murray Commission Report 86/1.

Pusey, B.J., Arthington, A.H. and Flanders, T.J. (1994) An outdoor replicated artificial stream system: design, operating conditions and initial invertebrate colonisation. *Ecotoxicology and Environmental Safety*, **27,** 177-91.

Quinn, J.M. and Hickey, C.W. (1990) Magnitude of effects of substrate particle size, recent flooding, and catchment development on benthic invertebrates in 88 New Zealand rivers. *New Zealand Journal of Marine and Freshwater Research*, **24,** 387-409.

Racke, K.D. (1973) Environmental fate of chlorpyrifos. *Review of Environmental Contaminate Toxicology*, **131,** 150.

Raddum, G.G., Fjellheim, A. and Hesthagen, T. (1988) Monitoring of acidification by the use of aquatic organisms. *Verhandlungen der Internationale Vereinigung für Theoretische und Angewandte Limnologie*, **23,** 2291-7.

Rae, J.G. (1985) A multivariate study of resource partioning in soft bottom lotic Chironomidae. *Hydrobiologia*, **126,** 275-85.

Rapport, D.J. (1991) Evaluating ecosystem health. *Journal of Aquatic Ecosystem Health*, **1,** 15-24.

Rasmuson, M. (1960) Frequency of morphological deviants as a criterion of developmental stability. *Hereditas*, **46,** 511-35.

Rasmussen, J.R. (1985) Effects of density and microdetritus enrichment on the growth of chironomid larvae in a small pond. *Canadian Journal of Fisheries and Aquatic Sciences*, **42,** 1418-22.

Rasmussen, K. and Lindegaard, C. (1988) Effects of iron compounds on macroinvertebrate communities in a Danish lowland river system. *Water Research*, **22,** 1101-8.

Rawn, G.P., Webster, G.R.B. and Findlay, G.M. (1978) Effect of pool bottom substrate on residues and bioactivity of chlorpyrifos, against larvae of *Culex tarsalis* (Diptera: Culicidae). *Canadian Entomologist*, **110,** 1269-76.

Reiff, N., Riss, H.W. and Orendt, C. (1992) Chironomiden, in *Informationsberichte des Bayersiche Landesamtes für Wasserwirtshaftschaft*, Heft 7192, München, pp.51-122.

Reiss, F. (1968) Ökologische und systematische Untersuchungen an Chironomiden (Diptera) des Bodensees. Ein Beitrag zur lakustrischen Chironomidenfauna des nördlichen Alpenvorlandes. *Archiv für Hydrobiologie*, **64,** 176-323.

Reiss, F. (1984a) Chironomiden (Diptera, Insecten) aus dem Ampertal bei Schöngeising, Oberbayern. *Mitteilungen aus dem zoologischen Gesellschaft Braunau*, **4,** (10-11), 211-20.

Reiss, F. (1984b) Die Chironomidenfauna (Diptera, Insecta) des Osterseengebietes in Oberbayern. *Bericht ANL*, **8,** 186-94.

Resh, V.H. (1979) Biomonitoring, species diversity indices, and taxonomy, in *Ecological Diversity in Theory and Practice*, (eds J.F. Grassle, G.P. Patil, W.K. Smith, and C. Taillie). International Cooperative Publishing House, Fairland, MD, pp.241-53.

Resh, V.H. and Grodhaus, G. (1983) Aquatic insects in urban environments, in *Urban Entomology: Interdisciplinary Perspectives* (eds G.W. Frankie and C.S. Koehler), Praeger Publishers, New York, New York, USA, pp.247-76.

Resh, V.M. and McElravy, E.P. (1993) Contemporary quantitative approaches to biomonitoring using benthic macroinvertebrates, in *Freshwater Biomonitoring and Benthic Macroinvertebrates*, (eds D.M. Rosenberg and V.H. Resh), Chapman and Hall, New York, pp.159-94.

Resh, V.H. and Unzicker, J.D. (1975) Water quality monitoring and aquatic organisms: the importance of species identification. *Journal of the Water Pollution Control Federation*, **47,** 9-19.

Reynoldson, T.B. (1987) Interactions between sediment contaminants and benthic organisms. *Hydrobiologia*, **149,** 53-66.

Richeux, C., Arias-Gonzalez, J.E. and Tourenq, J.N. (1993) A study of the diet of roach (Rutilus rutilus (L)) in the lake of Pareloup (Massif Central, France). *Annales de Limnologie*, **28**, 245-63.

Rieman, B., Christoffersen, K., Jensen, H.J., Muller, J.P., Lindegaard, C. and Bosselmann, S. (1990) Ecological consequences of a manual reduction of roach and bream in a eutrophic, temperate lake. *Hydrobiologia*, **200/201**, 241-50.

Rieradevall, M. (1991) *Ecology and Production of Benthos in Lake Banyoles* . Ph.D. Thesis, University of Barcelona, pp.242

Rieradevall, M. and Prat N. (1991) Benthic fauna of Banyoles Lake (NE Spain). *Verhandlungen der Internationalen Vereinigung für Theoretische und Angewandte Limnologie*, **24**, 1020-3.

Ringe, F. (1974) Chironomiden-Emergenz 1970 in Breitenbach und Rohrwiesenbach. Schlitzer produktionsbiologische Studien (10). *Archiv für Hydrobiologie Supplement*, **45**, 212-304.

Rist, S. (1979) The hydrology of River Laxa. *Oikos*, **32**, 271-80.

Roback, S.S. (1966) The Catherwood Foundation Peruvian-Amazon Expedition XII. Diptera, with some observations on the salivary glands of Tendipedidae. *Monographs of the Academy of Natural Sciences of Philadelphia*, **139**, 159-209.

Roback, S.S. (1975) A new subgenus and species of the genus *Tanytarsus* (Chironomidae: Chironominae: Tanytarsini). *Proceedings of the Academy of Natural Sciences of Philadelphia*, **127**, 71-80.

Roback, S.S. and Ferrington, L. (1983) The immature Stages of *Thienemannimyia barberi* (Coquillett) (Diptera: Chironomidae: Tanypodinae). *Freshwater Invertebrate Biology*, **2**(2), 107-11.

Rodova, R.A. (1969) Samki khironomid (Diptera, Chironomidae) IV (Chironomid females IV). *Trudy Instituta Biologii Vnutrennikh Vod. moskva*, **195**, 196-206.

Rodrigues-Capitulo, A., España, A., Ibañez, C. and Prat, N. (in press) Limnology of natural wells in the Ebro Delta (NE Spain). *Verhandlungen der Internationalen Vereinigung für Theoretische und Angewandte Limnologie*, **25**.

Rosenberg, D.M. (1993) Freshwater biomonitoring and Chironomidae. *Netherlands Journal of Aquatic Ecology*, **26**, 101-22.

Rosenberg, R. (1972) Benthic faunal recovery in a Swedish fjord following the closure of a sulphite pulp mill. *Oikos*, **23**, 92-108.

Rossaro, B. (1984) The chironomids of the Po River (Italy) between Trino Vercellese and Cremona. *Aquatic Insects*, **6**, 123-35.

Rossaro, B. (1992) Chironomids and water temperature. *Aquatic Insects*, **13**, 87-98.

Rossaro, B. (1993) Ordination methods and chironomid species in stony bottom streams. *Netherlands Journal of Aquatic Ecology*, **26**, 447-56.

Rothfels, K.H. and Fairlie, J.W. (1957) The non-random distribution of inversion breaks in the midge *Tendipes decorus*. *Canadian Journal of Zoology* **35**, 221-63.

Royal Society of Chemistry (1992) *Agrochemicals Handbook*. Cambridge, England.

Röser, B. (1980) Emergenz eines Mittelgebirgsbaches des Vorderwesterwaldes. *Archiv für Hydrobiologie Supplement*, **58**, 56-96.

Ruse, L.P. and Wilson, R.S. (1984) Monitoring water quality within the Great Ouse basin using the chironomid exuvial analysis technique. *Water Pollution Control*, **83**, 116-35.

Ryser, H.M., Scholl, A, and Wülker, W. (1983) Revision der Gattung *Chironomus* Meigen (Diptera) VII: *C. muratensis* n. sp. und *C. nudiventris* n. sp., Geschwisterarten aus der *plumosus*- Gruppe. *Revue suisse de Zoologie* **90**, 299-316.

Sæther, O.A. (1968) Chironomids of the Finse Area, Norway, with special reference to their distribution in a glacier brook. *Archiv für Hydrobiologie*, **64**, 426-83.

Sæther, O.A. (1969) Some Nearctic Podonominae, Diamesinae and Orthocladiinae (Diptera: Chironomidae). *Bulletin of the Fisheries Research Board of Canada*, **170**, 1-154.

Sæther, O.A. (1973) Four species of *Bryophaenocladius* Thien., with notes on other Orthocladiinae (Diptera: Chironomidae). *Canadian Entomologist*, **105**, 51-60.

Sæther, O.A. (1977a) Taxonomic studies on Chironomidae: *Nanocladius, Pseudochironomus*, and the *Harnischia*-complex. *Bulletin of the Fisheries Research Board of Canada*, **196**, 1-143.

Sæther, O.A. (1977b) Female genitalia in Chironomidae and other Nematocera: morphology, phylogenies, keys. *Bulletin of the Fisheries Research Board of Canada, 197*. 1-209.

Sæther, O.A. (1979a) Chironomid communities as water quality indicators. *Holarctic Ecology*, **2**, 65-74.

Sæther, O.A. (1979b) Hierarchy of the Chironomidae with special emphasis on the female genitalia (Diptera). *Entomologica Scandinavica Supplement*, **10**, 17-26.

Sæther, O.A. (1980) Glossary of chironomid morphology, terminology (Diptera, Chironomidae). *Entomologica Scandinavica Supplement*, **14**, 1-51.

Sæther, O.A. (1981) Orthocladiinae (Diptera: Chironomidae) from the British West Indies, with descriptions of *Antillocladius* n. gen., *Lipurometriocnemus* n. gen., *Compterosmittia* n. gen. and *Diplosmittia* n. gen. *Entomologica Scandinavica Supplement*, **16**, 1-46.

Sæther, O.A. (1982) Orthocladiinae (Chironomidae: Diptera) from the SE U.S.A. with descriptions of *Plhudsonia*, *Unniella* and *Platysmittia* n. genera and *Atelopodella* n. subgen. *Entomologica Scandinavica*, **13**, 465-510.

Sæther, O.A. (1983) The canalized evolutionary potential - inconsistencies in phylogenetic reasoning. *Systematic Zoology*, **32**, 343-59.

Sæther, O.A. (1985) *Trichochilus*, a new genus of Orthocladiinae (Diptera: Chironomidae). *Spixiana Supplement*, **11**, 31-6.

Sæther, O.A. (1986) The pupae of Buchonomyiinae (Diptera: Chironomidae) of the Holarctic region - Diagnoses. *Entomologica Scandinavica Supplement*, **28,** 115-7.

Sæther, O.A. (1989a) The adult males of Buchonomyiinae (Diptera: Chironomidae) of the Holarctic region - Diagnoses. *Entomologica Scandinavica Supplement* **34,** 125-7.

Sæther, O.A. (1989b) Phylogenetic trends and their evaluation in chironomids with special reference to orthoclads. *Acta Biologica Debrecina Oecologica Hungarica*, **2,** 53-75.

Sæther, O.A. (1990) A review of the genus *Limnophyes* Eaton from the Holarctic and Afrotropical regions (Diptera: Chironomidae, Orthocladiinae). *Entomologica Scandinavica Supplement*, **35,** 1-135.

Sæther, O.A. and Galloway (1980) Sexual anomalies in Chironomini (Chironomidae: Diptera) from Lake Winnipeg, Manitoba, with observations on mermithid (Nematoda) parasites. *Acta Universitatis Carolinae Biologica*, **1978,** 193-211.

Sæther, O A. and Halvorsen, G.A. (1981) Diagnoses of *Tvetenia* Kieff. emend., *Dratnalia* n. gen., and *Eukiefferiella* Thien. emend., with a phylogeny of the *Cardiocladius* group (Diptera: Chironomidae). *Entomologica Scandinavica Supplement*, **15,** 269-85.

Sagar, P.M. and Glova, G.J. (1988) Diel feeding periodicity, daily ration and prey selection of a riverine population of juvenile chinook salmon, *Oncorhynchus tshawytscha* (Walbaum). *Journal of Fish Biology*, **33,**: 643-53.

Sagar, P. M. and Glova, G.J. (1992) Invertebrate drift in a large, braided New Zealand river. *Freshwater Biology*, **27,** 405-16.

Sasa, M. and Kawai, K. (1985) Morphological accounts on selected chironomids collected in Toyama. *Bulletin of the Toyama Science Museum*, **7,** 7-22.

Sato, H. and Yasuno, M. (1979) Test on Chironomidae larvae susceptibility to various insecticides. *Eisei Dobutsu*, **30,** 361-6.

Saxena, S., Martin, J. and Dass, C.M.S. (1985) Polytene Chromosomes of *Nilodorum biroi* (Diptera: Chironomidae). *Genetica*, **66,** 213-21.

Schiemer, F. (1979) The benthic community of the open lake, in *Neusiedlersee - The Limnology of a Shallow Lake in Central Europe* (ed. by H. Löffler), Junk, The Hague, Boston, London.

Schleuter, A. (1986) Die Chironomiden-Besiedlung stehender Kleingewässer in Abhängigkeit von Wasserführung und Fallaubein trag. *Archiv für Hydrobiologie*, **105,** 471-87.

Schlosser, I.J. and Ebel, K.K. (1989) Effects of flow regime and cyprinid predation on a headwater stream. *Ecological Monographs*, **59,** 41-57.

Schlott, G. (1976) Die Chironomiden des Piburger Sees (Tirol, Österreich). *Dissertation aus der Abteilung für Limnologie des Instituts für Zoologie der Universität Innsbruck*, pp.1-77.

Schmid, P.E. (1986) The larvae of chironomids - The composition of species in the mountain brook Oberer Seebach (Ritrodat-Lunz). *Jahresbericht der Biologischen Station Lunz*, **9,** 66-74.

Schmid, P.E. (1992) Community structure of larval Chironomidae (Diptera) in a backwater area of the River Danube. *Freshwater Biology*, **27,** 151-67.

Schmincke, H.-U. (1976) The geology of the Canary Islands, in *Biogeography and Ecology in the Canary Islands* (ed. G. Kunkel), Junk, The Hague, pp.67-184.

Schreiber, E.S.G. (in press) Long-term patterns in invertebrate stream drift in an Australian temperate stream. *Freshwater Biology*,**35**.

Scott, K.M.F. (1959) Hydrobiological studies on the Great Berg River, Western Cape Province. Part 3: The Chironomidae. *Transactions of the Royal Society of South Africa*, **35,** 277-98.

Scullion, J. and Edwards, R.W. (1980) The effect of of coal industry pollutants on the macroinvertebrate fauna of a small river in the South Wales coal field. *Freshwater Biology*, **10,** 141-62.

Segraves, W.A. and Richards, G. (1990) Regulatory and developmental aspects of ecysone-regulated gene expression. *Invertebrate Reproduction and Development*, **18,** 67-76.

Sephton, T.W., Hicks, B.A., Fernando, C.H. and Paterson, C.G. (1983) Changes in the chironomid (Diptera: Chironomidae) fauna of Laurel Creek Reservoir, Waterloo, Ontario. *Journal of Freshwater Ecology*, **2,** 89-102.

Serra-Tosio, B. (1989) Revision des espèces oest-paléarctiques et néarctiques de **Boreoheptagyia** Brundin avec des clés pour les larves, les nymphes et les imagos. *Spixiana*, **11,** 133-73.

SETAC-EUROPE (1992) Guidance document on testing procedures for pesticides in freshwater mesocosms, from the workshop "*A meeting of experts on guidelines for static field mesocosm tests*", Monks Wood Experimental Station, Huntingdon, U.K., July 1991. Society of Environmental Toxicology and Chemistry, Europe. pp.46

SETAC-RESOLVE (1992) Proceedings of a *Workshop on Aquatic Microcosms for Ecological Assessment of Pesticides* Wintergreen, Virginia, USA, October 1991. SETAC Foundation for Environmental Education and the RESOLVE Program of the World Wildlife Fund. pp.56

Shannon, D.E. (1948) A mathematical theory of communication. *Bell System Technological Journal*, **27,** 379-423, 623-56.

Sharp, P.M. and Li, W.-H. (1989) On the rate of DNA sequence evolution in *Drosophila*. *Journal of Molecular Evolution*, **28,** 398-402.

Sheehan, P.J. and Knight, A.W. (1985) A multilevel approach to the assessment of ecotoxicological effects in a heavy metal polluted stream. *Verhandlungen der Internationalen Vereinigung für Theoretische und Angewandte Limnologie*, **22,** 2364-70.

Siebert, M. (1980) Die Emergenz der Chironomiden im Breitenbach 1969-1973. *Archiv für Hydrobiologie Supplement*, **58,** 310-355.

Simpson, K.W. (1983) Communities of Chironomidae (Diptera) from an acid-stressed headwater stream in the Adirondack Mountains, New York. *Memoirs of the American Entomological Society*, **34**, 315-27.

Singh, M.P. and Harrison, A.D. (1984) The chironomid community (Diptera: Chironomidae) in a Southern Ontario stream and the annual emergence patterns of common species. *Archiv für Hydrobiologie*, **99**, 221-53.

Smith, M.J. (1994) *Monitoring Recovery: The Effect of Acidity on Aquatic Invertebrate Communities*, Open File Record No. OFR 111, ARRRI: Office of the Supervising Scientist, Sydney.

Smits, A.J.M., de Lyon, M.J.H., van der Velde, G., Steentjes, P.L.M. and Roelofs, J.G.M. (1988) Distribution of three nymphaeid macrophytes (*Nymphaea alba* L., *Nuphar lutea* (L) Sm. and *Nymphoides peltata* (Gmel) O.Kuntze) in relation to alkalinity and uptake of inorganic carbon. *Aquatic Botany*, **32**(1-2), 45-62.

Smock, L.A., Gilinsky, E. and Stoneburner, D.L. (1985) Macroinvertebrate production in a southeastern United States blackwater stream. *Ecology*, **66**, 1491-503.

Snedecor, G.W. and Cochran, W.G.(1980) *Statistical Methods, 7th Edition*, The Iowa State University Press, Ames, Iowa.

Sokal, R.R. and Rohlf, F.J. (1981) *Biometry: The Principles and Practice of Statistics in Biological Research*, 2nd edition, W.H. Freeman & Co., New York.

Solomon, K.R. and Liber, K. (1988) Fate of pesticides in aquatic mesocosm studies - an overview of methodology. *Proceedings of the Brighton Crop Protection Conference - Pests and Diseases*, **1**, 139-48.

Soponis, A.R. (1977) A revision of the Nearctic species of *Orthocladius (Orthocladius)* van der Wulp (Diptera: Chironomidae). *Memoirs of the Entomological Society of Canada* **102**, 1-187.

Soponis, A.R. (1980) Taxonomic composition of Chironomidae (Diptera) in a sand-bottomed stream of Northern Florida, in *Chironomidae - Ecology, Systematics, Cytology and Physiology*, (ed. D.A. Murray), Pergamon Press, Oxford, pp.163-9.

Spiller, D. (1965) Methods used for chironomid larva surveys of sewage oxidation ponds and natural waters at Aukland, New Zealand. *California Vector Views*, **12**, 9-15.

Srokosz, K. (1980) Chironomidae communities of the River Nida and its tributaries. *Acta Hydrobiologia*, **22**, 191-215.

Stark, J.D. (1981) Chironomidae (nonbiting midges), in *Guide to the Aquatic Insects of New Zealand* (), *Bulletin of the Entomological Society of New Zealand* **5**, 60-7.

Statzner, B. and Higler, B. (1985) Questions and comments on the river continuum concept. *Canadian Journal of Fisheries and Aquatic Sciences*, **42**, 1038-44.

Steffan, A.W. (1967) Ectosymbiosis in aquatic insects, in *Symbiosis* (ed. S.M. Henry), Academic Press, New York, pp.207-89.

Stevens, M.M. (1991) Insecticide treatments used against a rice bloodworm, *Chironomus tepperi* (Diptera: Chironomidae): Toxicity and residual effects in water. *Journal of Economic Entomology*, **84**, 795-800.

Stevens, M.M. (1992) Toxicity of organophosphorus insecticides to the fourth instar larvae of *Chironomus tepperi* Skuse (Diptera: Chironomidae). *Journal of the Australian Entomological Society*, **31**, 335-7.

Stevens, M.M. (1993a) Larval development in *Chironomus tepperi* Skuse (Diptera: Chironomidae) under laboratory conditions. *Environmental Entomology*, **22**, 776-80.

Stevens, M.M. (1993b) Acute toxicity of synthetic pyrethroids to final instar larvae of the rice bloodworm, *Chironomus tepperi* Skuse (Diptera: Chironomidae). *General and Applied Entomology*, **25**, 61-4.

Stevens, M.M. and Warren, G.N. (1992) Insecticide treatments used against a rice bloodworm, *Chironomus tepperi* (Diptera: Chironomidae): Suppression of larval populations. *Journal of Economic Entomology*, **85**, 1606-13.

Stevens, M.M. and Warren, G.N. (1993) Ovicidal and early larvicidal activity of four insect growth regulators against the rice bloodworm, *Chironomus tepperi* Skuse (Diptera: Chironomidae). *General and Applied Entomology*, **24**, 47-52.

Storå, R. (1936) Chironomidae, in Die Dipteren fauna der Kanarischen Inseln unde ihre Probleme (ed. R. Frey), *Commentationes biologicae Helsingfors*, **6**, 21-30.

Storey, A.W. and Edward, D.H.D. (1989) Longitudinal variation in community structure of Chironomidae (Diptera) in two South-Western Australian river systems. *Acta Biologica Debrecina Supplementum Oecologica Hungarica*, **3**, 315-28.

Street, M. and Titmus, G. (1979) The colonization of experimental ponds by Chironomidae (Diptera). *Aquatic Insects*, **1**, 233-44.

Strixino S.T. and Strixino G. (1982) Ciclo de vida de *Chironomus sancticaroli* Strixino & Strixino (Diptera, Chironomidae). *Revista Brasileira de Entomologia*, **26**, 183-9

Sublette, J.E. (1970) Type specimens of Chironomidae (Diptera) in the Illinois Natural History Survey Collection, Urbana. *Journal of the Kansas Entomological Society*, **43**(1), 44-95.

Sublette, J.E. (1979) Scanning electron microscopy as a tool in taxonomy and phylogeny of Chironomidae (Diptera). *Entomologica Scandinavica Supplement*, **10**, 47-65.

Sublette, J.E. and Martin, J. (1995) Morphological variation in secondary sexual characteristics of gynandromorphic *Chironomus (Camptochironomus) tentans* Fabricius (Diptera, Chironomidae) in *Chironomids: from Genes to Ecosystems*, (ed. P.S. Cranston), CSIRO Publications, Melbourne, pp.343-54.

Sublette, J.E. and Sasa, M. (1994) Chironomidae collected in Onchocerciasis endemic areas of Guatemala (Insecta, Diptera). *Spixiana Supplement*, **20**, 1-60.

Sublette, J.E. and Sublette, M.S. (1965) Family Chironomidae (Tendipedidae) in *A Catalog of the Diptera of America north of Mexico*, (eds A. Stone *et al.*), U.S. Department of Agriculture Handbook, **276**, 142-81.

Sublette, J.E. and Sublette, M.F. (1974a) A review of the genus *Chironomus* (Diptera, Chironomidae) VI. The maturus-complex. *Studies in Natural Science (Portales, New Mexico)* **1**(8), 1-41.

Sublette, J.E. and Sublette, M.F. (1974b) A review of the genus *Chironomus* (Diptera, Chironomidae) VII. The morphology of *Chironomus stigmaterus* Say. *Studies in Natural Science (Portales, New Mexico)* **1**(10), 1-65.

Sublette, J.E. and Wirth, W.W. (1980) The Chironomidae and Ceratopogonidae (Diptera) of New Zealand's subantarctic islands. *New Zealand Journal of Zoology*, **7**, 299-378.

Sulc, K. and Zavr^{v}el, J. (1924) O epoikickych a parasitickyck larvach Chironomidu. *Acta Societatis scientarum naturalium moravo-silesiacae*, **1**, 353-91.

Suter, P.J. (1992) *Taxonomic Key to the Ephemeroptera (Mayflies) of the Alligator Rivers Region, Northern Territory*, Open File Record No. OFR 96, ARRRI: Office of the Supervising Scientist, Sydney.

Suter, P.J. and Williams, W.D. (1977) Effect of a total solar eclipse on stream drift. *Australian Journal of Marine and Freshwater Research*, **28**, 793-8.

Suter, P.J., Goonan, P.M., Beer, J.A. and Thompson, T.B. (1993) A Biological and Physico-chemical Monitoring Study of Wetlands from the River Murray Flood Plain in South Australia. *Australian Centre for Water Quality Research, Report* 7/93.

Syrjämäki J. (1965) Laboratory studies on the swarming behaviour of Chironomus strenzkei Fittkau in litt. (Chironomidae). *Annales Zoolici Fennici*, **2**, 143-52

SYSTAT Inc. (1992) *SYSTAT for Windows: Statistics, Version 5 Edition*. SYSTAT Inc., Evanston, IL.

Tabaru, Y. (1975) Outbreak of chironomid midges in a polluted river and chemical control of the larvae. *Japanese Journal of Sanitation and Zoology*, **26**, 247-51.

Tabaru, Y., Matsunaga, H. and Sato, A. (1978) Chemical control of the chironomid larvae (Diptera: Chironomidae) in a polluted river in reference to the controlled area and the mortality of the intoxicated larvae carried by the stream. *Japanese Journal of Sanitation and Zoology*, **29**, 87-91.

Tabaru, Y., Moriya, K. and Ali, A. (1987) Nuisance midges (Diptera: Chironomidae) and their control in Japan. *Journal of the American Mosquito Control Association*, **3**, 45-9.

ten Winkel, E.H., Foekema, E., and Timmermans, K.R. (1987) The impact of cyprinid fish on the benthos of the sandy littoral zone of Lake Maarsseveen, in Winkel, E.H. ten (1987) *Chironomid Larvae and Their Foodweb Relations in the Littoral Zone of Lake Maarsseveen*. Ph.D. Thesis, University of Amsterdam, pp.93-120.

ter Braak, C.J.F. (1988) *CANOCO - A FORTRAN Program for Canonical Community Ordination by [Partial] [Detrended] [Canonical] Correspondence Analysis, Principal Components Analysis and Redundancy Analysis (Version 2.1)*. Report LWA-88-02. Agricultural Mathematics Group, Wageningen, The Netherlands.

ter Braak, C.J.F. (1989) *CANOCO version 3.11*, Agricultural Mathematics Group DLO, AC Wageningen, the Netherlands.

ter Braak, C.J.F (1990a) *Update notes: CANOCO version 3.1*. Agricultural Mathematics Group, Wageningen, The Netherlands.

ter Braak, C.J.F. (1990b) Interpreting canonical correlation analysis through biplots of structure correlations and weights. *Psychometrika*, **55**, 519-31.

ter Braak, C.J.F. (1992a) *CANOCO - a FORTRAN program for Canonical Community Ordination (Version 3.1)*. Microcomputer Power, Ithaca, New York, USA.

ter Braak, C.J.F. (1992b) Permutation versus bootstrap significance tests in multiple regression and ANOVA, in *Bootstrapping and Related Techniques* (eds K.-H. Jöckel, G. Rothe and W. Sendler), Springer Verlag, Berlin, pp.79-86.

ter Braak, C.J.F. and Looman, C.W.N. (1994) Biplots in reduced-rank regression. *Applied Statistics* (in press).

ter Braak, C.J.F. and Prentice, I.C. (1988) A theory of gradient analysis. *Advances in Ecological Research*, **18**, 271-317.

Thienemann, A. (1913) Der Zusammenhang zwischen dem Sauerstoffgehalt des Tiefenwassers und der Zusammensetzung der Tierfauna unserer Seen. *Internationale Revue der Gesamten Hydrobiologie*, **12**, 1-65.

Thienemann, A. (1925) Die Binnengewässer Mitteleuropas. Eine Limnologische Einführung. *Binnengewässer*, **1**, 1-255.

Thienemann, A. (1932) Die Tierwelt der *Nepenthes*-Kannen. *Archiv für Hydrobiologie Supplement*, **11**, 1-54.

Thienemann, A. (1954) *Chironomus*. Leben, Verbreitung und wirtschafliche Bedeutung der Chironomiden. *Die Binnengewässer*, **20**, 1-834.

Thompson, P.E. (1971) Male and female heterogamety in populations of *Chironomus tentans* (Diptera: Chironomidae). *Canadian Entomologist* **103**, 743-7.

Thomson, J.A. (1975) Major patterns of gene activity during development in holometabolous insects. *Advances in Insect Physiology*, **11**, 321-98.

Tilley, L.J. (1989) Diel Drift of Chironomidae larvae in a pristine Idaho mountain stream. *Hydrobiologia*, **174**, 133-49.

Titmus, G. (1979) The emergence of midges (Diptera: Chironomidae) from a wet gravel-pit. *Freshwater Biology*, **9**, 165-79.

Tokunaga, M. and Kuroda, M. (1935) Unrecorded chironomid flies from Japan (Diptera) with a description of a new species. *Transactions of Kansai Entomological Society*, **6**, 1-8.

Tokunaga, M. and Kuroda, M. (1936) *Stenochironomus* midge from Japan (Diptera), with notes on controlling methods of a leaf mining midge. *Transactions of Kansai Entomological Society*, **7**, 1-6.

Topping, M.S. (1971) Ecology of larvae of *Chironomus tentans* (Diptera: Chironomidae) in saline lakes in central British Columbia. *Canadian Entomologist* **103**, 328-38.

Townes, H.K., Jr. (1945) The Nearctic species of Tendipedini [Diptera, Tendipedidae (= Chironomidae)]. *American Midland Naturalist*, **34**, 1-206.

Towns, D.R. (1978) Some little-known benthic insect taxa from a Northern New Zealand River and its tributaries. *The New Zealand Entomologist*, **6,** 409-19.

Towns, D.R. (1979) Composition and zonation of benthic invertebrate communities in a New Zealand kauri forest stream. *Freshwater Biology*, **9,** 251-62.

Towns, D.R. (1981) Effects of artificial shading on periphyton and invertebrates in a New Zealand stream. *New Zealand Journal of Marine and Freshwater Research*, **15,** 185-92.

Travis, J.L. and Schin, K. (1976) Evidence for haemoglobin uptake by oöcytes of *Chironomus thummi*. *Journal of Insect Physiology*, **22,** 1601-8.

Treverrow, N. (1985) Susceptibility of *Chironomus tepperi* (Diptera: Chironomidae) to *Bacillus thuringiensis* serovar *israelensis*. *Journal of the Australian Entomological Society*, **24,** 303-4.

Trewitt, P.M., Boyer, D.B. and Bertrom, G. (1986) Characterization of maternal haemoglobins in the eggs and embryos of *Chironomus thummi*. *Journal of Insect Physiology*, **32,** 963-9.

Tuiskunen, J. (1986) A new northern species of *Sympotthastia* (Diptera, Chironomidae, Diamesinae). *Annales Entomologici Fennici*, **52,** 78-80.

Underwood, A.J. (1993) The mechanics of spatially replicated sampling programmes to detect environmental impacts in a variable world. *Australian Journal of Ecology*, **18,** 99-116.

Urk, G. van, Kerkum, F.C.M. and Smit, H. (1992) Life cycle patterns, density, and frequency of deformities in *Chironomus* larvae (Diptera: Chironomidae) over a contaminated sediment gradient. *Canadian Journal of Fisheries and Aquatic Science*, **49** (11), 2291-9.

Urk, G. van, Kerkum, F.C.M. and Wiersma, S.M. (1985) Bodemfauna in verontreinigde onderwaterbodems. H_2O **18 (24),** 509-13.

van de Gutche, C. and Van Urk, G. (1989) Discrepancies in the effects of field and artificially heavy metal contaminated aquatic sediments upon midge larvae, in *Proceedings of Heavy Metals in the Environment, 7th International Conference, 12 - 15 September 1989, Geneva*.

van der Velde, G. and Hiddink R. (1987) Chironomidae mining in *Nuphar lutea* (L.) Sm. (Nymphaceae). *Entomologica Scandinavica Supplement*, **29,** 255-64.

Van Valen, L. (1962) A study of fluctuating asymmetry. *Evolution*, **16,** 125-42.

Vannote, R.L., Minshall, G.W., Cummins, K.W., Sedell, J.R. and Cushing, C.E. (1980) The river continuum concept. *Canadian Journal of Fisheries and Aquatic Sciences*, **37,** 130-7.

Varga, I. (1928) Ein interessanter Biotrop der Biocönose von Wasserorganismen. *Biologisches Zentralblatt*, **48,** 143-62.

Veevers, J.J. (1991) Phanerozoic Australia in the changing configuration of Proto-Pangea through Gondwanaland and Pangea to the present dispersed continents. *Australian Systematic Botany*, **4,** 1-11.

Verneaux, J. (1968) Contribution a l'étude d'une rivière de plaine les Doulonnes (Jura). Référence particulière aux diptères chironomides remarques écologiques. *Annales Scientifiques de l'Université de Besançon, 3ème Sér*, **4,** 29-40.

Verneaux, J. and Vergon, J.P. (1974) Fauna dulcaquicole de France - Compté, Sixième partie: les diptères chironomides. *Annales Scientifiques de l'Université de Besançon, 3ème Sér*, **11,** 179-98.

Verneaux, J. and G. Tuffery. (1976) Une métode zoologique pratique de détermination de la qualité biologique des eaux courantes. *Annales scientifiques de l'Université de Besançon. Zoologie*, **3,** 79-90.

Vinogradov, S.N., Walz, D.A. and Pohajdak, B. (1992) Organization of non-vertebrate globin genes. *Comparative Biochemistry and Physiology*, **103B,** 759-73.

Walker, E.D., Olds, K.J. and Merritt, R.W. (1988) Gut content analysis of mosquito larvae (Diptera: Culicidae) using DAPI stain and epifluorescence microscopy. *Journal of Medical Entomology*, **25,** 551-4.

Wallace, J.B. (1990) Recovery of lotic macroinvertebrate communities from disturbance. *Environmental Managment*, **14,** 605-20.

Walley, A.M. and Ross, M.I. (1991) Preliminary reconstructions for the Cretaceous to Cainozoic of the New Zealand - New Caledonia region. *Palaeogeography*, **31,** 1-43.

Walshe, B.M. (1947) Feeding Mechanisms of *Chironomus* Larvae. *Nature*, **160,** 474.

Walshe, B.M. (1951) The Feeding Habits of Certain Chironomid Larvae (Subfamily Tendipedinae). *Proceedings of the Zoological Society of London*, **121,** 63-79.

Walter, L. (1973) Syntheseprozesse an den Riesenchromosomen von *Glyptotendipes*. *Chromosoma*, **41,** 327-60.

Ward, A.F. and Williams, D.D. (1986) Longitudinal zonation and food of larval chironomids (Insecta: Diptera) along the course of a river in temperate Canada. *Holarctic Ecology*, **9,** 48-57.

Ward, S., Arthington, A.H. and Pusey, B.J (1994) The effects of chronic application of chlorpyrifos on the macroinvertebrate fauna in an outdoor artificial stream system: specific responses. *Ecotoxicology and Environmental Safety*, (in press).

Wartinbee, D.C. and Coffman, W.P. (1976) Quantitative determination of chironomid emergence from enclosed channels in a small lotic ecosystem. *American Midland Naturalist*, **95,** 479-85.

Warwick, W.F. (1980) Pasqua lake, Southeastern Saskatchewan: a preliminary assessment of trophic status and contamination based on the Chironomidae (Diptera), in *Chironomidae. Ecology, Systematics, Cytology and Physiology* (ed. D.E. Murray), Pergamon Press, Oxford, pp.255-67.

Warwick, W.F. (1985) Morphological abnormalities in Chironomidae (Diptera) larvae as measures of toxic stress in freshwater ecosystems: indexing antennal deformities in *Chironomus* Meigen. *Canadian Journal of Fisheries and Aquatic Science*, **42,** 1881-941.

Warwick, W.F. (1988) Morphological deformities in Chironomidae (Diptera) larvae as biological indicators of toxic stress, in *Toxic Contaminants and Ecosystem Health: A Great Lakes Focus* (ed. M.S. Evans), John Wiley & Sons, New York, pp.281-320.

Warwick, W.F. (1989) Morphological deformities in larvae of *Procladius* Skuse (Diptera: Chironomidae) and their biomonitoring potential. *Canadian Journal of Fisheries and Aquatic Science*, **46**, 1255-70.

Warwick, W.F. (1990) The use of morphological deformities in chironomid larvae for biological effects monitoring. National Hydrology Research Institute no. 43 IWD Sci. Series no. 173, pp.34.

Warwick, W.F. (1991) Indexing deformities in ligulae and antennae of *Procladius* larvae (Diptera: Chironomidae): application to contaminant-stressed environments. *Canadian Journal of Fisheries and Aquatic Science*, **48**, 1151-66.

Warwick, W.F and Tisdale, N.A. (1988) Morphological deformities in *Chironomus*, *Cryptochironomus* and *Procladius* larvae (Diptera: Chironomidae) from two differentially stressed sites in Tobin Lake, Saskatchawan. *Canadian Journal of Fisheries and Aquatic Science*, **45 (7)**, 1123-44.

Warwick, W.F., Fitchko, J., McKee, P.M., Hart, D.R. and Burt, A.J. (1987) The incidence of deformities in *Chironomus* spp. from Port Hope Harbour, Lake Ontario. *International Association of Great Lakes Research*, **13 (1)**, 88-92.

Waters, T.F. (1966) Production rate, population density and drift of a stream invertebrate. *Ecology*, **47**, 595-604.

Waters, T.F. (1969) The turnover ratio in production ecology of freshwater invertebrates. *American Naturalist*, **103**, 173-85.

Waters, T.F. (1972) The drift of stream insects. *Annual Review of Entomology*, **17**, 253-72.

Waters, T.F. (1977) Secondary production in inland waters. *Advances in Ecological Research*, **10**, 91-164.

Waters, T.F. (1979) Influence of benthos life history upon the estimation of secondary production. *Journal of Fisheries Research Board Canada*, **36**, 1425-30.

Waters, T.F. (1988) Fish production - benthos production relationships in trout streams. *Polskie Archiwum Hydrobiologiczne*, **35**, 545-61.

Wawrik, F. (1966) Die Waldviertler Fischteiche und ihre Entomostrakenfauna auf oekologischer Grundlage. *Hydrobiologia* **28**, 385-552.

Way, M.O. and Wallace, R.G. (1989) First record of midge damage to rice in Texas. *Southwestern Entomologist*, **14**, 27-33.

Welch, H.E., Jorgenson, J.K. and Curtis, M.F. (1988) Emergence of Chironomidae (Diptera) in fertilized and natural lakes at Saqvaqjuac, N.W.T. *Canadian Journal of Fisheries and Aquatic Sciences*, **45**, 731-7.

Wells, A. (1991) *A Guide to the Caddisflies (Trichoptera) of the Alligator Rivers Region, Northern Territory*, Open File Record No. OFR 84, ARRRI: Office of the Supervising Scientist, Sydney.

White, T.C.R. (1970) Airborne arthropods collected in South Australia with a drogue-net towed by a light aircraft. *Pacific Insects*, **12**, 251-9.

Wiederholm, T. (1980a) Chironomids as indicators of water quality in Swedish lakes. *Acta Universitatis Carolinae-Biologica*, **1978**, 275-83.

Wiederholm, T. (1980b) Use of benthos in lake monitoring. *Journal of the Water Pollution Control Federation*, **52**, 537-47.

Wiederholm T. (Ed.) (1983) Chironomidae of the Holarctic Region. Keys and diagnoses. Part 1. Larvae. *Entomologica Scandinavica Supplement*, **19**, 1-457.

Wiederholm, T. (1984a) Incidence of deformed chironomid larvae (Diptera: Chironomidae) in Swedish lakes. *Hydrobiologia*, **109**, 243-9.

Wiederholm, T. (1984b) Responses of aquatic insects to environmental pollution, in *The Ecology of Aquatic Insects* (eds V.H. Resh and D.M. Rosenberg), Praeger Publishers, New York, NY, pp.508-57.

Wiggins, G.B. (1977) *Larvae of the North American Caddisfly Genera*. University of Toronto Press, Ontario. pp.401

Wiggins, G.B., Mackay, R.J. and Smith, I. M. (1980) Evolutionary and ecological strategies of animals in annual temporary pools. *Archiv für Hydrobiologie*, **58 1/2**, 97-206.

Wiley, M. and Kohler, S.L. (1984) Behavioural adaptations of aquatic insects, in, *The Ecology of Aquatic Insects* (eds V.H. Resh and D.M. Rosenberg), Praeger, pp.101-33.

Willassen, E. and Cranston, P.S. (1986) Afrotropical montane midges (Diptera, Chironomidae, Diamesa). *Zoological Journal of the Linnean Society*, **87**, 91-123.

Williams, C.J. (1985) A comparison of net and pump sampling methods in the study of chironomid larval drift. *Hydrobiologia*, **124**, 243-50.

Williams, C.J. (1989) Downstream drift of the larvae of Chironomidae (Diptera) in the River Chew, SW England. *Hydrobiologia*, **183**, 59-72.

Williams, D.D. (1987) *The Ecology of Temporary Waters*. Croom Helm, London.

Williams, D.N. (1974) An infestation by a parthenogenetic chironomid. *Water Treatment and Examination*, **23**, 215-29.

Williams, K.A., Green, D.W.J., Pascoe, D. and Gower, D.E. (1986) The acute toxicity of cadmium to different larval stages of *Chironomus riparius* (Diptera: Chironomidae) and its ecological significance for pollution regulation. *Oecologica*, **70**, 362-66.

Williams, W.D. (1980) *Australian Freshwater Life*, Globe Press, Melbourne.

Wilson, R.S. (1977) Chironomid pupal exuviae in the River Chew. *Freshwater Biology*, **7**, 9-17.

Wilson, R.S. (1980) Classifying rivers using chironomid pupal skin exuviae, in *Chironomidae - Ecology, Systematics, Cytology and Physiology*, (ed. D.A. Murray), Pergamon Press, Oxford, pp.209-16.

Wilson, R.S. and McGill, J.D. (1977) A new method of monitoring water quality in a stream receiving sewage effluent, using chironomid pupal exuviae. *Water Research*, **11**, 959-62.

Wilson, R.S. and McGill, J.D. (1979) *The use of chironomid pupal exuviae for biological surveillance of water quality*. Department of the Environment, Water Data Unit, Technical Memorandum No. 18.

Wilson, R.S. and Wilson, S.E. (1985) A survey of the distribution of Chironomidae (Diptera, Insecta) of the river Rhine by sampling pupal exuviae. *Hydrobiological Bulletin*, **18**, 119-32.

Winberg, G.G. (1971) *Methods for the Estimation of Production of Aquatic Animals*. Academic Press, London, pp.1-175.

Winner, R.W., Boesel, M.W. and Farrel, M.P. (1980) Insect community structure as an index of heavy-metal pollution in lotic ecosystems. *Canadian Journal of Fisheries and Aquatic Sciences*, **37**, 647-55.

Winterbourn, M.J. (1987) Invertebrate Communities, in *Inland Waters of New Zealand* (ed. A.B. Viner), DSIR Bulletin 241, Wellington, pp.167-90.

Winterbourn, M.J., Rounick, J.S. and Cowie, B. (1981) Are New Zealand stream ecosystems really different? *New Zealand Journal of Marine and Freshwater Research*, **15**, 321-8.

Winterbourn, M.J. and Collier, K.J. (1987) Distribution of benthic invertebrates in acid brown water streams in the South Island of New Zealand. *Hydrobiologia*, **153**, 277-86.

Woodiwiss, F.S. (1964) The biological system of stream classification used by the Trent River Board. *Chemistry and Industry*, **83**, 443-7.

Wotton, R.S., Armitage, P.D., Aston, K., Blackburn, J.H., Hamburger, M. and Woodward, C.A. (1993) Colonization and emergence of midges (Chironomidae: Diptera) in slow sand filter beds. *Netherlands Journal of Aquatic Ecology*, **26**, 331-9.

Wright, I. (1992) *The Impact of Wentworth Falls Sewage Treatment Plant on Blue Mountain Creek 1989-1990*. Australian Water Technologies, Science and Environment, Sydney. Report No. 92/29.

Wright, J.F., Armitage, P.D., Furse, M.T. and Moss, D. (1988) A new approach to the biological surveillance of river quality using macroinvertebrates. *Verhandlungen der Internationale Vereinigung für Theoretische und Angewandte Limnologie*, **23**, 1548-52.

Wright, J.F., Armitage, P.D., Furse, M.T. and Moss, D. (1989) Prediction of invertebrate communities using stream measurements. *Regulated Rivers: Research and Management*, **4**, 147-55.

Wright, J.F., Moss, D., Armitage, P.D. and Furse, M.T. (1984) A preliminary classification of running-water sites in Great Britain based on macro-invertebrate species and the prediction of community type using environmental data. *Freshwater Biology*, **14**, 221-56.

Wülker, W. (1980) Basic patterns in the chromosome evolution in the genus *Chironomus* (Diptera). *Zeitschrift für Zoologische Systematik und Evolutionsforschung*, **18**, 112-23.

Wülker, W. (1991) Chromosomal, morphological, and biological differences between *Chironomus tenuistylus* Brundin and *C. longistylus* Goetghebuer (Diptera: Chironomidae) in FennoScandia and the USA. *Entomologica Scandinavica* **22**, 231-40.

Wülker, W. and Butler, M.G. (1983) Karyosystematics and morphology of northern *Chironomus* (Diptera: Chironomidae): Freshwater species with larvae of the *salinarius*-type. *Entomologica Scandinavica* **14**, 121-36.

Wülker, W. and Klotzli, A.M. (1973) Revision des Gattung *Chironomus* Meig. IV. Arten des *lacunarius* (*commutatus*) Komplexes. *Archiv für Hydrobiologie*, **72**, 474-89.

Wülker, W. and Martin, J. (1974) A review of the genus *Chironomus* (Diptera, Chironomidae) VI. Cytology of the maturus- complex. *Studies in Natural Science (Portales, New Mexico)* **1**(9), 1-21.

Wülker, W., Ryser, H.M. and Scholl, A, (1983) Revision der Gattung *Chironomus* Meigen (Diptera) VIII: Arten mit Larven des *fluviatilis*-Typs (*obtusidens*-Gruppe): *C. acutiventris* n. sp. und *C. obtusidens* Goetgh. *Revue Suisse de Zoologie*, **90**, 725-45.

Wülker, W., Sublette, J.E. and Martin, J. (1968) Zur cytotaxonomie der nordamerikanischer *Chironomus*-Arten. *Annales Zoologici Fennici*, **5**, 155-8.

Wülker, W., Sublette, J.E. and Martin, J. (1991) *Chironomus utahensis* Malloch and *Chironomus harpi* new species and their karyosystematic relationships to other species in the *decorus*- group of *Chironomus*. *Spixiana* **14**, 71-94.

Wülker, W., Sublette, J.E., Morath, E. and Martin, J. (1989) *Chironomus columbiensis* n. species in South America and *Chironomus anonymus* Williston in North America - closely related species. *Studies on Neotropical Fauna and Environment*, **24**, 121-36.

Wülker, W., Sublette, J.E., Sublette, M.F. and Martin, J. (1971) A review of the genus *Chironomus* (Diptera, Chironomidae) I. The *staegeri* group. *Studies in Natural Science (Portales, New Mexico)* **1**(1), 1-89.

Yano, K., Takayama, M. and Ueda, T. (1991) Biology of *Glyptotendipes tokunagai* Sasa (Diptera: Chironomidae).1. Development of immature stages. *Transactions of the Shikoku Entomological Society*, **19**, 177-88.

Yasuno, M., Hatakeyama, S. and Sugaya, Y. (1985) Characteristic distribution of chironomids in the rivers polluted with heavy metals. *Verhandlungen der Internationalen Vereinigung für Theoretische und Angewandte Limnologie*, **22**, 2371-7.

Yasuno, M., Iwakuma, T. Sugaya, Y. and Sasa M. (1983) Zoobenthos of Japanese lakes of different trophic status, with special reference to Chironomidae. *Research Report from Special Research Projects, Environmental Sciences*, **182-R12-17**, 21-48 (in Japanese).

Zakharov, V.M. (1989) Future prospects for population phenogenetics. *Soviet Scientific Reviews, F, Physiology and General Biology*, **4,** 1-79.

Zakharov, V.M. and Yablokov, A.V. (1990) Skull asymmetry in the Baltic grey seal: effects of environmental pollution. *Ambio*, **19,** 266-9.

Zar, J.H. (1984) *Biostatistical Analysis*, Prentice-Hall International, New Jersey.

Zelinka, M. and Working-Team (1977) Production conditions of the polluted trout brook. *Folia Facultatis Scientiarum Natauralium Universitatis Purkynianae Brunensis, Biologia*, **18, 7,** 1-105.

Zieba, J. (1971) Produktion of macrobenthos in fingerling ponds. *Polskie Archiwum Hydrobiologii*, **18,** 235-46.

Zischke, J.A., Arthur, J.W., Nordlie, K.J., Hermanutz, R.O., Standen, D.A. and Henry, T.P. (1983) Acidification effects on macroinvertebrates and fathead minnows (*Pimephales promelas*) in outdoor experimental channels. *Water Research*, **17,** 47-63.

Index

E

F

N

O